T0313790

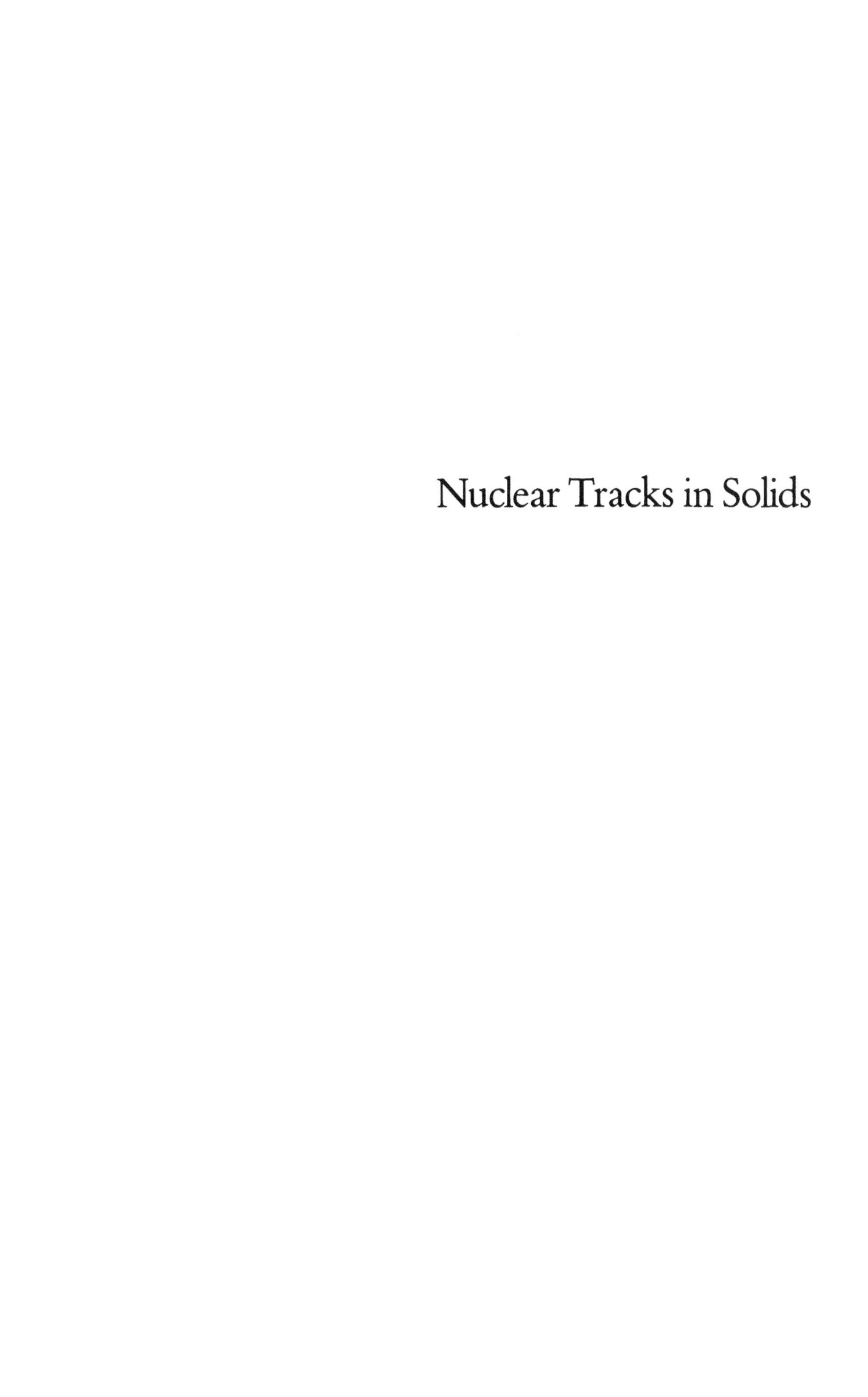

Nuclear Tracks in Solids

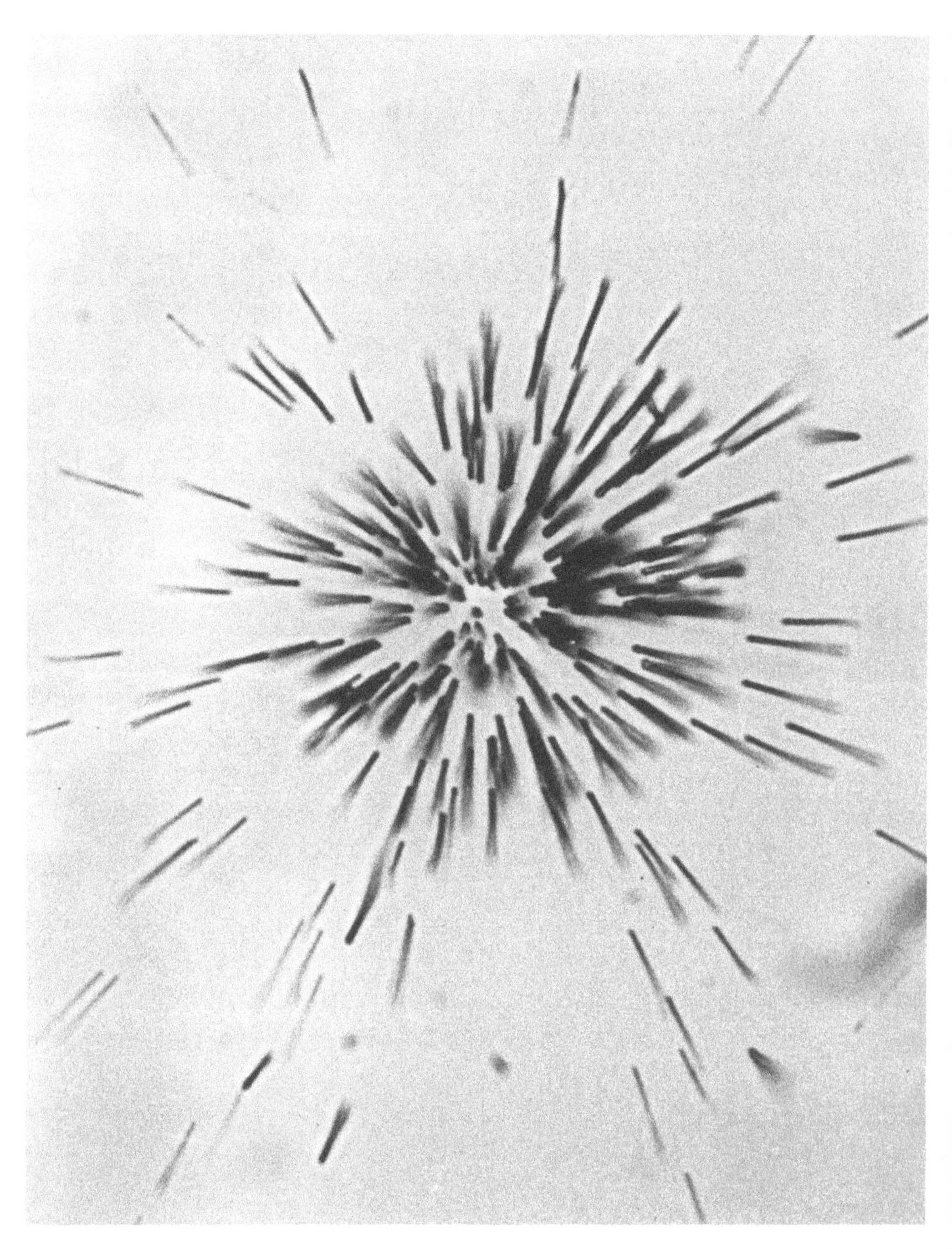

Portion of a sheet of mica that was exposed to 10^{17} thermal neutrons per cm and then etched 30 min in hydrofluoric acid. Fission fragment tracks (about 10^{-3} cm long) were emitted from a particle of ordinary dust containing 1 ppm of uranium.

Nuclear Tracks in Solids

Principles and Applications

Robert L. Fleischer
General Electric Research Laboratory, Schenectady

P. Buford Price
University of California, Berkeley

Robert M. Walker
Washington University, St. Louis

UNIVERSITY OF CALIFORNIA PRESS
BERKELEY · LOS ANGELES · LONDON

University of California Press
Berkeley and Los Angeles, California

University of California Press, Ltd.
London, England

Copyright © 1975, by
The Regents of the University of California

ISBN 0-520-02665-9
Library of Congress Catalog Card Number: 73-90670

Designed by Henry Bennett

Dedication

To those who guide the course of science and engineering in government, industry, and universities, who have the vision of basic science in uncharted areas as one of the great forces that serve the long range good of mankind.

"*The universe is not to be narrowed down to the limits of the understanding, which has been men's practice up to now, but the understanding must be stretched and enlarged to take in the image of the universe as it is discovered.*"

Francis Bacon
Parasceve, Aphorism 4

CONTENTS

PART II. EARTH AND SPACE SCIENCES

PART III. NUCLEAR SCIENCE AND TECHNOLOGY

PREFACE

The first direct photographs of damage trails created by fragments from the fission of ^{235}U were transmission electron micrographs of mica published by E. C. H. Silk and R. S. Barnes of Harwell in 1959. Their observations touched off a chain of ideas and experiments that soon led to the discovery of particle track etching and have helped to create an aura of excitement and fascination for particle tracks in solids. Part of this fascination lies in its basic simplicity; part comes from the diversity of scientific and technological areas where track etching proves to be useful. These range from nuclear science to botany and from arms control to beer. A brief description of the early development of this new field—its logical progression, the deliberate and chance interactions of people from different disciplines, and the frequent combats with preconceptions—may carry some lessons on the environments and mechanisms through which research progresses.

Our part of the story began at the General Electric Research Laboratory, where the three of us were working in very different areas. The report, at an international meeting on radiation effects in solids, of Silk and Barnes' observation caused one of us with a background in both high energy physics and point defects in solids (RMW) to evaluate quantitatively the possibility that tracks from nuclear interactions would be stored in mica in lunar rocks, forming a "fossil" record of the cosmic ray bombardment of the moon. Although nearly a decade passed before we found such tracks in lunar minerals and used them to elucidate surface processes on the moon, the search remained one of our long-term objectives over that time. Fortunately, at the beginning we were unaware of the absence of mica in any known extraterrestrial matter.

Another of us (PBP), whose interest had been in using electron microscopy to study crystal defects in thin crystals, made possible our early, brief, direct observations of tracks and the subsequent study of etched tracks. The first views of tracks were brief because of the frustrating tendency of the tracks in natural micas to fade in the electron beam of the microscope while they were being observed. This ailment received an unexpected cure as a result of a chance encounter in a corridor with Dr. Louis Navias, a ceramist whose collection happened to include synthetic fluoro-micas—which happily turned out to record tracks that were stable during electron-microscopic examination. This was the material that stimulated the whole line of investigation that otherwise might well have been dropped in its infancy. Instead it led to the observation in the electron microscope of the first deliberately etched, submicroscopic tracks. These holes, which unlike the tracks themselves do not fade, shortly thereafter allowed fossil spontaneous fission tracks to be seen by electron microscopy of etched mica crystals that contained uranium-rich inclusions. They also allowed us to begin seriously to pursue our early interest in nuclear physics.

The realization that tracks may be etched to sizes where they can be observed in ordinary microscopes is a conceptually simple advancement. But it was one that for us depended on a peripheral request from General Electric's Electronics Laboratory at Syracuse, New York by an engineer who wanted a controlled vacuum leak and happened to have heard of the etching of holes in mica at Schenectady. To make what he requested—a sheet of mica with a single, small, etched hole through it—required an initial, flaw-free sheet with no background of dislocations or other line defects which might etch into many holes. In order to identify such sheets, different samples were etched for much longer times than usual and examined for defects. What was found was that natural fission tracks were enlarged to sizes where they could conveniently be seen optically, and therefore studied at much lower concentrations than would be recognized at the high magnification inherent in transmission electron microscopy.

Realization of the generality of track etching and some of the diversity of the possibilities came about by our finding that mica was not unique; a great variety of qualitatively different materials also store tracks. This work started with the etching of tracks in ordinary glass by another of us (RLF)—again with a background in defects in crystals and also with experience in etching defects in solids. Our exploration of amorphous solids had for some time been stifled by a preconception—our simplified initial picture of a track in a crystal as a structureless region, i.e., a glass. Why should a glass within a glass etch preferentially? Etching tracks in glass was clear proof of what is now obvious—that not all glasses have the same chemical reactivity. This result led us quickly to organic glasses (plastics) and then back to crystals. At the present time we know of approximately one hundred and fifty insulating substances in which etched tracks can be observed.

Track techniques and our understanding of tracks themselves were developed along with quantitative methods for precise identification of individual particles. Part I of this book describes these subjects. The track techniques are also new tools that have been applied extensively in a number of fields, including geochronology, cosmic ray physics, meteoritic and lunar science, nuclear physics, chemical analysis and micro-chemical mapping, and radiation dosimetry; they have been applied less extensively in a series of other fields. These categories correspond in sequence to the chapters in Parts II and III of this book.

In asking ourselves what factors have allowed the recognition of the opportunities inherent in particle track etching and their rapid exploration, a few thoughts emerge. One is the importance of gaining new ways of looking at old problems by the simple expedient of changing from one field to another. Frequently the established ways of looking at a former field provide new insights in the context of the different set of problems of a fresh field. A good example is the problem of stabilizing and revealing tracks in various solids. At the time our work began several methods were known for revealing one kind of crystal defect—line dislocations—and it was natural that we with backgrounds in the study of defects in crystals might try them all. Fortunately the first one we tried—the chemical etching method—worked. Extending the track techniques to new fields has given us a great deal of satisfaction, not the least of which has been the personal friendships formed with scientists and engineers from many countries and in many disciplines. Working as a team with no inhibitions or strains introduced by questions of authorship priority—it was decided from the beginning to list our names alphabetically—had great advantages. We interacted freely and, by dividing our efforts, could afford to tackle problems that were potentially important but with seemingly low probabilities for success.

A second major ingredient in the rapid testing of the new scientific possibilities of particle tracks was a research environment in which divergent exploration was encouraged. In short, the philosophy that a significant fraction of qualitatively new science will lead to worthwhile practical discoveries made it easy to branch off into areas unconnected with the original focus on radiation damage and lunar history. Often, fresh understanding gained in one of the side branches of the work proved directly useful in solving stubborn problems that had resisted a direct frontal assault for some time. As one example, our electron microscope observation of fossil spontaneous fission tracks in mica suggested a new method of geochronology, but the idea was set aside as impractical as long as tracks had to be sought by tedious scanning of small, thin samples that would fit into the chamber of an electron microscope. With the later realization that tracks could be enlarged to optically visible size by prolonged etching, our effort was refocussed, and fission track dating quickly developed into a practical reality.

Luck is a precious ingredient in any work, and we have related how we benefited from a special ceramic donated by Louis Navias, from a request for a new vacuum leak, and from ignorance about where mica is found. We have had good fortune in our timing too, with track dating becoming possible just when new ideas about the Earth presented a means of providing a critical test of the concept of ocean bottom spreading, with the ability to understand the origins of tracks in extraterrestrial materials coming at a time when the first lunar samples were about to be retrieved and their histories studied, and with particle identification methods maturing as satellites, rockets, and balloons became capable of exposing and returning to Earth suitable detectors for examining the nature, history, and origin of particles from distant sources.

This brief historical view would be incomplete without a few words about D. A. Young at Harwell, England, who published a short but remarkable note in *Nature* in 1958. In it he described irradiating lithium fluoride with fission fragments and etching it to reveal shallow pits at the fission tracks. He also briefly described a possible mechanism of track formation whose principle is identical to what we later independently proposed for tracks in inorganic detectors. In one short paper he recognized the existence of tracks, demonstrated they could be etched, showed they could be made visible optically, and explained how they formed. This paper escaped us until later—after we had gone through our separate sequence of discovery. What is perhaps more surprising is that his work also apparently went unnoticed by Silk and Barnes who did their work at Harwell just one year later than Young.

PURPOSE OF THIS BOOK

We believe that particle track etching has reached a sufficiently mature stage of development that a comprehensive account of its methods and accomplishments is needed. Among the roughly 1500 published articles on track etching the student of the subject who seeks an overall view will find several short, tightly condensed papers that review scientific or technological track applications and a few isolated reviews that consider in greater depth individual specific areas of application. Even though the pace of new developments at the forefront of science and engineering remains high, there now exist extensive principles and methods for particle detection and identification and for data collection. We endeavor here to put such information into an accessible, self-consistent presentation in Part I—*Principles of Track Etching*, and then in Parts II and III discuss the more specialized methods and their results in the several fields of application in *Earth and Space Sciences* and in *Nuclear Science and Technology*.

We envision this book being useful for four groups of individuals. Research workers who use particle tracks should find in it useful compilations of data and up-to-date references along with overviews of areas of application that are different from those in which they may be active. For those who wish to begin track work this book is a logical base on which to build. Scientists that normally use other techniques within one of the disciplines on which tracks have been brought to bear can benefit from an understanding of what particle tracks have done or can do, and of what their limitations are. Finally, we believe that some of the chapters in Parts II and III are appropriate source matter for special topic or seminar courses at the advanced undergraduate or beginning graduate level.

The particle track literature scanned in the preparation of these manuscripts consisted primarily of the more than 1300 papers of which we were aware in mid-1973. These included published papers plus a number of preprints that we had received, a good many of which were in response to a letter sent out to track workers to locate current work. Although we have no adequate overall assessment of the thoroughness of our search, the existence of Georef, the computer-based search service of the American Geological Institute, gave us considerable encouragement that most of the relevant papers have been located. After assembling our lists in geochronology and geochemistry, with supplementary items from C. W. Naeser of the U. S. Geological Survey, whose help we are pleased to acknowledge, the listing under "fission tracks" was obtained from Georef. In each of the two fields five or fewer new references were added to our lists of roughly 200, and nearly half of the new items were Masters' theses or similar hard-to-retrieve items.

All papers for which we had a reference were sought by Mrs. J. Long in the listings of various abstracting services, including *Chemical Abstracts*, *Physics Abstracts*, and *Nuclear Science Abstracts*. She found that 87% of the references were listed in either *Chemical* or *Nuclear Science Abstracts*; only 3% more were found by consulting the *Physics Abstracts*.

ACKNOWLEDGMENTS

We are pleased to give thanks to the many people who have contributed in various ways to this work, particularly to the more than one thousand scientists who have done studies of particle track etching and who have kept us supplied with copies of their work. Although it is impossible to list here the names of all the students and colleagues with whom we have worked closely, we would like to single out for special mention H. R. Hart, Jr., D. Lal, and M. Maurette, who have been associated with us since the early days and who have made numerous fundamental contributions to the development of this field. The following have read individual chapters and offered constructive comments: C. P. Bean, R. W. Caputi, B. S. Carpenter, G. Crozaz, M. D. Fiske, S. C. Furman, C. Hohenberg, F. Hörz, D. Lal, D. Macdougall, C. W. Naeser, F. Podosek, G. M. Reimer, and D. Yuhas. We are especially indebted to Emilio Segrè for reading the entire manuscript and making numerous helpful suggestions for improvements. We are indebted to the following for permission to use illustrations: J. R. Arnold, F. Aumento, C. P. Bean, G. R. Bellavance, E. V. Benton, Big Bear Solar Observatory, D. S. Burnett, B. S. Carpenter, J. H. Chan, C. B. Childs, R. M. Chrenko, J. H. Davies, R. W. DeBlois, P. H. Fowler, D. Gault, F. Getsinger, J. E. Gingrich, F. Hörz, R. Katz, D. Lal, E. Lifshin, J. Lovering, D. B. Lovett, M. Maurette, Metropolitan Museum of Art, D. S. Miller, J. Mory, C. W. Naeser, National Geographic Society, A. Noonan, W. Z. Osborne, T. Plieninger, C. R. Porter, R. S. Rajan, H. S. Rosenbaum, R. Scanlon, J. Shirck, R. H. Stinson, D. Storzer, P. V. Tobias, G. A. Wagner, D. Yuhas, and D. Zimmerman. Permission to quote and discuss unpublished work was given by: C. P. Bean, D. Braddy, J. H. Chan, H. J. Crawford, D. Gault, J. E. Gingrich, D. S. Golibersuch, H. R. Hart, Jr., I. D. Hutcheon, D. B. Lovett, R. Scanlon, D. Storzer, J. D. Sullivan, J. R. M. Viertl, and E. Zinner. Major typing and secretarial assistance was given by: Cathy Ellison, Jan Haslam, Joan Kaczynski, Hilda Ketterer, Mary Klash, Laura Meier, Marge Miller, Bonnie Pasko, and Helen Tepedelen. We are extremely grateful to Mary Klash for assistance in assembling the manuscript into its final form. We are indebted to the General Electric Research and Development Center for its hospitality during a two-month period in early 1973, during which portions of this work came into existence. We also wish to thank the Physics Department of the University of California, Berkeley, for its hospitality during a two-week period in the summer of 1973 when several chapters were completed. R. L. Fleischer thanks the National Oceanic and Atmospheric Administration, Atmospheric Physics and Chemistry Laboratory and the National Center for Atmospheric Research for their hospitality during the preparation of Chapter 1. We thank Jean Long for carrying out the long and laborious job of obtaining abstracts for approximately one thousand papers and to P. K. Noaecker for extensive additional reference work. We thank C. P. Bean for the description of his work with the late Warren DeSorbo on the damage intensity distribution of tracks in polycarbonate detectors. Portions of Chapter 4 are adapted from an article by R. L. Fleischer and H. R. Hart, Jr. in *Calibration of Hominoid Evolution*, a conference held at Burg Wartenstein, Austria, July 1971. Special thanks should go to the track group at G.E.'s Vallecitos Nuclear Laboratory. Under the leadership of H. W. Alter and S. C. Furman, they carried out most of the earliest exploratory work in the practical applications of particle tracks—making filters; measuring radiation exposures of mine workers; doing nuclear fuel development, tracer work, neutron radiography, uranium exploration; and designing and building nuclear safeguard devices.

PART I

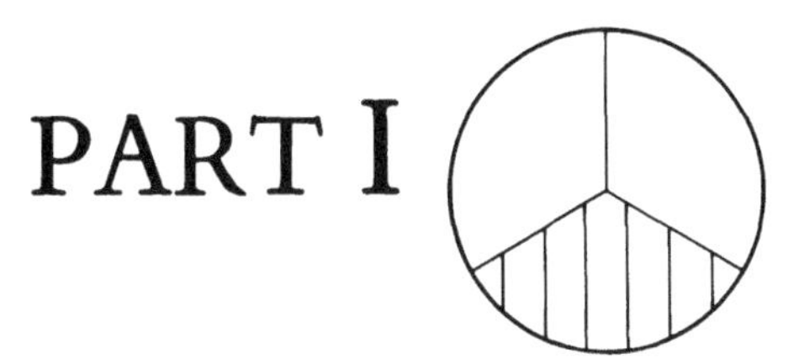

Principles of Track Etching

Chapter 1

Formation of Particle Tracks

1.1. INTRODUCTION

The passage of heavily ionizing, nuclear particles through most insulating solids creates narrow paths of intense damage on an atomic scale. These damage *tracks* may be revealed and made visible in an ordinary optical microscope by treatment with a properly chosen chemical reagent that rapidly and preferentially attacks the damaged material. It less rapidly removes the surrounding undamaged matrix in such a manner as to enlarge the etched holes that mark and characterize the sites of the original, individual, damaged regions. This simple technique of observing particles has been used in a wide variety of technical fields that range from nuclear science and engineering to cosmic ray astrophysics and from geology, archaeology, and suboceanic geophysics to lunar science and meteoritics. Fig. 1-1 shows the envisioned character of tracks in crystalline and polymeric solids and Fig. 1-2 shows etched tracks in a crystal, a glass, and a polymer.

In Part I of this book, we describe our present understanding of the nature of tracks and how they are developed and used quantitatively. In Parts II and III we show how tracks have been applied in a diversity of areas including those noted above.

This first chapter is concerned primarily with the formation of etchable tracks and with their nature, why ionizing particles form tracks only in dielectric solids, and why the tracks (unless heated) are normally permanent features of solids. In confining our attention to tracks that are usually observed after etching, we consider bulk samples and therefore deliberately exclude extensive observations and studies of tracks produced in thin films (for example, Bierlein and Mastel, 1960; Noggle and Stiegler, 1960; Kelsch et al., 1960; Izui and Fujita, 1961; Ruedl et al., 1961; Bowden and Chadderton, 1961; Pravdyuk and Golyanov, 1962). In these cases mechanisms of track formation are available that are peculiar

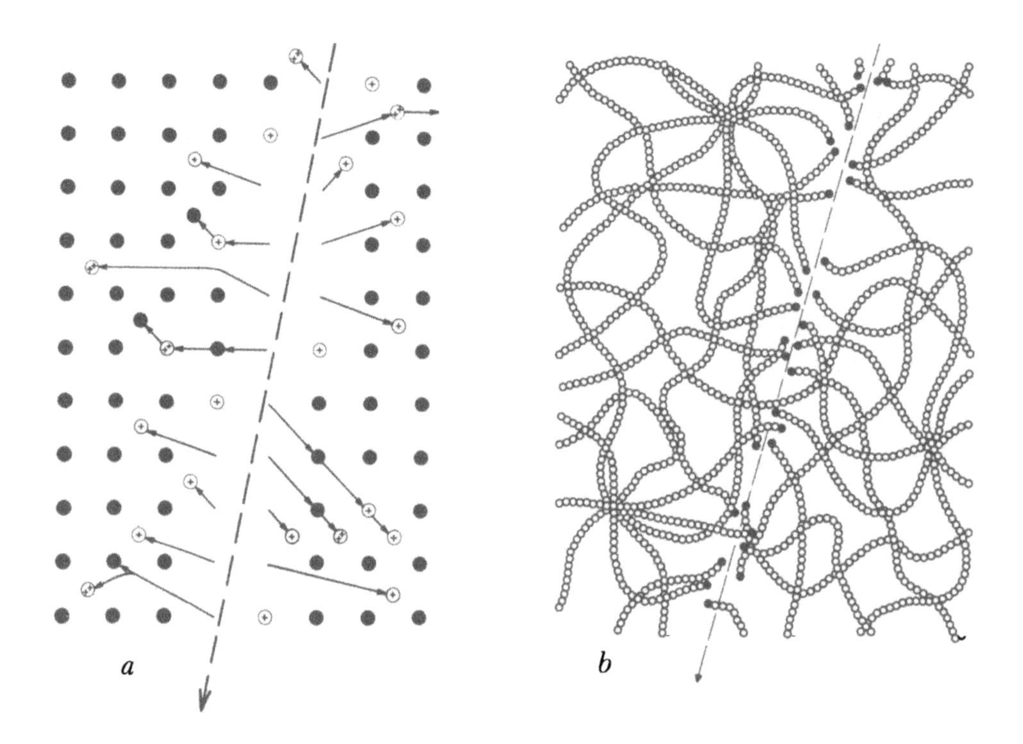

Fig. 1-1. *The atomic character of a particle track in (a) a crystal and (b) a polymer. In the crystal the damage consists of continuous disorder composed of vacant lattice sites and of interstitial ions or atoms. In the polymer new chain ends and other chemically reactive sites are formed. (After Fleischer et al., 1969.)*

to the proximity of a free surface (Kelsch et al., 1962; Merkle, 1962; Stiegler and Noggle, 1962; Whapham and Makin, 1962). For a brief history of the development of particle track etching—the direct observation of tracks by Silk and Barnes (1959), the discovery of track etching in mica by Price and Walker (1962a, b), the realization of the generality of track etching (Fleischer and Price, 1963a, b; 1964a), and the neglected paper by Young (1958)—the reader is referred to the Preface. This chapter, then, is concerned primarily with the nature of tracks. The detailed techniques for etching and particle identification are deferred until Chapters 2 and 3.

1.2. PROCEDURES FOR REVELATION OF TRACKS

The quality of the information we can acquire on the nature of tracks depends on our means of observation. For that reason we review here the several procedures used or attempted for seeing tracks. We will find that even the failure to observe tracks by a particular, potential method yields information about the structure of tracks.

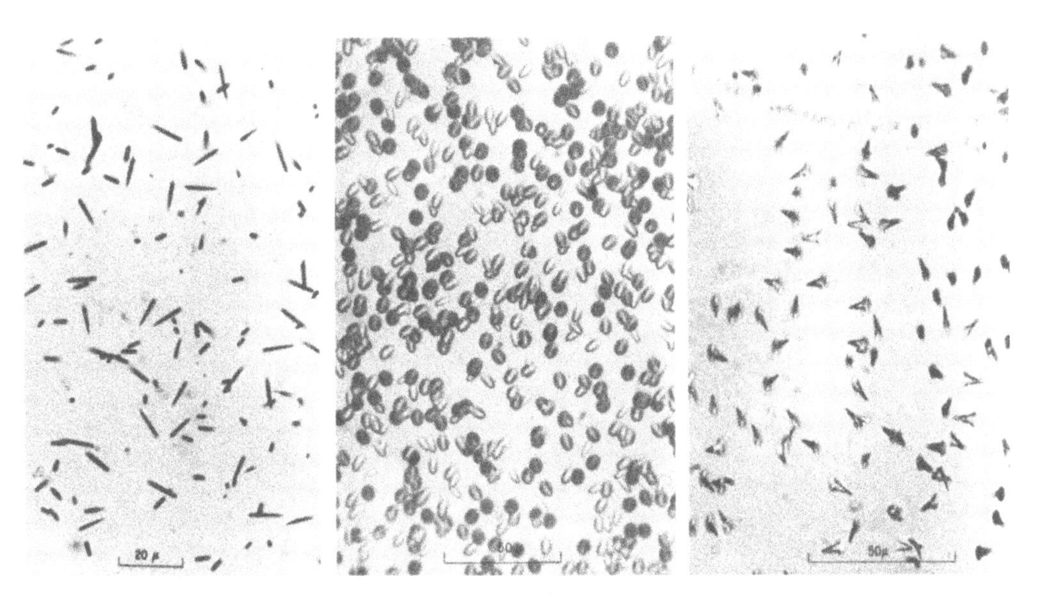

Fig. 1-2. *After chemical etching, particle tracks (in this case from ^{252}Cf fission fragments) can be viewed in a light microscope in a variety of materials: (right) a crystal, orthoclase, a feldspar mineral; (center) an ordinary, inorganic soda-lime glass; (left) an organic glass, Lexan polycarbonate, a high polymer. (After Fleischer et al., 1968.)*

1.2.1. Chemical Etching

By far the most general, useful means of observing particle tracks in solids has been the technique of preferential chemical attack to enlarge and display tracks, as illustrated in Fig. 1-2. The utility of the technique derives from its simplicity—only common chemicals being needed—and from the effective magnification that results from enlarging etched tracks to sizes where they can be viewed with an ordinary optical microscope. The procedure has allowed tracks to be observed in many dozens of substances, using the etching recipes listed in the next chapter. It is also the only procedure that has succeeded in revealing tracks of extremely low energy nuclei in solids—for example, solar wind ions (Burnett et al., 1972) and the recoil nuclei that result from the alpha decay of the heavy elements thorium and uranium (Huang and Walker, 1967). As we shall discuss later, it is believed that at these extremely low energies, $\sim$1 keV/amu (amu = atomic mass unit), the damage results from a different mechanism than is dominant for more heavily charged particles.

1.2.2. Transmission Microscopic Observation

The original observation of fission fragment tracks in micas was made by Silk and Barnes (1959) with a transmission electron microscope (TEM). They used

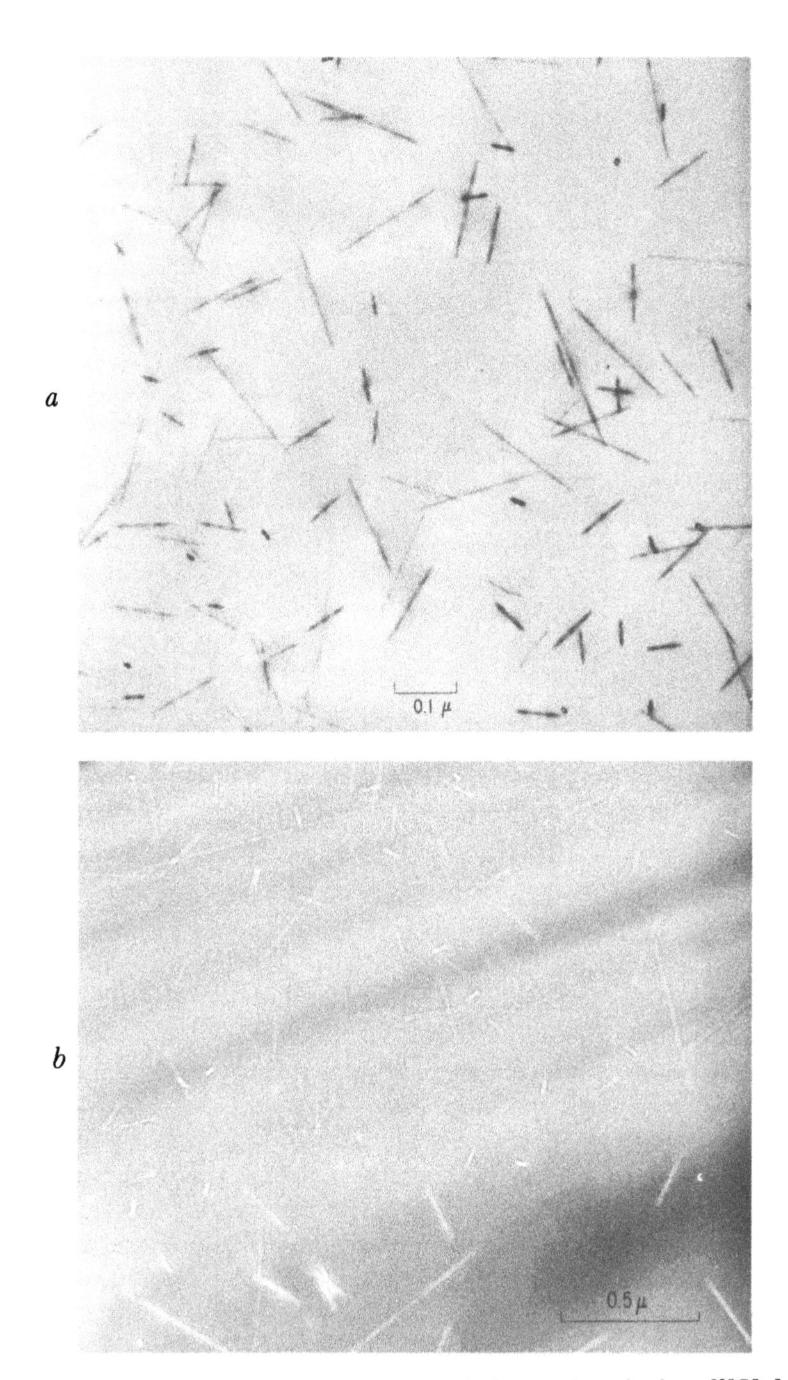

Fig. 1-3. *Transmission electron microscopic image of tracks from* ^{235}U *fission fragments in a synthetic fluor-phlogopite mica.* (*a*) *The periodic variations in darkness along tracks result from a diffraction contrast effect and are not fluctuations in the damage intensity.* (*After Price and Walker, 1962c.*) (*b*) *The light lines are the etched holes in mica viewed by thickness contrast. They show the diameter of the etched holes at their earliest visible stage.* (*After Price and Walker, 1962b.*)

the electron beam in a diffraction contrast mode to observe tracks as dark lines where crystal planes are bent sufficiently to scatter electrons out of the direction of the Bragg reflection that is imaged. The width of 100 to 150 Å for the dark lines in the upper portion of Fig. 1-3 depends on the magnitude of the strain distribution around the tracks and gives therefore an upper limit on the actual size of the disordered region. Tracks have been viewed in bulk materials in this manner by Silk and Barnes (1959), Bonfiglioli et al. (1961), and Price and Walker (1962d). The tracks may also be seen with the TEM by using thickness contrast in viewing etched samples. Here they appear as light lines such as are seen in the lower portion of Fig. 1-3. These give 60 Å as a more restrictive limit that is closer to the true width of the intense damage along fission tracks. TEM observations of unetched tracks by diffraction contrast are inherently limited to crystalline materials. They are conveniently scanned only when very high track densities are present, and in certain substances such as natural micas containing hydroxyl ions, track fading in the TEM makes observation difficult (Silk and Barnes, 1959), so that a cold stage is necessary for proper viewing (Price and Walker, 1962d).

1.2.3. Decoration Techniques

A variety of methods have been used to nucleate precipitates of a second phase along the damage tracks within the primary phase. In general the methods have the virtue that they allow tracks to be displayed throughout a large volume, so that the full lengths of tracks can be measured even though the tracks fail to reach a surface of the detector. Usually, however, the techniques suffer from an erratic variability that depends sensitively on the trace composition of the detector. Consequently, experimental reproducibility is difficult to achieve. In crystals and glasses it is thought that the primary driving force for precipitation at a track is supplied by its strain field. The strain creates sites that are energetically more favorable for oversized or undersized atoms to fit and therefore causes segregation, followed by precipitation that is closely analogous to that observed in the strain fields of dislocations, as has been observed for example in alkali halide crystals (Amelinckx, 1956).

Development of Silver Chloride

Probably the most successful of the decoration procedures to date is the development of tracks in AgCl in the manner shown in Fig. 1-4a. In this procedure, as described by Childs and Slifkin (1962; 1963), ultraviolet light creates free photoelectrons, which are pulsed through the crystal with applied electric fields, converting silver ions into interstitial, metallic silver atoms. These in turn rapidly diffuse through interstitial sites to the damage tracks, where they precipitate as narrow threads of silver. In the last few years Schopper et al. (1973) (also see references therein) have greatly improved the performance of AgCl crystals. By

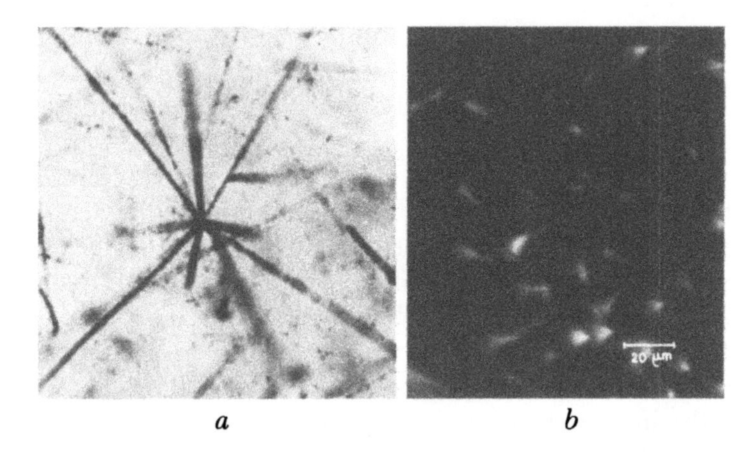

Fig. 1-4. *Examples of track decoration in different systems. (a) A proton-induced star is revealed by silver decoration in a silver chloride crystal. It is thought from the number of prongs in the star that 25 MeV alpha particles are observed. (Courtesy of C. B. Childs.) (b) Tracks of* ^{252}Cf *fission fragments are seen after propenoic acid has been polymerized by grafting onto chain ends along the tracks in cellulose triacetate. (After Monnin and Blanford, 1973.)*

adding 0.5% of Cd to otherwise highly pure AgCl, they achieve a reproducible response in which the mean track widths in AgCl and in K2 emulsion increase in the same way with ionization rate. Glass-backed crystals 200 to 300 μm thick can be grown with dimensions up to 4 × 15 cm. Tracks can be selectively stabilized by a simultaneous irradiation with yellow light and later made visible by an irradiation with light at ~4100 Å. An important advantage over nuclear emulsion is the absence of shrinkage and distortion during development. The sensitivity of AgCl crystals is greater than that of any of the etched detectors so far developed and is likely, therefore, to have some special uses in the future.

Silver Precipitation in Glass

At present the only inorganic, noncrystalline material in which precipitation on particle tracks has been observed is a phosphate glass containing silver oxide: 63 wt% P_2O_5:8% Al_2O_3:11% BaO:9% K_2O:9% Ag_2O (Fleischer and Price, 1963b). After irradiating with fission fragments, the glass was heated to 250°C in a hydrogen atmosphere to reduce the Ag_2O to metallic silver, which precipitated on tracks so as to form 50 Å diameter filaments of silver. This technique has not been utilized because of the small size of the decorated tracks, which must be viewed by TEM, and the associated difficulties in reproducing them.

Iron Precipitation in Mica

Linear filamentary precipitates have been observed in mica by Russell (1967; 1968). He interpreted these preferentially aligned features as tracks formed by showers of electrons that had been created by rare, high energy cosmic rays. For the precipitate atoms (which he believes are iron) to be sufficiently mobile that they segregate to tracks in the crystal, an elevated temperature is needed. Hence, for Russell's interpretation to be correct the electron showers would be ancient events recorded during the cooling of the mica after its formation.

However, Craig et al. (1968) have analyzed the orientation distribution of the supposed tracks and find that the observations are not quantitatively consistent with the known distribution of scattering angles of electrons in cosmic ray showers. In fact, the characteristics of these tracks are consistent with those described and observed for dislocations (Fleischer and Price, 1964a), as we illustrate in Fig. 4-8 and the accompanying discussion. Although the nature of the features in Russell's mica is by no means established, we believe that the dislocation hypothesis is the more plausible explanation.

POLYMERIZATION

Monnin and Blanford (1973) have utilized the high concentration of reactive chain ends and free radicals along tracks in polymer samples (schematically shown in Fig. 1-1b) to initiate polymerization along particle tracks. By using a monomer of a different polymer species they create a region of the new polymer by grafting onto the reactive sites at the track. Fig. 1-4b shows tracks that were made visible in cellulose triacetate by polymerizing propenoic (acrylic) acid at tracks from ^{252}Cf fission fragments. The polymerization can be thought of as producing a cellulose triacetate-propenoic acid co-polymer by grafting. Typically 15 to 20% propenoic acid is added; after irradiation the detector is heated to 55°C for 24 hours (or 70° for 6 hours), the excess monomer is removed and the precipitated phase is dyed with rhodamine B to improve optical contrast. Monnin and Blanford report a 100% detection efficiency for ^{252}Cf fission tracks. As yet no calibrations have been reported that would indicate what sensitivity can be achieved with this new technique. As it also remains to establish what variety of polymers can be utilized, we must regard this as a technique with high promise, but one which needs more extensive study before it can be widely applied.

1.2.4. Detection by Color Change

Ouseph (1973) suggested the use of color centers in a photochromic material to display particle tracks. He noted that light can be used to blacken originally yellow (chemically reduced) strontium titanate crystals containing 0.1% of iron. In his actual experimental work he applied a 1000 volt electric field between the detector and a uranium foil and observed 10 μm-diameter dots that he attributed to alpha particles from uranium oxide. As none were observed from unoxidized uranium, he reasoned that diffusion of oxygen ejected from the oxide played a critical role in the process. It appears that considerable development work would be required to transform Ouseph's idea into a working system for revealing tracks.

1.2.5. Detection with X-Rays

In principle it should be possible to image individual tracks in solids by means of their strain fields, in a manner similar to the x-ray topographic method of imaging dislocations (Lang, 1958). As yet this has not proved possible, possibly because the strain around a track is expected to decrease as r^{-2} with radial distance as compared to r^{-1} for the strain field of a dislocation. Possibly with increased sensitivity future work will allow such tracks to be observed.

On a statistical scale the effects of tracks have been recognized by low angle x-ray scattering techniques. Lambert et al. (1970) found that mica irradiated with $3 \times 10^{12}/\text{cm}^2$ of argon ions of 1 MeV/nucleon showed clear effects of the irradiation. They interpreted the scattering in terms of defects of diameter $\sim$30 Å, separated by $\sim$200 Å, coupled with many much smaller and more numerous defects of a few atomic volumes in size. Seitz et al. (1970) and Seitz (1972) similarly observed x-ray distortion in crystals of various minerals bombarded with heavy ions that produced tracks. The reader should be reminded, however, that even though composed of a heterogeneous distribution of atomic defects, the direct TEM observations on tracks show that the larger scale effects are those of simple cylindrical volume strains.

1.3. NATURE OF THE DAMAGE

Although we are far from knowing with assurance how each chemically reactive defect in the core of a particle track is produced, we can characterize in considerable detail the nature of the defects produced (at least in inorganic solids), the dimensions of tracks, and relative sensitivities and registration properties of different detector materials. We will find that these properties make clear the major fact that most tracks are ionization-produced defects—the result of the interaction of a charged particle with the electrons attached to atoms in the detector. As we will discuss in Section 1.4, there remains the significant question as to how one can best calculate the magnitude of the damage from the ionization and excitation processes along a particle trajectory.

1.3.1. Diameter of the Damaged Region

A fast, charged particle will eject electrons from atoms that were close to its path, giving them a distribution of energies that is strongly peaked toward low energies. Consequently, although the electrons carry a portion of the original energy of the incident particle to sites that are far from its path, most is concentrated close by; and all of the residual effects of the defects left or created where the electrons were removed are at, or near, the core of the track. Characterizing the extent of the most intensely damaged region of the core is an important step in understanding the nature of a particle track.

We have already noted that transmission electron microscopy gives upper limits on the diameter of the region of intense damage along fission tracks, with the most realistic number being the value of ~50 Å observed for micas in the early stages of etching (Price and Walker, 1962b). In fact, the most appropriate diameter to consider for track etching studies is clearly the minimum value that can be etched. The most detailed and quantitative evaluation of diameters comes from measurements by C. P. Bean and coworkers of electrical conductivity across thin detector membranes while transverse particle tracks are being etched through them. We are indebted to Bean for supplying the text from which the remainder of this section (1.3.1) was prepared. The results on muscovite mica are those of Bean et al. (1970). Those on polycarbonate are work by Bean and the late Warren DeSorbo; partial accounts of environmental effects observed in that work have been given by Crawford et al. (1968) and DeSorbo and Humphrey (1970).

Electrical Measurements of Track Diameter

Following a suggestion by R. M. Walker, a study was made of the process of etching of fission tracks in mica using the conductance through the pores as a measure of pore radius. The etchant is used as the conducting electrolyte. Briefly, the concept is that the conductance, Ω, of n pores of radius r and length l filled with a medium of conductivity k is

$$\Omega = n\pi r^2 k/l, \qquad l \gg r. \tag{1-1}$$

Hence, monitoring the conductance through pores that are being etched allows inference of the radius as a function of time. There are a number of factors that may interfere with this simple application of Ohm's law to find the true radius of a pore. Surface conductivity of the pore wall, internal heating, and electrolyte exhaustion all could cause complications. In addition, if not all the damaged regions etch identically or if the pore radius is a function of distance, the simple representation will not suffice. However, the measurements on mica showed that none of these possible sources of error was of importance.

Typical results on mica are shown in Fig. 1-5a (Bean et al., 1970). After a 13 sec delay time the conductivity abruptly jumps from zero (showing that there is no hole prior to etching) to that corresponding to a radius of 33 Å. After a subsequent slow rate of enlargement, the rate assumes the value corresponding to bulk attack at a constant rate v_G in the plane of the sheet of mica. In short, to a first approximation, the experiment showed the existence of a damaged core in which the radial etching rate was thousands of times faster than the radial attack rate in undamaged mica. The diameter of this region of high attack rate near the axis was quite reproducible. There was some evidence of a mild positive dependence of the

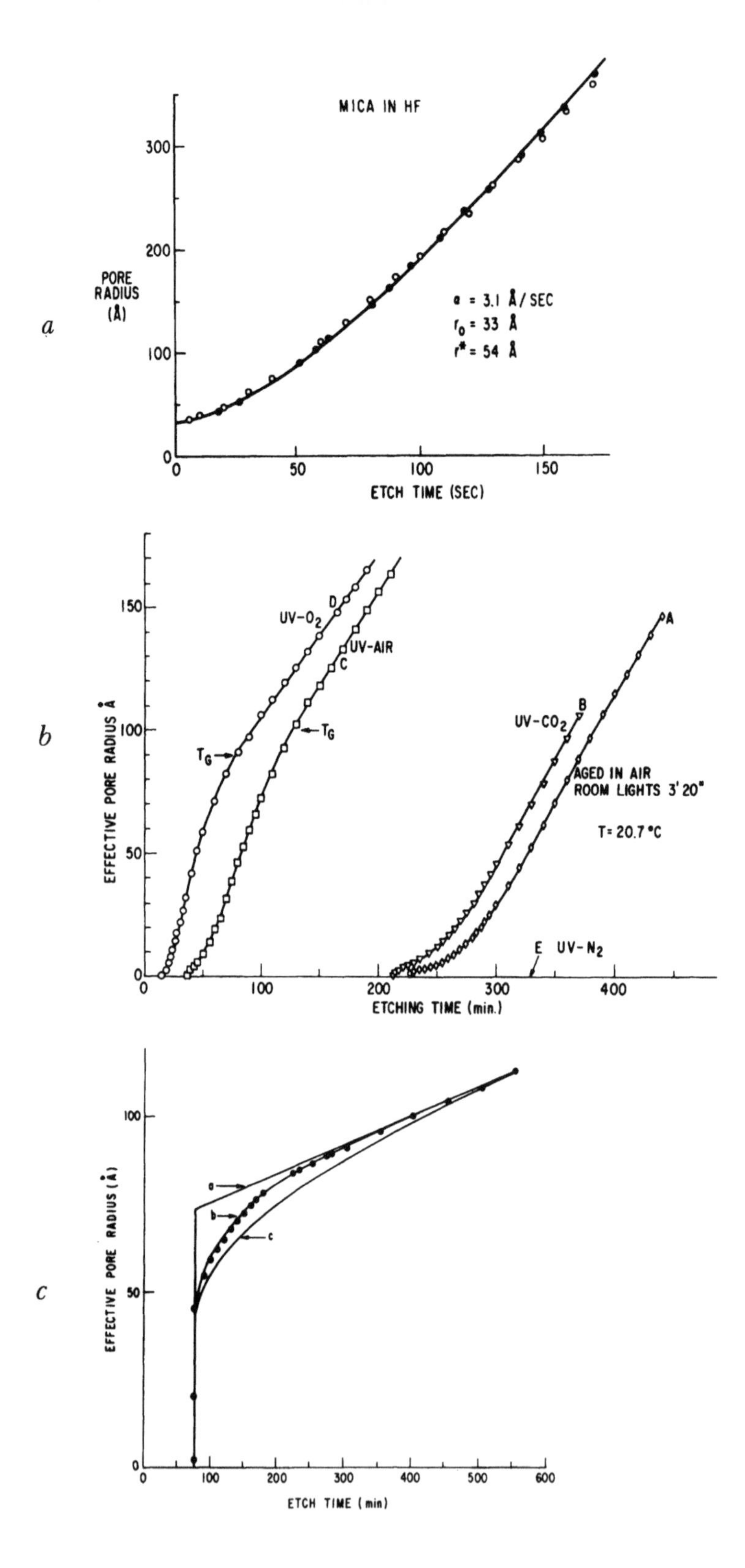

a
MICA IN HF
PORE RADIUS (Å)
a = 3.1 Å/SEC
r_0 = 33 Å
r^* = 54 Å
ETCH TIME (SEC)
b
EFFECTIVE PORE RADIUS Å
UV-O_2
D
UV-AIR
C
T_G
UV-CO_2
B
A
AGED IN AIR
ROOM LIGHTS 3' 20"
T= 20.7 °C
E UV-N_2
ETCHING TIME (min.)
c
EFFECTIVE PORE RADIUS (Å)
ETCH TIME (min)

radius on the energy of the fission fragment, a finding that is consistent with the higher ionization that occurs at higher energy.

The transient region of lower radial attack rate dr/dt (following the original breakthrough) is interpreted in terms of the extra work needed to create new surface. This work is relatively more important at small radii because the surface-to-volume ratio of the pore is highest. The variation of radius with time is fit by a solution of Bean's theoretical equation $dr/dt = \alpha \exp(-r^*/r)$, where r^* is a constant that depends on both the surface energy and temperature and α is a second constant.

Bean and DeSorbo used the same technique to characterize the etching process of fission tracks in polycarbonate plastic. Preliminary experiments showed a qualitative difference from the experience with mica in that the appearance of conductance (corresponding to the meeting of the regions etching from the opposite faces of the plastic) was not a sharp event and depended on the pretreatment of the polycarbonate film (Crawford et al., 1968; DeSorbo and Humphrey, 1970). To display this behavior we plot in Fig. 1-5b, as a function of time, the effective radius r_{eff} (which is proportional to the square root of the conductance) for the etching of about three thousand tracks in each case. The effective radius is defined by rearrangement of the previous equation as

$$r_{eff} = (l\Omega/n\pi k)^{1/2}. \tag{1-2}$$

The figure shows that a sample aged in air and room lights after irradiation has a delayed and prolonged pore breakthrough time, whereas one treated with ultra-

Fig. 1-5. (Facing page). *Growth of holes as measured by electrical conductivity through pores.* (*a*) *Inferred radius as a function of time for two irradiated muscovite mica samples, each of thickness 4.2 μm, etched in 34% HF at 25°C. Solid points are for one sample, open circles are for the other sample, and the solid curve is a theoretical result discussed in the text. The derived original radius when the etchant first penetrates is 33 Å.* (*After Bean et al., 1970.*) (*b*) *Effective pore radius in polycarbonate film as a function of etching time for various post-irradiation treatments. All etching was done at 20.7°C in 3.1 N NaOH with 0.5 volume percent Benax 2A1 surfactant. The UV source had a peak at 3700 Å. The various treatments were all for 3 hr 20 min. The time T_G denotes the time of last pore breakthrough of the 3000 tracks in each sample. Curve E shows there was no pore breakthrough for at least 320 minutes if the sample was exposed immediately after irradiation to UV in a pure nitrogen atmosphere.* (*Bean and DeSorbo, unpublished.*) (*c*) *Effective pore radius as a function of etching time for a single pore in polycarbonate. The radius is inferred from electrical measurements of pore resistivity during the etch process. The solid curves are for various assumed radial dependences of decreased activation energy for etchant attack. Curve a is a square well of depth 0.20 eV and radius 71 Å. Curve b is an arbitrary best fitting function,* $-0.20(1 - (r/97\,Å)^{1/3})$ *eV and curve c is* $0.20 \exp(-r/27Å)$ *eV.* (*Bean and DeSorbo, unpublished.*)

violet and oxygen has its average breakthrough time decreased by a factor of ten and, in addition, has a much more pronounced step in conductance at breakthrough. The residual dispersion in breakthrough times most probably arises from the fact that fission fragments have a spectrum of masses and energies. In any event, it is necessary to follow the etching of single pores to obtain a more accurate picture of the kinetics of etching. An example of such an experiment is shown in Fig. 1-5c. In this case, the polycarbonate film was etched in 3.1N NaOH at 7.2°C. It took 75 minutes for the track to etch through the 10.3 μm thickness of the film; therefore the maximum attack rate was 686 Å/min. This is designated as the track attack rate, v_T. The general attack rate, v_G, in undamaged material was 0.078 Å/min and hence v_T/v_G was almost 10^4. An examination of the temperature dependence of v_T and v_G over the range of 7°C to 55°C yields Van't Hoff dependences:

$$v_G = v_0 \exp(-0.87 \pm .03 \text{ eV}/kT) \quad \text{and} \quad v_T = v_0 \exp(-0.69 \pm .03 \text{ eV}/kT),$$

where v_0 is, within a factor of three, $2 \cdot 10^5$ cm/sec. (A first-order explanation of this pre-exponential factor is that it may be considered as the product of a vibration frequency of roughly 10^{12}/sec and an effective molecular thickness of tens of Ångstroms.) The effect of radiation damage, therefore, is primarily to lower the activation barrier for scission and removal of polymer molecules. Put another way, a fraction of the energy dissipated by a charged particle is stored in the material and puts the damaged material into a higher energy state where it is more susceptible to attack. A quantity of vital interest to both the question of the mechanics of radiation damage and the use of this track-etching process to make pores of extremely small diameter is that of the radial extent and distribution of the damage. Fig. 1-5c shows that once the effective pore radius exceeds 85 Å or so, the pore growth rate is constant and equal to v_G. Thus, this distance represents a limit to the radial extent of severe damage.

This observation may be made more quantitative by the assumption of specific models for the etching rate as a function of distance from the track and comparison of the results so derived with the data of Fig. 1-5c. In particular we assume arbitrary forms of an activation energy function, $E(r)$, such that

$$v_0 \exp(-E(0)/kT) = v_T \quad \text{and} \quad v_0 \exp(-E(\infty)/kT) = v_G. \tag{1-3}$$

With the assumption of any given function—say, a gaussian depression in activation energy—one can calculate pore radius as a function of local etch time. This result cannot be compared directly with the experimental results of Fig. 1-5c because local times of etching vary along the pore, i.e., it has a taper. As Maxwell (1904) has shown, for a smoothly tapering pore eq. (1-1) can be replaced by

$$\Omega^{-1} = \int_0^l dz/k\pi r^2(z) \tag{1-4}$$

where z measures distance along the pore. Thus, for a specific form of $E(r)$, $r(z, t)$ can be calculated and, through eq. (1-4), the conductance can be predicted. Lastly, the effective radius is calculated by the definition of eq. (1-2).

Fig. 1-5c shows the results of a best fit for various functional forms. For instance, the best fit for a square well model (a) occurs when it is assumed to have a range of 71 Å, but it is clearly not a close representation of the state of affairs. An exponential well (c) also differs from experiment. A quite close fit is obtained by the function used for curve (b):

$$\begin{aligned} E(r) &= E(\infty) - 0.2\ \text{eV}\,(1 - (r/97\ \text{Å})^{1/3}), & r &\leq 97\ \text{Å} \\ &= E(\infty), & r &> 97\ \text{Å} \end{aligned} \qquad (1\text{-}5)$$

a relation that has purely empirical significance. This function gives implicit values for the pore profile as a function of time. For instance, ten minutes after breakthrough the pore radius at the surface is predicted to be 66 Å and the constriction at the center is predicted to be 34 Å in radius.

Fig. 1-6 shows the local etching rates as a function of radial distance for the various models used in Fig. 1-5c. The attack rate is seen to be a strong function of distance—falling for model (c) by a factor of ten in 10 Å.

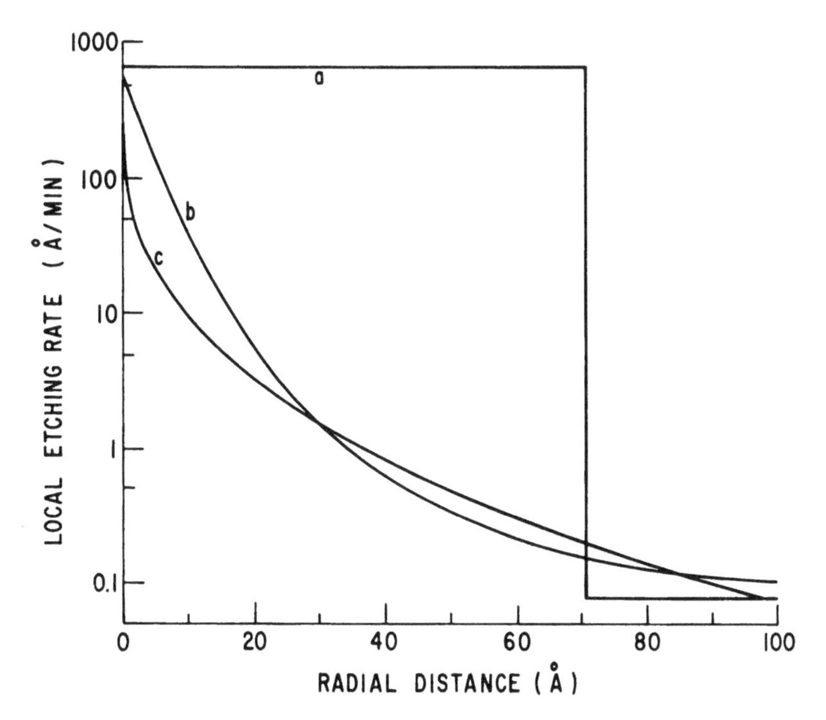

Fig. 1-6. *Radial attack rate as a function of radial distance for various models. Curve a assumes an exponential well of characteristic distance 28 Å. These are best fits to the data of Fig. 1-5c. (Courtesy of C. P. Bean.)*

STORED ENERGY

The total stored energy per unit length of track that is effective in reducing the activation energy for etching can be calculated with one additional assumption. If $\Delta E(r)$ is the reduction of activation enthalpy for each site and there are n activation sites/unit volume then the stored energy per unit length is $-n \int_0^\infty \Delta E(r) 2\pi r dr$. In turn, if ρ_d is the density, M is the molecular weight of the activating unit and N_A is Avogadro's number, then $n = \rho_d N_A/M$. Using the general form of activation energy reduction given in eq. (1-5), the effective stored energy/unit length is $\rho_d N_A E(0) \pi r_0^2/5M$. Using the appropriate values with an arbitrary assumption that M is 100, this is approximately 700 MeV/cm. The energy dissipation by a typical fission fragment is $\sim 0.5 - 1.0 \times 10^5$ MeV/cm. The result is that about one percent of the total energy dissipation is stored in such a way as to reduce the activation energy.

1.3.2. Relative Thresholds for Detection

Particle tracks may be formed in bulk samples of virtually any insulating material but not in metals or other good conductors (Fleischer et al., 1965a). Table 1-1 indicates the categories of track-storing and non-track-storing materials and shows that there appears to be a correlation with electrical resistivity, such that materials with values above about 2,000 ohm-cm generally store tracks. As we will discuss

Table 1-1. Relation of Track Formation to Electrical Resistivity (after Fleischer et al., 1965a)

Minerals			Resistivity Range (ohm-cm)
I.	Track-Forming		
	Insulators:	Silicate Minerals Alkali Halides Insulating Glasses Polymers	$10^6 - 10^{20}$
	Poor Insulators:	MoS_2	3,000 - 25,000
	Semiconductors*:	V_2O_5 glass	2,000 - 20,000
II.	Non Track-Forming		
	Semiconductors*:	Germanium Silicon	10 - 2,000
	Metals:	Aluminum Copper Gold Platinum Tungsten Zinc	$10^{-6} - 10^{-4}$

*In a series of thin film experiments Morgan and Chadderton (1968) found they could observe tracks in the semiconducting compounds MoS_2, $MoTe_2$ (α and β), WSe_2, $MoSe_2$, WS_2, and WTe_2 but not in $TiSe_2$, $TaSe_2$, $NbSe_2$, $TaTe_2$, and $NbTe_2$.

shortly, we believe that the mobility of positive current carriers is a more relevant quantity than resistivity, which we have quoted in Table 1-1 primarily because it is a simple, easily measurable, available number.

For irradiated thin films Morgan and Chadderton (1968) found a similar resistivity threshold for the appearance of tracks in a series of semiconducting compounds, but at a lower value, somewhere in the range 0.0004 to 0.003 ohm-cm. It is not yet known whether this difference is merely due to the different character of track formation in thin films alluded to earlier, whether it is a consequence of resistivity not being the real fundamental parameter of interest, or whether it is the special property of tracks that are nearly parallel to the planes of the layer crystals used. In addition, other measurements of the conductivities of several of the compounds involved (for example, Champion, 1965) give values that often differ by several orders of magnitude from those quoted by Morgan and Chadderton, and hence it is by no means clear whether quantitative significance can be given to their resistivity threshold for track registration.

For materials in which tracks can be formed, quantitative measurements can be made to establish whether particular particles produce tracks or not and consequently to decide the relative sensitivities of different detectors. For example, Fig. 1-7 shows that after etching, the tracks of a beam of ^{32}S ions of 139 MeV energy are clearly revealed in Lexan.

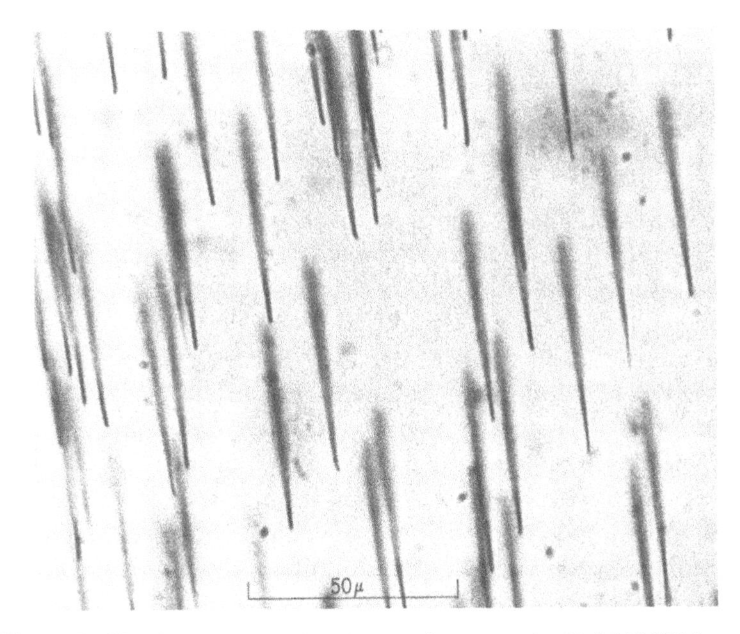

Fig. 1-7. *Tracks in Lexan polycarbonate of a beam of 139 MeV ^{32}S ions from the Berkeley heavy ion accelerator are revealed by etching. A series of such experiments with other ions and energies allows the damage threshold for registration to be defined for this detector material, as shown in Fig. 1-8. (After Fleischer et al., 1964a.)*

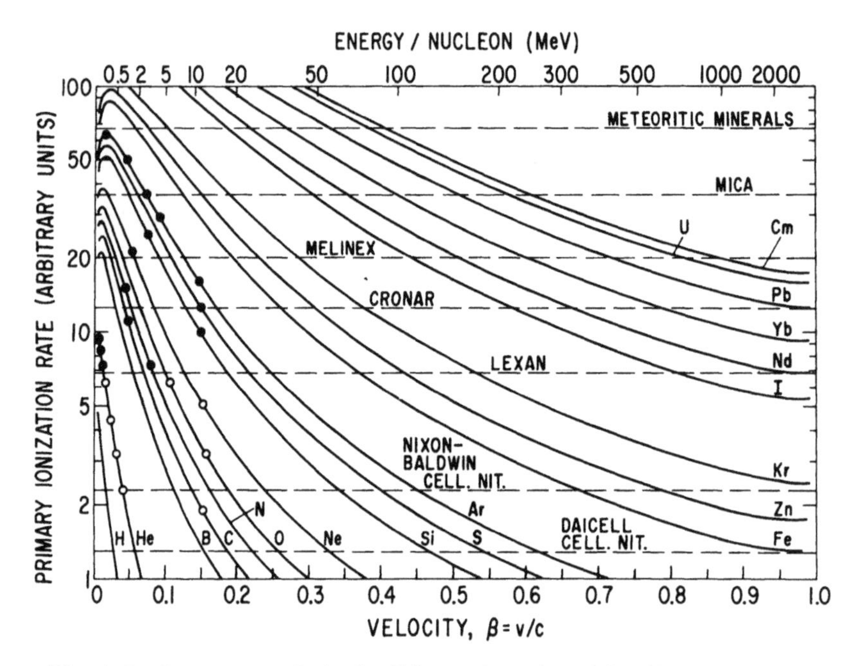

Fig. 1-8. *Damage vs. velocity for different charged particles. Each detector has a level below which no tracks are etched and one above which all particles create tracks. The experimental points for accelerator ions in Lexan polycarbonate are given as open circles for zero registration and as filled circles for 100% registration. Thresholds for other detectors are also indicated.*

A series of such measurements with different charged particles and different energies is used to define the registration properties of a substance (Fleischer et al., 1964a; 1967a). For example, Fig. 1-8 gives theoretical curves of the relative damage caused by different ions as a function of their velocities. Superimposed on these curves (whose meaning and derivation will be considered later) are experimental points for Lexan polycarbonate, where each point corresponds to a single track observation of the sort illustrated in Fig. 1-7. The dotted line labeled "Lexan" separates the region of low damage and zero registration from that of unit registration. The variety of responses of different materials is shown by the other threshold levels indicated in Fig. 1-8. The figure indicates the wide variation in response, which extends from that of cellulose nitrate, which will register low energy protons (Jones and Neidigh, 1967) to those of minerals so insensitive that they will not register argon even at its maximum ionization rate (Fleischer et al., 1966; 1967b).

Table 1-2 lists in approximate order of increasing sensitivity the materials for which we have relevant data (Fleischer et al., 1964a; 1965a, b; 1967a, b; 1969b; Debeauvais and Monnin, 1965; Monnin et al., 1966; Monnin, 1968; Price et al., 1968a; 1973; Blanford et al., 1970; Varnagy et al., 1970; Plieninger et al., 1972;

Lecerf and Peter, 1972; Endo and Doke, 1973; Shirk, 1974). Dotted lines separate detectors that are likely to have differing sensitivities; solid lines separate detectors for which clear differences have been observed. It must be emphasized that this table is of only qualitative value for two basic reasons. Firstly, the data come from a variety of sources with differing experimental conditions and choices of calibration particles used and are not intercomparable in a simple way. Secondly, the observed registration behavior of a material can depend on the particular etching conditions (Fleischer and Price, 1963a; Somogyi et al., 1968; Price et al., 1973), the particular formulation of a given type of plastic (Price et al., 1968b), and exposure to various environmental conditions (as described in Chapter 2). Therefore, the table should be regarded as giving the typical sequence of sensitivities

Table 1-2. Relative Sensitivities of Various Detectors

A. Inorganic Detectors

Detector	Atomic Composition	Least Ionizing Ion Seen
Hypersthene	$Mg_{1.5}Fe_{0.5}Si_2O_6$	100 MeV ^{56}Fe
Olivine	$MgFeSiO_4$	
Labradorite	$Na_2Ca_3Al_8Si_{12}O_{40}$	
Zircon	$ZrSiO_4$	
Bronzite	$Mg_{1.7}Fe_{0.3}Si_2O_6$	
Enstatite	$MgSiO_3$	
Diopside	$CaMg(SiO_3)_2$	170 MeV ^{56}Fe
Augite	$CaMg_3Fe_3Al_2Si_4O_{19}$	170 MeV ^{56}Fe
Oligoclase	$Na_4CaAl_6Si_{14}O_{40}$	4 MeV ^{28}Si
Bytownite	$NaCa_4Al_9Si_{11}O_{40}$	4 MeV ^{28}Si
Orthoclase	$KAlSi_3O_8$	100 MeV ^{40}Ar
Quartz	SiO_2	100 MeV ^{40}Ar
Phlogopite Mica	$KMg_2Al_2Si_3O_{10}(OH)_2$	
Muscovite Mica	$KAl_3Si_3O_{10}(OH)_2$	2 MeV ^{20}Ne
Silica Glass	SiO_2	16 MeV ^{40}Ar
Flint Glass	$18SiO_2:4PbO:1.5Na_2O:K_2O$	2-4 MeV ^{20}Ne
Tektite Glass (Obsidian similar)	$22SiO_2:2Al_2O_3:FeO$	
Soda Lime Glass	$23SiO_2:5Na_2O:5CaO:Al_2O_3$	20 MeV ^{20}Ne
Phosphate Glass	$10P_2O_5:1.6BaO:Ag_2O:2K_2O:2Al_2O_3$	

Table 1-2 (*continued*)

B. Organic Detectors

Detector	Atomic Composition	Least Ionizing Ion Seen
Amber	$C_2H_3O_2$	Full-energy fission fragments
Phenoplaste	C_7H_6O	
Polyethylene	CH_2	Fission fragments
Polystyrene	CH	
Polyvinylacetochloride	$C_6H_9O_2Cl$	42 MeV ^{32}S
Polyvinylchloride - Polyvinyledene chloride copolymer	$C_2H_3Cl + C_2H_2Cl_2$	42 MeV ^{32}S
Polyethylene Terephthalate (Cronar, Melinex)	$C_5H_4O_2$	
Polyimide	$C_{11}H_4O_4N_2$	36 MeV ^{16}O
Ionomeric Polyethylene (Surlyn)		36 MeV ^{16}O
Bisphenol A-polycarbonate (Lexan, Makrofol)	$C_{16}H_{14}O_3$	0.3 MeV ^4He
Polyoxymethylene (Delrin)	CH_2O	28 MeV ^{11}B
Polypropylene	CH_2	1 MeV ^4He
Polyvinylchloride	C_2H_3Cl	
Polymethylmethacralate (Plexiglas)	$C_5H_8O_2$	3 MeV ^4He
Cellulose Acetate Butyrate	$C_{12}H_{18}O_7$	
Cellulose Triacetate (Cellit, Triafol-T, Kodacel TA-401 unplasticized)	$C_3H_4O_2$	
Cellulose Nitrate (Daicell)	$C_6H_8O_9N_2$	0.55 MeV ^1H

Notes: Solid lines represent relatively clear separations; broken lines represent unclear but likely separations.

Many materials have different sensitivities depending on their exposure to oxygen, ultra-violet light, ..., and on the etchant used.

with differences between any two detectors being more reliable the farther apart in the table they are. Although the plastics in general are the most sensitive class of materials, there is some overlap with the sensitivities of crystals and inorganic glasses. For example, amber is less sensitive than diopside (Uzgiris and Fleischer, 1971). Because there are not sufficient data intercomparing the organic and the inorganic detectors, they are listed as separate subsections of the table.

1.3.3. Activation Processes in Track Annealing

The repair of the complicated atomic structure of a particle track is governed by a complicated series of atomic processes. Nevertheless, some simple inferences can be drawn from the results of measuring the kinetics of track repair in solids. In the simplest case (Fleischer and Price, 1964b) pieces of a solid containing tracks are heated at a series of temperatures (T), and the time (t) for total fading at each temperature is determined. Since the results normally fit a Boltzmann equation of the form $t = A \exp (E_{act}/kT)$, where k = Boltzmann's constant and A is another constant, an energy of activation (E_{act}) for total fading can be determined. By comparing this energy with known atomic and molecular processes it is hoped that the actual controlling mechanism or mechanisms can be identified.

Removal of tracks by annealing is frequently more complicated. Partial removal is often the result of heating for times that are less than times for complete erasure. If quantitative measurements of the fraction of tracks remaining are made (as Naeser and Faul (1969) were the first to show), a series of activation energies are inferred. Low activation energies are measured in the initial stages of annealing and monotonically higher values are found as more complete track removal occurs.

Significance of a Spectrum of Activation Energies

A simple, qualitative interpretation of the increasing activation energies follows from the idea that a track in an inorganic solid is a region with a continuous array of atoms in wrong positions (i.e., positions of high free energy). Not only are some atoms in sites from which they can be more easily moved by supplying thermal energy than are others, but most disturbed atoms are initially adjacent to atoms that are also not in normal sites. Consequently, the first to move and return to normal sites are aided by the high distortion energy of their surroundings and rearrange themselves rapidly with low activation energies. In the later stages of annealing, atoms that diffuse with higher activation energies have time to migrate and do so in a more nearly perfect matrix.

Significance of the Magnitude of Activation Energies

The actual values measured for annealing of tracks are given in Chapter 2, which discusses the effects of the environment on tracks in detectors. The values observed for complete erasure of tracks are usually at least two electron volts and are similar to the known or expected activation energies for atomic or ionic diffusion.

Fig. 1-9 gives a particularly informative example of annealing kinetics in an inorganic, semiconducting glass. The observed activation energy for track fading is a factor of five greater than that for the motion of electronic current carriers and strongly implies, then, that the lasting damage along a track is not electronic in nature. These data, along with the observed magnitude of activation energies in

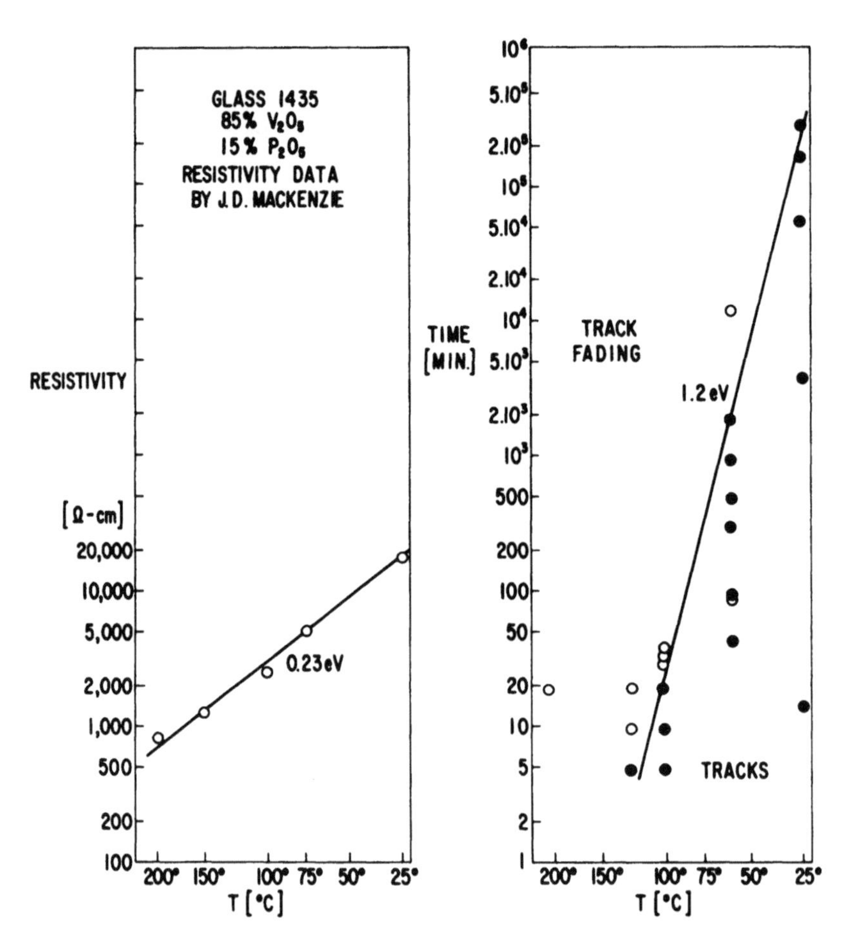

Fig. 1-9. *Variation of electrical conductivity and of track annealing time with temperature for $P_2O_5 \cdot 5V_2O_5$ glass. In the right hand graph solid dots indicate conditions for retention of tracks, open circles conditions for track fading.* (*After Fleischer et al., 1965a.*)

the inorganic materials, make clear that the lasting, chemically etchable damage along particle tracks consists of atomic disorder.

1.3.4. Range Deficits

One of the noteworthy properties of particle tracks is the fact that, in the materials with the highest thresholds for registration, the distance to the point where a track-forming particle comes to rest exceeds the length over which preferential etching is observed. The difference, called a *range deficit*, was recognized by Fleischer et al. (1964a) and Maurette (1966a) on the basis of studies with fission

Table 1-3. Range Deficits in Minerals of Different Sensitivities (data from Price et al., 1968a)

Mineral	Ion	Range Deficit (μm) (±0.5μm)
Mica	Fe	1.2
	Br	1.6
	I	1.2
Diopside	Fe	4.0
	Br	3.5
	I	3.4
Hypersthene	Fe	4.7
	Br	4.6
	I	4.1

fragments and later put on a firmer basis by Price et al. (1968a) using Fe, Br, and I ions of known energy. Although it is possible that a portion of the deficits observed is in fact caused by uncertainties in range-energy relations, the physically reasonable, systematic nature of the observations shown in Table 1-3 supports the conclusion that the deficit is real and that its magnitude increases as the sensitivity decreases. Other observations with lower starting energies obviate the need to use possibly imperfect range-energy relations. Thus Price (unpublished) observed tracks from 4 MeV ^{40}Ar in orthoclase but none in zircon, the less sensitive of the two; and Woods, Hart, and Fleischer (unpublished) observed tracks of 56 keV ^{56}Fe ions in Lexan polycarbonate, but not in phosphate glasses, which are less sensitive. All the above experiments are consistent with the existence of range deficits of increasing magnitude for detectors of decreasing sensitivity. These observations will be helpful in ruling out a mechanism of track formation that will be discussed in Section 1.4.

1.4. FORMATION MECHANISMS

Any detailed theory of how tracks form must fit the facts we have reviewed in the preceding sections: Particle tracks are narrow (<50 Å radius), stable, chemically reactive centers of strain that are composed mainly of displaced atoms rather than of electronic defects. They are not formed in good electronic conductors, and there is a particular sequence of sensitivities among the solids that do record tracks. The sensitivities are also correlated with a failure to record the last few microns of range in the least sensitive detectors.

1.4.1. General Description of Heavy Ion Energy Deposit in Solids

The means by which heavy ions lose energy as they slow down and come to rest in a solid is central to any attempt to understand track formation, and we therefore briefly review this subject.

A fast atom of atomic number Z moving through a solid would rapidly become an ion by being stripped of all or some portion of its orbital electrons. This stripping is a result of interaction of the electrons surrounding the moving atom and those around the atoms that make up the solid. From these interactions the ion acquires a net positive charge Z^* (in units of the electron's charge), an empirical form for which is given (Heckman et al., 1960) by

$$Z^* = Z[1 - \exp(-130\beta/Z^{2/3})], \qquad (1\text{-}6)$$

where β is the speed v of the ion relative to that of light. In moving in the solid the ion undergoes two types of collisions, the relative frequency of which is a strong function of velocity. At high velocities, where $Z^* \approx Z$, by far the dominant interaction is the electrical force between the ion and electrons attached to atoms within the solid. The effect of this force is either (1) to excite electrons to higher energy levels or (2) to loosen them from their atoms and eject them. In polymers de-excitation following process (1) can lead to breaks in the long chain molecules and to free radical production (Bovey, 1958). In any solid, process (2), ionization, creates charge centers. The ejected electron, called a *delta ray*, can produce further excitation and ionization if it carries enough energy. The original or *primary* ionization and excitation occur close to the path of the ion, while the secondary ionization and excitation are spread over larger radial distances from the core of the track. In the simplest case where the ion velocity is large compared to the electronic orbital velocity, the electrons can be treated as though they were originally at rest and the energy given the electrons is inversely proportional to the square of the impact parameter b (the distance between the ejected electron and the path of the ion). Since πb^2 is the area around a particle track within which electrons receive at least a given energy, it follows that relative to those electrons with an energy E only a fourth as many electrons receive energy $\geq 2E$, one sixteenth have energy $\geq 4E$, and so forth. In short, most of the ejected electrons move short distances from the core of the track region and only a very few move large distances corresponding to the kinetic limit in electron energy of $2mv^2/(1 - \beta^2)$, where m is the electron mass. Fig. 1-10, from calculations by Katz and Kobetich (1968), shows how close-in the energy lost by the delta rays is deposited and makes obvious the effect of the variation with velocity of the kinetic limit on electron energy.

When an ion slows down in passing through a solid, it eventually reacquires orbital electrons one-by-one as its velocity becomes comparable with the orbital velocities of less and less tightly bound electrons. Below ~50 keV/amu atomic

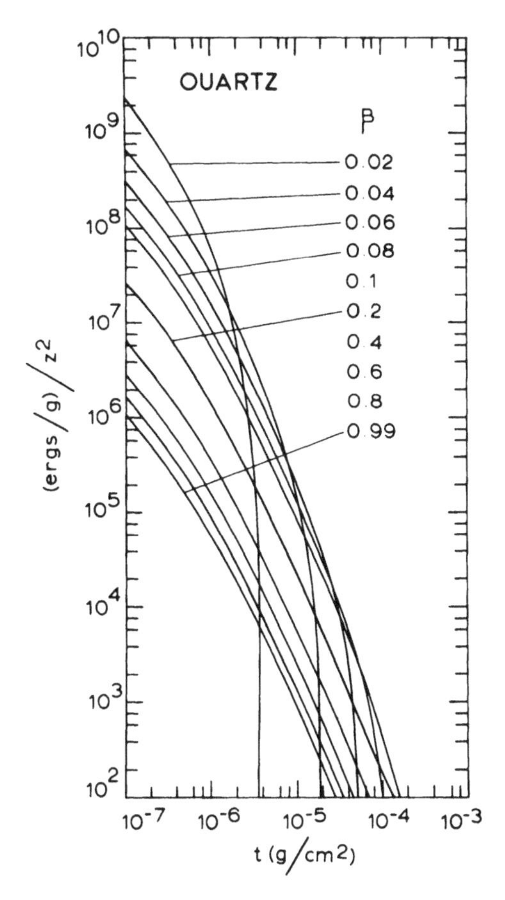

Fig. 1-10. *Computation by Katz and Kobetich (1968) of the spatial distribution of the energy deposited by delta rays around a particle track in silica.*

collisions (interactions of the moving ion with whole atoms or ions in the stopping medium) become the more dominant mode of energy loss. Lindhard and Scharff (1961), Lindhard and Thomsen (1962), and Lindhard et al. (1963) have considered the slowing-down of ions and the partition of energy into atomic and electronic stopping. The damage produced by atomic collisions consists of displaced atoms and the resultant vacancies.

An important question, whose resolution would be a major step toward quantitative calculation of the total damage along a track, is the relative importance (a) of the primary damage that results from the excitation and ionization caused directly by the heavy ion and (b) of that part of the secondary damage that is caused by the delta rays during their passage close to the path of the ion.

In attempting to answer this question we must consider separately the inorganic solids (crystals and glasses) and the organic solids (the high polymers). Experimental tests that we may make use of here are, firstly, measurements of the effects of electron irradiation on chemical dissolution rates and, secondly, measurements of the extent of the radial distribution of etchable damage, each of these being a measure of the effects of delta rays.

Crystals and Glasses

The inorganic solids make the simpler case because of their extreme insensitivity to electron bombardment. In contrast to the high polymers, which are severely damaged by $\sim 10^9$ erg/g of electron bombardment (Bovey, 1958; Charlesby, 1960), doses of 1.5 MeV electrons orders of magnitude greater than 10^9 erg/g applied to the several minerals that have been examined have no detectable effects on etching rates (Fleischer et al., 1965c), even though such high-energy electrons can cause atomic displacements. On silica glass, the one inorganic substance where we know of a positive effect, $\sim 10^{14}$ ergs/g were required to loosen the oxygen-silicon bonds (Sigsbee and Wilson, 1973) and increase the etching rate significantly (O'Keeffe and Handy, 1968; Krätschmer, 1971). Even then the rate is not as large as that along the tracks of heavy ions (Krätschmer, 1971). We should nevertheless note the effect of $\sim 3 \times 10^{12}$ ergs/g of laser light in producing etchable damaged regions in silica glass (Uzgiris and Fleischer, 1973). Here the etching rate is above that for fission tracks. It is believed, however, that the laser light creates a burst of sudden, dense ionization much like that along a particle track, so that *dispersed* ionization is not involved. Bean et al. (1970), in the high resolution experiments described earlier, saw no evidence for a radial variation of the etching rate along the track—such as one would expect from the pattern of the energy deposition by delta rays. Finally, we noted earlier in relation to Fig. 1-9 that the damage does not consist of electronic defects, which would be the most common product of irradiation by electrons. Thus the evidence is strong that the secondary effects of delta rays are unimportant in inorganic solids. By inference, the remaining effect, that of primary ionization, appears to be the major source of track damage. Maurette (1970), in a brief critical review, and Seitz (1972), in his thesis work, have come to the same conclusion. Possible mechanisms and their implications will be discussed shortly.

Polymers

For plastics the effects of delta rays can definitely not be neglected. Firstly, the work of Bean and DeSorbo presented in Section 1.3.1 shows that, for fission tracks in polycarbonate, accelerated chemical attack extends to a distance of ~ 86 Å, which is well outside the region of primary excitation and ionization. Secondly, the electron dose along a fission track exceeds that necessary to produce major damage at radial distances out to 100 to 200 Å according to the calculations of Kobetich and Katz (1968), Katz and Kobetich (1968), Baum (1969), and Fain et al. (1974). It is not known definitely what the relative importance of primary ionization is for polymers. However, since in the inorganic solids primary ionization is the major source of track damage, it is probable that both primary and secondary ionization (and excitation) contribute in the polymers and consequently will need to be considered in a complete theory. We include excitation as well as ionization in the above statement because it is known that excitation can lead to chain breaks and therefore to a reduced molecular weight (Bovey, 1958) and it is also known that the etching rate of a polymer increases with a decrease in its average molecular weight (Fleischer et al., 1965a).

We now consider some of the proposed mechanisms in more detail.

1.4.2. Unrealistic Mechanisms

DIRECT ATOMIC DISPLACEMENTS

Perhaps the first, most obvious thought about how particle tracks might come into existence is that direct atomic collisions produce interstitial atoms and vacant atomic sites either as a trail of nearby separate defects (Seitz, 1949; Lindhard and Scharff, 1961; Lindhard and Thomsen, 1962; Lindhard et al., 1963) or as a final dense clump of damage (a *displacement spike*) at the end of the trajectory where the mean free path for collision equals the atomic spacing (Brinkman, 1955). We know with certainty that these direct collisions with atoms are not the usual cause of tracks from charged particles since (a) they would be expected to occur equally in conductors and insulators, and (b) they become more prevalent near the end of the range of a charged particle (where we find that tracks often do not form). Point (b) has been further supported by detailed calculations by Maurette (1966b).

For one special class of tracks, however, it is likely that atomic displacements are important—very low-energy, heavy particles such as solar wind particles of mass $\gtrsim 50$ amu (Burnett et al., 1972; Walker et al., 1973) and heavy recoil fragments (mass $\sim$200 amu) such as result from the alpha decay of heavy nuclides (Huang and Walker, 1967). In each case the energies are $\sim$1 keV/amu and the track lengths of a few hundred Ångstroms that are observed are compatible with the dimensions expected for displacement spikes. An important experiment to test whether this is the relevant mechanism would be the observation of such tracks in metals.

Fig. 1-11 shows schematically the regimes of damage from direct displacements, from ionization, and from their sum; it compares them with the damage thresholds for etching in three detectors. For the entire range shown, damage density is above threshold in Lexan and therefore there is no range deficit. Below $\sim$4 μm residual range there is no etchable damage in hypersthene, which therefore has a sizable range deficit. For mica it appears possible that there is a damage gap between the region of decreasing displacement damage and increasing ionization damage, giving both range deficits for the more energetic particles and displacement spikes where they come to rest. Although theory does not always give a minimum in energy loss between electronic and atomic stopping regions (see Fig. 3-19), it is

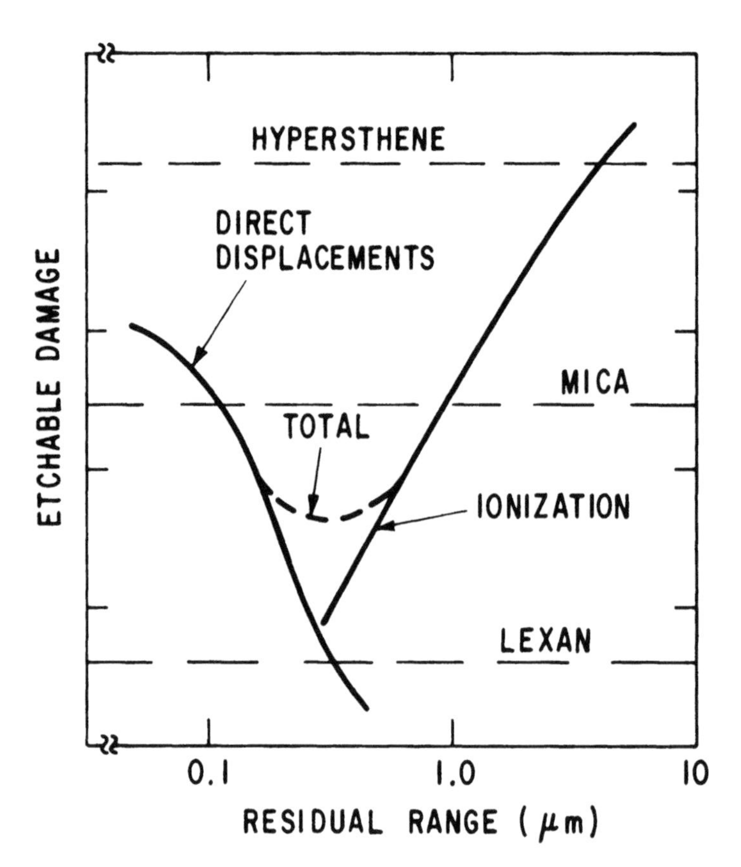

Fig. 1-11. *Etchable damage from a stopping heavy nucleus partitioned into the effects of ionization and of atomic displacements at low energies. Only for the detectors with low thresholds, such as Lexan polycarbonate, is a range deficit not observed. Mica may have a gap in registration between ~0.1 μm and 1 μm residual range. (Schematic.)*

still possible that there may be a minimum in etchable damage as hypothesized in Fig. 1-11 and suggested by experimental data.

Thermal Spike

The region of intense ionization and excitation along the path of a charged particle might be envisioned as a narrow cylinder of material that is rapidly heated to a high temperature and then rapidly quenched by thermal conduction into the surrounding matrix, possibly disordering the core material (Seitz, 1949; Bonfiglioli et al., 1961; Chadderton and Montagu-Pollock, 1963) or straining the matrix due to differential thermal expansion (Bullough and Gilman, 1966).

Bonfiglioli et al. (1961) reported a variation from one mica to another in the width of diffraction contrast images of tracks. They concluded that the width correlated with the relative thermal stability of the different materials. Their observation, however, is in disagreement with ours on micas (Price and Walker, 1962d) and other materials (Fleischer et al., 1965b). Further, as we noted earlier, electron microscopic observations of diffraction contrast do not give the true width of the region of damage and hence would be of doubtful relevance even if substantiated. Chadderton and Montagu-Pollock (1963) have used a thermal spike model to predict that intrinsic semiconductors should be the best track registering materials—in contradiction to the experimental results noted in Tables 1-1 and 1-2. Later Chadderton et al. (1966) discuss the transfer of electronic energy to the lattice of crystals and infer correctly that insulators are the best track recorders, as was known at the time.

The merit, however, of a successful theory is to make useful predictions, of which there is a dearth, both qualitatively and quantitatively, from the papers on the thermal spike. One clear expectation from the model is that the sensitivity of different materials would relate in some regular manner either to decomposition temperatures or to melting temperatures of the detectors or to their track annealing temperatures. None of these correlations exists. We therefore discard the thermal spike model as not leading to fruitful predictions.

Total Energy Loss

On the basis of limited, early experiments, we suggested that track formation was governed by the total energy loss rate, (dE/dx), of the track-forming particles (Fleischer et al., 1964a). Our more extensive experiments (Fleischer et al., 1967a, c), however, made it clear that this was not a satisfactory description and led us to suggest the primary ionization and excitation criterion (Fleischer et al., 1967a), which will be discussed shortly. One glaring violation of the dE/dx criterion was the prediction that relativistic Fe nuclei would leave etchable tracks in cellulose nitrate, in contrast to observation (Fleischer et al., 1967c).

It is not difficult to recognize the physical reason for the failure of the dE/dx criterion. As high energies are approached, an increasing fraction of the energy loss of a heavy nucleus goes into creating high energy delta rays, which leave most of their energy at distances far outside the 30 to 50 Å radius of the track.

Since the reader will shortly be exposed to a variety of proposed relations for damage-related quantities such as (dE/dx), $(dE/dx)_{E<E_0}$, E_v, J, and $I \cdot J + E_v$, it is appropriate to illustrate how they vary with energy and to note their differences. Fig. 1-12 gives, as a function of energy in MeV/amu, the total energy loss rate, dE/dx (Henke and Benton, 1966; Benton and Henke, 1969), the restricted energy loss rate $(dE/dx)_{E<350\,\mathrm{eV}}$ (Benton, 1967; Benton and Henke, 1972), the energy density, E_v, deposited at some standard distance (17 Å) from the center of the track (Katz and Kobetich, 1968), and the primary ionization J (Fleischer et al., 1969a), in each case calculated for ^{56}Fe ions passing through Lexan polycarbonate.

In general form each curve increases in energy loss toward lower energies and

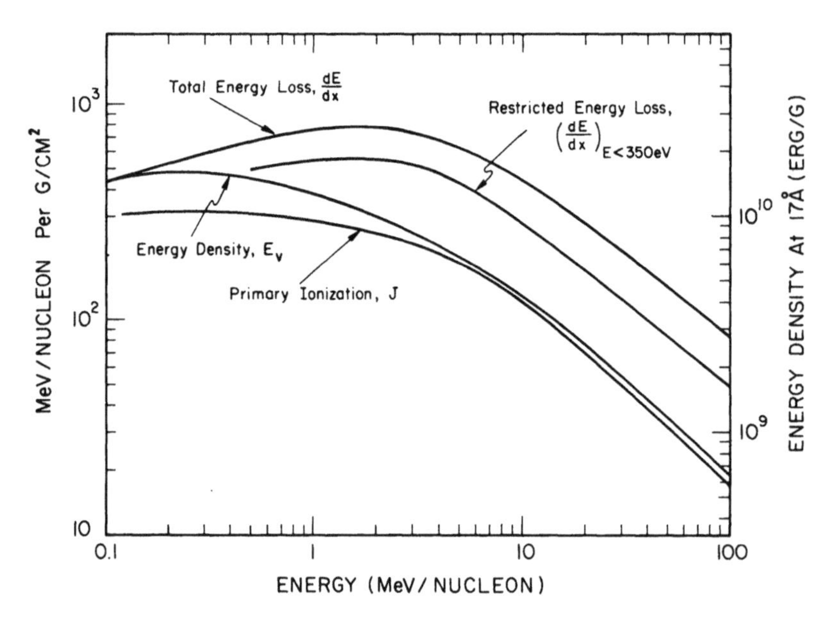

Fig. 1-12. *Various damage-related quantities as a function of energy. Calculations are given for* ^{56}Fe *in Lexan polycarbonate. The primary ionization data are in arbitrary units.*

then rather abruptly decreases over the last ~1 MeV/amu or less as zero energy is approached. The increase is simply due to the fact that in slowing down the ion spends more time near each electron upon which it exerts a force and therefore gives the electron a larger impulse. The drop-off at very low energy occurs because the ion picks up more and more orbital electrons as it slows down; as a result, it has a lower net charge and exerts a lower force on the electrons.

The restricted energy loss is the portion of the total energy loss that produces delta rays of less than some specified energy (in this case 350 eV). Because with faster ions a larger fraction of the delta rays have high energy, $(dE/dx)_{E<350\,\text{eV}}$ becomes a smaller fraction of the total energy loss at higher energies and has its maximum at a lower energy. The primary ionization, because it does not weight the delta rays by energy, has its maximum at still lower energies, whereas the energy deposited at 17 Å from the track center is intermediate between $(dE/dx)_{E<350\,\text{eV}}$ and J but drops more rapidly than the other three quantities at high energies. The equations that have been used for dE/dx, for $(dE/dx)_{E<E_0}$, and for J are the following:

$$\frac{dE}{dx} = \frac{C_1 Z^{*2}}{\beta^2}\left[\ln\left(\frac{W_{\max}{}^2}{I^2}\right) - 2\beta^2 - \delta - U\right] \tag{1-7}$$

$$\left(\frac{dE}{dx}\right)_{E<E_0} = \frac{C_1 Z^{*2}}{\beta^2}\left[\ln\left(\frac{W_{\max} E_0}{I^2}\right) - \beta^2 - \delta - U\right] \tag{1-8}$$

$$J = \frac{C_1 C_2 Z^{*2}}{I_0 \beta^2}\left[\ln\left(\frac{W_{\max}}{I_0}\right) - \beta^2 - \delta + K\right] \tag{1-9}$$

in which

$C_1 = 2\pi n_e e^4/mc^2$
n_e = number of electrons/cm³ in the detector
m = electron mass
$W_{\max} = 2mc^2\beta^2\gamma^2$
$\gamma = (1 - \beta^2)^{-1/2}$
δ = correction for effect of polarization of medium at relativistic velocities
U = low velocity correction for non-participation of inner electron shells
K = a constant that depends on the composition of the stopping medium
I = mean ionization potential of the detector
I_0 = ionization potential of the most loosely bound electrons in the detector
C_2 = effective fraction of the electrons in the detector in the most loosely bound state.

The original equations were derived by Bethe (1930) and improved by Bloch (1933) and others. The computation of E_v is less straightforward and the reader is referred to the paper by Kobetich and Katz (1968) for the method.

1.4.3. Possible Track Production Mechanisms

We have mentioned earlier that the nature of track production appears to be distinctly different in polymers from what it is in inorganic solids. We choose first to consider the inorganic solids as eighty percent of the track detectors that have been studied are inorganic and because they present the simpler case. For these solids we know that one of the two major factors in damage production may be neglected, the secondary damage by delta rays.

ION EXPLOSION SPIKE

From the discussion in Section 1.3 we find ourselves in the apparently paradoxical situation where we know that the damage along tracks in inorganic solids consists mainly of displaced atoms; yet the damage results from interactions with the electrons in the detector, not from direct atomic scattering. The escape from this dilemma that has been suggested, termed the *ion explosion spike* (Fleischer et al., 1965a), is to use the burst of ionization along the path of a charged particle to create an electrostatically unstable array of adjacent ions which eject one another

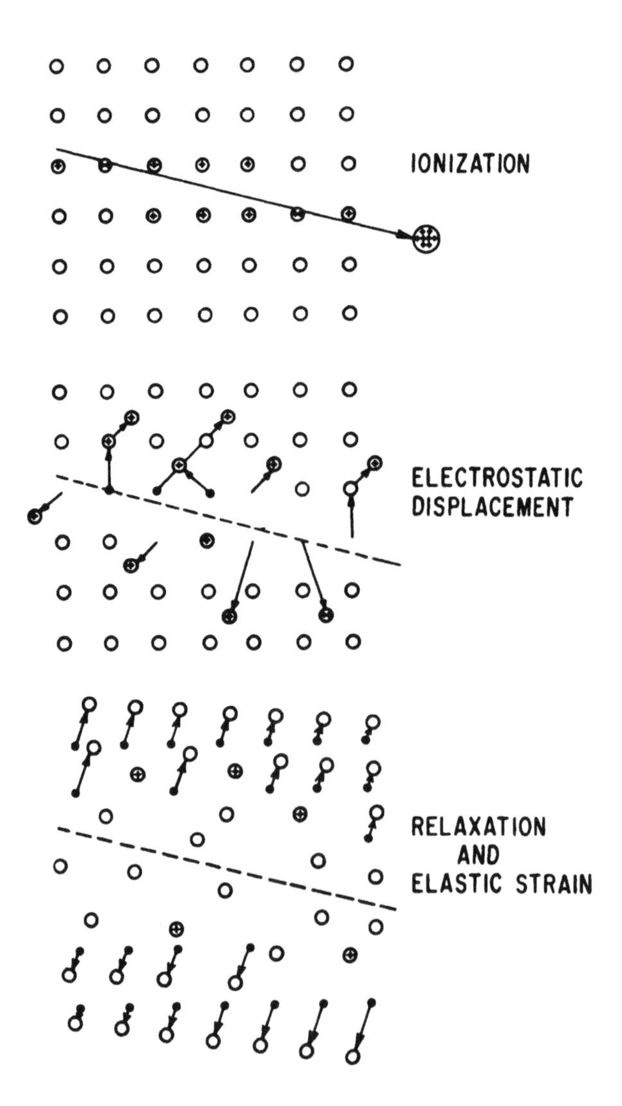

Fig. 1-13. *The ion explosion spike mechanism for track formation in inorganic solids. The original ionization left by passage of a charged particle (top) is unstable and ejects ions into the solid, creating vacancies and interstitials (middle). Later the stressed region relaxes elastically (bottom), straining the undamaged matrix.*

from their normal sites into interstitial positions. Fig. 1-13 illustrates this multiple-step process: Following the primary ionization, an array of interstitial ions and vacant lattice sites is produced by the Coulomb energy of the ions, after which elastic relaxation diminishes the acute, local stresses by spreading the strain more widely. It is the creation of long range strains in this third step that makes possible the direct observation of unetched tracks in crystals by transmission electron microscopy.

The idea that lattice instability could result from the creation of atomically adjacent ions of like sign is an old one, first suggested by Varley (1954a, b) as a means of producing individual interstitial-vacancy pairs in alkali halide crystals. Young (1958) mentions the possibility that track formation in lithium fluoride is basically a multiple Varley process. Separately, Stiegler and Noggle (1962) invoked a similar process as a means of producing fission track grooves at free surfaces of thin metal films. They imagined electrons to be ejected from the metal and to be followed by the ions thereby created. Later, we independently arrived at the same idea and used it to construct a specific, but approximate, semiquantitative model (Fleischer et al., 1965a).

Criterion for atomic displacements.— The physical picture given in Fig. 1-13 leads directly to this model for track formation. The model describes the condition that the Coulomb repulsive forces within the ionized region be sufficient to overcome the lattice bonding forces in terms of a local electrostatic "stress" being greater than the local "mechanical strength" or bonding strength. If two ions in material of dielectric constant ϵ and average atomic spacing a_0 have received an average ionization of n unit charges e, the force between them is $n^2e^2/\epsilon a_0^2$ or a local force per unit area (the electrostatic stress σ_e) of $n^2e^2/\epsilon a_0^4$. We find what is essentially the interatomic bonding force, in terms of a macroscopically measurable quantity, by noting that the theoretical mechanical tensile strength σ_M of material of Young's modulus Y is approximately $Y/10$ (Polanyi, 1921). We then see that the electrostatic stress will be larger than the mechanical strength if $n^2e^2/\epsilon a_0^4 > Y/10$ or if

$$n^2 > R \equiv Y\epsilon a_0^4/10e^2, \tag{1-10}$$

where the quantity R defined in (1-10) is called the stress ratio and should be a measure of the relative sensitivity of various track-storing materials.

Relation (1-10) indicates that tracks should be formed most easily in materials of low mechanical strength, low dielectric constant, and close interatomic spacing. Collection of data for these quantities yields the results given in Table 1-4, which uses improved, more extensive data than were presented earlier (Fleischer et al., 1965a). Comparison of Table 1-4 with Table 1-2 shows that the stress ratio averages for groups of detectors of closely similar sensitivities do a good job of ordering the groups in their proper sequence and of justifying why, on the average, plastics are more sensitive than glasses, which are more sensitive than most crystals.

Within each group there is considerable variability from one detector to another, so that the inverse relationship between R and sensitivity is by no means a complete story. The extreme of variability in R occurs for the plastics, with values that range between 0.04 and 0.002. As we noted, however, primary ionization is not the sole factor in producing tracks in plastics. It would in fact be astounding if the simple model presented explained in full detail the properties of even the crystals, ignoring

Table 1-4. Relation of Registration Thresholds to Calculated Stress Ratio

Detector Class	Detector Groups of Similar Registration Thresholds Highest Threshold at Top	Average Stress Ratio
Crystals	Hypersthene Olivine	4.5
	Labradorite Zircon	1.4
	Diopside Augite	1.3
	Oligoclase Bytownite Orthoclase Quartz Micas	0.5
Inorganic Glasses	Silica Flint Tektite	0.7
	Soda Lime Phosphate	0.5
Plastics	Group of 14 plastics	0.01

as it does any direct considerations of crystal structure, chemical composition, ordering, and so forth. The data of Table 1-4 do, however, support the major physical features of the model.

In certain detector materials, such as the micas, it is clear from experiments that tracks are not revealed unless the damage is atomically continuous (Fleischer et al., 1964b). We can translate that requirement into a condition that there be at least one ionization per atomic plane traversed by the charged particle. In short, $n > 1$ is a criterion for track formation in such materials. In other materials it is likely that n will be $\sim$ unity. By using such a criterion, we are limiting ourselves to tracks that can be revealed by the etching technique. For other methods of revealing tracks different criteria will no doubt apply. In our paper on the ion spike mechanism, we showed that the threshold for registration in muscovite in fact coincided with $n = 1$ (as calculated from a classical relation for ionization due to J. J. Thomson, 1912). Mott and Massey (1965) note that in a case where there was a test of the Thomson equation the ionization was in fact higher by a factor of nearly three, which if applicable to mica, would also be consistent with $n > 1$. The same reasoning when applied to the more sensitive plastics shows that tracks can be formed even when there are only 1 to 3 ionizations for each 20 atom distances along the track. In addition, however, excited electrons can lead to free radical formation and broken bonds. Excitations of 2 eV can produce bond breaks (Bovey, 1958) with a spacing approximately one per atom distance along a track for which the density of ionizations is only $\sim$1 per 20 atom distances. Hence, primary ionization and primary excitation together can give a continuous trail of damage. That there is adequate primary damage to explain track formation in polymers is further supported by calculations of Fain et al. (1971), who include effects from ionization of inner shells and conclude that ~0.6 ionizations per atom distance are produced at threshold in cellulose nitrate.

Specific questions have been raised concerning the validity of the Varley mechanism and therefore by inference concerning the ion spike process itself. Most of these have been specifically addressed to crystals with the rock salt structure and may not be relevant to other detector materials. We have noted that there are several significant differences between isolated charge pairs and the abundant array of ions that are the precursor to track formation, the most significant of which is the outward velocity given the ions of the detector that lie close to the path of the moving ion. This and other considerations are discussed by Fleischer et al. (1965a).

Katz and Kobetich (1968) have stated that adequate electrostatic energy is not available for atomic displacements. However, if we consider our minimum requirement of one ionization per atom plane crossed, a track of length l would have an electrostatic energy per atom of $\sim(e^2/a)$ $[\log(l/a) - 1]$, or 55 eV per atom for a 10 μm ionized track. Since this value is well above the 25 eV energy usually required (Seitz and Koehler, 1956) for atomic displacements, it would appear that the problem raised is not significant.

Restrictions on track formation.— Two conditions having to do with the concentration and mobility of current carriers explain why tracks are only found in insulators and the semiconductors of poorer conductivity.

The first of these requirements relates to the supply of electrons near an ionized track. If other electrons were able to replace those ejected by the energetic charged particle before the ionized atoms are forced into the adjacent material, no track would result. Thus, in order to suppress track formation it would be necessary to drain electrons from a cylindrical region around the ionized track in less time than the $\sim10^{-13}$ sec needed for the ions to be displaced from their sites. If the density of free electrons is n_n and the number of ionizations per atomic plane is n_a, the radius of the region to be drained is given by $\pi r^2 a_0 n_n = n_a$. The time for electrons to diffuse a distance r is r^2/D, where the diffusion constant is given by the Einstein relation $D = \mu_n kT/e$, μ_n being the electron mobility. Thus tracks may be formed only if $n_n < en_a/\pi a_0 \mu_n kTt$.

For a diffusion time t of $\sim10^{-13}$ sec (a lattice vibration time) we find that under typical conditions this relation is obeyed by semiconductors and insulators. It does, however, imply that tracks will not form in metals (unless the density of conduction electrons were less than $10^{20}/\text{cm}^3$).

A second requirement is set by the mobility of holes. Because the ionized region along a track is in essence a high concentration of holes, they might move away and thereby suppress permanent track formation. If we assume that in order to do this holes must diffuse at least half a track radius r_0, then the hole mobility μ_p must be more than $r_0^2e/4tkT$ for track formation to be prevented. This relation requires that at room temperature tracks will not appear in materials whose hole mobility is more than $\sim$10 cm^2/V-sec. It follows that metals and many semiconductors, including silicon and germanium, are normally not track-storing materials. On the other hand, in many other semiconductors, including the vanadium glass described earlier, it is thought (Van Houten, 1960) that the thermal activation occurs as a result of the intermittent overlap of ions due to thermal vibrations. For any such material the direct ejection of ions should be possible.

The quantitative validity of the estimated critical value of μ_p is doubtful though the level is roughly correct. Most of the semiconductors in which tracks have been sought and not found have mobilities above 150 cm^2/V-sec (Hannay, 1959), whereas $MoSe_2$, $MoTe_2$, and WTe_2, which do form tracks in thin films, have values of μ_p in the range 35 to 150 cm^2/V-sec (Wilson and Yoffe, 1969). In germanium the mobility is decreased strikingly in the presence of high electric fields (Conwell, 1967) such as must exist along the paths of charged particles. As a result, a rigorous derivation of a critical hole mobility may be difficult.

Quantitative etching rate relations.— Quantitative tests of a model are possible once we have adequate equations with which to describe the variation with an ion's energy and range of some measurable quantity that is related to the damage intensity along unetched tracks. As will be detailed in Chapter 3, the chemical attack rate v_T along the track is the best observable quantity we know of with which to measure the damage intensity. For the ion spike model the primary ionization (or primary ionization plus excitation for polymers) is the quantity which is postulated to determine the damage intensity. The problem arises of providing equations to describe the primary ionization and relate it to observable quantities such as v_T and range R. In practice Bethe's eq. (1-9) is the one that has been utilized to calculate damage vs. velocity. In addition a range-energy relation is needed to relate damage density, J, to range, and an empirical equation is inferred for v_T vs. J. Finally, the effective charge Z^* needs to be related to velocity and nuclear charge Z for use in eq. (1-9) and in the range-energy relations.

Unfortunately Bethe's equation has been derived rigorously only for ions moving through atomic hydrogen, and therefore the constants K and C_2/I_0 are unknown and in practice are treated as adjustable parameters for each detector. In Fig. 1-8 a series of damage curves is calculated for different ions. The criterion of success of the model then becomes whether the etching rates of different ions are equal at the same damage intensity on the graph, and, therefore, also whether the proper energies are predicted for the thresholds of detection of different ions. Fig. 1-8 shows that for Lexan polycarbonate the data are consistent with the model—that is, all thresholds are at the same level. More extensive data on cosmic ray particles at higher energies also fit the observed threshold. Above the threshold the etching rate data are also self-consistent (as will be described in more detail in Chapter 3). The exceptions and discrepancies that have appeared are associated with very low energies—where the shell corrections that enter into the quantity U in eqs. (1-7) and (1-8) and K in eq. (1-9) are of uncertain accuracy, as noted by Benton (1970). An additional uncertainty at low energy is the validity of the empirical relation given in eq. (1-6) for Z^* as a function of β and Z (Heckman et al., 1960). In a new range of Z the work of Cumming and Crespo (1967) gave data that did not agree with eq. (1-6). In addition, some of the observed shell effects along low index directions in crystals give warning that Z^* is unlikely to behave in any simple manner at very low energies (Cheshire and Poate, 1970). Further complications in the concept of an effective charge are discussed by Betz (1972) and extensive data are summarized.

Other Ionization-Based Models

Several models try to predict quantitatively the thresholds and etching rates for different ions in polymers. Each model that we discuss includes secondary effects of the primary ionization, and the first of these fails to specifically include the primary ionization and excitation.

Secondary energy loss.— A model proposed by Kobetich and Katz (1968) and Katz and Kobetich (1968) assumes the energy loss by secondary electrons at a specific radial distance near the path of the heavy ion is the critical quantity for track formation. In polymers they identify the critical dose (energy per unit volume, E_v) with that required to significantly alter the bulk properties, when imposed by dispersed electron or gamma ray irradiation. Thus, they are guided by experiments on effects of radiation on bulk properties to choose critical radii of 15 Å, 17 Å, and 19 Å and doses of 2.8×10^8 ergs/g, 7.5×10^8 ergs/g, and 3.5×10^9 ergs/g for cellulose nitrate, Lexan polycarbonate, and muscovite mica, respectively. Similar calculations of deposited energy density have been reported by Baum (1970) and Paretzke and Burger (1973).

Although the heavy ion registration data available to Katz and Kobetich were consistent with their model, subsequently accumulated information violates their predictions. The most clear-cut example is the prediction that in Lexan relativistic ions of atomic number less than 70 would not register, in contrast with the observation of tracks of ions down to 57 ± 2 (Price et al., 1970, 1971). In addition, there are conceptual difficulties (1) in the arbitrary selection of different critical radii for different materials and (2) in ignoring all of the primary effects, which must at the very least contribute importantly to the damage in the track core, and as we have noted previously, must be essentially the sole effect for inorganic materials such as mica. In particular, Katz and Kobetich's inferred critical dose for track formation in mica is only a small fraction (less than 2%) of electron doses that we have found experimentally to have negligible effect on track etching (Fleischer et al., 1965c). We also note that micas that have accumulated a fivefold higher dose over geological time show no accelerated general etching.

Additionally, it should be noted that Kobetich and Katz (1968) used assumed, interpolated range-energy relations at energies below 3,000 eV. In contrast, the calculations of Baum (1970) and Paretzke (Paretzke, 1973; Paretzke and Burger, 1973) did utilize the low energy measurements (Cole, 1969) of electron range and hence should be more realistic, particularly in the track core. Comparison with actual measurements in a gas of tissue-equivalent composition (Baum et al., 1973; Wingate and Baum, 1974), which is also similar to the composition of polymers, agreed best with the calculations of Paretzke (1973) and Paretzke and Burger (1973). Since the difference between Cole's observations and the range-energy relation for electrons assumed by Kobetich and Katz (1968) is such that the deposited energy lies closer to the track core than Katz and Kobetich (1968) inferred, it becomes even more difficult to make the distinction between primary and secondary effects in the core region than it appeared earlier.

Restricted energy loss.— It was suggested by Benton (1967) that, by considering only that part of the energy loss that goes into moderately low energy delta rays, a useful quantity to describe track damage intensities in plastics could be obtained. Originally, Benton (1967) restricted the energy loss to that expended in accelerating electrons to less than 1,000 eV; later he suggested that a lower value would fit the results better (Benton, 1970); and most recently quoted 350 eV as the preferred maximum energy (Benton and Henke, 1972). Empirically this criterion probably describes the data about as well as the criterion of primary-ionization-plus-excitation. For example, it predicts that relativistic nuclei of $Z = 55$ should register under specific etching conditions in Lexan, as compared to $Z = 60$ for primary ionization and 57 ± 2 observed. Reference to Fig. 1-12 shows why this should be—namely, the forms of eqs. (1-8) and (1-9) are very similar, particularly at the higher energies. The major difference comes at low energy where $(dE/dx)_{E<E_0}$, like dE/dx, predicts that ^{4}He nuclei should fail to register in Lexan polycarbonate, in contrast to the fact that they are observed at low energy, as expected under the criterion of primary-ionization-plus-excitation.

Conceptually, the $(dE/dx)_{E<E_0}$ model has its difficulties. Since the range of 350 eV electrons is a few hundred Ångstroms (Kobetich and Katz, 1968; Cole, 1969), energy deposited well outside the region of accelerated etching is included, whereas the fraction of their energy deposited within the core by higher energy delta rays is ignored. These two errors presumably partially compensate each other. The model also weights energy that is expended in primary ionization equally with that in secondary processes, which are less strategically concentrated at the core of the track.

Primary plus secondary damage.— Monnin (1970) and Fain et al. (1971, 1972, 1974) have embarked on a careful and promising set of calculations which are starting to provide a more complete theory about the energy distribution around a particle track. They have not yet, however, extended their work toward describing just what constitutes etchable damage around a path of a heavy ion. They have separately calculated the partitioning of energy at each radial distance into the primary processes of ionization and excitation and into the secondary processes of ionization, excitation, and vibration. Presumably, much of the excitation energy ends as thermal (vibrational) energy also. They find that the proportion of energy loss in primary processes is about 40% of the total at 1 MeV/amu and diminishes monotonically with increasing energy to 30%, 25%, and 20% at 10 MeV/amu, 100 MeV/amu, and 1,000 MeV/amu, respectively.

Although it is too early to draw definitive conclusions from the work of Fain, Monnin, and Montret, their results point out that in the core region a major fraction of the energy goes into the primary processes. There is a reasonable expectation that ultimately comparison of the results of their calculations with experimental measurements of the radial distribution of accelerated etching, using the method of Bean and DeSorbo (Section 1.3), will allow us to decide the relative contributions of primary and secondary processes in the cores of tracks.

Table 1-5 gives a summary of various models, the major criticisms that have been raised, and some general conclusions. For inorganic solids the ion explosion spike model appears to contain the right physics, even though a number of problems remain. For plastics both primary and secondary processes are probably significant. Even though the ion spike model omits secondary processes, it still empirically best describes observations of the existing models. However, the work of Fain et al. shows promise of allowing the missing physical processes to be added for a more realistic formulation.

1.5. PROBLEMS AND FUTURE WORK

The first, most immediate, and visible scientific problems of the particle track field were solid state physics questions: How does a track form? What atomic processes take place? What is the ultimate atomic configuration along and around a track? Curiously enough these problems remain as some of the least studied, presumably because of the intense interest in the many applications of track etching that has directed attention to the assortment of fields considered in Chapters 4 through 10.

In short, widespread technical use is being made of defects whose real nature is relatively incompletely known. There are several levels on which we would like to develop our understanding of particle tracks.

Most of the existing work has taken its incentives from the desire to *use* tracks

Table 1-5. Comparison of Models of Track Formation

Model	Major Criticisms	Conclusion
A. Thermal Spike	1. Threshold expected to increase with temperature required for phase change, contrary to fact. 2. Makes no useful predictions.	Not applicable.
B. Atomic Collisions	1. Predicts tracks are produced at lower energies than they are.	Not generally applicable; relevant at energies ~1 keV/amu
C. Total Energy Loss in Ionization	1. Predicts tracks are produced at higher energies than they are. 2. Includes energy deposited far from track.	Not applicable.
D. Energy deposited by Delta Rays in Track Core	1. Ignores defects produced by primary excitation and ionization. 2. Quantitatively wrong at low energy. 3. Does not predict relative thresholds nor absence of registration in conductors.	Incomplete; wrong for minerals.
E. Total Energy Loss in Track Core	1. Ignores difference between the qualitatively different defects produced by primary excitation and ionization and those produced by delta ray energy loss. 2. No useful predictions as yet. 3. Same as No. 3, Model D.	Promising for further study.
F. Restricted Energy Loss	1. Same as No. 1, Model E. 2. Assumes damage important at unrealistically large distances - contrary to observation. 3. Ignores interior dose deposited by delta rays that end outside of etchable region. 4. Predicts registration where not observed qualitatively like total energy loss but with smaller errors. 5. Same as No. 3, Model D.	Convenient but inexact analytic predictions.
G. Primary Ionization and Excitation	1. Ignores energy loss by delta rays within track region.	Most satisfactory at present for inorganic detectors; improvements needed for plastics where delta rays are more important.

Note: All attempts at analytic prediction encounter trouble at very low energy and require the use of arbitrary with more quantitative rigor, for example to identify individual particles and to measure their energies. Success here would lie in finding for each detector a precise function that relates etching rates to the velocity and nuclear species of the track-forming particle.

We have emphasized earlier that our description is incomplete in the low energy region where the net, average charge of a moving ion is decreasing as the ion slows and where the charge is fluctuating in a statistically controlled manner. A sustained effort will be required to transform our limited knowledge of effective charge in simple substances at intermediate velocities into quantitative information in complex detector materials at still lower velocities where the velocity changes rapidly over very short distances.

Similarly, there is a gap in present knowledge of track behavior at relativistic velocities. The dependence of Z^* on velocity indicated in eq. (1-6) has been tested only at low velocities. The necessity for a relativistic correction term δ in eqs. (1-8) and (1-9) has not yet been established in solid state track detectors and is included only because it is necessary in the total energy loss expression (1-7) for particles moving through a condensed medium. Eventually, when ions as heavy as Kr are accelerated to 2.5 GeV/amu at the Bevalac accelerator at Lawrence Berkeley Laboratory, it will be possible to determine both Z^* and δ experimentally.

The assumption that $(dE/dx)_{E<E_0}$, J, and E_v are proportional to Z^{*2} is based on the Born approximation, which is valid only if $Z/137\beta \ll 1$. Eby and Morgan (1972) have used Mott scattering theory to calculate dE/dx when the condition $Z/137\beta \ll 1$ is not met. The result deviates somewhat from a quadratic dependence on Z^*. No one has yet derived exact expressions for $(dE/dx)_{E<E_0}$, J, and E_v that hold for all values of $Z/137\beta$. Both distant and close collisions must be re-examined. It is likely that some correction for a deviation from a Z^{*2} dependence will be necessary in eqs. (1-8) and (1-9).

There is an intermediate scale of understanding of tracks where the attempt is to relate the fundamental atomic nature of the damage to macroscopically observable quantities. The major problem of interest is to understand the etching rate v_T in terms of the composition and structure of the detector and of the atomic disorder making up a track. The calculations of Bean reported in Section 1.3 comprise a first attempt in this direction.

Finally, the real scientific questions are the disorder itself, its nature and formation mechanism(s). Little has been done toward creating such high concentrations of tracks that many of the point defects present can be identified by such techniques as magnetic resonance and infrared absorption.

Although unaltered tracks are presumably composed of complicated, multi-atomic defects, recognition of what residual defects are present after partial annealing of tracks could be highly informative. Similarly, for polymers it should be possible to analyze the changes in the chain length distributions and free radical concentrations in polymers after track production.

Theoretical computer simulation of track formation with a lattice model into which an appropriately ionized region is entered as an initial condition would be useful. The formulations exist to do such studies on small model crystals (for example Beeler, 1966), but as yet no such work has been carried out on any realistic basis. Collection of pertinent experimental data on hole-mobility would allow a test of whether μ_p dictates the separation of solids in which charged particle tracks can be produced from those in which they cannot, as is predicted by the ion explosion spike model.

As yet there has been no experimental test of the assertion that tracks of <50 keV/amu particles in fact are produced by direct atomic scattering rather than by ionization processes. A successful experimental finding of such tracks in metals would be definitive. Although such a test is in preparation, no results are yet known.

Chapter 1 References

S. Amelinckx (1956), "The Direct Observation of Dislocation Nets in Rock Salt Crystals," *Phil. Mag.* **1,** 269–290.

J. W. Baum (1970), "Comparison of Distance- and Energy-Restricted Linear Energy Transfer for Heavy Particles with 0.25 to 1000 MeV/amu," *Proc. Second Symp. on Microdosimetry*, 20–24 Oct., Ispro, Italy, 653–666, Brussels: Euratom.

J. W. Baum, M. N. Varma, C. L. Wingate, H. G. Paretzke, and A. V. Kuehner (1973), "Nanometer Dosimetry of Heavy Ions," Brookhaven Natl. Lab. Rept. BNL 18219, Sept. 11.

C. P. Bean, M. V. Doyle, and G. Entine (1970), "Etching of Submicron Pores in Irradiated Mica," *J. Appl. Phys.* **41,** 1454–1459.

J. Beeler (1966), "Displacement Spikes in Cubic Metals. I. Alpha-Iron, Copper, and Tungsten," *Phys. Rev.* **150,** 470–487.

E. V. Benton (1967), "Charged Particle Tracks in Polymers No. 4: Criterion for Track Registration," USNRDL-TR-67-80, U. S. Nav. Rad. Def. Lab., San Francisco, Calif.

E. V. Benton (1970), "On Latent Track Formation in Organic Nuclear Charged Particle Track Detectors," *Rad. Effects* **2,** 273–280.

E. V. Benton and R. P. Henke (1969), "R(E) Calculations," *Nucl. Inst. Methods* **67,** 87–92.

E. V. Benton and R. P. Henke (1972), "On Charged Particle Tracks in Cellulose Nitrate and Lexan," Dept. of Physics, Univ. of San Francisco, Technical Rept. No. 19, July.

H. A. Bethe (1930), "Theory of the Passage of Rapid Corpuscular Rays Through Matter," *Ann. Physik* **5,** 325–400.

H.-D. Betz (1972), "Charge States and Charge-Changing Cross Sections of Fast Heavy Ions Penetrating Through Gaseous and Solid Media," *Rev. Mod. Phys.* **44,** 465–539.

T. K. Bierlein and B. Mastel (1960), "Damage in UO_2 films and Particles during Reactor Irradiation," *J. Appl. Phys.* **31,** 2314–2315.

G. E. Blanford, Jr., R. M. Walker, and J. P. Wefel (1970), "Calibration of Plastic Track Detectors for Use in Cosmic Ray Experiments," *Rad. Effects* **5,** 41–45.

F. Bloch (1933), "Stopping Power of Atoms with Many Electrons," *Z. Physik* **81,** 363–376.

G. Bonfiglioli, A. Ferro, and A. Mojoni (1961), "Electron Microscope Investigation on the Nature of Tracks of Fission Products in Mica," *J. Appl. Phys.* **32,** 2499–2503.

F. A. Bovey (1958), *The Effects of Ionizing Radiation on Natural and Synthetic High Polymers*. New York: Wiley-Interscience.

F. P. Bowden and L. T. Chadderton (1961), "Molecular Disarray in a Crystal Lattice Produced by a Fission Fragment," *Nature* **192,** 31–34.

J. A. Brinkman (1955), "Production of Atomic Displacements by High Energy Particles," *Am. J. Phys.* **24,** 246–267.

R. Bullough and J. J. Gilman (1966), "Elastic Explosions in Solids Caused by Radiation," *J. Appl. Phys.* **37,** 2283–2287.

D. S. Burnett, C. Hohenberg, M. Maurette, M. Monnin, R. Walker, and D. Woolum (1972), "Solar Cosmic Ray, Solar Wind, Solar Flare, and Neutron Albedo Measurements," *Apollo 16 Preliminary Science Report*, NASA Special Pub. 315, Chap. 15, pp. 19–32.

L. T. Chadderton and H. M. Montagu-Pollock (1963), "Fission Fragment Damage to Crystal Lattices: Heat Sensitive Crystals," *Proc. Roy. Soc.* **A274,** 239–252.

L. T. Chadderton, D. V. Morgan, I. McC. Torrens, and D. Van Vliet (1966), "On the Electron Microscopy of Fission Fragment Damage," *Phil. Mag.* **13,** 185–195.

J. A. Champion (1965), "Some Properties of (Mo, W) (Se, $Te)_2$," *Brit. J. Appl. Phys.* **16,** 1035–1037.

A. Charlesby (1960), *Atomic Radiation and Polymers*. London: Pergamon Press.

I. M. Cheshire and J. M. Poate (1970), "Shell Effects in Low Energy Atomic Collisions," in *Atomic Collision Phenomena in Solids*, D. W. Palmer, M. W. Thompson, and P. D. Townsend (eds.), Proc. Int. Conf., Univ. of Sussex, Brighton, England, 7 to 12 Sept. 1969. Amsterdam: North-Holland.

C. B. Childs and L. Slifkin (1962), "Detection of Nuclear Disintegrations Produced by 1.55-BeV Protons in Silver Chloride Single Crystals," *Phys. Rev. Letters* **9,** 354–355.

C. B. Childs and L. M. Slifkin (1963), "Delineating of Tracks of Heavy Cosmic Rays and Nuclear Processes Within Large Crystals of Silver Chloride," *Rev. Sci. Instr.* **34,** 101–104.

A. Cole (1969), "Absorption of 20 eV to 50,000 eV Electron Beams in Air and Plastic," *Rad. Research* **38,** 7–33.

E. M. Conwell (1967), "High Field Transport in Semiconductors," *Solid State Physics*, Suppl. 9. New York: Academic Press.

R. Craig, E. Mamidzhanian, and A. W. Wolfendale (1968), "Ancient Cosmic Ray Tracks in Mica," *Phys. Letters* **B26,** 468–470.

W. T. Crawford, W. DeSorbo, and J. S. Humphrey (1968), "Enhancement of Track Etching Rates in Charged Particle-irradiated Plastics by a Photo-oxidation Effect," *Nature* **220,** 1313–1314.

J. B. Cumming and V. P. Crespo (1967), "Energy Loss and Range of Fission Fragments on Solid Media," *Phys. Rev.* **161,** 287–293.

M. Debeauvais and M. Monnin (1965), "Measurement of the Registration Sensitivity of Various Ionizing Particles in Solid Polymers," *Compt. Rend.* **260,** 4728–4730.

W. DeSorbo and J. S. Humphrey, Jr. (1970), "Effects Upon Track Etching Rates in Charged Particle Irradiated Polycarbonate Film," *Rad. Effects* **3,** 281–282.

P. B. Eby and S. H. Morgan (1972), "Charge Dependence of Ionization Energy Loss for Relativistic Heavy Nuclei," *Phys. Rev.* **A5,** 2536–2541.

K. Endo and T. Doke (1973), "Calibration of Plastic Nuclear Track Detectors for Identification of Heavy Charged Nuclei Using Fission Fragments," *Nucl. Instr. Methods* **111,** 29–37.

J. Fain, M. Monnin, and M. Montret (1971), "Heavy Ion Effects in Polymers," Third Symposium on Radiation Chemistry, Tihany, Hungary, 10–15 May.

J. Fain, M. Monnin, and M. Montret (1972), "Spatial Energy-Density Distributions around Ion Paths in Polymers," *Proc. 8th Int. Conf. on Nuclear Photography and Solid State Track Detectors*, Bucharest, Rumania **1,** 34–72.

J. Fain, M. Monnin, and M. Montret (1974), "Spatial Energy Distribution around Heavy Ion Paths," *Rad. Research* **57,** 379–389.

R. L. Fleischer and P. B. Price (1963a), "Tracks of Charged Particles in High Polymers," *Science* **140,** 1221–1222.

R. L. Fleischer and P. B. Price (1963b), "Charged Particle Tracks in Glass," *J. Appl. Phys.* **34,** 2903–2904.

R. L. Fleischer and P. B. Price (1964a), "Techniques for Geological Dating of Minerals by Chemical Etching of Fission Fragment Tracks," *Geochim. Cosmochim. Acta* **28,** 1705–1714.

R. L. Fleischer and P. B. Price (1964b), "Glass Dating by Fission Fragment Tracks," *J. Geophys. Res.* **69,** 331–339.

R. L. Fleischer, P. B. Price, R. M. Walker, and E. L. Hubbard (1964a), "Track Registration in Various Solid State Nuclear Track Detectors," *Phys. Rev.* **133A,** 1443–1449.

R. L. Fleischer, P. B. Price, E. M. Symes, and D. S. Miller (1964b), "Fission Track Ages and Track Annealing Behavior of Some Micas," *Science* **143,** 349–351.

R. L. Fleischer, P. B. Price, and R. M. Walker (1965a), "The Ion Explosion Spike

Mechanism for Formation of Charged Particle Tracks in Solids," *J. Appl. Phys.* **36,** 3645–3652.
R. L. Fleischer, P. B. Price, and R. M. Walker (1965b), "Tracks of Charged Particles in Solids," *Science* **149,** 383–393.
R. L. Fleischer, P. B. Price, and R. M. Walker (1965c), "Effects of Temperature, Pressure, Ionization on the Formation and Stability of Fission Tracks in Minerals and Glasses," *J. Geophys. Res.* **70,** 1497–1502.
R. L. Fleischer, P. B. Price, R. M. Walker, and E. L. Hubbard (1966), "Ternary Fission of Heavy Compound Nuclei in Thorite Track Detectors," *Phys. Rev.* **143,** 943–946.
R. L. Fleischer, P. B. Price, R. M. Walker, and E. L. Hubbard (1967a), "Criterion for Registration in Dielectric Track Detectors," *Phys. Rev.*, **156,** 353–355.
R. L. Fleischer, P. B. Price, R. M. Walker, and M. Maurette (1967b), "Origins of Fossil Charged Particle Tracks in Meteorites," *J. Geophys. Res.* **72,** 333–353.
R. L. Fleischer, P. B. Price, R. M. Walker, R. C. Filz, F. Fukui, E. Holeman, M. W. Friedlander, R. S. Rajan, and A. S. Tahmane (1967c), "Tracks of Cosmic Rays in Plastics," *Science* **155,** 187–189.
R. L. Fleischer, P. B. Price, and R. M. Walker (1968), "Charged Particle Tracks: Tools for Geochronology and Meteor Studies," *Radiometric Dating for Geologists*, E. Hamilton and R. M. Farquhar (eds.), 417–435. London: Wiley-Interscience.
R. L. Fleischer, P. B. Price, and R. M. Walker (1969a), "Nuclear Tracks in Solids," *Sci. Amer.* **220,** 30–39, June.
R. L. Fleischer, P. B. Price, and R. T. Woods (1969b), "Nuclear Particle Track Identification in Inorganic Solids," *Phys. Rev.* **88,** 563–567.
N. B. Hannay (ed.) (1959), *Semiconductors.* New York: Reinhold Press.
H. H. Heckman, B. L. Perkins, W. G. Simon, F. M. Smith, and W. Barkas (1960), "Ranges and Energy Loss Processes of Heavy Ions in Emulsion," *Phys. Rev.* **117,** 544–556.
R. P. Henke and E. V. Benton (1966), "Range-Energy and Range-Energy Loss Tables," U. S. Naval Radiological Defense Laboratory, Rept. No. TR-1102, unpublished.
W. H. Huang and R. M. Walker (1967), "Fossil Alpha-particle Recoil Tracks: A New Method of Age Determination," *Science* **155,** 1103–1106.
K. Izui and F. E. Fujita (1961), "Observation of the Tracks of Fission Fragments in Molybdenite," *J. Phys. Soc. Japan* **16,** 1779.
W. D. Jones and R. V. Neidigh (1967), "Detection of Light Nuclei with Cellulose Nitrate," *Appl. Phys. Letters* **10,** 18–19.
R. Katz and E. J. Kobetich (1968), "Formation of Etchable Tracks in Dielectrics," *Phys. Rev.* **170,** 401–405.
J. J. Kelsch, O. F. Kammerer, and P. A. Buhl (1960), "New Technique for the Direct Investigation of Fission Events," *Brit. J. Appl. Phys.* **11,** 555.
J. J. Kelsch, O. F. Kammerer, A. N. Goland, and P. A. Buhl (1962), "Observation

of Fission Fragment Damage in Thin Films of Metal," *J. Appl. Phys.* **33,** 1475-1482.

E. J. Kobetich and R. Katz (1968), "Energy Deposition by Electron Beams and Delta Rays," *Phys. Rev.* **170,** 391–396.

W. Krätschmer (1971), *Etching of Heavy Ion Tracks in Quartz Glass,* Ph.D. Dissertation, Heidelberg.

M. Lambert, A. M. Levelut, M. Maurette, and H. Heckman (1970), "Etude par diffusion des Rayons X aux petits angles de mica muscovite irradie par des ions d'Argon," *Rad. Effects* **3,** 155–160.

A. R. Lang (1958), "Direct Observation of Individual Dislocations by X-Ray Diffraction," *J. Appl. Phys.* **29,** 597–598.

M. Lecerf and J. Peter (1972), "Detecteurs en Verre," *Nucl. Instr. Methods* **104,** 189–195.

J. Lindhard and M. Scharff (1961), "Energy Dissipation by Ions in the keV Region," *Phys. Rev.* **124,** 128–130.

J. Lindhard and P. V. Thomsen (1962), "Sharing of Energy Dissipation between Electronic and Atomic Motion," in *Radiation Damage in Solids* **1,** 66–76, Vienna: Int. Atomic Energy Agency.

J. Lindhard, M. Scharff, and H. E. Schiott (1963), "Range Concepts and Heavy Ion Ranges (Notes on Atomic Collisions, II)," *Kgl. Danske Videnskab. Selskab., Mat.-Fys. Medd.* **33,** No. 14, 1–42.

M. Maurette (1966a), "Study of the Registration of Fission Fragment Tracks in Certain Substances," *J. Physique* **27,** 505–512.

M. Maurette (1966b), "Study of Heavy Ion Tracks in Terrestrial and Extraterrestrial Natural Minerals," *Bull. Soc. Franc. Min. Crist.* **89,** 41–79.

M. Maurette (1970), "Track Formation Mechanisms in Minerals," *Rad. Effects* **3,** 149–154.

J. C. Maxwell (1904), *A Treatise on Electricity and Magnetism,* 3rd Ed., Vol. 1, Section 306. Oxford: Oxford University Press.

K. L. Merkle (1962), "Fission-Fragment Tracks in Metal and Oxide Films," *Phys. Letters* **9,** 150–152.

M. Monnin (1968), *Etude de l'interaction des particules lourdes ionisantes avec les macromolecules a l'état solide,* Ph.D. thesis, June, Report, PNCF 68-RI 9, Univ. de Clermont.

M. Monnin (1970), "Mechanism of the Formation of Tracks in Polymers," *Rad. Effects* **5,** 69–73.

M. Monnin and G. E. Blanford, Jr. (1973), "Detection of Charged Particles by Polymer Grafting," *Science* **181,** 743–744.

M. Monnin, H. Besson, S. Sanzelle, and L. Avan (1966), "Quelque nouveaux detecteurs ionographiques de la famille des detecteurs solides de traces," *Compt. Rend.* **263B,** 1337–1338.

D. V. Morgan and L. T. Chadderton (1968), "Fission Fragment Tracks in Semiconducting Layer Structures," *Phil. Mag.* **17,** 1135–1143.

N. F. Mott and H. S. W. Massey (1965), *The Theory of Atomic Collisions*, 3rd Ed., Clarendon Press, Oxford.
C. W. Naeser and H. Faul (1969), "Fission Track Annealing in Apatite and Sphene," *J. Geophys. Res.* **74,** 705–710.
T. S. Noggle and J. O. Stiegler (1960), "Electron Microscope Observations of Fission Fragment Tracks in Thin Films of UO_2," *J. Appl. Phys.* **31,** 2199–2208.
T. W. O'Keeffe and R. M. Handy (1968), "Fabrication of Planar Silicon Transistors Without Photoresist," *Solid-State Electronics* **11,** 261–266.
P. J. Ouseph (1973), "Energetic Heavy Particle Detection by Photochromic Material," *Phys. Rev. Letters* **30,** 1162–1165.
H. G. Paretzke (1973), "Comparison of Track Structure Calculations with Experimental Results," *4th Symp. on Microdosimetry*, Verbania Pollanza, Italy, 24–28 Sept.
H. Paretzke and G. Burger (1973), "Spatial Distribution of Deposited Energy Along the Path of Heavy Charged Particles," *Second Symposium on Microdosimetry*, H. G. Ebert (ed.), Stresa, Italy, 20–24 Oct., 615–627.
T. Plieninger, W. Krätschmer, and W. Gentner (1972), "Charge Assignment to Cosmic Ray Heavy Ion Tracks in Lunar Pyroxenes," *Proc. Third Lunar Sci. Conf.* **3,** 2933–2939. Cambridge: MIT Press.
M. Polanyi (1921), "Nature of the Breaking Process," *Z. Physik* **7,** 323–327.
N. F. Pravdyuk and V. M. Golyanov (1962), "Examination of Uranium Fission with an Electron Microscope," International Conference, Berkeley Castle, Gloucestershire, England, May 30–June 2, 1961. *Properties of Reactor Materials and the Effects of Radiation Damage*, P. J. Littler (ed.), 160–175. London: Butterworths.
P. B. Price and R. M. Walker (1962a), "Electron Microscope Observation of Etched Tracks from Spallation Recoils in Mica," *Phys. Rev. Letters* **8,** 217–219.
P. B. Price and R. M. Walker (1962b), "Chemical Etching of Charged Particle Tracks," *J. Appl. Phys.* **33,** 3407–3412.
P. B. Price and R. M. Walker (1962c), "Electron Microscope Observation of a Radiation-Nucleated Phase Transformation in Mica," *J. Appl. Phys.* **33,** 2625–2628.
P. B. Price and R. M. Walker (1962d), "Observation of Charged Particle Tracks in Solids," *J. Appl. Phys.* **33,** 3400–3406.
P. B. Price, R. L. Fleischer, and C. D. Moak (1968a), "On the Identification of Very Heavy Cosmic Ray Tracks in Meteorites," *Phys. Rev.* **167,** 277–282.
P. B. Price, R. L. Fleischer, D. D. Peterson, C. O'Ceallaigh, D. O'Sullivan, and A. Thompson (1968b), "Plastic Track Detectors for Identifying Cosmic Rays," *Can. J. Phys.* **46,** S1149–S1153.
P. B. Price, R. L. Fleischer, and G. E. Nichols (1970), "Identification of Tracks of Super Heavy Cosmic Rays in Plastics," *Acta Phys. Acad. Sci. Hung.* **29,** Suppl. 1, 411–416.
P. B. Price, P. H. Fowler, J. M. Kidd, E. J. Kobetich, R. L. Fleischer, and G. E.

Nichols (1971), "Study of the Charge Spectrum of Extremely Heavy Cosmic Rays Using Combined Plastic Detectors and Nuclear Emulsions," *Phys. Rev.* **D3,** 815–823.

P. B. Price, D. Lal, A. S. Tamhane, and V. P. Perelygin (1973), "Characteristics of Tracks of Ions of $14 \leq Z \leq 36$ in Common Rock Silicates," *Earth Planet. Sci. Lett.* **19,** 377–395.

E. Ruedl, P. Delavignette, and S. Amelinckx (1961), "Fission Fragment Damage to Platinum," *J. Appl. Phys.* **32,** 2492–2493.

F. M. Russell (1967), "Tracks in Mica Caused by Electron Showers," *Nature* **216,** 907–909.

F. M. Russell (1968), "Mica as a Neutrino Detector," *Science Journal* 57–63, February.

E. Schopper, F. Granzer, G. Haase, D. Haseganu, G. Henig, J. U. Schott, F. Wendnagel, and F. Zörgiebel (1973), "AgCl Particle Track Detectors," *Proc. 13th Inter. Cosmic Ray Conf.*, Univ. of Denver, **4,** 2855–2860.

F. Seitz (1949), "The Disordering of Solids by the Action of Fast Massive Particles," *Disc. Faraday Soc.* **5,** 271–282.

M. G. Seitz (1972), "Heavy Ion Irradiation Studies in Terrestrial Materials," Ph.D. Thesis, Washington University.

F. Seitz and J. S. Koehler (1956), "Displacement of Atoms During Irradiation," *Solid State Physics* **2,** 305–448.

M. Seitz, M. C. Wittels, M. Maurette, and R. M. Walker (1970), "Accelerator Irradiations of Minerals: Implications for Track Formation Mechanisms and for Studies of Lunar and Meteoritic Materials," *Rad. Effects* **5,** 143–148.

E. K. Shirk (1974), "Observation of Trans-Iron Solar Flare Nuclei in an Apollo 16 Command Module Window," *Astrophys. J.*, **190,** 695–702.

R. A. Sigsbee and R. H. Wilson (1973), "Electron Irradiation Dilation in SiO_2," *Appl. Phys. Letters* **23,** 541–542.

E. C. H. Silk and R. S. Barnes (1959), "Examination of Fission Fragment Tracks with an Electron Microscope," *Phil. Mag.* **4,** 970–971.

G. Somogyi, M. Varnagy, and G. Petö (1968), "Application of Plastic Track Detectors for Detection of Light Nuclei," *Nucl. Instr. Methods* **59,** 299–304.

J. O. Stiegler and T. S. Noggle (1962), "Nitrogen Ion Bombardment of Thin Pt Films," *J. Appl. Phys.* **33,** 1894–1895.

J. J. Thomson (1912), "Ionization by Moving Charged Particles," *Phil. Mag.* **23,** 449–457.

E. E. Uzgiris and R. L. Fleischer (1971), "Amber: Charged Particle Track Registration, Track Stability, and Uranium Content," *Nature* **234,** 28–30.

E. E. Uzgiris and R. L. Fleischer (1973), "Etched Laser Filament Tracks in Glasses and Polymers," *Phys. Rev.* **A7,** 734–740.

S. Van Houten (1960), "Semiconduction in $Li_xNi_{(1-x)}O$," *Phys. Chem. Solids* **17,** 7–17.

J. H. O. Varley (1954a), "A Mechanism for the Displacement of Ions in an Ionic Lattice," *Nature* **174**, 886–887.

J. H. O. Varley (1954b), "New Interpretation of Irradiation-Induced Phenomena in Alkali Halides," *J. Nucl. Energy* **1**, 130–143.

M. Varnagy, J. Csikai, S. Szegedi, and S. Nagy (1970), "Observations of Proton Tracks by a Plastic Detector," *Nucl. Instr. Meth.* **89**, 27–28.

R. M. Walker, E. Zinner, and M. Maurette (1973), "Measurements of Heavy Solar Wind and Higher Energy Solar Particles During the Apollo 17 Mission," *Apollo 17 Preliminary Science Report*, NASA Special Pub. 330, Chap. 19, pp. 2–11.

A. D. Whapham and M. J. Makin (1962), "The Nature of Fission Fragment Tracks in Uranium Dioxide," *Phil. Mag.* **7**, 1441–1455.

J. A. Wilson and A. D. Yoffe (1969), "The Transition Metal Dichalcogenides," *Adv. Phys.* **18**, 193–335.

C. L. Wingate and J. W. Baum (1974), "Measured Radial Distributions of Dose and LET for Alpha and Proton Beams in Hydrogen and Tissue Equivalent Gas," Brookhaven Natl. Lab. Rept. BNL 18187, submitted to *Rad. Research.*

D. A. Young (1958), "Etching of Radiation Damage in Lithium Fluoride," *Nature* **182**, 375–377.

Chapter 2

Basics of Track Etching

The diversity of solids that record the passage of charged particles was illustrated in Fig. 1-2 (Fleischer et al., 1968a), which also displays the remarkable variety of etching responses that exist. In each of the three solids pictured—a crystal, an ordinary glass, and a high polymer—the particles were full-energy fission fragments from the same fission source, so that the marked difference in the resulting etched tracks relates to the registration properties of the different detectors and to the etching processes. In this chapter we examine the geometry of track etching and infer how this geometry quantitatively affects observable track parameters, and through them the track densities. Then we describe how track etching treatments are chosen and how the environment of a sample affects track revelation. Finally we discuss optical scanning and other special techniques for track detection and measurement.

2.1. TRACK GEOMETRY

2.1.1. Track Geometry for Constant v_T

The geometry of track etching is dictated in the simplest case by the simultaneous action of two etching processes: chemical dissolution along the particle track at a linear rate v_T and general attack on the etched surface and on the interior surface of the etched track at a lesser rate v_G (Fleischer and Price, 1963a,b). Fig. 2-1 illustrates how this process creates a cone that has the original track as its axis. Here we have assumed that (1) v_T is constant along the track and (2) v_G is constant and isotropic. Shortly we will consider the consequences of violations of these assumptions. For the present, we note that the assumption that v_G/v_T is constant is approximately true for short etching distances $v_T t$ [t = etching time] and that v_G is normally isotropic for noncrystalline solids.

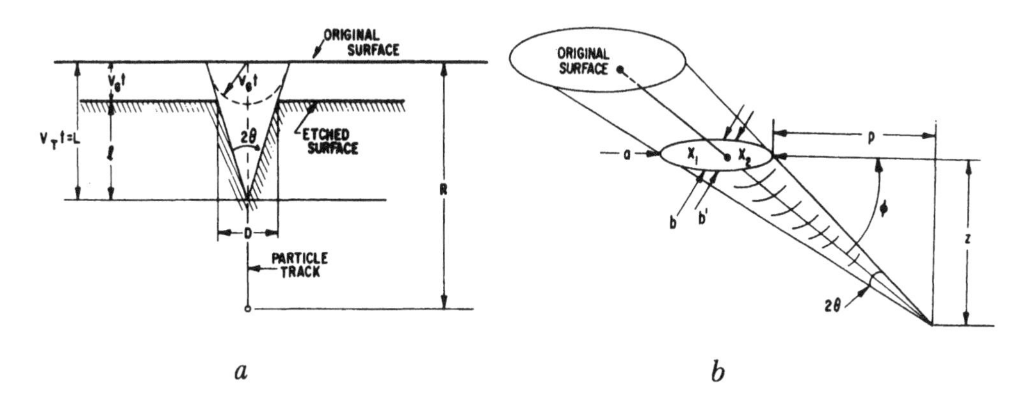

a *b*

Fig. 2-1. *Track geometry with v_T and v_G constant and (a) a vertically incident particle or (b) a particle incident at dip angle ϕ. (After Price and Fleischer, 1971.)*

Consider first the simplest case of normal incidence. Fig. 2-1a shows that surface material is removed at the same time that the etched track is developing, so that both of the directly observable quantities, the track diameter D and the visible track length l, are the result of the competition between the effects of v_G and v_T, both D and l being less, the smaller the excess of v_T over v_G (or equivalently the larger the cone angle θ [= arcsin $(v_G/v_T]$ (Fleischer and Price, 1963a). To illustrate this behavior, we note that simple geometry gives the measurable quantities in terms of etching parameters:

$$l = (v_T - v_G)t \tag{2-1}$$

$$D = 2v_G t \sqrt{(v_T - v_G)/(v_T + v_G)}. \tag{2-2}$$

Both l and D vanish for $v_T = v_G$. Eq. (2-1) applies when $v_T t$ is less than the range R over which θ is effectively constant and eq. (2-2) applies for a longer time given by

$$t < (R/v_G) \cdot \{1 + [(1 - \sin\theta)(\cos\theta/2 + \sin\theta/2)/(\cos\theta/2 - \sin\theta/2)\cos\theta]\}.$$

We can also give the etching parameters in terms of measurable quantities:

$$\begin{aligned} v_T/v_G &= 2\sqrt{(D/2)^2 + l^2}/D = \csc\theta \\ v_G t &= (D/2)(D/2l + \sqrt{(D/2)^2 + l^2}/l) = (D/2)(\tan\theta + \sec\theta) \\ v_T t &= \sqrt{(D/2)^2 + l^2} \\ &\quad \cdot (D/2l + \sqrt{(D/2)^2 + l^2}/l) = D\csc\theta\,(\tan\theta + \sec\theta)/2. \end{aligned}$$

The sketch in Fig. 2-1b allows one to derive geometrical relations applying to a track tilted at an angle ϕ to the surface. The intersection of the conical track with the etched surface is an ellipse, with the track position displaced from the center, as shown in the figure. The major axis of the ellipse approaches infinity as ϕ approaches the cone angle θ, which will lead us in Section 2.2 to the concept of an etching efficiency. The relations for the angles and track length in terms of the measurable quantities Z, p, a, and v_G are

$$\phi = [\arctan (Z/(a+p)) + \arctan (Z/p)]/2$$
$$= \arcsin \left(\frac{\sqrt{Z^2+p^2}\sqrt{(a+p)^2+Z^2} - p^2 - ap + Z^2}{2\sqrt{Z^2+p^2}\sqrt{(a+p)^2+Z^2}} \right)^{1/2}$$
$$\theta = [-\arctan (Z/(a+p)) + \arctan (Z/p)]/2$$
$$= \arcsin \left(\frac{\sqrt{Z^2+p^2}\sqrt{(a+p)^2+Z^2} - p^2 - ap - Z^2}{2\sqrt{Z^2+p^2}\sqrt{(a+p)^2+Z^2}} \right)^{1/2}$$
$$L = (Z + v_G t)/\sin \phi.$$

Relations for the dimensions of the elliptical intersection with the etched surface are

$$b'/b = [1 - (\sin^2 \theta/\sin^2 \phi)]/(1 - \sin^2 \theta)$$
$$b = 2v_G t\,(\sin \phi - \sin \theta)/\sqrt{\sin \phi + \sin \theta}$$
$$a = 2v_G t\,(\sin \phi - \sin \theta) \cos \theta/[\sin (\phi + \theta) \sin (\phi - \theta)]$$
$$X_2 = v_G t\,(1 - \sin \theta/\sin \phi)/\sin (\phi + \theta)$$
$$X_1 = v_G t\,(1 - \sin \theta/\sin \phi)/\sin (\phi - \theta)$$
$$b' = 2v_G t\,(\sin \phi - \sin \theta)/\cos \theta \sin \phi.$$

The above relations come from Price and Fleischer (1971); other geometrical relations are given by Henke and Benton (1971).

2.1.2. Track Geometry for Varying v_T

The reader may well conclude that happiness is believing v_T to be constant, as now we are about to see that etched track geometry is appreciably more complicated when v_T increases or decreases along the length of a track. The bright side of this cloud, however, shows itself in Chapter 3, where we see how individual particles are identified by measuring v_T along their tracks.

It is now experimentally clear that in most materials, whether plastic (Somogyi, 1966a,b; Price et al., 1967a, 1968a,b; Blanford et al., 1970c), glass (Höppner et al., 1969; Fleischer et al., 1969b), or crystalline (Fleischer et al., 1964d; Price et al., 1973), v_T in general increases with ionization rate. Only for short etching times can we usefully approximate v_T (or θ) as constant, and therefore, we must face the question of how etching geometry is to be related to etching rate.

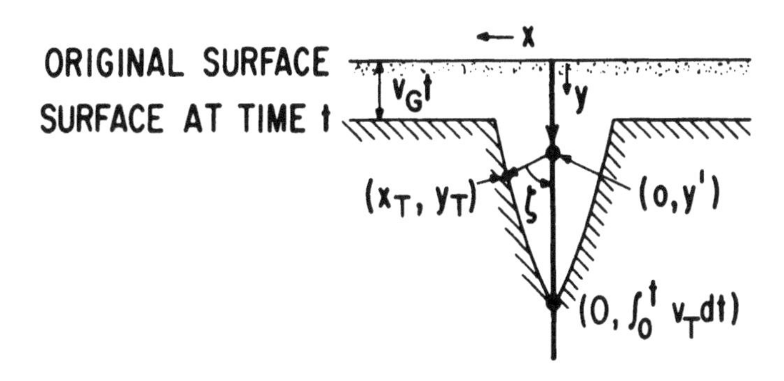

Fig. 2-2. *Track geometry with v_T decreasing along the track is convex as viewed from the outside.* (*After Fleischer et al., 1969b.*)

The problem is, given $v_T(y)$ (the variation of v_T with position y along the track), what is the track length L, diameter D, and detailed profile (x_t, y_t) as a function of the etching time, x_t and y_t being defined in Fig. 2-2. The reverse problem—of inferring $v_T(y)$ from measurements of L, D, or x_t and y_t—allows particle identification, as we shall see in Chapter 3. The way in which the convex outward shape sketched in Fig. 2-2 arises can be understood simply by the motorboat analogy. In this picture, the two dimensional track profile is similar to the bow wave from a motorboat (where the boat velocity is v_T and the wave propagates across the water surface at a velocity v_G). For constant v_T the wave angle to the "boat" is arcsin (v_G/v_T). As the "boat" slows down, its decreasing velocity rounds the track outward, as in Fig. 2-2. When the "boat" accelerates, a concave outward wave shape results. These shapes are illustrated by the tracks shown in Fig. 2-3. The shapes allow the direction of motion to be inferred for these two cosmic ray nuclei, which passed into an Apollo space helmet. As a particle slows, its ionization rate increases and the taper of its etched track diminishes.

The value of L can be calculated from the integral $t = \int_0^L dy/v_T(y)$. If this equation cannot be solved in closed form for L, computer calculation of this integral can be used to find the value of L that sets $t - \int_0^L dy/v_T(y)$ equal to zero. To calculate the track profile (x_t, y_t) we use the idea of Fleischer et al. (1969b) and later Paretzke et al. (1973b) that each point on the profile of an etched track can be found by Huygens (or least-time) construction such as is sketched in Fig. 2-2. First, using the coordinate system indicated there we calculate the shape of the etch pit $y_t(x_t)$ for an arbitrary $v_T(y)$ describing the attack rate for a track along the y axis. Twice the value of x_t when $y_t = v_G t$ gives the diameter as a function of time. The etching time to reach the point (x_t, y_t) is given by

$$t = \int_0^{y'} dy/v_T(y) + [(y_t - y')^2 + x_t^2]^{1/2}/v_G. \qquad (2\text{-}3)$$

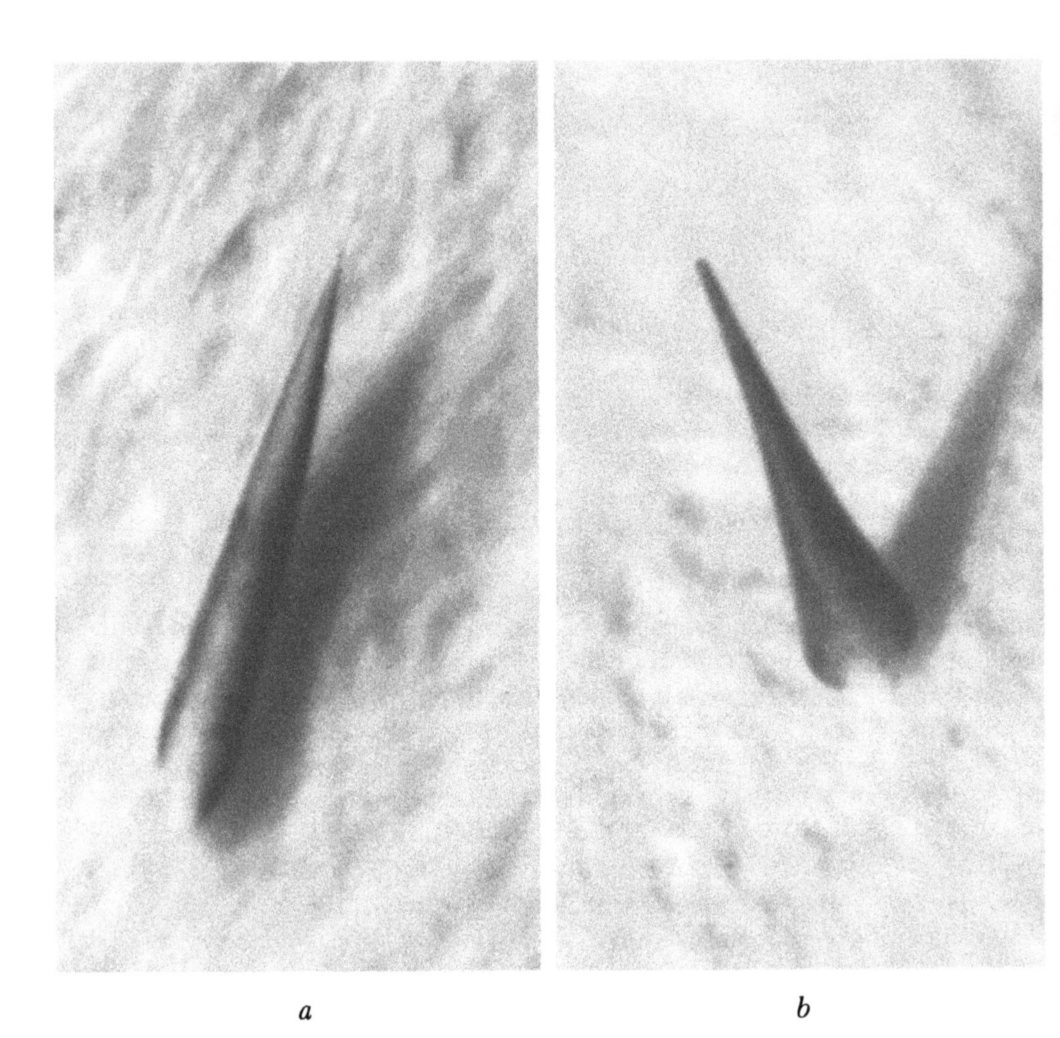

Fig. 2-3. *Photographs of etched cosmic ray tracks on the inside of an Apollo space helmet made of Lexan polycarbonate. The curvature of the sides of a track shows how v_T varies along it. (a) A track that is more tapered at the tip than at the base. (b) An ending track from a particle that passed through the interior of the helmet and came to rest in the opposite side. The tracks are 0.5 mm and 0.7 mm in length. (After Comstock et al., 1971.)*

The first term gives the time to etch along the track from (0,0) to $(0, y')$ and the second term gives the time to etch from $(0, y')$ to (x_t, y_t). Since the actual path from (0,0) to (x_t, y_t) is that corresponding to least time, we take the derivative of time with respect to y' and set it equal to zero, giving the relation

$$y' = y_t - x_t/[(v_T(y')/v_G)^2 - 1]^{1/2}, \qquad (2\text{-}4)$$

which, together with eq. (2-3), defines x_t and y_t for a given t, y', and $v_T(y)$. By noting that the angle ζ in Fig. 2-2 is arctan $[x_t/(y_t - y')]$, we can write eqs. (2-3) and (2-4) in a more useful parametric form in y' as follows:

$$x_t = [(y_t - y')^2 + x_t^2]^{1/2} \sin(\zeta),$$
$$y_t = y' + [(y_t - y')^2 + x_t^2]^{1/2} \cos(\zeta),$$

which can be rewritten

$$x_t = v_G\left[t - \int_0^{y'} dy/v_T(y)\right][1 - v_G^2/v_T^2(y')]^{1/2} \qquad (2\text{-}5)$$

$$y_t = y' + v_G^2/v_T(y')\left[t - \int_0^{y'} dy/v_T(y)\right]. \qquad (2\text{-}6)$$

These equations are valid for all y' and t.

Although in general these two equations cannot be put into a closed expression for $D(t)$, the procedure for calculating $D(t)$ is straightforward: Values of y' are substituted in (2-6) until that value of y' for which $y_t = v_G t$ is obtained; this value of y' in eq. (2-5) gives $D/2$ directly. It can be readily confirmed that for the simple case where v_T is constant the result quoted as eq. 2-2 obtains.

As an example of the use of these equations, Fig. 2-4 shows the profiles calcu-

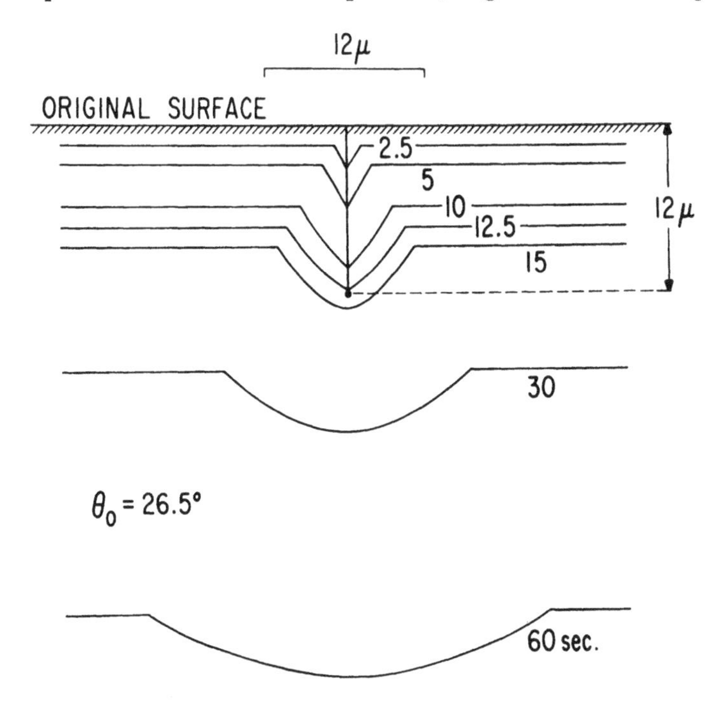

Fig. 2-4. *Calculated etching sequence of a fission fragment track with initial cone angle 26.5° in an obsidian-like glass. It was assumed that $v_G = 0.58$ μm/sec, $v_T = 1.36 - 0.0657y'$ μm/sec, and $R = 12$ μm. (After Fleischer et al., 1969b.)*

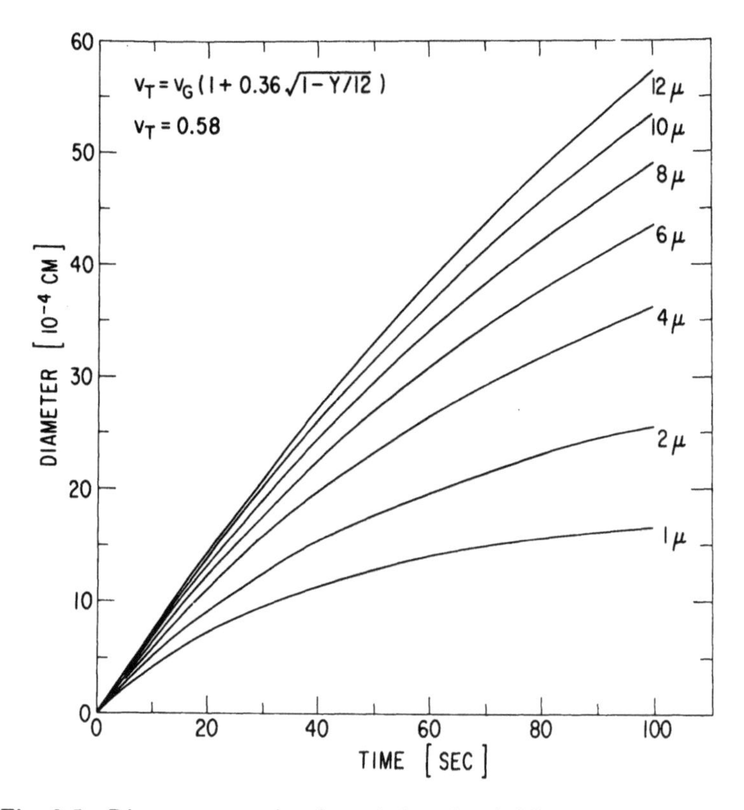

Fig. 2-5. *Diameter versus time for etched tracks of different lengths in glass. In this calculation, it was assumed that* $v_T = v_G \cdot [1 + 0.36(1 - y'/12)^{1/2}]$, *where* y' *is the distance along the etchable particle range, and starting values of* y' *of 0, 2, 4, 6, 8, 10, and 11 were used corresponding to fission track lengths of 12, 10, 8, 6, 4, 2, and 1 μm.* $v_G = 0.58$ *μm/sec. The conditions considered are appropriate to ordinary soda glass (microscope slides, etc.). (After Fleischer et al., 1969b.)*

lated for an etched track for various etching times and a specific, simple choice of v_T (decreasing linearly to v_G to the end of the track). The track remains pointed until very close to its end and then rounds out. The diameter increases monotonically at a decreasing rate. Although larger diameter differences appear at long etching times, the unique, pointed character of the tracks is lost and they become difficult to discriminate from background pits. Fig. 2-5 shows the calculated diameter increase for an ordinary soda-lime glass.

2.1.3. Track Geometry for Anisotropic v_G

So far we have assumed that v_G is isotropic, which is true for glasses but untrue for most crystals. In high polymers chain molecules sometimes are preferentially oriented during processing into sheets, with the result that differences in chemical attack rate are produced between directions that are dominantly along the pre-

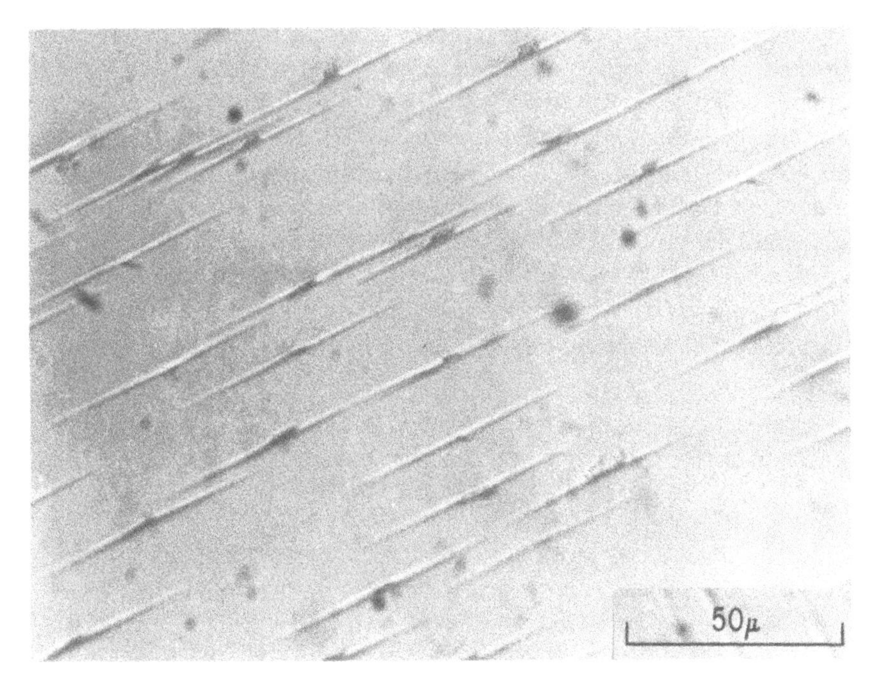

Fig. 2-6. *Etched tracks in gypsum are dark, spike-like features centered at the bottoms of aligned grooves whose direction is dictated by the crystal structure, which causes v_G to be highly anisotropic.* (*After Fleischer and Price, 1964b.*)

ferred direction and those normal to the molecular alignment. Occasionally the processing sequence will cause opposite faces of the same sheet to have different registration characteristics. Hence specific tests must be made of each individual plastic for each processing history.

For crystals the rate of general attack as a function of crystal surface orientation can be elaborately complex. For example, in gypsum etching gives rise to slitlike tracks (Fig. 2-6). Similarly, Maurette's (1966) work on olivine showed v_G to be different for the three principal crystal directions, and the ratios of the three v_G's to vary according to the etchant used. Although in principle the track geometry can be calculated if v_G is known as a function of orientation and v_T as a function of orientation and range, it is usually more useful to seek a simpler detector. For this reason we shall not consider the case of general anisotropy further, but content ourselves with noting that this gap in our understanding of crystal chemistry offers a potentially fruitful area of future research.

2.2. ETCHING EFFICIENCY

Tracks inclined at less than the cone angle θ to a surface are not etched (Fleischer and Price, 1963a; 1964a). It follows that for nonzero θ we must consider an etching efficiency η, where η is defined as the fraction of tracks intersecting a given surface that are etched on the surface under specified etching conditions. It is important here to remember that etching efficiency has to do with purely geometrical requirements for revealing tracks that are present. Registration efficiency is a different concept: the fraction of particles passing through a surface that produces radiation damage tracks as continuous damage with $v_T > v_G$.

To understand why tracks inclined at less than θ are not revealed by etching, consider Fig. 2-7. Part (a) reminds the reader that the angle θ depends on v_T and v_G; part (b) shows that if the component normal to the etched surface of preferential etching along the track ($v_T \sin \phi$) is less than v_G (the general rate of surface

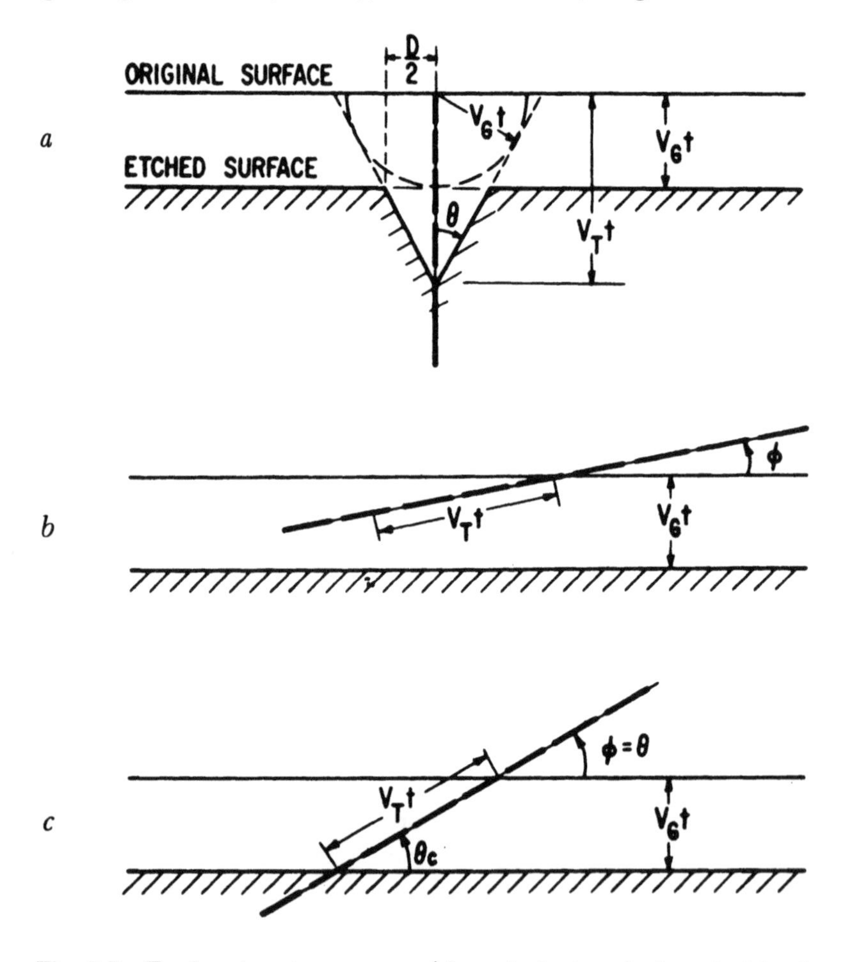

Fig. 2-7. *Track registration geometry: (a) track pit shape is determined by the general rate of attack v_G and the preferential attack rate v_T along damage trail; (b) for angle of incidence less than arcsin (v_G/v_T), the surface is removed at a greater rate than the normal component of v_T and therefore no track is observed; (c) arcsin (v_G/v_T) is the critical angle, θ_c, above which tracks are registered.*

removal), no track is revealed (i.e., the material is removed so rapidly that the preferential etching fails to keep ahead). The critical condition is shown in 2-7c where $\phi = \theta$, and, therefore, the track just fails to be developed. When θ is not constant along a track the situation is more complicated, as will be discussed shortly.

The fraction of solid angle over which tracks are not observed is given by

$$\frac{1}{2\pi}\int_0^{\theta} 2\pi \cos\theta'\, d\theta' = \sin\theta \tag{2-7}$$

so that $\eta = 1 - \sin\theta$ for the case where equal numbers of tracks occur in each solid angle interval. This would, therefore, be the case for an external irradiation of a surface from a thin, infinite source of fission fragments parallel to or at the surface of interest. Here "thin" means much less than the etchable range of the particles in the detector.

In the case where particle tracks originate throughout the volume of the detector itself, tracks that may be revealed by etching originate at depths ranging from 0 to H, where H is the maximum depth from which a particle may start and still produce an etchable track. Those from depth H must all have traveled normal to the surface to have reached it, and those at a depth δ can arrive at angles between arcsin (δ/H) and 90°. Because in order to be etched at the surface the angle to the surface must be greater than θ, all of the tracks from a given depth will be etched as they cross the surface provided $\delta > H \sin\theta$. For $\delta < H \sin\theta$ some of the tracks are inclined at less than θ to the surface and are, therefore, not revealed by etching. The situation is illustrated in Fig. 2-8, in which the line BD represents the fraction

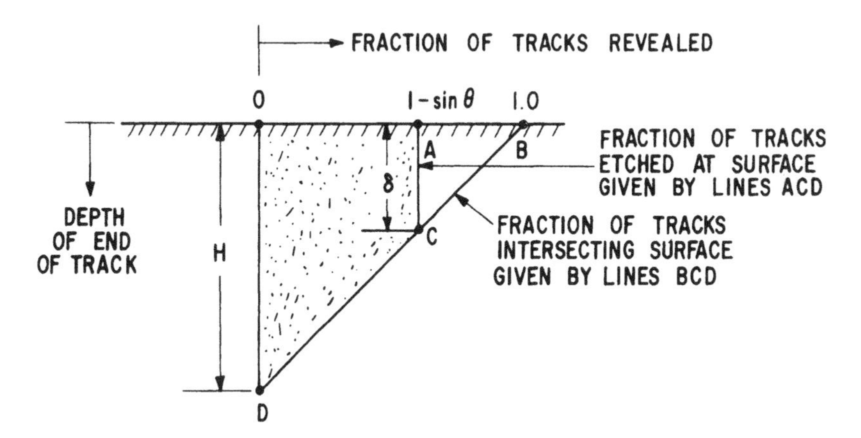

Fig. 2-8. *The fraction of a random distribution of tracks of length L that is revealed by a short etch as a function of the depth of the lower end of the track. The fraction is constant to a depth δ and then decreases linearly to zero at depth L. A fraction $\cos^2\theta$ of tracks that intersect the surface is revealed by etching.*

of the tracks in the upward hemisphere of solid angle that reach the surface and *ACD* (which bounds the shaded area) represents the fraction that is etchable. Simple geometry given in Fig. 2-8 shows that the fraction of tracks reaching the surface but not revealed by etching is just the ratio of the areas of the two triangles *ABC*/*OBD*, which is $\sin^2 \theta$. Alternatively, the fraction (η) revealed is $1 - \sin^2 \theta$ ($= \cos^2 \theta$) (Fleischer and Price, 1964a). Spatially the case just discussed is equivalent to that where tracks are dispersed homogeneously and isotropically through a volume and we ask how many will be revealed at a particular surface.

2.2.1. Etching Efficiency vs. Registration Efficiency

The distinction was made between etching efficiency and registration efficiency. Since tracks inclined within an angle θ of the surface are not revealed by etching, there is only one safe, decisive way to show whether registration occurs for a given particle, and that is to etch after bombarding at normal incidence. If tracks from particles at less than normal incidence are sought, but not found, it is possible that tracks were present but that θ was greater than the angle of incidence. Before we realized this, we were led incorrectly to conclude that mica was more sensitive than glasses because certain particles at 30° incidence in mica would produce etchable tracks but would not do so in glass (Fleischer et al., 1965d). More recent results, with a better understanding of the etching process and use of irradiations at normal incidence (Fleischer et al., 1969b) make it clear that some glasses are more sensitive than muscovite mica.

2.2.2. Etching Efficiency for Anisotropic v_G

Fig. 2-7 demonstrates that there is a critical angle for revealing tracks such that $\theta = \arcsin (v_G/v_T)$. This same relation applies also for anisotropic v_G so long as the value used in this relation is the rate of attack normal to the etched surface, v_G(normal), i.e., $\theta = \arcsin [v_G(\text{normal})/v_T]$. One case of special interest occurs where v_G(normal) is effectively zero, as exemplified by the micas when etched on the layer planes by hydrofluoric acid. Since $\theta_c \approx 0$ for $v_G(\text{normal}) \approx 0$, the etching efficiency is close to unity. Consequently mica is a useful material for making absolute measures of etching efficiencies in other substances, by exposing a control piece of mica to the same integrated isotropic particle flux as the material to be examined.

As an aside it should be noted that since v_G(parallel)—the rate(s) of attack along the layer plane—is finite, tracks in mica can have a "cone" angle greater than zero yet show unit etching efficiency since it is still true that $\arcsin (v_G(\text{normal})/v_T) = 0$. In general tracks in crystals will not be simple cones, but because of crystal anisotropy can assume shapes such as daggers, pyramids, or rectangular parallelopipeds.

2.2.3. Etching Efficiency when v_T Varies

Etching efficiency with a varying θ (or v_T) is naturally more complicated than for fixed θ. Because with sufficiently short etching time θ is essentially constant at the etched surface, the critical angle is again arcsin (v_G/v_T) for an individual track. For a random array of tracks with v_T/v_G decreasing linearly with range, averaging over the different positions at which a track may be intersected by an etched surface leads to an efficiency $\eta = 1 - \sin\theta$ (Fleischer and Hart, 1972a). For non-linear variations of v_T/v_G with range, numerical solutions are necessary as described in further detail by Fleischer and Hart (1972a). Fig. 2-9 gives examples of measured variations of v_T/v_G with range in soda-lime glass and fused quartz. θ is not a constant for either, so that $\cos^2\theta$ is not a good approximation to the true result. For fused quartz a linear decrease of v_T/v_G with residual range is close to the true result—giving $1 - \sin\theta = \eta = 0.67$ as compared to a numerically determined value of 0.69. This value is clearly less than the 0.89 which would be calculated for a

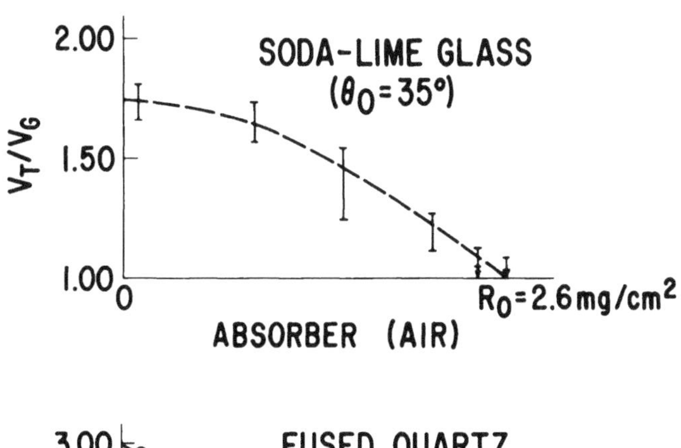

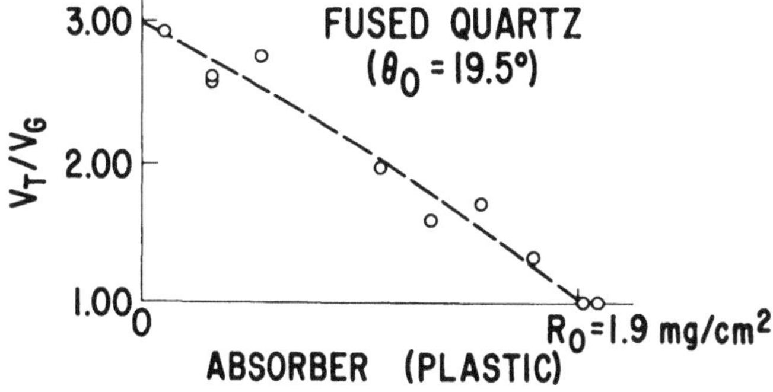

Fig. 2-9. *Etch rate ratio v_T/v_G vs. absorber thickness for two glasses irradiated with Cf^{252} fission fragments (Fleischer and Hart, 1972a). These data illustrate the decrease in track etching rate v_T with decreasing residual range. The data have been approximated by the dashed lines and used to calculate etching efficiencies of 0.69 for the fused quartz and 0.50 for the soda-lime glass.*

Table 2-1. Etching Efficiencies for Short Etching Times

$$\left[\eta = \frac{\text{number of tracks revealed on surface}}{\text{number of tracks intersecting surface}}\right]$$

Case		Relation for η
I.	Tracks random through volume; θ constant along length of track.	$\eta = \cos^2\theta$
II.	Tracks random in orientation from external thin source.	$\eta = 1 - \sin\theta$
III.	Tracks random through volume; θ decreases linearly to zero along range of the particles; short etching time.	$\eta = 1 - \sin\theta$
IV.	Tracks random through volume; θ an arbitrary function of R.	Numerical calculation, with some examples intermediate between Case I and Case III given by Fleischer and Hart (1972a).

constant cone angle of 19.5° appropriate to full-energy fission fragments. For soda-lime glass neither a constant value of 35° for θ, which would give $\eta = 0.67$, nor a linear variation with residual range ($\eta = 0.43$) describes the observed variation adequately, although the two values do bracket the numerically determined value, which is 0.50. Table 2-1 summarizes the etching efficiencies for short etching times.

2.2.4. Prolonged Etching Effects

When etching times are so long that it is not valid to assume that θ is constant over the rather short portion of each track that has been etched, the etching efficiency is increased by an amount that must be calculated numerically or measured directly. The effect arises because it is the highest value of v_T/v_G over the length of track removed during the etch that determines whether a visible etch pit results. If, at the beginning of etching, θ is less than the dip angle, the track is revealed, and continued etching will retain the effects of this accelerated etching. Conversely, if, at the start of etching, θ is greater than the dip angle, no track is being revealed; but as time goes on the general attack may be able to reach a position along the track where θ allows preferential etching, and the track will begin to be evident. These possibilities are illustrated in Fig. 2-10, which points out how tracks that do not preferentially etch originally (such as number 4 at time t_1) will be revealed by a longer etch. The figure also points out that tracks such as 1 and 2, which are etched from the start, remain pointed as long as a track is present, regardless of its angle to the original surface.

Longer etches reveal more pits for three major reasons: (1) very small features that were not originally perceptible are enlarged; (2) new tracks that began and ended beneath the original surface are encountered as material is removed; and

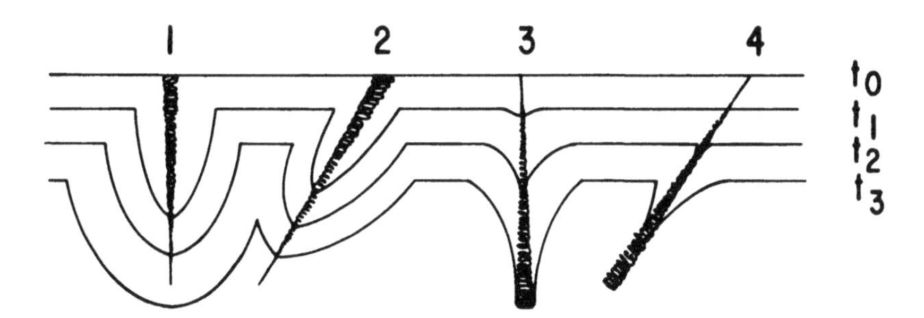

Fig. 2-10. *For tracks with damage intensity that varies along their lengths the conical etch pits are convex or concave outward depending on whether the damage decreases or increases with depth. The drawn width is meant to describe the track etching rate* v_T *along the track and has no relation to the actual diameter of atomic disorder. Once preferential etching has begun, it persists to the end of the track.*

(3) the effect noted above, that etching along a track having a low dip angle and damage that increases downward reveals tracks part way along their lengths. Fig. 2-4 makes clear that a pit, once started, remains and grows for all time but becomes less clear and finally imperceptible as it rounds out and its edges become closer and closer to being coplanar with the original surface. The pits do not get shallower, as Khan and Durrani (1972c) suggest, but they do become less distinct. Except in materials such as the micas, where v_G normal to the layer planes is negligible, use of fixed etching conditions and time(t) is advisable so that the thickness $v_G t$ of removed material is standardized. Otherwise the total number of pits will increase monotonically (see for example Khan and Durrani, 1972c).

2.2.5. Counting Conventions

A convenient practical choice of $\sim 1 \mu m$ for $v_G t$ can be made so that fission tracks are large enough to be visible in the microscope but such that few new tracks have been uncovered by mechanisms (2) and (3) of the last paragraph. A second alternative is to make $v_G t$ large relative to a track length and count only sharp-bottomed pits. Since for long etching times the rate of display of new tracks will be balanced by the rounding out of old tracks after their full lengths have been dissolved away, the procedure will give a reproducible, steady state value, as long as the pit density is not so large that overlapping pits are common. A more laborious but well-defined procedure (Khan and Durrani, 1972c) is to count all tracks as a function of etching time and extrapolate back to the track density at zero etching time.

2.2.6. Effect of Heating

Severe heating, as we will discuss further in Chapter 4 and as is documented in Section 2.4, can erase tracks; less intense heating can reduce the number of tracks that are revealed by etching. In the latter case we can quantitatively understand the decrease in etched tracks if we heat samples containing tracks and observe progressive changes in the relation between v_T/v_G and range. For fission tracks it is found to consist primarily of a reduction in v_T/v_G (which could equivalently be described as an increase in cone angle θ) and a decrease in etchable range (Fleischer and Hart, 1972a).

An increase in θ means that η is reduced, so that a smaller fraction of the tracks is revealed even though *none* have been removed by heating. At the same time eq. (2-2) makes it clear that the average track diameter must be reduced, an effect first recognized by Berzina et al. (1966) and later used by Storzer and Wagner (1969) to improve the process of fission track dating (see Chapter 4).

Table 2-2. Etching Conditions for Fission Fragments

Note: Because of chemical variations within most minerals, glasses, or plastics of a given type, optimum conditions may vary somewhat from those given. Only the preferred etchant is listed.

A. ETCHANTS FOR MINERALS

Mineral	Etching Conditions	Reference
allanite $[H_2O \cdot 4(Ca,Fe)O \cdot 3(Al,Fe)_2O_3 \cdot 6SiO_2]$	50N NaOH, 2-60 min, 140°C	Naeser and Dodge (1969)
apatite $[Ca_5(F,Cl)(PO_4)_3]$	0.25% HNO_3, ~1 min, 23°C, or olivine etch without oxalic acid	Bhandari et al. (1971b) Lal (unpub.)
aragonite ($CaCO_3$)	same as for calcite	
autunite $[Ca(UO_2)_2P_2O_8 \cdot 8H_2O]$	10% HCl, 10-30 sec, 23°C	Fleischer and Price (1964b)
barite ($BaSO_4$)	70% HNO_3, 3 h, 100°C	Fleischer and Price (1964b)
barysilite ($Pb_3Si_2O_7$)	glacial acetic acid, 5-70 sec, 23°C	Haack (1973)
bastnäsite ($CeFCO_3$)	20% HCl, 20-150 min, 155°C	Fleischer and Naeser (1972)
benitoite ($CaTiSi_3O_9$)	1 ml 40% HF:1 ml 65% HNO_3, 5 min, 230°C	Haack (1973)
beryl ($Be_3Al_2Si_6O_{18}$)	19N KOH, 9 h, 150°C	Fleischer and Price (1964b)
bismutite ($Bi_2O_3 \cdot CO_2 \cdot H_2O$)	1g NaOH:1g H_2O, 50 min, 140°C	Fleischer, Price and Woods (unpub.)
brewsterite $[(Sr,Ba,Ca)O \cdot Al_2O_3 \cdot 6SiO \cdot 5H_2O]$	2 ml 48% HF:1 ml 95% H_2SO_4: 4 ml H_2O, 3 sec, 230°C	Haack (1973)
calcite ($CaCO_3$)	olivine etch with NaOH added for pH 12, 30 min, 23°C	Lal (unpub.)
celsian ($BaAl_2Si_2O_8$)	19N NaOH, 20 min, boiling	Haack (1973)
cerussite ($PbCO_3$)	glacial acetic acid, 10-30 min, 23°C	Fleischer et al. (1965a)
chlorite $[(Mg,Fe)_5(Al,Fe)_2Si_3O_{10}(OH)_8]$	48% HF, 10 min, 23°C	Debeauvais et al. (1964)

Table 2-2 (continued)

Mineral	Etching Conditions	Reference
clinopyroxene [augite, $Ca(Mg,Fe,Al)(Al,Si)_2O_6$]	2 ml 48% HF: 1 ml 80% H_2SO_4: 4 ml H_2O, 5-20 min, 23°C	Crozaz et al. (1970)
	3 g NaOH: 2 g H_2O, 90 min, boiling	Lal et al. (1968)
clinopyroxene [diopside, $CaMgSi_2O_6$]	3 g NaOH: 2 g H_2O, 80 min, boiling	Lal et al. (1968)
clinopyroxene [pigeonite, $((Mg,Fe)SiO_3)_x \cdot (CaMg(SiO_3)_2)_{1-x}$]	3 g NaOH: 2 g H_2O, 80 min, boiling	Lal et al. (1968)
cryolite (Na_3AlF_6)	50% HI, 3 min, 23°C	Haack (1973)
epidote [$Ca_2(Al,Fe)_3(SiO_4)_3OH$	50N NaOH, 0.5-2 h, 140°C	Naeser and Dodge (1969)
eulytite [$Bi_4(SiO_4)_3$]	5% HCl, 60 sec, 23°C	Fleischer, Price and Woods (unpub.)
	33% HNO_3, 15 min, 23°C	Haack (1973)
feldspar [albite, $NaAlSi_3O_8$]	3 g NaOH: 4 g H_2O, 85 min, boiling	Lal et al. (1968)
feldspar [anorthite, $CaAl_2Si_2O_8$]	3 g NaOH: 4 g H_2O, 14 min, boiling	Lal et al. (1968)
feldspar [bytownite, An_8Ab_2]	3 g NaOH: 4 g H_2O, 19 min, boiling	Lal et al. (1968)
feldspar [labradorite, An_6Ab_4]	3 g NaOH: 4 g H_2O, 40 min, boiling	Lal et al. (1968)
feldspar [microcline; orthoclase, $KAlSi_3O_8$]	5 g KOH: 1 g H_2O, 80 min, 190°C	Fleischer, Price and Woods (unpub.)
feldspar [oligoclase, An_2Ab_8]	3 g NaOH: 4 g H_2O, 75 min, boiling	Lal et al. (1968)
fluorite (CaF_2)	98% H_2SO_4, 10 min, 23°C	Fleischer and Price (1964b)
garnet [pyrope, $Mg_3Al_2(SiO_4)_3$]	50N NaOH, 0.5-2 hr, 140°C	Naeser and Dodge (1969)
garnet [almandine-pyrope, $(Fe_{1-x}Mg_x)_3 \cdot Al_2(SiO_4)_3$]	5 to 30 min 50N NaOH, boiling	Haack and Gramse (1972)
garnet [spessartine, $Mn_3Al_2(SiO_4)_3$]	10 min, 50N NaOH, boiling	Haack and Gramse (1972)
garnet [andradite-grossular; $Ca_3(Fe_{1-x}Al_x)_2 \cdot (SiO_4)_3$]	1 to 6 h, 75N NaOH, boiling	Haack and Gramse (1972)
gillespite ($FeO \cdot BaO \cdot 4SiO_2$)	19 N NaOH, 12 min, boiling	Haack (1973)
glass (see separate list of etchants for glasses)		
gypsum ($CaSO_4 \cdot 2H_2O$)	5% HF, 5-10 sec, 23°C	Fleischer and Price (1964b)
halite (NaCl)	1 g/ℓ $HgCl_2$ in ethanol, 30 sec, 23°C†	Komarov (unpub.)

Table 2-2 (continued)

Mineral	Etching Conditions	Reference
harmotome [$Ba_5(NaK)Al_{11}Si_{29}O_{80}\cdot 25H_2O$]	2.4% HF, 30 sec, 23°C	Haack (1973)
hardystonite ($Ca_{1.99}Pb_{.01}ZnSi_2O_7$)	1 g NaOH: 1 g H_2O, 20-70 min, 140°C	Price et al. (1970b)
heulandite [a zeolite, $(Ca,Na_2)O\cdot Al_2O_3\cdot 9SiO_2\cdot 6H_2O$]	10 ml aqua regia: 1 ml 48% HF, 30 sec, 23°C	Fleischer (unpub.)
huebnerite ($MnWO_4$)	5 g NH_4Cl: 5 g $Na_4P_2O_7$: 5 ml H_3PO_4: 20 ml H_2O, 110 min, boiling	Haack (1973)
kleinite ($Hg_2N(Cl,SO_4)\cdot nH_2O$]	37% HCl, 7 min 23°C	Fleischer, Price and Woods (unpub.)
leuchtenbergite (low-iron clinochlore)	49% HF, 10 min, 23°C	Debeauvais et al. (1964)
lithium fluoride (LiF)	H_2O + .13 gm/liter LiF + 0.5 ppm Fe, ~1 min, 23°C†	Johnston (unpub.)
margarite [$CaAl_4Si_2O_{10}(OH)_2$]	48% HF, 2 min, 23°C	Fleischer and Price (1964b)
mica [biotite, $K(Mg,Fe)_3AlSi_3O_{10}(OH)_2$]	20% HF, 1-2 min, 23°C	Price and Walker (1962a)
mica [lepidolite; zinnwaldite, $K_2Li_3Al_4Si_7O_{21}(OH,F)_3$]	48% HF, 3-70 sec, 23°C	Fleischer and Price (1964b)
mica [muscovite, $KAl_3Si_3O_{10}(OH)_2$]	48% HF, 10-40 min, 23°C	Price and Walker (1962a)
mica (phlogopite; lepidomelane, $KMg_2Al_2Si_3O_{10}(OH)_2$]	48% HF, 1-5 min, 23°C	Price and Walker (1962a)
microlite ($Ca_2Ta_2O_7$)	1 ml 48% HF; 1 ml 65% HNO_3, 6 min, 23°C	Haack (1973)
mimetite [$Pb_5Cl(AsO_4)_3$]	33% HNO_3, 3 sec, 23°C (10$\bar{1}$0 planes)	Haack (1973)
monazite [$(Ce,La,Th)(PO_4, SiO_4)$]	98% H_2SO_4, 6-8 min, 23°C †	Muralli and Rajan (unpub.)
mullite ($Al_6Si_2O_{13}$)	25N NaOH, 100°C, 4-5 h	Fleischer (unpub.)
nasonite [$Pb_4(PbOH)_2Ca_4(Si_2O_7)_3$]	1 g NaOH: 1 g H_2O, 10 min, 137°C	Fleischer, Price and Woods (unpub.)
nickel chloride; nickel bromide ($NiCl_2$, $NiBr_2$)	air (40% humidity), 10 min, 20°C (submicroscopic)	Caspar (1964)

Mineral	Etching Conditions	Reference
olivine [$(Mg,Fe)_2SiO_4$]	1 ml H_3PO_4: 1 g oxalic acid: 40 g EDTA: 100 ml H_2O: ~4.5 g NaOH*, 2–3 h, 125°C (or 6 h, 95°C)	Krishnaswami et al. (1971)
orpiment (As_2S_3)	0.25N NaOH, 10–15 min, 23°C	Perelygin and Otgonsuren (unpub.)
orthopyroxene [bronzite, $Mg_{1-f}Fe_fSiO_3$, (.1<f<.2)]	3 g NaOH: 2 g H_2O, 40 min, boiling	Lal et al. (1968)
orthopyroxene [enstatite, $MgSiO_3$]	3 g NaOH: 2 g H_2O, 35 min, boiling	Lal et al. (1968)
orthopyroxene [ferrohypersthene, $Mg_{1-f}Fe_fSiO_3$, f>.5]	3 g NaOH: 2 g H_2O, 70 min, boiling	Lal et al. (1968)
orthopyroxene [hypersthene, $Mg_{1-f}Fe_fSiO_3$, (.2<f<.5)]	3 g NaOH: 2 g H_2O, 42 min, boiling	Lal et al. (1968)
pennine (a chlorite)	48% HF, 5 min, 23°C	Fleischer and Price (1964b)
pollucite ($H_2O\cdot 2Cs_2O\cdot 2Al_2O_3\cdot 9SiO_2$)	5–8% HF, 55 sec, 23°C	Haack (1973)
pucherite ($BiVO_4$)	5% HCl, 90 sec, 23°C	Fleischer, Price and Woods (unpub.)
pyromorphite ($Pb_5Cl(PO_4)_3$)	33% HNO_3, 5 sec, 23°C, (on $(10\bar{1}0)$ planes)	Haack (1973)
quartz (SiO_2)	KOH(aq), 3 h, 150°C, or 48% HF, 24 h, 23°C	Fleischer and Price (1964b)
raspite ($PbWO_4$)	6.25N NaOH, 4 min, 23°C	Fleischer, Price, and Woods (unpub.)
sanbornite ($BaSi_2O_5$)	19N NaOH, 60 min, boiling	Haack (1973)
scheelite ($CaWO_4$)	6.25N NaOH, 90 min, 95°C	Fleischer, Price, and Woods (unpub.)
sphene ($CaTiSiO_5$)	conc HCl, 0.5–1.5 h, 90°C 6N NaOH, 20–30 min, 130°C	Naeser and Faul (1969) Calk and Naeser (1973)
spodumene ($LiAlSi_2O_6$)	48% HF, 24 h, 23°C	Fleischer and Price (1964b)
stilbite [$(Ca,Na_2)O\cdot Al_2O_3\cdot 6SiO_2\cdot 6H_2O$), a zeolite]	1% HF, 60 sec, 23°C	Fleischer (unpub.)

Table 2-2 (continued)

Mineral	Etching Conditions	Reference
stibiotantalite $[(SbO)_2(Ta,Nb)_2O_6]$	1 ml 48% HF: 1 ml 65% HNO_3, 6 min, 23°C	Haack (1970)
talc $[Mg_3Si_4O_{10}(OH)_2]$	48% HF, 15 min, 23°C	Walker (1963)
thorite $(ThSiO_4)$	H_3PO_4, 1 min, 250°C	Fleischer et al. (1965a; 1966b)
torbernite $[Cu(UO_2)_2P_2O_8 \cdot 12H_2O]$	10% HCl, 10 min, 23°C	Fleischer et al. (1965a)
tridymite (SiO_2)	10% HF, 1 h, 23°C	Fleischer et al. (1965a)
topaz $[Al_2SiO_4(F,OH)_2]$	KOH(aq), 100 min, 150°C	Fleischer and Price (1964b)
tourmaline (complex silicate)	KOH(aq), 20 min, 220°C	Fleischer and Price (1964b)
vanadinite $[Pb_5Cl(VO_4)_3]$	33% HNO_3; 1 sec, 23°C	Haack (1973)
vermiculite (biotite-derived)	48% HF, 5-10 sec, 23°C	Fleischer and Hart (unpub.)
whitlockite $[Ca_3(PO_4)_2]$	0.25% HNO_3, 10 sec to 2 min, 23°C, or olivine etch without oxalic acid	Fleischer et al. (1965a) Lal (unpub.)
willemite (Zn_2SiO_4)	33% HNO_3, 10 sec, 23°C	Haack (1973)
zircon $(ZrSiO_4)$	H_3PO_4, few sec, 375-500°C, or NaOH(aq), .25 to 5 h, 220°C, or 2 ml 48% HF: 1 ml 80% H_2SO_4 at 180°C in a pressure bomb	Fleischer et al. (1964a) Naeser (1969) Krishnaswami et al. (1973)

*to adjust pH to 8.0; † shallow pits

B. ETCHANTS FOR GLASSES

Type	Etching Conditions (at 23°C if not otherwise noted)	Reference
alumino-silicate (Corning 1720)	5.7% HF, 6 min	Fleischer and Hart (1972c)
andesitic glass ($Ab_{60}An_{40}$)	5% HF, 3-5 min 29% HBF_4: 5% HNO_3: 0.5% acetic acid, 50 min	Fleischer et al. (1969a) Macdougall (1971)
basaltic glass	20% HF, 1 min 25% HBF_4; 5% HNO_3; 0.5% acetic acid, 10 min	Fleischer et al. (1968b) Macdougall (1971)
borate glass	H_2O, 1 min	Fleischer and Price (1963b)
flint (lead-silicate) glass	5.7% HF, 3 min	Fleischer et al. (1971a)
germania glass (GeO_2)	48% HF, 6 sec	Fleischer (unpub.)
lead phosphate glass	1 ml 70% HNO_3: 3 ml H_2O, 2-20 min	Lal (unpub.)
obsidian	48% HF, 30 sec	Fleischer and Price (1964c)
phosphate glass	48% HF, 5-20 min	Fleischer and Price (1963b)
pumice	5% HF, 500 sec	Fleischer et al. (1965e)
silica glass (fused quartz; Vycor; Libyan Desert Glass)	48% HF, 1 min	Fleischer and Price (1963b)
soda-lime (microscope slide; cover slip; window glass)	48% HF, 5 sec (better: 5% HF, 2 min) 24% HBF_4: 5% HNO_3: 0.5% acetic acid, 1 h	Fleischer and Price (1963b) Macdougall (1971)
tektite	48% HF, 30 sec 24% HBF_4: 5% HNO_3: 0.5% acetic acid, 90 min	Fleischer and Price (1964a) Macdougall (1971)
uranium-soda glass	48% HF, 5 sec	Brill et al. (1964)
uranium phosphate glass	50% HF, 30 min	Hart (unpub.)
$V_2O_5 \cdot (P_2O_5)_5$ (semiconducting)	48% HF, 10 sec	Fleischer et al. (1965d)

Table 2-2 (continued)

C. ETCHANTS FOR PLASTICS

Plastic (Trade Names)	Etching Conditions	Reference
amber	30 g $K_2Cr_2O_7$: 50 ml conc. H_2SO_4, 40 h, 28°C	Uzgiris and Fleischer (1971)
cellulose acetate (Kodacel; Triafol T; Cellit)†	1 ml 15% NaClO: 2 ml 6.25N NaOH, 1 h, 40°C	Price et al. (1971)
	25 g NaOH: 20 g KOH: 4.5 g $KMnO_4$: 90 g H_2O, 2-30 min, 50°C	Somogyi et al. (1968)
cellulose acetate butyrate†	6.25N NaOH, 12 min, 70°C	Fleischer et al. (1965d)
cellulose nitrate (Diacell; Nixon-Baldwin)†	6.25N NaOH, 2-4 h, 23°C	Fleischer et al. (1965b)
cellulose propionate (Cellidor)	28% KOH, 100 min, 60°C	Becker (1969)
cellulose triacetate (Kodacel TA401, unplasticized; Bayer TN)†	1 ml 15% NaClO: 2 ml 6.25N NaOH, 1 h, 40°C	Price et al. (1970a)
dimethyl siloxane (crosslinked)	25N NaOH, 3 min, 115°C	Fleischer and Bergeron (unpub.)
formophenol (ambrolithe, phenoplaste)	6N NaOH, 1 h, 40°C; 48% HF, 30 sec, 40°C, in sequence	Monnin et al. (1966)
HBpaIT (polyester, $C_{17}H_9O_2$)	6.25N NaOH, 8 min, 70°C	Fleischer et al. (1965a)
ionomeric polyethylene (Surlyn)†	10 g $K_2Cr_2O_7$: 35 ml 30% H_2SO_4, 1 h 50°C	Besson et al. (1967)
polyamide (H-Film)	$KMnO_4$ (25% aq), 1.5 h, 100°C	Besson et al. (1967)
	6N NaOH solution	Fleischer (unpub.)
polyimide	$KMnO_4$ in H_2O	Monnin and Isabelle (1970)
poly 1-4 butylene terephthalate	1 ml 6.25N NaOH: 1 ml ethanol, 24 h, 23°C	Fleischer (unpub.)
polycarbonate (Lexan; Makrofol; Merlon; Kimfol)†	6.25N NaOH, 20 min, 50°C	Fleischer and Price (1963a)
	6.25N NaOH + 0.4% Benax*, 20 min, 70°C	Price et al. (1968a)
polyethylene	10 g $K_2Cr_2O_7$: 35 ml 30% H_2SO_4, 30 min, 85°C	Monnin et al. (1967)

Plastic (Trade Names)	Etching Conditions	Reference
polyethylene terephthalate (Mylar; Chronar; Melinex; Terphane)	6.25N NaOH, 10 min, 70°C $KMnO_4$ (25%, aq), 1 h, 55°C	Fleischer and Price (1963a) Monnin et al. (1967)
polymethyl methacralate (Plexiglas; Lucite)†	sat. $KMnO_4$, 8 min, 85°C	Monnin et al. (1966)
polyoxymethylene (Delrin)	5% $KMnO_4$, 10 h, 60°C	Monnin et al. (1966)
polyphenoxide	$KMnO_4$, (25% aq), 4 min, 100°C	Besson et al. (1967)
polyphenylene oxide (PPO)	$KMnO_4$ aq., sat., 24 h, 93°C	Fleischer (unpub.)
polypropylene (Cryovac-Y)†	35 ml 30% H_2SO_4: 10 g, $Cr_2K_2O_7$, 5 min, 94°C	Besson et al. (1967)
polystyrene	sat. $KMnO_4$, 2.5 h, 85°C	Monnin et al. (1966)
	10 g $K_2Cr_2O_7$: 35 ml, 30% H_2SO_4, 3 h, 85°C	Monnin et al. (1967)
polyvinyl acetate (Formvar)	6.25N NaOH, 200 h, 23°C	Fleischer (unpub.)
polyvinylaceto-chloride	$KMnO_4$ (25% aq.) 30 min, 100°C	Besson et al. (1967)
polyvinylchloride	sat. $KMnO_4$, 2.5 h, 85°C	Monnin et al. (1966)
polyvinylidene chloride (Saran)	$KMnO_4$ (25% aq.), 2 h, 55°C	Besson et al. (1967)
polyvinyl toluene	$KMnO_4$, sat., aq., 30 min, 100°C	Fleischer and Price (unpub.)
silicone-polycarbonate copolymer	6.25N NaOH 20 min, 50°C	Fleischer et al. (1972b)
siloxane-cellulose copolymer	8N NaOH + ~0.1% Dowfax, 3 h, 85°C	Fleischer, Viertl and Holub (unpub.)

*Dow surfactant 2A1, Dowfax, Dow Corning; presaturated with etch products (Peterson, 1970).
†Tracks of low-energy alpha particles can be revealed by etching this plastic.

2.3. ETCHING TECHNIQUES

In this section we give some guidelines that may be of use in deciding how to etch a given sample. The simplest approach is to use etchants and etching conditions that have previously been used successfully to reveal fission tracks as listed in Table 2-2. The references in the table are not exhaustive nor are they meant to establish credit. In many cases successful etchants were in use by one laboratory for several years before being discovered and published by someone at another laboratory. We have omitted reference to unpublished work unless no published reference exists. It is appropriate here to give credit to Lal and his coworkers for their major role in developing improved etchants for producing tracks with very small cone angles in many minerals.

It is instructive to ask what procedures and principles have been followed to develop the etchants in Table 2-2. Although trial and error has played and will continue to play a major role in the selection of etching conditions, certain general principles can be recommended:

(1) Use a sample known to have tracks at the surface to be etched. A weak ^{252}Cf source (with, for example, $\sim 10^5$ fissions per minute from a 0.2 cm^2 area) is a convenient means of being sure that tracks will be present (so long as the solid records and stores tracks).

(2) Use a chemical reagent that slowly dissolves the material to be etched at a constant rate. Since the tracks are normally to be enlarged to optically visible size, bulk attack is necessary. Chemical polishes which tend to round corners are not suitable since they will obscure etched tracks. The following comment should be made concerning nomenclature relevant to high polymers: A "solvent" in the language of polymer scientists is a liquid that disperses the long chain molecules into solution, thus producing what a track scientist would term a "gummy mess." "Degrading agents" which break polymer chains at random along their lengths are what are desired for etchants. For a number of the polymers oxidizing agents provide the needed degradation and hence are suitable etchants.

(3) Temperature and concentration of chemicals are variables that allow control of the attack rate so that etching occurs, but not too rapidly, as illustrated by Fig. 2-11. For example, Blanford et al. (1970a) observed that changing the concentration of sodium hydroxide solutions from 1N to 12N increased the general etching rate v_G by factors of 100, 1.6, and 5.1 for Lexan polycarbonate, Nixon-Baldwin cellulose nitrate, and Bayer cellulose acetate butyrate, respectively. Above 6N the cellulose nitrate is almost totally insensitive to concentration, while v_G for the polycarbonate is still rapidly rising. Somogyi et al. (1970) found a minimum in cone angle occurred for cellulose acetate and polycarbonate

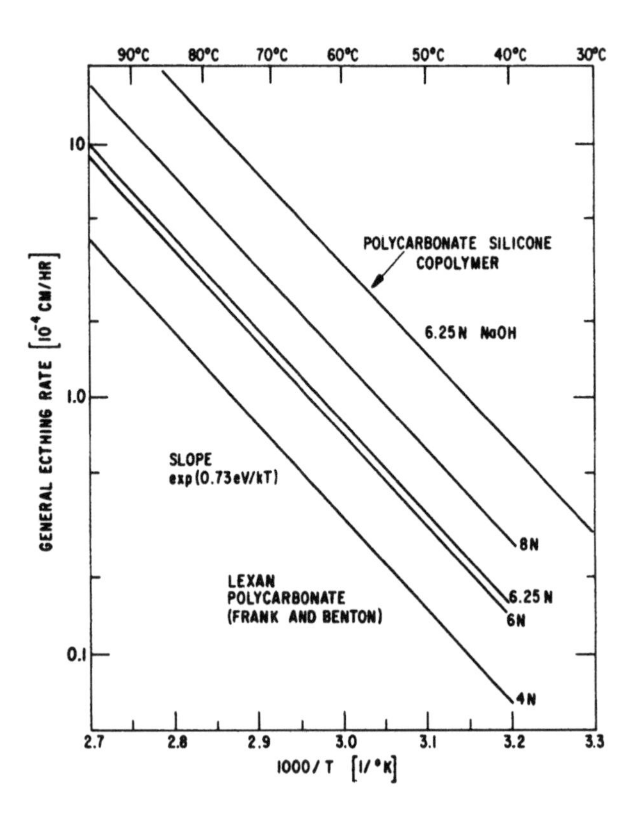

Fig. 2-11. *Etching rates are typically exponential functions of inverse temperature and increase with concentration. Lexan results from Frank and Benton (1970); copolymer results from Fleischer et al. (1972b).*

at 20% concentration of NaOH. A different type of alteration of the etching solution—adding alcohol—can speed up attack considerably. For example, a hundredfold increase in v_G for polycarbonate results from increasing the ethanol content of $KOH:C_2H_5OH:H_2O$ solutions used for etching at 50°C (Somogyi, 1972b). Ethanol increases the sensitivity and the etching rate but has the unfortunate effect of causing brittleness. Similarly, etching can be accelerated by increasing the temperature. For example, in 6.25N NaOH, a 30°C increase will speed the attack on Lexan by a factor of 10 (Peterson, 1969).

(4) Etchants that reveal dislocations—the other type of line defects commonly found in insulating crystals—normally will reveal tracks. An extensive list is given by Johnston (1961).

(5) The surfaces to be etched should be smooth on an optical scale. Cleavage of cleavable crystals or fracture of glass or noncleavable crystals will usually produce suitable surfaces. If a polished surface is used, a high quality finish is necessary in order to avoid the presence of

Table 2-3. Typical Cone Angles θ for Different Materials Irradiated with Full-Energy Fission Fragments

Material	θ
Most Minerals	
(with some etches, minerals such as LiF and zircon give angles (≳70°)	0°–5°
Glasses	
(soda, lime, borate, flint, borosilicate, tektite, obsidian, basaltic, lunar)	20°–70°
silica	13°–19°
phosphate glasses	1°–5°
Most Plastics	
(with some etches larger angles can be obtained)	0°–5°

numerous submicroscopic cracks or polishing scratches which can etch to visible dimensions and thereby obscure etched tracks or on occasion be confused with tracks by an unpracticed observer. Chapter 4 gives a more extensive discussion of sample preparation and the methods of distinguishing tracks from other linear defects in natural materials.

In examining a material for etched tracks it is helpful to have some idea what they will look like. Since their shape and appearance are primarily dependent on the cone angle, we list in Table 2-3 typical values of cone angles that are commonly observed after irradiating with fission fragments and etching. Most minerals and plastics have small cone angles and glasses large ones, but there are exceptions in each group. Often, different etches will give strikingly different cone angles in a particular substance. For a substance with a small cone angle the length of the track as given in eq. (2-1) will determine its visibility, and for a large cone angle the diameter, eq. (2-2), is the more relevant parameter. Khan and Durrani (1972a) give measurements of cone angles for a number of substances and Debeauvais et al. (1964) and Kapustsik et al. (1964) give etching efficiencies that can be used to derive cone angles.

2.3.1. Etches to Be Used for Particle Identification

The list of etchants given in Table 2-2 is intended primarily for use in revealing the presence of heavily ionizing particles such as fission fragments. More sophisticated, highly controlled treatments are necessary when the purpose is to identify individual particles by making quantitative microscopic measurements of the dimensions of their etched tracks. For some detectors, for example cellulose nitrate, commercially available 6.25N NaOH (volumetric standard solution, treated to pH 8.6, available from Anachemia Chemicals Ltd., Toronto, Montreal and

Champlain, N.Y.) at 23°C is satisfactory (Price et al., 1967a). For cellulose acetate or cellulose triacetate two of the options that are available which have proven adequate for particle identification are NaOH-NaClO solutions (Price and Fleischer, 1970) and the more complicated etch given by Varnagy (1970), described as follows:

> "Dissolve 15 g of NaOH in 50 ml water and 12 g of KOH in 40 ml water; mix the two solutions after about an hour (when they are clarified). Add 4.5 g of $KMnO_4$ after half an hour. After an hour of standing, the solution will be ready for use. Hanging on a thin plastic-coated wire, the plastics are etched at 50°C. After the development, remove the precipitated MnO_2 with 20% HCl, wash the detectors in distilled water and dry them."

The former etch gives cleaner tracks and allows easier handling; the latter yields a higher sensitivity for particle detection but embrittles and discolors the detectors.

Lexan polycarbonate has been the most widely used plastic for particle identification, with a greater variety of etches having consequently been utilized. We believe the most satisfactory etch to be the 40° NaOH etch that is presaturated with etch products (Peterson, 1970) so as to stabilize the bulk attack rate, v_G. Paretzke et al. (1973a) have shown that this procedure is necessary for Lexan polycarbonate but not for Makrofol polycarbonate.

Another procedure that has been used (Price et al., 1968a) is etching in NaOH at 70°C. The response function, like that for the saturated etch just described, has a simple analytic character (as will be discussed in Chapter 3). Since some track annealing will occur at 70°C there is danger that the particle resolution that is available could be somewhat altered. Whether the resolution is in practice significantly degraded is not presently known from published data.

Of the remaining two etches the sodium hydroxide-ethanol mixture (Price et al., 1970a) has the convenience of giving a useful attack rate at room temperature but leads to troublesome brittleness and has the undesirable property that the etching rate depends on the dip angle of the particle track—presumably an effect of diffusion of some of the components of the etch into the plastic from the surface being etched.

Finally, the unsaturated 40° NaOH etch gives a complicated response function that defies simple analytical description. This function causes many of the ending segments of tracks to be flared out into thermometer-bulb-like shapes due to an instability in v_G caused by the etch products not diffusing out of the etching track rapidly enough (Peterson, 1970).

In etching phosphate glass with hydrofluoric acid a "poisoning" effect lowers v_G if the same solution is used for too long a time, either continuously or for a series of repeated etches (Woods, 1973). Repeated use of fresh batches of reagent or of a large volume of solution relative to the surface area being etched is recommended.

2.4. ENVIRONMENTAL EFFECTS

In using particle tracks it is vital to know the extent to which either the prior history of a sample or its processing in the laboratory may have altered tracks. This knowledge leads to useful cautions in the choice of laboratory procedures and new opportunities for doing science. For example, ultraviolet radiation from fluorescent lighting will alter tracks, and consequently quantitative studies must either be done without UV exposure or with carefully chosen, controlled doses. If tracks in a material are altered by heating to 100°C, etching in boiling sodium hydroxide solution would not be a sensible choice. On the other hand, a quantitative understanding of thermal effects on tracks in a mineral is a tool that allows us to learn about thermal events that occurred in the distant past, as we will show in Chapter 4. Similarly, selective heating can be used to erase unwanted tracks in order to study those of primary interest.

The types of effects we consider here are thermal, mechanical, chemical irradiation, electrical, photochemical, and temporal.

2.4.1. Thermal Effects

By far the most pervasive and useful environmental effect on particle tracks has been the thermal alteration of tracks. The most useful type of information comes from heating a series of duplicate samples, each for a different time (t) — temperature (T) combination, and plotting the results on a diagram of log t vs. T^{-1} such as Fig. 2-12. Here solid circles indicate retention of some specific fraction of the tracks and open circles indicate fading of more than that fraction. As is usually (but not always) true, a single straight line separates the fading and no-fading regions on the diagram, indicating that a relation of the form $t = A \exp(U/kT)$, where k is Boltzmann's constant and A is another constant, describes the time and temperature for fading. Here U is the activation energy for fading. In most cases where quantitative counts have been made, U was found to be higher for a given, small fraction of tracks retained than for a large fraction of tracks retained. Fig. 4-10 shows an example of a case where the activation energies for the start and end of fading differ by a factor of nearly two.

Typically the observed activation energies have the same magnitude as those for atomic diffusion, and therefore have been interpreted (Fleischer et al., 1965d) as implying that in crystals they correspond to the diffusional motion of displaced atoms back to proper lattice positions.

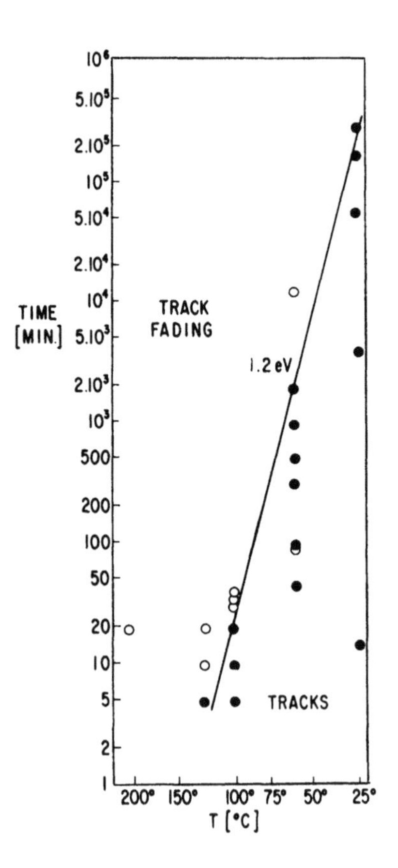

Fig. 2-12. *Variation of track annealing time with temperature for a $V_2O_5 \cdot 5\,P_2O_5$ glass. Open circles represent time-temperature combinations for which tracks disappear; closed circles are for conditions where tracks are retained. (After Fleischer et al., 1965d.)*

Detailed fading kinetics including activation energies have been measured in several dozen materials, and temperatures that allow track fading after one hour have been measured for ~20 others. These results are summarized in Table 2-4. Most of the data are for annealing of fission fragment tracks, except for a few that have been collected for iron-group cosmic ray tracks and alpha-recoil nuclei. These are less intensely ionizing particles. Their tracks being of correspondingly less severe damage, they anneal out at lower temperatures than do fission fragments; and alpha-recoil nuclei fade at lower temperatures than do iron group cosmic ray nuclei. Extensive heavy-ion annealing data that defy concise presentation are given by Price et al. (1973) for Cl, Ti, Fe, Zn, and Kr in an assortment of common olivines, feldspars, and pyroxenes. The decreases in v_T with position along the tracks are used to characterize the results.

In general, the high-energy ends of heavy-ion tracks, where the atomic damage is least, are the most readily affected by heating. Fission tracks are notable exceptions since the initial energy of each fission fragment is below that of the Bragg peak, and the damage decreases toward each end of the track.

Few systematics are recognized for track annealing properties. For example, there appears to be no correlation with melting temperature. Polycarbonate retains tracks to where it becomes a viscous fluid, whereas mullite loses all tracks at 0.5 of its melting temperature. In some cases phase changes can erase tracks, as occurs in autunite at ~60°C; in other cases phase changes do not erase them, as in quartz at 723°C. Hydrated minerals such as the micas, which may lose water, often have complicated annealing behavior (Price and Walker, 1962b; Maurette et al., 1964). Among certain crystal classes Haack (1972) has recognized some reasonably regular variations with bonding strength. These patterns could be helpful in identifying compositions that possess particular desired annealing properties. That the patterns are far from rigorous is, however, evident by the results on sphene, which Naeser has shown (Naeser and Faul, 1969; Calk and Naeser, 1973) depend on the etch used.

The extrapolation into the past of the annealing kinetics measured in the laboratory is a necessary and useful exercise in evaluating the meaning of natural tracks in solids. Such extensions of the annealing results are therefore considered in Chapter 4, where natural tracks in solids are discussed.

2.4.2. Mechanical Effects

The effects of stress or strain on fission tracks in solids have been evaluated for a limited number of substances that have undergone applied hydrostatic stresses and shear stresses, both static and dynamic.

The effects of hydrostatic stresses are typically much less important than that of temperature (Fleischer et al., 1965c). In olivine and zircon a pressure of 80 kilobars produced an effect equivalent to an increase in temperature of only ~50°C. Tektite glass was more sensitive, 10 kb being equivalent to a temperature shift of ~150°C.

We can compare these properties with typical pressure and temperature increases with depth in the earth. Pressure increases at roughly 0.3 kb/km and temperature at ~15°/km (Roy et al., 1968), so that for deeply buried tektite glass the effect of pressure is minor but not negligible, whereas for the crystals that have been studied the effects of hydrostatic pressure are negligible.

Nonhydrostatic stresses have been evaluated quantitatively only for shock-loaded samples. In applying shock to individual minerals under laboratory conditions Ahrens et al. (1970) found that tracks were partially erased at 112 kb pressure in tektite glass and 42 kb in soda-lime glass, that they survived 45 kb in apatite but that the crystals were fragmented at higher stresses, and that tracks were retained in biotite even at 173 kb. In examining minerals from rocks shocked in nuclear explosions Fleischer et al. (1974) found the start of fading in apatite to

Table 2-4. Track Annealing Characteristics

Note: Data given here pertain (except as noted) to fission fragment tracks, which in general will have different annealing characteristics from other tracks. Unless noted differently, all data are for annealing in air at atmospheric pressure. For extensive annealing data for tracks of Kr, Zn, Fe and Cℓ ions in various silicate minerals, see Price et al. (1973).

	Activation Energy (Electron Volts)			1 Hour Annealing Temperature [°C]			
Material	Total Fading	50% Track Loss	Start of Track Loss	Total Fading	50% Track Loss	Start of Track Loss	Reference
Amber (Baltic)	-	3.0	-	-	110	-	Uzgiris and Fleischer (1971)
Apatite	-	-	-	530	-	400	Fleischer and Price (1964b)
	-	2.2	-	-	336	-	Wagner (1968)
	2.8	2.1	1.62	375	322	275	Naeser and Faul (1969)
Aragonite ($CaCO_3$)	-	-	-	150	-	130	Fleischer et al. (1968a)
Autunite	-	-	-	60	-	40	Fleischer and Price (1964b)
Barysilite	-	3.3	-	-	440	-	Fleischer, Price and Woods (unpub.)
Calcite ($CaCO_3$)	-	-	-	-	200(g)	-	Macdougall (unpub.)
Cellulose acetate (Cellit-T)				165	~160	>100	Somogyi (1972b)
Cellulose nitrate							
(Nixon-Baldwin)	-	-	-	-	85	-	Fleischer et al. (1965d)
(Daicel; fission frag.)	-	-	-	147	~140	>110	Somogyi (1972b)
(Daicel; alpha particles)	-	-	-	138	~130	<110	Somogyi (1972b)
Epidote	18.2(a)	12.5	6.0	715	680	575	Naeser and Dodge (1969)
	-	8	-	-	650	-	Wagner (1972)
Feldspar (Albite)	~5	-	-	775(b)	-	-	Fleischer et al. (1967)
Feldspar (Anorthite)	-	-	-	~680(b)	550(b)	350(b)	Crozaz et al. (1970)
Feldspar (Bytownite)	-	6.6	-	790	750	690	Fleischer et al. (1965d)
Garnet ($Ca_{2.91}Fe_{.09}$) $\cdot(Fe_{.82}Al_{1.18})\cdot(SiO_4)_3$	8.4	5.4	2.1	690	665	560	Haack and Potts (1972)
Glass (Aluminosilicate Corning 1720)	-	-	-	>500	320	100	Fleischer and Hart (1972c)
Glass (Andesitic impact glass)	2.5	-	1.1	250	-	190	Fleischer et al. (1969a)

Glass (Basaltic, tachylitic impact glasses)	-	2.6	-	-	190 to 225	-	Fleischer et al. (1969a)
Glass (Basaltic)	1.9	1.6	1.3	240	190	140	Macdougall (1973)
Glass (Basaltic, tholeitic)	-	1.74	-	300	-	280	Fleischer et al. (1968b)
	-	2.2	-	-	275	-	Fleischer et al. (1971c)
	-	2.3	-	-	275	-	Aumento (1969)
Glass (Borosilicate, pyrex)	>2.5	1.2	-	380	302	-	Fleischer (unpub.)
	-	-	-	-	275	-	Fleischer et al. (1965d)
Glass (Lunar black, Apollo 12)	-	1.3	-	-	250	-	Fleischer et al. (1971b)
Glass (Feldspathic)	-	-	-	235	-	200	Fleischer et al. (1968c)
Glass (Lunar green, Apollo 15, >2.6 g/cc)	-	3.5(b)	-	417(b)	337(b)	321(b)	Fleischer and Hart (1972b)
Glass (Lunar green, Apollo 15, <2.6 g/cc)	-	5.3(b)	-	372(b)	307(b)	285(b)	Fleischer and Hart (1972b)
Glass (Libyan desert, 98% SiO_2)	3.6(c)	1.9	1.2(d)	570(c)	460	330(d)	Storzer and Wagner (1971)
	-	-	1.3(e)	-	-	335(e)	Gentner et al. (1969)
Glass (Obsidian)	-	1.8	-	-	390	-	Suzuki (1970)
Glass (Obsidian, .3% H_2O)	-	2.3(f)	-	-	262(f)	-	Lakatos and Miller (1972b)
Glass (Obsidian, .8% H_2O)	-	2.0(f)	-	-	250(f)	-	Lakatos and Miller (1972b)
Glass (Obsidian, 2.2%, H_2O)	-	1.5(f)	-	-	190(f)	-	Lakatos and Miller (1972b)
Glass (Ryolite)	-	1.9	-	-	262	-	Aumento and Souther (1973)
Glass (Phosphate)	-	-	-	300	-	150	Perelygin et al. (1969)
(Phosphate)	-	-	-	190	130	90	Fleischer and Hart (1972c)
Glass (Pitchstone)	-	1.6	-	-	450	-	Storzer (1970)
Glass (Silica)	-	-	-	700	-	-	Fleischer and Price (1963b)
	-	-	-	>500	290	100	Fleischer and Hart (1972c)
Glass (Soda lime)	-	-	-	370	-	-	Fleischer and Price (1963b)
	-	-	-	-	190	-	Fleischer and Hart (1972c)
Glass (NBS soda lime) (4 glasses)		1.5-2.0	0.7-1.2		230	110	Reimer et al. (1972)
Glass (Australian tektite)	2.8	1.7	1.05	510	380	225	Storzer and Wagner (1969)

Material	Activation Energy (Electron Volts)			1 Hour Annealing Temperature [°C]			Reference
	Total Fading	50% Track Loss	Start of Track Loss	Total Fading	50% Track Loss	Start of Track Loss	
Glass (Bediasite tektite)	3.5(c)	1.6	.94(d)	510(c)	370	220(d)	Storzer and Wagner (1971)
	2.8(g)	1.6	1.15(e)	490(g)	360	235(e)	Durrani and Khan (1970)
Glass (Indochina tektite)							
1 atm	-	2.51	-	-	500	-	Fleischer and Price (1964a)
10 kbar	-	1.7	-	-	360	-	Fleischer et al. (1965c)
60 kbar	-	.75	-	-	100	-	Fleischer et al. (1965c)
Glass ($V_2O_5 \cdot 5P_2O_5$)	-	1.2	-	-	95	-	Fleischer et al. (1965d)
Hardystonite	-	5.7	-	-	450	-	Fleischer, Price and Woods (unpub.)
Hornblende	-	-	-	630	590	530	Fleischer et al. (1968a)
	-	-	-	750	720	650	Maurette (1970); Crozaz et al. (1969)
	-	-	-	500(h)	315(h)	100(h)	Maurette (1970); Crozaz et al. (1969)
Mica (Biotite)	-	1.7	-	-	-	-	Shukolyukov and Komarov (1966)
Mica (Lepidolite)	-	2.0	-	-	-	-	Shukolyukov and Komarov (1966)
Mica (Muscovite)	-	-	-	540	510	450	Fleischer et al. (1964b)
	-	2.8	-	680	-	510	Widell (1969)
	-	-	-	550	-	275	Perelygin et al. (1969)
	-	-	-	700	670	550	Maurette (1970)
	-	-	-	600(h)	520(h)	290(h)	Maurette (1970)
	-	1.0	-	-	-	-	Shukolyukov et al. (1965)
	-	-	-	>600	540	400	Mehta and Rama (1969)
Mica (Phlogopite)	-	1.8	-	550	-	350	Maurette et al. (1964); Fleischer et al. (1964b)
Monazite	-	0.3-0.7	-	-	300	-	Shukolyukov and Komarov (1970)
Mullite				270(±130)	525(±25)	625(±75)	Fleischer, Hart and Giard (unpub.)
Nasonite	-	5.3	-	450	-	-	Fleischer, Price and Woods (unpub.)
Olivine (1 atm and at 80 kbar)	3.3	-	-	500	-	-	Fleischer et al. (1965c)
Pollucite	-	1.3	-	-	670	-	Fleischer, Price and Woods (unpub.)

Polycarbonate (Lexan)	-	-	-	>185	-	-	Fleischer and Price (1963a)
	-	-	-	-	-	<40	Hart, Giard and Fleischer (unpub.)
				-	170	110	Khan and Durrani (1972a)
(Makrofol)				200	~190	100	Somogyi (1972b)
Pyroxene (Augite)				560(b)	480(b)	300(b)	Crozaz et al. (1970)
Pyroxene (Diopside)	-	~10	-	-	875	-	Fleischer et al. (1967)
	-	-	-	880	850	820	Fleischer et al. (1968c)
Pyroxene (Enstatite)	-	~5(b)	-	-	450(b)	-	Fleischer et al. (1967)
Pyroxene (Hypersthene)	-	-	-	475(b)	-	-	Fleischer et al. (1968c)
	-	-	-	600	-	525	Maurette (1970)
	-	-	-	500(b)	330(b)	300(b)	Maurette (1970)
Pyroxene (Pigeonite)	-	-	-	530	-	500	Fleischer et al. (1968a)
Quartz (SiO_2)(b)	-	-	-	1050	-	1000	Fleischer et al. (1968a)
	4.0	-	-	-	-	-	Kosanke (1972)
Sphene (HCl etch)	5.5	4.0	3.0	637	620	520	Naeser and Faul (1969)
(NaOH etch)	5.5	4.8	4.3	780	740	680	Calk and Naeser (1973)
Zircon (1 atm)	-	3.6	-	-	700	-	Fleischer et al. (1964a)
(80 kbar)	-	3.6	-	-	675	-	Fleischer et al. (1965c)

(a) No fading observed at T < 600°C. (b) Fe-group cosmic ray tracks. (c) 95% loss. (d) 5% loss. (e) 10% loss. (f) 30% loss. (g) 90% loss. (h) Alpha interaction tracks. (i) Synthetic quartz has lower track retentivity.

occur at ~25 kb, 95% fading at 100 (±25) kb, and total erasure above 400 kb. Tracks in sphene were retained at 30 kb, 95% erased at 100 kb, and erased totally at ~350 kb. Considering that there are considerable differences in the state of stress and duration of the pressure pulse, the two sets of results on the one common mineral (apatite) are regarded as self-consistent.

Track disappearance in the presence of shear deformation is believed to be caused by individual tracks being cut into shorter tracks where active slip planes intersect the tracks, as shown schematically in Fig. 2-13. When tracks are shortened suddenly by this mechanism, they cease to be visible in the microscope and hence appear to have been erased. This phenomenon (Fleischer et al., 1972a; Fleischer and Hart, 1973b) is discussed briefly in Chapter 4 in relation to Fig. 4-17 and further in Chapter 6 in noting how this phenomenon is useful in lunar applications.

Static shear stresses have also been found to produce track fragmentation of the sort depicted in Fig. 2-13. The observation was made in mica by Fleischer et al. (1965c), who also noted that a 1 kb shear stress had no effects on tracks in a tektite glass.

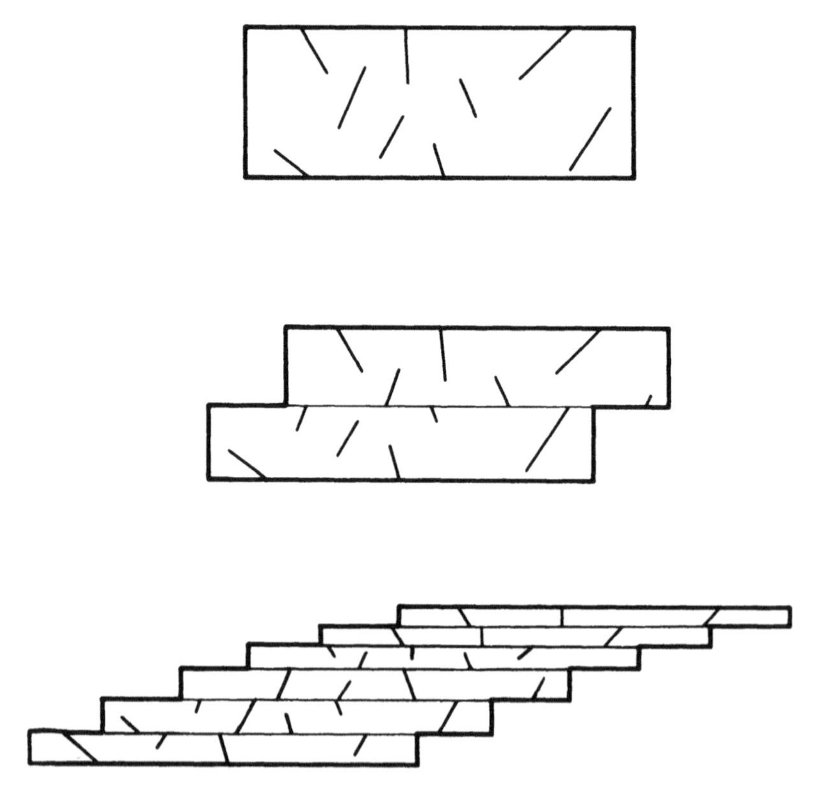

Fig. 2-13. *Intersecting slip cuts tracks into sub-portions whose median lengths are twice the spacing of the slip planes. Short lengths may not be visible when viewed in the optical microscope. (After Fleischer and Hart, 1973b.)*

If fragmented tracks from natural static deformation could be identified in minerals that are appropriate for geochronology, they would be useful for dating of ancient earth movements. Whether such cases can be found is a question yet to be resolved and a promising area for further exploration.

2.4.3. Chemical Effects

Chemical effects on tracks can arise from pretreatments that alter etching rates, and from effects of water on track annealing properties. Table 2-5 summarizes effects that have been observed.

Treatments during irradiation with active agents such as O_2, O_3, H_2O_2, and even H_2O have been observed to increase the track etching rate of polymers, whereas neutral environments such as a vacuum or an N_2 atmosphere decrease v_T, presumably by excluding oxygen. For a qualitative physical picture of the positive effects, imagine the active species as diffusing along the radiation-damaged core of a track and attaching to the active species that make up the track (free radicals, broken bonds, displaced atoms, . . .), thereby preventing broken polymer chains from reuniting and repairing part of the damage.

An interesting and useful effect that has been observed by M. Monnin (personal communication) is resensitization of cellulose triacetate that was irradiated in a vacuum. Even when the desensitization prevented alpha particle tracks from being seen after a normal etch, the detector could be resensitized by treatment in oxygen at pressures above 100 atmospheres.

No atmospheric environmental effects have as yet been demonstrated for crystals or glasses. Effects of water content in the structure have, however, been examined for mica by Lakatos and Miller (1970; 1973) and for volcanic glass by Lakatos and Miller (1972a,b). They have performed annealing experiments which demonstrate that for both minerals annealing proceeds more rapidly the higher the water content. The quantitative results are included in Table 2-4.

2.4.4. Electrical Effects

Two effects which were first thought to be of electrical origin were observed by Crannell et al. (1969) and Blanford et al. (1970b). Crannell and coworkers exposed cellulose nitrate to alpha particles and found that the fraction that registered was enhanced when an electrical field was applied across the plastic sheet, independent of the sign of the field. In the light of our present knowledge that ozone enhances track-etching parameters in cellulose nitrate, it may be that ozone is the agent responsible for the observed effect.

Blanford et al. (1970b) observed a sensitivity enhancement of Lexan polycarbonate in the presence of a spark discharge. Although ozone was also produced here, their experiments eliminated ozone and the direct electrical effects of the

Table 2-5. Chemical Effects on Particle Tracks

Track Detector	Chemical	Environmental Effect	Reference
Plastics			
cellulose acetate	ozone	v_G and v_T increased	Somogyi (1972a)
cellulose acetate	oxygen	v_G and v_T increased	Kartuzhanskii et al. (1970)
cellulose triacetate	humid air	v_T/v_G increased	Becker (1968)
cellulose triacetate	nitrogen	v_T/v_G decreased	Becker (1968)
cellulose triacetate	H_2O_2	v_T/v_G increased	Becker (1968)
cellulose nitrate	high electric field (ozone?)	v_T/v_G increased	Crannell et al. (1969)
cellulose nitrate	oxygen	v_T/v_G increased	Boyett et al. (1970)
cellulose nitrate	oxygen	v_T or v_G increased	Kartuzhanskii et al. (1970)
cellulose nitrate	camphor added	v_T/v_G increased	Veprik et al. (1970)
cellulose nitrate	ozone	v_G increased	Somogyi (1972a)
cellulose nitrate	water at 50^oC	v_G and v_T increased equally	Enge et al. (1970)
polycarbonate	nitrogen	v_T decreased	Crawford et al. (1968)
polycarbonate	products of spark discharge (a nitrogen oxide?)	v_T increased	Blanford et al. (1970b)
polycarbonate	Cl replacement of H	v_T/v_G decreased	Pai (1973)
polycarbonate	oxygen	v_T increased	Monnin (1968)
polycarbonate	oxygen	v_T increased	Henke et al. (1970)
polycarbonate	ozone	v_G and v_T increased	Somogyi (1972a)
Minerals			
muscovite mica	H_2O pressure	increases fading rate	Lakatos and Miller (1970, 1973); Miller and Lakatos (1970)
volcanic glass	H_2O pressure	increases fading rate	Lakatos and Miller (1972a,b)

discharge itself. The authors suggested that one of the nitrogen oxides produced by the discharge was the most likely source of the effect. Since nitrous oxide has been observed to have an opposite effect, one of the other oxides presumably was responsible.

2.4.5. Photochemical and Irradiation Effects

Irradiation with ultraviolet and higher energy photons and with electrons or other particles that do not themselves make tracks can have significant and sometimes profound effects on the properties of track detectors. Except at extremely high doses the observed effects are confined to plastics.

Crawford et al. (1968) and, independently, Benton and Henke (1969) discovered that ultraviolet light in the presence of oxygen can greatly increase the etching rate of tracks in Lexan without significantly changing v_G. Unirradiated Lexan is optically transparent at wavelengths longer than ~2750 Å (but see section 2.6.4 for an exception.) Lexan that has been subjected to a massive dose of radiation has a strong absorption peak at ~3050 Å. Individual tracks also absorb strongly at about the same wavelength if the Lexan contains oxygen. Presumably, the energy deposited breaks additional chemical bonds along the track and these are stabilized by oxidation. Further work has been done by DeSorbo and Humphrey (1970), by Henke et al. (1970), and by Stern and Price (1972). The latter showed that with suitable cooling, Lexan could be subjected to very high doses of UV at ~3100 Å so that alpha particles of energies up to ~3 MeV could be revealed, whereas normal, untreated Lexan does not record alpha particles at all. In Chapter 3, we discuss the processing of Lexan with UV for the purpose of particle identification. As noted in Table 2-6, NO and H_2O_2 are other agents in addition to O_2 that have been shown to be active in promoting the ultraviolet enhancement of v_T in Lexan. In no other plastics have track-etching rates been increased by photo-oxidation reactions to nearly the extent observed in Lexan.

Irradiation by lightly ionizing radiation can either promote cross linking of a polymer and hence decrease v_G, as we have observed in dimethyl siloxane, or can cut the long chain molecules, converting them into shorter, more rapidly etchable materials. The latter behavior, as Table 2-7 indicates, has been the more usual result with irradiated plastics. Tens to hundreds of Megarads (1 Megarad = 10^8 ergs/cm^3) are typically needed to produce noticeable effects. In some cases, as after bombardment with alpha particles (Stone, 1969) or fast neutrons (Medveczky and Somogyi, 1966), tracks from recoiling carbon, nitrogen, or oxygen can form a background of shallow pits which in effect elevate v_G. With electron or gamma ray irradiation the damage is dispersed on an atomic scale, so that it is not possible to see the individual etched features resulting from the defects responsible for accelerated etching.

Table 2-6. Photo-Oxidation Effects on Particle Tracks

Material	Exposure	Effect	Reference
cellulose nitrate	O_2 + UV	sensitivity increased (plus bulk damage)	Henke et al. (1970)
polycarbonate	O_2 + UV	v_T increased (max. increase at 3040 (±100) Å	Crawford et al. (1968); DeSorbo and Humphrey (1970)
polycarbonate	O_2 + UV	v_T increased (max. increase at 3130Å)	Henke et al. (1970)
polycarbonate	N_2O + UV	v_T decreased	DeSorbo and Humphrey (1970)
polycarbonate	NO + UV	v_T increased	DeSorbo and Humphrey (1970)
polycarbonate	H_2O_2 + UV	v_T increased; sensitivity increased	Caputi and Crawford (1970); Henke et al. (1970)
polyethylene terephthalate	O_2 + UV	sensitivity increased	Henke et al. (1970)

Table 2-7. Irradiation Effects on Particle Track Etching

Material	Irradiation	Effect	Reference
cellulose nitrate (Daicell)	1.5 MeV electrons, 16 MR (Megarads)	v_G doubled	Price, Fleischer and Miller (1969, unpub.)
cellulose nitrate (Daicell)	thermal neutrons, 10^{14} nvt	v_G increased factor of 6	R. Scanlon (unpub.)
cellulose nitrate (Hercules)	5 MR, γ-rays	10% increase in v_G	Boyett and Becker (1970)
cellulose nitrate (Hercules)	.006 MR, fast neutrons	10% increase in v_G	Boyett and Becker (1970)
cellulose triacetate (Triafol T)	6 MR, γ-rays	1% increase in v_G	Boyett and Becker (1970)
cellulose triacetate (Triafol T)	.008 MR, fast neutrons	0.5% increase in v_G	Boyett and Becker (1970)
cellulose triacetate (Kodacel)	1.5 MeV electrons, 12 MR	v_G doubled	Price, Fleischer and Miller (1969, unpub.)
dimethyl siloxane	1.5 MeV electrons, 20 MR	v_G decreased	Fleischer, Hart and Nichols (1972, unpub.)
polycarbonate (Lexan)	1.3 MeV γ-rays, 200 MR	v_G doubled	Frank and Benton (1970)
polycarbonate (Lexan)	1.5 MeV electrons, 25 MR	v_T increased	DeSorbo (1968, unpub.)
polycarbonate (Lexan)	alpha particles, 100 MR	v_G increased by recoil pits	Stone (1969)
polycarbonate (Kimfol)	.006 MR, fast neutrons	10% increase in v_G	Boyett and Becker (1970)
polycarbonate (Kimfol)	8 MR, γ-rays	<0.1% increase in v_G	Boyett and Becker (1970)
polyethylene terephthalate (Mylar)	.007 MR, fast neutrons	1% increase in v_G	Boyett and Becker (1970)
polyethylene terephthalate (Mylar)	7 MR, γ-rays	1% increase in v_G	Boyett and Becker (1970)
phosphate glass	UV, 2×10^{11} ergs/cm^2	v_G or v_T decreased	Hart et al. (1973, unpub.)
natural micas	100 keV electrons	track fading	Silk and Barnes (1959)
olivine; muscovite; biotite; phlogopite; tektite glass	1.5 MeV electrons, 2×10^3 MR	no effect	Fleischer et al. (1965c)
silica glass	20 keV electrons, 3×10^5 MR	v_G doubled	Krätschmer (1971)
silica glass	UV, 2×10^{11} ergs/cm^2	no effect	Hart et al. (1973, unpub.)
soda glass	UV, 2×10^{11} ergs/cm^2	no effect	Hart et al. (1973, unpub.)
soda glass	1.4×10^{17} protons of 3 MeV	v_G doubled; v_T increased 50%	Durrani (1973, unpub.)

In the case of crystalline minerals high doses of lightly ionizing radiation have so far failed to alter measured etching parameters.

Irradiation effects have been observed for glasses. Krätschmer (1971) showed that 3×10^5 Megarads of 20 keV electrons would double the general rate of etching of silica glass. This dose is more than three orders of magnitude greater than that required to produce an effect of similar magnitude in polymers. If one uses intense ultraviolet irradiation equivalent to a one- or two-day exposure to unattenuated solar UV, phosphate glasses (but not silicate glasses) can be affected in the opposite sense, i.e., etching is slowed. The physical mechanisms that cause the above effects in glass are presently unknown.

2.4.6. Temporal Effects

After Lexan polycarbonate has been irradiated with track-forming particles, the reactivity (as measured by v_T) increases with time at room temperature even when the sample is protected from ultraviolet illumination. For neon ions the effect increases rapidly during the first two days but slows to a negligible rate after two to three weeks (Peterson, 1969; Price and Fleischer, 1970). Henke et al. (1970) found for ^{40}Ar ions that v_T increases logarithmically with time for at least six months. The aging must occur in the presence of oxygen and is speeded by ultraviolet irradiation. Polycarbonate film that was irradiated and then stored in vacuum for up to 6 months did not show tracks after chemical etching (personal communication from R. Caputi), unless the normal waiting time in air was allowed after the exposure to vacuum.

The existence of this delay time for slow chemical reactions along tracks implies that when quantitative track etching rate measurements are to be made, a waiting time must be specified. Although the effect was noted earlier by workers who had irradiated polycarbonate with fission fragments in nuclear reactors (Engelkemeier and Walton, 1964), it was not made quantitative until Peterson's work.

2.5. SCANNING TECHNIQUES

The basic information needed for many track studies consists simply of measurements of track densities. While often a total track count is needed, sometimes tracks with certain distinctive characteristics need to be selectively counted. The most direct, commonly used counting method is manual observation through the microscope using an eyepiece with a calibrated grid and counting the number of tracks in a series of fields of view. Details of techniques for track recognition in the presence of other defects are presented in Chapter 4, where we discuss natural tracks.

Microscopes that many workers have found suitable for this purpose are the Leitz Ortholux and the Zeiss Ultraphot or Photomicroscope. The Leitz turret head has the special convenience of allowing rapid changes from one magnification to another. For prompt photographic work it is helpful to have a Polaroid camera attachment. An adapter to allow such an attachment for the Leitz Ortholux can be obtained from Monroe Microscope Service, 969 S. Clinton, Rochester, N.Y., 14620. It is also important to equip the microscope for viewing samples in both reflected and transmitted light simultaneously.

Often it is desirable or even necessary to enhance the ease of recognition of tracks or to use automated or analogue readout techniques, either because extensive statistics are needed, the track density is low, the samples to be examined are numerous, or the contrast between the tracks and the background is low. In this section we first consider methods of improving contrast, then methods for enlarging the track or marking its location clearly, and finally, more automated, but less direct, readout methods.

2.5.1. Enhancing Track Contrast

Contrast enhancement can be achieved by interference contrast microscopy (Piesch, 1971), by filling etched tracks with various opaque substances such as metals (Possin, 1970), dye (Cross and Tommasino, 1967), or ink (Lal et al., 1968), or by coating the track with a thin opaque metal film (Macdougall et al., 1971; Seitz et al., 1973). For minerals and glasses the best method is probably that of Macdougall et al. (1971), who deposit silver inside etched tracks by means of the silvering solution traditionally employed for coating vacuum flasks. After the layer of silver on the surface of the sample has been wiped away, the silvered tracks can be viewed with the high resolution attainable with oil immersion lenses and yet can later be etched further, if desired, after cleaning out the silver using nitric acid. Seitz et al. (1973) have shown that different observers can make track counts that are reproducible to within $\pm 3\%$ if they use reflected light to observe etch pits on mica surfaces onto which silver has been evaporated. Other procedures involve inserting a fluorescent dye (for example Zyglo, available from Magnaflux Corp., 7309 W. Ainslie Ave., Chicago, Ill.) in the tracks and viewing under ultraviolet light (Parizet and Monnin, 1969; Morley, 1972); using polarized light (Medveczky, 1969) through a crossed polarizer to see only light scattered by track surfaces (Morley, 1972), or light that is directly transmitted through tracks (Fleischer, 1971, unpublished); and the use of colored detector films viewed through filters that are opaque to the wavelengths transmitted (Fleischer and Walker, 1971). Fig. 2-14(a) shows an example of contrast obtained using Kodak Pathé LR-115 red-dyed cellulose nitrate and the particular etch recommended (Table 2-2). The 8 μm-thick dyed layer is removed along the etched tracks; when viewed through a green filter, light is transmitted only where it has passed through the holes.

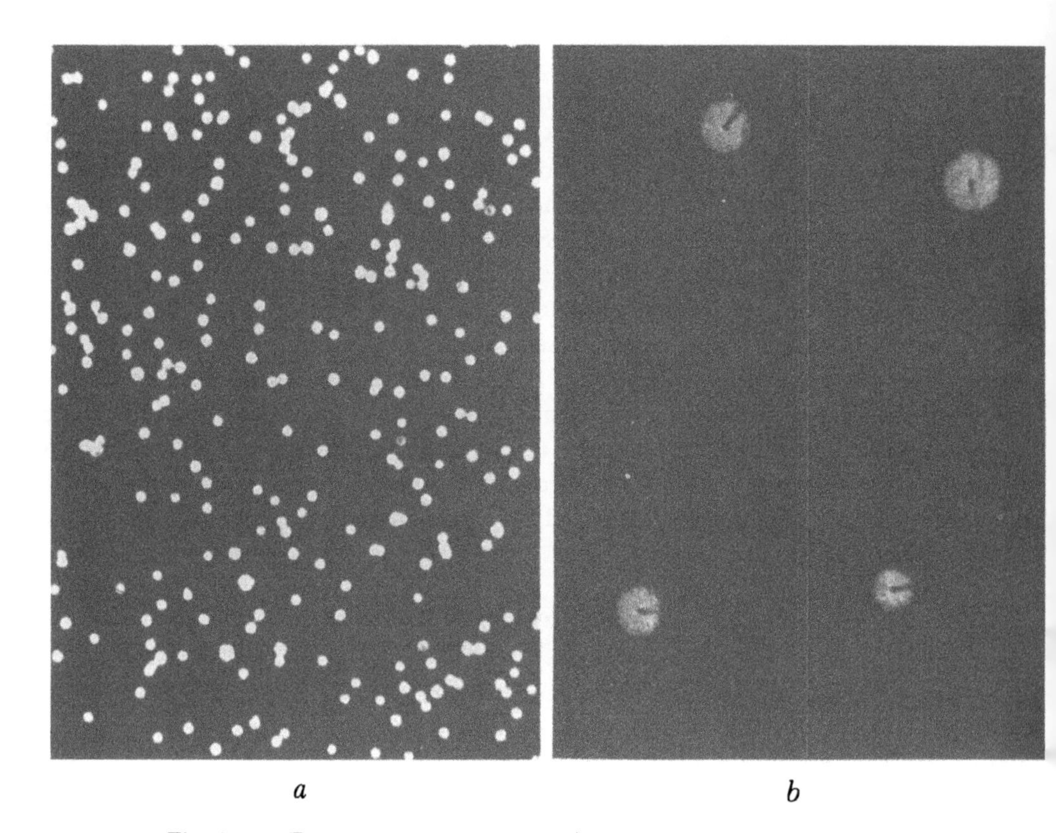

a *b*

Fig. 2-14. *Track contrast enhancement: (a) Track-etched holes in a special red Kodak Pathé LR-115 film show high contrast when viewed with green light (after Fleischer et al., 1972c). (b) A Lexan-silver-Lexan sandwich structure reveals track locations by dissolving silver to allow light transmission. Tracks are etched through one layer of Lexan to the silver. A separate reagent dissolves silver around the tracks (Viertl and Fleischer, unpublished).*

Having a defect-free surface can be a great help. One simple method of producing a scratch-free background on which tracks will be readily visible is the use of flame polishing, i.e., melting a thin surface layer and allowing it to resolidify as it cools. This procedure is generally useful on glasses (Fleischer and Price, 1963b) and with gentle heating can even be used on some plastics (Fleischer and Price, 1963a; Gilliam and Knoll, 1970). Casting of plastics from solution onto water, mercury, or a clean plate is another alternative.

Surfaces are often produced by grinding and polishing, and considerable care (as described in detail in Chapter 4) is necessary to avoid leaving polishing defects that can increase scanning time considerably or can in extreme cases obscure tracks.

In some cases it is desired to count fission tracks in the midst of an intense background of short recoil tracks, such as fast neutrons induce in plastics. Because the etching rate (v_T) along the fission tracks is greater than that along recoil tracks, a short etch will produce fine, long holes along these longer tracks before the recoil tracks are etched to visible size. Stone (1969) and Somogyi (1972b) have noted that if, after a short, preliminary etch the plastic is heated to remove the recoil tracks, subsequent re-etching will enlarge the fission tracks to a convenient size without the troublesome recoil track background.

Both optical scanning and specifying the location of an observed track for later re-examination can be greatly facilitated by having a regular indexed grid on the sample being scanned. Birabeau et al. (1972) have described detailed procedures which can be used to photoengrave useful grids on cellulose nitrate or polycarbonate detectors.

2.5.2. Track Location and Enlargement

Where track densities are unusually low (which may mean 10^{+4} to $10^{-5}/cm^2$), it is useful to have ways of locating and/or enlarging individual tracks. Table 2-8 lists a number of the techniques that have been developed. Each method makes use of the transmission of something through holes that cross or nearly cross the detector sheet. The transmitted substance may be dye or a gas that acts as a dye, an etchant that dissolves an opaque metal film, or charge carriers that supply an electrical current. A case where the track needs only to "nearly cross" the detector occurs when a high voltage is applied across a film with a hole that ends nearly all the way through the film. Dielectric breakdown can open and enlarge such a track so that it can be located quickly at low magnification. Measurements of background counts on etched but unirradiated films must be made since the track shapes are altered almost unrecognizably in this type of breakdown. Use of lower voltages (where dead-ended holes are not punched through) allows through holes to be located by spark-scanning without altering the etched track (Blanford et al., 1969). A second case where nearly crossing the sheet is adequate is the UV technique of Pretre et al. (1968).

In the aluminum-backed plastic technique (Fleischer et al., 1966a) listed in Table 2-8, etching with NaOH from one side of an aluminized detector creates holes through to the aluminum, which is then dissolved away to form a hole in the otherwise opaque film. The procedure suffers from two defects: the etched sheet is fragile, since it must be thin to allow tracks to cross it; the holes in the aluminum are nonuniform in size, since the tracks at normal incidence etch through the sheet before those at lower angles do. By sandwiching a thin, metal-backed Lexan detector against a thick piece of Lexan at elevated temperature a more rugged assembly can be made (and one that is less susceptible to catastrophic leaks). If the aluminum-backed side is kept in dry air, the NaOH solution dries as it emerges through the tracks, thus limiting the diameter of the holes to a few μm. By using a metal such as silver, which is not attacked by NaOH, the holes can be etched in

Table 2-8. Methods of Locating or Enlarging Tracks

Technique	Track Density Where Useful	Limitations or Special Features	Reference
Etch aluminum-backed plastic from front; transmits light where metal removed.	<100/cm^2	Oblique tracks require longer etch; non-uniform hole sizes (but see text).	Fleischer et al. (1966a)
Apply potential; (a) observe spark location; (b) scan enlarged holes.	 <10/cm^2 <10^4/cm^2	High voltage can enlarge holes, including ones that were not quite open.	Flemons (1968); Cross and Tommasino (1968); Lark (1969)
Apply potential; scan holes evaporated in thin aluminum coating by discharge.	<3000/cm^2	Alters etched tracks.	Cross and Tommasino (1970, 1972).
Press dye through holes onto filter paper.	<10/cm^2	Separate sheet shows track location.	Price (1963, unpub.); Alter (1968); Sherwood (1970); Khan (1971).
Blow NH_3 through holes, thus staining sensitive paper.	<1/cm^2	Gives separate sheet showing track location; rapid scanning rate (10^3 cm^2/min).	Blok et al. (1969).
Measure resistance of track etched across membrane in electrolyte.	[1 to 10 tracks total]	Can give etching rates v_T and v_G	Walker et al. (1963); DeSorbo and Humphrey (1970); Bean et al. (1970); DeBlois and Bean (1970).
Enlarge images using slide projector.	10^4 to $5x10^6$	Individually counted on projection screen.	Khan and Durrani (1972b)
Measure UV light transmission in Mylar, which strongly absorbs.	10^3 to 10^7	Detects light when hole not quite through sheet.	Pretre et al. (1968).

the detector without removing any of the metal film; and when a second etch is done with nitric acid, which dissolves the silver, holes of nearly constant diameter can be made in the silver, as shown in Fig. 2-14b.

The fastest of the methods listed in Table 2-8 is the ammonia scanning technique. This procedure has the additional merit of not altering the shapes of the etched tracks, so that the individual locations where ammonia has penetrated the sheet can be examined microscopically and the flaws separated from the tracks. In the original form Blok et al. (1969) laid an etched plastic sheet against a photosensitive copying paper, taped the two around the edges with a gas-tight seal, and passed the pair through the final stage of an Ozalid copying machine so that ammonia gas passed through holes in the plastic and produced blue spots on the sensitive paper. A more controllable method is to tape the copying paper and plastic tightly onto a cylindrical metal backing with a large radius of curvature and to place the assembly into a transparent box into which ammonia from a gas cylinder is passed. The assembly can quickly be removed after the blue spots are visible and can be replaced with a second assembly. For best results the location of the spots should be marked within a few minutes, since the smallest spots tend to fade somewhat with time.

In a novel procedure of somewhat similar principle, Geisler (1972) etched holes through a sheet and then exposed it simultaneously to hydrogen chloride gas on one side and ammonia on the other. When the two gases meet at a hole, ammonium chloride nucleates and grows into a cluster of readily seen, columnar crystals at each hole.

2.5.3. Integrating or Automated Scanning Methods

Numerous procedures that have been developed for rapid counting are outlined in Table 2-9 and Griffith (1973) lists laboratories that are using various techniques for scanning. The techniques include spark counting, automated microscopes, electrical conductivity measurements through tracks containing an electrolyte, pattern recognition systems, and measurement of light scattered by the tracks or transmitted through an opaque substance that the tracks cross. Becker (1972) and Medveczky (1969) have reviewed a number of these techniques.

At present the spark-scanning technique is by far the simplest, most widely used one for track densities up to $\sim 3000/cm^2$. The following description of the procedure, which commonly is carried out on 10 μm polycarbonate film (Kimfol or Makrofol) through which holes have been etched to diameters of 2 to 5 μm diameter, is adapted from Becker (1972). After etching, rinsing, and drying, the etched foil is placed on a flat electrode plate. Clamping of the thin foils in retainer rings, or attaching them to plastic rings by rubber cement, simplifies their handling. The film is then covered with a piece of aluminized Mylar (McCordi Metallizing Corp., Mamaroneck, N.Y.) which is pressed down with the aluminized side facing

Table 2-9. Automated or Integrating Scanning Methods

Technique	Scanning Rate cm^2/min	Track Density cm^{-2}	Comments	Reference
Optical scanning	.15		Pulse height discrimination.	Bitter et al. (1967).
Optical scanning (Quantimet)			Signal characteristic can be selected.	Besant and Ipson (1970); Abmayr et al. (1969); Ruegger et al. (1969); Jowitt (1971).
Optical scanning (with automatic focusing)	.002	10^3–10^5	Works well on Makrofol polycarbonate, less well on mica.	Cohn et al. (1969); Cohn and Gold (1972); Gold and Cohn (1972); Goel and Eggmann (1972).
D.C. spark counting with scaler	10	0–3000	D.C. voltage develops some flaws.	Lark (1969); Cross and Tommasino (1968, 1970, 1972); Abmayr et al. (1969); Congel et al. (1970); Johnson et al. (1970); Somogyi and Gulyas (1972); Paretzke (1972).
A.C. spark counting with scaler	100 to 200 (up to 100 counts/sec)		Avoids most flaws.	Geisler and Phillips (1972).
Light scattering		10^4 to 10^6		Becker (1966); Tuyn (1967); Schultz (1968); Khan (1971).
Light transmission through opaque plastic		$4x10^4$ to $3x10^6$	Ultraviolet light through Mylar.	Pretre et al. (1968).
Light transmission through opaque plastic	100 – 1500		Blue light through red (Kodak LR-115) cellulose nitrate.	Sohrabi and Becker (1971).
Electrical conductivity through membrane		1 to 10^{11}		Walker et al. (1963); DeSorbo and Humphrey (1970).
Surface barrier detector plus membrane detector		10^2–10^5	Detects alpha particles transmitted through holes in membrane.	Khan and Durrani (1972b).
Intensity of scattered coherent light		$5x10^4$ to 10^6	Linear for scattering at 54°.	Platzer et al. (1972).

the etched film and making contact to an outer, grounded electrode. A positive voltage of about 500 V causes sparks to occur through the perforations in the etched film. Each spark evaporates aluminum from the aluminized Mylar near the hole in an area several orders of magnitude larger than the original hole in the detector film. This removal opens the circuit so that no multiple sparking occurs through individual holes and a plainly visible "replica" of each hole remains in the aluminum layer. The replica provides a map of the track distribution for possible radiographic applications, and is a permanent record of the results. Several such replicas can be made from the same detector film.

The counter is coupled to a scaler through a simple Geiger-Müller-type quenching circuit so that each spark is recorded in the scaler. The spark counter is a simple device which requires only an RC pulse-shaping circuit, a high-voltage supply, and a scaler for auxiliary apparatus. The simple RC quenching circuit operates the scaler reliably and accurately.

If the spark counting is done as a function of sparking voltage, a "plateau" around 500 V to 800 V is obtained (the exact voltage depending on the counting circuit, foil thickness, etching conditions (Abmayr et al., 1969), and the aluminum thickness). The sparking operation apparently completes the penetration of the holes by spark-punching etched tracks that did not quite cross the foil because of their low angle of incidence. For this reason, in the preferred procedure foils are sparked at least twice, first at a higher voltage to expose a larger fraction of the etched tracks and secondly (for data-taking) at a somewhat reduced voltage to avoid the multiple sparking that begins to occur at about 600 to 800 V. The evaporated aluminum is partially deposited on the detector film and the positive electrode, making it necessary to clean both occasionally by wiping the electrode with cotton and rinsing the film in an NaOH or KOH solution. If foils are "overetched" or thinner than 3 to 4 μm, electrical breakdown occurs even in unperforated areas. In fission-fragment spark counting with polycarbonate films, only a very small "background" count ($\sim$0.01 count/cm^2) from unexposed, etched films is normally observed.

We note that, although full-energy fission fragments at normal incidence will be revealed with unit efficiency by this technique, only a fraction $1 - t/R$ of an isotropic dose of fission fragments is revealed. Here t is the depth in the foil of etched tracks that just fail to be punched through by the high voltage and R is the etchable range of the fission fragments. The punching-through reduces t from what otherwise would be the thickness of the etched foil and hence increases the efficiency.

Geisler and Phillips (1972) have improved the technique by sweeping a roller electrode across the foil and using repeated brief electrical pulses rather than applying a d.c. voltage. This procedure avoids most of the breakdowns that occur at occasional flaws under d.c. conditions, and consequently has a lower background of spurious counts and higher reliability at very low track densities.

The spark-counting technique, which is usually used for the counting of heavily ionizing fission fragments, has also been applied to the counting of alpha particle tracks in cast cellulose nitrate foils with a high registration efficiency (Johnson et al., 1970). Alpha particle track spark-counting is, however, not a routine matter because the necessary thin detector foils are not presently available commercially, the etching parameters need to be more carefully controlled, and environmental alpha radiation or artifacts in the foil may cause significant background counts. As long as alpha particle registering material is available with a plane surface of good quality, the Quantimet microscope system with human supervision appears to be the best present solution for automatic scanning of alpha particle tracks.

2.6. MISCELLANEOUS SPECIAL TECHNIQUES

There are a number of helpful techniques of various types which we discuss separately in this section.

2.6.1. Measurement of Track Dimensions

The total etchable lengths of tracks in minerals after a long etch can be useful indications of the atomic numbers of the nuclei that formed the tracks. (This will be discussed in the next chapter.) In the usual case where a crystal is cut and polished, a portion of each track that intersects the surface has been removed so that the total track is not available for etching. Lal et al. (1968, 1969) and Lal (1969) have described two simple techniques that avoid this difficulty.

The track-in-track ("TINT") procedure shown in Fig. 2-15a depends on there being occasional long tracks that etch rapidly to well below the polished surface and provide access to other tracks of interest. When the hole along such a track is enlarged sufficiently by etching, it will intersect other tracks and etch them toward *both* ends, so that the full information in the intersected tracks can be made available for measurement in the optical microscope. Fig. 2-16 gives an example of such tracks in a meteoritic crystal that accumulated cosmic ray tracks during its long exposure in space.

In a variation of the above procedure, the track-in-cleavage ("TINCLE") technique utilizes natural or induced fractures in crystals or glasses as a means of allowing the etchant to reach and etch the tracks that intersect the fracture plane (Fig. 2-15b).

In another variation of the TINT technique, Benton (1971) has irradiated a detector with heavy ions to produce access tracks normal to the surface. When these tracks are enlarged by etching, they intersect tracks within the body of the detector as in Lal's case. This procedure is helpful in supplying the needed access

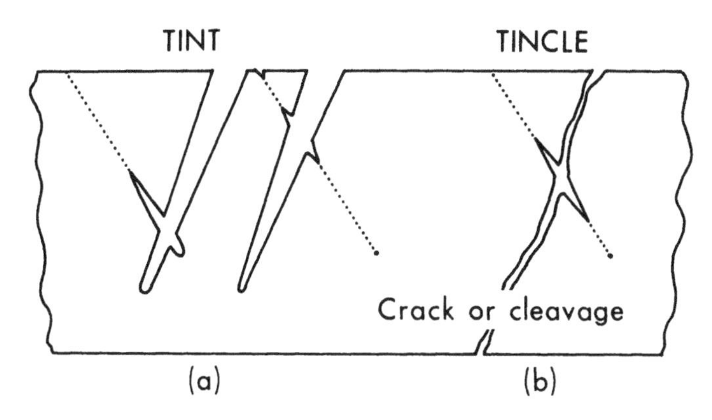

Fig. 2-15. *Techniques for measuring etchable track lengths within a mineral or glass: (a) In the track-in-track (TINT) method the occasional long "host" tracks that are produced by more heavily ionizing particles than are the interior intersecting short tracks give the etchant access within the crystal, allowing the short tracks to be etched toward their ends in both directions. (b) Tracks-in-cleavage (TINCLE) are revealed if the etchant can run into a natural or deliberately produced crack (adapted from Lal et al., 1969).*

tracks in an orientation where they give a minimum of optical interference with the microscopic observation of the tracks to be studied.

In a more sophisticated version of the TINCLE method, Bastin et al. (1974) create a narrow groove or planar hole in a crystal by bombarding its surface through a slit with a very intense krypton beam. Etching for a long enough time that the etched tracks overlap creates a groove of the desired orientation and with a width that can be made large enough to ensure simultaneous access by the etchant to all of the tracks that intersect the groove.

Straightforward microscopic measurements (Peterson, 1969) of track length, diameter, and angle of incidence of etched tracks are conventionally used in the particle identification procedures to be described in the next chapter. For some subtleties, see O'Ceallaigh and Daly (1970). We note here some more unusual optical techniques in which lasers are proving useful for measuring track diameters (Varnagy et al., 1973) and cone angles (Hart and Fleischer, unpublished) where arrays of nominally identical tracks are obtained—as, for example, when a beam of ions from an accelerator is used for calibration purposes. Imagine shining a laser beam through an etched glass plate containing tracks at normal incidence and then viewing the light on a projection screen. Light traveling through the tracks is scattered out of each area $\pi D^2/4$ occupied by a track, leaving the unscattered light to interfere with itself—giving a series of concentric light and dark rings on the screen. The track diameter D in terms of the ring radius r_i for the ith ring is given by $D = C_i \lambda S/r_i$ where λ is the wavelength of light and S the spacing of the

Fig. 2-16. *An example of TINT's in a meteoritic crystal (courtesy of D. Lal).*

tracks from the projection screen. The constants C_i are 1.22, 2.23, and 3.24 for the first three dark rings and 1.64, 2.69, and 3.72 for the first three light rings. This sort of measurement can be done on a transparent etched sample or on a transparent replica made from an opaque sample. If there is a large dispersion of diameters the rings will be washed out.

The cone angle θ of tracks can be determined from the refracted light. The refraction angle ϕ depends on θ and the indices at refraction of the media through which the light passes. In air the angle ϕ is given by $\sin \phi = \cos \theta\, (\sqrt{n^2 - \cos^2 \theta} - \sin \theta)$ where n is the refractive index of the detector. Fig. 2-17 allows θ to be inferred from measured values of ϕ and n.

For large values of θ the simple reflection technique depicted in Fig. 2-18 is useful. In this case the sample is placed on a horizontal rotating table, and the light is reflected from the inside of the cones, forming a vertical line on the screen. By noting the difference in angle β between the two sketched positions where this

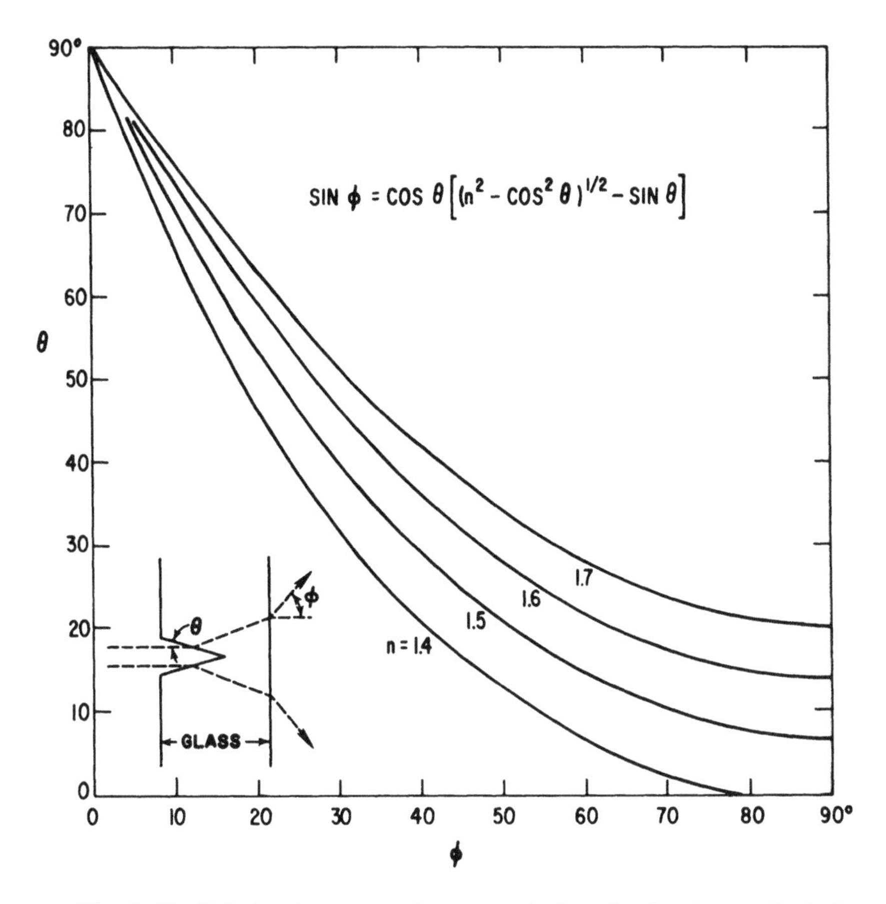

Fig. 2-17. *Relation between track cone angle θ and refraction angle ϕ for different values of index of refraction n (Hart and Fleischer, unpublished).*

line passes through the beam direction the cone angle can be calculated to be $\pi/2 - \beta/2$.

Each of the procedures given is particularly useful where an average value of D or θ is desired from an array of closely similar tracks.

2.6.2. Preparation of Replicas for Scanning Electron Microscopy (SEM)

Direct observation of an etched surface by SEM permits one to determine track densities provided there is no ambiguity between tracks and other etched defects. All one needs to do following etching is to evaporate a conducting coating onto the surface. Many of the studies of extraterrestrial samples to be discussed in Chapter 6 were made by direct observation of rock surfaces in the SEM. The small depth of penetration of the electron beam prevents one from seeing the full

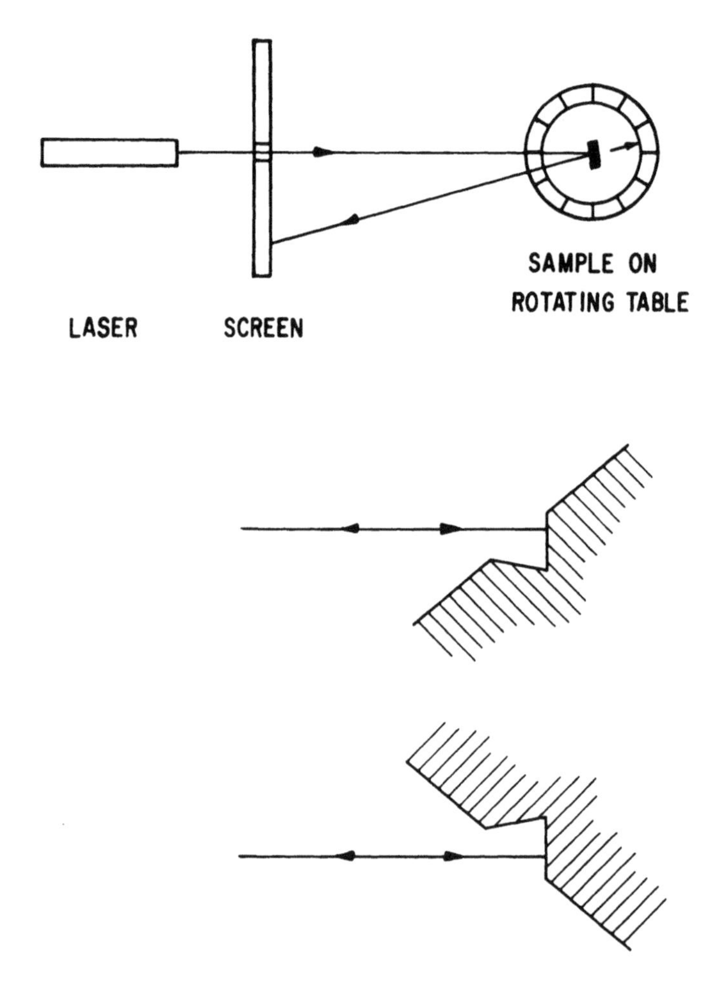

Fig. 2-18. *Alternate procedure for measuring cone angles (Hart and Fleischer, unpublished).*

lengths of etched tracks, and it is often extremely valuable to make a replica of the etched surface, which can be peeled away, coated with a conducting film, and examined in the SEM. If the replica completely filled the etched tracks, one can make accurate measurements of track lengths because the electron beam does not have to penetrate into the interior of the replica. Because the delicate portions of the replica that filled the tracks will usually be twisted when the replica is peeled away, it is possible to determine directionality only if the cone angle is large, so that the replica of the track is self-supporting.

We have tried cellulose acetate, silicone rubber, and formvar replicas. With rock surfaces cellulose acetate is quite satisfactory, the only drawback being that at very high magnification it "withers" in the intense SEM beam unless great skill is used. Formvar is far more resistant to distortion in the electron beam but has the disadvantage that in porous rocks it makes too good a replica. When softened in a suitable solvent such as acetic acid, it runs deeply into crevices and sometimes cannot be peeled away. With glass surfaces all three materials are adequate and we recommend formvar because of its stability in the electron beam. With plastic surfaces one must choose a solvent that softens the replica without attacking the plastic, and one must later be able to peel the replica off. J. H. Chan (University of California, Berkeley), who took the SEM photographs of replicas of tracks in Lexan shown in Fig. 3-19, has found that a saturated solution of formvar in glacial acetic acid makes an excellent replica of a Lexan surface, provided the Lexan has first been coated with a thin film of gold ($\sim$100 to 200 Å), which prevents the formvar from sticking to the Lexan. To replicate a plastic other than Lexan one must find a solvent that dissolves formvar (or some other plastic that is to serve as the replica) without attacking the plastic to be replicated.

2.6.3. Sample Preparation for 4π Solid Angle Observations

Often for fission experiments of various types, including cross-section measurements and particle dosimetry, it is desirable to observe events through a full 4π solid angle either in order to measure angular distributions or to ensure that there is a high efficiency for seeing fission events. A simple technique is to evaporate a thin film of the element to be examined on one side of a detector of high efficiency and press a second sheet of the same detector material against the film to form a sandwich structure. If a plastic such as Lexan polycarbonate is used, combined pressure and heat will form a tight bond and produce a structure that is easy to handle. After irradiation the evaporated layer can be dissolved away and the detector then etched.

A similar sandwich using mica (Cieslak et al., 1966) has the advantage that it can withstand higher irradiation doses and has nearly unit efficiency of registration. Hudis and Katcoff (1969) developed an effective sandwiching technique for mica. They enclosed the mica layers between 100 μm sheets of Mylar, evacuated the space between the Mylar sheets and then sealed the edges of the plastic. In air the atmospheric pressure on the Mylar presses the enclosed mica sheets firmly together so that the evaporated layer is in intimate contact with both sheets. After irradiation, alignment holes are drilled through the stack. In some cases it is desirable to produce tracks where both fission fragments are etched in the same sheet of mica. Boos et al. (1970) showed that a convenient way to do this is to produce fission tracks at normal incidence, etch them, deposit into the etched tracks a solution containing the element of interest (for example uranyl nitrate to deposit uranium), wipe away the excess solution, and allow evaporation to deposit the element of interest near the bottom of the holes. After irradiation new tracks will emerge from near the tips of the original, head-on tracks and allow the angular distribution to be examined in a single piece of mica without the annoyance and experimental difficulty of needing to match up adjacent sheets. Background damage from intense irradiations (up to 10^{13} alpha particles/cm^2), which in some cases might obscure optical examination, can be removed by a brief 600°C anneal of muscovite mica without removing fission tracks (Zhagrov et al., 1968; Rumyantsev et al., 1968).

Angular distributions of nuclear interaction products or scattering events can readily be recorded in a simple form by wrapping a plastic sheet about the interaction site and recording the distribution over 360° (see, for example, Pilcher et al., 1972).

2.6.4. Commercially Available Detectors

The firm of Bayer A.g., in Leverkusen, W. Germany, sells cellulose acetate butyrate (Bayer BN), cellulose triacetate (Triafol TN), and polycarbonate (Makrofol) in a great variety of thicknesses. The BN contains a purple dye, the TN contains a blue dye, and both suffer the drawback that opposite sides of the sheet have different sensitivity. Their polycarbonate is distributed in the U.S. by Peter J. Schweitzer, Lee, Mass., under the trade name Kimfol.

Daicel, the most sensitive cellulose nitrate, is distributed in the U.S. by Kanematsu Gosho, Inc., 1 Whitehall St., New York, N.Y., or is available from Daicel, Ltd., 8 Kawaramachi, 3 Chome Higashi-Ku, Osaka. The minimum order unfortunately is 1,000 pounds! A thin coating of cellulose nitrate, either clear or dyed red, on a polyester backing, is made by Kodak Pathé, Vincennes, France, and is also distributed by Eastman Kodak, 343 State St., Rochester, N.Y. For high-resolution neutron radiographic use, Kodak-Pathé is supplying sheets of their film CA 8015, type B, coated with a thin neutron-converter film made of lithium borate. CA 8015 is also available as clear, uncoated sheets.

Cellulose nitrate in the form of powder containing 30% ethyl alcohol and suitable for casting can be bought from Hercules Powder Corp., Wilmington, Del. Films cast from highly nitrated, high-viscosity cellulose nitrate have, in general, superior etching and spark-counting characteristics compared with other grades such as Parlodion. We reproduce here the recipe used by Benton (1968) for casting such films: To 24.3 g of the cellulose nitrate powder containing 30% ethyl alcohol, solvents, co-solvents and plasticizer are added sequentially: 124.7 g ethyl acetate, 4 g isopropyl alcohol, 5 g butyl alcohol, 8 g cellosolve acetate (trade name for ethylene glycol monoethyl ether acetate), and 4 g dioctylphthalate. This results in a 10% cellulose nitrate solution, which should be stirred thoroughly until clear. After about 4 days at room temperature, a 10–20 g aliquot (depending on the desired film thickness of about 10 to 20 μm) of the solution is diluted to 240 ml with ethyl acetate. Then 40 ml aliquots of this diluted solution are transferred to glass dishes 15 cm in diameter and placed on a leveled surface in a clean area. The dishes are partially covered by a glass plate to obtain slow evaporation rate and to help prevent dust contamination. After one to two days' drying of the foils at room temperature, they are removed by water flotation and dried between layers of blotting paper. Before use, the films are annealed at 80°C overnight to remove residual solvents and background tracks.

The cellulose triacetate normally available from Eastman Chemical Products, P.O. Box 847, Framingham, Mass., contains a plasticizer that makes it completely unsuitable as a particle detector, but on special request we have been able to obtain limited quantities made without plasticizer. V. E. B. Filmfabriki, Wolfen, E. Germany, makes a brand of cellulose triacetate called Cellit-T. The General Electric Plastics Dept., Mt. Vernon, Indiana, makes Lexan polycarbonate. If one is planning to use the UV sensitization technique, one should specify the non-UV-stabilized Lexan. Frequently we have bought Lexan labeled as non-UV-stabilized and discovered when we checked its transmission with a spectrophotometer that it was actually the UV-stabilized type, opaque to wavelengths shorter than ~3500 Å. *Caveat emptor!*

Polyethylene terephthalate is distributed as Cronar by E. I. duPont de Nemours and Co., Wilmington, Del., as Hostaphan by Kalle A.g., Wiesbaden-Biebrich, W. Germany, as Melinex by Imperial Chemical Industries, Ltd., England, and as Lavsan in the USSR.

As a final comment, we point out that additives put into plastics to improve their mechanical properties can also grossly alter their etching characteristics. For example, Schultz (1970) cast cellulose nitrate films combining boron and three different plasticizers. He found that for samples containing 20% dibutyl phthalate, 20% camphor, 20% dioctyl phthalate, and no additives, the general etching rates v_G were in the ratios 1 to 2.5 to 12 to 35.

Chapter 2 References

W. Abmayr, F. Grunauer, and G. Burger (1969), "Automatic Optical Registration and Computer Analysis of Etched Ion-Tracks in Polymer Foils," *Proc. Inter. Conf. Nucl. Track Registration in Insulating Solids*, Clermont-Ferrand, Sect. III, 46–55.

T. J. Ahrens, R. L. Fleischer, P. B. Price, and R. T. Woods (1970), "Erasure of Fission Tracks in Glasses and Silicates by Shock Waves," *Earth Planet. Sci. Lett.* **8,** 420–426.

H. W. Alter (1968), "Visual Imaging of Track-Etched Patterns," U.S. Patent 3,373,683.

F. Aumento (1969), "The Mid-Atlantic Ridge Near 45° N. Fission Track and Manganese Chronology," *Can. J. Earth Sci.* **6,** 1431–1440.

F. Aumento and J. G. Souther (1973), "Fission Track Dating of Late Tertiary and Quaternary Volcanic Glass from Mt. Edziza Volcano, British Columbia," *Can. J. Earth Sci.* **10,** 1156–1163.

G. Bastin, G. M. Comstock, J. C. Dran, J. P. Duraud, M. Maurette, and C. Thibaut (1974), "Lunar Soil Maturation, Part III: Short-Term and Long-Term Aging of Radiation Damage Features in the Regolith," *Lunar Science V*, 44–46, The Lunar Sci. Inst., Houston.

C. P. Bean, M. V. Doyle, and G. Entine (1970), "Etching of Submicron Pores in Irradiated Mica," *J. Appl. Phys.* **41,** 1454–1459.

K. Becker (1966), "Photographic, Glass or Thermoluminescence Dosimetry?" *Health Phys.* **12,** 955–964.

K. Becker (1968), "The Effect of Oxygen and Humidity on Charged Particle Registration in Organic Foils," *Rad. Research* **36,** 107–118.

K. Becker (1969), "Alpha Particle Registration in Plastics and Its Applications for Radon, Thermal, and Fast Neutron Dosimetry," *Health Phys.* **16,** 113–123.

K. Becker (1972), "Dosimetric Applications of Track Etching," *Topics in Radiation Dosimetry*, Suppl. **1,** 79–142.

E. V. Benton (1968), *A Study of Charged Particle Tracks in Cellulose Nitrate*, USNRDL-TR-68-14, U.S. Nav. Rad. Def. Lab., San Francisco, Calif.

E. V. Benton (1971), "A Method for Development of Volume Tracks in Dielectric Nuclear Track Detectors," *Nucl. Instr. Methods* **92,** 97–99.

E. V. Benton and R. P. Henke (1969), "Sensitivity Enhancement of Lexan Nuclear Track Detector," *Nucl. Instr. Methods* **70,** 183–184.

I. G. Berzina, I. V. Vorob'eva, Ya. E. Geguzin, and I. M. Zlotova (1966), "Annealing of Tracks of Fragments from Spontaneous Fission of Uranium in Glasses and Mica Crystals," *Sov. Phys. Doklady* **11,** 1105–1107.

C. B. Besant and S. S. Ipson (1970), "Measurement of Fission Ratios in Zero Power Reactors Using Solid-State Track Recorders," *J. Nucl. Energy* **24,** 59–69.

H. Besson, M. Monnin, and S. Sanzell (1967), "New Solid State Track Detectors and Chemical Etching Methods," *Compt. rend.*, Paris **264B,** 1751–1752.

N. Bhandari, S. Bhat, D. Lal, G. Rajagopalan, A. S. Tamhane, and V. S. Venkatavaradan (1971a), "Spontaneous Fission Record of Uranium and Extinct Transuranic Elements in Apollo Samples," *Proc. Second Lunar Sci. Conf.* **3,** 2599–2609. Cambridge: MIT Press.

N. Bhandari, S. G. Bhat, D. Lal, G. Rajagopalan, A. S. Tamhane, and V. S. Venkatavaradan (1971b), "Fission Fragment Tracks in Apatite: Recordable Track Lengths," *Earth Planet. Sci. Lett.* **13,** 191–199.

J. P. Birabeau, A. Cordaillat, and O. Mendola (1972), "Réalisation de Grilles sur des Détecteurs en Matière Plastique," *Nucl. Instr. Methods* **99,** 355–357.

F. Bitter, G. Fiedler, and J. Wollnik (1967), "Automatic Scanning of Nuclear

Track and Solid Detector Plates (German)," *Nucl. Instr. Methods* **51,** 241–244.

G. E. Blanford, Jr., R. L. Fleischer, M. W. Friedlander, J. Klarmann, G. E. Nichols, P. B. Price, R. M. Walker, J. P. Wefel, and W. C. Wells (1969), "Observation of Trans-Iron Nuclei in the Primary Cosmic Radiation," *Phys. Rev. Letters* **23,** 338–342.

G. E. Blanford, Jr., R. M. Walker, and J. P. Wefel (1970a), "Track Etching Parameters for Plastics," *Rad. Effects* **3,** 267–270.

G. E. Blanford, Jr., R. M. Walker, and J. P. Wefel (1970b), "Enhancement of Track Etching by a Spark Discharge," *Rad. Effects* **3,** 263–266.

G. E. Blanford, Jr., R. M. Walker, and J. P. Wefel (1970c), "Calibration of Plastic Track Detectors for Use in Cosmic Ray Experiments," *Rad. Effects* **5,** 41–45.

J. Blok, J. S. Humphrey, and G. E. Nichols (1969), "Hole Detection in Polymer Films and in Plastic Track Detectors," *Rev. Sci. Instr.* **40,** 509.

A. H. Boos, R. Brandt, and K. Starke (1970), "Mica Targets with Nearly a 4π Geometry for Fission Studies," *Rad. Effects* **3,** 235–238.

R. H. Boyett and K. Becker (1970), "LET Effects on the Chemical Resistance of Irradiated Polymers," *J. Appl. Polymer Sci.* **14,** 1654.

R. H. Boyett, D. R. Johnson, and K. Becker (1970), "Some Studies on the Chemical Damage Mechanism Along Charged Particle Tracks in Polymers," *Rad. Research* **42,** 1–12.

R. H. Brill, R. L. Fleischer, P. B. Price, and R. M. Walker (1964), "The Fission Track Dating of Man Made Glasses: Part 1, Preliminary Results," *J. Glass Studies* **6,** 151–155.

L. C. Calk and C. W. Naeser (1973), "The Thermal Effect of a Basalt Intrusion on Fission Tracks in Quartz Monzonite," *J. Geol.* **81,** 189–198.

R. W. Caputi and W. T. Crawford (1970), "Track Registration Process," U.S. Patent 3,662,178, issued 1972.

P. E. Caspar (1964), "Etching and Enlarging of Fission Fragment Tracks by Atmospheric Moisture," *Nature* **201,** 1203–1204.

E. Cieslak, J. Piekarz, J. Zakrzewski, M. Dakowski, H. Piekarz, and M. Sowinski (1966), "Observation of Fission Events in Mica Sandwiches," *Nucl. Instr. Methods* **39,** 224–231.

C. E. Cohn and R. Gold (1972), "A Computer-Controlled Microscope for Automatic Scanning of Solid State Nuclear Track Recorders," *Rev. Sci. Instr.* **43,** 12–17.

C. E. Cohn, R. Gold, and T. W. Pienias (1969), "A Computer-Controlled Microscope for Scanning Fission Track Plates," *Am. Nucl. Soc. Trans.* **12,** 68.

G. M. Comstock, R. L. Fleischer, W. R. Giard, H. R. Hart, Jr., G. E. Nichols, and P. B. Price (1971), "Cosmic Ray Tracks in Plastics: The Apollo Helmet Dosimetry Experiment," *Science* **172,** 154–157.

F. J. Congel, J. H. Roberts, R. J. Armani, R. Gold, and B. G. Oltman (1970), "Absolute Slow Neutron Measurements with Solid State Track Recorders," *Proc. Seventh Inter. Colloq. Corpuscular Photography and Visual Solid Detectors*, Barcelona, 469–479.

F. J. Congel, J. H. Roberts, E. N. Strait, J. Kastner, B. J. Oltman, R. Gold, and R. Armani (1972), "Automatic System for Counting Etched Holes in Thin Dielectric Foils," *Nucl. Instr. Methods* **100,** 247–252.

H. Crannell, C. J. Crannell, F. J. Kline, and L. Battist (1969), "Particle Track Enhancement in Cellulose Nitrate by Application of an Electric Field," *Science* **166,** 606–607.

W. T. Crawford, W. DeSorbo, and J. S. Humphrey (1968), "Enhancement of Track Etching Rates in Charged Particle-Irradiated Plastics by a Photo-Oxidation Effect," *Nature* **220,** 1313–1314.

W. G. Cross and L. Tommasino (1967), "Detection of Low Doses of Fast Neutrons with Fission Track Detectors," *Health Phys.* **13,** 932.

W. G. Cross and L. Tommasino (1968), "Electrical Detection of Fission Fragment Tracks for Fast Neutron Dosimetry," *Health Phys.* **15,** 196.

W. G. Cross and L. Tommasino (1970), "Rapid Reading Technique for Nuclear Particle Damage Tracks in Thin Foils," *Rad. Effects* **5,** 85.

W. G. Cross and L. Tommasino (1972), "Improvements in the Spark Counting Technique for Damage Track Neutron Dosimeters," *First Int. Symp. on Neutron Dosimetry in Biology and Medicine*, 15–18 May, Neuherberg, Germany.

G. Crozaz, M. Hair, M. Maurette, and R. M. Walker (1969), "Nuclear Interaction Tracks in Minerals and Their Implications for Extraterrestrial Materials," *Proc. Inter. Conf. Nucl. Track Registration in Insulating Solids*, Clermont-Ferrand, Sect. VII, 41–53.

G. Crozaz, U. Haack, M. Hair, M. Maurette, R. Walker, and D. Woolum (1970), "Nuclear Track Studies of Ancient Solar Radiations and Dynamic Lunar Surface Processes," *Proc. Apollo 11 Lunar Sci. Conf.* **3,** 2051–2080. New York: Pergamon Press.

M. Debeauvais, M. Maurette, J. Mory, and R. Walker (1964), "Registration of Fission Fragment Tracks in Several Substances and Their Use in Neutron Detection," *Int. J. Appl. Rad. Isotopes* **15,** 289–299.

R. W. DeBlois and C. P. Bean (1970), "Counting and Sizing of Submicron Particles by the Resistive Pulse Technique," *Rev. Sci. Instr.* **41,** 909–916.

W. DeSorbo and J. S. Humphrey, Jr. (1970), "Studies of Environmental Effects Upon Track Etching Rates in Charged Particle Irradiated Polycarbonate Film," *Rad. Effects* **3,** 281–282.

S. A. Durrani and H. A. Khan (1970), "Annealing of Fission Tracks in Tektites: Corrected Ages of Bediasites," *Earth Planet. Sci. Lett.* **9,** 431–445.

W. Enge, R. Beaujean and H-P. Nicken (1970), "The Influence of Humidity on Track Registration in Cellulose Nitrate Plastic Detectors," *Proc. Seventh Inter.*

Colloq. Corpuscular Photography and Visual Solid Detectors, Barcelona, 175–179.

D. Engelkemeier and B. N. Walton (1964), "Experimental Study of Fission Fragment Scattering," AERE-R 4716, Harwell, England.

R. L. Fleischer and H. R. Hart, Jr. (1972a), "Fission Track Dating: Techniques and Problems," in *Proceedings Burg Wartenstein Conference on Calibration of Hominoid Evolution*, W. W. Bishop, J. A. Miller, and S. Cole (eds.), 135–170. Edinburgh: Scottish Academic Press.

R. L. Fleischer and H. R. Hart, Jr. (1972b), "Particle Track Record of Apollo 15 Green Soil and Rock," in *The Apollo 15 Lunar Samples*, J. W. Chamberlain and C. Watkins (eds.), 368–370, Lunar Sci. Inst., Houston; also in *Earth Planet. Sci. Lett.* **18,** 357–364 (1973).

R. L. Fleischer and H. R. Hart, Jr. (1972c), "The Apollo 16 Cosmic Ray Experiment," *Apollo 16 Preliminary Science Report*, NASA Special Pub. 315, Chapter 15, 1–11.

R. L. Fleischer and H. R. Hart, Jr. (1973b), "Mechanical Erasure of Particle Tracks, a Tool for Lunar Microstratigraphic Chronology," *J. Geophys. Res.* **78,** 4841–4851.

R. L. Fleischer and C. W. Naeser (1972), "Tracks in Mountain Pass Bastnäsite," *Nature* **40,** 465.

R. L. Fleischer and P. B. Price (1963a), "Tracks of Charged Particles in High Polymers," *Science* **140,** 1221–1222.

R. L. Fleischer and P. B. Price (1963b), "Charged Particle Tracks in Glass," *J. Appl. Phys.* **34,** 2903–2904.

R. L. Fleischer and P. B. Price (1964a), "Glass Dating by Fission Fragment Tracks," *J. Geophys. Res.* **69,** 331–339.

R. L. Fleischer and P. B. Price (1964b), "Techniques for Geological Dating of Minerals by Chemical Etching of Fission Fragment Tracks," *Geochim. Cosmochim. Acta* **28,** 1705–1714.

R. L. Fleischer and P. B. Price (1964c), "Fission Track Evidence for the Simultaneous Origin of Tektites and Other Natural Glasses," *Geochim. Cosmochim. Acta* **28,** 755–760.

R. L. Fleischer and R. M. Walker (1971), "Radiation Detection Method," U.S. Patent 3,770,762, issued 1973.

R. L. Fleischer, P. B. Price, and R. M. Walker (1964a), "Fission Track Ages of Zircons," *J. Geophys. Res.* **69,** 4885–4888.

R. L. Fleischer, P. B. Price, E. M. Symes, and D. S. Miller (1964b), "Fission Track Ages and Track-Annealing Behavior of Some Micas," *Science* **143,** 349–351.

R. L. Fleischer, P. B. Price, and E. M. Symes (1964c), "On the Origin of Anomalous Etch Figures in Minerals," *Am. Mineralogist* **49,** 794–800.

R. L. Fleischer, P. B. Price, R. M. Walker, and E. L. Hubbard (1964d), "Track Registration in Various Solid State Nuclear Track Detectors," *Phys. Rev.* **133A,** 1443–1449.

R. L. Fleischer, P. B. Price, and R. M. Walker (1965a), "Solid State Track Detectors: Applications to Nuclear Science and Geophysics," *Ann. Rev. Nuc. Sci.* **15,** 1–28.

R. L. Fleischer, P. B. Price, and R. M. Walker (1965b), "Tracks of Charged Particles in Solids," *Science* **149,** 383–393.

R. L. Fleischer, P. B. Price, and R. M. Walker (1965c), "Effects of Temperature, Pressure, and Ionization on the Formation and Stability of Fission Tracks in Minerals and Glasses," *J. Geophys. Res.* **70,** 1497–1502.

R. L. Fleischer, P. B. Price, and R. M. Walker (1965d), "The Ion Explosion Spike Mechanism for Formation of Charged Particle Tracks in Solids," *J. Appl. Phys.* **36,** 3645–3652.

R. L. Fleischer, P. B. Price, R. M. Walker, and L. S. B. Leakey (1965e), "Fission Track Dating of Bed I, Olduvai Gorge," *Science* **148,** 72–74.

R. L. Fleischer, P. B. Price, and R. M. Walker (1966a), "Simple Detectors for Neutrons or Heavy Cosmic Ray Nuclei," *Rev. Sci. Instr.* **37,** 525–527.

R. L. Fleischer, P. B. Price, R. M. Walker, and E. L. Hubbard (1966b), "Ternary Fission of Heavy Compound Nuclei in Thorite Track Detectors," *Phys. Rev.* **143,** 943–946.

R. L. Fleischer, P. B. Price, R. M. Walker, and M. Maurette (1967), "Origins of Fossil Charged Particle Tracks in Meteorites," *J. Geophys. Res.* **72,** 333–353.

R. L. Fleischer, P. B. Price, and R. M. Walker (1968a), "Charged Particle Tracks: Tools for Geochronology and Meteorite Studies," in *Radiometric Dating for Geologists,* E. Hamilton and R. M. Farquhar (eds.), 417–435. New York: Wiley-Interscience.

R. L. Fleischer, J. R. M. Viertl, P. B. Price, and F. Aumento (1968b), "Mid-Atlantic Ridge: Age and Spreading Rates," *Science* **161,** 1339–1342.

R. L. Fleischer, P. B. Price, and R. M. Walker (1968c), "Identification of Pu^{244} Fission Tracks and the Cooling of the Parent Body of the Toluca Meteorite," *Geochim. Cosmochim. Acta* **32,** 21–31.

R. L. Fleischer, J. R. M. Viertl, and P. B. Price (1969a), "Age of the Manicouagan and Clearwater Lakes Craters," *Geochim. Cosmochim. Acta* **33,** 523–527.

R. L. Fleischer, P. B. Price, and R. T. Woods (1969b), "Nuclear Particle Track Identification in Inorganic Solids," *Phys. Rev.* **88,** 563–567.

R. L. Fleischer, H. R. Hart, Jr., and G. M. Comstock (1971a), "Very Heavy Solar Cosmic Rays: Energy Spectrum and Implications for Lunar Erosion," *Science* **171,** 1240–1242.

R. L. Fleischer, H. R. Hart, Jr., G. M. Comstock, and A. O. Evwaraye (1971b), "The Particle Track Record of the Ocean of Storms," *Proc. Second Lunar Sci. Conf.* **3,** 2559–2568. Cambridge: MIT Press.

R. L. Fleischer, J. R. M. Viertl, P. B. Price, and F. Aumento (1971c), "A Chronological Test of Ocean Bottom Spreading in the North Atlantic," *Rad. Effects* **11,** 193–194.

R. L. Fleischer, G. M. Comstock, and H. R. Hart, Jr. (1972a), "Dating of Mechanical Events by Deformation-Induced Erasure of Particle Tracks," *J. Geophys. Res.* **77**, 5050–5053.

R. L. Fleischer, J. R. M. Viertl, and P. B. Price (1972b), "Biological Filters with Continuously Adjustable Hole Size," *Rev. Sci. Instr.* **43**, 1708–1709.

R. L. Fleischer, H. W. Alter, S. C. Furman, P. B. Price, and R. M. Walker (1972c), "Particle Track Etching: Technological Applications of Science," *Science* **178**, 255–263.

R. L. Fleischer, R. T. Woods, H. R. Hart, Jr., P. B. Price, and N. M. Short (1974), "Effect of Shock on Fission Track Dating of Apatite and Sphene Crystals from the Hardhat and Sedan Underground Nuclear Explosions," *J. Geophys. Res.* **79**, 339–342.

R. S. Flemons (1968), "Hole Detection," Canadian Patent No. 82510, issued 1969.

A. L. Frank and E. V. Benton (1970), "Dielectric Plastics as High Exposure X-Ray Detectors," *Rad. Effects* **2**, 269–272.

Yu. E. Geguzin, I. G. Berzina, and I. V. Vorob'eva (1967), "Features of Kinetics of Disappearance of Tracks Left by Uranium Fission Fragments in Thin Mica Plates," *Doklady Akad. Nauk USSR* **175**, 807–809.

Yu. E. Geguzin, I. G. Berzina, and I. V. Vorob'eva (1968), "On Thermally Stable Tracks Derived from the Impacts of Fragments of Uranium Fission in Single Crystals of Muscovite," *Izvestia Akad. Nauk SSSR, Geological Series* **31**, 21–28.

F. H. Geisler (1972), *Search for Superheavy Elements in Terrestrial Minerals and Cosmic Ray Induced Fission of Heavy Elements*, Ph.D. Thesis, Washington University.

F. Geisler and P. R. Phillips (1972), "An Improved Method for Locating Charged Particle Tracks in Thin Plastic Sheets," *Rev. Sci. Instr.* **43**, 283–284.

W. Gentner, D. Storzer, and G. A. Wagner (1969), "New Fission Track Ages of Tektites and Related Glasses," *Geochim. Cosmochim. Acta* **33**, 1075–1081.

D. M. Gilliam and G. F. Knoll (1970), "Counting of Heavily Etched Fission Fragment Tracks in Polycarbonate," *Trans. Am. Nucl. Soc.* **13**, 526.

J. J. Gilman and W. G. Johnston (1956), "Observation of Dislocation Climb and Glide in LiF Crystals," *J. Appl. Phys.* **27**, 1018–1022.

B. Goel and R. Eggmann (1972), "Automatic Focusing of Scanning Microscopes," *Nucl. Instr. Methods* **104**, 225–226.

R. Gold and C. E. Cohn (1972), "Analysis of Automatic Fission Track Scanning in Solid State Nuclear Track Recorders," *Rev. Sci. Instr.* **43**, 18–28.

R. V. Griffith (1973), "Results of the 1972 Survey on Track Registration," Lawrence Livermore Lab. Rept. UCRL-51362, March 15.

U. Haack (1970), "Fission Track Age of Stibiotantalite from Alto Ligonha, Mozambique," *Contr. Mineral. Petrol.* **29**, 183–185.

U. Haack (1972), "Systematics in the Fission Track Annealing of Minerals," *Contr. Mineral. Petrol.* **35**, 303–312.

U. Haack (1973), "Search for Superheavy Transuranic Elements," *Naturwiss.* **60,** 65–70.

U. K. Haack and M. Gramse (1972), "Survey of Garnets for Fossil Fission Tracks," *Contr. Mineral. Petrol.* **34,** 258–260.

U. K. Haack and M. J. Potts (1972), "Fission Track Annealing in Garnet," *Contr. Mineral. Petrol.* **34,** 343–345.

R. Henke and E. Benton (1971), "On Geometry of Tracks in Dielectric Nuclear Track Detectors," *Nucl. Instr. Methods* **97,** 483–489.

R. P. Henke, E. V. Benton, and H. H. Heckman (1970), "Sensitivity Enhancement of Plastic Nuclear Track Detectors Through Photo-Oxidation," *Rad. Effects* **3,** 43–49.

U. Höppner, E. Konecny, and G. Fiedler (1969), "Diameter of Etched Fission Fragment Tracks in Solid State Detectors as a Function of Particle Energy," *Nucl. Instr. Methods* **74,** 285–290.

J. Hudis and S. Katcoff (1969), "High-Energy-Proton Fission Cross Sections of U, Bi, Au, and Ag Measured with Mica Track Detectors," *Phys. Rev.* **180,** 1122–1130.

D. R. Johnson, R. H. Boyett, and K. Becker (1970), "Sensitive Automatic Counting of Alpha Particle Tracks in Polymers and Its Applications in Dosimetry," *Health Phys.* **18,** 424–427.

W. G. Johnston (1961), "Dislocation Etch Pits in Non-Metallic Crystals," *Progress in Ceramic Science,* **2,** 1–75. New York: Pergamon Press.

D. Jowitt (1971), "Measurement of $^{238}U/^{235}U$ Fission Ratios in Zebra Using Solid State Track Recorders," *Nucl. Instr. Methods* **92,** 37–44.

A. Kapustsik, V. P. Perelygin, and S. P. Tretiakova (1964), "Efficiency of Registration of Fission Fragments with the Aid of Glass," *Pribory i. Tekh. Eksp.* **5,** 72–75.

A. L. Kartuzhanskii, V. E. Privalova, E. S. Sorokin, and V. Ya. Fedyukin (1970), "Recording of Tracks of Ionizing Particles on Cellulose Films in Various Gaseous Media," *Zh. Nauch. Prikl. Fotogr. Kinematogr.* **15,** 59–61.

H. A. Khan (1971), "Semi-Automatic Scanning of Tracks in Plastics," *Rad. Effects* **8,** 135–138.

H. A. Khan and S. A. Durrani (1972a), "Efficiency Calibration of Solid State Nuclear Track Detectors," *Nucl. Instr. Methods* **98,** 229–236.

H. A. Khan and S. A. Durrani (1972b), "Electronic Counting and Projection of Etched Tracks in Solid State Nuclear Track Detectors (SSNTD)," *Nucl. Instr. Methods* **101,** 583–587.

H. A. Khan and S. A. Durrani (1972c), "Prolonged Etching Factor in Solid State Track Detection and Its Applications," *Rad. Effects* **13,** 257–266.

H. D. Kosanke (1972), "Track Etch Registrants for High Temperature Applications," *Trans. Am. Nucl. Soc.* **15,** 124.

W. Krätschmer (1971), *Etchable Tracks of Artificially Accelerated Heavy Ions in Quartz Glass*, Ph.D. Thesis, Heidelberg.

S. Krishnaswami, D. Lal, N. Prabhu, and A. S. Tamhane (1971), "Olivines: Revelation of Tracks of Charged Particles," *Science* **174**, 287–291.

S. Krishnaswami, D. Lal, N. Prabhu, and D. Macdougall (1973), "Characteristics of Fission Tracks in Zircon: Applications to Geochronology and Cosmology," *Earth Planet. Sci. Lett.* **22**, 51–59.

S. Lakatos and D. S. Miller (1970), "Water-Pressure Effect on Fission-Track Annealing in an Alpine Muscovite," *Earth Planet. Sci. Lett.* **9**, 77–81.

S. Lakatos and D. S. Miller (1972a), "Evidence for the Effect of Water Content on Fission-Track Annealing in Volcanic Glass," *Earth Planet. Sci. Lett.* **14**, 128–130.

S. Lakatos and D. S. Miller (1972b), "Fission-Track Stability in Volcanic Glass of Different Water Contents," *J. Geophys. Res.* **77**, 6990–6993.

S. Lakatos and D. S. Miller (1973), "Problems of Dating Mica by the Fission-Track Method," *Can. J. Earth Sci.* **10**, 403–407.

D. Lal (1969), "Recent Advances in the Study of Fossil Tracks in Meteorites Due to Heavy Nuclei of the Cosmic Radiation," *Space Sci. Rev.* **9**, 623–650.

D. Lal, A. V. Muralli, R. S. Rajan, A. S. Tamhane, J. C. Lorin, and P. Pellas (1968), "Techniques for Proper Revelation and Viewing of Etch-Tracks in Meteoritic and Terrestrial Minerals," *Earth Planet. Sci. Lett.* **5**, 111–119.

D. Lal, R. S. Rajan, and A. S. Tamhane (1969), "Chemical Composition of Nuclei of $Z > 22$ in Cosmic Rays Using Meteoritic Minerals as Detectors," *Nature* **221**, 33–37.

N. L. Lark (1969), "Spark Scanning for Fission-Fragment Tracks in Plastic Foils," *Nucl. Instr. Methods* **67**, 137.

D. Macdougall (1971), "Fission Track Dating of Volcanic Glass Shards in Marine Sediments," *Earth Planet. Sci. Lett.* **10**, 403–406.

D. Macdougall (1973), "Fission Track Dating of Oceanic Basalts," *Trans. Amer. Geophys. Union* **54**, 987.

D. Macdougall, D. Lal, L. L. Wilkening, S. G. Bhat, S. S. Liang, G. Arrhenius, and A. S. Tamhane (1971), "Techniques for the Study of Fossil Tracks in Extraterrestrial Samples. I: Methods of High Contrast and High Resolution Study," *Geochem. J.* **5**, 95–112.

M. Maurette (1966), "Etude des Traces d'Ions Lourds dans les Mineraux Naturels d'Origine Terrestre et Extra-Terrestre," *Bull. Soc. Franc. Min. Crist.* **89**, 41–79.

M. Maurette (1970), "On Some Annealing Characteristics of Heavy Ion Tracks in Silicate Minerals," *Rad. Effects* **5**, 15–19.

M. Maurette, P. Pellas, and R. M. Walker (1964), "Etude des Traces de Fission Fossiles dans le Mica," *Bull. Soc. Franc. Min. Crist.* **87**, 6-17.

L. Medveczky (1969), "Track Revealing and Visualization," *Proc. Inter. Conf. Nucl. Track Registration in Insulating Solids*, Clermont-Ferrand, Sect. III, 2–13.

L. Medveczky and G. Somogyi (1966), "The Possibility of Neutron Personnel Dosimetry with Solid State Track Detectors," *Proc. 2nd Symp. on Health Phys.*, Budapest, **1,** 61–64.

P. P. Mehta and Rama (1969), "Annealing Effects in Muscovite and Their Influence on Dating by Fission Track Method," *Earth Planet. Sci. Lett.* **7,** 82–86.

D. S. Miller and S. Lakatos (1970), "Fission Track Stability of Alpine Micas," *Ecologae Geol. Helv.* **63/1,** 229.

M. Monnin (1968), *Etude de l'Interaction des Particules Lourdes Ionisantes avec les Macromolecules a l'Etat Solide*, Ph.D. Thesis, Univ. of Clermont, Clermont-Ferrand, France.

M. Monnin and D. B. Isabelle (1970), "Solid Track Detectors and Their Applications in Biology," *Ann. Phys. Biol. Med.* **4,** 95–113.

M. Monnin, H. Besson, S. Sanzelle, and L. Avan (1966), "Physique Corpusculaire," *Compt. rend.* **263B,** 1337–1938.

M. Monnin, H. Besson, S. Sanzelle, and L. Avan (1967), "Nouveaux Detecteurs Solides de Traces Nucleaire et Nouvelles Methodes de Developpement Chimique des Detecteurs deja Connus," *Compt. rend.* **B264,** 1751–1752.

J. A. Morley (1972), "Two Techniques to Increase Contrast of Track Etch Neutron Radiographs," *Am. Nucl. Soc. Trans.* **15,** 120.

C. W. Naeser (1969), "Etching Fission Tracks in Zircons," *Science* **165,** 388.

C. W. Naeser and F. C. W. Dodge (1969), "Fission-Track Ages of Accessory Minerals from Granitic Rocks of the Central Sierra Nevada Batholith, California," *Geol. Soc. of Amer. Bull.* **80,** 2201–2212.

C. W. Naeser and H. Faul (1969), "Fission Track Annealing in Apatite and Sphene," *J. Geophys. Res.* **74,** 705–710.

C. O'Ceallaigh and J. Daly (1970), "The Measurement of Cone Length and Range in Etched Plastics," *Rad. Effects* **5,** 135–138.

H. L. Pai (1973), "The Development of a Special Polycarbonate Detector for Heavy Element Fission Study," Paper at Eastern Region Nuclear Physics Conference, March 17.

H. G. Paretzke (1972), unpublished report.

H. G. Paretzke, T. A. Gruhn and E. V. Benton (1973a), "The Etching of Polycarbonate Charged Particle Detectors by Aqueous Sodium Hydroxide," *Nucl. Instr. Methods* **107,** 597–600.

H. G. Paretzke, E. V. Benton, and R. P. Henke (1973b), "On Particle Track Evolution in Dielectric Track Detectors and Charge Identification Through Track Radius Measurement," *Nucl. Instr. Methods* **108,** 73–80.

M. J. Parizet and M. Monnin (1969), "Observation des Traces de Fluorescence," *Proc. Inter. Conf. Nucl. Track Registration in Insulating Solids*, Clermont-Ferrand, Sect. IX, 6–8.

V. P. Perelygin, S. P. Tretiakova, N. H. Shadieva, and E. Cieslak (1969), "The Discrimination of Heavy Particles in Phosphate Glass and Muscovite Mica," *Proc. Inter. Conf. Nucl. Track Registration in Insulating Solids*, Clermont-Ferrand, Sect. I, 28–39.

D. D. Peterson (1969), *A Study of Low Energy, Heavy Cosmic Rays Using a Plastic Charged Particle Detector*, Ph.D. Thesis, Rensselaer Polytechnic Institute.

D. D. Peterson (1970), "Improvement in Particle Track Etching in Lexan Polycarbonate Film," *Rev. Sci. Instr.* **41,** 1214–1255.

E. Piesch (1971), "The Use of Interference-Contrast Microscopy for Nuclear-Particle Recording in Solids," *Zeiss Inform.* **76,** 58–60.

V. E. Pilcher, C. C. Jones, and G. R. Ellmers (1972), "Particle Tracks in Cellulose Nitrate," *Am. J. Phys.* **40,** 679–683.

H. Platzer, W. Abmayr, and H. G. Paretzke (1972), "Evaluation of Dielectric Track Detectors with Diffracted Laser Light," *Atomkernenergie* **20,** 162.

G. E. Possin (1970), "A Method for Forming Very Small Diameter Wires," *Rev. Sci. Instr.* **41,** 772–774.

S. Prêtre, E. Tochilin, and N. Goldstein (1968), "A Standardized Method for Making Neutron Fluence Measurements by Fission Fragment Tracks in Plastics. A Suggestion for an Emergency Neutron Dosimeter with Rad-Response," *Proceedings of the First International Congress of Radiation Protection*, 491–505. New York: Pergamon Press.

P. B. Price and R. L. Fleischer (1970), "Particle Identification by Dielectric Track Detectors," *Rad. Effects* **2,** 291–298.

P. B. Price and R. L. Fleischer (1971), "Identification of Energetic Heavy Nuclei with Solid Dielectric Track Detectors: Applications to Astrophysical and Planetary Studies," *Ann. Rev. Nuc. Sci.* **21,** 295–334.

P. B. Price and R. M. Walker (1962a), "Chemical Etching of Charged Particle Tracks," *J. Appl. Phys.* **33,** 3407–3412.

P. B. Price and R. M. Walker (1962b), "Electron Microscope Observation of a Radiation-Nucleated Phase Transformation in Mica," *J. Appl. Phys.* **33,** 2625–2628.

P. B. Price and R. M. Walker (1963a), "A Simple Method of Measuring Low Uranium Concentrations in Natural Crystals," *Appl. Phys. Letters* **2,** 23–25.

P. B. Price and R. M. Walker (1963b), "Fossil Tracks of Charged Particles in Mica and the Age of Minerals," *J. Geophys. Res.* **68,** 4847–4862.

P. B. Price, R. L. Fleischer, D. D. Peterson, C. O'Ceallaigh, D. O'Sullivan, and A. Thompson (1967a), "Identification of Isotopes of Energetic Particles with Dielectric Track Detectors," *Phys. Rev.* **164,** 1618–1620.

P. B. Price, R. S. Rajan, and A. Tamhane (1967b), "On the Pre-Atmospheric Size and Maximum Space Erosion Rate of the Patwar Stony-Iron Meteorite," *J. Geophys. Res.* **72,** 1377–1388.

P. B. Price, R. L. Fleischer, D. D. Peterson, C. O'Ceallaigh, D. O'Sullivan, and A. Thompson (1968a), "High Resolution of Low Energy Cosmic Rays with Lexan Track Detectors," *Phys. Rev. Letters* **21,** 630–633.

P. B. Price, R. L. Fleischer, D. D. Peterson, C. O'Ceallaigh, D. O'Sullivan, and A. Thompson (1968b), "Plastic Track Detectors for Identifying Cosmic Rays," *Can. J. Phys.* **46,** S1149–S1153.

P. B. Price, D. D. Peterson, R. L. Fleischer, C. O'Ceallaigh, D. O'Sullivan, and A. Thompson (1970a), "Composition of Cosmic Rays of Atomic Number 12 to 30," *Acta Phys. Acad. Sci. Hung.* **29,** Suppl. 1, 417–422.

P. B. Price, R. L. Fleischer, and R. T. Woods (1970b), "Search for Spontaneously Fissioning Trans-Uranic Elements in Nature," *Phys. Rev.* **1,** 1819–1821.

P. B. Price, P. H. Fowler, J. M. Kidd, E. J. Kobetich, R. L. Fleischer, and G. E. Nichols (1971), "Study of the Charge Spectrum of Extremely Heavy Cosmic Rays Using Combined Plastic Detectors and Nuclear Emulsions," *Phys. Rev.* **D3,** 815–823.

P. B. Price, D. Lal, A. S. Tamhane, and V. P. Perelygin (1973), "Characteristics of Tracks of Ions with $14 \leq Z \leq 36$ in Common Rock Silicates," *Earth Planet. Sci. Lett.* **19,** 377–395.

G. M. Reimer, G. A. Wagner, and B. S. Carpenter (1972), "The Thermal Stability of Fission Tracks in the Standard Reference Material Glass Standard (National Bureau of Standards)," *Rad. Effects* **15,** 273–274.

R. F. Roy, E. R. Decker, and F. Birch (1968), "Heat Flow in the United States," *J. Geophys. Res.* **73,** 5207–5221.

B. Ruegger, R. Richmond, and W. Zürcher (1969), "Les Detecteurs Solid Visuals: Comptage Automatique des Traces et Loi des Pertes par Chevauchment," Swiss Federal Institute of Research and Reactor Materials, Würenlingen, unpublished report.

O. V. Rumyantsev, Y. A. Selitskii, and V. B. Funshtein (1968), "Recording of Fission Fragments by Means of Mica Under Intensive Detector Irradiation with Charged Particles," *Pribory i. Tech. Eksp.* **1,** 51–52.

W. W. Schultz (1968), "Track Density Measurement in Dielectric Track Detectors with Scattered Light," *Rev. Sci. Instr.* **39,** 1893–1896.

W. W. Schultz (1970), "$1/v$ Dielectric Track Detectors for Slow Neutrons," *J. Appl. Phys.* **41,** 5260–5269.

M. G. Seitz, R. M. Walker, and B. S. Carpenter (1973), "Improved Methods for Measurement of Thermal Neutron Dose by the Fission Track Technique," *J. Appl. Phys.* **44,** 510–512.

H. F. Sherwood (1970), "Enhancing Radiation Damage for Nuclear Particle Detection," U.S. Patent 3,501,636.

Yu. A. Shukolyukov and A. N. Komarov (1966), "Possible Paleothermometry on the Grounds of Fission Tracks of Uranium in Minerals," *Izvestia Akad. Nauk SSSR* **9,** Geol. Series **31,** 137–141.

Yu. A. Shukolyukov and A. N. Komarov (1970), "Tracks of Fission in Monazites," *Akad. Sauk SSSR* **9,** 20–26.
Yu. A. Shukolyukov, I. N. Krylov, I. N. Tolstikhin, and G. V. Ovchinnikova (1965), "Tracks of Uranium Fission Fragments in Muscovite," *Geochem. Intern'l.* **2,** 202–213.
E. C. H. Silk and R. S. Barnes (1959), "Examination of Fission Fragment Tracks with an Electron Microscope," *Phil. Mag.* **4,** 970–972.
R. F. Sippel and E. D. Glover (1964), "Fission Damage in Calcite and the Dating of Carbonates," *Science* **144,** 409–410.
M. Sohrabi and K. Becker (1971), "Some Studies on the Application of Track Etching in Personnel Fast Neutron Dosimetry," Oak Ridge National Lab., ORNL-TM-3605.
G. Somogyi (1966a), "A New Possibility for Determination of Energy Distributions of Charged Particles in Solid State Track Detectors," *Nucl. Instr. Methods* **42,** 312–314.
G. Somogyi (1966b), "New Method for Measuring the Particle Energy According to the Diameters of the Holes in Solid State Nuclear Track Detectors," *Atomki Kozl.* **8,** 218–224.
G. Somogyi (1972a), "Effects of Ozone Atmosphere on the Detecting Properties of Plastic Track Recorders," *Rad. Effects.* **16,** 233–243.
G. Somogyi (1972b), "Influence of Thermal Effects on the Track Registration Characteristics of Plastics," *Rad. Effects.* **16,** 245–251.
G. Somogyi and J. Gulyás (1972), "Methods for Improving Radiograms in Plastic Track Detectors," *Radioisotopy* **13,** 549–568.
G. Somogyi, M. Varnagy, and G. Pëto (1968), "Application of Plastic Track Detectors for Detection of Light Nuclei," *Nucl. Instr. Methods* **59,** 299–304.
G. Somogyi, M. Varnagy, and L. Medveczky (1970), "The Influence of Etching Parameters on the Sensitivity of Plastics," *Rad. Effects* **5,** 111–116.
R. A. Stern and P. B. Price (1972), "Charge and Energy Information from Heavy Ion Tracks in 'Lexan,'" *Nature Phys. Sci.* **240,** 82–83.
D. R. Stone (1969), "Identification of Fission Fragment Tracks in Lexan After Pretreatment to High Doses of Alpha Particles," *Health Phys.* **16,** 772–773.
D. Storzer (1970), "Fission Track Dating of Volcanic Glasses and the Thermal History of Rocks," *Earth Planet. Sci. Lett.* **8,** 55–60.
D. Storzer and G. A. Wagner (1969), "Correction of Thermally Lowered Fission Track Ages of Tektites," *Earth Planet. Sci. Lett.* **5,** 463–468.
D. Storzer and G. A. Wagner (1971), "Fission Track Ages of North American Tektites," *Earth Planet. Sci. Lett.* **10,** 435–440.
M. Suzuki (1970), "Fission Track Ages and Uranium Contents of Obsidians," *Zinruigaku Zassi, J. Anthrop. Soc. Nippon* **78,** 50–58.
J. W. N. Tuyn (1967), "Solid-State Nuclear Track Detectors in Reactor Physics Experiments," *Nucl. Appl.* **3,** 372–374.

E. E. Uzgiris and R. L. Fleischer (1971), "Amber: Charged Particle Track Registration, Track Stability, and Uranium Content," *Nature* **234,** 28–30.

M. Varnagy (1970), *A Few Properties of the Cellulose Acetate (T-Cellit) Detector and Its Applications*, Ph.D. Thesis, Kossuth University, Debrecen.

M. Varnagy, J. Szabo, S. Juhasz, and J. Csikai (1973), "Determination of Track Parameters by Diffraction Method Using Laser Light," *Nucl. Instr. Methods* **106,** 301–305.

Ya. M. Veprik, V. P. Perelygin, V. P. Romanenki, S. P. Tretiakova, and Yu. Z. Vinogradov (1970), "Effect of Chemical Factors on the Sensitivity Drift of Polymeric Detectors," *Pribory i. Tekh. Eksp.* **4,** 51–53.

G. A. Wagner (1968), "Fission Track Dating of Apatites," *Earth Planet. Sci. Lett.* **4,** 411–415.

G. A. Wagner (1972), "Spaltspurenalter von Mineralen und Naturlichen Glasern: eine Ubersicht," *Fortschr. Miner.* **49,** 114–145.

R. M. Walker (1963), "Characteristics and Applications of Solid State Detectors," *Proc. Strasbourg Conference on New Methods of Track Detection*, Centre de Recherches Nucleaires, Strasbourg, France.

R. M. Walker, P. B. Price, and R. L. Fleischer (1963), "A Versatile, Disposable Dosimeter for Slow and Fast Neutrons," *Appl. Phys. Letters* **3,** 28–29.

C. O. Widell (1969), "Stability of Latent Tracks in Mica," *Proc. Inter. Conf. Track Registration in Insulating Solids*, Clermont-Ferrand, Sect. I, 24–27.

R. T. Woods (1973), *The Nature of Low Energy Interplanetary Nuclei Down to Energies of 50 keV/amu from an Analysis of Solid State Etched-Track Detectors from the Surveyor III Spacecraft and the Apollo 16 and 17 Cosmic Ray Experiments*, Ph.D. Thesis, State University of New York at Albany.

D. A. Young (1958), "Etching of Radiation Damage in Lithium Fluoride," *Nature* **182,** 375–377.

E. A. Zhagrov, Yu. A. Nemilov, and Yu. A. Selitskii (1968), "Yield and Angular Anistropy of ^{226}Ra Photofission Fragments," *Sov. J. Nucl. Phys.* **7,** 183–185.

Chapter 3

Methods of Nuclear Particle Identification

3.1. INTRODUCTION

In a microscopic region around the trajectory of an energetic particle, the structure of a track-recording solid and its chemical reactivity are modified by amounts that depend on the charge Z and velocity β of the particle. We shall discuss several methods of particle identification that depend upon the quantitative relationship between track etch rate v_T and ionization rate J. The most highly developed method and the one that has been applied to all classes of dielectric solids involves measuring track length L at one or more values of residual range R. It is sometimes called the LR-plot method. The track-profile method involves detailed measurements of the change of shape of the etch pit and is particularly useful for identifying very low-energy particles. In certain circumstances the change of etch pit diameter with etch time is a better measure of Z and E than is etch pit length. The maximum etchable track length is also related to Z, though caution must be used in interpreting such measurements.

3.2. EXPERIMENTAL PROBLEMS

Certain kinds of solid track detectors are capable of high resolution; others give only a rough measure of nuclear charge but may have other unique advantages (e.g., sampling particles in space 4.5 billion years ago). In all cases to get the best results it is necessary to control etching conditions carefully and to calibrate the detectors with beams of ions with Z and E as close as possible to those of interest. Rarely can we prescribe an exact equation that the reader may confidently use to identify ions with his own detector. The reason is not hard to find. No commercial firm makes dielectric detectors for particle identification. We usually acquire plastics, glass, or minerals intended for some other purpose. Polymers vary in composition from batch to batch. Those with volatile plasticizers change in composition with time. Most glasses contain a variety of oxides whose composition is not rigorously controlled. Minerals are notoriously variable in composition. Etchants and environmental conditions such as ambient humidity and temperature will vary from one laboratory to another. A similar difficulty applies to nuclear emulsions and in fact to practically all detectors of which exacting performance is expected.

Having faced up to the necessity of acquiring and calibrating his own detector, the researcher should buy a large stock of the particular solid he wants to use and then calibrate portions of it to check for uniformity of response.

3.2.1. Plastic Detectors

One must keep in mind the adverse environmental effects discussed in Chapter 2. After irradiation Lexan needs to be protected against UV from such sources as sunlight and fluorescent lights, and it should be aged for a sufficient time to permit slow chemical reactions along the track to reach completion. Two weeks is reasonably safe. Because of the major effect of O_2 on track formation (see Chapter 2), cellulosics should not be left in a vacuum system any longer than necessary. After irradiation they should be thoroughly aerated at atmospheric pressure, or better yet, at $\sim 10^2$ atm (M. Monnin, private communication). They should be processed promptly or kept refrigerated until use, because in contrast to the situation with Lexan the etch rate v_T in cellulose nitrate slowly decreases with time and the plastic gradually decomposes at room temperature.

In the method of particle identification to be discussed in Section 3.3, the etch pit length L is measured at a known residual range R. Fig. 3-1 shows two methods of doing this. If the particle produces etch pits in several sheets, followed by a final, completely etched portion, there is no difficulty in measuring L and R at several points (Fig. 3-1a). If the final portion is not etched to the end of the range, or if the particle does not cross a single sheet, it is necessary to re-etch the sheet in order to determine the total range. To avoid the extra labor of having to relocate each such event and make a second measurement after a second stage of etching, Stern and Price (1972) developed a three-stage process consisting of an initial etch, an intense ultraviolet irradiation which greatly increases the reactivity of the unetched portion, and a short final etch which completely reveals the remainder of the track (Fig. 3-1b). The initial and final etched portions can easily be distinguished by the abrupt change of slope. Fig. 3-2 shows photomicrographs of some tracks processed in this way. The method only works on plastics such as Lexan, for which v_T can be strongly increased by a UV irradiation. A four-stage variation of the method can be used to identify particles with ionization rate too small to produce an initial cone: a short initial UV exposure, an initial etch,

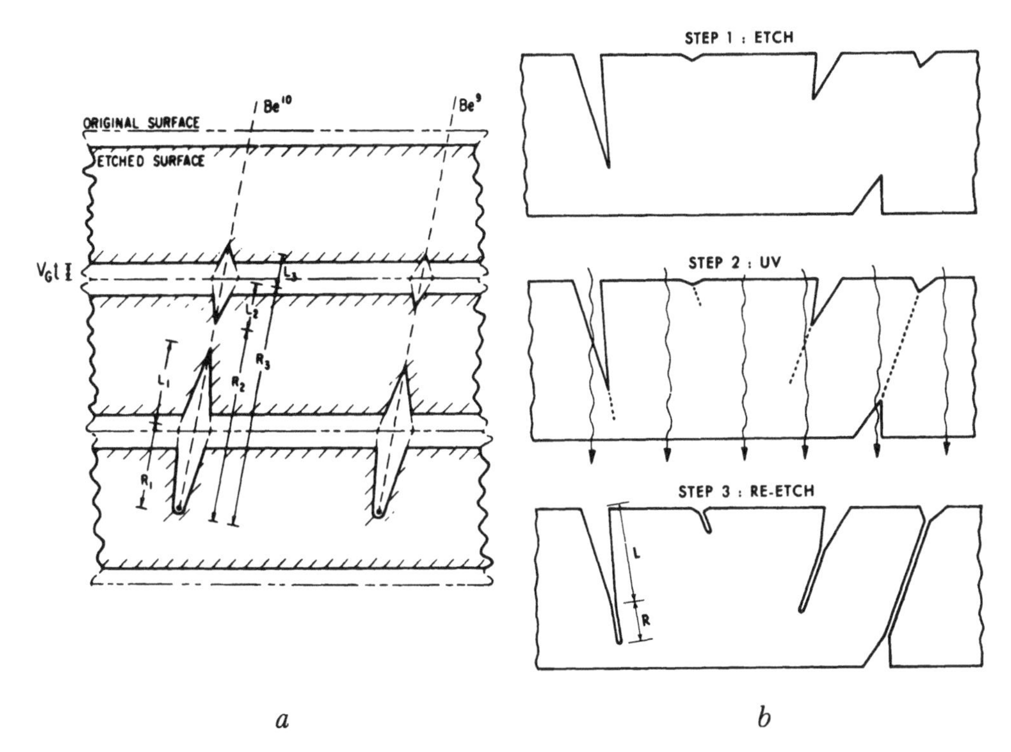

Fig. 3-1. *Methods of particle identification by measuring etch pit lengths L_i at residual ranges R_i: (a) Particle produces etchable cones in several sheets of a stack. Isotopes of the same element should produce distinguishable signals; different elements produce larger signal differences. (After Price et al., 1967.) (b) Particle stops in a single sheet; both L and R can be measured at the same time if detector is a plastic such as Lexan for which v_T can be increased by a UV irradiation; three-stage process requires an initial etch, a UV exposure, and a short final etch of the same sheet. (After Stern and Price, 1972.)*

a long UV exposure, and a final etch lead to an etched track with a sharp change of slope. A cautionary remark is in order. Some batches of Lexan contain a small amount of an additive that increases their stability against ultraviolet radiation. Such Lexan is virtually opaque to UV and does not respond well to a multistage process that includes a UV treatment. It can be identified by a spectrophotometer as being strongly absorbing at wavelengths shorter than ~3500 Å.

At energies so low that the range of the particle is less than a few microns, the resolution of visible light is insufficient to permit measurements of L and R. Scanning electron microscopes (SEM) are easy to use and increasingly common in well-equipped laboratories. With an SEM one can still use the L, R technique or use the track-profile method to be discussed in Section 3.8. It is necessary to make a plastic replica of the etched surface, using a solvent that softens the repli-

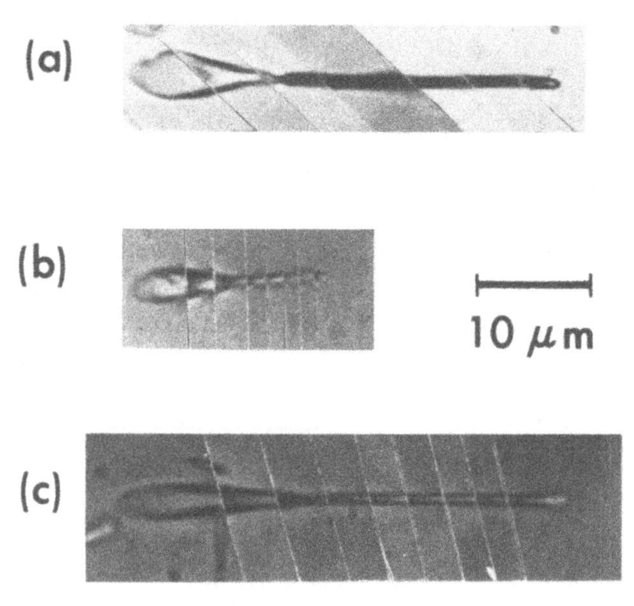

Fig. 3-2. *Examples of tracks in Lexan processed in three stages. The photomicrographs are composites made by focusing at different points along the tracks and then assembling the portions in focus. The change of slope between initial cone and final cylinder is evident. (After Stern and Price, 1972.)*

cating plastic but does not affect the etched plastic detector. With Lexan one can use either a strip of formvar, softened with acetic acid, which does not attack Lexan, or one can use RTV silicone rubber, following the prescription of Fleischer et al. (1970). In either case the replica must be made conducting by coating it with an evaporated metal layer, Au being a very satisfactory coating. Formvar is particularly resistant to the damaging effects of an SEM beam.

3.2.2. Glass and Mineral Detectors

Fig. 3-3 shows a method of calibrating brittle solids such as glass or minerals, using a beam of monoenergetic heavy ions (Price et al., 1973). After irradiation, the detector is embedded in epoxy at a shallow angle and then the epoxy is ground away until a wedge-shaped portion of the detector is removed. When this new surface is polished and etched, one has an array of tracks with all possible residual ranges, increasing from zero to the full range linearly with distance from the edge of the ground-off region. The shapes and sizes of etch pits with a large cone angle in a glass such as SiO_2 can be determined much better from a replica than by viewing the glass itself. Lengths as short as ~1 μm can be estimated reasonably well by optical microscopy if the replica is shadowed at a shallow angle by a pin-

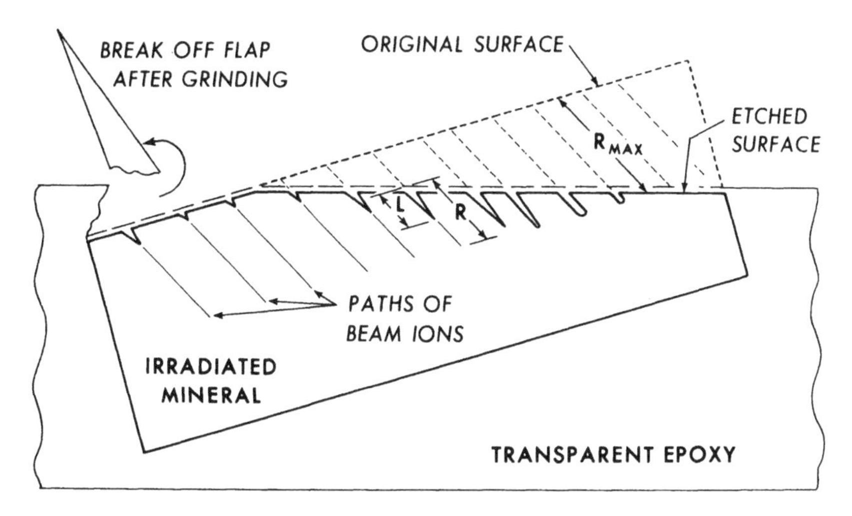

Fig. 3-3. *Method of preparing a mineral or glass detector for determination of L vs. R. The solid is irradiated with monoenergetic ions of range* R_{max}*, imbedded at an angle to the surface of an epoxy disc, and ground and polished so that, in a single operation, tracks with a distribution of ranges from 0 to* R_{max} *are etched at the same time.* (*After Price et al., 1973.*)

hole source of metal in a vacuum evaporator. The height of the replica of an etch pit is determined from the length of the shadow and a knowledge of the shadowing angle.

Tracks in the interior of a mineral or a glass with a small cone angle for etched tracks can be observed with the TINT or TINCLE method described in Chapter 2 (Lal et al., 1969; Bhandari et al., 1971). These techniques are useful when one is trying to identify particles by measuring maximum etchable track lengths (Section 3.10).

3.3. PARTICLE IDENTIFICATION BY MEASUREMENTS OF ETCH RATE VS. RESIDUAL RANGE

This method was developed in a series of papers (Price et al., 1967; 1968a,b; 1969; Fleischer et al., 1969) and reviewed by Price and Fleischer (1970; 1971). It is useful to go step by step through the procedure for establishing the response of a solid detector to ions of different Z and E. We will then give a few examples that indicate the kind of resolution that can be expected and that illustrate some of the practical problems that can arise. Chapters 4 and 5 contain examples of the contributions that dielectric track identifiers have made to nuclear physics, cosmic ray physics, and solar particle physics.

3.3.1. Calibration with Beams of Known Ions

Table 3-1 lists some of the beams available now or in the near future at various heavy ion accelerators. If at all possible one should rely on direct calibrations with ions whose Z and E are similar to those being studied, but in studies of cosmic rays and solar particles there is useful information in the natural beam itself. For example, Fe and Si are known to be much more abundant than the intervening elements, and even-Z elements are more abundant than their odd-Z neighbors. Among the heaviest elements the short-lived actinides $84 \leq Z \leq 89$ are missing, which helps to establish the location of Pb($Z = 82$) and Th($Z = 90$).

Fig. 3-4 shows accelerator data on an LR-plot for the first material studied—cellulose nitrate (Price et al., 1967). In addition to demonstrating the feasibility of particle identification by etch rate measurements, this work showed that in favorable conditions it is possible to resolve isotopes of the same element—in this case ^{10}B and ^{11}B.

Fig. 3-5 shows accelerator data for the minerals oligoclase and augite (Price et al., 1973). To achieve a spread of beam energies the samples were ground as shown in Fig. 3-3. In these calibrations the residual range was not observed but the total range $R_T = L + R$ of each particle was calculated from a knowledge of the angle of entry of the beam, the grinding angle, and the distance of each track from the edge of the ground-off region. Note that these LR_T-plots differ from the LR-plot in Fig. 3-4 in that the abscissa gives total range instead of residual range. Obviously no data can appear in the region where $L > R_T$.

In crystalline solids such as minerals, v_T does not vary as smoothly with ioniza-

Table 3-1. Heavy Ion Accelerators

Machine	Energy	Ions
Bevatron (LBL)	2.1 GeV/nuc	^{12}C, ^{14}N, ^{16}O, ^{20}Ne
Superhilac (LBL)	up to ~8 MeV/nuc	Practically all ions up to U
88" Cyclotron (LBL)	up to ~20 MeV/nuc for ^{16}O; ~3.2 MeV/nuc for ^{56}Fe	Most ions up to ^{56}Fe
Hilac (Yale)	10.3 MeV/nuc	^{10}B, ^{11}B, ^{12}C, ^{14}N, ^{16}O, ^{20}Ne, ^{19}F, ^{32}S
Linac (Manchester)	10 MeV/nuc	Cu, Fe, Ca and lighter ions
Cyclotron (Dubna)	~6 MeV/nuc	Xe, Zn, Ca and lighter ions
Cyclotron (Orsay)	~7 MeV/nuc	Kr and lighter ions
Tandem van de Graaffs at many locations	Usually ≲2 MeV/nuc	U, I, Br, As, Fe, Cℓ, S and lighter ions
Electrostatic accelerators at many locations	~200 keV/nuc	Many species
Unilac (Darmstadt)	~8 MeV/nuc	Practically all ions (~1975)
Bevalac (LBL; uses Superhilac as injector to Bevatron)	2.1 GeV/nuc	Any ions up to ~Fe (currents decreasing with Z; late 1974)

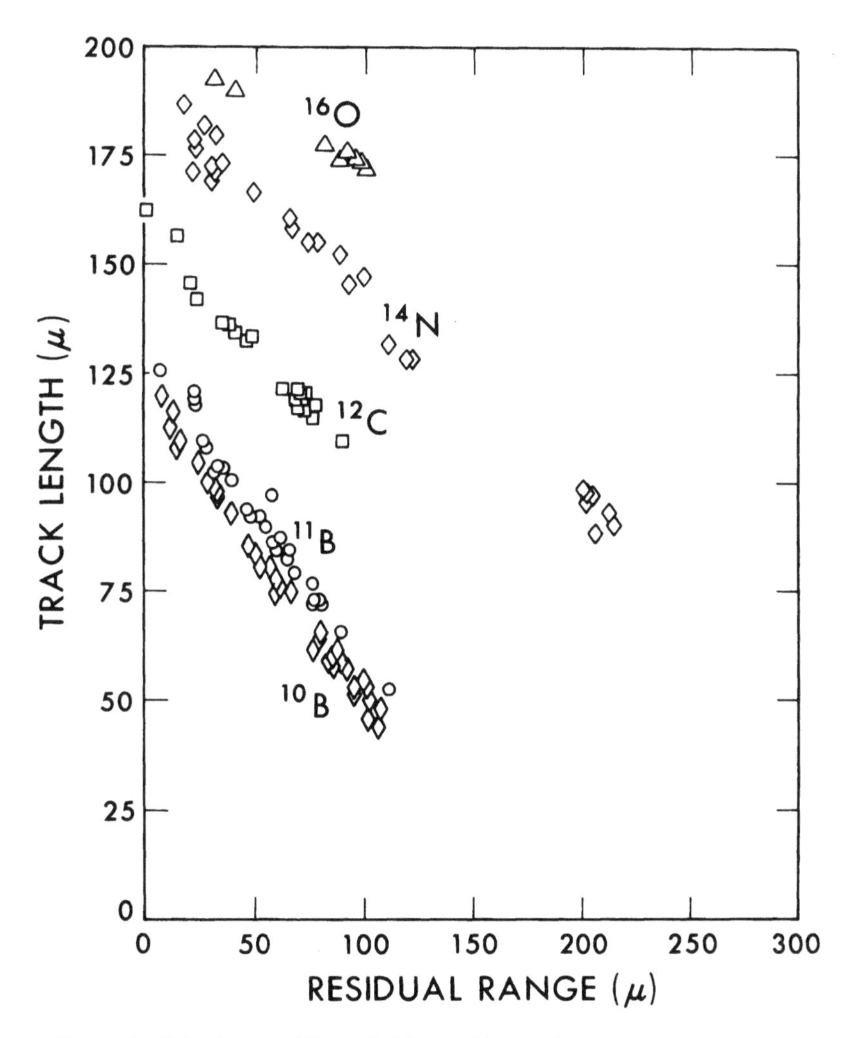

Fig. 3-4. *LR-plots for Nixon-Baldwin cellulose nitrate bombarded with various ions having energies from 0 to 10.3 MeV/nucleon and etched 9 hours in 6.25N NaOH at 20°C. (After Price et al., 1967.)*

tion rate as it does in amorphous solids such as plastics and glasses, possibly because the radiation damage is distributed in a nonuniform way as a result of anisotropies and periodicities of the crystal structure. At very low ionization rates the spread in values of L on an LR_T-plot may be quite large. The visibility of etched tracks may be better for some directions than others, because the cross section of the track is not circular but has some other geometric shape. At high ionization rates, in a few minerals such as augite (Fig. 3-5c and d), v_T saturates and thus does not provide a measure of ionization rate for high-Z particles. It is possible, however, if one is especially interested in high-Z particles, to anneal the

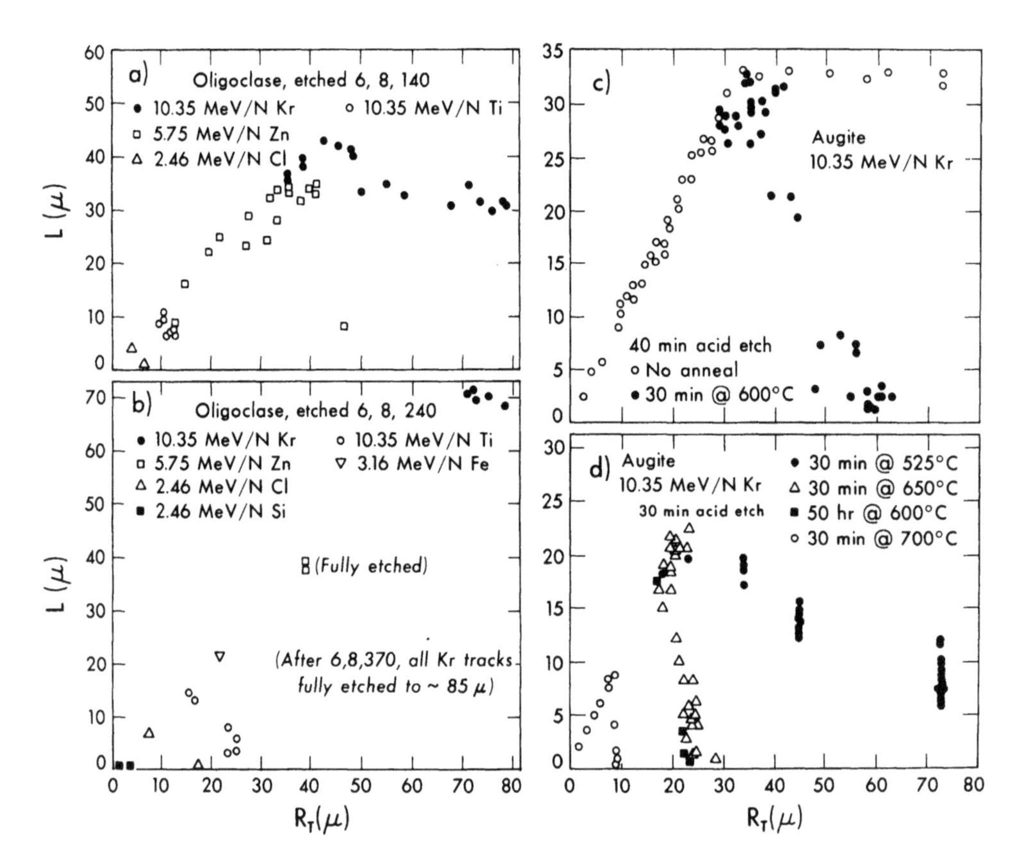

Fig. 3-5. *Plots of L vs.* $R_T(\equiv L + R)$ *for ions of various initial energies (as labeled) in oligoclase and augite crystals ground as shown in Fig. 3-3. The points along the line* $L = R_T$ *represent tracks that became cylinders during etching; points falling off to the right of that line are for etch pits whose lengths provide a measure of charge and energy. In oligoclase L is an increasing function of charge and decreases with energy. In augite the etch rate for Kr ions saturates at a high value but can be decreased so as to change with energy if the crystal is annealed before etching. For details of the etching recipes and further data, see Price et al. (1973).*

mineral before etching, so that the density of radiation damage is reduced and v_T is decreased. In Fig. 3-5c and d we see that the LR_T curve for Kr ions in augite can be lowered and its slope made steeply negative by a high-temperature anneal. Price et al. (1973) have given an exhaustive treatment of the problems of identifying particles with mineral detectors. We refer the interested reader to that paper for further details and for calibration data for a variety of minerals.

3.3.2. Range-Energy Relations

From calibrations with a limited set of ion species, the goal is to derive a single relation $v_T = f(J)$ from which LR-curves can be computed for ions of any Z and A.

We first convert LR-curves to curves of $\bar{v}_T$ vs. $\bar{R}$, where $\bar{v}_T \equiv L/t$ is the average etch rate and $\bar{R} \equiv (L/2) + R$ is the range to a point halfway along the etched portion. To a good approximation $\bar{v}_T(\bar{R})$ is equal to the instantaneous value of v_T at that point. To test whether the curves of $\bar{v}_T$ vs. $\bar{R}$ for several ion species transform into a universal curve of $\bar{v}_T$ vs. J, we need to know range-energy relations for each species, so that the energy and velocity of each ion for which a cone-length measurement was made can be determined from its R, Z, and A. Then J can be determined from the relation

$$J = a(Z^{*2}/\beta^2)[\ln(\beta^2\gamma^2) + K - \beta^2] \tag{3-1}$$

and a knowledge of its Z and β. This expression is the same as eq. (1-9) with appropriate choice of the constants a and K. Recall from Chapter 1 that Z^*, the effective charge, can be approximated by the semi-empirical expression

$$Z^* = Z[1 - \exp(-130\,\beta/Z^{2/3})]. \tag{3-2}$$

Range-energy relations for heavy ions have been calculated by Henke and Benton (1967), by Steward (1968), and by Northcliffe and Schilling (1970), using different input data and different approaches. For most ions Steward's ranges appear to be systematically greater than those of the other two. The tables of Northcliffe and Schilling extend only up to 10 MeV/nucleon, and the tables of Henke and Benton are not claimed to be reliable below 0.1 MeV/nucleon. From our experience, and from measurements of Benton and Henke (1969), we recommend that the tables of Henke and Benton be used at energies above ~1 MeV/nucleon. At lower energies the data are scanty. J. H. Chan (1973, unpublished) has used a scanning electron microscope to look at formvar replicas of etched tracks in Lexan with lengths from ~0.2 μm to ~2 μm produced by ions of ^{19}F, ^{23}Na, ^{40}Ar, and ^{56}Fe with energies from ~13 to ~250 keV/nucleon. P. B. Price has used an optical microscope to study etched tracks of Fe, Br, and I ions with energies up to ~1 MeV/nucleon. Fig. 3-6 compares track-length measurements for Fe ions with ranges predicted by the three groups. For most ions the observations agree better with the ranges calculated by Northcliffe and Schilling than with those calculated by Steward, though the scatter is large. The observations of Chan indicate that in Lexan the damage along the entire range of the particle appears capable of being etched.

3.3.3. Response Curves v_T vs. J for Various Solids

For each etchant and detector we must choose the optimum value of K, which determines the spacing between the curves of J vs. R for different charges. Fig. 3-7 gives a set of ionization curves for $K = 12$, indicating the four different energy

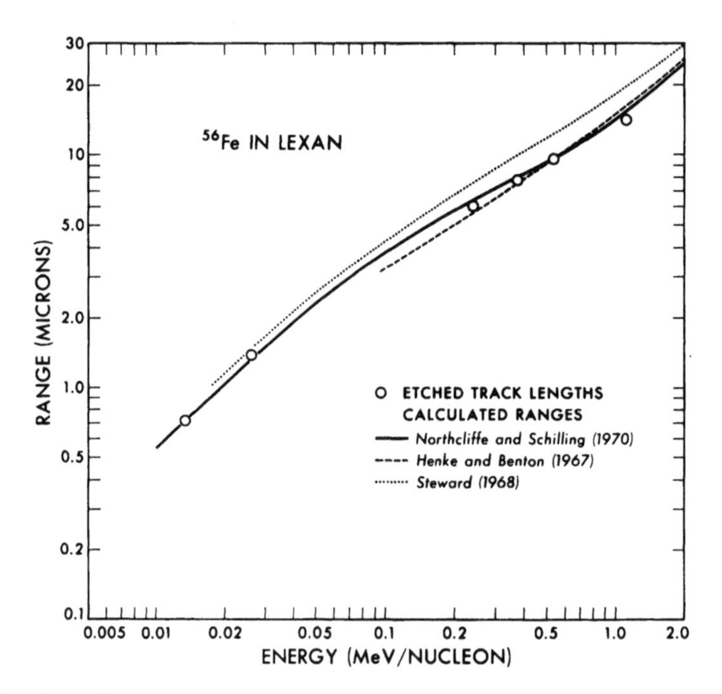

Fig. 3-6. *Comparison of observed lengths of etched tracks in Lexan with range-energy relations calculated by several authors. The energy of the* ^{56}*Fe ions was adjusted at the accelerator.* (*Courtesy of J. H. Chan.*)

regimes that will be discussed in later sections of this chapter. For most plastics it is an adequate approximation to express the response curve as a power function.

$$v_T = BJ^{\zeta} \tag{3-3}$$

where B and ζ must be evaluated empirically. In the energy regime $10 \lesssim E \lesssim 1000$ MeV/nucleon we can also approximate the relation between J and R by a power law

$$J \approx C(A, Z)R^{-\eta} \tag{3-4}$$

which is often convenient when one is doing rough calculations of v_T over a limited energy interval. The procedure is to start with a trial value of K and compute J for all measured points on a plot of v_T vs. R, thus giving v_T vs. J. If all points, independent of Z or A, lie closely on a single curve, the value of K is satisfactory. If the scatter is wide, the procedure should be repeated with a different value of K.

Lexan has been the most extensively studied solid. For different batches and etching conditions, values of K have ranged from about 16 to ∞ (the latter being the limiting case when J is found to depend only on Z^{*2}/β^2), and values of ζ have

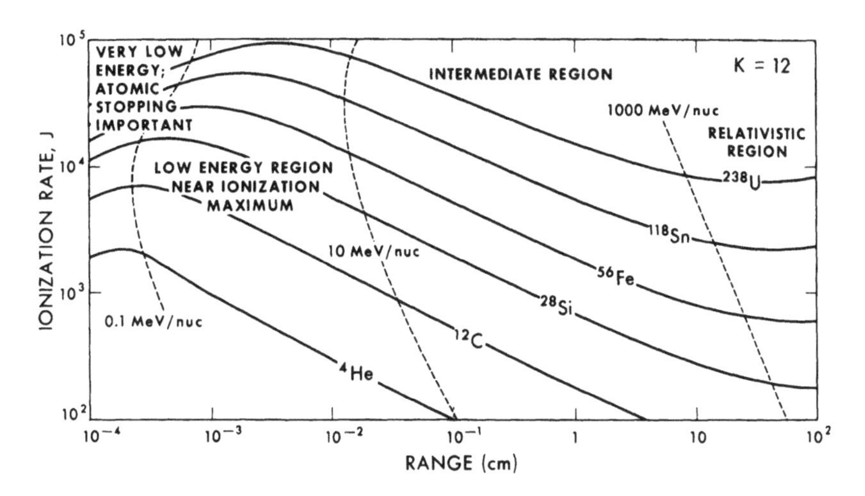

Fig. 3-7. *Curves of radiation damage vs. range, showing the different energy regimes discussed in the text.*

ranged from 1.5 to ~2.4 (Price et al., 1968b; 1971; O'Sullivan et al., 1971a; Wefel, 1971; Shirk et al., 1973; Blanford et al., 1973; Benton and Henke, 1972a,b).

Wefel (1971) has obtained values of ζ from ~1.7 to ~2.1 for CAB (cellulose acetate butyrate) and CTA (cellulose triacetate) made by Bayer and for unplasticized Kodacel CTA made by Eastman. Benton and Henke (1972a,b) have approximated the response curve of cellulose nitrate by two power laws with ζ changing from ~8 at low ionization to ~4.2 at high ionization.

Fig. 3-8 shows response curves for several polymers, glasses, and minerals.

3.3.4. Generation of Curves of v_T vs. R for All Ions

Having found a satisfactory response curve for a particular etchant and solid, we can calculate LR-curves for any ions of interest by integrating the expression

$$t = \int_{R_0}^{R_0+L} dR[v_T(J)]^{-1} = \int_{R_0}^{R_0+L} dR[v_T(R, Z, A)]^{-1}, \tag{3-5}$$

where R_0 is the range at the low-energy end of the etch pit and t is the etching time. If v_T is a complicated function of J the equation must be integrated numerically. In the special case where $v_T \approx BJ^\zeta$, with $\zeta \approx 2$, v_T can be replaced with its average value $\bar{v}_T$ and the integral simplifies to the expression

$$L = t \cdot v_T(R_0 + L/2, Z, A) = t \cdot B[J(R_0 + L/2, Z, A)]^\zeta. \tag{3-6}$$

Fig. 3-9 shows LR-curves generated in this manner for ultraheavy cosmic rays in Lexan.

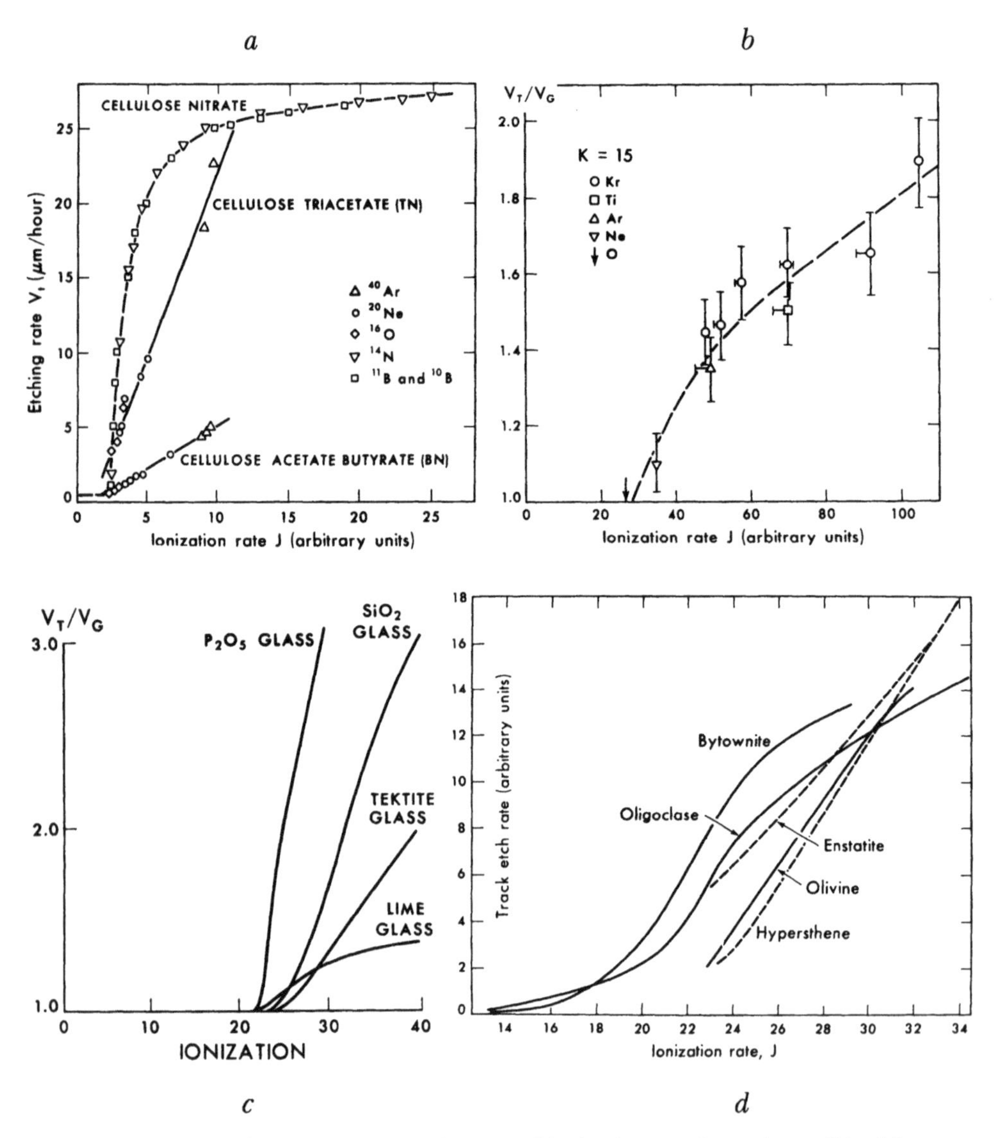

Fig. 3-8. *Track etch rate as a function of ionization rate for various solids. (a) Cellulosic polymers irradiated with heavy ions. Data for CN are adapted from Price et al. (1967); data for TN and BN are from Blanford et al. (1970). (b) Flint glass irradiated with heavy ions. (Unpublished results of I. D. Hutcheon, University of California, Berkeley, and of Fleischer et al., 1971.) (c) Various glasses irradiated with fission fragments. (After Fleischer et al., 1969.) (d) Silicate minerals. Because etch rate depends on such factors as temperature and concentration of etchant, as well as on the composition of the specific mineral sample used, the units of etch rate were arbitrarily chosen to position all the curves in a convenient region of the graph. Note that the response curves increase gradually rather than abruptly with ionization rate. (After Price et al., 1973.)*

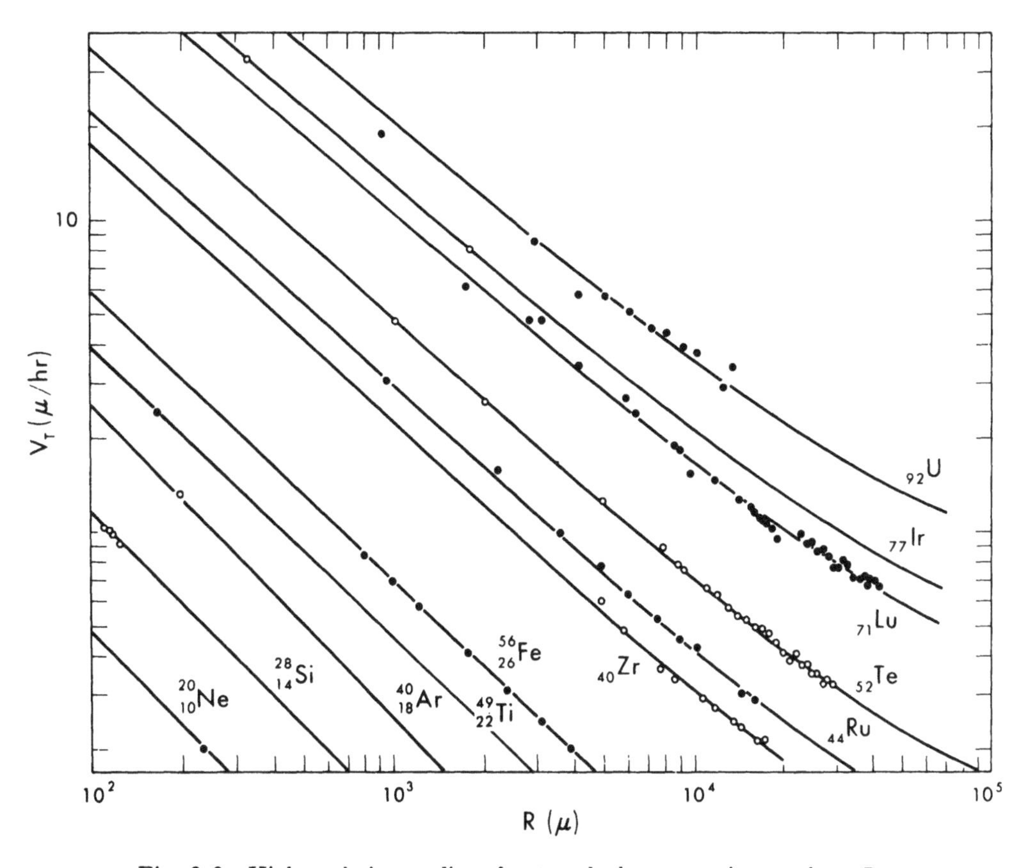

Fig. 3-9. *High-resolution studies of extremely heavy cosmic rays in a Lexan stack by the etch rate method. The Ne, Si, Ar, and Ti data are averages of measurements of many tracks from accelerator bombardments. The Fe data represent the spread in measurements of about 50 cosmic rays that stopped in the stack. Data for other, less abundant elements are omitted for clarity. The data points for the six nuclei assigned charges of 40, 44, 52, 71, 77, and 92 give etch rate values at many positions along their trajectories in the stack. The curves were calculated from eq. (3-5) with $K = 62$, normalized to fit ^{56}Fe and ^{28}Si. (After O'Sullivan et al., 1971a.)*

A number of questions and difficulties arise in the use of solid track detectors for particle identification. It is convenient to categorize the problems according to the four different energy regimes indicated by dashed lines in Fig. 3-7, starting at the highest energies and working our way down.

3.4. RELATIVISTIC ENERGIES, $E \gtrsim 1$ GeV/NUCLEON

Several difficulties are encountered at high energies. First of all, for each element the ionization rate has approximately its minimum value, so that only extremely heavy particles can produce detectable tracks. We are aware of no minerals whose sensitivity is sufficient to detect minimum-ionizing particles even with Z as high as 114 (the region of charge relevant to hypothetical superheavy elements). This insensitivity never ceases to amaze the typical high-energy experimentalist, who is accustomed to study minimum-ionizing tracks of singly charged particles in bubble chambers and spark chambers. Some of the special phosphate glasses have adequate sensitivity to detect the heaviest nuclei in the Periodic Table at relativistic energies, but for most practical purposes one is restricted to plastics, which are available in large areas and are useful in studying relativistic cosmic rays with $Z \gtrsim 30$.

The second difficulty is that relativistic particles travel great distances in a solid before coming to rest, so that it becomes impractical to try to identify them by determining v_T vs. R. Stacks of plastic as thick as 10 cm have been flown in balloons (O'Sullivan et al., 1971a; Blanford, 1971; Wefel, 1971). This thickness would be adequate to stop an 800 MeV/nucleon Sn nucleus or a 1200 MeV/nucleon U nucleus, but the probability of a nuclear disintegration within this distance is $\sim$0.7 for Sn and $\sim$0.85 for U, so that these heavy particles tend to break up more often than come to rest. Three methods have been used to avoid the necessity for a thick stack:

(1) In cosmic ray experiments the earth's magnetic field can be used to screen out all but those particles with rigidity exceeding a cutoff value for that latitude. We will discuss the problems associated with this method in Chapter 5.

(2) In a stack of intermediate thickness, $\sim$1 to 2 g/cm^2, subrelativistic particles will show a detectable change of ionization rate as they pass through the stack, whereas particles with ultra-high energy will show no detectable change. From curves of v_T vs. R such as those in Fig. 3-9, and a knowledge of the experimental spread in etch rates due to variations in the homogeneity of the detector and in the energy deposition rate of an ion at fixed energy, one can estimate the uncertainty in charge determination as a function of Z and E for a given stack thickness. Under favorable conditions such as existed for the 1 g/cm^2 stack designed by Price and Shirk (1973) and exposed inside the orbiting Skylab in a controlled atmosphere and temperature, it has proven possible to detect a systematic increase of v_T through the stack for particles with $Z \gtrsim 70$ and E up to $\sim$1500 MeV/nucleon.

(3) In a very thin stack consisting of as few as two or three sheets but including a layer of a fast photographic film that detects Cerenkov light from an adjoining transparent plastic radiator, it is possible to infer β by measurements of the dimensions of the elliptical image resulting from the cone of Cerenkov light (Pinsky, 1972; Shirk et al., 1973). The range

of velocities that can be determined by this scheme depends on the refractive index, n, of the plastic radiator through the relation $\cos \theta = (n\beta)^{-1}$, where θ is the angle at which the light is emitted with respect to the particle momentum. For most plastics $n \approx 1.6$. The lowest velocity for which Cerenkov light is generated is given by $\beta_{min} = n^{-1} \approx 0.65$, which corresponds to a minimum energy of ~300 MeV/nucleon. As β approaches β_{min}, θ goes to 0, whereas as β approaches 1, θ goes to arccos $(n^{-1}) \approx 50°$. In practice it becomes difficult to determine velocities $\beta \gtrsim 0.9$ accurately, because θ is close to its maximum value. This corresponds to an energy of about 1200 MeV/nucleon. By means of either the Cerenkov technique or by using a stack ~1 to 2 g/cm² thick, it is then possible to establish whether the energy of a heavy nucleus (one that leaves a detectable etch cone) exceeds ~1000 to ~1500 MeV/nucleon.

The third difficulty is that we do not know accurately how the response of a particle with $E \gtrsim 1000$ MeV/nucleon varies with energy in the ultrarelativistic region where the polarization of the medium due to the proximity of other atoms becomes important. Using an expression due to Sternheimer (1956) for the relativistic rise in ionization rate in a plastic leads to a velocity-dependent correction $\delta(\beta)$ to the equation for J:

$$J = a(Z^{*2}/\beta^2)[\ln(\beta^2\gamma^2) + K - \beta^2 - \delta(\beta)]. \tag{3-7}$$

The polarization term counteracts the logarithmic rise in J with increasing γ so that in the range $0.9 \leq \beta \leq 1$ the quantity in brackets changes only from $K + 0.45$ to $K + 2.21$, which amounts to less than a 3% change for $K \gtrsim 60$. Thus, if it can be established that $E \gtrsim 1000$ MeV/nucleon, the simple expression

$$J \approx aKZ^2/\beta^2 \tag{3-8}$$

is accurate to within better than 3%.

The fourth problem, which arises when the quantity $Z/137\beta$ becomes comparable to or greater than unity, is the breakdown of the first-order Born approximation for the cross section for scattering of electrons due to Coulomb interactions with the incoming heavy ion. The consequence is that J may not have a simple quadratic dependence on Z. The papers by Fowler et al. (1970) and by Eby and Morgan (1972) show that corrections as large as ~15% may have to be applied to eqs. (1-7) and (1-8) for total energy loss and for restricted energy loss when particles are identified by measurements on delta rays of ~50 keV in nuclear emulsion. The paper by Price et al. (1971) points out that in plastics most of the radiation damage within a few tens of Ångstroms of the trajectory is caused

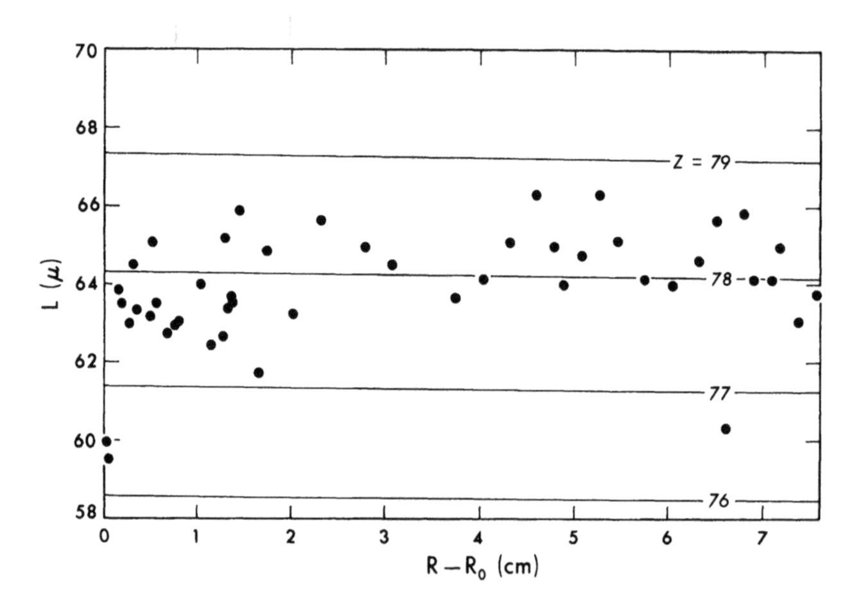

Fig. 3-10. *Etch pit length measurements as a function of position ($R - R_0$) along the trajectory of a cosmic ray nucleus with $Z \approx 78$. The curves were calculated from eq. (3-5) with $K = 62$, assuming the nucleus had an energy of 8 GeV/nucleon at entry and had a residual range of R_0 at exit. (After O'Sullivan et al., 1971b.)*

by low-energy delta rays, whose scattering cross section is given to an excellent approximation by the simple Rutherford scattering law. The necessary corrections to eq. (3-1) for J are thus likely to be small.

Fig. 3-10 gives an indication of the uniformity of response of an 8 g/cm² stack of commercial Lexan to a cosmic ray with $Z \approx 78$ that penetrated the entire stack without perceptible change of ionization rate (O'Sullivan et al., 1971b). The uncertainty in charge was $\sim \pm 0.6$ units, when account was taken both of the magnitude and slope of the points. The same particle also passed through a layer of nuclear emulsion and was assigned a charge $Z = 76 \pm 2$.

3.5. INTERMEDIATE ENERGIES, $10 \lesssim E \lesssim 1000$ MeV/NUCLEON

In this interval the relations between J and R for various ions are approximated reasonably well by power laws. Furthermore, the range is long enough that the particle penetrates more than one sheet of a typical stack of plastics (thickness usually 50 to 250 μm per sheet). Measurements of etching rates can be made at several points along the trajectory, as shown in Fig. 3-1a, and the particle can be identified rather accurately. Figs. 3-9, 3-11, and 3-12 illustrate the kind of resolution that can be expected for particles that come to rest in Lexan and cellulose nitrate. For ultraheavy nuclei ($Z \gtrsim 30$) that pass through several sheets before

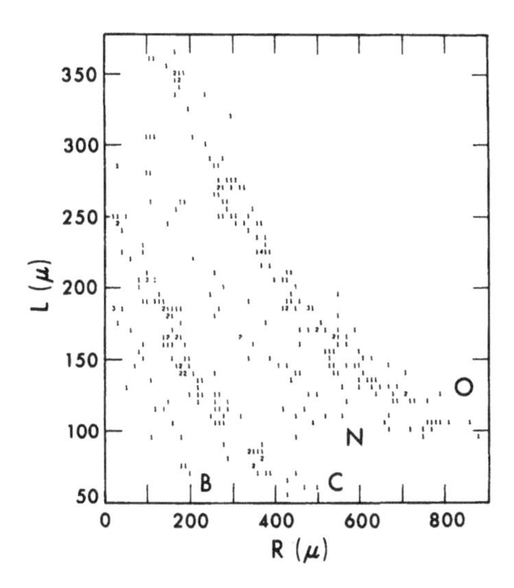

Fig. 3-11. *LR-plot (with suppressed zero) for low-energy cosmic rays that stopped in a stack of cellulose nitrate sheets. Etchant was 6N NaOH at 40°C for 6 hours. The B, C, N, and O are clearly separated and the mean mass of each element can be determined to within ±0.1 amu. (After Beaujean and Enge, 1972.)*

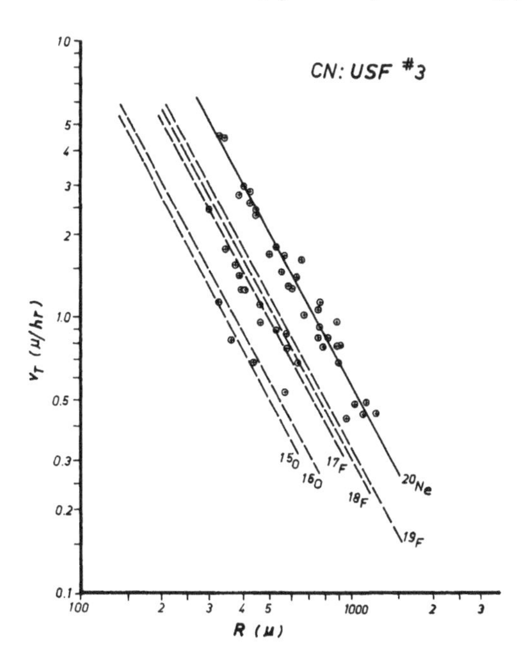

Fig. 3-12. *Etch rate vs. $\bar{R}$ for heavy ions stopping in a cellulose nitrate stack at the Princeton-Penn accelerator. Etchant was 6.25N NaOH solution at 40°C for 20 hours. The original beam consisted of 290 MeV/nucleon ^{20}Ne ions, which came to rest in the stack. The F and O isotopes are interaction products of Ne. (After Benton and Henke, 1972b.)*

coming to rest, the resolution is $\Delta Z \approx \pm 1$, part of the uncertainty arising from the large number of stable isotopes of the heavy elements (O'Sullivan et al., 1971a). For nuclei with $14 \lesssim Z \lesssim 30$ in Lexan, the resolution is $\Delta Z \approx \pm 0.5$ but should in principle be as good as $\Delta Z \approx \pm 0.3$ (Price, 1971). For nuclei with $5 \leq Z \leq 8$, Beaujean and Enge (1972) found that the mass resolution in Daicel cellulose nitrate is $\Delta A = \pm 0.4$ amu for beams of ions from an accelerator and $\Delta A \approx \pm 0.7$ amu for cosmic rays. The charge separation is exceedingly good, as shown in Fig. 3-11. For nuclei with $6 \leq Z \lesssim 14$ in Lexan (Sullivan et al., 1973) and for nuclei with $6 \leq Z \lesssim 30$ in cellulose triacetate (O'Sullivan et al., 1973), the charge resolution is $\Delta Z \approx \pm 0.5$.

Even if the particle completely penetrates the stack without coming to rest, it may still be possible to identify it if the change in etch rate from one side of the stack to the other is detectable. One simply adds a constant amount to the residual range for each etch pit so that the resulting *LR*-curve gives the best fit to one of the family of possible *LR*-curves. The procedure is discussed in detail by Benton and Henke (1972a) and by Shirk et al. (1973).

3.6. LOW ENERGIES, $0.1 \lesssim E \lesssim 10$ MeV/NUCLEON

This region is important for the diversity of its applications, which include the study of products of nuclear reactions; the composition of solar flare particles; quiet-time interplanetary particles and particles in the radiation belts; and ancient tracks in extraterrestrial bodies. It is relatively easy to obtain suitable heavy-ion beams (Table 3-1) with which to calibrate detectors of plastics, glasses, and minerals. At the low end of this energy interval, ranges are only a few microns, which limits charge resolution and may necessitate the use of a scanning electron microscope instead of an optical microscope.

One difficulty in this energy interval is that heavy ions are incompletely stripped and capture electrons as they slow down—as indicated in eq. (3-2)—so that their ionization rate goes through a maximum (Fig. 3-7), with the result that curves of J vs. R are hard to fit analytically. As Fig. 3-13 shows, the chemical reactivity also goes through a maximum (Diamond et al., 1973). Since L is the integral $\int v_T dt$, the functional form of v_T is not apparent if etching is allowed to proceed through the region where v_T changes rapidly with R. In their method of charge identification, Diamond et al. (1973) measured the time for the etchant to penetrate a single very thin Makrofol sheet, and inferred $\bar{v}_T$ = (sheet thickness) $\div$ (breakthrough time). They floated aluminized sheets on the surface of a NaOH solution and with time-sequence photography watched bright points of light develop as the solution etched through and dissolved the aluminum. By determining the breakthrough times for the same event in several sheets of an original stack, they could

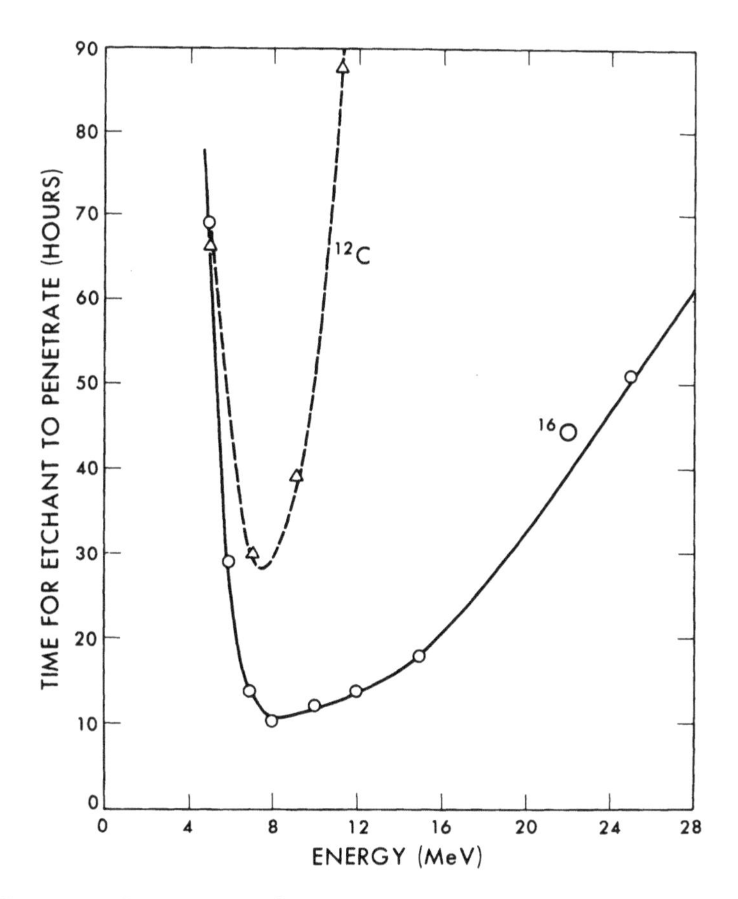

Fig. 3-13. *The time, t = (thickness) ÷ v_T, for etchant to penetrate a thin, aluminized Makrofol sheet from one side can be monitored by placing a movie camera on one side and a light source on the other side of the sheet, which floats on the surface of NaOH solution. The ionization maximum shows up as an etch time minimum. (After Diamond et al., 1973.)*

map out the curves of v_T vs. R and resolve charges of ions with energies down to ~0.5 MeV/nucleon. This is an interesting example of a nonmicroscopic method of data acquisition.

Fig. 3-14 illustrates the charge separation attainable with Daicel cellulose nitrate (Price and Fleischer, 1970). Various isotopes of He, Li, and Be with energies up to several MeV/nucleon are occasionally emitted during the ternary fission of ^{252}Cf. The fact that most of the ^{8}Li fragments (identifiable by virtue of their T-shaped appearance resulting from the reaction $^{8}\text{Li} \rightarrow {}^{8}\text{Be} \rightarrow 2\,{}^{4}\text{He}$) lie at the top of the Li distribution, which consists mainly of ^{7}Li and ^{6}Li tracks, suggests that mass separation is also attainable.

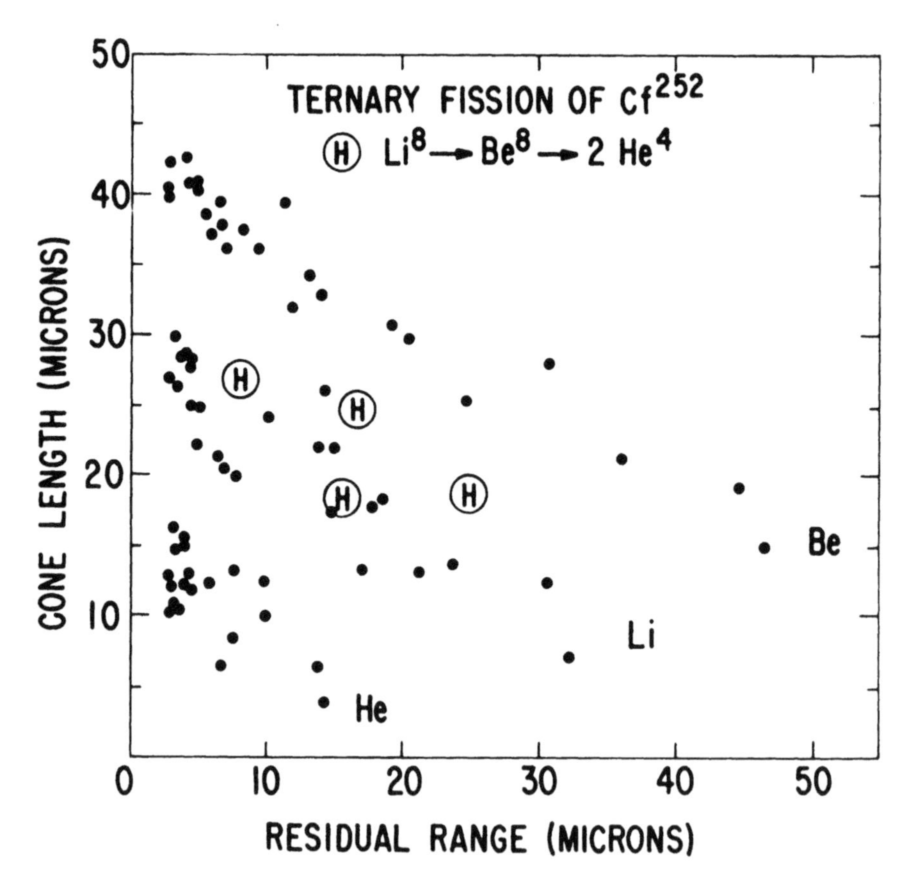

Fig. 3-14. *LR-plot for light fragments emitted during rare ternary fission of* ^{252}Cf. *Points labeled H are identified as* ^{8}Li *from their T-shaped appearance resulting from the decay* $^{8}Li \rightarrow {}^{8}Be \rightarrow 2\,{}^{4}He$. (*After Price and Fleischer, 1970.*)

Endo and Doke (1973) have used Lexan detectors to identify fission fragments, which have atomic numbers ranging from ~30 to ~60 and typical energies of ~1 MeV/nucleon. Fig. 3-15 is from their work. They claim a resolution $\Delta Z \approx 1.5$ for light fragments and $\Delta Z \approx 2$ for heavy fragments.

At energies of only a few tenths MeV/nucleon, charge resolution is difficult for any detector, but solid track detectors have definite advantages over detectors that respond to total deposited energy. Fig. 3-16 compares the response of a semiconductor detector, which senses ΔE deposited in a crystal of thickness ΔX, and the response of a silica glass track detector, which senses only that portion of the energy deposited within a narrow cylinder around the particle's trajectory (see discussion in Chapter 1). At residual ranges of only a few microns the responses are the same, because none of the energy is deposited outside the narrow cylinder, but at ranges

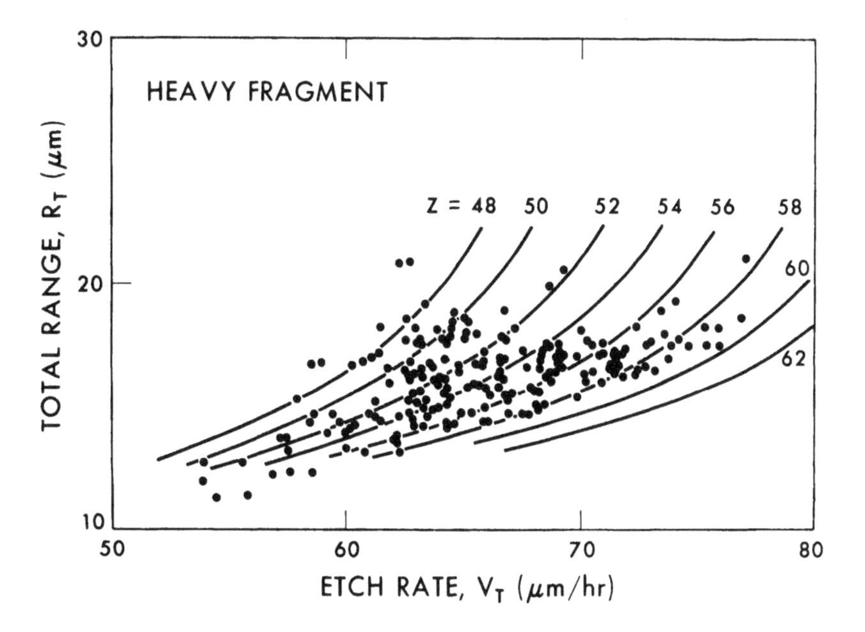

Fig. 3-15. *Etch rate as a function of total range for the heavier of the two fragments emitted in the neutron-induced fission of ^{235}U. Etchant is 6.4N NaOH at 50°C. Tracks of both fragments from each event are observed in a Lexan sandwich enclosing a thin ^{235}U target. (After Endo and Doke, 1973.)*

greater than a few microns the response curves of various ions are more widely separated in glass and other solid track detectors than are curves of total dE/dx. At a few tenths MeV/nucleon, with the LR-plot method one can resolve individual charges of the lightest elements (up to $Z \approx 6$) using a suitable plastic detector, and one can easily determine relative abundances of the major heavy elements in solar flare particles—Fe, Si, and Ne—because of the large difference in their charges. In Sections 3.8 and 3.9 we will discuss other methods for identifying particles in the interval $0.1 \lesssim E \lesssim 10$ MeV/nucleon.

3.7. VERY LOW ENERGIES, $E \lesssim 0.1$ MeV/NUCLEON

This regime is important for studies of heavy ions in the solar wind and very low-energy nuclear reaction products.

As Fig. 3-17 shows, the total energy loss rate of an ion is the sum of two contributions, an "electronic" part which arises from energy transfer to the electrons of the medium and an "atomic" (or "elastic") part arising from energy transfer to screened nuclei of the medium. Lindhard et al. (1963) give formulas and curves

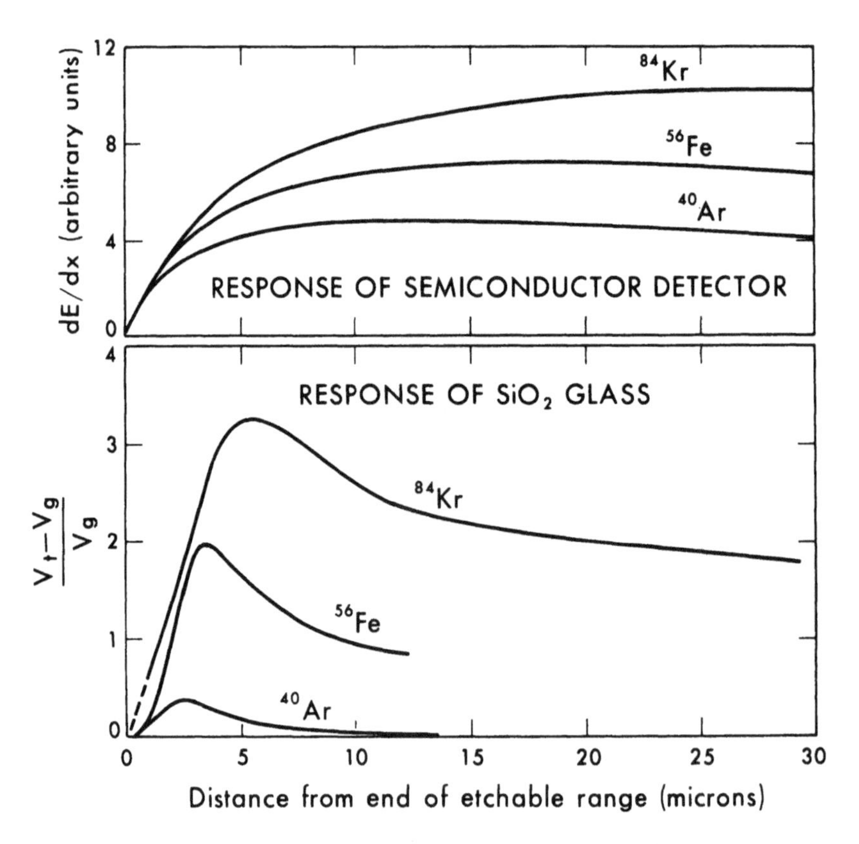

Fig. 3-16. *Response of SiO_2 glass compared with total dE/dx. The complete $v_T(R)$ functions were determined by SEM measurements of track profiles using replicas of glass surface. (Unpublished data of D. Braddy.)*

for these contributions. At very low energy the electronic part is proportional to ion velocity as shown by the dashed lines. With decreasing energy below ~0.1 MeV/nucleon the atomic part becomes increasingly important. Whereas the range straggling due to the stochastic nature of electronic stopping is usually less than ~1%, the range straggling due to the stochastic nature of atomic stopping is quite large, typically more than 20% and largest for an ion moving through a medium of the same atomic number, as in that case all of its energy can be given up in a single head-on collision. Thus, in addition to the experimental difficulty of dealing with extremely short tracks, the large magnitude of the range straggling limits our ability to resolve charges by means of an *LR*-plot, and because the nature of the energy loss process is changing from electronic to atomic, we can expect that solids with rather different structures may respond quite differently, perhaps even not at all, to very low-energy particles. Calibrations with particles of interest are essential, and to this date few have been done.

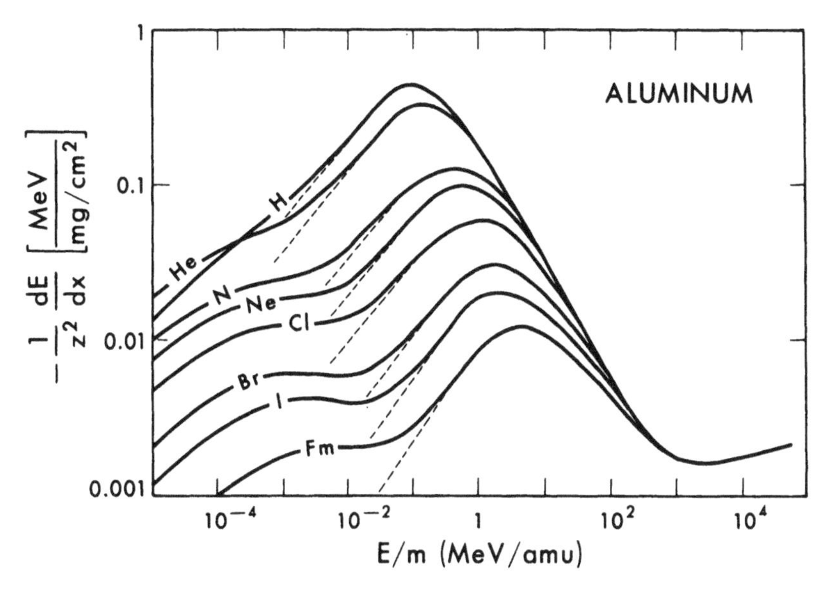

Fig. 3-17. *Total energy losses (divided by Z^2) of heavy ions in Al. The dashed lines indicate the "electronic" contribution at low energies. At very low energies the "atomic" contribution dominates. (After Northcliffe and Schilling, 1970.)*

Bibring et al. (1972) and Zinner (Washington University) have bombarded mica with various ions from F up to Bi, at energies from ~0.2 to ~200 keV/nucleon, and have etched the mica in concentrated HF for various times. In this energy interval they find that in favorable cases and for selected samples the depths of the etch pits (determined by interference microscopy and by electron microscope observations of shadowed replicas) are roughly independent of mass and increase monotonically with energy/nucleon, whereas the average pit diameter after a given etch time is an increasing function of mass. Fig. 3-18 compares the appearance of etch pits due to 0.2 keV/nucleon ions of Fe, Sb, Bi, and the molecular compound SbBi after the same etching treatment. Many experimental difficulties need to be resolved. Background pits due to recoiling daughter nuclei from spontaneous alpha decay of U and Th impurities must be removed by annealing before the mica can be used, but the annealing treatment seems to damage the structure of the mica. Different batches of mica give quite disparate results. Worst of all, the scatter in pit depths for monoenergetic ions is far larger than can be attributed to range straggling and seems to be an intrinsic feature of a crystalline detector. They conclude that energy losses due to nuclear collisions can produce etchable damage in mica and that further development effort is warranted, since mica seems to be the only solid capable of detecting heavy ions with energy as low as a few hundred eV/nucleon.

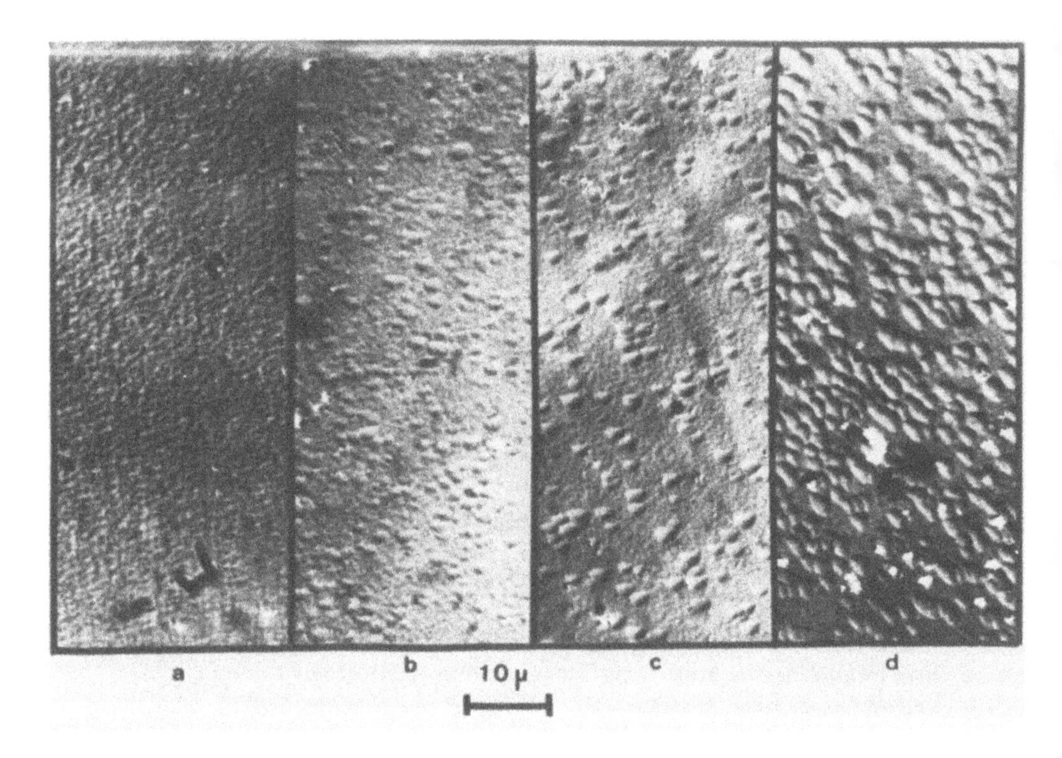

Fig. 3-18. *Etch pits in mica bombarded with 0.2 keV/nucleon ions of (a) Fe, (b) Sb, (c) Bi, and (d) the compound SbBi, then etched 2 h in 40% HF at 30°C and viewed by reflected light, phase-contrast microscopy. (After Bibring et al., 1972.)*

Woods, Hart, and Fleischer (1973, unpublished) have found that pits can be etched at tracks of Fe ions with energies only above ~50 keV/nucleon in phosphate glasses.

J. H. Chan (University of California, Berkeley) has used an SEM to study formvar replicas of etch pits in Lexan irradiated with ions of ^{19}F, ^{23}Na, ^{40}Ar and ^{56}Fe, at energies from 3 to 250 keV/nucleon. He finds that even at the lowest energies the lengths of etch pits agree well with calculated ranges, indicating that any range deficit is less than ~0.1 μm and that in Lexan atomic stopping processes contribute to a local increase in chemical reactivity. Some of his data for Fe ions are shown in Fig. 3-6. Fig. 3-19 shows SEM pictures of formvar replicas of F, Na, Ar, and Fe tracks. In 3-19d the Fe ions had such low energy (13 keV/nucleon) that the range straggling is quite noticeable. Lindhard et al. (1963) have shown that at very low energies, where atomic stopping is dominant, the range straggling is given approximately by

$$\Delta R/\overline{R} \approx 2A_1A_2/3(A_1 + A_2)^2 \tag{3-9}$$

which is ~0.3 for Fe ions in Lexan. The spread in etch pit lengths measured with

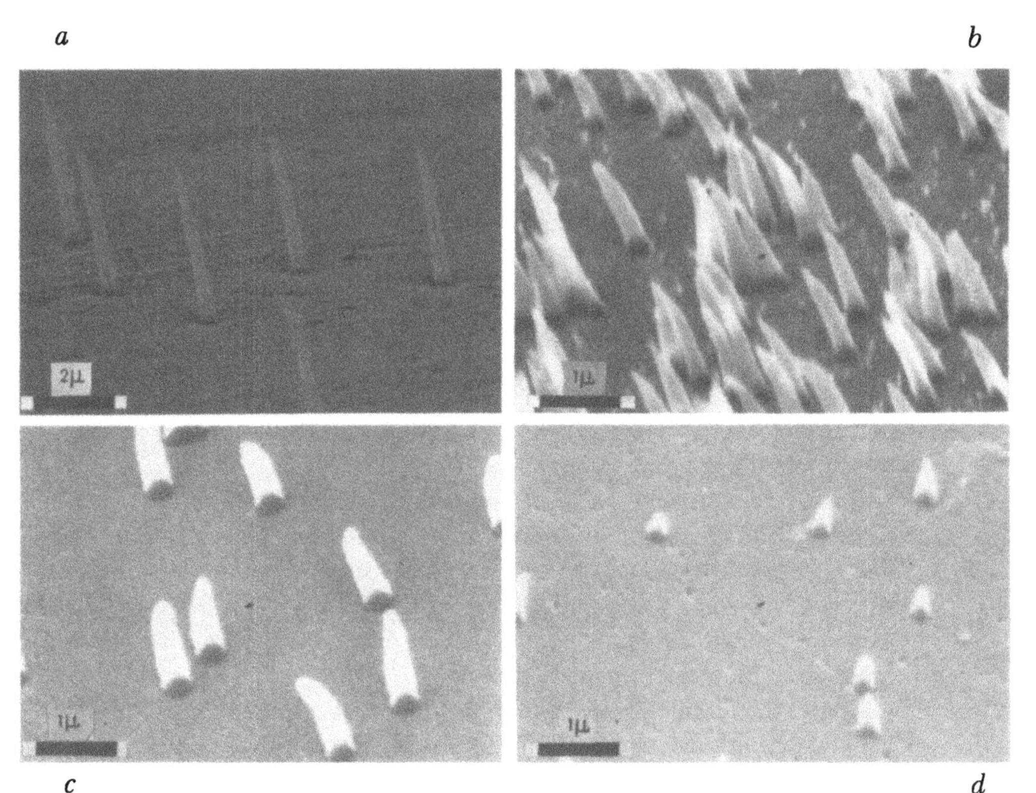

Fig. 3-19. *Formvar replicas of etch pits of very low-energy ions in Lexan, observed with an SEM after a 1.5 h etch in 6.25N NaOH at 40°C: (a) 125 keV/nucleon ^{19}F ions; (b) 41 keV/nucleon ^{23}Na ions; (c) 21 keV/nucleon ^{40}Ar ions; (d) 13 keV/nucleon ^{56}Fe ions. (Courtesy of J. H. Chan.)*

the SEM appeared to be consistent with the calculated value. The lowest energy for which etch pits in Lexan have been detected so far is ~3 keV/nucleon for Fe ions.

Though the Lexan replica technique for detecting very low energy ions looks quite promising, it remains to be seen whether atomic numbers can be determined by application of the methods of this chapter. The response of Lexan to monoenergetic ions is much more uniform than the response of mica, probably because of its random structure, which contrasts with the remarkably anisotropic layer structure of mica. The reader should recall from our earlier discussion of data at higher energies that *LR*-plots constructed for heavy ions in crystalline solids (Fig. 3-5) show far more scatter than those for heavy ions in plastics and glasses.

Several other techniques have been devised for identifying particles with $E \lesssim 0.1$ MeV/nucleon, but all have drawbacks:

(1) On the IMP-G satellite Fan et al. (1971) have an electrostatic deflection instrument that measures charge per unit kinetic energy down to ~100 keV/charge, but it has a very small geometric factor, is capable

of resolving only groups of nuclei such as C, N, O, and cannot distinguish incompletely stripped nuclei with different Z but the same Z^*/(total energy).

(2) On the same satellite the gas-flow proportional counter of Hovestadt and Vollmer (1971) successfully detected nuclei with energies down to ~0.4 MeV/nucleon but ceased to operate after about one month due to a gas leak (Hovestadt et al., 1973). The energy threshold is not particularly low, the charge resolution is only modest, and the window is easily punctured.

(3) Thin solid state detectors have limited charge resolution (see Fig. 3-16) and a low-energy threshold of ~0.2 MeV/nucleon (for a 3-μm Si detector recording alpha particles).

(4) Secondary emission detectors record the time-of-flight of a particle from its time of impact and generation of secondary electrons in a thin film to its time of arrival at an energy detector. The cumbersome electron optics system, with the necessary high voltages (~30 kV) for rapid focussing, make the system unattractive for space applications, though acceptable in accelerator experiments.

(5) Muga (1971) has shown that thin, laminated plastic scintillator films, used to record ΔE and time-of-flight, are potentially capable of good energy and charge resolution. S. M. Krimigis (Johns Hopkins University) has developed a thin-film detector capable of identifying charges at energies down to ~30 keV/nucleon, but it has not been tested yet in an actual experiment.

There is thus considerable incentive for exploring the capabilities of Lexan and other solids for particle identification at energies of a few keV/nucleon.

3.8. PARTICLE IDENTIFICATION BY THE TRACK PROFILE METHOD

The shapes of etched tracks are not simple cones with straight walls as might be surmised from the oversimplified drawings in Fig. 3-1, but contain complete information about the variation of v_T with R. As Fig. 3-20 illustrates, the sine of the half cone angle θ at a distance x from the end of the range equals v_G/v_T at the distance R located where the normal to the wall meets the particle's trajectory. Fleischer et al. (1970) pointed out that the track profile can be transformed into a curve of v_T vs. R and used the method to identify tracks of heavy cosmic rays that stopped in the astronauts' Lexan helmets. If the etchant has reached the end of the particle's range, the LR-plot method will not work but the track profile

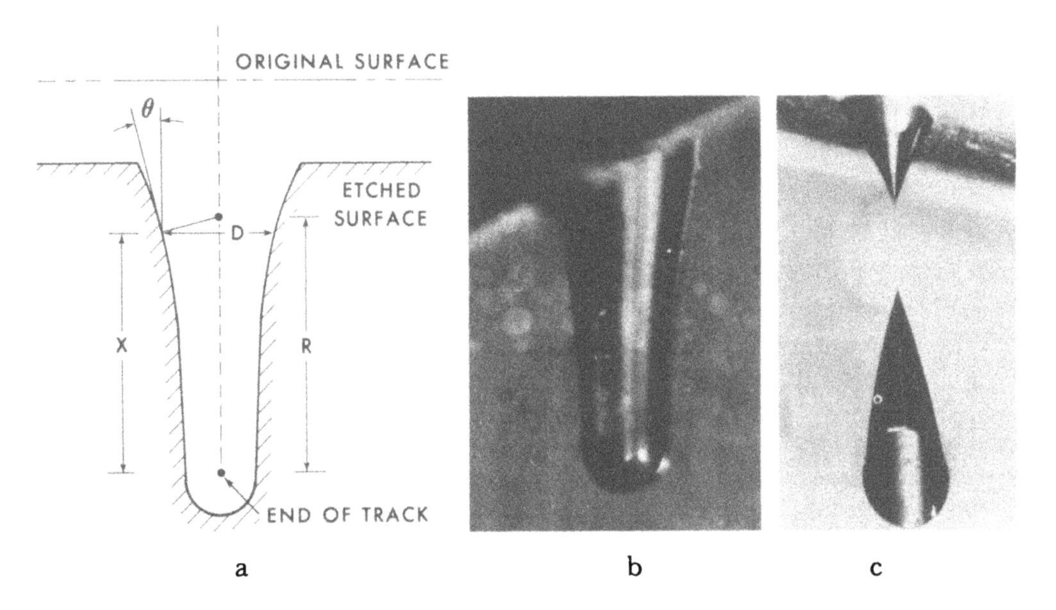

Fig. 3-20. *Track profile method of particle identification: (a) $\theta(x)$ can be transformed into $v_T(R)$. (After Fleischer et al., 1970.) Micrographs (Fleischer et al., 1973) show heavy cosmic rays (b) stopping and (c) passing through electrophoresis cell made of Lexan and flown on one of the Apollo missions. Curvature of walls indicates that v_T is changing along the trajectory.*

method does. Another feature is that the ability to map out instantaneous values of v_T enables one to determine the function $v_T(R)$ accurately in the region where its curvature is rapidly changing and makes it possible to identify very low-energy particles from SEM observations of track replicas. The response curves for Ar, Fe, and Kr ions in SiO_2 glass, shown in Fig. 3-16, were determined by D. Braddy (University of California, Berkeley), who photographed cellulose acetate replicas with an SEM and measured the profiles. This method has been used very successfully to determine the energy spectrum and composition of low-energy ions emitted during a solar flare (Braddy et al., 1973) and to detect low-energy, rare, trans-iron nuclei in the same flare (Shirk and Price, 1973). At energies of $\sim$0.5 to $\sim$1 MeV/nucleon the resolution is better than one unit of charge for ions with $18 \leq Z \lesssim 30$ in SiO_2 glass. The separation of different species is not nearly as good in this glass if only the length or diameter is used instead of the profile (Krätschmer, 1971). Because of its attractive materials properties, further studies should be done to see how well ions of very high Z can be resolved at low energies in SiO_2 glass.

It should be emphasized that the track profile can be measured much more accurately from a replica than from the track itself. The added resolution attainable with an SEM is essential if one wishes to use the track-profile method at energies below ~1 MeV/nucleon.

3.9. PARTICLE IDENTIFICATION BY MEASUREMENTS OF ETCH PIT DIAMETER

Though this method is not as sensitive as the track-profile method, it does not require the use of an SEM and is extremely simple to use. It was first suggested by Somogyi (1966), who has recently discussed the kinetics of track-diameter growth in considerable detail (Somogyi and Szalay, 1973). In principle the method should work with particles incident at arbitrary angles on a solid surface, but in practice it is much simpler if the detector can be positioned such that particles are nearly normally incident. For tracks with large cone angles, such as are encountered in most glasses, and with very light particles in plastics, the diameter is a more sensitive function of ionization rate than is track length. Etch pits with large cone angles are not visible at shallow angles, which is another reason for working at nearly normal incidence.

In Section 2.1 we described the least-time method (Fleischer et al., 1969) for calculating track profiles if the dependence of v_T on R was known. In the previous section we saw how the function $v_T(R)$ could be determined directly from observations of the track profile. A measurement of diameter, D, after only one etching time is not sufficient to allow one to determine both Z and E, but in certain cases may be adequate to determine E if the charge is known. Measurements of the change of diameter with time (such as are shown in Fig. 2-4) can give both Z and E.

The Debrecen group has made extensive empirical studies of the growth of pit diameters with time for ^{3}He, ^{4}He, and ^{3}H of various energies in cellulose acetate etched in various solutions (Somogyi, 1966; Somogyi et al., 1968a,b; 1970; Varnagy, 1970). With suitable calibrations they found it possible to distinguish the various isotopes of H and He and to determine alpha-particle energies with a resolution of 300 to 400 keV at ~2 to ~4 MeV and a resolution of about 100 keV at ~6 to ~9 MeV. Empirical data for D vs. time agree extremely well with calculated curves based on the least-time method. Fig. 3-21 compares data for alpha particles with initial energies from 0.5 to 5.48 MeV normally incident on cellulose nitrate with curves calculated from the experimentally determined function $v_T(R)$ (Paretzke et al., 1973). It can be seen from this figure that, in general, a single measurement of D is inadequate to provide an unambiguous determination of E even if it is known that the particles are all alpha particles. A different family of curves can be generated for other charged particles. The measurable quantity most closely related to charge, Z, is the second derivative, d^2D/dt^2, which is greatest for particles of lowest Z because their ionization rate has the fastest rate

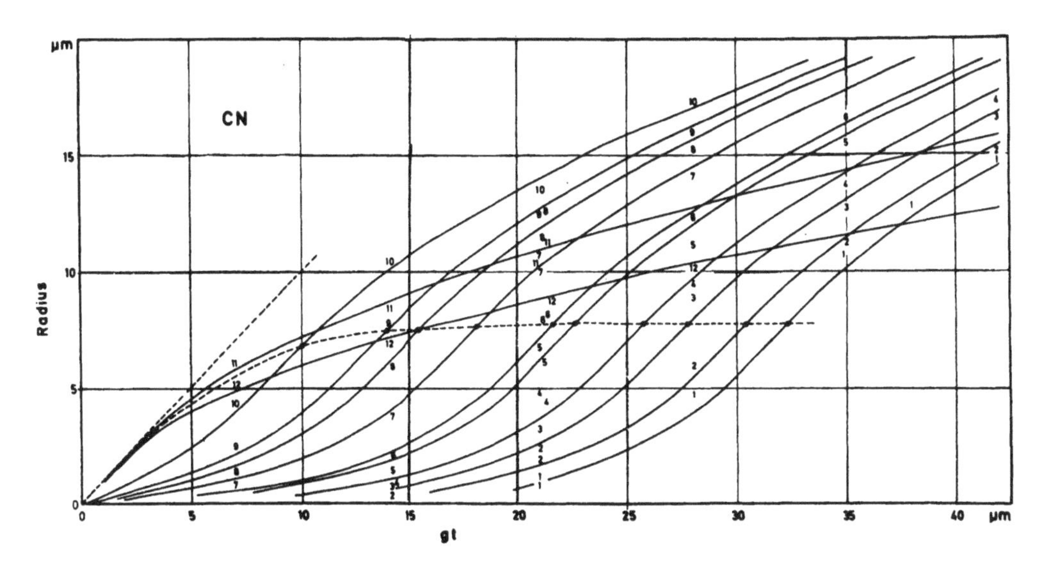

Fig. 3-21. *Experimental measurements (symbolized by a number increasing with decreasing energy) of track radii as a function of surface removed* ($v_G t$) *for 0.5, 0.8, 2.4, 3.1, 3.3, 3.75, 4.2, 4.3, 4.7, 5.0, 5.25, and 5.48 MeV alpha-particles in cellulose nitrate detectors. Solid lines are responses calculated from measured* $v_T(R)$ *function for alpha-particles. Dotted line shows response for infinite* v_T. *Dashed line is locus of inflection points. (After Paretzke et al., 1973.)*

of change of curvature with distance. Paretzke et al. (1973) give a comprehensive discussion of the behavior of $D(t)$ curves. Beaujean et al. (1970) have used the least-time method to study the response of cellulose nitrate to lightly ionizing alpha particles. Blanford et al. (1970) have shown that measurements of pit diameters in Lexan with an SEM can be used to identify relativistic ultraheavy nuclei. When $Z \lesssim 70$, the cone angle is large and consequently D is a more sensitive measure of Z than is L.

Diameter measurements of tracks in glasses have been used to study very heavy particles. Fission fragment energies can be determined to within 1 or 2 MeV (Höppner et al., 1969; Fiedler and Höppner, 1970) and the data agree well with the least-time calculations of Fleischer et al. (1969). The effect of fission fragment mass on the response $D(E)$ is surprisingly small. Phosphate glass, with its small cone angle, gives the best resolution (Aschenbach et al., 1974). Braddy et al. (1973) found that particles of different charge in the vicinity of Fe could be separated on the basis of curves of D vs. t, but abandoned the method because they were able to achieve higher resolution with the track-profile method, using an SEM. Lecerf and Peter (1972) have reported measurements of etch pit diameters in microscope slide glass irradiated with C, O, Ne, Ar, Ni, and Kr ions with energies up to 1.1 MeV/nucleon.

3.10. PARTICLE IDENTIFICATION BY MEASUREMENT OF MAXIMUM ETCHABLE TRACK LENGTH

The earliest approach to the problem of particle identification was to assume the existence of a critical ionization rate, J_c, below which $v_T = v_G$ and above which v_T abruptly rises to a large value (Fleischer et al., 1965; Maurette, 1966). It then followed that, after a sufficiently long etching time, the portion of the trajectory of a particle for which $J > J_c$ would be completely etched out. Light particles could never satisfy this inequality and could not produce etchable tracks. Heavy particles would have two values of range, R_1 and R_2, at which $J = J_c$ and a well-defined maximum recordable range $R_{max} = R_2 - R_1$ for which $J > J_c$. This R_{max} would then be a rapidly increasing function of Z. Based on limited calibration data, several theoretical curves of R_{max} vs. Z have been published for minerals (Fleischer et al., 1965, 1967a; Maurette, 1966; Price and Fleischer, 1971) and cellulose nitrate (Benton and Henke, 1968). The discovery of trans-iron elements in the cosmic radiation was based on the application of this concept to track lengths in meteorites (Fleischer et al., 1965, 1967b). The problem of observing the entire etchable range of a particle that entered a solid with $J < J_c$ and had J exceed J_c within the solid seemed to be solved with the development of the TINT and TINCLE methods illustrated in Fig. 2-15.

Later, however, careful studies of response curves showed that, usually, v_T approaches v_G gradually rather than abruptly and that the concept of a critical ionization rate is not, strictly speaking, valid (Beaujean et al., 1970; Price et al., 1973; Paretzke et al., 1973). The asymptotic approach of v_T to v_G means that the full, recordable range is very difficult to measure. But in practice the track density is often so high, particularly in the case of extraterrestrial minerals, that the etch time must be kept short to prevent the crystal from disintegrating.

Price et al. (1973) and the Heidelberg group (Plieninger, 1972; Plieninger et al., 1972, 1973) have compared TINT and TINCLE distributions from accelerated heavy ion beams and from fossil tracks in lunar and meteoritic crystals. Selected aspects of their results are shown in Figs. 3-22 and 3-23. Fig. 3-22, from the work of Price et al. (1973), illustrates several of the experimental limitations on the R_{max} method. In (a) we see that a basic etchant (6, 4, 120 means 6g NaOH + 4g H_2O for 120 min at its boiling point) gives longer etchable tracks than does an acidic etchant (2 HF:1 H_2SO_4:4 H_2O). In (b) we see that the peak in a TINCLE distribution shifts to increasing length with increasing etch time. Thus, if the track density is so high that one can only etch for 60 minutes, which often happens with extraterrestrial samples, the tracks will be incompletely developed. In (c) we see that a 600 hour anneal of olivine, a typical lunar mineral, at a temperature as low as 175°C, produces a detectable decrease in the TINCLE lengths. At maximum lunar surface temperatures ($\sim$130°C) we expect that tracks in olivine will be shortened. In (d) we again see the increase in TINCLE lengths with etch time.

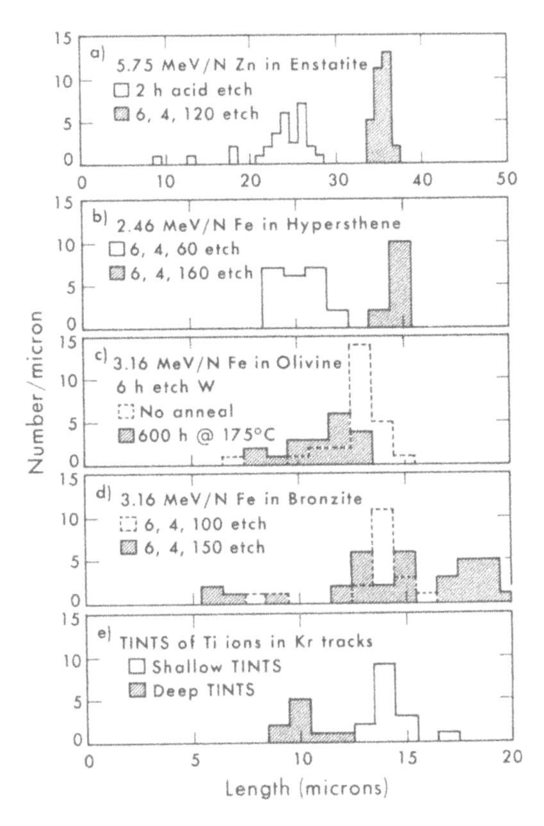

Fig. 3-22. *Problems associated with the method of maximum etchable track lengths. See text for discussion. (After Price et al., 1973.)*

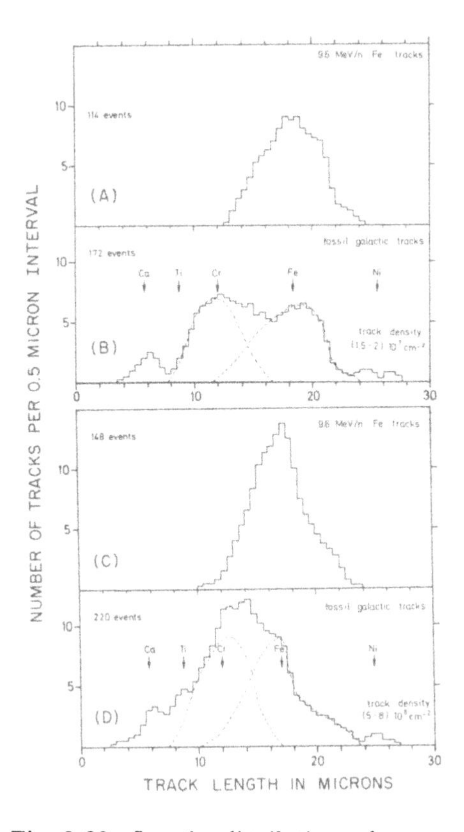

Fig. 3-23. *Length distributions for artificially generated Fe and galactic VH ion tracks measured in the same pyroxene crystals under identical conditions. (After Plieninger et al., 1973.) The crystals are from lunar rock 10049 and results are shown for crystals with different track densities corresponding to different amounts of shielding. From the shape and position of the distribution of the calibration ions, the amount of Fe and of elements belonging to the Cr group (V + Cr + Mn) was evaluated for the galactic VH ions (dashed lines). The charge assignments are based on calibrations with Ca and Cu ions.*

In (e) we see another difficulty, that deep TINTS are shorter than shallow TINTS because of the delay time for the etchant to extend down the host track until it meets the interior track of interest. Price et al. found that, as the beam energy is raised, the etch rate decreases (as we saw in Fig. 3-5) and becomes intermittent, apparently reflecting local, statistical variations in radiation damage distribution in the track. The existence of a single, well-defined R_{max} is thus an idealization that is not attained in practice.

The work of Plieninger et al. (1972, 1973) suggests that, with internal calibrations of the same samples one wishes to study, it may still be possible, on a statistical basis, to do particle identification by a TINT or TINCLE method. Fig. 3-23 compares TINT distributions of Fe ions and of unknown cosmic rays in two samples of lunar pyroxene with different track densities corresponding to exposures at different depths at the lunar surface. The Fe length distribution is rather broad, because the Fe tracks were cut by host tracks at random points along their lengths corresponding to different values of v_T. Even though the peak is at a much shorter length than the expected R_{max}, the width of the peak is sufficiently small that it can be used to distinguish Fe from other ions, with a resolution that we estimate at $\Delta Z \approx \pm 2$. The interpretation of the distribution of fossil cosmic ray tracks is a problem that will be addressed in Chapter 6.

Chapter 3 References

J. Aschenbach, G. Fiedler, H. Schreck-Köllner, and G. Siegert (1974), "Special Glasses as Energy Detectors for Fission Fragments," *Nucl. Instr. Methods* **116,** 389–395.

R. Beaujean and W. Enge (1972), "Study on the Isotopic Composition of Low Energy Boron, Carbon, Nitrogen and Oxygen Cosmic Ray Particles," *Z. Physik* **256,** 416–440.

R. Beaujean, W. Enge, and H. Meyerbroecker (1970), "Determination of the Sensitivity of Various Plastic Detectors," *Acta Phys. Acad. Sci. Hung.* **29,** Suppl. 4, 387–390.

E. V. Benton and R. P. Henke (1968), "A New Method for Charge Determination of Heavy, Cosmic Ray Particles," *Nucl. Instr. Methods* **58,** 241–244.

E. V. Benton and R. P. Henke (1969), "Heavy Particle Range-Energy Relations for Dielectric Nuclear Track Detectors," *Nucl. Instr. Methods* **67,** 87–92.

E. V. Benton and R. P. Henke (1972a), "High-*Z* Particle Cosmic-Ray Exposure of Apollo 8-14 Astronauts," Air Force Weapons Laboratory TR-72-5.

E. V. Benton and R. P. Henke (1972b), "On Charged Particle Tracks in Cellulose Nitrate and Lexan," Univ. of San Francisco Tech. Rept. No. 14.

N. Bhandari, S. Bhat, D. Lal, G. Rajagopalan, A. S. Tamhane, and V. S. Venkatavaradan (1971), "Spontaneous Fission Record of Uranium and Extinct Transuranic Elements in Apollo Samples," *Proc. Second Lunar Sci. Conf.* **3,** 2599–2609, Cambridge: MIT Press.

J. Bibring, J. Borg, J. Dran, M. Maurette, R. Meunier, J. Peters, and R. Walker (1972), "Detecteurs Mineraux et Traces d'Ions Lourds de Très Faible Energie ($0.2 < E < 200$ keV/uma)," *Proc. 8th Inter. Conf. Nuclear Photography and Solid State Track Detectors*, Bucharest, **1,** 485–499.

G. E. Blanford (1971), *Calibration and Use of Plastic Track Detectors in the Study of Extremely Heavy Cosmic Rays*, Ph.D. Thesis, Washington University.

G. E. Blanford, R. M. Walker, and J. P. Wefel (1970), "Calibration of Plastic Track Detectors for Use in Cosmic Ray Experiments," *Rad. Effects* **5,** 41–43.

G. E. Blanford, M. W. Friedlander, J. Klarmann, S. S. Pomeroy, R. M. Walker, J. P. Wefel, P. H. Fowler, J. M. Kidd, E. J. Kobetich, R. T. Moses, and R. T. Thorne (1973), "Observation of Cosmic Ray Particles with $Z > 35$," *Phys. Rev.* **D8,** 1707–1722.

D. Braddy, J. Chan, and P. B. Price (1973), "Charge States and Energy-Dependent Composition of Solar-Flare Particles," *Phys. Rev. Letters* **30,** 669–671.

W. T. Diamond, A. E. Litherland, J. Goldemberg, H. L. Pai, and A. H. Chung (1973), "Development of Polycarbonate Damage Track Detectors (PDT) for Heavy Ions," presented at International Conference on Photonuclear Reactions and Applications, Asilomar, California, March 26–30.

P. B. Eby and S. H. Morgan (1972), "Charge Dependence of Ionization Energy Loss for Relativistic Heavy Nuclei," *Phys. Rev.* **A5,** 2536–2541.

K. Endo and T. Doke (1973), "Calibration of Plastic Nuclear Track Detectors for Identification of Heavy Charged Nuclei Using Fission Fragments," *Nucl. Instr. Methods* **111,** 29–37.

C. Y. Fan, G. Gloeckler, and E. Tums (1971), "An Electrostatic Deflection vs. Energy Instrument for Measuring Interplanetary Particles in the Range 0.1 to $\sim$3 MeV/charge," *Proc. 12th Inter. Conf. on Cosmic Rays*, Hobart, Australia, **4,** 1602–1607.

G. Fiedler and V. Höppner (1970), "The Diameter of Etched Fission Fragment Tracks in Solid State Nuclear Track Detectors as a Function of the Particle Energy," *Nucl. Instr. Methods* **74,** 285–290.

R. L. Fleischer, P. B. Price, and R. M. Walker (1965), "Solid-State Track Detectors: Applications to Nuclear Science and Geophysics," *Ann. Rev. Nuc. Sci.* **15,** 1–28.

R. L. Fleischer, P. B. Price, R. M. Walker, and M. Maurette (1967a), "Origins of Fossil Charged-Particle Tracks in Meteorites," *J. Geophys. Res.* **72,** 331–353.

R. L. Fleischer, P. B. Price, R. M. Walker, M. Maurette, and G. Morgan (1967b), "Tracks of Heavy Primary Cosmic Rays in Meteorites," *J. Geophys. Res.* **72,** 355–366.

R. L. Fleischer, P. B. Price, and R. T. Woods (1969), "Nuclear-Particle-Track Identification in Inorganic Solids," *Phys. Rev.* **188,** 563–567.

R. L. Fleischer, H. R. Hart, and W. R. Giard (1970), "Particle Track Identification: Application of a New Technique to Apollo Helmets," *Science* **170,** 1189–1191.

R. L. Fleischer, H. R. Hart, and G. M. Comstock (1971), "Very Heavy Solar Cosmic Rays: Energy Spectrum and Implications for Lunar Erosion," *Science* **171,** 1240–1242.

R. L. Fleischer, H. R. Hart, G. M. Comstock, M. Carter, A. Renshaw, and A. Hardy (1973), "Apollo 14 and 16 Heavy Particle Dosimetry Experiments," *Science* **181,** 436–438.

P. H. Fowler, V. M. Clapham, V. G. Cowen, J. M. Kidd, and R. T. Moses (1970), "The Charge Spectrum of Very Heavy Cosmic Ray Nuclei," *Proc. Roy. Soc.*, London **A318,** 1–43.

R. P. Henke and E. V. Benton (1967), "A Computer Code for the Computation of Heavy-ion Range-energy Relationships in any Stopping Material," U.S. Naval Rad. Defense Lab. TR-67-122.

V. Höppner, E. Konecny, and G. Fiedler (1969), "The Diameter of Etched Fission Fragment Tracks in Solid State Nuclear Track Detectors as a Function of the Particle Energy," *Nucl. Instr. Methods* **74,** 285–290.

D. Hovestadt and O. Vollmer (1971), "A Satellite Experiment for Detecting Low Energy Heavy Cosmic Rays," *Proc. 12th Inter. Conf. on Cosmic Rays*, Hobart, Australia, **4,** 1608–1613.

D. Hovestadt, O. Vollmer, G. Gloeckler, and C. Y. Fan (1973), "Measurement of Elemental Abundance of Very Low Energy Solar Cosmic Rays," *Proc. 13th Inter. Cosmic Ray Conf.*, Denver, **2,** 1498–1503.

W. Krätschmer (1971), *Die Anätzbaren Spuren Künstlich Beschleunigter Schwerer Ionen in Quarzglas*, Ph.D. Thesis, Max-Planck-Institut für Kernphysik, Heidelberg.

D. Lal, R. S. Rajan, and A. S. Tamhane (1969), "Chemical Composition of Nuclei of $Z > 22$ in Cosmic Rays Using Meteoritic Minerals as Detectors," *Nature* **221,** 33–37.

M. Lecerf and J. Peter (1972), "Detecteurs en Verre: Influence de la Masse et de l'Energie des Particules sur le Diametre des Traces," *Nucl. Instr. Methods* **104,** 189–195.

J. Lindhard, M. Scharff, and H. E. Schiott (1963), "Range Concepts and Heavy Ion Ranges," *Kgl. Danske Videnskab. Selskab, Mat.-Fys. Medd.* **33,** No. 14, 1–42.

M. Maurette (1966), "Etude des Traces d'Ions Lourds dans les Mineraux Naturels d'Origine Terrestre et Extra-terrestre," *Bull. Soc. Franc. Min. Crist.* **89,** 41–79.

J. C. D. Milton and J. S. Fraser (1962), "Time of Flight Fission Studies on U^{233}, U^{235}, and Pu^{239}," *Can. J. Phys.* **40,** 1626–1663.

M. L. Muga (1971), "A Versatile dE/dx Detector for Heavy Mass Nuclear Particles," *Nucl. Instr. Methods* **95,** 349–359.

L. C. Northcliffe and R. F. Schilling (1970), "Range and Stopping Power Tables for Heavy Ions," *Nuclear Data Tables* **A7,** 233–263.

D. O'Sullivan, P. B. Price, E. K. Shirk, P. H. Fowler, J. M. Kidd, E. J. Kobetich, and R. Thorne (1971a), "High Resolution Measurements of Slowing Cosmic Rays from Fe to U," *Phys. Rev. Letters* **26,** 463–466.

D. O'Sullivan, E. J. Kobetich, E. K. Shirk, and P. B. Price (1971b), "Resolution of High Energy Extremely Heavy Cosmic Rays in Plastic Detectors," *Phys. Letters* **34B,** 49–50.

D. O'Sullivan, A. Thompson, and P. B. Price (1973), "Composition of Galactic Cosmic Rays with $30 < E < 130$ MeV/Nucleon," *Nature Phys. Sci.* **243,** 8–9.

H. G. Paretzke, E. V. Benton, and R. P. Henke (1973), "On Particle Track Evolution in Dielectric Track Detectors and Charge Identification through Track Radius Measurement," *Nucl. Instr. Methods* **108,** 73–80.

L. S. Pinsky (1972), "A Study of Heavy Trans-Iron Primary Cosmic Rays ($Z \geq 55$) with a Fast Film Cerenkov Detector," NASA Tech. Memo X-58102, Manned Spacecraft Center, Houston.

T. Plieninger (1972), *Mineral als Spurdetektor zur Untersuchung der Schweren Komponente der Kosmischen Primarstrahlung*, Ph.D. Thesis, Max-Planck-Institut für Kernphysik, Heidelberg.

T. Plieninger, W. Krätschmer, and W. Gentner (1972), "Charge Assignment to Cosmic Ray Heavy Ion Tracks in Lunar Pyroxenes," *Proc. Third Lunar Sci. Conf.*, **3,** 2933–2939. Cambridge: MIT Press.

T. Plieninger, W. Krätschmer, and W. Gentner (1973), "Indications for Long-Time Variations in the Galactic Cosmic Ray Composition Derived from Track Studies on Lunar Samples," *Proc. Fourth Lunar Sci. Conf.*, **3,** 2337–2346. New York: Pergamon Press.

P. B. Price (1971), "The Study of Isotopes of Heavy Cosmic Ray Nuclei by Means of Tracks in Plastics," *Isotopic Composition of the Primary Cosmic Radiation*, P. M. Dauber (ed.), Danish Space Research Institute, Lyngby, Denmark, 63–74.

P. B. Price and R. L. Fleischer (1970), "Particle Identification by Dielectric Track Detectors," *Rad. Effects* **2,** 291–298.

P. B. Price and R. L. Fleischer (1971), "Identification of Energetic Heavy Nuclei with Solid Dielectric Track Detectors: Applications to Astrophysical and Planetary Studies," *Ann. Rev. Nuc. Sci.* **21,** 295–334.

P. B. Price and E. K. Shirk (1973), "Search for Trans-uranic Cosmic Rays on Skylab," paper at Symposium on High-Energy Astrophysics, Tucson, December 7.

P. B. Price, R. L. Fleischer, D. D. Peterson, C. O'Ceallaigh, D. O'Sullivan, and A. Thompson (1967), "Identification of Isotopes of Energetic Particles with Dielectric Track Detectors," *Phys. Rev.* **164,** 1618–1620.

P. B. Price, D. D. Peterson, R. L. Fleischer, C. O'Ceallaigh, D. O'Sullivan, and A. Thompson (1968a), "Plastic Track Detectors for Identifying Cosmic Rays," *Can. J. Phys.* **46,** S1149–S1153.

P. B. Price, R. L. Fleischer, D. D. Peterson, C. O'Ceallaigh, D. O'Sullivan, and A. Thompson (1968b), "High Resolution Study of Low Energy Cosmic Rays with Lexan Track Detectors," *Phys. Rev. Letters* **21,** 630–633.

P. B. Price, R. L. Fleischer, and D. D. Peterson (1969), "Identification of Nuclear Particles by Measurements of Etching Rates Along Their Tracks in Dielectric Solids," in *Reactivity of Solids*, J. W. Mitchell, R. C. DeVries, R. W. Roberts, and P. Cannon (eds.), 735–741. New York: Wiley.

P. B. Price, P. H. Fowler, J. M. Kidd, E. J. Kobetich, R. L. Fleischer, and G. E. Nichols (1971), "Study of the Charge Spectrum of Extremely Heavy Cosmic Rays Using Combined Plastic Detectors and Nuclear Emulsions," *Phys. Rev.* **D3,** 815–823.

P. B. Price, D. Lal, A. S. Tamhane, and V. P. Perelygin (1973), "Characteristics of Tracks of Ions of $14 \leq Z \leq 36$ in Common Rock Silicates," *Earth Planet. Sci. Lett.* **19,** 377–395.

E. K. Shirk and P. B. Price (1973), "Observation of Trans-Ion Solar Flare Nuclei in an Apollo 16 Command Module Window," *Proc. 13th Inter. Cosmic Ray Conf.*, Denver, **2,** 1474–1478.

E. K. Shirk, P. B. Price, E. J. Kobetich, W. Z. Osborne, L. S. Pinsky, R. D. Eandi, and R. B. Rushing (1973), "Charge and Energy Spectra of Trans-Iron Cosmic Rays," *Phys. Rev. D*, **7,** 3220–3232.

G. Somogyi (1966), "A New Possibility for the Determination of Energy Distributions of Charged Particles in Solid State Nuclear Track Detectors," *Nucl. Instr. Methods* **42,** 312–314.

G. Somogyi and S. A. Szalay (1973), "Track-Diameter Kinetics in Dielectric Track Detectors," *Nucl. Instr. Methods* **109,** 211–232.

G. Somogyi, B. Schlenk, M. Varnagy, L. Mesko, and A. Valek (1968a), "Application of Solid-State Track Detectors for Measuring Angular Distributions of Alpha-Particle Groups from Nuclear Reactions," *Nucl. Instr. Methods* **63,** 189–194.

G. Somogyi, M. Varnagy, and G. Pëto (1968b), "Application of Plastic Track Detectors for Detection of Light Nuclei," *Nucl. Instr. Methods* **59,** 299–304.

G. Somogyi, M. Varnagy, and L. Medveczky (1970), "The Influence of Etching Parameters on the Sensitivity of Plastics," *Rad. Effects* **5,** 111–116.

R. A. Stern and P. B. Price (1972), "Charge and Energy Information from Heavy Ion Tracks in Lexan," *Nature Phys. Sci.* **240,** 82–83.

R. M. Sternheimer (1956), "Density Effect for the Ionization Loss in Various Materials," *Phys. Rev.* **103,** 511–515.

P. G. Steward (1968), "Stopping Power and Range for Any Nucleus in the Specific Energy Interval 0.01- to 500-MeV/amu in Any Nongaseous Material," Lawrence Radiation Lab. Rept. UCRL-18127.

J. D. Sullivan, H. J. Crawford, and P. B. Price (1973), "Relative Abundances of

Nuclei with $2 \leq Z \leq 36$ in Solar Flares," *Proc. 13th Inter. Cosmic Ray Conf.*, Denver, **2,** 1522–1525.

M. Varnagy (1970), *Few Properties of the Cellulose Acetate (T-Cellit) Detector and Its Applications*, Ph.D. Thesis, Kossuth University, Debrecen, Hungary.

J. P. Wefel (1971), *Measurements of the Relative Abundances of Extremely Heavy Cosmic Rays*, Ph.D. Thesis, Washington University.

PART II

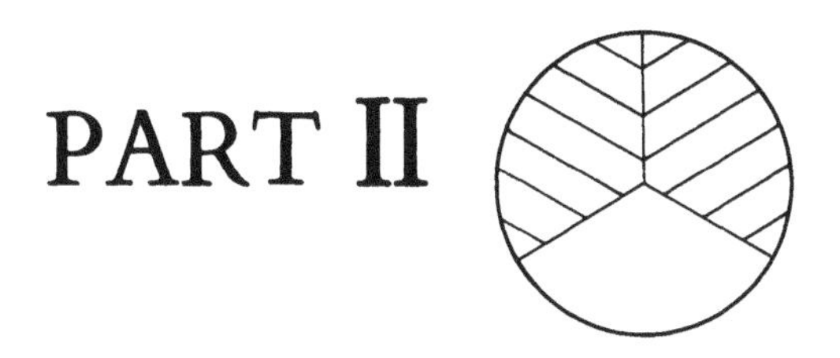

Earth and Space Sciences

Chapter 4

Fission Track Dating (with Applications to Geochronology, Geophysics and Archaeology)

4.1. INTRODUCTION

Since the period of extreme excitement when we found that virtually all natural micas contained a background of spontaneous fission tracks (Price and Walker, 1962), natural tracks in solids have provided new insights into geology, archaeology, and the space sciences. The types of natural tracks that were first recognized are the basis for fission track dating, a new method that has by now been applied to a remarkable diversity of significant and interesting problems. These range from a confirmation of how early man's hominid ancestors existed, to the first direct confirmation by radiometric age dating of the recent geophysical model that describes how the continents and ocean floors evolved, to extensive work on the times of creation, flight, and fall of those mysterious glass objects—the tektites—whose origin remains a puzzle. Recent advances in the understanding of environmental effects have made possible internal checks on the rigor and meaning of fission track ages and allow the dating of thermal events of the past—ranging in scale from regional metamorphism, to local igneous intrusions, to the dying out of the fire in a neolithic hearth. Over the six year period from 1964 when the first track age determinations were reported (Fleischer and Price, 1964b; Fleischer et al., 1964b; Maurette et al., 1964), fission track dating has developed and is now an expanding, widely used technique. The number of papers in track dating has increased to the point that in the three years 1970–1972 they make up 6% of all publications in geochronology.

In this chapter we consider the possible types of internal track sources in natural materials and the age measurements that they make possible. Because fission track dating is in an advanced stage of development and is the only widely used track dating method, the major emphasis here will be given to techniques and assumptions of fission track dating, to describing ways in which these assumptions can become invalid, and to presenting examples of the application of the method to the variety of geological, geophysical, and anthropological problems that were mentioned in the preceding paragraph.

Most radiometric dating processes depend on the statistical regularity of the decay of one parent radionuclide into a daughter nuclide, for example ^{40}K into ^{40}Ar, ^{87}Rb into ^{87}Sr, or ^{238}U, ^{235}U, and ^{232}Th into ^{206}Pb, ^{207}Pb, and ^{208}Pb, respectively. By measuring the relative abundance of the parent and the daughter in any pair an age can be derived. [See Faul (1966a) for a lucid review of radiometric dating methods.] Fission track dating is analogous except that the counting of radiation damage tracks replaces isotopic measurements.

Radiometric dating techniques are in general complementary to one another in that each method produces an age with a special meaning—the last outgassing, the last melting, the last melting accompanied by mixing with isotopically separate material, the last heating to remove tracks, and so forth. Different methods are in general appropriate to different materials. For example, fission tracks provide the best method for dating the chilled glass that rims young, deep ocean basalts, while older, fine-grained crystalline basalts are better examined by K-Ar dating (Aumento et al., 1968; Fleischer et al., 1968b). Where the various other methods do overlap with fission track dating, the techniques that use isotopic analysis are capable of higher analytical precision because automatic scanning of natural tracks has yet to be developed to the extent needed to accumulate extensive statistical track counts. The special merits of fission track dating are simplicity, low cost, wide time span over which it has been applied (from less than a year to the age of the solar system), and applicability to a wide range of substances—both crystalline and glassy. It is especially valuable in the awkward time span from $\sim 4 \times 10^4$ years, where carbon dating ceases to be accurate, to $\sim 10^6$ years, where K-Ar dating begins to become less laborious. The technique has the special property that it allows ages of tiny samples to be measured—less than a microgram of old, accessory minerals being ample in some cases. This quality is especially helpful in dating precious samples or samples with complicated histories where there is some reason to expect ages to vary with position. Finally, in measuring thermal events in the past, fission track dating is uniquely flexible and is receiving considerable use.

4.2. ORIGIN OF NATURAL TRACKS: INTERNAL SOURCES

The number of possible track origins in natural samples is sharply limited by the low sensitivities of natural crystals and glasses (Fleischer et al., 1964d, 1967a,c;

Price et al., 1968), none of which record tracks of alpha particles or of any less intensely ionizing radiation such as is produced in α- or β-decay. This insensitivity is a unique feature of track-recording solids. Thus the internal tracks are due entirely to either fission (Price and Walker 1963, Fleischer et al., 1965e, 1968c) or to recoil nuclei that are much heavier than alpha particles (Huang and Walker, 1967). By excluding cosmic ray effects we are implicitly considering only samples that are well shielded—located beneath a thick atmosphere or at a depth of more than 600 gm/cm^2 within the interior of a body without an atmosphere.

At present there are four varieties of natural tracks of internal origin that have been positively identified: tracks from spontaneous fission of ^{238}U (Price and Walker, 1963); tracks from spontaneous fission of ^{244}Pu (Fleischer et al., 1965e, 1968c); tracks of the recoiling heavy nuclei resulting from the alpha decay of ^{232}Th, ^{235}U, and ^{238}U (Huang and Walker, 1967); and tracks produced by the scattering of alpha particles from the normal constituents of silicate minerals (Crozaz et al., 1969). In the last two cases it should be emphasized that alpha particles themselves do not create tracks in natural minerals; only the short range ($<5 \times 10^{-5}$ cm), heavy, residual, recoil nuclei leave etchable tracks. Alpha scattered nuclei have not proved useful and will not be discussed further here; ^{244}Pu tracks will be considered in Chapter 6.

Although in principle many other varieties of natural tracks can exist, detailed consideration (Price and Walker, 1963) shows that normally on Earth these alternative sources are so rare as to be unimportant. Specific alternatives that have been considered include the following:

1. Spontaneous fission of other naturally existing heavy elements or nuclides (^{235}U, Th, Bi, Pb, and lighter elements). (Extensive searches have produced no evidence of spontaneous fission of any element that is lighter than uranium.)
2. Fission of heavy elements induced by natural α, β, or γ activity.
3. Fission of ^{235}U induced by thermalized neutrons from spontaneous fission, (γ, n) reactions, and (α, n) reactions.
4. Fission of other heavy nuclides induced by fast neutrons from spontaneous fission, (γ, n), and (α, n) reactions.
5. Fission induced by cosmic ray primaries.
6. Fission induced by cosmic ray secondaries, including neutrons and μ-mesons.
7. Spallation caused by cosmic ray secondaries.

Under certain very special conditions, such as those for a mineral imbedded in a uranium ore, induced fission events can be comparable to spontaneous events (Price and Walker, 1963). Such cases are obviously rare. The recent inference that the Oklo, Gabbon uranium mine was once the site of a naturally occurring, water-moderated reactor (Bodu et al., 1972) is an example of such a special condition. Although several reports have been published (Flerov and Perelygin, 1969; Flerov et al., 1972) purporting to show the existence of naturally fissioning "superheavy elements" in terrestrial minerals, we have examined a large number of minerals that are geochemically favorable locations of superheavy elements (Price et al., 1970; Fleischer and Naeser, 1972; Haack, 1973a) without finding any anomalously high fission track densities, i.e., tracks possibly caused by the spontaneous fission of any isotope other than ^{238}U.

Although the effects of cosmic rays on terrestrial samples generally can be neglected, it is possible to produce detectable amounts of induced fission in minerals that are both low in uranium and rich in other very heavy elements such as Pb and Bi (Fleischer et al., 1967a; Maurette, 1966; Geisler et al., 1973; Flerov et al., 1972). Such induced fission events are now agreed (Flerov et al., 1972; Geisler, 1972) to be almost certainly responsible for the fission tracks that had been described as possibly due to the decay of superheavy elements (Flerov and Perelygin, 1969; Otgonsuren et al., 1969; 1972).

The only other identified "unusual" example of fission in natural materials was the observation of fission tracks induced by neutrons from an atomic weapon test (Walker, 1963). Hopefully these should not be properly regarded as natural tracks.

4.3. ^{238}U SPONTANEOUS FISSION: FISSION TRACK DATING

Fission track dating is conceptually the simplest of the several dating techniques that provide absolute measures of time from the slow but statistically steady decay of radioactive nuclides. The major isotope of uranium (^{238}U) decays by spontaneous fission at a rate of $\sim 10^{-16}$ per year, leaving tracks such as we see in the mica shown in Fig. 4-1, so that it is necessary merely to count the fraction of uranium atoms that have fissioned within a sample in order to tell its age. We now describe the procedures, including a discussion in some detail of problems that, if not understood, can result in confusion, or when carefully approached, can lead to significant insights into geological history.

Since the absolute rate at which spontaneous fission tracks accumulate in minerals is proportional to the uranium content, it is necessary to measure both the natural track abundance and the uranium content. Fortunately this latter quantity can be determined simply by inducing a measurable fraction of the uranium to fission, making use of the fact that thermal (slow) neutrons cause fission of the less abundant isotope, ^{235}U. After the natural tracks are counted, the sample is exposed to a specified dose of neutrons in a nuclear reactor and the newly induced tracks are counted. The ratio of the two track densities, together with a measure of the neutron dose, allows an age to be calculated. We now describe this calculation and the assumptions behind it. The basic ideas here are those of Price and Walker (1963) with refinements given by Fleischer and Price (1964b), Fleischer et al. (1965c), and Fleischer and Hart (1972).

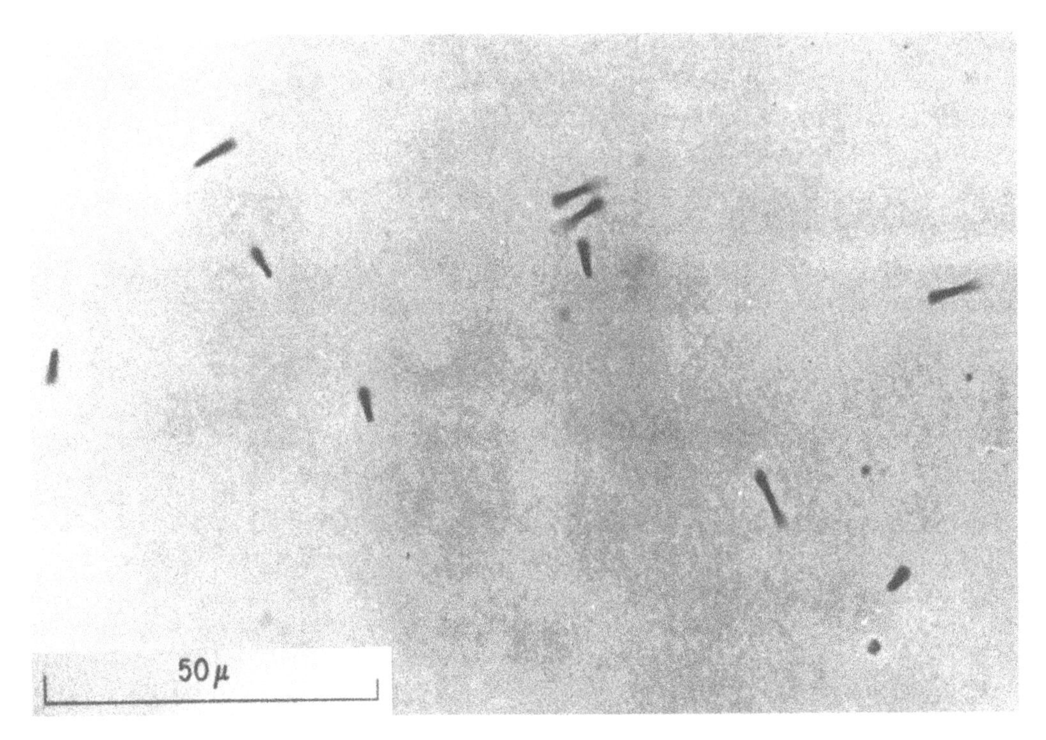

Fig. 4-1. *Mica was the first substance in which ^{238}U spontaneous fission tracks were identified. The tracks in this Madagascar phlogopite were revealed by etching in hydrofluoric acid.* (*After Price and Walker, 1963.*)

4.3.1. Age Calculation and Assumptions

Fission track ages are determined by three track densities ρ_s, ρ_i, and ρ_D. For simplicity and clarity of presentation we neglect for the present the fact that the concentration of ^{238}U changes slowly with time (half-life 4.51×10^9 years) due to alpha decay. For ages of less than 10^8 years the fractional change is negligible.

The spontaneous track density ρ_s (through an internal surface) is then given by the following expression, which is linear in time and uranium content:

$$\rho_s = (A\lambda)(N_v c^{238} R^{238} \eta^{238}) \tag{4-1}$$

where

A = time over which spontaneous fission tracks have been stored,
λ = decay rate by fission of ^{238}U,
N_v = number of atoms per unit volume in the material,
c^{238} = fraction of those atoms that are ^{238}U,

R^{238} = length of the etchable track of a ^{238}U fission fragment,
η^{238} = etching efficiency, the fraction of the tracks crossing a surface that are revealed by etching.

The above equation can be understood as simply the product of two terms, $A\lambda$ being the fraction of ^{238}U atoms that fission in the time A, and $N_v c^{238} R\eta$ being the number per unit area of ^{238}U atoms that are within a range R of the etched surface and can lead to etched tracks at that surface.

After neutron irradiation with Φ thermal neutrons per unit area, the induced track density ρ_i measured on a freshly cut internal surface is given by

$$\rho_i = (\sigma\Phi)(N_v c^{235} R^{235} \eta^{235}) \quad (4\text{-}2)$$

where σ = cross section for inducing fission of ^{235}U with thermal neutrons, and the quantities in the second parenthesis are defined analogously to those in the first equation. Again there is a first term ($\sigma\Phi$) which represents the fraction of uranium (^{235}U in this case) caused to fission and a remaining term giving the atoms per unit area that can lead to etchable tracks at the etched surface.

Use of eq. (4-2) in element measurement and mapping is described in Chapter 8. We should point out that, although in principle c^{238} could be determined by any suitably sensitive analytic technique, the method of inducing tracks by neutron irradiation, as originally worked out, has important advantages. It can take into account spatial heterogeneities in uranium concentration and it permits geometric factors and other constants to cancel.

To measure the neutron dose Φ we include with each group of samples to be neutron-irradiated a piece of standard track detector (a *dosimeter glass*) of fixed uranium content. The density of fissions induced in this glass, ρ_D, will be a measure of the neutron dose

$$\rho_D = B\Phi \quad (4\text{-}3)$$

where the constant B is given formally by the expression $\sigma N_v R_D C_D{}^U \eta^D$—quantities defined for the dosimeter analogously to those in eq. (4-2). It is not necessary to measure B explicitly as long as it is constant. The same expression with different value of B applies if the fissions from the glass are registered in a separate Lexan or mica detector pressed against the glass.

Three sources of standard glass are available: Dr. R. H. Brill, Corning Museum of Glass, Corning, N.Y. (Schreurs et al., 1971); Dr. R. L. Fleischer, General Electric Research Laboratory, Schenectady, N.Y. (Fleischer et al., 1965c); and Office of Reference Materials, National Bureau of Standards, Washington, D.C.

(certificates 610 to 617 and 961 to 964, Carpenter, 1972; Carpenter and Reimer, 1973). Chapter 9 discusses dosimetry more generally.

The above equations can be combined to give the age in terms of the three track densities and a constant:

$$A = \zeta(\rho_s/\rho_i)\rho_D \tag{4-4}$$

where

$$\zeta = \frac{\sigma}{B\lambda}\frac{c^{235}}{c^{238}}\frac{R^{235}}{R^{238}}\frac{\eta^{235}}{\eta^{238}}.$$

Once this constant is known, three track counts allow the age to be found, for samples $\lesssim 10^8$ years old.

As we noted, because the ^{238}U abundance more than 10^9 years ago was significantly greater than at present, eq. (4-1) must be altered for old samples to include the time variation of c^{238}. If the total decay rate for ^{238}U is λ_D, the ancient concentration is given by exp $(\lambda_D A)$ times the present concentration, and the relation for ρ_s becomes

$$\rho_s = [\exp(\lambda_D A) - 1](N_v c^{238} R^{238} \eta^{238} \lambda/\lambda_D) \tag{4-5}$$

so that eq. (4-1) would be rewritten

$$\exp(\lambda_D A) - 1 = \zeta(\rho_s \rho_D \lambda_D/\rho_i), \tag{4-6}$$

which reduces to eq. (4-1) for $A \lesssim 10^8$ years.

The assumptions in the use of the age equations (4-4 or 4-6) are threefold. Firstly, the natural tracks counted must be only tracks from spontaneous fission of uranium. Secondly, the same etching and track counting methods are to be used in counting natural and induced tracks in the sample, so that a correct value of (ρ_s/ρ_i) results. Finally, ζ is assumed to be a constant. Precautions for obtaining reliable (ρ_s/ρ_i) measurements will be given in the section on techniques.

Turning to the third requirement of the method, the question of the constancy of the proportionality factor ζ, we can consider individually the various factors. σ and λ describe nuclear properties and are clearly constant. For a given dosimeter glass, B is a fixed property as long as the same glass is used with the same etching and counting procedure. c^{238}/c^{235} is an isotopic ratio whose variations in nature have frequently been sought but have only once been observed (Bodu et al., 1972). That case (the Oklo mine), as noted earlier, involved a uranium ore, which is not a material that would normally be used for geochronology. The ranges R^{238} and

R^{235} are very nearly equal (Togliatti, 1965), and since they apply to the same detector material, their ratio will be constant. Although for fresh, unaltered tracks the same statement applies to the ratio η^{235}/η^{238}, heating can cause tracks to fade and produce changes in this ratio. Recent advances in fission track dating (Storzer and Wagner, 1969; Gentner et al., 1969a; Storzer, 1970) and an understanding of thermal effects on etching efficiency revolve about such possible environmental effects. These effects, along with fundamental considerations that determine track geometry, will be discussed after describing how the constant ζ is measured.

In principle we can measure ρ_s, ρ_i, and ρ_D for one sample whose age A is independently known. ζ is then given by $(A\rho_i/\rho_s\rho_d)$, a number that can be used thereafter to determine previously unknown A's for all other samples. Since any particular sample might in principle have suffered from track fading, a more certain and statistically valid procedure is to measure ζ for many samples, so that the occasional high value that would result from track fading (giving a low ρ_s) can be discarded. Fig. 4-2 shows how this has been done, following the procedure of Fleischer and Price (1964b) by plotting the known age against the fission track age for a number of samples. Occasional samples fall off the line (statistically significant deviations are always to the left, corresponding to a lowered value of η^{238}): a best 45° line fit to the other samples determines the proper value of ζ. This value is 0.0138 (million years/track per cm^2) using our original dosimeter (Fleischer et al., 1965c) for which $B = 2.26 \times 10^{11}$ neutrons/track.

Although λ is clearly a constant, direct measurements have given a wide range of values from $0.53 \times 10^{-16}\ y^{-1}$ (Petrzhak and Flerov, 1940) to $1.2 \times 10^{-16}\ y^{-1}$ (Gerling et al., 1959). If one wished to calculate age values by measuring all of the terms making up ζ separately, the track methods of measuring ζ are the most directly relevant values. Such measurements of λ have been reported by Fleischer and Price (1964a) [$6.85\ (\pm.20) \times 10^{-17}$], Roberts et al. (1968) [$7.03\ (\pm0.11) \times 10^{-17}$], Kleeman and Lovering (1971) [$6.8\ (\pm0.6) \times 10^{-17}$], Leme et al. (1971) [$7.30\ (\pm.16) \times 10^{-17}$], Khan and Durrani (1973) [$6.82\ (\pm0.55) \times 10^{-17}$], and Suzuki (1973a) [$7.5\ (\pm.5) \times 10^{-17}$], all of which are consistent with $7.03\ (\pm.11) \times 10^{-17}$, the value with the smallest quoted error (a value that is remarkably close to the result of the early measurement of Segré (1952)—$6.82 \times 10^{-17}\ y^{-1}$). It should be emphasized, however, that measuring ζ directly using a standard or set of standards of known age is the simpler, more precise, and more reliable procedure that is recommended.

A material that has been proposed as a fission track age standard consists of apatite and zircon crystals from the Fish Canyon, volcanic welded tuff (San Juan Mountains, South Fork, Colorado), which has a 27.2 (±0.7) My K-Ar age (Steven et al., 1967) based on concordant biotite, hornblende, plagioclase, and sanidine and has been used for an extensive interlaboratory comparison. These samples, designated No. 70L-126 or (later) No. 72N8, are available from Dr. C. W. Naeser, U. S. Geological Survey, Federal Center, Denver, Colorado.

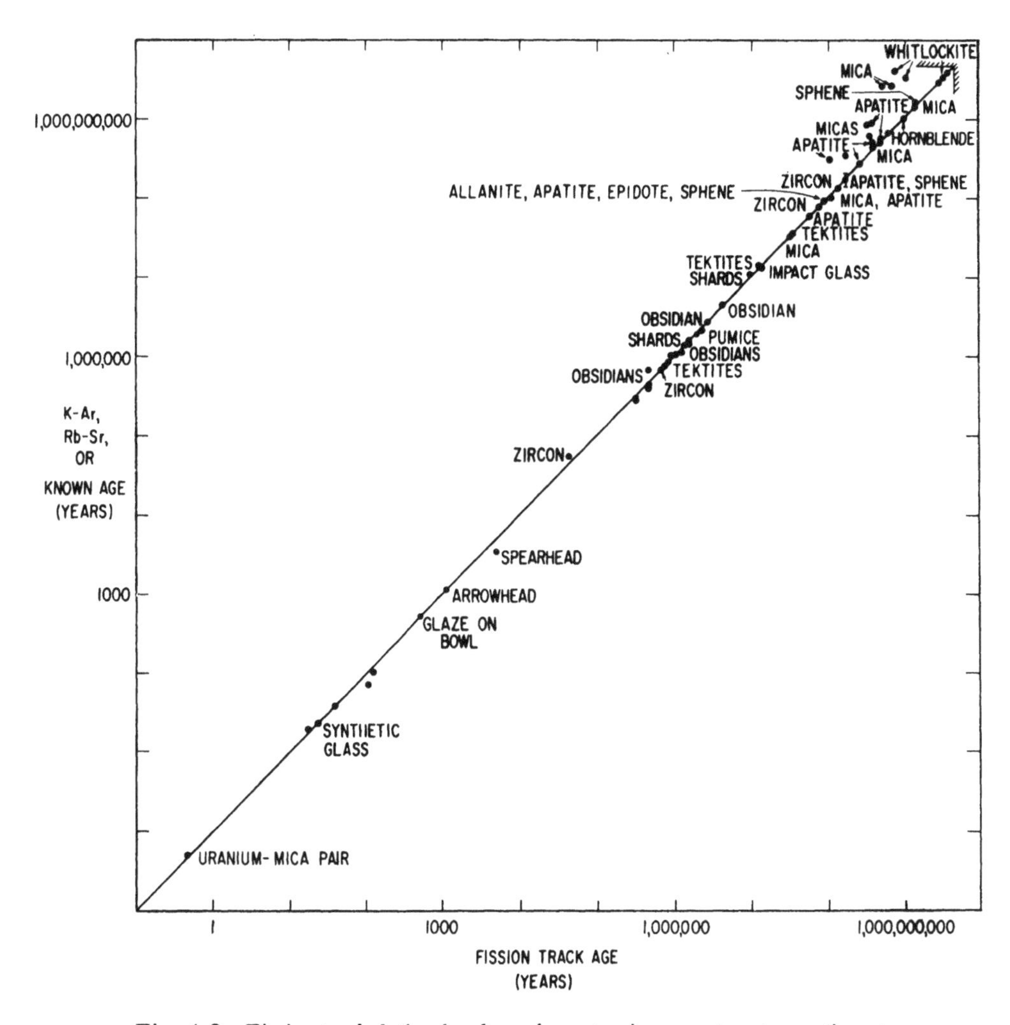

Fig. 4-2. *Fission track dating has been shown to give correct ages over time spans ranging from 1/2 year to more than a billion years. The graph compares fission track ages, measured by counting natural tracks formed by spontaneous fission of* ^{238}U, *with ages known by other means, either documented ages for man-made samples which appear in the lower left or ages measured by other radioactive decay techniques for the geological samples indicated in the upper right. Occasionally fission track ages of natural samples are lowered by thermal effects, allowing the fission track technique to serve as a thermometer for past events.*

4.3.2. Experimental Procedures

SELECTION OF SURFACES TO BE COUNTED

The preferred method for measurement of fission track ages is the "internal surface" method just described—in which the induced tracks are measured on an interior surface that was prepared after the neutron irradiation of the material being dated. Because the same detector is used to measure ρ_s and ρ_i (with, therefore, a fixed etching efficiency), the ratio ρ_s/ρ_i can be determined without the need for additional corrections.

One possible procedure when working with separated minerals is to divide the grains into two groups. The first group is mounted, polished, and etched to determine the fossil track density. The second group is heated to a temperature that will cause complete annealing of the fossil tracks. These grains are then irradiated, mounted, polished, etched, and counted for the induced track density. This is a useful method only if uranium concentration of all grains is similar. For the most part apatite is the only mineral that is reliably dated this way.

In a variation on the above technique, low-uranium crystals within a high uranium matrix may be used to date their time of deposition. Bonadonna and Bigazzi (1969, 1970) used biotite from within a pyroclastic tuff. The uranium content of the tuff is subsequently determined by irradiating it adjacent to a biotite detector.

If the surface to be etched was not an interior surface during irradiation, eqs. (4-4) and (4-6) must be altered in different ways depending on whether the "external surface" method or the "external detector" method is used. In the first of these the induced track density is measured on an external surface of the material to be dated; in the latter it is measured on a detector (usually a muscovite mica or Lexan polycarbonate with a low uranium content) that was pressed against the sample during neutron irradiation. For the "external surface" method, the surface must be protected during irradiation by being placed against a material containing much less uranium than the sample. For example, one useful material that normally has a negligibly small concentration of uranium is quartz.

The equations are then modified by replacing ρ_i by $f\rho_i$, where in the simplest case $f = 2$. This change is needed because the contribution to the spontaneous track density comes from both sides of the plane produced by cleaving or polishing, whereas the contribution to the induced track density at the exposed surface comes only from one side. The replacement of ρ_i by $2\rho_i$ is rigorous only for the case of an etch in which negligible material is removed relative to the etchable range of a fission fragment. An additional correction must be made if the sectioned surface dissolves at an appreciable rate, so that new tracks ending below the original surface appear as etching progresses. The need for this correction was noted in relation to dating of hornblende (Fleischer and Price, 1964d) and quantitative corrections have been derived as a function of etching time for glasses of three different etching efficiencies by Reimer et al. (1970). They have also measured the appropriate factors for several minerals and glasses for what they consider to be convenient etching conditions. For the etches they specify, f would be 1.45 to 1.59 for apatite, 1.75 for muscovite, 1.61 for sphene, 1.59 to 1.67 for zircon, and 1.13 to 1.39 for some natural glasses. Edwards (1967) has done similar work on an Icelandic obsidian. Since different samples may respond differently even to the same etch, it is necessary for the fission track dater to measure this factor separately under his own etching conditions for each mineral sample.

In the "external detector" technique it is necessary to know the etching efficiency of the natural tracks on an internal surface of the sample and the etching efficiency for an external detector used to register the induced fission fragments from the sample. For an external mica detector this factor is close to unity (Debeauvais et al., 1964; Kleeman and Lovering, 1970; Seitz et al., 1973), although inefficiency in scanning short tracks when the silvering technique of Seitz et al. is not employed may reduce the value (as in the observations of Reimer et al., 1970). We will discuss this technique shortly.

In many natural samples, the uranium occurs in highly localized regions, either in discrete mineral phases or in disordered areas such as grain boundaries. This gives rise to localized fission stars such as are shown in Fig. 4-3a. In principle such fission stars can also be used for dating. Although little work with inclusions has

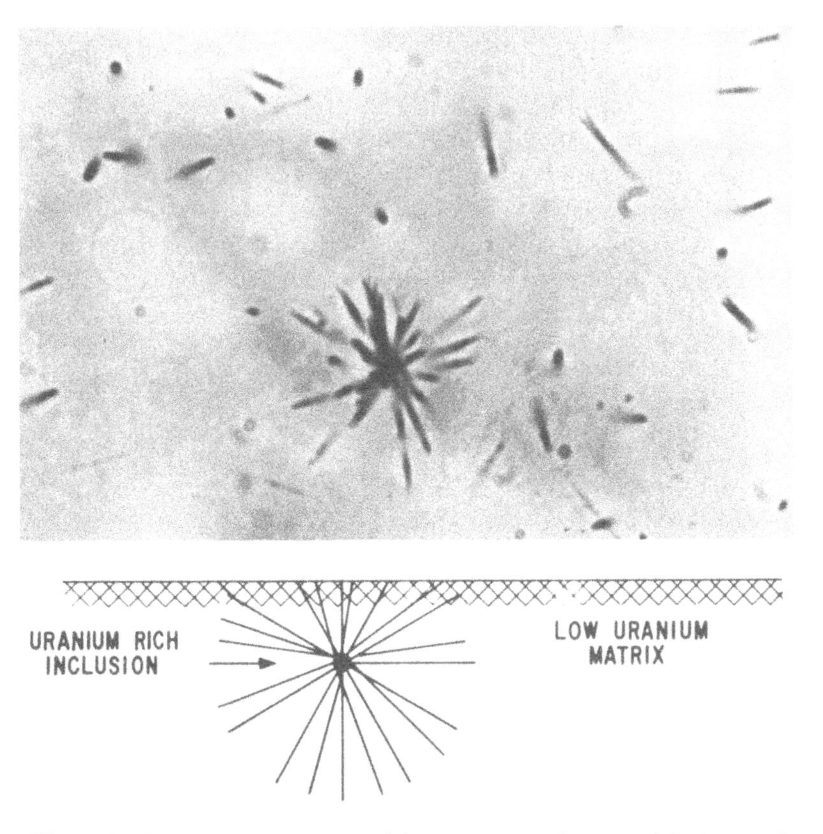

Fig. 4-3. *Fission stars in crystals: (a) Micrograph of a natural fission star in a mica. The cluster of tracks emerges from a small inclusion that is rich in uranium. (After Price and Walker, 1963.) (b) Sketch of the geometry of such a sunburst of tracks.*

been reported, we include a discussion here for the sake of completeness. This method is basically similar to that for dating the time at which a uranium-poor material was placed in a uranium-rich matrix, as we described in an earlier paragraph.

Fig. 4-3b shows a diagram of a fission star such as would be produced by a uranium-rich apatite crystal buried in a larger feldspar crystal. The number of fission tracks that reach the surface of the feldspar is clearly proportional to the total uranium in the apatite and the age. In principle, the number of uranium atoms in the inclusion capable of producing tracks that reach the surface can be determined by placing an external detector on the surface of the feldspar and irradiating with thermal neutrons. If n_f is the number of fossil tracks in the host feldspar and n_i is the number counted in the external detector after irradiation with a dose, Φ, of thermal neutrons, the age is given by eq. (4-6) with ρ_s replaced by n_f and ρ_i with n_i.

However, the number and energy spectrum of the fission fragments that passed through the surface depend on the burial depth of the inclusion and introduce special problems in this method of dating. Consider, for example, the case where the external detector (e.g., plastic or mica) is more sensitive than the feldspar. In this case fission fragments that would emerge from the feldspar surface at very shallow angles to the surface (i.e., at low energies) would still be registered in the external detector though no longer capable of producing tracks in the feldspar. The value of n_i would thus be artificially increased with respect to n_f, thereby giving an artificially low age. As the depth of the inclusion increases, the effect becomes more pronounced and, in the limit where the inclusion is buried at a depth equal to the registered range of fission tracks in feldspar, the apparent age would be zero! Similar difficulties of the opposite sign exist if the external detector is less sensitive than the feldspar. The conclusion is clear that the external detector should be composed of the same material as that in which the fossil tracks are being registered, and ideally should have the same crystallographic surface exposed as does the natural sample.

Even in this event the method has potential problems. The track length distributions are perforce different in the external detector and in the crystal being studied and this may lead to errors in track counts, depending on the track counting criteria that are used. Additional corrections, dependent on the depth of the inclusion, have to be made if an appreciable depth of the surface of the host crystal (and identical external detector) is removed in the etch used to reveal tracks. From these various considerations buried inclusions in mica, studied with a mica external detector, would be the simplest system to study. The rate of surface removal of the mica is negligible and the reflection method of track counting (Seitz et al., 1973) gives essentially 100% track counting efficiency regardless of the total length of the etched fission tracks.

The inclusion method also differs from other fission track dating methods in that the etching for fossil tracks should generally be done *after* neutron irradiation. This is to prevent the uranium in the inclusion from being leached out by the solution used to reveal tracks in the host mineral. The basic track counts in the host mineral will then consist of a combination of fossil and induced tracks. Although the induced track number n_i can be simply measured in the external detector and then subtracted from the total track number in the host crystal to give the fossil contribution n_f, it is clear that the induced track number cannot be made too high without producing errors in the determination of n_f. An optimum choice would generally give $n_i \approx n_f$.

In spite of these limitations, it is likely that inclusion dating will be utilized in the future. It should be particularly useful in studies of the thermal histories of rocks where the radioactive inclusions themselves and the host minerals have different track fading characteristics.

Track Identification

Although only fission events produce genuine particle tracks, all linear features displayed by etching are not necessarily tracks. Examples of spurious, somewhat tracklike features are rod-shaped crystalline phases in glass and dislocations in crystals (Fleischer et al., 1964c; Fleischer and Walker, 1966). Care in preparation of glass samples (discussed later in this chapter) allows such features to be identified on the relatively rare occasions when they are present. Dislocations in crystals can be identified by unique properties that distinguish them from fission tracks (length, distribution, preferred orientation, etc.). These properties will be reviewed and illustrated in detail shortly. Both varieties of defects are recognizable; only occasionally will their presence preclude age determinations on samples.

Sample Handling

The sequence normally followed consists of sample selection, preparation, etching, scanning, and irradiation, followed by a second sequence of preparation, etching, and scanning of the sample and dosimeter.

Selection.— Samples must both retain tracks over long periods of time and have a countable track density. We have discussed thermal effects on tracks in Chapter 2. Table 2-4 gives track thermal fading characteristics that have been observed on an assortment of minerals and glasses. The higher the fading temperature the more likely a mineral is to have retained tracks since its formation time; conversely, samples with lower annealing temperatures will be more sensitive detectors of past thermal effects such as campfires, adjacent igneous intrusions, or slow cooling.

Since the number of fissions depends both upon age and uranium concentration, there is some minimum product $A \cdot c^{238}$ which allows convenient track counting. Since minerals are characterized by certain "typical" uranium contents, for each there is a minimum age that normally can be measured without investing unreasonable time in scanning for tracks. For an easy track-counting job $AC > 300{,}000$, where A is in years and C is in weight parts per million; for what we describe as a labored determination, a 40 hour stint at track counting, $AC > 8{,}000$ is a rough requirement. For a typical obsidian (3-6 ppm of uranium) ages down to 50,000–100,000 years are therefore easy and 1,000–3,000 years possible, but their determination is difficult. By comparison, a zircon with an unusually high uranium concentration (1,000 ppm uranium) would allow ages as short as 3,000 years to be measured with ease, but if the age were greater than $\sim 10^8$ years one would have to use an electron microscope to resolve the tracks. For typical small zircon grains from rocks, 200,000 years is a convenient lower limit. Fig. 4-4 indicates age ranges accessible to various minerals, assuming typical uranium contents. Materials that have been used for fission track dating include allanite, anthophillite, apatite, bastnäsite, epidote, garnet, hornblende, micas, quartz, sphene, stibiotantalite, topaz, whitlockite, zircon, and a wide variety of glasses—both natural and man-made. The Appendix to this chapter gives references to dating of specific materials.

Preparation.— Many different sample preparation techniques have been used to obtain suitable surfaces for etching. The simplest procedure is cleavage to produce

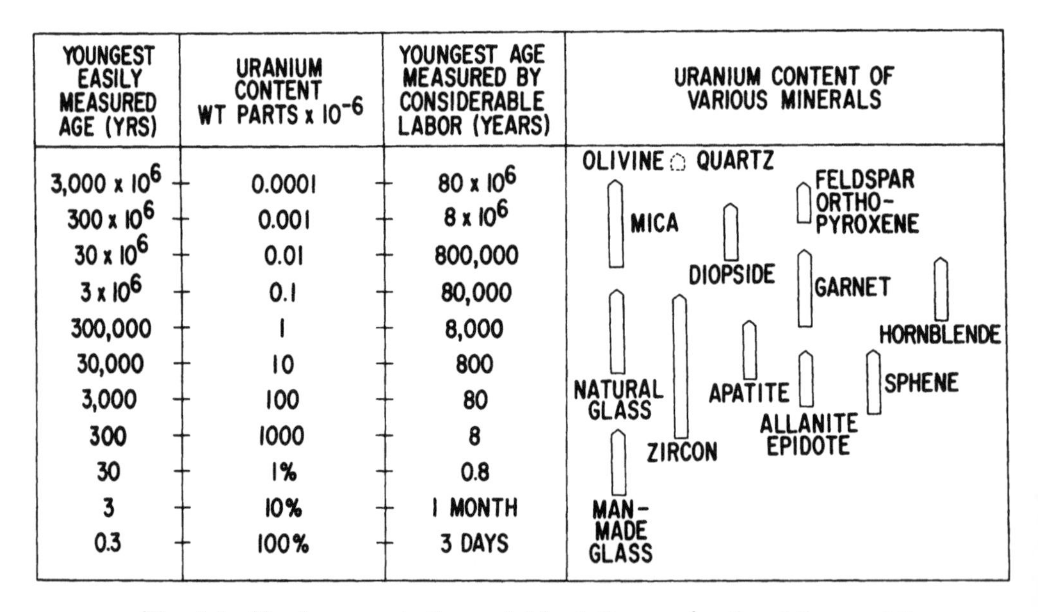

Fig. 4-4. *Uranium concentration needed for dating as a function of the age of the sample. For an "easily" measured age, it is assumed that the observer will spend one hour at his microscope counting fission-track etch pits. For a determination "by considerable labor" nearly 40 hours of such work are assumed. Material that is free from inclusions and bubbles is considered for estimating the time necessary for counting tracks. On the right are indicated the lowest ages which should be measurable in various minerals on the basis of typical uranium contents, or as in the man-made glasses, available uranium contents. (After Fleischer and Price, 1964d.)*

a fresh surface that was previously interior to the sample. For micas this is direct and easy; many other crystals can be cleaved if care is taken (see, for example, Gilman, 1959). Some materials (such as glass, olivine, and quartz) do not fracture along crystal planes. Even so, many glasses have high enough track densities that a fracture surface can be used by occasionally shifting the vertical position of the microscope stage while it is being translated during scanning. For other samples the usual technique is to imbed them in plastic (epoxy or glycol phthalate) and then polish. If proper care is taken during mounting, cutting, and polishing, the local heating will not significantly affect tracks. It is vital, however, to obtain an optical quality polish (optically smooth but not necessarily optically flat) in order that surface defects be minimized. Otherwise, microcracks and scratches from polishing will etch into distinctive but numerous figures that obscure fission tracks and make counting arduous. Conventional thin sections for petrographic work have inadequate finishes for subsequent etching. In our work the final finish is obtained using 0.3 μm Al_2O_3 polishing powder or diamond paste.

For most minerals—those in which tracks etch as long, narrow, spikelike features—transmitted light is needed for track identification. Samples must therefore be thin enough to transmit light. For polycrystalline rock samples a section is desired whose thickness is not greater than the typical diameter of crystals making up the rock; for nearly opaque minerals (hornblende, augite) the necessary thickness may be less. Glass, on the other hand, is normally best scanned using reflected light, which better displays wider, cone-shaped tracks. Occasionally a glass will contain inclusions or elongated pores that give rise to defects that are difficult to distinguish from etched particle tracks. In these cases the use of thin slabs of glass that can be viewed with both reflected and transmitted light usually allows a decisive track count to be made.

In minerals such as mica, where the etch pits have a distinctive geometry, it is often useful to silver the surface and observe in reflected light. Over-etching to give very large fission pits whose diameter is greater than the range of fission fragments can also be used to advantage in measuring low track densities.

Techniques for distinguishing tracks from other etched defects have been described in two articles (Fleischer and Price, 1964b,d) which form the basis for the following discussion. The important requirement is that no features be counted that are not fission tracks, even (as we shall discuss in the section on *Repreparation and Counting* on p. 183) at the expense of ignoring a fraction of the true fission tracks.

Scanning procedure for glass.— The dating process depends on proper identification of fission fragment tracks and accurate counts. We remind the reader that, as shown in Fig. 4-5, the etched tracks seen in most glasses have a distinctive appearance. They are cones similar to those which may be formed in irradiated commercial soda-lime or borate glasses (Fleischer and Price, 1963). The sample shown

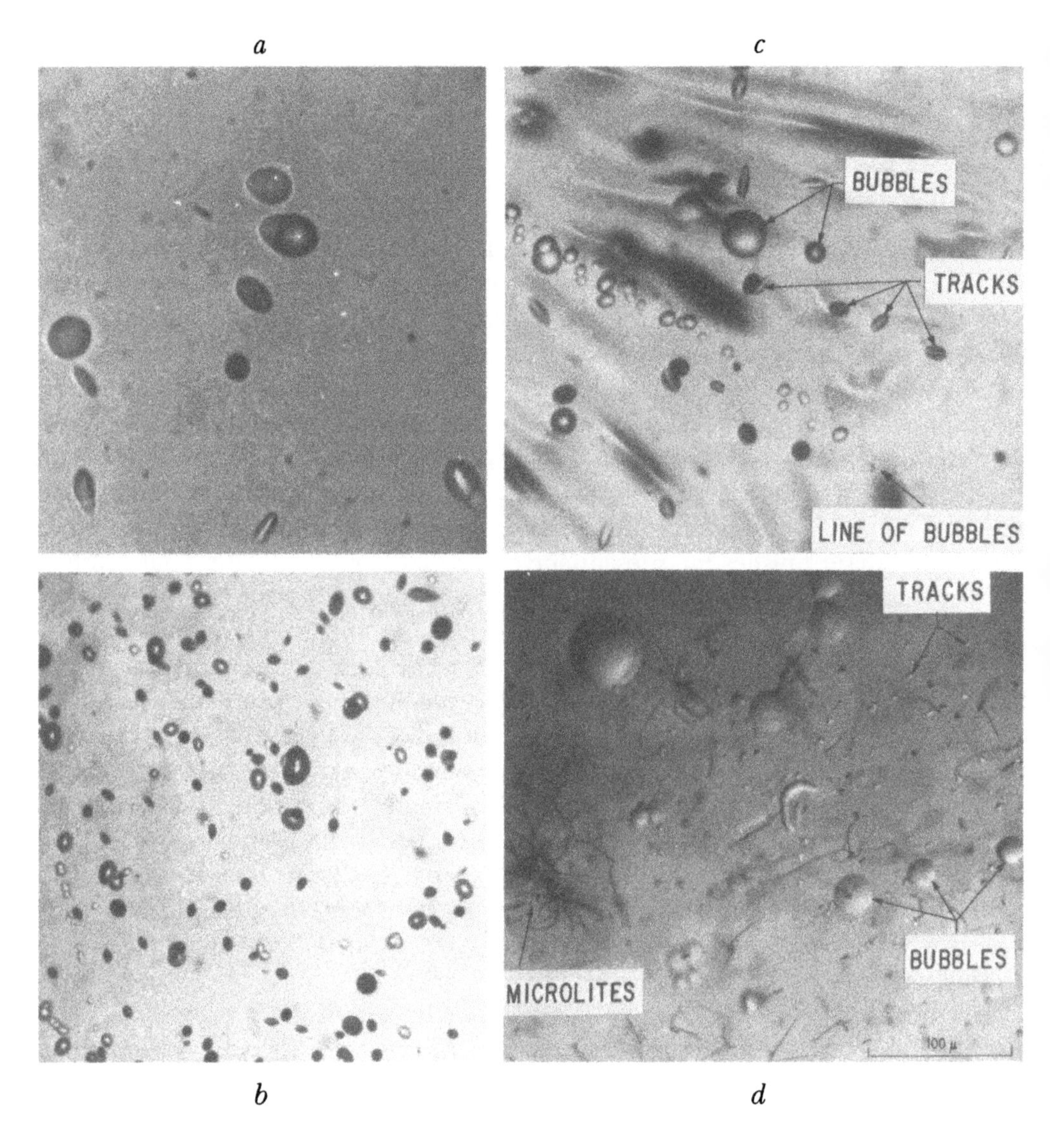

Fig. 4-5. *Tracks in natural glass. (a) Fossil fission tracks in a tektite. (b) Fission tracks in a tektite that was irradiated in a fluence of* 5×10^{15} *thermal neutrons/cm*^{2}. *From the ratio of the track densities an age can be computed. (c) and (d) Tracks and artifacts that have to be discriminated against when dating natural glass. (After Fleischer and Price, 1964b.)*

in Fig. 4-5a is etched to reveal fossil tracks; the one in Fig. 4-5b was neutron-irradiated and then fractured and etched. As long as the entire track has not been etched out, the cone remains sharp. But once the end of a track is reached, the general attack of the etchant causes a rounding of the pit bottom. Because such etched-out tracks are easily confused with etched pores, a counting procedure must be chosen which includes only sharp-bottomed pits. There are two possibilities:

(1) One procedure is to form only very tiny pits by a very brief etch. In this manner a negligible fraction of the tracks will have gone flat. This technique may suffer from one or both of two limitations. The worker may inadvertently count small, irrelevant imperfections or inclusions, as for example can be caused by imperfect polishing (Durrani and Hancock, 1970) or by microporosity. The other limitation is that very small pits are difficult to count rapidly when their density is low. For example, it would be inconvenient to scan young samples such as the Far Eastern tektites or ocean-bottom basaltic glass unless reasonably large, easily identifiable pits were present.

(2) A more convenient procedure, and the one most frequently used, is to form pits large enough to identify, but small enough that they do not overlap appreciably, and then to count only sharp-bottomed pits. Some workers prefer to use a plastic surface replica viewed in reflected light. For accuracy it is necessary to examine freshly created surfaces, rather than old, irregular external surfaces, where stray fission fragments from external sources will almost certainly be found in numbers that are different from those that occur well within the samples.

Examples are given in Fig. 4-5c and d of some of the stray defects that are sometimes present in glasses but can be distinguished from tracks that are still sharp-bottomed. When viewed in reflected light, bubbles, in particular, are distinctive in having bright reflections from their centers in contrast to the black character of head-on (circular) etched tracks. If microlites are short and straight, they can occasionally etch to a tracklike appearance, but are readily seen and identified using transmitted light. Other examples of defects in obsidian are given by Suzuki (1973a).

In counting only the pointed pits, some tracks will not be counted even though they crossed the original fracture surface that was removed by etching. On the other hand, new tracks will be reached as the fracture surface is dissolved away—the two effects partially balancing. Since the age depends only on the track ratios, it is only necessary that the same procedure be used for finding ρ_s and ρ_i.

Track-counting can conveniently be done at a magnification of 300X to 600X by viewing a sample systematically with a microscope having a rectangularly outlined field of view of known area. Since glass fracture surfaces, though smooth, are not plane, it is necessary with such surfaces to run the plane of focus up and down for each field of view to establish whether pits are present and are sharp-bottomed. Fig. 4-5a illustrates the differences between sharp- and flat-bottomed pits and the usual need for adjusting the plane of focus.

Ordinarily, at least 400 pits should be counted for each sample to obtain adequate statistics, although for some samples with low track densities this will not be practical. When a pit is found inside a second one, it should not be counted since it would not have been revealed without the first pit. Other pits that overlap are both counted if sharp.

Scanning procedure for crystals.— In order to count fission tracks in crystals, it is necessary to know which etch pits are to be excluded. The process of distinguishing pits due to spontaneous fission tracks from other pits is done on the basis of the known properties of tracks created by spontaneous fission:

(1) They form line defects.
(2) They are straight.
(3) They are randomly oriented.
(4) They are of limited length (typically on the order of 5-20 μm).
(5) They can be caused to disappear by suitable heating.

Property (1) allows elimination of other defects besides fission tracks and dislocations, while properties (2)–(5) may be used to distinguish between dislocations and fission tracks. Further discussion and examples are given by Fleischer and Price (1964d), Fleischer et al. (1964c), and Fleischer and Walker (1966).

Fission track pits which are acutely pointed or, in the limit, perfect cylinders are easier to identify than those which are shallow. Fortunately, most minerals develop

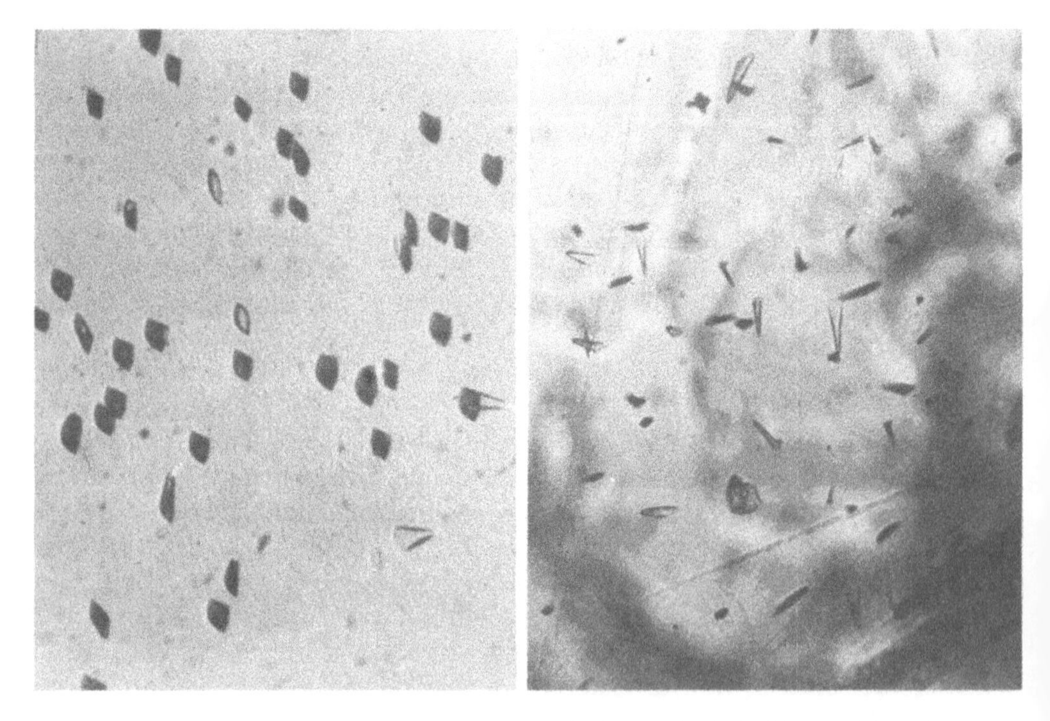

Fig. 4-6. *Fission track etch pits in a hornblende from Froland, Norway. Most of the fission tracks were produced by fission of ^{235}U impurities induced by $\sim 10^{17}$ thermal neutrons/cm^2. Left: Reflected light shows the spikelike character of only the few pits lying nearly parallel to the crystal surface. Right: Recognition of the tracks is considerably easier when a thin section is prepared and transmitted light is used. (After Fleischer and Price, 1964d.)*

such pits. We discuss these first. If the mineral is nearly opaque, the details of pit shape, length, and direction can be revealed only by viewing a thin section in transmitted light. For example, compare the two sides of Fig. 4-6 showing tracks in hornblende as viewed in reflected (left) and transmitted light (right). Since sharply tapered pits automatically satisfy property (1), they must correspond to either tracks or dislocations. Etch pits which are curved—property (2)—or which are regularly arrayed or lined up—property (3)—or which are much longer than $\sim 20\ \mu m$—property (4)—can be eliminated from consideration. Fig. 4-7 shows fission tracks in a variety of crystals and Fig. 4-8 illustrates a variety of defects that are not tracks.

In some crystals the true lengths of etched tracks are not evident from a single etch. Often channels along dislocations will lengthen with continued chemical attack. In contrast, individual fission track channels will only lengthen to $\sim 20\ \mu m$ at most. During further etching, they will widen, the ends becoming blunted in the process.

When the uranium is nonuniformly distributed in a crystal, the resulting fission tracks will have a unique appearance that is distinct from that of possible dislocation configurations. Fig. 4-9 shows examples of fission tracks emerging from small, uranium-rich particles within a tourmaline crystal. Photographs of fission tracks produced by irregular uranium distributions in micas have been given by Price and Walker (1963). Such concentrations of uranium should allow use of the previously discussed method for dating inclusions.

Under favorable conditions a treatment at elevated temperature will erase fission tracks but will not remove dislocations. In such a case the difference between the local pit count before and after heating gives the density of fission tracks. Sippel and Glover (1964) have used this procedure to count fission tracks in calcite. In crystals such as orthoclase, where we have seen etch pits arrayed in regular patterns at dislocations generated by plastic deformation, it is likely that dislocation rearrangement and mutual annihilation during heating will change the fraction of the etch pit count that is due to dislocations. In this case a reliable fission track count cannot be made by the annealing procedure. More often in refractory minerals, however, most of the dislocations were formed during crystallization and have been firmly anchored in place by impurities either in solution or present as precipitate particles strung along the dislocations (see Fig. 4-8, lower right). Finally, the simplest condition for counting fission tracks occurs when they are vastly more numerous than other defects, as for example, in the apatite crystal in Fig. 4-7.

Irradiation.— As noted earlier, to obtain the ratio ρ_s/ρ_i, fission of uranium is induced by irradiation with a known dose, Φ, of thermal neutrons. In the simple procedure outlined a standard particle track dosimeter is included with each irradiation, the subsequent counting of etched tracks being a measure of Φ. The

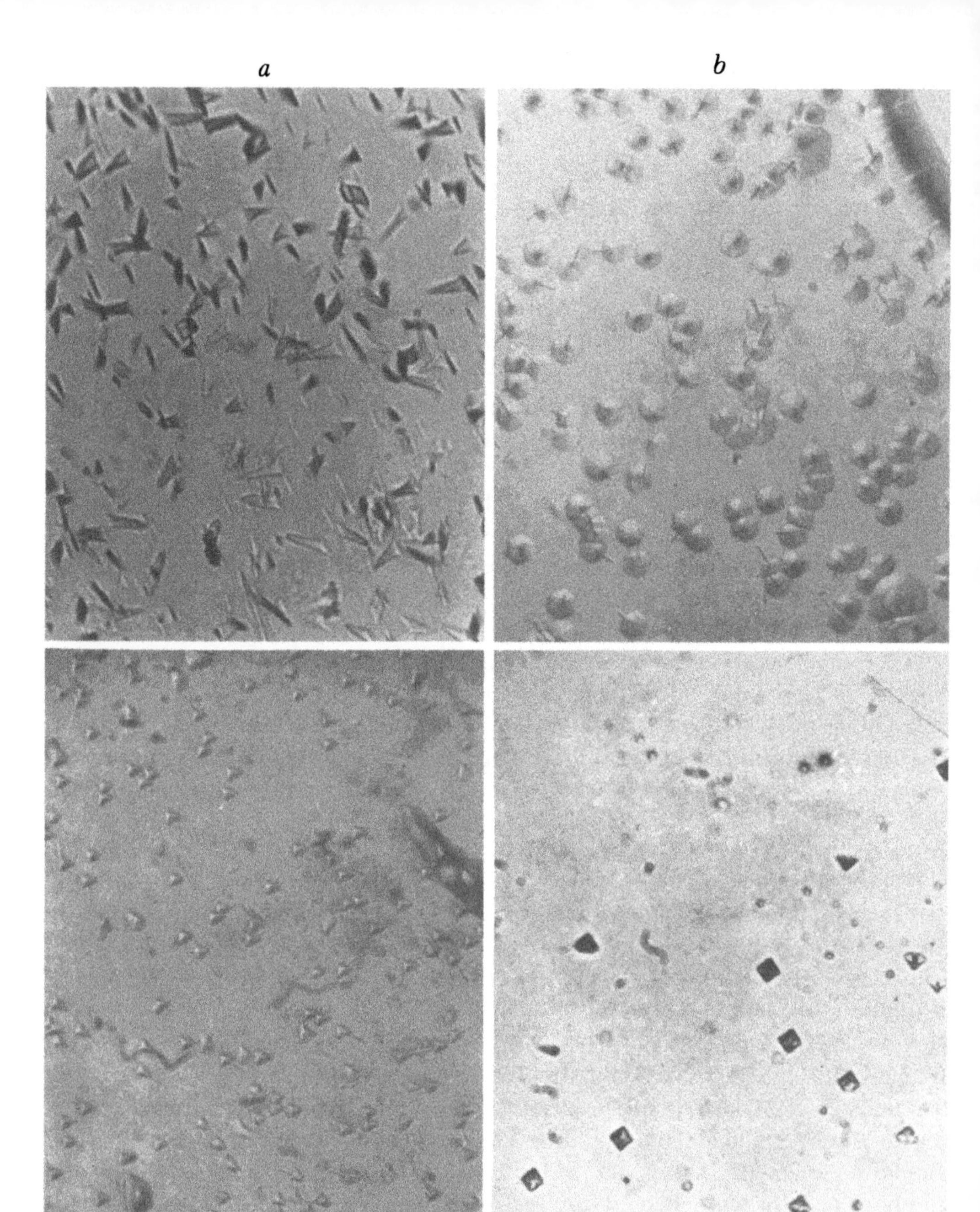

Fig. 4-7. *Etched fission tracks of undesirable shapes* (*a*) *on the* ($10\bar{1}0$) *plane of apatite*, (*b*) *on the* (*0001*) *plane of apatite*, (*c*) *on the* (*0001*) *plane of quartz, and* (*d*) *in zircon. By choosing the optimal etchants listed in Table 2-2, nearly hairlike tracks similar to those in the mica in Fig. 4-1 can be produced in these and most other minerals.* (*After Fleischer and Price, 1964d.*)

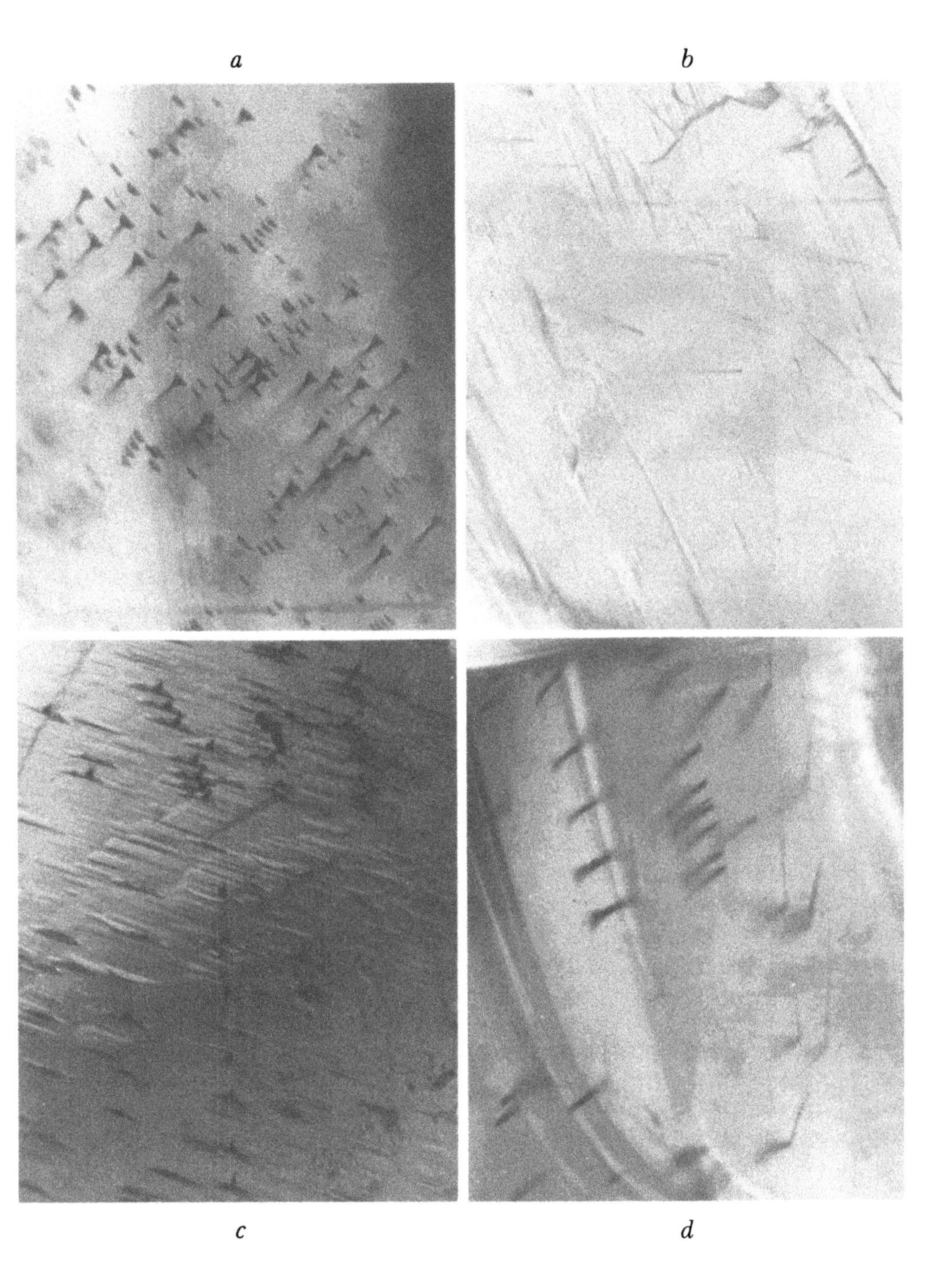

Fig. 4-8. *Examples of linear features that are not tracks in etched crystals. Dislocations may be distinguished by having preferred directions (as in (a) and (b)), by being arranged in arrays (as in (a) and (c)), by being longer than fission tracks (as in (a), (b), and (d)), or by not being straight (as in (d)). (a) shows etch pits in beryl etched 35 h in KOH at 165°C to produce two types of pits, neither of which is due to fission tracks. Those with channels attached are aligned and hence probably represent grown-in dislocations. The others, which are present in imperfectly aligned rows, are*

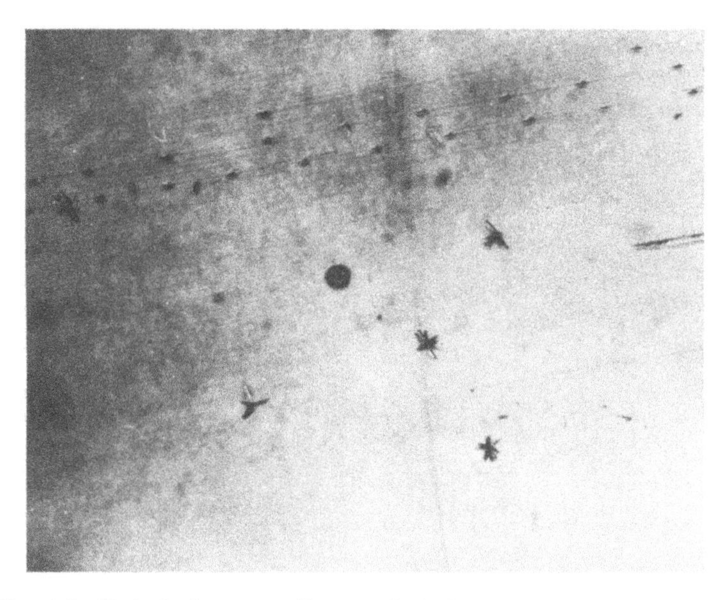

Fig. 4-9. *Etch pits in a tourmaline crystal which was irradiated with 1.2 × 10^{17} thermal neutrons/cm² and etched 20 min in KOH at 220°C. There are five clusters of fission tracks emanating from small, uranium-rich particles inside the tourmaline. Lined-up features in a double row are probably small-angle grain boundaries. (After Fleischer and Price, 1964d.)*

specific glass dosimeter which we use is either a soda-lime General Electric-Fisher glass MSA described by Fleischer et al. (1965c), or the Corning glass (Schreurs et al., 1971), or the National Bureau of Standards glass (certificates 610 to 617). See Table 4-1. Tracks can be counted directly in the glass dosimeter under standardized etching conditions (5 sec in 48% HF at 23°C for the General Electric-Fisher glass)

---

probably other dislocations which have been positioned by plastic deformation subsequent to crystal growth. (b) shows etch channels close to the basal plane of a topaz crystal which was etched for 6 h in KOH at 150°C. Because of their length of 100 μm or more and their alignment, these channels are almost surely dislocations, which tend to line up along the crystal growth direction. (c) shows etch pits in a barite crystal which was irradiated with fission fragments from ^{252}Cf and etched 7 h in HNO_3 at 100°C. The regularly-arrayed pits with connecting channels are almost certainly dislocations in small-angle boundaries. Most of the pits without connecting channels mark the ends of fission tracks. (d) shows channels in topaz which has been etched 6 h in KOH at 150°C. The channels are long and not straight, and hence are not fission track pits. The alignment and occasional discontinuous appearance suggest that these are dislocations onto which a second phase has been precipitated. The dotted regions are probably precipitates and therefore are not a result of the etching treatment. (After Fleischer and Price, 1964d.)

Table 4-1
Uranium Content of Standard Glasses
for Dosimetry and Uranium Measurements

Glass	Uranium Concentration (10^{-6}g/g)	U^{235}/U^{238}	References
General Electric-Fisher	0.35 ± .02		Fleischer et al. (1965c)
General Electric 1497	10,000 ± 300		Fleischer et al. (1965c)
Corning U-1	40.9 ± 0.7	.00360	Schreurs et al. (1971)
Corning U-2	43 ± 1	.00360	Schreurs et al. (1971)
Corning U-3	52 ± 1†	.00715	Schreurs et al. (1971)
Corning U-4	142 ± 7	.00715	Schreurs et al. (1971)
Corning U-5	4 (±1) x 10^{-3}		Schreurs et al. (1971)
Corning U-6	500 ± 25	.00435	Schreurs et al. (1971)
Corning U-7	675 ± 35	.00435	Schreurs et al. (1971)
Corning T-1	0.0457 ± .00003 (41 ± 1 Thorium)*	.00528	Schreurs et al. (1971)
NBS 610‡, 611	461.5 ± 1.1 (455 ± 3.6 Thorium)	.002376	NBS Certificate of Analysis; Carpenter (1972)
NBS 612‡, 613	37.38 ± .08 (37.55 ± 0.09 Thorium)	.002392	NBS Certificate of Analysis; Carpenter (1972)
NBS 614‡, 615	0.823 ± .002 (0.746 ± 0.007 Thorium)	.002792	NBS Certificate of Analysis; Carpenter (1972)
NBS 616‡, 617	0.0721 ± .0013 (0.025 ± 0.004 Thorium)	.00616	NBS Certificate of Analysis; Carpenter (1972)

* 30.94 ± 0.16 x 10^{-6}g/g by W.R. Shields, Isotope Dilution Mass Spectrometry, NBS; and 31.5 ± 0.9 x 10^{-6}g/g by B.S. Carpenter, Fission tracks using fast and thermal neutrons.

† 55.889 ± 0.0003 x 10^{-6}g/g by W.R. Shields and B.S. Carpenter.

‡ Portions irradiated with known doses of thermal neutrons are available as NBS Standard Reference Materials 961 through 964 respectively (Carpenter and Reimer, 1973).

or can be read from an adjacent detector such as Lexan (Lovering, 1968, personal communication) or mica (Seitz et al., 1973).

The dosimetry procedure using mica is to place an annealed piece of mica next to the standard glass and measure the induced track density in the mica. As described by Seitz et al. (1973), if, after etching, the mica is covered with a thin reflecting film of aluminum or silver and then studied by reflected light, essentially all the fission tracks can be counted. As a result differences between different observers are negligible. The constant relating track density in the mica to neutron dose is the same for different laboratories using the same standard glass. High-quality muscovite mica samples with ~5 ppb of uranium are available as grade V-1 from Essex International, IWI Division, Lynfield St., Peabody, Mass.

The external detector methods using mica or Lexan have the advantage of requiring less decision-making on whether an etched feature is a track, but have the disadvantage of a possible danger of uranium contamination at the dosimeter-detector interface, which would give spuriously high inferred fluences. The normal etching conditions for the General Electric-Fisher MSA dosimeter (5 sec with 48% HF at 23°C) should be followed for reproducible results, since the cone angle θ, and thus η, have been found to vary with both concentration of etchant (Hart et al., 1970) and temperature. For example, the etching efficiency of a similar soda-lime glass has been found to change from 0.53 to 0.43 as the etchant is changed from 10% HF to 48% HF. Scanning is most reproducibly done in air with reflected light at a standardized magnification, typically ~500 times.

The choice of dose is dictated by the usual need to have the final track density $\rho_s + \rho_i$ much greater than ρ_s so that a statistically useful value of $\rho_i = (\rho_s + \rho_i) - \rho_s$ can be deduced. If, for example, we desire ρ_i to be at least ten times ρ_s, the equations given earlier yield $\Phi > 10A\lambda c^{238}/c^{235}\,\sigma$. It follows that if A is 10^6 years, then Φ should be at least $\sim 1.5 \times 10^{14}$ neutrons/cm^2; for $A = 4.5 \times 10^9$ years (the approximate age of the earth), Φ must exceed $\sim 7 \times 10^{17}$ neutrons/cm^2. It is therefore useful in deciding the neutron dose to have some idea of the likely age of the sample. An alternative to making the subtraction $(\rho_s + \rho_i) - \rho_s$ is to heat the sample before neutron irradiation in order to remove the original track density ρ_s. The data in Table 2-4 are useful guides to choosing appropriate heating conditions.

It is important to use a predominantly thermal neutron source in order to induce fission of ^{235}U only. Fast neutrons cause fission of ^{238}U and ^{232}Th, so that, if irradiations that include fast neutrons are used, σc^{235} in the dating equation constant is replaced by $\sigma_{235}c^{235} + \sigma_{238}c^{238} + \sigma_{232}c^{232}$. In short, the constant would vary with the Th/U ratio and with the fast/thermal ratio in the reactor. The problem can be dealt with by doing cadmium difference experiments, as described by Price and Walker (1963), or by using a location within a reactor where the fissions induced by fast neutrons are negligible. Since the cross section σ_{235} for fission by thermal neutrons is several hundred times larger than those for fast fission of ^{238}U or ^{232}Th, thermal-to-fast ratios of greater than 20 to 1 normally will ensure that fast fission introduces errors in calculated ages that are less than the other errors—statistical and systematic. The exception occurs if thorium is more abundant relative to uranium than the usual ratio, which is about 3.6/1 to 4/1. Zircon, whitlockite, and apatite are known to have Th/U ratios that may deviate considerably from the usual value and therefore require special care in fission track dating.

As we have previously indicated, the preferred method of dating uses the "internal" detector, where an internal surface is first counted, and then a fresh internal surface is exposed by cleaving or polishing after irradiation. This method suffers the sole disadvantage that variations in uranium concentration over short distances may lead to incorrect ages. This can be checked by making replicate determinations on a number of samples.

In those cases where the external detector method, in which the *same* surface is counted before and after irradiation, is deemed necessary or desirable, the major problem encountered (in addition to the correction for surface dissolution previously discussed) is uranium contamination of the counted surface. In this method great care must be used to keep the surfaces clean. A laminar flow clean hood provides a highly desirable environment for sample preparation prior to irradiation. In practice it is difficult to avoid contamination below the 50 ppb level, though with great care samples with 5 ppb of uranium can be studied. Normal laboratory detergents such as Alkonox are satisfactory at this level as are reagent-grade acetone and alcohol. Decontaminant soaps normally used to clean radioactive spills are to be especially avoided. Even when fresh they generally contain unacceptable levels of uranium.

The neutron flux in reactors can have a surprisingly large spatial variation (up to ~10%/cm). For careful work it is therefore essential that the dosimeter glass be placed directly adjacent to the sample. In the case where several samples are being measured an appropriate number of separate dosimeters should be included and the entire package should be made as compact as possible.

Lexan or various other plastics can be used in place of mica if the external detector dosimeter is used; however, in this case great care must be taken to ensure that thermal fading of tracks does not occur in the reactor. Thermal effects in particular reactors may also cause problems with specific samples and we caution the reader to be aware of this possibility.

Repreparation and counting.— Because uranium concentrations may vary with position in natural samples, eq. (4-1) or (4-5) yields rigorously accurate ages only (1) if the induced track density ρ_i is produced with the same uranium distribution on both sides of the etched surface as existed in the natural sample and (2) if tracks are counted on the same surface as used for the natural track count. Because in all minerals except micas some material is removed from the exposed surface by the etchant, this condition seldom obtains. It is possible in general to etch a fresh surface that is within 25 μm of the original surface, and normally this is adequate to avoid wide variations in uranium content. Even in the micas where the same surface can be re-etched (Fleischer et al., 1964b) there is the continual danger that in the handling and etching prior to irradiation some uranium has been inadvertently introduced onto the surface of interest. In general the optimum surface is nearby, but one that was interior during irradiation. Merely repolishing a mounted sample sufficiently (>12 μm) to remove the previously etched natural tracks is adequate.

In re-etching and scanning after irradiation it is vital to (1) use the same etching procedure as before and (2) use the same criteria for identifying tracks. Fortunately the quantity that is needed is the ratio of ρ_s/ρ_i so that if, for example, 25% of genuine tracks are not counted in deciding ρ_s there is no problem as long as 25% are also not counted for ρ_i. Suppose that in a particular glass sample there are vesicles along a particular flow direction that, when etched, resemble tracks of that orientation. An appropriate procedure would be to count no tracks that are aligned within say 30° of the flow axis. As long as this same procedure is followed on the same surface for $\rho_i + \rho_s$ after irradiation, no error is introduced.

It is usual and informative to measure the uranium content explicitly by placing a detector of unit (or near unit) etching efficiency adjacent to each sample to be dated. Most plastics are suitable (list of etchants in Table 2-2) as is muscovite mica, provided the uranium content of the sample is much higher than that of the mica (which typically is $\sim 5 \times 10^{-9}$ g/g). Fragments from fission induced in the sample enter the plastic or mica and are counted there. A sample of known uranium content provides the needed reference value.

Table 4-1 lists uranium contents of several glass standards and where known, U^{235}/U^{238} ratios and thorium contents. Four of these standards are also available with known preirradiations. Chapter 8 describes in further detail the mapping of uranium distributions.

4.3.3. Environmental Effects

ETCHING EFFICIENCY: EFFECTS OF ANNEALING

A specimen which has been exposed to an elevated temperature or a gross mechanical shock can have a reduced etching efficiency (η^{238}) for fossil fission tracks. This loss or fading of tracks leads to a young apparent age of the specimen. Although such young ages often may be corrected to give the true age of formation of a given sample, they are frequently interesting in themselves since they give a measure of the thermal history of a specimen, difficult to obtain in any other way. In recognizing and assessing track fading, it is helpful to have an understanding of the way in which tracks are formed by etching. We shall develop this background after pointing out that the primary track fading process of importance in archaeological and geological dating is thermal in origin; natural effects on fission tracks from mechanical deformation or hydrostatic pressure have yet to be detected. Shock effects are unimportant except for specimens obtained near impact craters or nuclear explosion sites. (Ahrens et al., 1970; Fleischer et al., 1972a; 1974).

In this section we calculate the etching efficiency as a function of the etch pit diameter for samples (such as the glasses) that etch isotropically. This relation (Hart et al., 1970, unpublished) is the physical basis of the empirical track fading correction devised by Storzer and Wagner (1969). Finally, experimental laboratory annealing results and examples of their use are given.

Track fading.— The possibility always exists that track fading has occurred. Such fading can be caused by a high-temperature event of short duration or by a heating of extended duration at lower temperatures. It is important to be able to estimate the track fading which may occur over archaeological or geological times. As a consequence, laboratory track annealing data have been obtained for many substances. We emphasize that annealing can be a complicated process (as first noted by Maurette et al., 1964) and is often therefore not describable by a single activation energy or by simple annealing kinetics. In cases where extensive studies have been made, results similar to those of Fig. 4-10 (Naeser and Faul, 1969) are often obtained. Here we see that track annealing, measured by the fraction of tracks that are lost as a function of time and temperature, yields so-called Arrhenius plots in which a given percentage loss of tracks has a characteristic slope or activa-

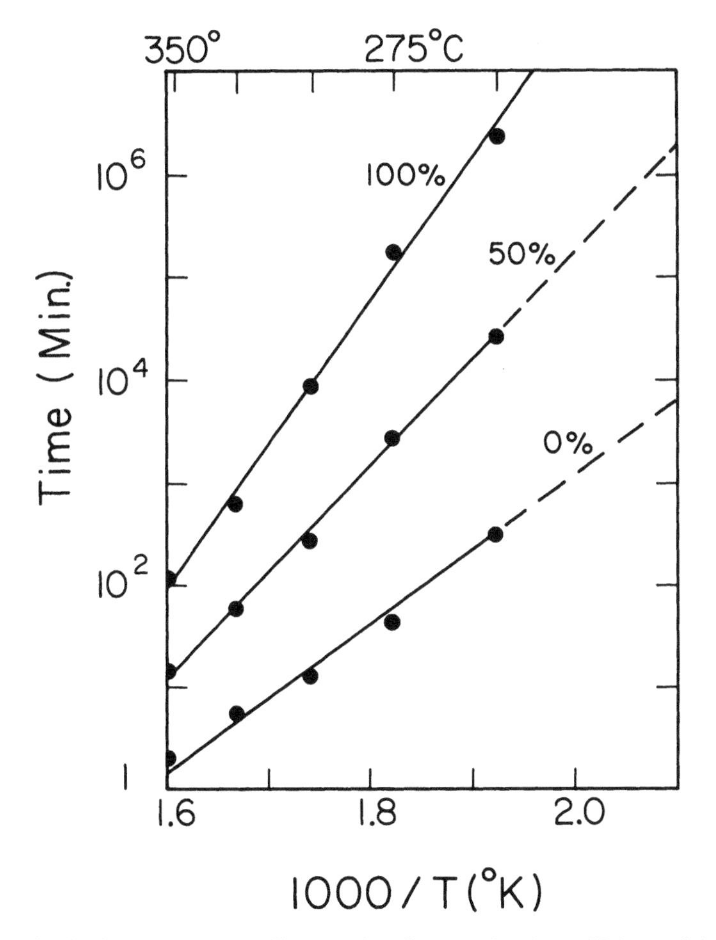

Fig. 4-10. *Laboratory annealing results for apatite from Eldora, Colorado amphibolite. (After Naeser and Faul, 1969.) Points with a given percent loss in track density lie along a straight line.*

tion energy, U. Thus, over the range of the laboratory experiments, the effects of annealing time τ_a and temperature T can be expressed accurately for a given track density reduction by the relation $\tau_a = a \exp(U/kT)$, where k is Boltzmann's constant and a is another constant.

Table 2-4 presents in tabular form the results of laboratory track annealing measurements for several minerals, glasses, and plastics. As an example of the range in sensitivity of minerals to track fading, we note that annealing temperatures that produce complete loss of tracks in one hour are 60°C for autunite and 1050°C for quartz. It should be noted that the use of different etchants can give rise to different apparent track fading rates (Calk and Naeser, 1973, Price et al., 1973), so that from a single mineral diverse information can be extracted.

Our best estimate of track annealing for geological or archaeological times is the extrapolation of the straight lines of the laboratory Arrhenius plots, i.e., the use of the above equation assuming a and U to be independent of T. As we shall see in discussing Fig. 4-11, the extrapolation is indeed extreme, data taken over four orders of magnitude being extrapolated over an additional eleven. This assumption that fading rate depends on the Boltzmann factor appears to be justified as it is consistent with the retention or loss of tracks observed in a number of systems of geological interest. Several such examples of the use of extrapolated laboratory annealing curves in fission track dating follow.

A material that laboratory experiments show to be very susceptible to track fading may well give track ages which are too low, as in a study of impact glasses from the Manicouagan and Clearwater Lakes craters (Fleischer et al., 1969d). Two Manicouagan glass samples were dated. One gave a track age of 208 ± 25 My, in agreement with a K-Ar age of 225 (±30) My. The other gave a track age of 36 ± 3.5 My. Laboratory annealing data for the two glasses indicated striking differences; in one glass (36 My) the tracks would fade in 30–40 My at 60°C while in the other glass (208 My), several thousand million years would be needed for fading at 60°C. The Clearwater sample yielded an apparent track age of 33 My, much younger than the K-Ar ages of 285–300 My. Laboratory annealing studies indicated that track fading would be detectable after a few hundred years at ambient temperatures and that 70–80% of the tracks would be gone after 300 My. We see that the fading of fossil tracks in these glasses is roughly consistent with the extrapolated laboratory annealing experiments.

In some cases the environment of the sample can assure us that even a temperature-sensitive material is yielding accurate ages. In the study of ocean-bottom spreading Fleischer et al. (1968b, 1971b) and Aumento (1969) measured the fission track ages of glass veneers taken from exposed pillow lavas. Even though tracks in the glass are inferred to begin fading at relatively low temperature (~1 My at 80°C as shown in Fig. 4-11a), the knowledge that the temperature of the glass had been very close to that of the water at the ocean bottom (~3°C) gives us confidence in the accuracy of the fission track age.

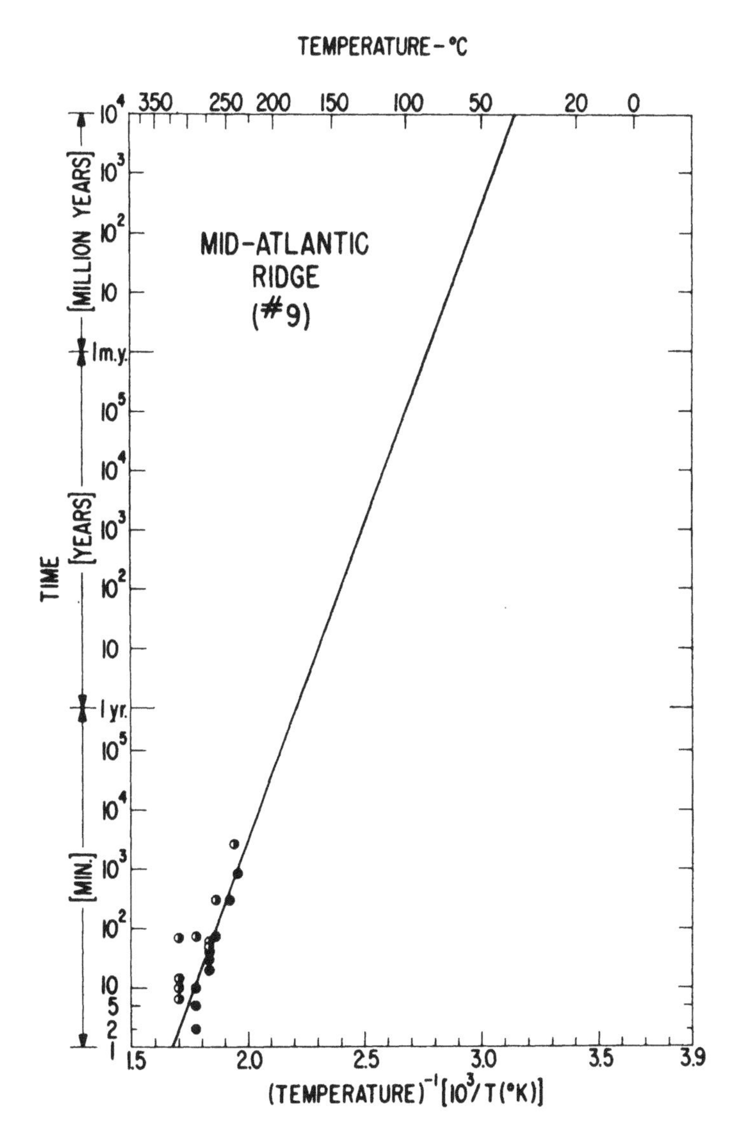

Fig. 4-11. *Track-fading conditions for a Mid-Atlantic ocean-bottom basaltic glass sample from 7.6 km west of the Median Valley. Solid dots indicate time and temperature conditions for which tracks are unaffected. Open dots indicate conditions when tracks are reduced in number or are absent. (After Fleischer et al., 1971b.)*

If several materials of differing fading sensitivities are found together, the dates for the different materials can tell much about not only the original cooling age of the assembly but also its cooling rate and subsequent thermal history. Naeser and Dodge (1969) found the *same* age, 87 ± 3 My, in four minerals—apatite, sphene, epidote, and allanite—all from the same sample of Lamarck granodiorite. In such a case we can be sure that the track age is accurate and that the rock cooled from 600°C to below 100°C in 10 My or less. In rocks from other regions, the ages measured in the different minerals do not agree. In all these cases the minerals that are more sensitive to annealing yield the younger ages; either there has been a heating episode following the initial cooling or the rock cooled slowly after formation or during long periods of orogenic uplift.

Lakatos and Miller (1973) obtained clear evidence that fossil fission tracks in muscovite are sometimes altered over time. Their measurements of track densities as a function of etching time showed that fossil tracks in selected (low fission track age) micas were revealed more slowly than are fresh tracks induced by neutron irradiation (Fig. 4-12). These particular micas were chosen for study because they had given fission track ages that were lower than the ages inferred by other radiometric techniques. It is clear that some environmental effect—most probably

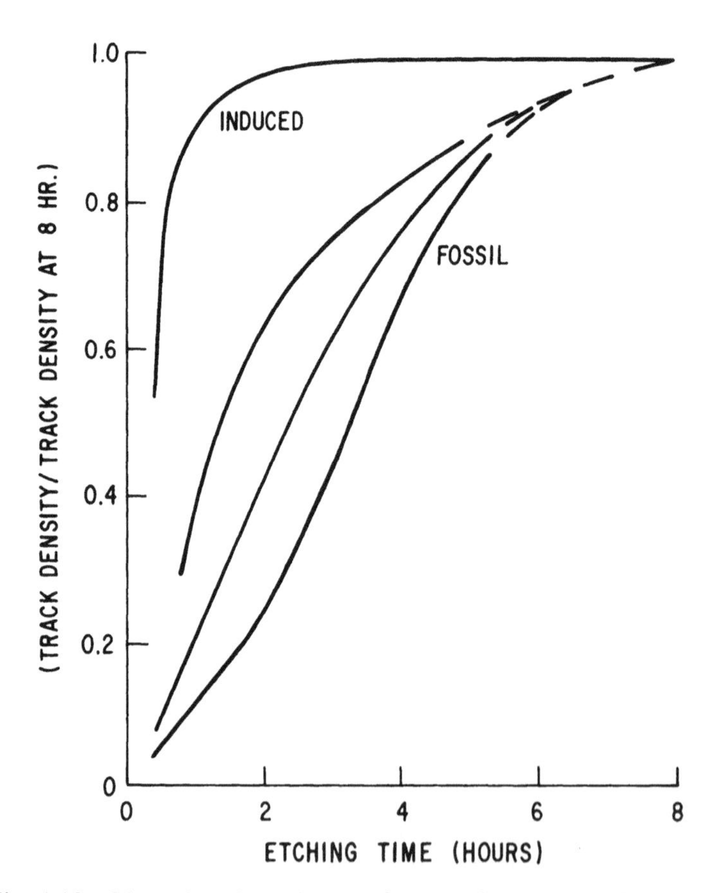

Fig. 4-12. *Observed track density as a function of etching time for natural and freshly induced tracks in muscovite micas that gave unexpectedly low fission track ages. The effect of natural heating or deformation is to give low ages for normal etching times. (After Lakatos and Miller, 1973.)*

low-grade heating, but possibly deformation—has so altered tracks that etching for the conventional etching time would reveal only a fraction of the natural track density and would therefore give low ages. The observation suggests that a measurement of ρ_s as a function of etching time provides a useful procedure for recognizing when there will be special problems in determining fission track ages of micas.

These results emphasize that one of the major contributions of fission track dating arises from the temperature-sensitive nature of the track damage, which allows real thermal events in the past to be recognized. The possibility of dating thermal events that could occur long after the original solidification of a substance was realized once the temperature sensitivity was measured (Fleischer and Price, 1964d; Shukolyukov and Komarov, 1966) and has been used to date contact zones (Naeser, 1967a; Calk and Naeser, 1973) and measure effects of orogenic uplift (Storzer, 1970; Sun, 1971; Burchart, 1972; Wagner and Reimer, 1972) in addition to the examples already cited.

Correction for thermally affected tracks by etch pit size measurements.— We showed in Chapter 2, on the basis of an etching geometry model (Hart et al., 1970, unpublished), that track fading or annealing in a glass yields undersized etch pits, an effect first recognized experimentally by Berzina et al. (1966). Storzer and Wagner (1969), Gentner et al. (1969a), Durrani and Khan (1970), Storzer (1970), and Suzuki (1973a) have used etch pit dimensions as the basis of a correction for ages determined with partially annealed tracks. Maurette et al. (1964) suggested that alterations in lengths of tracks in minerals might be used to correct ages, a suggestion that Wagner and Storzer (1970) have recently used to construct a simple correction procedure. Unfortunately, it failed on the one mineral to which it has been applied.

In their procedure for glass the etch pit size distributions are determined for both fossil and reactor-induced fission tracks. Only if the two distributions are indistinguishable is fading insignificant and the directly calculated track age accurate. The track age then gives the last time the sample was substantially heated. This may, of course, be more recent than the age of formation itself. If the average size of the fossil tracks is smaller, then information concerning the time-temperature history may be obtained. In particular, if the fading has proceeded at a constant rate, the data may be corrected to give the true age of formation as well as to infer the average temperatures of the sample during its lifetime. Episodic heating (leading to a bimodal track population of partially faded and nonfaded tracks) can also be detected and measured. Of course, if the thermal history is very complicated these various effects are difficult to unravel. In the case of partial fading at a constant temperature the age may be corrected using a correlation curve relating etch pit size to track density as determined in laboratory annealing experiments. A laboratory annealing correlation curve for an australite glass sample (Storzer and

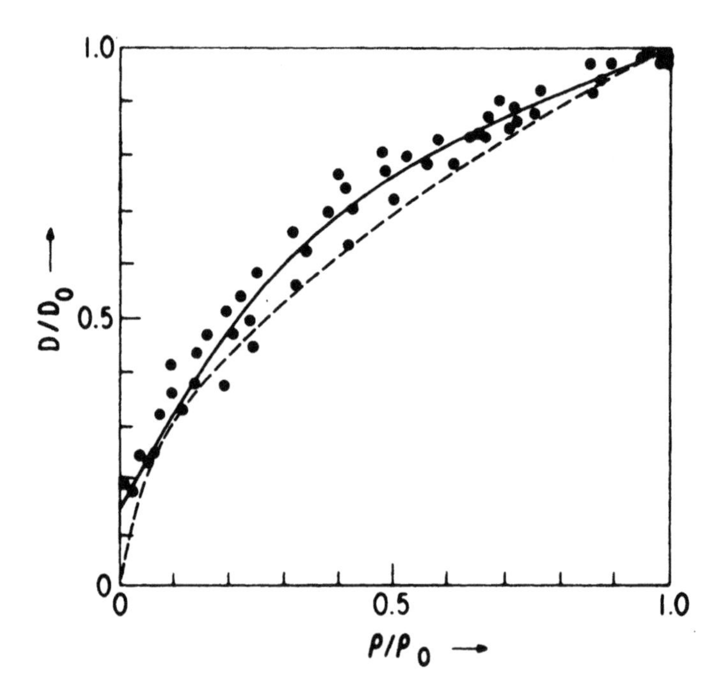

Fig. 4-13. *Etch pit diameter vs. track density for an australite as determined in laboratory annealing experiments (Storzer and Wagner, 1969). The mean track diameter in an annealed sample is plotted against the track density for that sample; the data are normalized to the mean diameter and track density for an unannealed sample. The solid curve is smoothed through the points. The dashed curve was calculated using parameters measured on a synthetic tektite glass furnished by D. R. Chapman. (After Fleischer and Hart, 1972a.)*

Wagner, 1969) is shown as the solid curve in Fig. 4-13. (We shall shortly discuss the dashed curve.) Here the average diameter (in practice the major dimension of the ellipse) measured after annealing is plotted against the track density after annealing, in both cases normalized to the unannealed average diameter and track density. This plot then relates the reduced etching efficiency to the reduced etch pit size. While the direct application of the correction curve using the average etch pit size is valid only for a single relatively recent event, a careful analysis of the etch pit size distribution (Storzer, 1970) can sometimes define the date of a relatively old thermal event as well as the date of the original cooling. In principle, the analysis of an etch pit size distribution can also yield information about lengthy, relatively low-temperature periods in the history of the sample.

The application of this type of analysis, though rich in information, requires extensive laboratory heating experiments. For this reason an etching geometry model (Fleischer et al., 1970, unpublished; summarized by Fleischer and Hart,

1972) was applied, along with further annealing studies, to the analysis and understanding of the correction curve of etch pit size as a function of track density. The findings were: (1) The major reduction of both the etch pit size and the track density occurs through a decrease in v_T, i.e., an increase in θ. Only a small part of the loss is caused by the decrease in the etchable range R. (2) The shape of the correction curve is determined rather well by the unannealed value of the cone angle θ, a parameter rather easily measured (Fleischer et al., 1969g). Subsequently, Somogyi and Nagy (1972) have done a somewhat similar calculation in which they neglected the variation of v_T along the tracks and assumed that all tracks are counted rather than only those which are sharp. Although the first assumption is known not to be true, their calculation is in reasonable accord with experiment.

The dashed curve in Fig. 4-12 was calculated using parameters measured on a synthetic tektite glass supplied by D. R. Chapman (Fleischer et al., 1970, unpublished), having a composition roughly similar to the australite for which Storzer and Wagner (1969) determined the empirical correlation shown by the data points. The calculated curve leads to correction factors within 5% of those obtained from the empirical curve. Future work along these lines may simplify the correction of fission track ages for partial track fading. Somogyi and Nagy (1972) have discussed this problem.

The literature on track fading corrections has been applied primarily to glasses; it has been noted (Fleischer and Hart, 1972) that a similar age correction should be possible for minerals by making use of track length reduction, an indicator that was first suggested by Maurette et al. (1964). For most minerals, in contrast with most glasses, the cone angle is small and the etch pit diameter is insensitive to partial annealing. The pertinent indicator of annealing is the track etching rate, which for short etching times can be measured by the length of the track; the distribution of track lengths is thus a record of thermal events subsequent to the initial cooling. Length reductions caused by heating were measured by Fleischer et al. (1964b) and Berzina et al. (1967) in mica and by Bhandari et al. (1971) in apatite, and Bigazzi (1967) has observed that older micas tend to have shorter tracks. The specific form of the curve of l/l_0 vs. ρ/ρ_0 shown in Fig. 4-14 is qualitatively what the model of Hart et al. would predict for a material of an initially small cone angle. Mehta and Rama (1969) were the first workers to indicate how such age corrections could be made on crystals. They applied their technique to numerous muscovites (Gupta et al., 1971a,b; Mehta and Nagpaul, 1971). Wagner and Storzer (1970, 1972) and Storzer and Wagner (1973) have determined a curve of reduced length vs. reduced track density from their annealing results on apatite (Fig. 4-14). The age correction is analogous to that shown in Fig. 4-13. The procedure has been further discussed by Wagner (1972) and later by Naeser and Fleischer (1975), who note that it has failed with the one mineral to which it has been applied.

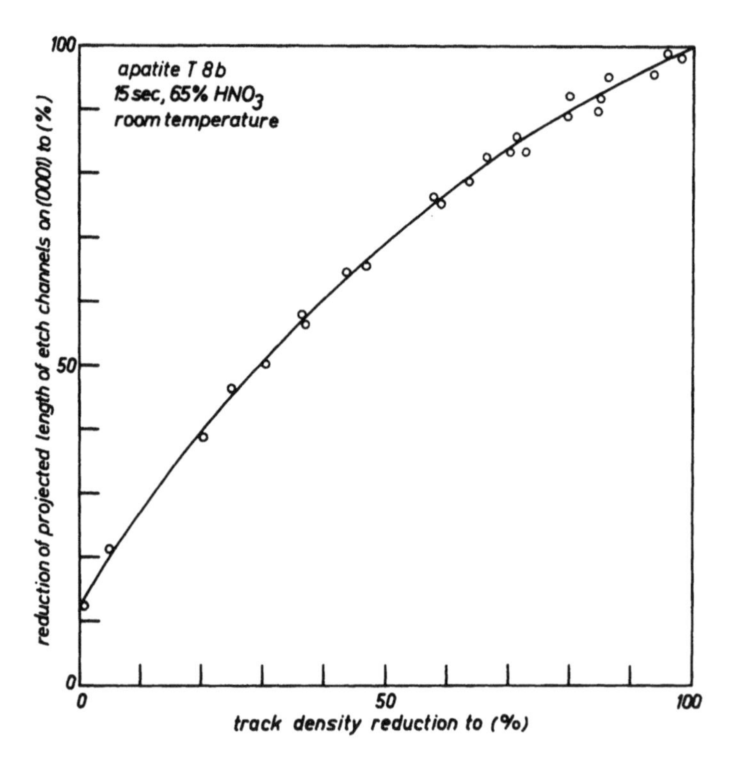

Fig. 4-14. *Track-fading data for age correction in apatite. After different heatings the fractional track length retention is related to the relative track density. Etching for 15 sec in 65% HNO_3 on (0001) at 23°C. (After Wagner and Storzer, 1970.) Corrections using such data have not been successful.*

Correction for thermally affected tracks by the age plateau method.— A second, more recently devised technique for recognizing thermal effects on tracks and correcting fission track ages has been outlined by Storzer et al. (1973b), Storzer (1973), and Burchart et al. (1973). Storzer et al. anneal separate pairs of samples at each of a series of increasing temperatures. One sample contains only fossil tracks, which have been thermally affected, and the other contains only fresh, neutron-induced tracks. As is shown in Fig. 4-15, in both samples the track density decreases with increasing temperature, but in the sample containing only fresh tracks the rate of decrease is at first faster than in the sample containing fossil tracks. For the pair of samples a fission track age is calculated for each temperature. The apparent age increases at low temperatures but reaches a plateau at temperatures above that corresponding to a previous heating event or at temperatures at which tracks fade naturally. This procedure, which is analogous to that conventionally employed in $^{40}Ar-^{39}Ar$ dating (Merrihue and Turner, 1966), is claimed to give concordant results with ^{40}Ar dating to a quoted precision of $\pm3\%$. The method exploits the fact that the resistance to thermal fading is an increasing function of radiation

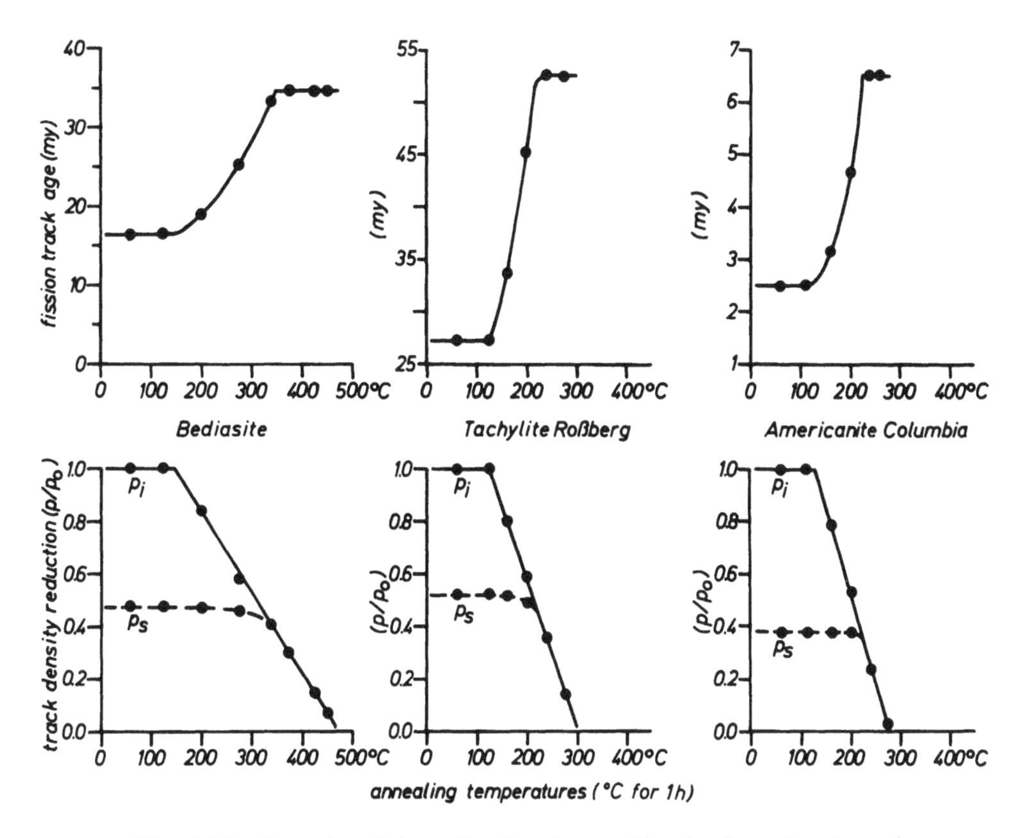

Fig. 4-15. *Correction of thermally affected ages. After heating pairs of samples with natural and induced tracks to successively increasing temperatures, an age can be computed at each temperature. In the bottom row are curves showing fraction of spontaneous* (ρ_s) *and induced* (ρ_i) *tracks retained as a function of annealing temperature for glasses from three locations; in the top row are the corresponding "age plateaus."* (*Courtesy of D. Storzer.*) *This procedure has been found to give meaningful corrections for the glasses to which it has been applied but not for the one crystalline mineral that has been examined carefully.*

damage density. As a fission fragment slows down it captures electrons and produces less radiation damage. The ends of tracks are more readily annealed than the high-energy portions. Thus, partly annealed tracks are shorter than fresh tracks and a smaller fraction of them are revealed at a given surface. Curves like those in Fig. 4-10 show implicitly that it is harder to erase the more heavily damaged ends of tracks than the more lightly damaged ends. In the plateau method the sequence of increasing temperatures simply reproduces in the induced fission tracks the same amount of thermal relaxation experienced by the fossil tracks.

Although this method has been applied with good success to a number of glasses, a discrepancy has appeared in the dating of the one crystalline mineral to which it has been applied. For Durango apatite the "corrected" age of 39 My found by Storzer is nearly 25% higher than a well-documented true age of 31 My. Recent work by Naeser and Fleischer (1975) gave obviously incorrect, "corrected" ages up to 2,400 My. Such results would arise if tracks are stabilized or intensified rather than partially annealed by long-time storage at low ambient temperatures. Such effects have been observed in epidote by Naeser et al. (1970) and in polycarbonate by Peterson (1969), as noted by Price and Fleischer (1970).

Erasure of Tracks by Plastic Deformation: Dating Mechanical Events

Mechanical deformation of crystals can cut particle tracks (Fleischer et al., 1965a, 1969f) and fragment them into such short, disconnected subsegments that they are no longer optically visible (Fleischer et al., 1972a, 1974; Fleischer and Hart, 1973). The "erasure" of tracks in this way effectively resets the fission track clock and allows the time interval after deformation to be measured by the track density produced by the subsequent accumulation of fresh tracks.

Deformation erasure of tracks by shock has been studied in terrestrial minerals exposed to laboratory shock conditions (Ahrens et al., 1970) and in natural rocks exposed to shock conditions in nuclear explosions (Fleischer et al., 1974, and Fig. 4-16). Fragmented, shortened tracks such as those shown in Fig. 4-17 have been identified in lunar minerals that also display extensive other shock features (Fleischer et al., 1972a; Fleischer and Hart, 1973).

By using electron microscope replicas of etched surfaces of deformed crystals the density of short tracks may often be measured. When this is possible, the number of tracks is a measure of the time interval during which tracks accumulated prior to deformation. If thermal effects are absent, the cutting of tracks does not alter the number per unit area, since the increase in the number of tracks per unit

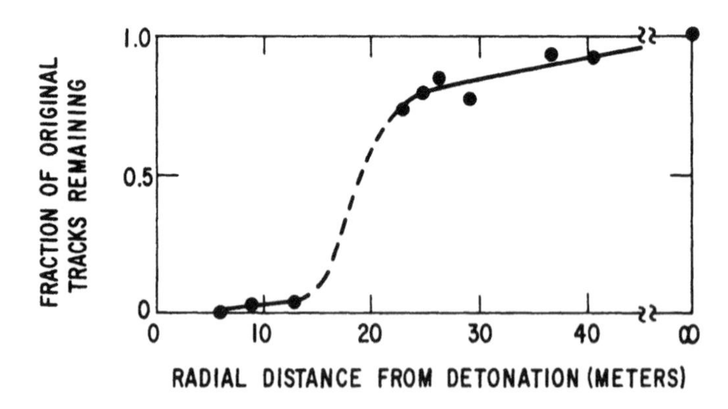

Fig. 4-16. *Track erasure in apatite from rocks that were shocked in the Hardhat nuclear explosion. Tracks are erased in samples taken close to the detonation point. Data of Fleischer et al. (1974).*

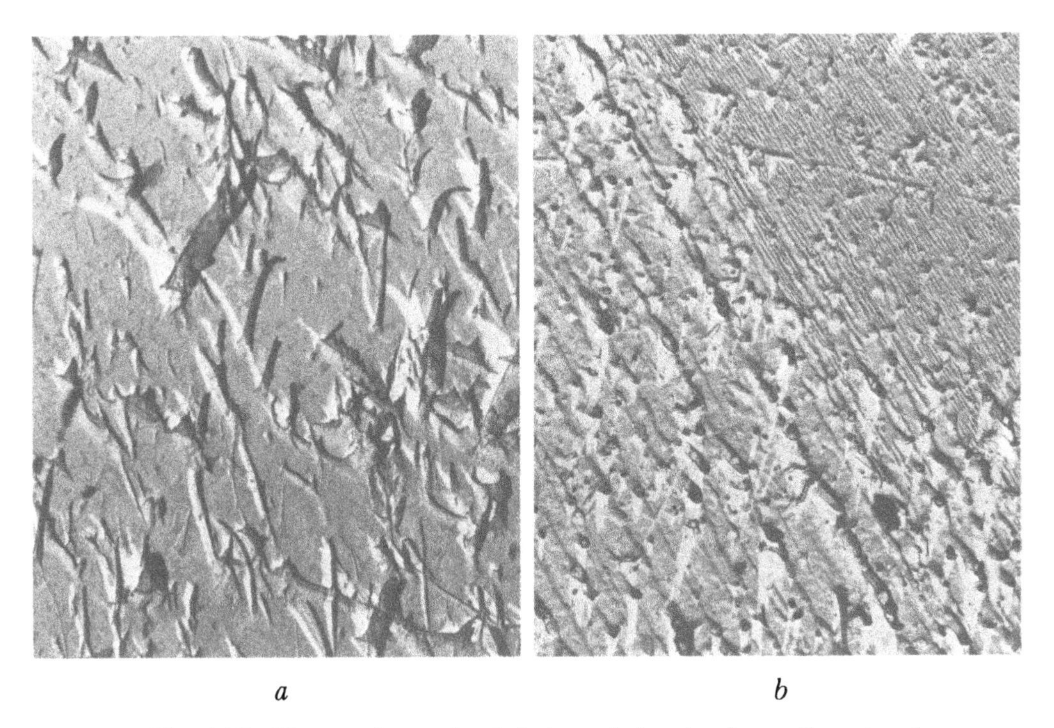

Fig. 4-17. *Electron micrographs of platinum-shadowed carbon replicas of tracks in lunar pyroxenes illustrate shortening of tracks by mechanical deformation.* (*a*) *Cosmic-ray tracks in an undeformed pyroxene from lunar rock 12028;* (*b*) *portion of a crystal 600 μm in diameter from the same rock with regions of fine slip* (*slip spacing ~0.07 μm*) *and coarser slip* (*spacing ~0.7 μm*). *The tracks are shortened to a length that is approximately the spacing of the slip lines. Each photo is approximately 30 μm high.* (*After Fleischer et al., 1972a.*)

volume is exactly balanced by a decrease in the average length. However, as a practical matter the efficiency of observing very short tracks is normally lower and therefore the density of short tracks would give a lower limit for the time prior to deformation.

So far, naturally deformed crystals in which tracks have been erased have been identified only in lunar samples where cosmic ray tracks predominate, and they have been useful in deriving soil histories (Fleischer and Hart, 1973) as described in Chapter 6. As yet, terrestrial samples have not been searched for deformation effects on tracks, so that the dating of times of mechanical deformation by the fission track method remains an interesting area for future research.

Possibility of Uranium Exchange

The success of the fission track dating method depends on the assumption that the uranium concentration of the sample has not changed in the time interval during which the tracks are stored. Although variations in the isotopic composition of other elements are sometimes found in natural samples, indicating ion exchange, most materials that have been measured do not appear to have lost uranium to the environment. For an exception, see Grauert and Seitz (1973). If they had, the apparent fission track ages would be larger than the formation age of the samples. In spite of extensive searches for anomalously high fission track ages by Price et al. (1970) and Haack (1973a) in attempts to find superheavy elements (see Chapter 7), none have been found, which supports the view that uranium is seldom lost over time. An exceptional case appears to be in mica, where Gupta et al. (1971a,b) have found evidence that uranium is frequently added, presumably by ground water, to "loose" cleavage planes in mica. When a thick sample of mica is immersed in acid, the normal cleavage structure is such that the acid normally enters into a large number of cleavage cracks, thus revealing tracks on many internal surfaces. Mehta and Nagpaul (1971) showed that more concordant track ages were obtained if these surfaces are *eliminated* and studies made only on surfaces that have to be physically cleaved to open them up. In contrast, the "easy cleavage" planes produced by the first cleaving frequently gave young ages, suggesting that uranium was added to these particular planes during the lifetime of the sample.

4.3.4. Examples of Fission Track Dating

In this section we give specific examples that illustrate some of the special abilities and some special techniques of fission track dating. Because it is beyond the scope of this work to review in detail all of the literature on fission track dating, we have added, in an appendix to this chapter, a listing of dating work by mineral, glass type, or rock type for the convenience of workers who have particular problems they wish to attack.

Ocean Bottom Spreading

In what constitutes a genuine revolution in thinking about our planet, modern geophysicists have come to the realization that our "solid earth" is by no means rigid on a long time scale but is rather a plastic, evolving body. The primary reasons for this revolution may be understood by considering two striking features in the map shown in Fig. 4-18. The first is the apparent match between the margins of the continental shelves on the opposite sides of the Atlantic. This feature long ago suggested the idea that the continents have reached their present positions as the result of the separation of an original, larger land mass into the present grouping by a fracturing of the single continental plate and subsequent *continental drift.*

The second feature is the crinkly set of parallel submarine ridges and valleys down the Mid-Atlantic, shown in a closer view in Fig. 4-19. This relief, together with extensive data on the magnetic polarity of ocean-bottom rocks (for a review see Cox et al., 1967), suggested that a process of *ocean-bottom spreading* had taken

Fig. 4-18. *This map of the N. Atlantic Ocean shows the matching of the continental margins at opposite sides and the ridge system centered about the median rift valley along the Mid-Atlantic ridge. (Courtesy National Geographic Society.)*

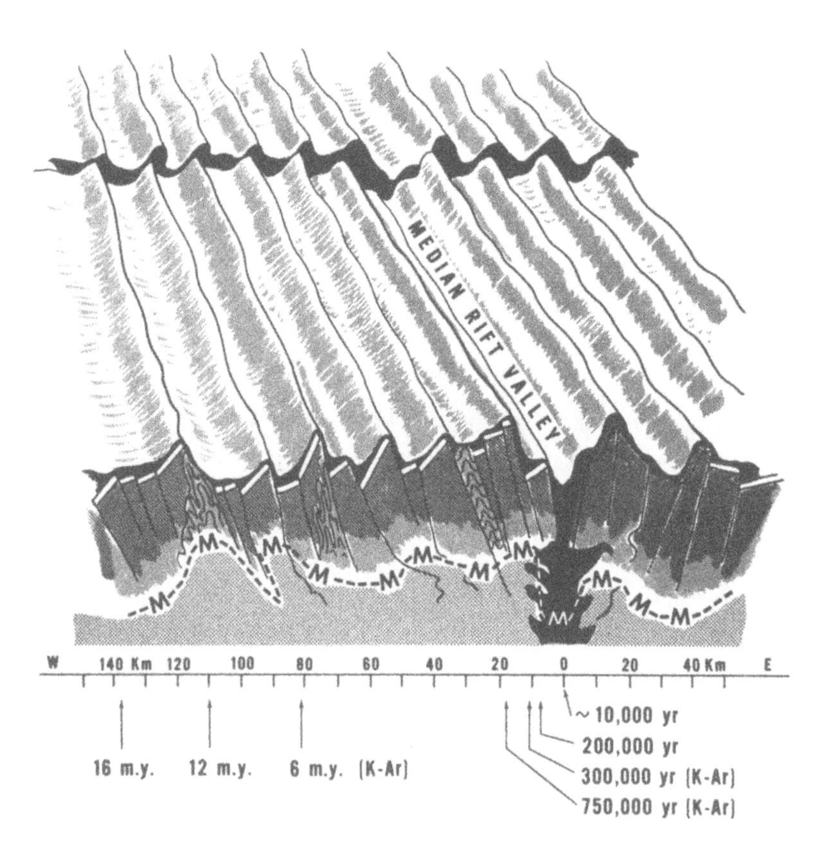

Fig. 4-19. *Schematic of the Mid-Atlantic Ocean floor, where fresh lava intrudes at the Median Valley, cools and is moved horizontally in the process called ocean-bottom spreading. A critical test that this process occurs is whether ages increase outward from the Median Valley. The rounded-off fission track ages from Fleischer et al. (1968b, 1971b) and Aumento (1969) and the K-Ar ages from Aumento et al. (1968), shown at the bottom of the figure, provide the proof. The full data are given in Fig. 4-20.*

place. This primary model for the evolution of the floors of our great ocean basins describes the mid-ocean ridges and their associated median valleys as sites where fresh, warm rock and lava are upwelling from deep within the earth, after which they flow away from the ridge—forming part of the circulation responsible for continental drift.

The most direct and critical test of such a hypothesis is a measurement at positions on either side of the mid-ocean ridges of the time at which ocean-bottom rock was last hot. Ocean-bottom spreading would give rise to ages increasing monotonically with distance from the source, corresponding to lava that had risen to the ocean floor, had been cooled, had begun to retain particle tracks, and subsequently had moved slowly away from the ridge.

K-Ar dating has been applied to numerous ocean-bottom rocks with erratic results, whose explanation only became obvious with the realization that pre-existing ^{40}Ar can be retained in lava that solidifies at the high pressures found at the ocean bottom, leading to spuriously high ages (Dalrymple and Moore, 1968; Funkhouser et al., 1968). Fission track dating does not encounter this problem and has been used to give clear results that are fully consistent with the ideas of ocean-bottom spreading (Fleischer et al., 1968b, 1971b; Aumento et al., 1968; Aumento, 1969).

The glassy skins of pillow basalts from the ocean bottom, because they are kept cool by the ocean, are highly retentive of tracks. Such basalts were used for track dating in one series of studies, along with K-Ar dating of low-viscosity lavas (in which retained ^{40}Ar was not a problem), to date a sequence across the Mid-Atlantic at 45° N. Lat. giving 11,000 y (F.T.), 230,000 y (F.T.), 310,000 y (K-Ar), 750,000 y (K-Ar), 8,000,000 y (K-Ar), and 16,000,000 y (F.T.), showing the continuous increase expected and providing strong support for the ocean-bottom spreading model (Fleischer et al., 1968b, 1971b; Aumento et al., 1968). Aumento (1969) has greatly extended this work and, in fact, the data plotted in Fig. 4-20 make it clear that the spreading rate has changed at least once in the past, increasing from 0.6 cm/y more than 2 My ago to 3 cm/y in more recent times. A date of 35 My obtained recently by Luyendyk and Fisher (1969) on an East Pacific basalt is also consistent with the ocean-bottom spreading model, using a spreading rate inferred from paleomagnetic information.

The above are cases where fission track dating was uniquely useful, since other methods either could not be applied (because of inherited ^{40}Ar) or were not sensitive enough for dating such geologically young material (e.g., at age $\sim$11,000 y). Special note should be made of the technique used, which was not that described earlier of scanning a section and then irradiating and rescanning. On the samples that were both young and low in uranium it was necessary, in order to locate an adequate number of tracks, to follow the original scan with a re-etch (removing 2×10^{-3} cm) sufficient to "peel" off a skin of material and reveal tracks at a different depth in the sample, then to scan, re-etch, scan, re-etch, until enough spontaneous fission tracks were found. Finally the piece was irradiated and scanned once more. Fortunately, in all these samples the uranium was found to be uniform in the remaining (i.e., irradiated) portion of the sample; hence, it was assumed to have been uniform throughout.

Ancient Man

East Africa.— One of the earlier accomplishments of fission track dating in the field of archaeology was the dating of volcanic pumice from Bed I, Olduvai Gorge, providing a useful cross check for the previous K-Ar dating that had produced surprise and unearned skepticism on the part of the anthropological community.

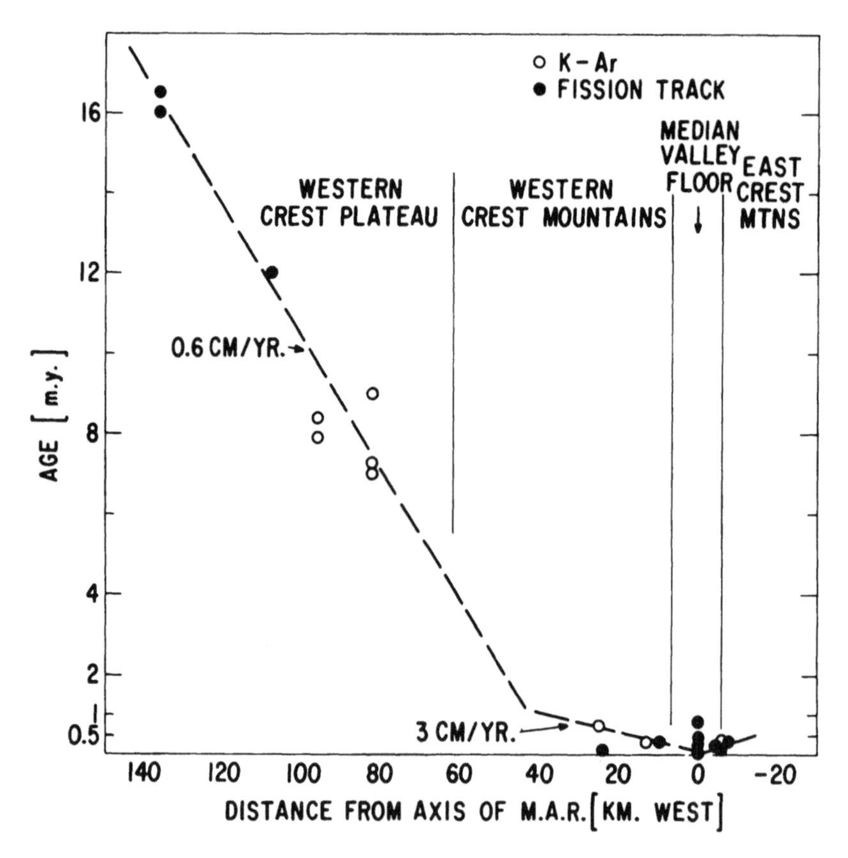

Fig. 4-20. *Age of basalts as a function of the distance from the Median Valley of the Mid-Atlantic Ridge at 45° North latitude as dated by fission-track and K-Ar methods. The rate of ocean-bottom spreading during the last million years is ~5 times that between 6 and 16 million years ago—a result that has not been explained by theories of the earth.* (*Aumento et al., 1968; Fleischer et al., 1968b, 1971b; Aumento, 1969.*)

Bed I has been shown (Hay, 1963) to be contemporaneous with the existence of two varieties of early man, *Zinjanthropus*, whose skull is shown in Fig. 4-21, and *Homo Habilis* (Leakey, 1959; Leakey and Leakey, 1964). The fission track age of 2.0 (±0.3) My (Fleischer et al., 1965f,g) was consistent with the K-Ar age of 1.75 (±.05) My found by Leakey et al. (1961), and the agreement has caused this site to be regarded as one of the best-dated of the anthropological sites that are too ancient for carbon dating.

From the point of view of dating technique, the pumice presented a unique problem because of its porosity, as shown in Fig. 4-22. If the pumice is etched (as it must be) between two track counts, the flat area upon which tracks may be distinguished is reduced by the general attack of the etchant working on the interiors of the pores. It was observed that etching may cause the projected area

Fig. 4-21. *Skull and model of jaw of the more ancient of the two varieties of ancient man found in Bed I, Olduvai Gorge. This variety of man was first named Zinjanthropus by its discoverer, the late Louis Leakey; it has more recently been regarded as a hyper-robust australopithecine and designated A. boisei. Potassium-argon and fission track dating both indicate that A. boisei lived nearly two million years ago.* (*Courtesy P. V. Tobias and Witwatersrand Univ. Press; photo by R. Campbell and A. R. Hughes.*)

of a given cross-section of pumice to be reduced by as much as 50%, so that a correction for the fraction of area where tracks can be seen is necessary. This fraction was measured in this case by a point counting method: A transparent plastic with a square grid of dots was laid over a photograph of the area of interest. The dots overlying areas where tracks could be distinguished if they were present

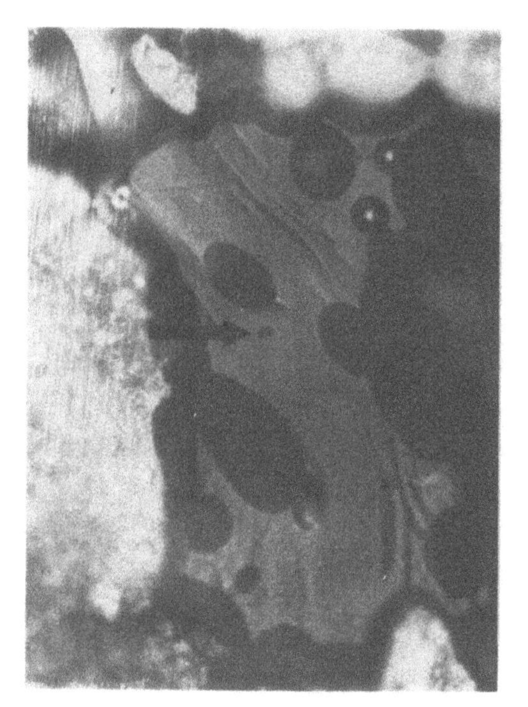

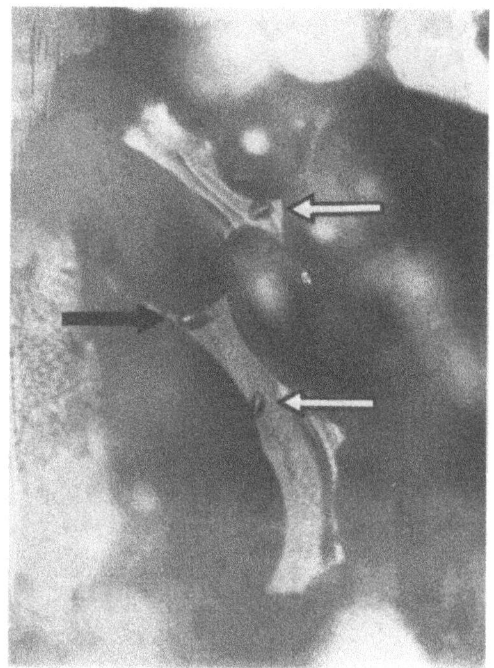

Fig. 4-22. *Spontaneous and induced tracks in pumice from the Olduvai Gorge. The view at the left shows a single spontaneous fission track (solid arrow); the view at the right shows the same area after neutron irradiation and a 500 second immersion in 5% HF. The pit from the spontaneous fission track is now flat-bottomed and two new tracks (open arrows) have been induced by the irradiation. (After Fleischer et al., 1965g.)*

were counted and used to calculate the fractional area actually used for track counting.

Western Hemisphere.— One of the more fascinating and recent archaeological controversies concerns the antiquity of man in the Americas, long believed to be not more than 10,000 years. Advanced man-made tools have been found at Valsequillo, Mexico, in clear association with the bones of animals they had been used to kill. A number of age determinations and estimates are each of low precision but in substantial agreement that the beds are at least 200,000 years old (Szabo et al., 1969). Probably the most telling bits of age data are the fission track dates found by C. W. Naeser on zircon from pumice and volcanic ash in the beds just above and below the deposit. The values (Naeser, personal communication) are 370,000 ± 100,000 years for the ash and 600,000 ± 170,000 years for the pumice. Of these, the age for the ash is considered to be geologically more reliable, because the pumice could have been influenced by addition of older components during its more extended deposition interval. The large statistical uncertainties which arise from the small samples used for separation of the zircons could be markedly reduced in the future by applying mineral separation techniques to larger samples. Nevertheless the agreement between these ages, the uranium-series disequilibrium ages and age limits, and other geological data makes it clear that there is a profound disagreement between the geological age of this site and the pre-existing archaeological data that showed man's presence in the Western Hemisphere only over a much shorter time span. One wonders whether, as in the case of Olduvai Gorge, the results from the Valsequillo site mark the beginning of a realization of the tenuousness of man's discoveries of his own earliest existence on Earth.

Thermal Dating in Archaeology

Because a sample of natural glass, such as an obsidian, when heated to a high enough temperature, loses all previous fossil tracks, the fission track age measured for such a glass will be the age of the heating event. In several cases the heating occurred as a result of man's activities and thus the track ages of the glasses are of archaeological interest.

This approach was first applied (Fleischer et al., 1965b) to the dating of an obsidian knife blade found in 1927 at Gamble's Cave II, Elmenteita, Kenya, by Dr. L. S. B. Leakey and his expedition. The wilted shape of the knife blade, Fig. 4-23, indicates that it was heated to a high temperature at some time after it had been struck from an obsidian core. A 0.1g piece was cut from the knife, polished, and etched for 15 sec in HF. After track counts were made, the sample was re-etched to expose a fresh surface and another track count was made. This procedure was repeated 36 times, yielding 17 tracks on 5.6 cm^2 of surface. The sample was then exposed to thermal neutrons for comparison, yielding a thermal age of 3,700 $\pm$ 900 years B.P. (before the present). The blade was apparently heated by a Neolithic hearth overlying an earlier culture level. The date agrees rather well with a carbon-14 date from a Neolithic cremation at the nearby Njoro River Rock Shelter just 20 km distant. The quoted error of nearly 25 percent is the statistical error resulting from the small number (17) of tracks counted. Since the scanning

Fig. 4-23. *Burnt obsidian blade, 7.3 cm long, from Gamble's Cave II, Elmenteita, Kenya. The blade was resoftened by heating about 4000 years ago as determined by counting the tracks formed since that heating removed the pre-existing tracks. (After Fleischer et al., 1965b.)*

required about 2 hours/cm^2, 60 hours would have been required to count 90 tracks and reduce the error to 400 years.

Watanabe and Suzuki (1969) have applied much the same procedure to several samples of glass from Japan. The first group, all obsidian, was a spearhead baked into a pottery vase and an arrowhead and flake collected from a dwelling pit destroyed by fire. The latter two specimens were wilted, indicating that they had been heated above 850–900°C. The last specimen was the glass glaze on a fragment of a bowl.

Each specimen was ground to expose a fresh surface and searched for fission tracks. Multiple repolishing, etching, and counting were used to increase the effective surface area. Annealing studies yielded track erasure temperatures of 350 to 400°C for the obsidians and 170 to 230°C for the glaze.

The track count for the spearhead from the vase was 175 tracks in 37.9 cm^2, yielding 5,080 ± 400 years B.P., an age about 1,500 years older than radiocarbon dates from a nearby dwelling pit associated with the same kind of pottery. The ages for the arrowhead and flake were obtained from 44 counts in 47.6 cm^2 and 7 counts in 7.0 cm^2, respectively. The dates were 1,060 ± 160 years B.P. and 1,150 ± 440 years B.P. These dates are to be compared with radiocarbon dates ranging from 990 to 1,420 years B.P. In the bowl glaze there were 25 counts in 71.0 cm^2, yielding 520 ± 110 years B.P., an age lying within the time of operation of the kiln where the fragment was found.

These experiments show that it is possible to date thermal events 500 to 5,000 years ago in samples containing 3–6 ppm of uranium. Quite clearly, such measurements require considerable patience and perseverance on the part of the investigator.

A similarly impressive study was done by Nishimura (1971) in dating archaeological objects mostly by removing tiny zircons from pottery, mounting and etching large numbers of small zircons for each sample dated. Nine zircon-containing archaeological materials were dated in the age range 700 to 2,300 years B.P. and seven glass objects in the range 400 to 1,500 years B.P.

In order to trace out ancient trade routes by characterizing obsidian sources by their geological ages Suzuki (1969a,b; 1970a,b) and Kaneoka and Suzuki (1970) dated an extensive suite of obsidians from Japan. Of 48 samples, 17 had ages between 1 and 10 My, 25 had ages between 0.1 and 1 My, and 6 were less than 100,000 years old, the two youngest definite ages being 20,000 years. Durrani et al. (1971) have carried out studies of European and Near Eastern obsidians with similar objectives in mind.

Thermal Dating in Geological and Space Sciences

Fossil fission tracks have been useful in studying impacts of extraterrestrial objects and in studying geological processes. Two examples are the dating of impact glasses and tektites and the recent study by Storzer (1970) of the geological thermal history of the South Tyrol. In both examples the recognition and correction for thermal track fading are important.

Two distinct types of glasses are found on the surface of the earth. The first of these, called *impact glasses*, are usually associated with craters and are thought to be local rock and soil that was melted by hypervelocity impacts. The second group are called *tektites*, which are rounded glass objects that often show definite evidence of having passed through the earth's atmosphere and fallen on certain limited areas of the earth. In the absence of track fading, fission track ages for these glasses indicate original solidification dates. In the case of the impact glasses this is clearly the date of the major impact that caused the crater. In the case of the tektites the meaning is not as clear, though Fleischer and Price (1964c) have shown by dating both the surface portion, or flange, which was melted upon passing downward through the atmosphere, and the interior, which was not significantly heated, that the date of original solidification is very nearly the same as the date of the tektite fall. Fission track dating, together with K-Ar dating, has allowed us to distinguish several distinct major families of tektites and to associate each of these families with a major event yielding an impact glass. A number of the key papers on tektites and tektite dating have been collected by Barnes and Barnes (1973).

Tektites have been found in four large but rather well defined regions of the earth. It used to be said that each region yields tektites having a single, discrete age: 0.7 My, ~1 My, 15 My, and 34 My. More recently it has been shown that a small, chemically distinct subgroup of tektites has an age of ~4 My (Fleischer et al., 1969b), even though they are found in the region where most tektites are 0.7 My old; so there is evidence for five distinct tektite falls. Discussion of tektites purported to have been found elsewhere is given by Faul and Wagner (1972).

For each of the first four tektite falls there is an impact glass thought to have the same age (Fleischer and Price, 1964b,c; Fleischer et al., 1965d, 1969c; Gentner et al., 1967, 1969a,b, 1970, 1973; Wagner, 1966a). Storzer and Wagner (1971) dispute one of the four cases, in contradiction to previous work (Fleischer and Price, 1964b; Durrani and Khan, 1970). Nevertheless, the set of remarkable coincidences makes it highly likely that each tektite family with an age-associated impact glass results from a single common impact event in which objects in a wide range of sizes fell onto the earth. It is not clear whether these many objects originated from a primary impact on the moon or on the earth. It should, however, be noted that tektite compositions do not match any of the material returned from the eight lunar sites that have been sampled.

Although the ages of the impact events are rather well established at this time, the effects of annealing or track fading have complicated their determination. In some regions a range of apparent ages is found, the younger ages being associated with samples having obviously subnormal etch pit sizes that undoubtedly resulted from annealing. When several of the oldest samples give the same age, especially when the oldest age agrees with the K-Ar age, that age can be taken with confidence as the age of the event (Fleischer and Price, 1964c). The spurious younger apparent ages may have resulted from chance grass fires, which have partially annealed the specimens.

Storzer and Wagner (1969) have applied their fading correction technique (described in Section 4.3.3) to a series of tektites from Australia similar to those for which Fleischer and Price (1964c) had obtained a wide range of ages. The track-fading correlation curve developed for these australites is shown in Fig. 4-13. The success of this procedure can be seen in Table 4-2, wherein apparent ages ranging from 0.13 My to 0.7 My are corrected to a common formation age of 0.7 My. Similar successful corrections have been made by Gentner et al. (1969a) and by Durrani and Khan (1970).

Storzer (1970) has measured fission track ages for a series of nine volcanic glasses from a Permian quartz porphyry shield in the southern Alps. He obtained apparent ages varying from 61 to 186 My, all younger than expected from the stratigraphy of the region. In preparation for a fading correction he measured the etch pit diameter distribution, finding a doubly peaked distribution in which the upper peak corresponded to the distribution of pit diameters expected for unannealed fission tracks. He thus had evidence for a complicated thermal history: an original cooling, a later thermal event, and, finally, an extended period with essentially no track fading. Here two ages are of interest, the age of the original cooling or solidification and the age of the last thermal event. Storzer separated the two etch pit distributions and applied a laboratory-developed annealing correction to the portion of the etch pit population showing the effects of annealing.

Table 4-2. Measured and Corrected Fission Track Ages of Australites (Storzer and Wagner, 1969)

Sample	Measured Track Age (My)	Normalized Mean Diameter of Fossil Tracks	Corrected Track Age (My)
A 1	0.66 ± 0.02*	1.00	0.66
A 2	0.71 ± 0.03	1.00	0.71
A 3	0.24 ± 0.02	0.63	0.71
A 4	0.54 ± 0.02	0.91	0.71
A 5	0.33 ± 0.04	0.77	0.66
A 6 core	0.34 ± 0.03	0.75	0.70
flange	0.13 ± 0.02	0.43	0.74
A 7 core I	0.40 ± 0.05	0.79	0.74
core II	0.63 ± 0.04	0.96	0.72
flange	0.48 ± 0.06	0.87	0.71
A 8 core	0.31 ± 0.04	0.73	0.71
flange	0.33 ± 0.03	0.76	0.68
A 9 core	0.71 ± 0.04	1.00	0.71
flange	0.70 ± 0.04	1.00	0.70

*statistical counting error

The sums of the ages obtained for the two distributions (one corrected for fading) should yield the time of the original cooling. These combined ages range between 231 My and 276 My, in good agreement with the Permian age of the quartz porphyry. The ages of the last thermal event range from 21 My in the north to 149 My in the south. Storzer offers two alternative explanations for this regional variation: 1) The volcanic glasses, after solidification, sank deep enough to experience temperatures greater than 130°C. They were then uplifted at different times, starting in the south. 2) There were regional and temporal heat flow variations in the geological history of the Alps.

In a recent example that illustrates the special power of fission tracks in dating extremely young samples (where K-Ar dating is difficult), Aumento and Souther (1973) have dated a stratigraphic series of volcanic glass from basalts, ryolites, and welded ashflows extending from 3.1 My ago to ~1,000 years ago for the most recent samples, which were also dated at 1,340 years by carbon-14. The results mesh with magnetic data on the polarity of the earth's magnetic field and with a single K-Ar date that is available at ~1.1 My.

One of the most graphic examples that demonstrates how a thermal event can reset the fission track "clock" is the recent set of results by Calk and Naeser (1973). The Cathedral Peak granite at Yosemite National Park is intruded by a basalt plug. Fig. 4-24 shows the variation in fission track ages of apatite and sphene

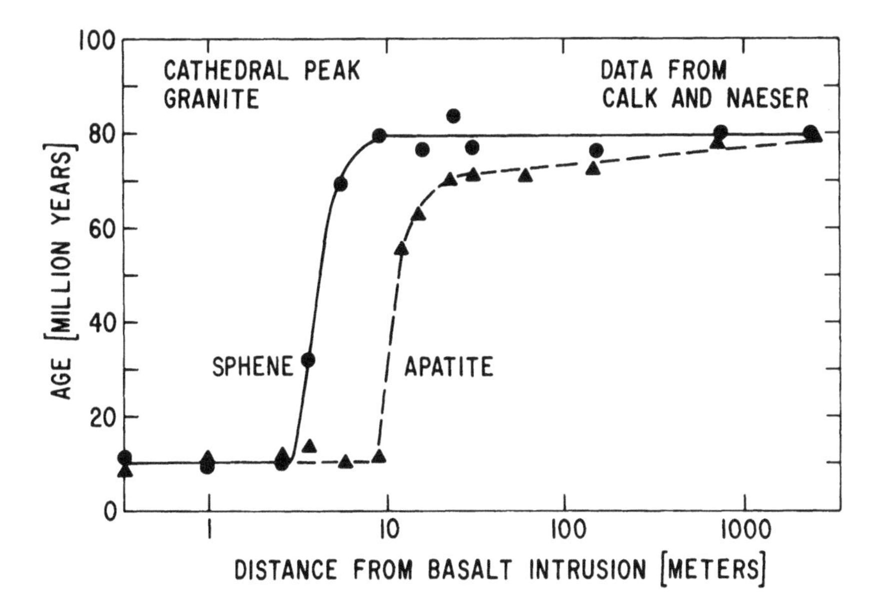

Fig. 4-24. *Effect of an igneous intrusion on fission track ages. Ages are lowered where heating from the intrusion of a basalt plug into Cathedral Peak, Yosemite granite caused track fading in apatite and sphene. The more retentive sphene crystals allowed fading only up to 10 meters from the contact as opposed to > 100 meters for the less retentive apatite. (Data of Calk and Naeser, 1973.)*

crystals in the granite as a function of the distance from the intrusion. The age of the granite, ~80 My, is obtained for both minerals at a large distance; but because of heating, they fall to the age of the intrusion, 9.4 My, close to the basalt. Since the tracks in the apatite are more sensitive to thermal effects than are those in the sphene, the apatite ages are influenced to greater distances from the basalt. The two age profiles and the knowledge of the track fading kinetics of the minerals (Naeser and Faul, 1969) allow a model to be constructed of the ancient thermal conditions around the hot intrusion. We showed earlier (Fig. 4-16) that a similar resetting of fission track ages occurs close to shock events.

The importance of this work to dating techniques is that it indicates how thermal track annealing, admittedly often a nuisance in age determinations, can be used as a recording thermometer for geological studies.

4.3.5. Future Applications of Fission Track Dating

Predicting the future is inherently a risky undertaking, but present trends are in some cases clear and can be given reasonable extrapolations. It is clear, for example, that fission track dating is a tool that is being applied by laboratories around the world and is likely to continue for routine age measurements because of its simplicity (in most cases), its flexibility, and its low cost. The techniques which lead to the unraveling of the thermal history of geological materials will almost certainly expand. Etch pit size distributions and track densities together reveal records of intervening thermal events in addition to the original cooling date. Future etch pit size (length as well as diameter) studies will include minerals as well as glasses and will hopefully become simpler to do as our understanding of the annealing and etching processes improves. The age plateau method of correcting for thermally altered ages is very promising for samples having fossil track densities high enough so that the necessary laboratory heating sequence does not reduce the final track density to an unacceptably low value. D. Macdougall (private communication) believes that many samples in ocean cores from the Deep Sea Drilling Project have thermally reduced ages that can be successfully corrected by the age plateau method. This makes fission track dating the only method presently capable of studying the absolute chronology of deep ocean sediments. Dating of mechanical events by their effects on track length is an opportunity still to be explored.

One trend for the future is increased standardization. As noted earlier in this chapter, glass standards have been prepared by the U.S. National Bureau of Standards and Corning Glass for use in dosimetry and uranium analysis. More widespread use of these dosimetry standards plus their intercalibration with the General Electric standard (Fleischer et al., 1965c) will facilitate placing all fission track ages on a common basis for proper intercomparison, as also will the availability of the standard, well-dated apatite and zircon described earlier.

4.4. ALPHA-RECOIL TRACKS

The primary mode of decay of ^{238}U and the sole decay mode of ^{232}Th are by emission of alpha particles. Such decays are followed by a series of beta decays and further alpha decays that lead ultimately to Pb isotopes. In each such alpha decay in a solid there exists a heavy, recoil nucleus of mass 206 to 234 that can create a track of length $\sim 2 \times 10^{-6}$ cm. Although they are very short, etched alpha-recoil tracks are easily observed in mica by several techniques (Huang and Walker, 1967; Huang et al., 1967). The simplest and most useful of these is illustrated in Fig. 4-25a which shows a Nomarski phase contrast picture taken of an etched, silvered mica surface. Fig. 4-25b shows similar alpha-recoil tracks as seen in a scanning electron microscope. The Nomarski system is an exclusive feature of Zeiss microscopes; the Smith interference system used by Leitz gives equally good viewing. Ordinary phase-contrast microscopy is also usable but less

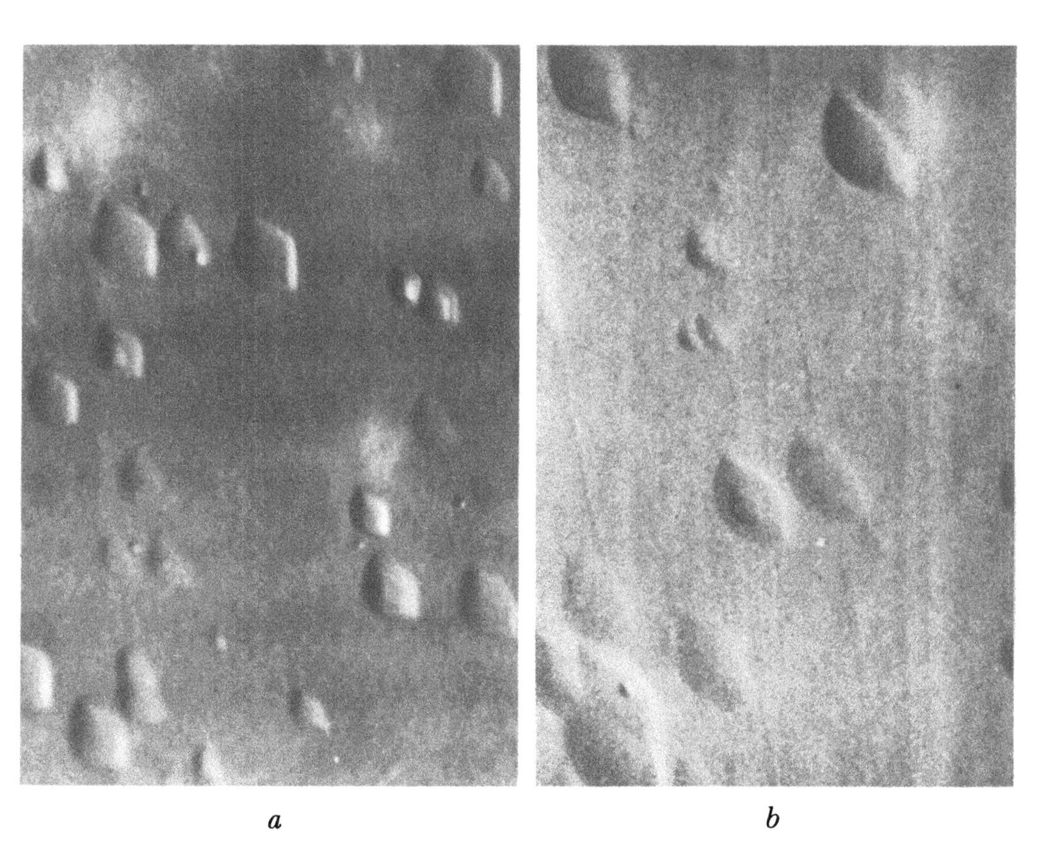

a *b*

Fig. 4-25. *Etched alpha-recoil tracks in muscovite mica as viewed (a) in the optical microscope using the Nomarski interference contrast technique and (b) in the scanning electron microscope.*

good than the above mentioned systems. Scanning electron microscopy of surface replicas also gives good alpha-recoil images.

The ^{238}U alpha-decay rate is $\sim 2 \times 10^6$ times the fission-decay rate, producing tracks that are only 10^{-3} times as long, so that the track density should be higher by a factor of roughly 2,000 for samples with the same uranium content. In addition ^{232}Th contributes recoil tracks, further amplifying the time sensitivity, typically by a factor of 2 for the most common Th/U ratio of 3.8. There do exist other natural alpha emitters; of these ^{147}Sm appears to be a potential complication if enriched relative to uranium and thorium.

An age equation similar to eq. (4-5) can be written for the density ρ of alpha-recoil tracks

$$\rho_\alpha = \sum_i \{[\exp(\lambda_{\alpha i} A) - 1](N_v C_i R_i \eta_i)\} \tag{4-7}$$

where the subscript i denotes the different alpha emitters, normally ^{238}U, ^{235}U, and ^{232}U.

Empirically it has been shown by Huang and Walker (1967) that the alpha-recoil density is $\sim$3,500 times higher than the fission track densities in micas of different ages. Whether the variation of this ratio by $\pm 30\%$ among different samples was caused by variations in the Th/U ratios or in the temperature histories of the different samples is not known. Although Huang and Walker originally reported that alpha recoils were more thermally stable than fission tracks, subsequent work (Crozaz et al., 1969) indicates that this was in error and that alpha-recoil tracks fade more easily than do fission tracks. The alpha-to-fission ratio of 3,500 can be used as a crude measure of the age of young muscovite mica samples in which the fossil fission track densities are too low to be easily measurable. In using this ratio it is important to reproduce the etching time of 2 hours at 20°C in 48% HF for which it was measured since etched alpha-recoil track densities increase with etching time. Unfortunately this equation contains several features that make alpha-recoil dating inherently more complicated than fission track dating. Firstly, it is obvious that we now need to know the thorium content as well as the uranium content. Although this measurement can be done by using irradiations with two different types of particles (a description of this technique is given in Chapter 8), it makes the method more cumbersome. A second complicating feature is encountered for samples that are comparable in age to the longest lived isotopes in the uranium and thorium decay chains ($T_{1/2} = 2.5 \times 10^5$ y (^{234}U), 3.2×10^4 y (^{231}Pa), and 5.7 y (^{228}Ra), respectively for ^{238}U, ^{235}U, and ^{232}Th). In this case the *number* of recoils that have occurred at the site of a single, initially decaying parent nucleus is time-dependent. Thus R_i is not constant with time. A related complication is the question of whether or not the daughter products of prior U and Th decay are separated from the substance at the time of its formation. If the removal is not perfect, then the decays of the daughter product will also contribute alpha-recoil tracks. Although for old samples these problems are unimportant, for young samples, where the promise of alpha-recoil track dating is highest, they must be explicitly answered in each case.

Although alpha-recoil tracks have been observed in feldspar (Turkowsky, 1969) using transmission electron microscopy, they have not been observed using the simpler and more useful technique of interference microscopy except in mica. In particular, unpublished experiments by one of us (RMW) have failed to find alpha-recoil tracks in either zircon or apatite, minerals which are of great interest because of their usually high uranium content and their widespread occurrence in archaeological ceramics (Nishimura, 1971; Zimmerman et al., 1973). The lack of a simple method of observing alpha-recoil tracks in substances other than mica, which is a substance for which other dating methods are generally adequate, is the largest single impediment to the development of this dating technique. Further work is required to tell whether the alpha-recoil method will be useful.

APPENDIX. REFERENCES TO DATING WORK ON SPECIFIC MINERALS, GLASS TYPES, AND ROCK TYPES

Asterisk indicates papers where no actual dates are given but fossil track identification is made.

Minerals

Allanite: Engels and Crowder (1971); Naeser and Dodge (1969); Naeser et al. (1970).

Amber: Uzgiris and Fleischer (1971)*.

Anthophillite: Nishimura (1969).

Apatite: Banks and Stuckless (1973); Berzina et al. (1971); Brookins and Naeser (1971); Burchart (1972); Christopher (1969); Church and Bickford (1971); Fleischer and Naeser (1972); Fleischer and Price (1964d); Fleischer et al. (1974); Hurford (1973); Kolodny et al. (1971); Märk et al. (1971); Mehta and Nagpaul (1970); Menzer (1970); Milton and Naeser (1970); Naeser (1967a,b); Naeser and Dodge (1969); Naeser and McKee (1970); Naeser et al. (1970, 1971); Nishimura (1969); Patel et al. (1967)*; Poupeau (1969); Storzer and Wagner (1973); Stuckless and Naeser (1972); Stuckless and Sheridan (1971); Sun (1971); Wagner (1968, 1969); Wagner and Reimer (1972, 1973); Wagner and Storzer (1970, 1972*); Zimmermann and Reimer (1973).

Bastnäsite: Fleischer and Naeser (1972).

Biotite: Abdullaev et al. (1966); Berzina et al. (1966); Bigazzi et al. (1971b); Bonadonna and Bigazzi (1969, 1970); Fleischer et al. (1964b); Gupta et al.

(1971b); Kere (1966); Miller (1968); Miller and Jaeger (1968); Miller et al. (1968); Nishimura (1969); Shima et al. (1969); Suzuki and Yamanoi (1970); Syromyatnikov et al. (1971)*; Welin et al. (1972).

Calcite: Sippel and Glover (1964)*.

Epidote: Engels and Crowder (1971); Naeser and Dodge (1969); Naeser et al. (1970).

Garnet: Haack (1973b)*; Haack and Gramse (1972)*; Hrichova (1966)*; Naeser and Dodge (1969).

Hornblende: Fleischer and Price (1964d); Nishimura (1969); Suzuki and Chinzei (1973)*; Welin et al. (1972).

Lepidolite and/or *Zinnwaldite:* Fleischer et al. (1964b); Maurette et al. (1964).

Micas: Berzina et al. (1966); Bigazzi et al. (1971b); Kachoukeev et al. (1970); Price and Walker (1962, 1963)*.

Monazite: Shukolyukov and Komarov (1970).

Muscovite: Abdullaev et al. (1966); Bigazzi (1967); Bigazzi et al. (1971b); Fleischer et al. (1964b, 1969h); Gupta et al. (1971a); Kachoukeev et al. (1970); Kere (1966); Lakatos and Miller (1973)*; Manecki (1968)*; Manecki and Mochnacka (1969)*; Maurette et al. (1964); Mehta and Nagpaul (1971); Mehta and Rama (1969)*; Miller (1968); Miller and Jaeger (1968); Miller et al. (1968); Shukolyukov and Komarov (1966); Shukolyukov et al. (1965).

Phlogopite and *Lepidomelane:* Abdullaev et al. (1966); Bigazzi et al. (1971b); Fleischer et al. (1964b); Maurette et al. (1964); Welin et al. (1972).

Quartz (amethyst): Suzuki and Yamanoi (1970).

Sphene: Banks and Stuckless (1973); Fleischer and Naeser (1972); Fleischer et al. (1974); Hurford (1973); Menzer (1970); Naeser (1967b, 1971); Naeser and Dodge (1969); Naeser and McKee (1970); Naeser et al. (1970); Stuckless and Naeser (1972); Stuckless and Sheridan (1971).

Stibiotantalite: Haack (1970).

Tanzanite: Naeser and Saul (1973).

Topaz: Patel and Patel (1969)*.

Zircon: Banks and Stuckless (1973); Bigazzi and Ferrara (1971); Bigazzi et al. (1972); Carbonnel and Poupeau (1969); Fleischer et al. (1964e); Machida and Suzuki (1971); McKee and Burke (1972); Menzer (1970); Milton et al. (1972); Naeser (1969*, 1971); Naeser and McKee (1970); Naeser et al. (1971, 1973); Nagai (1968); Nishimura (1971); Stuckless and Sheridan (1971); Sun (1971); Turner et al. (1973); Wagner and Storzer (1970); Yabuki and Shima (1970).

Mineral intercomparisons (multiple minerals from the same rock): Banks and Stuckless (1973): apatite, sphene, zircon; Calk and Naeser (1973): apatite, sphene; Engels and Crowder (1971): allanite, epidote; Fleischer and Naeser (1972): apatite, bastnäsite, sphene; Fleischer et al. (1974): apatite, sphene; Menzer (1970): apatite, sphene, zircon; Milton and Naeser (1970): apatite, bastnäsite, zircon; Naeser (1967b): apatite, sphene; Naeser (1971): sphene, zircon; Naeser and Dodge

(1969): allanite, apatite, epidote, sphene, zircon; Naeser and McKee (1970): apatite, sphene, zircon; Naeser et al. (1970): apatite, allanite, epidote, sphene; Naeser et al. (1971): apatite, zircon; Stuckless and Sheridan (1971): apatite, sphene, zircon.

Glasses

Archaeological: Bigazzi and Bonadonna (1973); Brill (1964); Brill et al. (1964); Durrani et al. (1971); Fleischer et al. (1965b,f,g); Kaneoka and Suzuki (1970); Kaufhold and Herr (1967); Kobayashi et al. (1971); Machida et al. (1971); Nishimura (1969); Suzuki (1969a,b, 1970a,b, 1972, 1973a,b); Uzgiris and Fleischer (1971); Watanabe and Suzuki (1969).

Impact glasses and structures: Fleischer and Price (1964b,c); Fleischer et al. (1965d, 1969c,d); Fredriksson et al. (1973); Gentner and Wagner (1969); Gentner et al. (1967, 1969a,b, 1973); Kaufhold and Herr (1967); Milton and Naeser (1970); Milton et al. (1972); Storzer (1971); Storzer and Gentner (1970); Storzer et al. (1971); Wagner (1966a, 1969).

Man-made: Brill (1964); Brill et al. (1964); Nishimura (1971); Watanabe and Suzuki (1969).

Obsidians: Aumento and Souther (1973); Bigazzi and Bonadonna (1973); Bigazzi et al. (1971a); Durrani et al. (1971); Fleischer and Price (1964c); Fleischer et al. (1965b); Kaneoka and Suzuki (1970); Kobayashi et al. (1971); Komarov et al. (1972); Lipman et al. (1970); Machida and Suzuki (1971); Machida et al. (1971); Sato et al. (1972); Storzer (1970); Suzuki (1969a,b, 1970a,b, 1972, 1973a,b); Suzuki and Chinzei (1973); Suzuki and Yamanoi (1970); Wagner and Storzer (1970); Watanabe and Suzuki (1969).

Glass shards: Bigazzi et al. (1973); Fleischer et al. (1965g); Hurford (1974); Macdougall (1971a); Shima et al. (1967).

Submarine basalts: Aumento (1969); Aumento et al. (1971); Fisher (1969a,b, 1971, 1972); Fleischer et al. (1968b, 1971b); Hawkins et al. (1971); Luyendyk and Fisher (1969); Macdougall (1971b, 1973).

Tektites: Durrani and Khan (1970, 1971); Faure and Fontaine (1969); Fleischer and Price (1964b,c); Fleischer et al. (1965d, 1969b,c); Garlick et al. (1971); Gentner and Wagner (1969); Gentner et al. (1967, 1969a,b, 1970); Glass et al. (1973); Kashkarov and Genaeva (1966); Kaufhold and Herr (1967); Komarov et al. (1972); Storzer and Gentner (1970); Storzer and Wagner (1969, 1970, 1971); Storzer et al. (1973a); Wagner (1966a,b, 1969).

Unknown glass: Vand et al. (1964).

Rocks

Alemellite: Hurford (1973).

Andesite: Shima and Yabuki (1969).

Ash flows, pumice, tuffs: Aumento and Souther (1973); Bonadonna and Bigazzi (1969, 1970); Fleischer et al. (1965f,g); Hurford (1974); Machida and Suzuki (1971); McKee and Burke (1972); Naeser and McKee (1970); Naeser et al. (1973); Nishimura (1969); Nishimura and Sasajima (1970); Sato et al. (1972); Stuckless and Sheridan (1971); Suzuki and Yamanoi (1970); Yabuki and Shima (1970).

Basalt: Carbonnel and Poupeau (1969); Shima and Yabuki (1969); Turner et al. (1973).

Calcitic ("Mottled zone"): Kolodny et al. (1971).

Carbonatite: Fleischer and Naeser (1972); Milton and Naeser (1970); Naeser (1967b).

Clay: Bigazzi et al. (1973).

Diabase: Naeser (1967b).

Diatremes and dikes: Naeser (1971).

Diorite: Banks and Stuckless (1973); Stuckless and Naeser (1972).

Gabbro: Bigazzi et al. (1972).

Gneiss: Manecki (1968)*; Manecki and Mochnacka (1969)*; Nagai (1968).

Granites: Berzina et al. (1971); Burchart (1972); Church and Bickford (1971); Engels and Crowder (1971); Fleischer et al. (1974); Hurford (1973); Menzer (1970); Naeser (1967b); Naeser and Dodge (1969); Nagai (1968); Nagpaul et al. (1973); Shima et al. (1969); Stuckless and Naeser (1972); Stuckless and Sheridan (1971).

Iron ore: Sun (1971).

Kimberlites: Brookins and Naeser (1971); Naeser (1971).

Manganese crust and nodules: Aumento (1969); Fleischer et al. (1969a); Otgonsuren et al. (1969); Shima and Okada (1968).

Schist: Manecki and Mochnacka (1969)*.

Syenite: Christopher (1969).

Alpha-recoil tracks: Huang and Walker (1967); Huang et al. (1967); Katcoff (1969); Sakagami et al. (1967); Turkowsky (1969).

Review papers on fission track dating: Fleischer and Hart (1972); Klose (1968) (Italian); Lehtovaara (1972) (Finnish); Nagai (1968) (French); Nishimura (1972) (Japanese); Shima et al. (1969) (Japanese); Shukolyukov (1970) (Russian); Wagner (1972) (German).

Short summaries of fission track dating: Berzina and Stolyarova (1968); Faul (1966a,b); Faul and Wagner (1971); Fleischer et al. (1964a, 1965a,f,h, 1967b, 1968a, 1969e,f, 1971a, 1972b); Kachoukeev et al. (1970); Komarov et al. (1972); Hamilton (1965); Oncescu et al. (1972); Pellas (1969); Stolz and Dörschel (1968); Suzuki (1973a); Wagner (1966b); Walker et al. (1968); Woods et al. (1971).

Chapter 4 References

K. H. Abdullaev, S. K. Gorbachev, V. P. Perelygin, and S. P. Tretiakova (1966), "Determination of the Geological Age of Mica by the Tracks of Uranium Fission Fragments," Dubna preprint 3-2961.

T. J. Ahrens, R. L. Fleischer, P. B. Price, and R. T. Woods (1970), "Erasure of Fission Tracks in Glasses and Silicates by Shock Waves," *Earth Planet. Sci. Lett.* **8,** 420–426.

F. Aumento (1969), "The Mid-Atlantic Ridge Near 45° N. Fission Track and Manganese Chronology," *Can. J. Earth Sci.* **6,** 1431–1440.

F. Aumento and J. G. Souther (1973), "Fission Track Dating of Late Tertiary and Quaternary Volcanic Glass from Mt. Edziza Volcano, British Columbia," *Can. J. Earth Sci.* **10,** 1156–1163.

F. Aumento, R. K. Wanless, and R. D. Stevens (1968), "Potassium-Argon Ages and Spreading Rates on the Mid-Atlantic at 45° N," *Science* **161,** 1338–1339.

F. Aumento, B. D. Longarevic, and D. I. Ross (1971), "Hudson Geotraverse: Geology of the Mid-Atlantic Ridge at 45° N," *Phil. Trans. Roy. Soc.*, London, **A268,** 623–750.

G. Banks and J. S. Stuckless (1973), "Chronology of Intrusion and Ore Deposition at Ray, Arizona: Part II, Fission-Track Ages," *Economic Geology* **68,** 657–664.

V. E. Barnes and M. A. Barnes (1973), *Tektites.* Strondsburg, Pa.: Dowden, Hutchinson, and Ross.

I. G. Berzina (1966), "Some Data on Micas from Tracks of U Fission Fragments," *Doklady Akad. Nauk SSSR* **170,** 681–683.

I. G. Berzina and A. N. Stolyarova (1968), "Sources of Error in Dating Minerals by the Fission-Track Method," *Izv. Akad. Nauk SSSR, Ser. Geol.* **4,** 114–119.

I. G. Berzina, I. B. Berman, and I. M. Zlotova (1966), "Age Determination of Micas Using Fission Tracks of Uranium," *Izv. Akad. Nauk SSSR, Ser. Geol.* **31,** 10–25.

I. G. Berzina, I. V. Vorob'eva, Ya. E. Geguzin, and I. M. Zlotova (1967), "Annealing of Tracks of Fragments from Spontaneous Fission of Uranium in Glasses and Mica Crystals," *Sov. Phys. Doklady* **11,** 1105–1107.

I. G. Berzina, O. P. Eliseeva, and A. N. Stolyarova (1971), "Uranium in Accessory Apatites," *Izv. Akad. Nauk SSSR, Ser. Geol.* **7,** 79–86.

N. Bhandari, S. G. Bhat, D. Lal, G. Rajagopalan, A. S. Tamhane, and V. S. Venkatavaradan (1971), "Fission Fragment Tracks in Apatite: Recordable Track Lengths," *Earth Planet. Sci. Lett.* **13,** 191–199.

G. Bigazzi (1967), "Length of Fission Tracks and Age of Muscovite Samples," *Earth Planet. Sci. Lett.* **3,** 434–438.

G. Bigazzi and F. Bonadonna (1973), "Fission Track Dating of the Obsidian of Lipari Island (Italy)," *Nature* **242,** 322–323.

G. Bigazzi and G. Ferrara (1971), "Fission Tracks in the Determination of the Age of Zircons," *Rend. Soc. Ital. Min. Petrol.* **27,** 295–304.

G. Bigazzi, F. Bonadonna, G. Bulluomini, and L. Malpieri (1971a), "Studi sulle Ossidiane Italiane. IV. Datazione con il Metodo delle Tracce di Fissione," *Bull. Soc. Geol. It.* **90,** 469–480.

G. Bigazzi, M. Cattani, U. G. Cordani, and K. Kawashita (1971b), "Comparison Between Radiometric and Fission Track Ages of Micas," *Ann. Acad. Brasil. Cien.* **43,** 633–638.

G. Bigazzi, G. Ferrara, and F. Innocenti (1972), "Fission Track Ages of Gabbros from Northern Apennines Ophiolites," *Earth Planet. Sci. Lett.* **14,** 242–244.

G. Bigazzi, F. P. Bonadonna, and F. Radicati de Brozolo (1973), "Fission-Track Dating of Volcanic Glasses in Neogene Sediments and Quaternary Chronostratigraphic Problems," Geochronology Conference, "ECOG II," Oxford, 3–8 Sept.

R. Bodu, H. Bouzigues, N. Morin, and J. P. Pfiffelman (1972), "On the Existence of Anomalous Isotopic Abundances in Uranium from Gabon," *Compt. rend.*, Paris, **275,** 1731–1732.

F. Bonadonna and G. Bigazzi (1969), "Studi sul Pleistocene del Lazio VII—Eta di un Livello Tufaceo del Bacino Diatomitico di Riano Stabilita con il Metodo delle Tracce di Fissione," *Bull. Soc. Geol. It.* **88,** 439–444.

F. P. Bonadonna and G. Bigazzi (1970), "Studi sul Pleistocene del Lazio VIII—Datazione di Tufi Intertirreniani della Zona di Cerveteri (Roma) Mediante il Metodo delle Tracce di Fissione," *Bull. Soc. Geol. It.* **89,** 463–473.

R. H. Brill (1964), "Applications of Fission-Track Dating to Historic and Prehistoric Glasses," *Archaeometry* **7,** 51–57.

R. H. Brill, R. L. Fleischer, P. B. Price, and R. M. Walker (1964), "The Fission Track Dating of Man-Made Glasses: Part I, Preliminary Results," *J. Glass Studies* **6,** 151–155.

D. Brookins and C. Naeser (1971), "Age of Emplacement of Riley County, Kansas, Kimberlites and a Possible Minimum Age for the Dakota Sandstone," *Bull. Geol. Soc. Am.* **82,** 1723–1726.

J. Burchart (1972), "Fission-Track Age Determinations of Accessory Apatite from the Tatra Mountains, Poland," *Earth Planet. Sci. Lett.* **15,** 418–422.

J. Burchart, M. Dakowski, and J. Galazka (1973), "A Method for Fission-Track Dating of Minerals with Very High Track Densities," Geochronology Conference, "ECOG II," Oxford, 3–8 Sept.

L. C. Calk and C. W. Naeser (1973), "The Thermal Effect of a Basalt Intrusion on Fission Tracks in Quartz Monzonite," *J. Geol.* **81,** 189–198.

J. P. Carbonnel and G. Poupeau (1969), "Premiers Elements de Datation Absolue par Traces de Fission des Basaltes de l'Indochine Meridionale," *Earth Planet. Sci. Lett.* **6,** 26–30.

B. S. Carpenter (1972), "Determination of Trace Concentration of Boron and

Uranium in Glass by the Nuclear Track Technique," *Anal. Chem.* **44,** 600–602.
B. S. Carpenter and G. M. Reimer (1973), "Calibrated Glasses as Neutron Monitors for Fission Track Studies," Geochronology Conference, "ECOG II," Oxford, 3–8 Sept., 58.
P. A. Christopher (1969), "Fission Track Ages of Younger Intrusions in Southern Maine," *Bull. Geol. Soc. Am.* **80,** 1809–1814.
S. E. Church and M. E. Bickford (1971), "Spontaneous Fission Track Studies of Accessory Apatite from Granitic Rocks of the Sawatch Range, Colorado," *Bull. Geol. Soc. Am.* **82,** 1727–1732.
A. Cox, G. B. Dalrymple, and R. R. Doel (1967), "Reversals of the Earth's Magnetic Field," *Sci. Amer.* **216** (2), 44–54.
G. Crozaz, M. Hair, M. Maurette, and R. Walker (1969), "Nuclear Interaction Tracks in Minerals and Their Implications for Extra-terrestrial Materials," *Proc. Inter. Conf. Nucl. Track Registration in Insulating Solids*, Clermont-Ferrand, Sect. VII, 41–53.
G. B. Dalrymple and J. G. Moore (1968), "Ar-40: Excess in Submarine Pillow Basalts from Kilauea Volcano, Hawaii," *Science* **161,** 1132–1135.
M. Debeauvais, M. Maurette, J. Mory, and R. Walker (1964), "Registration of Fission Fragment Tracks in Several Substances and Their Use in Neutron Detection," *Int. J. Appl. Rad. Isotopes* **15,** 289–299.
S. A. Durrani and D. A. Hancock (1970), "Effect of Strain on Fission-Track Ages of Tektites," *Earth Planet. Sci. Lett.* **8,** 157–162.
S. A. Durrani and H. A. Khan (1970), "Annealing of Fission Tracks in Tektites: Corrected Ages of Bediasites," *Earth Planet. Sci. Lett.* **9,** 431–445.
S. A. Durrani and H. A. Khan (1971), "Ivory Coast Microtektites: Fission Track Age and Geomagnetic Reversals," *Nature* **232,** 320–323.
S. Durrani, H. Khan, M. Taj, and C. Renfrew (1971), "Obsidian Source Identification by Fission Track Analysis," *Nature* **233,** 242–245.
J. Edwards (1967), "A Comparison Between Etched Fission Track Densities on Internal and External Glass Surfaces after Neutron Irradiation," *Geophys. J. Roy. Astron. Soc.* **13,** 541–543.
J. C. Engels and D. F. Crowder (1971), "Late Cretaceous Fission-Track and Potassium-Argon Ages of the Mount Stuart Granodiorite and Beckler Peak Stock, N. Cascades, Washington," *U.S. Geol. Surv. Prof. Papers* **750D,** 39–43.
H. Faul (1966a), *Nuclear Clocks*, U.S. Atomic Energy Commission, Div. of Tech. Inf., P.O. Box 62, Oak Ridge, Tenn.
H. Faul (1966b), *Ages of Rocks, Planets, and Stars*. New York: McGraw-Hill.
H. Faul and G. A. Wagner (1971), "Fission Track Dating," in *Dating Techniques for the Archaeologist*, H. N. Michael and E. K. Ralph (eds.), 152–156. Cambridge: MIT Press.
H. Faul and G. A. Wagner (1972), "Vagabond Tektites," *Earth Planet. Sci. Lett.* **14,** 357–359.

C. Faure and H. Fontaine (1969), "Geochronologie du Viet-Nam Meridional," *Viet-Nam Archives Geol.* **12,** 213–222.

D. E. Fisher (1969a), "Fission Track Ages of Deep Sea Glasses," *Nature* **221,** 549–550.

D. Fisher (1969b), "Dating the Spreading Sea Floor," *New Scientist*, 23 Oct., 185–187.

D. E. Fisher (1971), "Excess Rare Gases in a Subaerial Basalt from Nigeria," *Nature Phys. Sci.* **232,** 60–61.

D. E. Fisher (1972), "U/He Ages as Indicators of Excess Argon in Deep Sea Basalts," *Earth Planet. Sci. Lett.* **14,** 255–258.

R. L. Fleischer and H. R. Hart, Jr. (1972), "Fission Track Dating: Techniques and Problems," *Calibration of Hominoid Evolution*, W. W. Bishop, D. A. Miller, and S. Cole (eds.), 135–170. Edinburgh: Scottish Academic Press.

R. L. Fleischer and H. R. Hart, Jr. (1973), "Mechanical Erasure of Particle Tracks, a Tool for Lunar Microstratigraphic Chronology," *J. Geophys. Res.* **78,** 4841–4851.

R. L. Fleischer and C. W. Naeser (1972), "Search for ^{244}Pu Tracks in Mountain Pass Bastnaesite," *Nature* **240,** 465.

R. L. Fleischer and P. B. Price (1963), "Charged Particle Tracks in Glass," *J. Appl. Phys.* **34,** 2903–2904.

R. L. Fleischer and P. B. Price (1964a), "Decay Constant for Spontaneous Fission of U^{238}," *Phys. Rev.* **133B,** 63–64.

R. L. Fleischer and P. B. Price (1964b), "Glass Dating by Fission Fragment Tracks," *J. Geophys. Res.* **69,** 331–339.

R. L. Fleischer and P. B. Price (1964c), "Fission Track Evidence for the Simultaneous Origin of Tektites and Other Natural Glasses," *Geochim. Cosmochim. Acta* **28,** 755–760.

R. L. Fleischer and P. B. Price (1964d), "Techniques for Geological Dating of Minerals by Chemical Etching of Fission Fragment Tracks," *Geochim. Cosmochim. Acta* **28,** 1705–1714.

R. L. Fleischer and R. M. Walker (1966), "Etchable Line Defects in Crystals Are Not Necessarily Dislocations," *Phil. Mag.* **13,** 1083–1084.

R. L. Fleischer, P. B. Price, and R. M. Walker (1964a), "Fossil Records of Nuclear Fission," *New Scientist* **21,** 406–408.

R. L. Fleischer, P. B. Price, E. M. Symes, and D. S. Miller (1964b), "Fission Track Ages and Track-Annealing Behavior of Some Micas," *Science* **143,** 349–351.

R. L. Fleischer, P. B. Price, and E. M. Symes (1964c), "On the Origin of Anomalous Etch Figures in Minerals," *Am. Mineralogist* **49,** 794–800.

R. L. Fleischer, P. B. Price, R. M. Walker, and E. L. Hubbard (1964d), "Track Registration in Various Solid State Nuclear Track Detectors," *Phys. Rev.* **133A,** 1443–1449.

R. L. Fleischer, P. B. Price, and R. M. Walker (1964e), "Fission Track Ages of Zircons," *J. Geophys. Res.* **69,** 4885–4888.
R. L. Fleischer, P. B. Price, and R. M. Walker (1965a), "Solid State Track Detectors: Applications to Nuclear Science and Geophysics," *Ann. Rev. Nuc. Sci.* **15,** 1–28.
R. L. Fleischer, P. B. Price, R. M. Walker, and L. S. B. Leakey (1965b), "Fission Track Dating of a Mesolithic Knife," *Nature* **205,** 1138.
R. L. Fleischer, P. B. Price, and R. M. Walker (1965c), "Neutron Flux Measurements by Fission Tracks in Solids," *Nucl. Sci. Eng.* **22,** 153–156.
R. L. Fleischer, P. B. Price, and R. M. Walker (1965d), "On the Simultaneous Origin of Tektites and Other Natural Glasses," *Geochim. Cosmochim. Acta* **29,** 161–166.
R. L. Fleischer, P. B. Price, and R. M. Walker (1965e), "Spontaneous Fission Tracks from Extinct Pu^{244} in Meteorites and the Early History of the Solar System," *J. Geophys. Res.* **70,** 2703–2707.
R. L. Fleischer, P. B. Price, and R. M. Walker (1965f), "Applications of Fission Tracks and Fission Track Dating to Anthropology," *Proc. Seventh Int. Congress on Glass*, Brussels, **224,** 1–7.
R. L. Fleischer, P. B. Price, R. M. Walker, and L. S. B. Leakey (1965g), "Fission Track Dating of Bed I, Olduvai Gorge," *Science* **148,** 72–74.
R. L. Fleischer, P. B. Price, and R. M. Walker (1965h), "Tracks of Charged Particles in Solids," *Science* **149,** 383–393.
R. L. Fleischer, P. B. Price, R. M. Walker, and M. Maurette (1967a), "Origins of Fossil Charged Particle Tracks in Meteorites," *J. Geophys. Res.* **72,** 333–353.
R. L. Fleischer, P. B. Price, and R. M. Walker (1967b), "Fission Track Dating," *Int. Dictionary of Geophysics*, S. K. Runcorn (ed.), **1,** 529–532.
R. L. Fleischer, P. B. Price, R. M. Walker, and E. L. Hubbard (1967c), "Criterion for Registration in Dielectric Track Detectors," *Phys. Rev.* **156,** 331–355.
R. L. Fleischer, P. B. Price, and R. M. Walker (1968a), "Charged Particle Tracks: Tools for Geochronology and Meteorite Studies," in *Radiometric Dating for Geologists*, E. Hamilton and R. M. Farquhar (eds.), 417–435. London: Wiley-Interscience.
R. L. Fleischer, J. R. M. Viertl, P. B. Price, and F. Aumento (1968b), "Mid-Atlantic Ridge: Age and Spreading Rates," *Science* **161,** 1339–1342.
R. L. Fleischer, P. B. Price, and R. M. Walker (1968c), "Identification of Pu^{244} Fission Tracks and the Cooling of the Parent Body of the Toluca Meteorite," *Geochim. Cosmochim. Acta* **32,** 21–31.
R. L. Fleischer, H. R. Hart, Jr., I. S. Jacobs, P. B. Price, W. M. Schwarz, and F. Aumento (1969a), "Search for Magnetic Monopoles in Deep Ocean Deposits," *Phys. Rev.* **184,** 1393–1397.
R. L. Fleischer, P. B. Price, and R. T. Woods (1969b), "A Second Tektite Fall in Australia," *Earth Planet. Sci. Lett.* **7,** 51–52.

R. L. Fleischer, P. B. Price, J. R. M. Viertl, and R. T. Woods (1969c), "Ages of Darwin Glass, Macedon Glass, and Far Eastern Tektites," *Geochim. Cosmochim. Acta* **33,** 1071–1074.
R. L. Fleischer, J. R. M. Viertl, and P. B. Price (1969d), "Age of the Manicouagan and Clearwater Lakes Craters," *Geochim. Cosmochim. Acta* **33,** 523–527.
R. L. Fleischer, P. B. Price, and R. M. Walker (1969e), "Quaternary Dating by the Fission-Track Technique," in *Science in Archaeology*, 2nd ed., D. R. Brothwell & H. Higgs (eds.), 58–61. London: Thames and Hudson.
R. L. Fleischer, P. B. Price, and R. M. Walker (1969f), "Fission Track Dating and Processes in the Earth's Interior," in *The Application of Modern Physics to the Earth and Planetary Interiors*, Part 6, Chapter 1. London: Wiley-Interscience.
R. L. Fleischer, P. B. Price, and R. T. Woods (1969g), "Nuclear Particle Track Identification in Inorganic Solids," *Phys. Rev.* **88,** 563–567.
R. L. Fleischer, P. B. Price, and R. T. Woods (1969h), "Search for Tracks of Massive, Multiply Charged Magnetic Poles," *Phys. Rev.* **184,** 1398–1401.
R. L. Fleischer, P. B. Price, and R. M. Walker (1971a), "Fission Track Dating," in *Encyclopedia of Science and Technology* **5,** 307–308. New York: McGraw-Hill.
R. L. Fleischer, J. R. M. Viertl, P. B. Price, and F. Aumento (1971b), "A Chronological Test of Ocean Bottom Spreading in the North Atlantic," *Rad. Effects* **11,** 193–194.
R. L. Fleischer, G. M. Comstock, and H. R. Hart, Jr. (1972a), "Dating of Mechanical Events by Deformation-Induced Erasure of Particle Tracks," *J. Geophys. Res.* **77,** 5050–5053.
R. L. Fleischer, P. B. Price, and R. M. Walker (1972b), "Fission Track Dating," in *Encyclopedia of Geochemistry and Environmental Sciences*, IV A of *Encyclopedia of Earth Sciences* Series, 366–367. New York: Van Nostrand Reinhold.
R. L. Fleischer, R. T. Woods, H. R. Hart, Jr., P. B. Price, and N. M. Short (1974), "Effect of Shock on Fission Track Dating of Apatite and Sphene Crystals from the Hardhat and Sedan Underground Nuclear Explosions," *J. Geophys. Res.* **79,** 339–342.
G. N. Flerov and V. P. Perelygin (1969), "On Spontaneous Fission of Lead—Search for Very Far Transuranium Elements," preprint D7-4205, Joint Institute for Nuclear Research, Dubna.
G. N. Flerov, V. P. Perelygin and O. Otgonsuren (1972), "On the Origin of Tracks of Fission Fragments in Lead Glasses," *Atomnaya Energiya* **33,** 979–984.
K. Fredriksson, A. Dube, D. J. Milton, and M. S. Balasundaram (1973), "Lonar Lake, India: An Impact Crater in Basalt," *Science* **180,** 862–864.
J. G. Funkhouser, D. E. Fisher, and E. Bonatti (1968), "Excess Argon in Deep Sea Rocks," *Earth Planet. Sci. Lett.* **5,** 95–100.
G. D. Garlick, C. W. Naeser, and J. R. O'Neil (1971), "A Cuban Tektite," *Geochim. Cosmochim. Acta* **35,** 731–734.

F. H. Geisler (1972), *Search for Superheavy Elements in Terrestrial Minerals and Cosmic Ray Induced Fission of Heavy Elements*, Ph.D. Thesis, Washington University.

F. H. Geisler, P. R. Phillips, and R. M. Walker (1973), "Search for Superheavy Elements in Natural and Proton-irradiated Materials," *Nature* **244,** 428–429.

W. Gentner and G. A. Wagner (1969), "Altersbestimmungen an Riesgläsern und Moldaviten," *Geol. Bavarica* **61,** 296–303.

W. Gentner, B. Kleinmann, and G. A. Wagner (1967), "New K-Ar and Fission Track Ages of Impact Glasses and Tektites," *Earth Planet. Sci. Lett.* **2,** 83–86.

W. Gentner, D. Storzer, and G. A. Wagner (1969a), "New Fission Track Ages of Tektites and Related Glasses," *Geochim. Cosmochim. Acta* **33,** 1075–1081.

W. Gentner, D. Storzer, and G. A. Wagner (1969b), "Das Alter von Tektiten und verwandten Gläsern," *Naturwiss.* **56,** 255–260.

W. Gentner, B. P. Glass, D. Storzer, and G. A. Wagner (1970), "Fission Track Ages and Ages of Deposition of Deep-Sea Microtektites," *Science* **168,** 359–361.

W. Gentner, T. Kirsten, D. Storzer, and G. A. Wagner (1973), "K-Ar and Fission Track Dating of Darwin Crater Glass," *Earth Planet. Sci. Lett.* **20,** 204–210.

E. K. Gerling, Yu. A. Shukolyukov, and B. A. Makarochkin (1959), "Determination of the Half Life of the Spontaneous Uranium-238 Disintegration from the Xenon Content in Uranium Minerals," *Radiokhimiya* **1,** 223–226.

J. J. Gilman (1959), "Cleavage and Ductility in Crystals," *Conference on Fracture*, B. L. Averbach, D. K. Felbeck, G. T. Hahn, and D. A. Thomas (eds.), Chapter 8, 1–27. New York: Wiley.

B. P. Glass, R. N. Baker, D. Storzer, and G. A. Wagner (1973), "North American Microtektites from the Caribbean Sea and Their Fission Track Age," *Earth Planet. Sci. Lett.* **19,** 184–192.

B. Grauert and M. G. Seitz (1973), "Uranium Gain of Detrital Zircons Studied by Isotopic Analyses and Fission Track Mapping," *Trans. Amer. Geophys. Union* **54,** 495.

M. L. Gupta, K. K. Nagpaul, and P. P. Mehta (1971a), "Fission Track Ages of Some Indian Muscovites," *Can. J. Earth Sci.* **8,** 1491–1495.

M. L. Gupta, P. P. Mehta, and K. K. Nagpaul (1971b), "Fission Track Ages of Some Indian Biotites," *Indian J. Pure Appl. Phys.* **9,** 466–469.

U. Haack (1970), "Fission Track Age of Stibiotantalite from Alto Ligonha, Mozambique," *Contr. Mineral. Petrol.* **29,** 183–185.

U. K. Haack (1973a), "Search for Superheavy Transuranic Elements," *Naturwiss.* **60,** 65–70.

U. K. Haack (1973b), "The Influence of Uranium Distribution on the Error of Fission Track Ages of Garnets," Geochronology Conference, "ECOG II," Oxford, 3–8 Sept.

U. K. Haack and M. Gramse (1972), "Survey of Garnets for Fossil Fission Tracks," *Contr. Mineral. Petrol.* **34,** 258–260.

E. I. Hamilton (1965), *Applied Geochronology*, 152–158. London: Academic Press.

J. W. Hawkins, E. C. Allison, and D. Macdougall (1971), "Volcanic Petrology and Geologic History of Northeast Bank, Southern Calif. Borderland," *Bull. Geol. Soc. Am.* **82,** 219–227.

R. L. Hay (1963), "Stratigraphy of Beds I Through IV, Olduvai Gorge, Tanganyika," *Science* **139,** 829–833.

R. Hrichova (1966), "Study of Corrosion on Garnets," *Scientific Papers of the Inst. of Chemical Tech.*, Prague, **G8.**

W. H. Huang and R. M. Walker (1967), "Fossil α-particle Recoil Tracks: A New Method of Age Determination," *Science* **155,** 1103–1106.

W. H. Huang, M. Maurette, and R. M. Walker (1967), "α-particle Recoil Tracks and Their Implications for Dating Measurements," in *Radioactive Dating and Methods of Low-Level Counting*, Monaco, IAEC pub. no. 4135, pp. 415–429.

A. J. Hurford (1973), "A Fission Track Dating Study of Some Scottish Caledonian Intrusive Rocks," Geochronology Conference, "ECOG II," Oxford, 3–8 Sept., 47.

A. J. Hurford (1974), "Fission Track Dating of a Vitric Tuff from East Rudoff, North Kenya," *Nature* **249,** 236–237.

N. T. Kachoukeev (1969), "Recherches Effectuees aupres du Reacteur 'IRT' a Sofia a l'Aide de Detecteurs Solides Isolants des Fragments de Fission," *Proc. Inter. Conf. Nucl. Track Registration in Insulating Solids*, Clermont-Ferrand, Sect. IX, 56–60.

N. T. Kachoukeev, T. Nikifor, T. Taneva, and R. Ignatova (1970), "Determination of the Geological Age of Some Bulgarian Micas from Tracks Left by Fragments of Uranium Fission," *Dokl. Bolg. Akad. Nauk* **23,** 559–562.

I. Kaneoka and Ozima (1970), "On the Radiometric Ages of Volcanic Rocks from Japan," *Bull. Volc. Soc. Jap.* **15,** 10–12.

I. Kaneoka and M. Suzuki (1970), "K-Ar and Fission Track Ages of Some Obsidians from Japan," *J. Geol. Soc. Japan* **76,** 309–314.

L. L. Kashkarov and L. I. Genaeva (1966), "On the Possibility of Evaluating the Intensity of Primary Cosmic Radiation from the Tracks of Nuclear Particles in Meteorites and Tektites," *Bull. Acad. Sci. USSR Phys. Series* **30,** 1799–1801.

S. Katcoff (1969), "Alpha-Recoil Tracks in Mica: Registration Efficiency," *Science* **166,** 382–384.

J. Kaufhold and W. Herr (1967), "Influence of Experimental Factors on Dating by the Fission-Track Method," in *Radioactive Dating and Methods of Low-Level Counting*, Monaco, IAEC, pub. no. 4135, pp. 403–413.

S. S. Kere (1966), "Fission-Track Ages of Indian Micas," *Current Sci.*, India, **35,** 509–510.

H. A. Khan and S. A. Durrani (1973), "Measurement of Spontaneous-Fission Decay Constant of ^{238}U with a Mica Solid State Track Detector," *Rad. Effects* **17,** 133–135.

J. D. Kleeman and J. F. Lovering (1970), "The Evaluation of the Spatial Parameters in Equations for Fission Track Uranium Analysis," *Rad. Effects* **5,** 233–237.
J. D. Kleeman and J. F. Lovering (1971), "A Determination of the Decay Constant for Spontaneous Fission of Natural Uranium Using Fission Track Accumulation," *Geochim. Cosmochim. Acta* **35,** 637–640.
A. Klose (1968), "Tracce di Fissioni Nucleari e Loyo Applicagione alla Datizione ed Alla Microcrialisi dell Uranio in Reperti Geologici," Conferenze e Seminari dell Instituto di Geochemica: Universita Degli Studi di Roma, 1–22.
T. Kobayashi, S. Oda, K. Hatori, and M. Suzuki (1971), "A Study of the Preceramic Site, Nogawa," *Daiyonki Kenkyu* **10,** 231–270.
Y. Kolodny, M. Bar, and E. Sass (1971), "Fission Track Age of the 'Mottled Zone Event' in Israel," *Earth Planet. Sci. Lett.* **11,** 269–272.
A. N. Komarov, N. V. Skovorodkin, and S. G. Karapetyan (1972), "Determination of the Age of Natural Glasses According to Tracks of Uranium Fission Fragments," *Geochimiya* **6,** 693–698.
S. Lakatos and D. S. Miller (1973), "Problems of Dating Mica by the Fission-Track Method," *Can. J. Earth Sci.* **10,** 403–407.
L. S. B. Leakey (1959), "A New Fossil Skull from Olduvai," *Nature* **84,** 491–493.
L. S. B. Leakey and M. B. Leakey (1964), "Recent Discoveries of Fossil Hominids in Tanganyika at Olduvai and Near Lake Natron," *Nature* **202,** 5–7.
L. S. B. Leakey, J. F. Evernden, and G. H. Curtis (1961), "Age of Bed I, Olduvai Gorge, Tanganyika," *Nature* **191,** 478–479.
J. Lehtovaara (1972), "Fission Track Dating," *Geologi* **24,** 27–28.
M. P. T. Leme, C. Renner, and M. Cattani (1971), "Determination of the Decay Constant for Spontaneous Fission of Uranium-238," *Nucl. Instr. Methods* **91,** 577–579.
P. W. Lipman, T. A. Steven, and H. H. Mehnert (1970), "Volcanic History of the San Juan Mountains, Colorado as Indicated by Potassium-Argon Dating," *Bull. Geol. Soc. Am.* **81,** 2329–2352.
B. P. Luyendyk and D. E. Fisher (1969), "Fission Track Age of Magnetic Anomaly 10: A New Point on the Sea-Floor Spreading Curve," *Science* **164,** 1516–1517.
D. Macdougall (1971a), "Fission Track Dating of Volcanic Glass Shards in Marine Sediments," *Earth Planet. Sci. Lett.* **10,** 403–406.
D. Macdougall (1971b), "Deep Sea Drilling: Age and Composition of an Atlantic Basaltic Intrusion," *Science* **171,** 1244–1245.
D. Macdougall (1973), "Fission Track Dating of Oceanic Basalts," *Trans. Amer. Geophys. Union* **54,** 987.
H. Machida and M. Suzuki (1971), "Geological Dating of the Late Quaternary by Using the Absolute Age of Volcanic Ashes Based on the Fission-Track Method," *Kagaku (Science) Tokyo* **41,** 263–270.
H. Machida, M. Suzuki, and A. Miyazaki (1971), "Chronology of the Preceramic

Age in South Kanto, with Special Reference to Tephrochronology, Radiocarbon Dating and Obsidian Dating," *Daiyonki Kenkyu* **10,** 290–316.
A. Manecki (1968), "Significance of Spontaneous Fission of Uranium Atom Impurities in Minerals for Metasomatic Processes," *Bull. Acad. Pol. Sci., Ser. Sci. Geol. Geogr.* **16,** 169–170.
A. Manecki and K. Mochnacka (1969), "Preliminary Investigations of Fission Fragment Tracks in Micas of Some Polish Rocks," *Pol. Akad. Nauk, Oddzial W. Krakowie, Kom. Nauk Mineral. Pr. Mineral.* **20,** 71–82.
E. Märk, M. Paul, and T. D. Märk (1971), "Fission-Track-Alter von Durango-Apatit, Mexiko," *Contr. Mineral. Petrol.* **32,** 147–148.
M. Maurette (1966), "Study of the Registration of Fission Fragment Tracks in Certain Substances," *J. de Physique* **27,** 505–512.
P. Maurette, P. Pellas, and R. M. Walker (1964), "Etude des Traces de Fission Fossiles dans le Mica," *Bull. Soc. Franc. Min. Crist.* **87,** 6–17.
E. H. McKee and D. B. Burke (1972), "Fission-Track Age Bearing on the Permian-Triassic Boundary and Time of the Sonoma Orogeny in North-Central Nevada," *Bull. Geol. Soc. Am.* **83,** 1949–1952.
P. P. Mehta and K. K. Nagpaul (1970), "Fission Track Ages of Some Indian Apatites," *Indian J. Pure Appl. Phys.* **8,** 397–400.
P. P. Mehta and K. K. Nagpaul (1971), "Fission Track Ages of Mica Belts and Some Other Precambrian Deposits," *Rev. Pure Appl. Geophys.* **87,** 174–191.
P. P. Mehta and Rama (1969), "Annealing Effects in Muscovite and Their Influence on Dating by Fission Track Method," *Earth Planet. Sci. Lett.* **7,** 82–86.
F. J. Menzer, Jr. (1970), "Geochronologic Study of Granitic Rocks from the Okanogan Range, North-Central Washington," *Bull. Geol. Soc. Am.* **81,** 573–578.
C. Merrihue and G. Turner (1966), "Potassium-Argon Dating by Activation with Fast Neutrons," *J. Geophys. Res.* **71,** 2852–2857.
D. S. Miller (1968), "Fission Track Ages on 250 and 2500 M.Y. Micas," *Earth Planet. Sci. Lett.* **4,** 379–383.
D. S. Miller and E. Jaeger (1968), "Fission Track Ages of Some Alpine Micas," *Earth Planet. Sci. Lett.* **4,** 375–378.
D. S. Miller, M. Bagley and M. N. Rao (1968), "Fission Track Ages of Mica Minerals," *Z. Naturf.* **23A,** 1093–1094.
D. Milton and C. Naeser (1970), "Evidence for an Impact Origin of the Pretoria Salt Pan, South Africa," *Nature Phys. Sci.* **229,** 211–212.
D. J. Milton, B. C. Barlow, R. Brett, A. R. Brown, A. Y. Glikson, E. A. Manwaring, F. J. Moss, E. C. E. Sedmik, J. Van Son, and G. A. Young (1972), "Grosses Bluff Impact Structure, Australia," *Science* **175,** 1199–1207.
C. W. Naeser (1967a), *Fission Track Age Relationships in a Contact Zone*, Ph.D. Thesis, Southern Methodist University.

C. W. Naeser (1967b), "The Use of Apatite and Sphene for Fission Track Age Determinations," *Bull. Geol. Soc. Am.* **78,** 1523–1526.
C. W. Naeser (1969), "Etching Tracks in Zircons," *Science* **165,** 388.
C. W. Naeser (1971), "Geochronology of the Navajo-Hopi Diatremes, Four Corners Area," *J. Geophys. Res.* **76,** 4978–4985.
C. W. Naeser and F. C. W. Dodge (1969), "Fission-Track Ages of Accessory Minerals from Granitic Rocks of the Central Sierra Nevada Batholith, California," *Bull. Geol. Soc. Am.* **80,** 2201–2212.
C. W. Naeser and H. Faul (1969), "Fission-Track Annealing in Apatite and Sphene," *J. Geophys. Res.* **74,** 705–710.
C. W. Naeser and R. L. Fleischer (1975), "The Age of the Apatite at Cerro de Mercado, Mexico: A Problem for Fission-Track Annealing Corrections," *Geophys. Res. Letters,* in press.
C. W. Naeser and E. H. McKee (1970), "Fission-Track and K-Ar Ages of Tertiary Ash-Flow Tuffs, North-Central Nevada," *Bull. Geol. Soc. Am.* **81,** 3375–3384.
C. W. Naeser and J. M. Saul (1973), "Fission Track Dating of Tanzanite," *Am. Mineralogist* **59,** 613–614.
C. W. Naeser, J. C. Engels, and F. C. W. Dodge (1970), "Fission Track Annealing and Age Determination of Epidote Minerals," *J. Geophys. Res.* **75,** 1579–1584.
C. W. Naeser, R. Kistler, and F. Dodge (1971), "Ages of Coexisting Minerals from Heat-Flow Borehole Sites, Central Sierra Nevada Batholith," *J. Geophys. Res.* **76,** 6462–6463.
C. W. Naeser, G. A. Izett, and R. E. Wilcox (1973), "Zircon Fission-Track Ages of Pearlette Family Ash Beds in Meade County, Kansas," *Geology* **1,** 93–95.
J. Nagai (1968), "Fission Tracks in Zircon: Age, Distribution and Concentration of Uranium," *Mem. Fac. Sci. Kyoto Univ. Ser. Geol. Miner.* **35,** 119–130.
K. K. Nagpaul, M. K. Nagpal, P. P. Mehta, and M. L. Gupta (1973), "Fission Track Ages of Some Indian Pegmatite and Granites," Geochronology Conference, "ECOG II," Oxford, 3–8 Sept., 55.
S. Nishimura (1969), "The Determination of Ages and Uranium Content of Some Minerals and Glasses by Fission Track Registration," *Proc. 9th Jap. Radioisotope Conf.*, Tokyo, 223–225.
S. Nishimura (1971), "Fission Track Dating of Archaeological Materials from Japan," *Nature* **230,** 242–243.
S. Nishimura (1972), "Absolute Age Determination Methods," *Tampakushitsu Kakusan Koso, Bessatsu,* 49–59.
S. Nishimura and S. Sasajima (1970), "Fission-Track Age of Volcanic Ash-Layers of the Plio-Pleisto Series in Kinki District, Japan," *Earth Sci.* (*Chikyu Kagaku*) **24,** 222–224.
M. Oncescu, P. Sandru, A. Danis, and S. Minazatu (1972), "Influence of Experimental Factors on Age Results Obtained by the Fission Track Dating Method," *Rev. Roum. Phys.* **17,** 63–69.

O. Otgonsuren, V. P. Perelygin, and G. N. Flerov (1969), "Search for Very Far Transuranium Elements in Iron-Manganese Nodules," *Doklady Akad. Nauk USSR* **189,** 1200–1203.

O. Otgonsuren, V. P. Perelygin, S. P. Tretyakova, and Yu. A. Vinogradov (1972), "Search for Tracks of Spontaneous Fission Fragments from the Heavier Transuranium Elements in Naturally-Occurring Minerals," *Atomnaya Energiya*, USSR, **32,** 344–347.

A. R. Patel and M. M. Patel (1969), "Fission Tracks on (001) Cleavages of Topaz," *Indian J. Pure Appl. Phys.* **7,** 491–493.

A. R. Patel, M. K. Agarwal, and C. C. Desai (1967), "Fission Tracks on $(10\bar{1}0)$ Cleavages of Natural Apatite Crystals," *J. Phys. Soc. Japan* **23,** 553–556.

P. Pellas (1969), "Etude des Traces Fossiles dans les Materiaux Terrestres: Geochronologie, Geophysique et Archeologie," *Proc. Inter. Conf. Nucl. Track Registration in Insulating Solids*, Clermont-Ferrand, Sect. VI, 2–4.

D. D. Peterson (1969), *A Study of Low Energy, Heavy Cosmic Rays Using a Plastic Charged Particle Detector*, Ph.D. Thesis, Rensselaer Polytechnic Institute.

K. A. Petrzhak and G. N. Flerov (1940), "Spontaneous Fission of Uranium," *Phys. Rev.* **58,** 89.

G. Poupeau (1969), "Le Sandia Granite. Measures d'Ages 'Traces de Fission,' " *Proc. Inter. Conf. Nucl. Track Registration in Insulating Solids*, Clermont-Ferrand, Sect. VI, 53–54.

P. B. Price and R. L. Fleischer (1970), "Particle Identification by Dielectric Track Detectors," *Rad. Effects* **2,** 291–298.

P. B. Price and R. M. Walker (1962), "Observation of Fossil Particle Tracks in Natural Micas," *Nature* **196,** 732–734.

P. B. Price and R. M. Walker (1963), "Fossil Tracks of Charged Particles in Mica and the Age of Minerals," *J. Geophys. Res.* **68,** 4847–4862.

P. B. Price, R. L. Fleischer, and C. D. Moak (1968), "Identification of Very Heavy Cosmic Ray Tracks in Meteorites," *Phys. Rev.* **167,** 277–282.

P. B. Price, R. L. Fleischer, and R. T. Woods (1970), "Search for Spontaneously Fissioning Trans-Uranic Elements in Nature," *Phys. Rev.* **C1,** 1819–1821.

P. B. Price, D. Lal, A. S. Tamhane, and V. P. Perelygin (1973), "Characteristics of Tracks of Ions of $14 \leq Z \leq 36$ in Common Rock Silicates," *Earth Planet. Sci. Lett.* **19,** 377–395.

G. M. Reimer, D. Storzer, and G. A. Wagner (1970), "Geometry Factor in Fission Track Counting," *Earth Planet. Sci. Lett.* **9,** 401–404.

J. H. Roberts, R. Gold, and R. J. Armani (1968), "Spontaneous-Fission Decay Constant of ^{238}U," *Phys. Rev.* **174,** 1482–1484.

M. Sakagami, T. Nakanishi, and T. Nakagi (1967), "Studies on the Observation of Particle Tracks (fission tracks, alpha recoils) in Solid Detectors," *11th Japan Symp. on Radiochem.*, Tokyo, 63–64.

K. Sato, S. Aramaki, and J. Sato (1972), "Discrepant Results of Carbon-14 and Fission Track Datings for Some Volcanic Products in Southern Kyushu," *Geochem. J.* **6,** 11–16.

J. W. H. Schreurs, A. M. Friedman, D. J. Rokop, M. W. Hair, and R. M. Walker (1971), "Calibrated U-Th Glasses for Neutron Dosimetry and Determination of Uranium and Thorium Concentrations by the Fission Track Method," *Rad. Effects* **7,** 231–233.

E. Segré (1952), "Spontaneous Fission," *Phys. Rev.* **86,** 21–28.

M. G. Seitz, R. M. Walker, and B. S. Carpenter (1973), "Improved Methods for Measurement of Thermal Neutron Dose by the Fission Track Technique," *J. Appl. Phys.* **44,** 510–512.

M. Shima and A. Okada (1968), "Study on the Manganese Nodule. (I) Manganese Nodules Collected from a Long Deep-Sea Core on the Mid-Pacific Ocean Floor," *Ganseki Kobutsu Kosho Gakkaishi* **60,** 47–56.

M. Shima and S. Yabuki (1969), "Cross-Checking of Fission Track Method and Potassium-Argon Method for Age Dating, II: Volcanic Rocks," *Ganseki Kobutsu Kosho Gakkaishi* **62,** 299–303.

M. Shima, S. Amano, and A. Okada (1966), "Analytical Method for Uranium and Age-Determination Procedure Using Tracks of Charged Particles," *Rika Gaku Kenkyusho Hokoku* **42,** 205–210.

M. Shima, S. Amano, and A. Okada (1967), "Age Dating of Volcanic Glass Collected from the Mid-Pacific Ocean Floor," *Bull. Nat. Sci. Mus. Tokyo* **10,** 467–470.

M. Shima, A. Okada, and H. Yabuki (1969), "Cross-Checking of the Fission Track Method and Potassium-Argon Method for Age Dating," *Ganseki Kobutsu Kosho Gakkaishi* **61,** 100–105.

Yu. A. Shukolyukov (1970), "Fission of Uranium Nuclei in Nature," *Atomizdat*, Moscow, *Appl. in Geochim., Cosmochemistry of Radiation Defects Induced by Fission Fragments of Uranium*, 168–226 (bibliography on pp. 256–266).

Yu. A. Shukolyukov and A. N. Komarov (1966), "Possible Paleothermometry on the Grounds of Fission Tracks of Uranium in Minerals," *Izv. Akad. Nauk SSSR* **9,** 137–141, Ser. Geol. **31.**

Yu. A. Shukolyukov and A. N. Komarov (1970), "Tracks of Fission Fragments in Monazites," *Izv. Akad. Nauk SSSR* **9,** 20–26, *Ser. Geol.* **31.**

Yu. A. Shukolyukov, I. N. Krylov, I. N. Tolstikhin, and G. V. Ovchinnikova (1965), "Tracks of Uranium Fission Fragments in Muscovite," *Geochem. Inter.* **2,** 203–213, *Geokhimiya 1965*, **3,** 291–301.

R. F. Sippel and E. D. Glover (1964), "Fission Damage in Calcite and the Dating of Carbonates," *Science* **144,** 409–410.

G. Somogyi and M. Nagy (1972), "Remarks on Fission-Track Dating in Dielectric Solids," *Rad. Effects* **16,** 223–231.

T. A. Steven, H. H. Mehnert, and J. D. Obradovich (1967), "Age of Volcanic Activity in the San Juan Mountains, Colorado," U.S. Geol. Surv. Prof. Paper 575-D, 47–55.

W. Stolz and B. Dörschel (1968), "Solid State Track Detectors," *Kernenergie* **11,** 137–147.

D. Storzer (1970), "Fission Track Dating of Volcanic Glasses and the Thermal History of Rocks," *Earth Planet. Sci. Lett.* **8,** 55–60.

D. Storzer (1971), "Fission Track Dating of Some Impact Craters in the Age Range Between 6,000 y and 300 My," *Meteoritics* **6,** 319.

D. Storzer (1973), "Fission Track Plateau-Ages: A New Method for Correcting Thermally Lowered Track Ages," Geochronology Conference, "ECOG II," Oxford, 3–8 Sept., 67.

D. Storzer and W. Gentner (1970), "Spaltspuren-Alter von Riesgläsern, Moldaviten und Bentoniten," *Jber. Mitt. Oberrh. Geol. Ver.* **52,** 97–111.

D. Storzer and G. Poupeau (1973), "Fission Track Dating of Lunar Glass Spherules," *Meteoritics* **8,** 444–445.

D. Storzer and G. A. Wagner (1969), "Correction of Thermally Lowered Fission Track Ages of Tektites," *Earth Planet. Sci. Lett.* **5,** 463–468.

D. Storzer and G. A. Wagner (1970), "A Correction Method for Thermally Lowered Fission Track Ages," *Rad. Effects* **5,** 129–131.

D. Storzer and G. A. Wagner (1971), "Fission Track Ages of North American Tektites," *Earth Planet. Sci. Lett.* **10,** 435–440.

D. Storzer and G. A. Wagner (1973), "Längenanalysen von Spaltspuren in Apatiten aus dem Odenwald und ihre Bedeutung für die thermische und tektonische Geschichte von Gesteinen," *Fortschr. Min.* **50,** 22–23.

D. Storzer, P. Horn, and B. Kleinmann (1971), "The Age and the Origin of Köfels Structure, Austria," *Earth Planet. Sci. Lett.* **12,** 238–244.

D. Storzer, G. A. Wagner, and E. A. King (1973a), "Fission Track Ages and Stratigraphic Occurrence of Georgia Tektites," *J. Geophys. Res.* **78,** 4915–4919.

D. Storzer, G. Poupeau, and M. J. Orcel (1973b), "Ages-plateaux de Mineraux et Verres par la Methode des Traces de Fission," *Compt. rend.*, Paris, **276,** 137–139.

J. S. Stuckless and C. W. Naeser (1972), "Rb-Sr and Fission-Track Age Determinations in the Precambrian Plutonic Basement Around the Superstition Volcanic Field, Arizona," *U.S. Geol. Surv. Prof. Paper* **800B,** B191–B194.

J. S. Stuckless and M. F. Sheridan (1971), "Tertiary Volcanic Stratigraphy with Goldfield and Superstition Mountains, Arizona," *Bull. Geol. Soc. Am.* **82,** 3235–3240.

S. S. S. Sun (1971), *Fission Track Study of the Cheney Pond Titaniferous Iron Ore Deposit, Tahawus, New York*, Ph.D. Thesis, Washington University.

M. Suzuki (1969a), "Fission Track Dating and Uranium Contents of Obsidian (I)," *Daiyonki Kenkyu* (*Quaternary Research*) **8,** 123–130.

M. Suzuki (1969b), "Fission Track Identification of Geologic Source of Obsidian Artifacts Collected from the Sone Site," *Archaeological J.* **36,** 12–15.
M. Suzuki (1970a), "Fission Track Ages and Uranium Contents of Obsidians," *J. Anthro. Soc. Nippon* **78,** 50–58.
M. Suzuki (1970b), "Fission Track Dating and Uranium Contents of Obsidian II," *Daiyonki Kenkyu* **9,** 1–6.
M. Suzuki (1972), "Chronology of the Tachikawa Loam as Established by Fission Track and Obsidian Hydration Dating," *Daiyonki Kenkyu* **11,** 281–288.
M. Suzuki (1973a), "Chronology of Prehistoric Human Activity in Kanto, Japan, Part I—Framework for Reconstructing Prehistoric Human Activity in Obsidian," *J. Fac. Sci.*, Univ. of Tokyo, Sec. V, **3,** 241–318.
M. Suzuki (1973b), "Discussion of ^{14}C and Fission Track Obsidian Hydration Dates for the Ceramic and Preceramic Periods in Japan," *Archaeological J.* **81,** 88–93.
M. Suzuki and K. Chinzei (1973), "The Use of Obsidian for Fission Track Dating with Special Reference to the Fading of Spontaneous Fission Tracks Observed in Samples from the Niigata Oil Field," *Mem. Geol. Soc. Japan* **8,** 173–182.
M. Suzuki and N. Watanabe (1968), "Fission Track Dating of Archaeological Materials from Japan," VIIIth Congress of Anthropological and Ethnological Sciences, Tokyo, Sept. 3–10, 169–171.
M. Suzuki and N. Watanabe (1969), "Fission Track Dating of Archaeological Materials from Japan," *Nature* **222,** 1057–1058.
M. Suzuki and T. Yamanoi (1970), "Fission Track Dating of the Uonuma Group," *J. Geol. Soc. Japan* **76,** 317–318.
N. G. Syromyatnikov, L. A. Trofimova, and A. I. Ivanov (1971), "Determination of the Age of Minerals According to the Tracks from the Spontaneous Fission of Uranium," *Izv. Akad. Nauk Kaz. SSR* **27,** 52–57.
B. J. Szabo, H. E. Malde, and C. Irwin-Williams (1969), "Dilemma Posed by Uranium Series Dates on Archaeologically Significant Bones from Valsequillo, Puebla, Mexico," *Earth Planet. Sci. Lett.* **6,** 237–244.
V. Togliatti (1965), "Distribuzioni dei Ranges e Distribuzioni Angolari delle Tracce di Fissioni Fossili di U^{235} in Mica," *Boll. Geofis. Teor. Appl.* **7,** 326–335.
C. Turkowsky (1969), "Electron-Microscopic Observation of Artificially Produced Alpha-Recoil Tracks in Albite," *Earth Planet. Sci. Lett.* **5,** 492–496.
L. Turner, B. Forbes and C. W. Naeser (1973), "Radiometric Ages of Kodiak Seamount and Giacomini Guyot, Gulf of Alaska: Implications for Circum-Pacific Tectonics," *Science* **182,** 579–581.
E. E. Uzgiris and R. L. Fleischer (1971), "Amber: Charged Particle Track Registration, Track Stability, and Uranium Content," *Nature* **234,** 28–30.
V. Vand, F. Dachille, and P. Y. Simons (1964), "Qualitative Dating of Glasses—Applied to Tektite-Like Objects from the Rieskessel Meteoritic Crater," *Nature* **201,** 597–598.

G. A. Wagner (1966a), "Age Measurements on Tektites and Other Natural Glasses by Tracks from Spontaneous Fission of U^{238}," *Z. Naturf.* **21A**, 733–745.

G. A. Wagner (1966b), "Altersbestimmungen an Gläsern und Mineralien mit der Spaltungsspurenmethode (unter dem Mikroskop)," *Z. Deutsch Geol. Ges. Jahrgang* **118,** 209–216.

G. A. Wagner (1968), "Fission Track Dating of Apatites," *Earth Planet. Sci. Lett.* **4,** 411–415.

G. A. Wagner (1969), "Spuren der Spontanen Kernspaltung des ^{238}Urans als Mittel zur Datierung von Apatiten und ein Beitrag zur Geochronologie des Odenwaldes," *N. Jb. Miner. Abh.* **110,** 252–286.

G. A. Wagner (1972), "Spaltspurenalter von Mineralen und Naturlichen Gläsern: eine Ubersicht," *Fortschr. Miner.* **49,** 114–145.

G. A. Wagner and G. M. Reimer (1972), "Fission Track Tectonics: The Tectonic Interpretation of Fission Track Apatite Ages," *Earth Planet. Sci. Lett.* **14,** 263–268.

G. A. Wagner and M. Reimer (1973), "Fission Track Ages of Apatites from the Alps and Their Tectonic Interpretation," Geochronology Conference, "ECOG II," Oxford, 3–8 Sept., 73.

G. A. Wagner and D. Storzer (1970), "Die Interpretation von Spaltspurenaltern (fission track ages) am Beispiel von Naturlichen Gläsern, Apatiten und Zirkonen," *Ecologae Geologicae Helvetiae,* **63/1,** 335–344.

G. A. Wagner and D. Storzer (1972), "Fission Track Length Reductions in Minerals and the Thermal History of Rocks," *Trans. Amer. Nucl. Soc.* **15,** 127.

R. M. Walker (1963), "Characteristics and Applications of Solid State Track Detectors," *Proc. Strasbourg Conf. on New Methods of Track Detection,* Centre de Recherches Nucleaires, Strasbourg, France.

R. M. Walker, M. Maurette, R. L. Fleischer, and P. B. Price (1968), "Applications of Solid State Nuclear Track Detectors to Archeology," in *Science and Archaeology,* R. H. Brill (ed.), M.I.T. Press: Cambridge, Chap. 21, pp. 271–283.

N. Watanabe and M. Suzuki (1969), "Fission Track Dating of Archaelogical Glass Materials from Japan," *Nature* **222,** 1057–1058.

E. Welin, I. Lundström, and G. Åberg (1972), "Fission Track Studies on Hornblende, Biotite and Phlogopite from Sweden," *Bull. Geol. Soc. Finl.* **44,** 35–46.

R. T. Woods, R. L. Fleischer, and P. B. Price (1971), "Application of Particle Track Detectors to Geochemistry and Geochronology Including the Discovery of a Second Tektite Fall in Australia and the Search for Spontaneously Fissioning Transuranic Elements in Nature," *International Geochemical Congress,* A. I. Tubarinov (ed.), 203, Moscow. Acad. Sci. USSR.

H. Yabuki and M. Shima (1970), "Fission Track Dating of Zircon in the Quaternary Period," *Rika Gaku Kenkyusho Hokoku* (*Reports of the Inst. of Special and Chemical Research*) **46,** 59–62.

D. W. Zimmerman, M. P. Yuhas, and P. Meyers (1973), "Thermoluminescence Authenticity Measurements on Core Material from the Bronze Horse at the New York Metropolitan Museum of Art," submitted to Archaeometry.
R. A. Zimmermann and G. M. Reimer (1973), "Fission Track Dates of Apatites from the Northern Appalachians of Eastern North America," Geochronology Conference, "ECOG II," Oxford, 3–8 Sept., 74.

Chapter 5

Modern Energetic Particles in Space

5.1. INTRODUCTION

In astronomy texts written as recently as 1959 interplanetary space was depicted as an evacuated medium containing planets, comets, asteroids, and dust. The magnetic fields of the earth and sun were thought to be approximately dipolar (Fig. 5-1a). Cosmic rays were known to be high-energy charged particles, mostly protons, that entered the solar system from all directions. Those with sufficiently high magnetic rigidity (momentum/charge) could penetrate the earth's field and secondaries resulting from nuclear interactions in the earth's atmosphere could be detected in balloons and at ground level. In a historic balloon flight in 1948 Freier et al. (1948) found tracks of primary cosmic rays with Z up to $\sim$26 in nuclear photographic emulsions. In the 1940's and 1950's ground-level ion chambers and neutron monitors established that particles up to a few GeV were emitted once every few years during solar flares.

Since Sputnik our concept of interplanetary space has undergone revolutionary changes. We now know that a tenuous, magnetized plasma streams out from the sun at all times, forming a solar wind that distorts and limits the earth's magnetic field to a region called the geomagnetosphere (see Fig. 5-1b). This region extends only to a distance of about 10 Earth radii in the sunward direction but stretches out far beyond the orbit of the moon in the antisunward direction, forming a long, cometlike tail. Within the magnetosphere are a complex variety of particles and fields, the most famous being the Van Allen belts of trapped radiation. The combined action of the radially streaming, ionized solar wind and the rotating sun cause the solar magnetic field lines to assume roughly the shapes of Archimedes spirals that extend out many AU beyond Earth (1 AU = mean distance from sun to Earth) as sketched in Fig. 5-2. The solar wind expands spherically at supersonic speeds ($\sim$400 km/sec on the average) until it reaches a distance from the sun

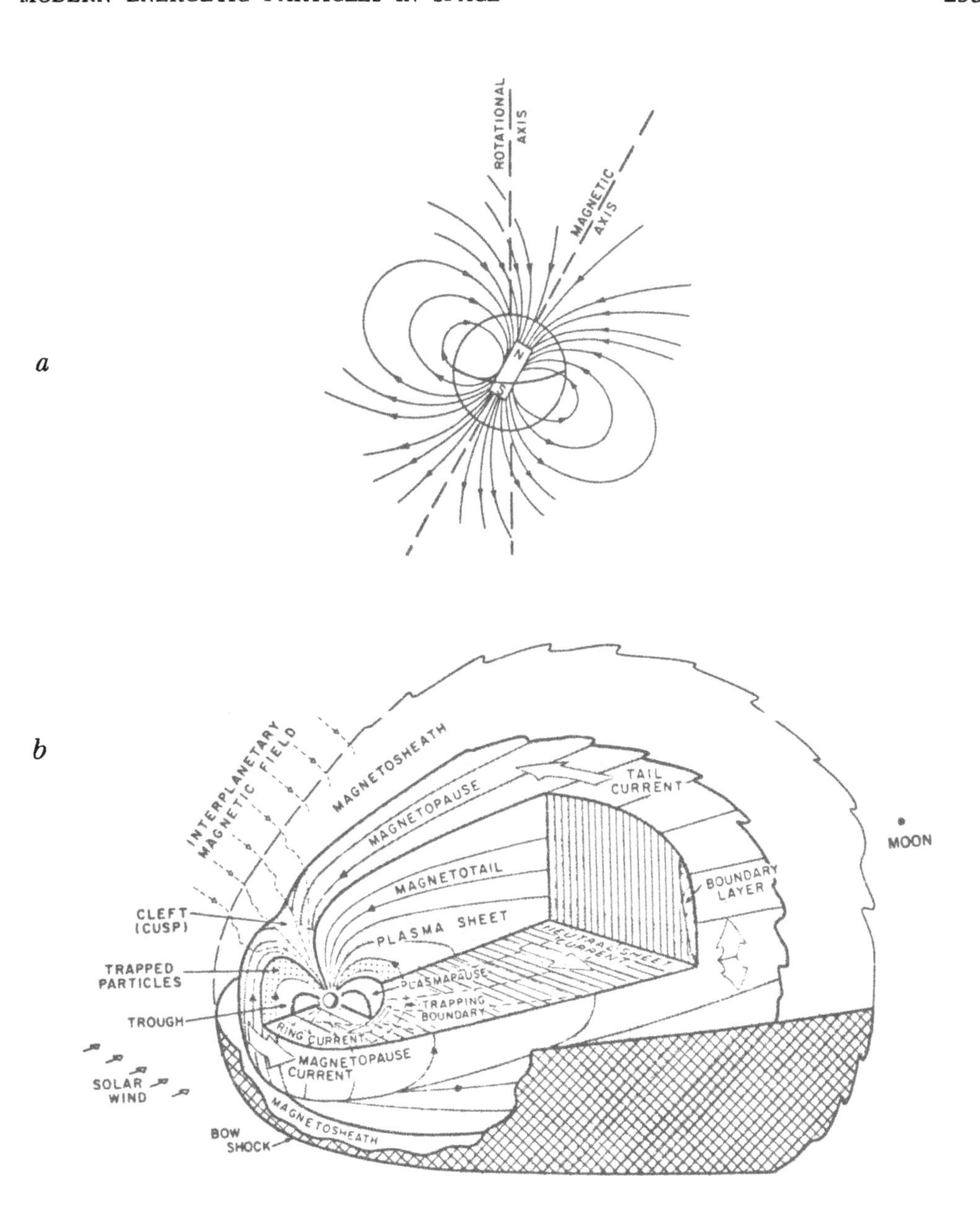

Fig. 5-1. (*a*) *Concept of Earth's magnetic field prior to the space age.* (*After Struve et al., 1959.*) (*b*) *Modern concept of Earth's magnetic field.* (*After Space Science Board, 1973.*)

where its pressure is balanced by the pressure of the interstellar medium. There it probably undergoes a shock transition to subsonic flow; the subsonic plasma beyond the shock forms a boundary shell. Because the solar system is moving with respect to the interstellar medium, the "heliomagnetosphere" is expected to be asymmetric with a tail extending in the direction opposite to that of motion of the

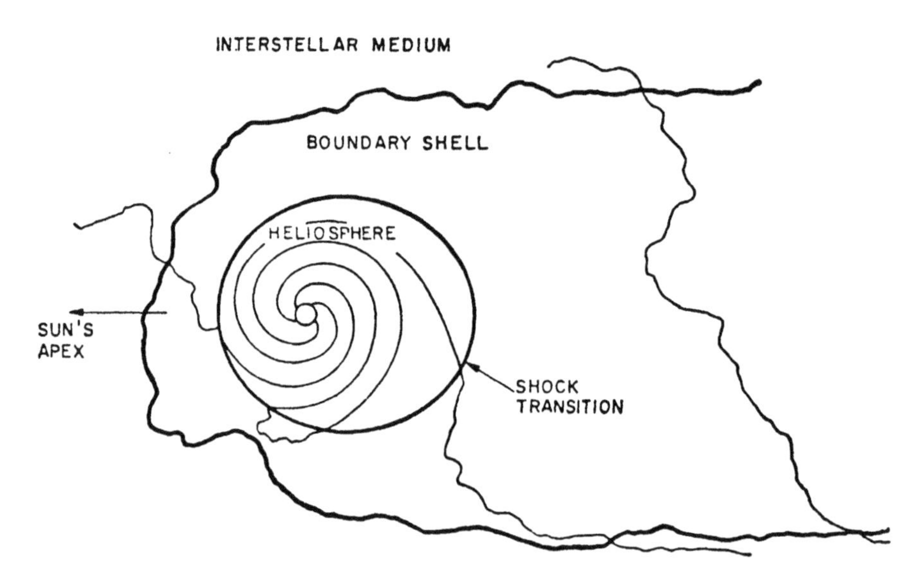

Fig. 5-2. *Modern concept of spiral magnetic field structure viewed in the plane of the ecliptic.* (*After Kovar and Dessler, 1967.*)

sun and analogous to the Earth's tail. For an elementary but more detailed description of interplanetary space we suggest the book by Haymes (1971).

The main subject of this chapter is the energetic particles that are found within interplanetary space at the present day, as distinguished from the study of records of ancient energetic particles in the meteorites and lunar samples, which will be covered in the next chapter. The emphasis will, of course, be on those aspects of energetic particles that can be studied with nuclear tracks in solids. We must, therefore, narrow the discussion to include only charged particles with ionization rates high enough to leave detectable tracks. In so doing, we exclude in this chapter several kinds of radiations—x rays and gamma rays; electrons and positrons; solar neutrinos; and neutrons—all of which are currently of great interest and importance in space science and astrophysics. We are left with energetic nuclei, which can be classified according to their origin and charge and energy distribution. Fig. 5-3 gives our best estimate of the average energy spectrum in interplanetary space of Fe nuclei—the most abundant of the very heavy cosmic rays. At the end of the chapter we will discuss how the curves were derived. In the next three sections we shall discuss each of the categories of energetic particles in turn, beginning with the most energetic—the galactic cosmic rays—and ending with the solar wind ions.

Another category of energetic particles of interest to geophysics and plasma physics is found within the magnetospheres of magnetized planets such as the Earth and Jupiter. Consisting mainly of electrons and protons, they can be trapped in quasi-stable orbits within the radiation belts and are also detected in other

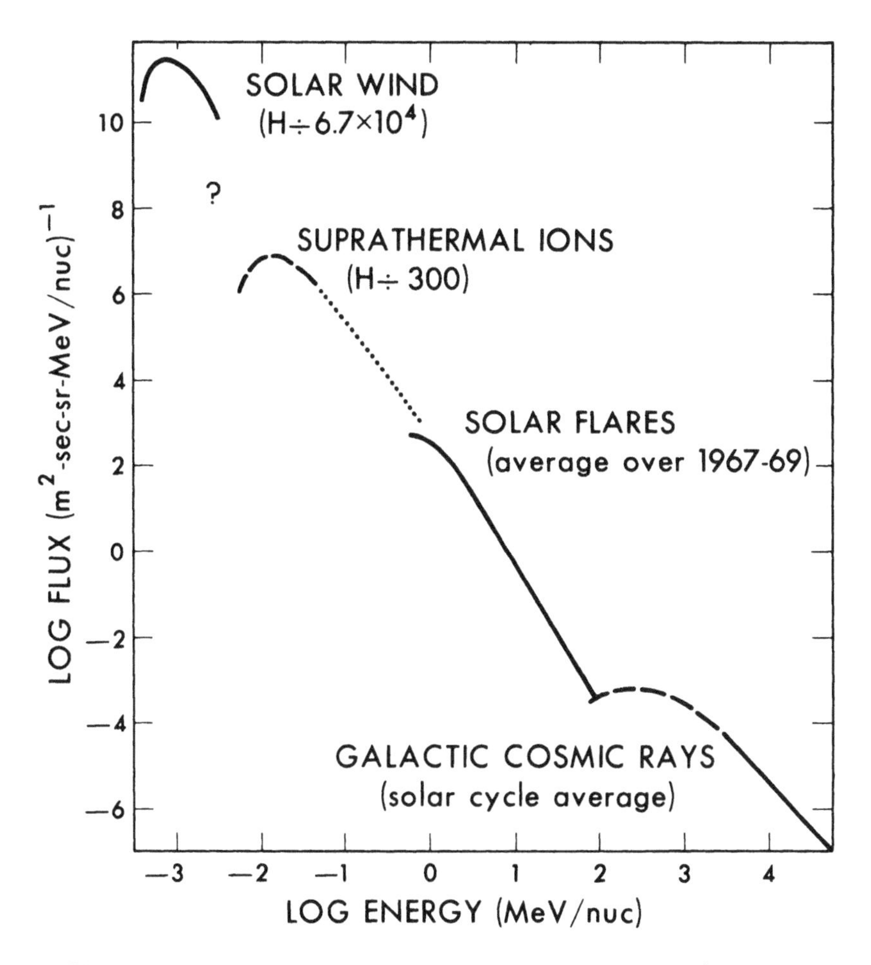

Fig. 5-3. *Average present-day energy spectrum of nuclei with $Z \geq 24$ in interplanetary space. See discussion in Section 5.5.*

regions within the magnetosphere, giving rise to large-scale currents sketched in Fig. 5-1b. Magnetospheric particles with $Z \geq 2$ exist but are quite rare (Mogro-Campero and Simpson, 1970; Krimigis et al., 1970). Solid track detectors judiciously deployed in Earth-orbiting satellites would offer an attractive means for determining the composition of heavy trapped particles in a background of singly charged trapped particles, but have not so far been used in this way.

The basic data on interplanetary charged particles are their composition, energy spectra, time variation, spatial variation, and angular distribution. Much of our knowledge comes from instruments on spacecrafts located outside the magnetosphere, but there are also many occasions when it is possible, even sometimes preferable, to make measurements in high-altitude balloons or in sounding rockets. A case in point is the study of the rare, ultraheavy elements in the cosmic radiation. Giant balloons containing 2×10^7 ft^3 of helium can maintain one-ton payloads at a residual atmospheric depth of ~3 g/cm^2 for at least 40 hours, exposing active areas up to perhaps 40 m^2 to cosmic rays. Electronic detectors on satellites have an active area of only ~1 cm^2. All of our knowledge of the present-day trans-iron cosmic rays has come from balloon-borne detectors.

Once a detector is above the blanket of air around the Earth, the problem is to take into account the complex interrelationship of particles and fields in the magnetosphere and in the interplanetary space so that the characteristics of the particles themselves can be determined. We can see from Figs. 5-1b and 5-2 that a detector at the top of the Earth's atmosphere is cloaked within the geomagnetosphere, which in turn is cloaked within the heliomagnetosphere. When galactic cosmic rays enter the heliosphere their energy spectrum is modulated by the solar wind. The magnetic field lines carried out from the sun by the solar wind do not have a perfectly uniform spiral configuration but contain small-scale irregularities which act as scattering centers through which the charged cosmic rays must diffuse. The steady state between the outward *convection* and inward *diffusion* establishes the density of cosmic rays within interplanetary space as some fraction of their density in interstellar space outside the heliosphere. In addition, because of the radial divergence of the solar wind, cosmic rays are *adiabatically decelerated* when they begin moving in the magnetic fields carried outward by the wind. The extent of solar modulation varies with the 11-year activity cycle of the sun. During periods of high solar activity the cosmic ray activity near earth is at a minimum, while during solar minimum one observes a maximum intensity. Modulation phenomena have been described with considerable success by the transport model of Parker (1965). More recently Goldstein et al. (1970) have emphasized that, as a result of adiabatic deceleration, very low-energy cosmic rays cannot reach the vicinity of Earth and cosmic rays with energies less than a few hundred MeV/nucleon near Earth had energies higher by several hundred MeV/nucleon in interstellar space. Recently Urch and Gleeson (1973) have shown that, on the average, particles with $Z \geq 2$ entering the solar system lose about 150 MeV/nucleon at solar minimum or about 400 MeV/nucleon at solar maximum, with large dispersion about this average value.

The motion of charged particles in the geomagnetic field and the resulting geomagnetic cutoff as a function of latitude and longitude are reasonably well understood in terms of electromagnetic theory. To detect low-energy particles from balloons or rockets, one must choose a launch site near one of the geomagnetic poles, where the horizontal component of the earth's field is too weak to exclude those particles. To study ultraheavy nuclei at a balloon altitude of ~3 g/cm^2 in the northern hemisphere, there is little point in going further north than Minnesota, because the minimum energy detectable would be determined at more northern latitudes by residual air pressure rather than by the geomagnetic cutoff. On the other hand, rockets can carry detectors to altitudes of ~150 km, corresponding to a residual atmospheric thickness of only $\sim 10^{-5}$ g/cm^2, which makes it possible to detect particles practically down to solar wind energies if the cutoff is negligible. Most of the solar flare studies reported in Section 5.3 were made possible by firing rockets from a base located at Fort Churchill, Manitoba, where the cutoff is extremely low. Smart and Shea (1972) have calculated that the cutoff rigidity there is ~ 150 MV during the daytime and less than 20 MV at night, the difference resulting from the asymmetric shape of the magnetosphere (Fig. 5-1b). Since we will need to use these cutoffs in our discussion of rocket data, we point out here that the relation between magnetic rigidity R_m (usually expressed in MV) and kinetic energy E_n (usually expressed in MeV/nucleon) is

$$R_m = (A/Z)m_n c^2 \beta\gamma = (A/Z)\sqrt{E_n{}^2 + 1876\, E_n} \tag{5-1}$$

where $m_n c^2$ (≈ 938 MeV) is the rest energy of a nucleon, βc is the velocity and $\gamma = (1 - \beta^2)^{-1/2}$.

During a magnetic storm which often accompanies a solar flare, a plasma blast wave from the sun hits the magnetosphere and so strongly distorts the field lines at high latitude that the cutoff may temporarily drop to a negligible value. Other interesting consequences such as auroras, power and communication failures, and the disorientation of homing pigeons need not concern us.

5.2. GALACTIC COSMIC RAYS

We will apply the term "cosmic rays" only to the energetic nuclei that stream through the solar system from sources distributed through our Galaxy and perhaps beyond. To avoid confusion we will denote energetic particles emitted from the sun as "solar particles" or "solar flare particles" and avoid the term "solar cosmic rays" sometimes used by others. Fig. 5-4 shows the approximate location of our solar system in the Galaxy. The presence in the flattish, disclike portion of the Galaxy of a several microgauss magnetic field with numerous irregularities along it causes cosmic rays to be repeatedly scattered, and, even before they enter our solar system and interact with the solar magnetic field, their directions of travel are remarkably isotropic. We shall see in the next chapter that their presence is not a transient phenomenon, and they appear to be rather uniformly distributed in space and time. In energy they extend from as low as a few MeV (possibly even down to thermal energies) to maximum energies of at least $\sim 10^{20}$ eV with a strongly nonthermal spectrum, one that can be represented over many orders of magnitude as a power law in energy with an exponent ~ 2.6. Only a small portion of the energy spectrum of galactic cosmic rays is shown in Fig. 5-3; below a few GeV/nucleon solar modulation depresses the spectrum and we are ignorant of the true nature of the spectrum outside the heliosphere. If it continues to rise as

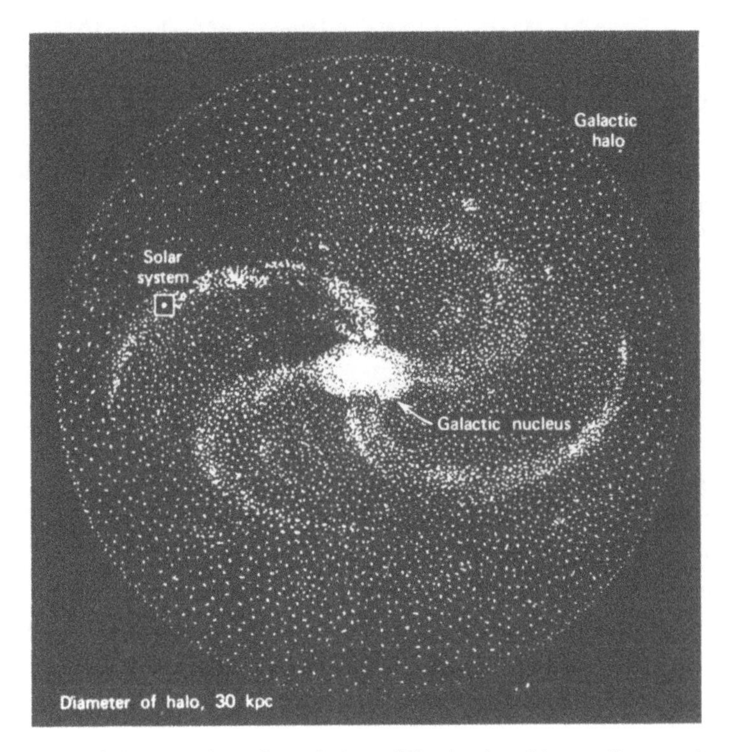

Fig. 5-4. *Schematic view of our Galaxy. The density of interstellar gas decreases from several atoms/cm³ in the densest regions of the spiral arms to less than* 10^{-2} *atom/cm³ in the halo.*

a power law, then the total energy density in low-energy cosmic rays is enough to explain the heating of interstellar gas clouds, which is a subject of considerable controversy among astrophysicists. Even at energies higher than a GeV/nucleon where their spectrum is fairly well-known, the energy density in cosmic rays is comparable to that of all other forms of energy in the Galaxy, and their dynamical interaction with galactic magnetic fields is believed to play an important role in the condensation of gas into stars.

As we shall see when we discuss Fig. 5-5, cosmic rays consist mostly of hydrogen, with about 10% helium and 1% consisting of heavier elements extending all the way through the Periodic Table. Because they are actually material particles from other parts of our Galaxy, their detailed elemental and isotopic composition may bear on several astrophysical problems including galactic cosmochemistry, stellar nucleosynthesis, nuclear reactions at high energy, and the origin and propagation of cosmic rays through the Galaxy. The task of particle identifiers such as solid track detectors is to provide compositional data, preferably at a variety of energies, which will make it possible to shed light on some of these astrophysical problems. To supplement the ensuing discussion of cosmic ray composition, we refer the reader to recent review articles by Meyer (1969), Shapiro and Silberberg (1970), and Cowsik and Price (1971).

Though the emphasis today is on the astrophysical role of cosmic rays, we should point out that cosmic rays first attracted the attention of the particle physicists, who discovered the positron, several kinds of mesons, and strange particles among their interaction products. Today the particle physicists use high-energy cosmic rays to test theories of both strong and weak interactions at energies beyond those accessible with accelerators and to search for hypothetical particles such as the quark, the intermediate boson, and the magnetic monopole. Through the production of radioisotopes in the earth's atmosphere and in meteorites and the moon's surface, cosmic rays are also useful in geophysics, planetology, archaeology, and meteorology. These more general aspects of cosmic rays are discussed by Pal (1967) and Hayakawa (1969). A historical introduction to cosmic ray research is given by Rossi (1964).

5.2.1. Chemical and Isotopic Composition of Cosmic Ray Nuclei up to Iron

Since the discovery of heavy cosmic ray nuclei by Freier et al. (1948), many attempts have been made to resolve individual elements, using first nuclear emulsions flown in balloons and later plastic detectors in balloons and electronic detectors on satellites. At first it was possible only to determine abundances of groups of charges, and cosmic rays were classified as hydrogen, helium, light nuclei ($3 \leq Z \leq 5$), medium nuclei ($6 \leq Z \leq 9$), heavy nuclei ($10 \leq Z \leq 19$), and very heavy nuclei ($Z \geq 20$). Resolution of individual charges is especially difficult when the ionization rate is different from that at the relativistic minimum, because then two parameters are required—a measure of ionization rate and a measure of velocity or range. A few careful measurements of the relative abundances of minimum-ionizing cosmic rays were made in the early 1960's (Kristiansson et al., 1963) but these studies were of limited value because very few particles were studied.

It was recognized very early (Bradt and Peters, 1950) that the relative abundance of naturally rare nuclei such as Li, Be, and B with respect to heavier nuclei could be used to determine the average amount of matter (predominantly gas) traversed by cosmic rays between their sources and the solar system. These nuclei, even if totally absent in the source, would be produced by spallation of heavier nuclei in collisions with interstellar gas. Taking into account the various cross sections for nuclear interactions, the relative abundance of "light" nuclei observed many years ago indicated that cosmic rays on the average have traversed several g/cm^2 of gas, and more recent measurements of relative abundances of other elements that are rare in nature but are observed in cosmic rays have confirmed and elaborated this picture. One of the motivations for trying to resolve individual isotopes is that additional information about the history of cosmic rays could be derived from measurements of relative abundances of long-lived radioactive isotopes such as ^{10}Be (mean life = 2.3×10^6 y). A knowledge of the mean age of cosmic rays from isotopic studies, together with a knowledge of the mean path-length traversed, would allow one to determine the density of matter in the region of the Galaxy in which cosmic rays propagate and thus to decide whether they are largely confined to the high-density, disclike region with its spiral arm structure or are somehow loosely contained in a much larger, more tenuous halo region (see Fig. 5-4). From a glance at Fig. 5-5 we see that there are several regions of the Periodic Table where elements are quite rare in nature but are present in much higher proportions in cosmic rays. We must keep in mind the necessity to

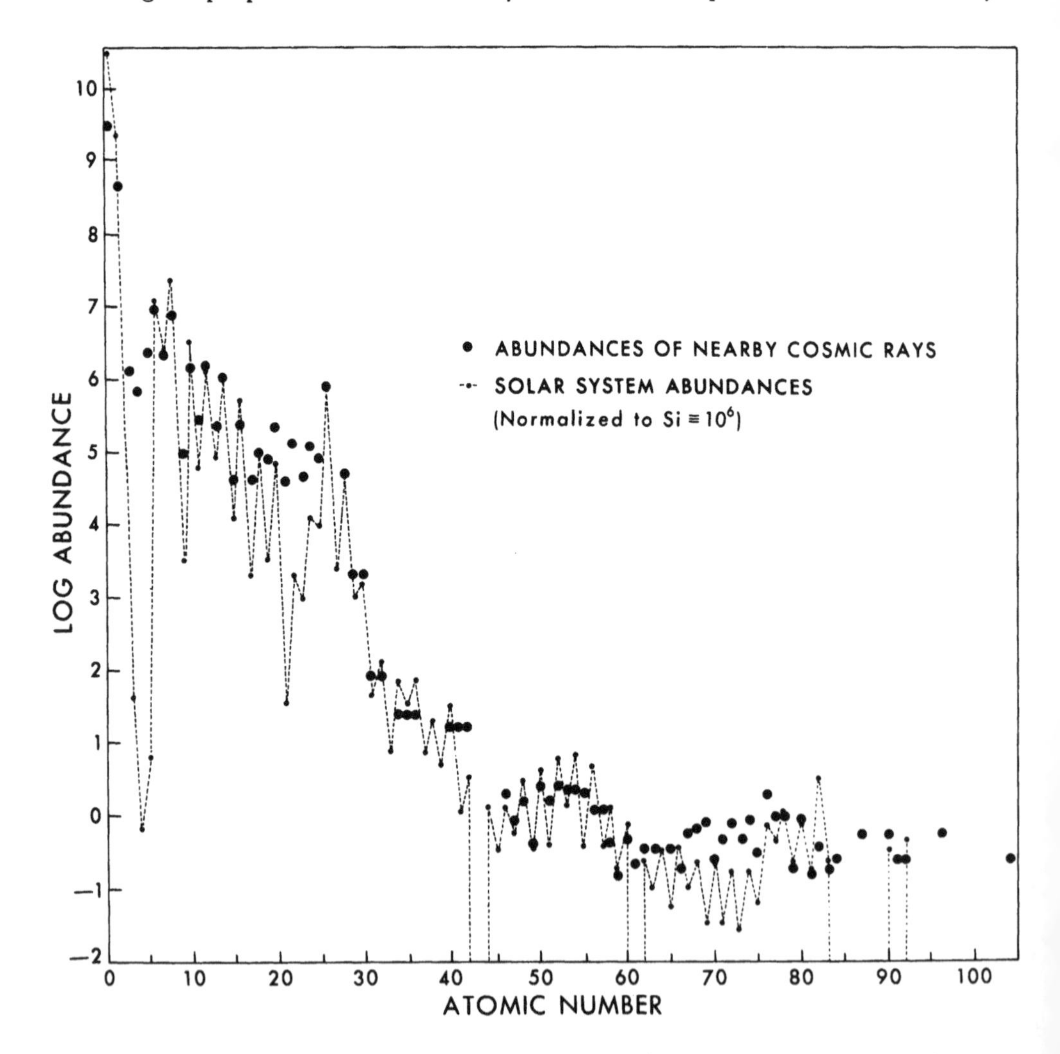

Fig. 5-5. *Composition of cosmic rays at the top of the atmosphere compared with solar system abundances. (After Price, 1973a.)*

correct the observed cosmic ray charge spectrum for products of nuclear reactions when we attempt later in this chapter to determine their abundances at sources.

Shortly before electronic detectors with good resolution began to operate on satellites, Lexan stacks flown in a balloon at Ft. Churchill produced the first charge spectrum of low-energy cosmic rays with good statistics extending from $Z = 12$ to 28 (Price et al., 1968). This experiment was the first application of the then new technique of particle identification by etch rate measurements. Particles with initial energies of ~200 to ~400 MeV/nucleon at the top of the atmosphere came to rest in a stack ~1 g/cm² thick. Measurements were made of etched cone length versus residual range after a standard etch time as described in the chapter on particle identification. The results, displayed in Fig. 5-6 with the symbols X, showed a preference for even-Z nuclei. The novelty of this information was not long-lasting, however, for at the Budapest Cosmic Ray Conference Garcia-Munoz and Simpson (1970) reported abundances for about the same energy interval, obtained with their satellite-based semiconductor detector, that confirmed the even-odd features and relative abundances found with the Lexan detector. Both the original Lexan data and additional measurements (Price et al., 1970) from a

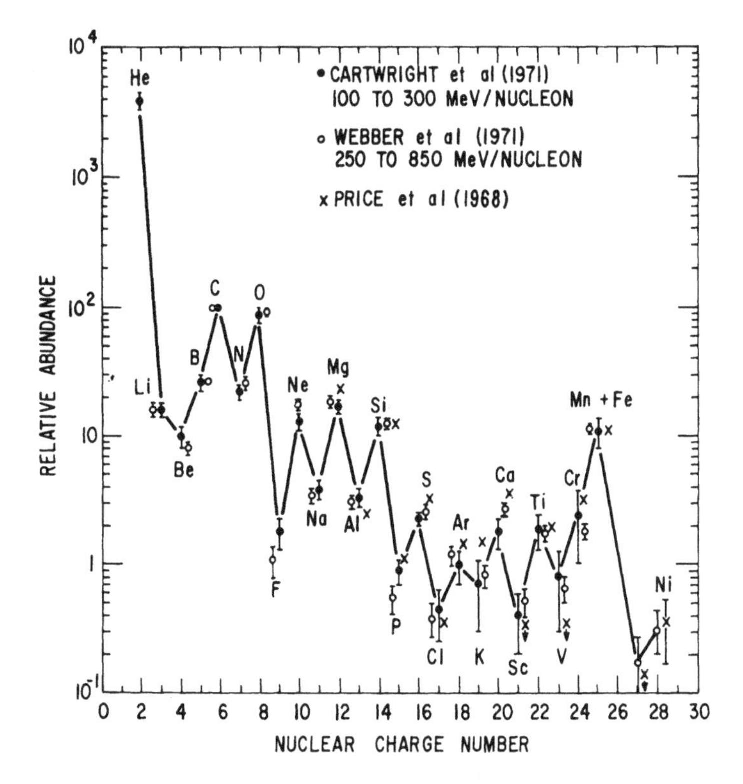

Fig. 5-6. *Relative abundances of elements in the galactic cosmic rays at low energies, normalized to carbon = 100.*

second balloon flight, presented at the same conference, indicated the presence of a surprisingly high abundance of Cr and Mn, each having ~40% the abundance of Fe. Price (1971) has discussed several possibilities: (1) the Mn and Cr might have been synthesized during a supernova explosion and accelerated to cosmic ray energies; (2) the Mn and Cr might be low-energy spallation products of Fe, since the cross-section reaches a strong maximum at ~200 MeV/nucleon; (3) the Fe might be dominantly ^{54}Fe, which would cause its signal (etch rate vs. residual range; see Chapter 3) to be more nearly like ^{55}Mn and make it more difficult to separate Mn and Fe than if it were dominantly ^{56}Fe.

Recent high-resolution results from Cartwright et al. (1971) and from Webber et al. (1972) are shown as closed and open circles in Fig. 5-6. The newer electronic data have better statistics and appear to have better resolution than Lexan, but the general agreement in the chemical composition serves the valuable function of demonstrating that the Lexan particle identification technique is on a firm footing.

One of the problems still not completely settled is whether the abundances of Cr and Mn are as high as measured in Lexan. The Chicago detector (Cartwright et al., 1971) does not attempt to separate Mn from Fe, but Webber et al. (1972) and Israel et al. (1973) report a very low Mn/Fe ratio in strong disagreement with the results from Lexan. Further experiments with Lexan detectors (Enge et al., 1971; Bartholomä et al., 1973) indicated a distribution of Cr, Mn, and Fe that was very similar to that of Price et al. (1968, 1970), whereas electronic measurements of the composition of very high-energy cosmic rays have failed to find a high Cr and Mn abundance (Cassé et al., 1971; Juliusson et al., 1972).

Except at the very lowest charges (Z = 1, 2, and 4), definitive isotopic resolution has proved to be an elusive goal. Improvements in the design of semiconductor particle identifiers have made it possible to identify isotopes of elements up to about nitrogen among the products of high-energy reactions in accelerator experiments (see, for example, Poskanzer et al., 1971), but the collecting power of such instruments has been too small for them to see use in cosmic ray studies. Webber et al. (1973a) are attempting to resolve isotopes of elements up to Fe, using a large system consisting of a Cerenkov radiation detector on top of a system of plastic scintillators, but the results are not yet convincing. Garcia-Munoz et al. (1973) have succeeded in separating the ^{7}Be isotope from the sum of ^{9}Be + ^{10}Be, a task that is facilitated because ^{8}Be is unstable. As yet they have not succeeded in separating the signals from ^{9}Be and ^{10}Be.

Relative abundances of the isotopes ^{1}H, ^{2}H, ^{3}He, and ^{4}He at energies from ~20 to ~300 MeV/nucleon have been reported by several groups, who used satellite-based electronic detectors to measure the changes of energy ΔE during passage through thin detectors and the total energy deposited in a thick detector. At the same energy/nucleon different isotopes of the same element have ranges and energy deposition rates proportional to their masses and can, in principle, be distinguished. Simpson (1972) has summarized the experimental and theoretical situation. The ratio $^3He/^4He$ appears to be $\sim$0.1 at all energies whereas the ratio $^2H/^4He$ appears to be less than $\sim$0.2, varying with energy, and still rather uncertain in value. When solar modulation (including adiabatic deceleration) and nuclear reactions leading to the production of ^2H and ^3He are taken into account, it is possible to start with no ^2H and ^3He in cosmic ray sources and to make the observed quantities by allowing ^1H and ^4He to pass through a distribution of pathlengths in interstellar gas with a mean value of $\sim$4 g/cm^2. This pathlength can also account for the relative abundances of Li, Be, and B and of heavier secondary nuclei such as the ones with $17 \leq Z \leq 25$ as well, provided there is a wide distribution of pathlengths ranging from very small to very large values. The simplest assumption consistent with the data is that the cosmic rays are in a steady state, being lost through nuclear interactions, radioactive decay, ionization loss, and leakage from their confinement regions in the Galaxy and being replaced by fresh production in a uniform distribution of sources (Cowsik et al., 1967). This assumption leads to an exponential distribution of leakage pathlengths with a mean value $\Lambda_l \approx 4$ g/cm^2, to which effective pathlengths for nuclear interaction, radioactive decay, and ionization loss are added harmonically:

$$\Lambda_l^{-1} + \Lambda_{\text{int}}^{-1} + \Lambda_{\text{decay}}^{-1} + \Lambda_{\text{ioniz}}^{-1} = \Lambda_{\text{total}}^{-1} \qquad (5\text{-}2)$$

As a first approximation Λ_l can be assumed independent of energy. In a later section we will discuss important recent evidence for a gradual change of Λ_l with energy. At high energies and for stable nuclei only the first two terms are important. Their relative importance varies with the mass A of the cosmic ray, since

$$\Lambda_{\text{int}} = M_g/\sigma N_0 \propto M_g/N_0 A^{2/3} \qquad (5\text{-}3)$$

where M_g is the average atomic weight of interstellar gas and N_0 is Avogadro's number. For the light and medium nuclei $\Lambda_{\text{int}} \gg \Lambda_l$ so that Λ_{total} is about the same for the nuclei that produce ^2H and ^3He (predominantly ^1H and ^4He) and for the nuclei that produce Li, Be, and B (predominantly C, N, and O). For the heavier nuclei Λ_{int} becomes shorter than Λ_l so that their propagation tends to be governed by nuclear interactions rather than by leakage. For Fe, $\Lambda_{\text{int}} \approx 2.5$ g/cm^2 in interstellar hydrogen, and for U, $\Lambda_{\text{int}} \approx 0.7$ g/cm^2. The value of Λ_{total} thus gradually decreases with Z, so that the typical heaviest nuclei cannot have reached us from very distant sources, even though their probability of leakage is formally the same as that for the light nuclei. The steady state model of cosmic ray transport receives strong support from observations of ultraheavy nuclei, to be discussed in a later section. We will come back to this model in our discussions of energy-dependent composition and of cosmic ray origin. The reader interested in learning more details of cosmic ray transport theory should refer to recent papers by Shapiro and Silberberg (1970), Webber et al. (1972), Cowsik and Wilson (1973), and Meneguzzi (1973), and to the book by Ginzburg and Syrovatskii (1964).

Information about the isotopic distribution in nuclei with $Z > 3$ has been obtained by indirect methods. Detectors that measure Z and velocity (by means of a Cerenkov counter) can be used in conjunction with the geomagnetic cutoff rigidity to make statistical determinations of the mean A/Z of various nuclei. The technique, which is described by Lund et al. (1970), is based on the definition of magnetic rigidity—see eq. (5-1). In an exposure at a given latitude (i.e., at a given cutoff rigidity), of the nuclei with a particular Z that penetrate the magnetic field the spectrum of a high-mass isotope will extend to a lower velocity than the spectrum of a low-mass isotope. Many thousands of events with a given charge must be measured in order to estimate the mean mass of that charge, and the spectral shapes of all the elements compared must be the same. Lund et al. (1970) concluded that the dominant isotopes of the elements B, C, N, and O are ^{11}B, ^{12}C, ^{15}N, and ^{16}O and that for Be the dominant isotopes are ^{9}Be or ^{10}Be rather than ^{7}Be. (Note, however, that Garcia-Munoz et al. (1973) conclude that, in the energy interval 50 to 100 MeV/nucleon, ^{7}Be and $^{9}Be + ^{10}Be$ are of comparable abundance.)

Beaujean and Enge (1971, 1972) have used a cellulose nitrate stack to study the charge and isotope composition of cosmic rays with $5 \leq Z \leq 8$. Though they were not able to separate individual isotopes, they observed displacements of the positions of their charge peaks from precise agreement with the positions of peaks at the most abundant isotopes determined from accelerator calibration bombardments. They reported values of mean masses shown in Table 5-1. Their values agree rather well with the mean masses obtained by Webber with an electronic detector and with Lund et al. (1970), whose work we have just discussed. The results of Beaujean and Enge, using ordinary commercial cellulose nitrate, offer encouragement that the isotopes ^{7}Be, ^{9}Be, and ^{10}Be may be resolvable in future measurements to be made in stacks of cellulose nitrate manufactured under more stringent conditions. Abundance ratios $^{10}Be/^{9}Be$ and $^{7}Be/^{9}Be$ are of astrophysical importance because ^{10}Be, with its 2.3×10^{6} year mean life, plays the role of a clock, and ^{7}Be, being a radioisotope that decays only by K-capture, has an effective half-life that depends on its velocity and on the density of the interstellar medium through which it moves. Good measurements of the relative abundances of the

Table 5-1. Mean Isotopic Masses of Low-Energy Cosmic Rays

	Mean mass (amu)	
Element	(Beaujean and Enge)	Webber (unpublished)
B	10.88 ± 0.16	10.82 ± 0.10
C	12.22 ± 0.08	12.08 ± 0.07
N	14.62 ± 0.13	14.65 ± 0.08
O	≡16.00	≡16.00

beryllium isotopes at several different energies would thus enable limits to be placed on the mean age of the cosmic rays and on the mean density of the confining medium. The behavior of ^{7}Be and other isotopes that decay by electron capture is discussed by Raisbeck (1971). Other interesting astrophysical aspects of cosmic ray isotopes are discussed in the proceedings of a conference on Isotopic Composition of the Primary Cosmic Radiation (Dauber, 1971) and in papers by Cassé (1973a,b) and Stone (1973). The detailed study of isotopic abundances of cosmic rays, together with observations of nuclear gamma ray lines, are two areas of future research that will be most useful and exciting in the eventual elucidation of nuclear processes in distant reaches of the Galaxy.

As a by-product of their studies of ultraheavy cosmic rays, using plastic detectors, Blanford et al. (1972) have detected low-energy heavy cosmic rays at a geomagnetic latitude at which such particles in the primary radiation are excluded by the earth's field. They found the flux of particles with $6 \lesssim Z \lesssim 26$ and kinetic energies of < 150 MeV/nucleon at 4 g/cm^2 residual atmosphere over Texas to be $\sim 2 \times 10^{-5}$ particle/m^2 ster sec. This flux poses a serious obstacle to the study of ultraheavy cosmic rays because it turns out to be about half the intensity of geomagnetically accepted relativistic primary cosmic rays with $Z \gtrsim 36$. Some of the heaviest of these low-energy particles have an ionization rate comparable to that of the relativistic, much heavier particles and thus produce an event background that must be discriminated against in searches for ultraheavy cosmic rays in balloon flights at intermediate latitude. They can be recognized if there are several sheets of plastic through which the tracks can be traced, because the ionization rate will be seen to increase rapidly. Blanford et al. (1972) intrepreted these slow particles in terms of return albedo arising from the nuclear interactions of relativistic primary Fe-group cosmic rays in the earth's atmosphere.

5.2.2. Energy Dependence of Cosmic Ray Composition

Until quite recently there was no decisive evidence for any significant variation of cosmic ray composition with energy. Since 1972 the situation has changed in two ways. Several large electronic detectors have provided compositional information at energies up to ~ 100 GeV/nucleon, an order of magnitude beyond the highest energies previously explored. The results show that the leakage pathlength Λ_l is a decreasing function of energy and place strong constraints on models of cosmic ray transport. In 1973 both dielectric track detectors and a new generation of electronic detectors on satellites have opened up the previously unexplored region between ~ 0.1 and ~ 100 MeV/nucleon. Quite unexpectedly, several investigators have reported that in the interval between ~ 1 and ~ 10 MeV/nucleon the composition is quite different from that at either lower or at higher energies, suggesting that a new source of charged particles has been discovered. We will consider both the low-energy and the high-energy data in turn.

The Changing Composition at Energies Below 100 MeV/Nucleon

Glass detectors and stacks of Lexan and cellulose triacetate were exposed outside the lunar modules on both the Apollo 16 and 17 missions (Fleischer and Hart, 1972; Price et al., 1972; Burnett et al., 1972; Price and Chan, 1973; Walker and Zinner, 1973; Woods et al., 1973). The hope was to be able to determine the composition and energy spectra of heavy nuclei at energies below ~100 MeV/nucleon and from these measurements to establish the relative contribution of (1) solar particles, either as a residue from flares or as a continual quiet-time component, and (2) galactic cosmic rays that enter the solar system in spite of solar modulation. The relative abundance of secondary nuclei such as LiBeB and the elements $17 \leq Z \leq 25$ as a function of energy should be directly related to the galactic cosmic ray contribution since they are extremely rare in the sun and in solar flare particles.

During both missions the sun was uncooperative. During Apollo 16 a solar flare produced a temporary background of energetic particles that exceeded the quiet-time flux at energies out to ~30 MeV/nucleon. Although during Apollo 17 the sun was considerably quieter, there was still a significant solar contribution at energies out to ~1 MeV/nucleon. The low-energy particles of solar origin proved to be interesting in themselves and are discussed later in this chapter. Here we consider only the particles that appear to have originated outside the solar system.

From a cellulose triacetate stack on the Apollo 16 mission, O'Sullivan et al. (1973) determined the relative abundances of the nuclei from carbon to nickel, at energies from ~30 to ~130 MeV/nucleon. A shifting mechanism enabled them to distinguish tracks of nuclei that entered the stack when it was mounted outside the spacecraft from undesired tracks of nuclei that penetrated the spacecraft and entered the stack after it was brought inside. They found the important result that the abundances of individual elements at energies below 130 MeV/nucleon were indistinguishable from the abundance distributions of Cartwright et al. (1971) and of Webber et al. (1972) shown in Fig. 5-6 for energies up to 850 MeV/nucleon. Table 5-2 compares the results of O'Sullivan et al. and Webber et al. for certain charge groupings. The ratios are independent of energy within counting statistics. At first sight, it might be expected that the abundance ratio $[17 \leq Z \leq 25]/[\mathrm{Fe} + \mathrm{Co} + \mathrm{Ni}]$ should increase at low energy due to the much larger cross section for breakup of Fe into Mn and Cr at low energy than at high energy. However, such an increase would be masked for observers situated inside the heliosphere because cosmic rays typically lose from ~150 to ~400 MeV/nucleon of energy by adiabatic deceleration in the process of moving into the inner solar system (Urch and Gleeson, 1973).

From their small Lexan stack on the Apollo 17 mission Price et al. (1973a) were able to make crude abundance measurements in the interval 17 to 40 MeV/nucleon. There they found that the abundance ratio $[17 \leq Z \leq 25]/[\mathrm{Fe} + \mathrm{Co} +$

Table 5-2. Abundance Ratios

Ratio	Our result	Energy Interval (MeV/N)	Results of Webber et al. (1972) 250-850 MeV/N	>850 MeV/N
CNO/(23≤Z≤28)	17.3 ± 5.8	60-90	16.6	17.56
CNO/(FeCoNi)	25.0 ± 8.0	60-90	22.8	21.3
10≤Z≤14 / Fe+Co+Ni	5.6 ± 1.5	60-100	5.76	5.07
Ne+Mg+Si / Fe+Co+Ni	4.2 ± 1.1	60-100	5.07	4.58
17≤Z≤25 / Fe+Co+Ni	1.50 ± 0.5 1.52 ± 0.6	60-100 100-130	1.15	0.87

Ni] was comparable to that observed at higher energies (Table 5-2) and concluded that, during times when the sun is quiet, galactic cosmic rays comprise the majority of interplanetary particles at energies as low as ~17 MeV/nucleon.

With an electronic detector the Chicago group (Hsieh and Simpson, 1969) had earlier shown that at least 85% of the particles with $Z = 1$ at 20 MeV/nucleon were of galactic origin. (Both ^{2}H and the nuclei with $17 \leq Z \leq 25$ are produced primarily by nuclear reactions of galactic cosmic rays.) Recently the Chicago group (Mogro-Campero et al., 1973) has been able to determine the composition of individual elements at low energies that range from ~13 to 18 MeV/nucleon for C to ~25 to 60 MeV/nucleon for the Fe-group. Within their statistics they find no evidence for any significant difference in the composition at low and at higher energies. We conclude that the composition of cosmic rays with $Z \gtrsim 4$ appears to be independent of energy from about 25 MeV/nucleon to about 2000 MeV/nucleon.

Completely unexpected compositional anomalies begin to show up at energies below ~10 MeV/nucleon, in a region of the cosmic ray energy spectrum that is extremely difficult to study because the flux is about at a minimum. In his rapporteur paper at the Denver Cosmic Ray Conference, Webber reviewed evidence presented in post-deadline papers by McDonald et al. (1974) and by Hovestadt et al. (1973) that the ratio O/C changes rapidly with energy in the interval ~1 to ~10 MeV/nucleon. Fig. 5-7 shows their data, which were taken only during days when the sun was quiet. Because the data at energies below ~8 MeV/nucleon were obtained on IMP-7 at ~1 AU from the sun, whereas the data at energies above ~8 MeV/nucleon were obtained on Pioneer 10 at distances of ~3 to 5 AU from the sun, it would be useful to have some independent evidence for the reality of the huge effect. Such evidence comes from the Apollo 17 Lexan experiment of Chan and Price (1974), the results of which are shown in Fig. 5-8. Both the shape and magnitude of the oxygen spectrum agree well with the composite oxygen

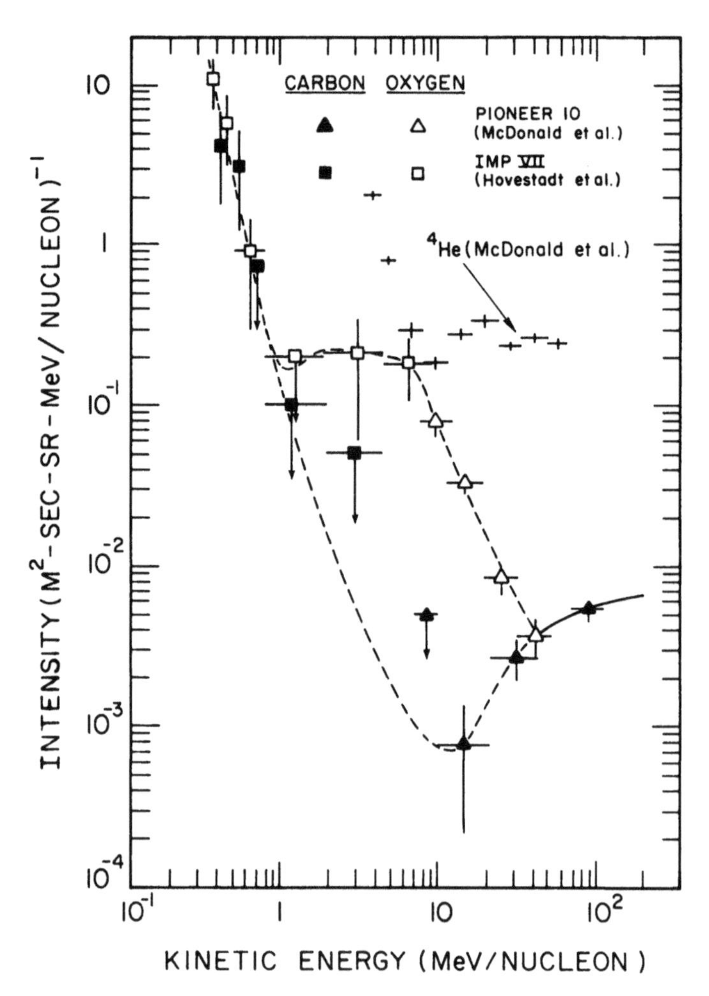

Fig. 5-7. *Anomalous composition at energies between ~1 and ~20 MeV/nucleon during solar quiet times. The O/C ratio increases by more than an order of magnitude.*

spectrum in Fig. 5-7, except that at very low energies (below ~0.5 MeV/nucleon) the oxygen flux during Apollo 17 was somewhat elevated due to a background of solar particles.

Table 5-3 compares the anomalous abundance ratios obtained by McDonald et al. (1974) and by Chan and Price (1974) with ratios for the sun and for galactic cosmic rays at energies where the composition is normal. The anomalous composition was still poorly characterized as of early 1974, but the data in Table 5-3 place strong constraints on the origin of the particles in the interval ~1 to 10 MeV/nucleon. The increase in the O/C and N/C ratios by an order of magnitude over either galactic or solar composition is especially striking. Some strong selection

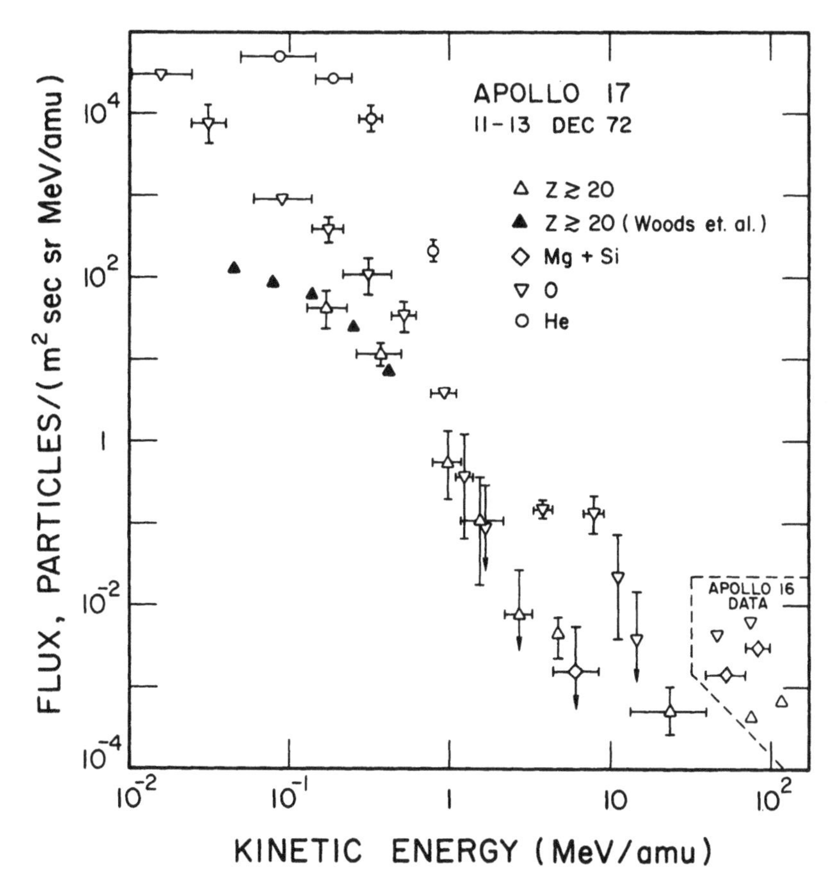

Fig. 5-8. *Measurement of the quiet-time low-energy spectra of various heavy ions by means of the Lexan stacks on Apollo 16 and 17 (data from Chan and Price, 1974). The "hump" at ~1 to 10 MeV/nucleon inferred from measurements on Pioneer 10 and IMP-7 is confirmed.*

mechanism appears to be enhancing the flux of O and N and perhaps depressing the flux of C, Mg, Si, and Fe. The absence of Li, Be, and B suggests that the particles have not passed through much matter. It is difficult to see how such low-energy particles could have entered the solar system without severe adiabatic deceleration, nor how they could have passed through much interstellar gas without preferential depletion of heavy elements due to ionization loss. Fisk et al. (1974) recently suggested that we might be seeing interstellar gas that enters the helio-magnetosphere as neutral atoms, which become singly ionized by charge exchange with the solar wind or by photo-ionization, and then some fraction of which are accelerated by moving magnetic fields. Elements such as He, O, N, and Ne, which have a high first ionization potential, can freely enter the solar system whereas elements with a low ionization potential are excluded. A critical test of the model

Table 5-3. Anomalous Composition of Interplanetary Particles at Low Energies

Ratio	Observed Value	Energy Interval (MeV/nucleon)	Solar System Cameron (1973)	Cosmic Rays (>850MeV/nuc) Webber et al. (1972)
LiBeB/CNO	<0.059*	3.4-4.4	1.08×10^{-5}	0.56
(C+N)/O	<0.63*	3.4-4.4	0.72	1.36
(Mg+Si)/O	<0.045*	4.5-8.6	0.096	0.33
(Z≥20)/O	0.018*	3.7-11	0.046	0.2
B/He	0.0026±0.0013†	8.3-30	1.59×10^{-7}	0.0097±0.006
C/He	0.0067±0.0022†	9.0-30	0.0054	0.031±0.002
N/He	0.089±0.024† 0.025±0.007†	7.3-10.6 10.6-17.4	0.0017	0.008±0.001
O/He	0.42±0.06† 0.109±0.015†	7.3-10.6 10.6-17.4	0.0098	0.028±0.001
Ne/He	0.0038±0.0017†	12-30	0.0016	0.0049±0.0004
Mg/He	<0.0017†	13.4-30	0.0005	0.0057±0.0004
Si/He	<0.0018†	14.5-30	0.00045	0.0038±0.0003

*Chan and Price (1974); †McDonald et al. (1974).

is the large mass to charge ratio of the ions resulting from their being singly charged. Using the earth's field as a magnetic rigidity analyzer, J. H. Chan (unpublished) has found high densities of 10 MeV/nucleon oxygen ion tracks in a Lexan stack exposed 72 days outside the Skylab. The flux level is consistent with the ions being singly charged and thus having a sixteen times higher rigidity than that of protons with the same velocity.

The Changing Composition at Energies Above 3 GeV/Nucleon

Recent measurements of cosmic ray composition at energies up to almost 100 GeV/nucleon (Juliusson et al., 1972; Ormes and Balasubrahmanyan, 1973; Smith et al., 1973; Webber et al., 1973b) have shown that with increasing energy above ~2 GeV/nucleon two changes occur: (1) among the primary nuclei (i.e., those that have not undergone nuclear interactions) the proportion of heavy elements increases; and (2) the proportion of secondary nuclei relative to primary nuclei decreases. Some of the data are shown in Fig. 5-9, which also includes the low-energy data obtained with plastic detectors. To understand the implications of the changing composition we must refer back to eq. (5-2). Cowsik and Wilson (1973) and Meneguzzi (1973) have suggested that Λ_l is a decreasing function of energy, so that at very high energy $\Lambda_l < \Lambda_{\text{int}}$ even for elements as heavy as Fe. In fact, they infer that at ~100 GeV/nucleon $\Lambda_l \lesssim 0.5$ g/cm². With such a short leakage pathlength interactions are unimportant, very few secondaries are produced, and a larger fraction of the heavy primary nuclei like Fe are seen relative to lighter primary nuclei than at low energies. To account for the detailed shapes of the energy spectra, which must, according to Fig. 5-9, have different slopes for different nuclei, both Cowsik and Wilson (1973) and Meneguzzi (1973) are led to a startling conclusion: cosmic rays leak out of uniformly distributed source regions

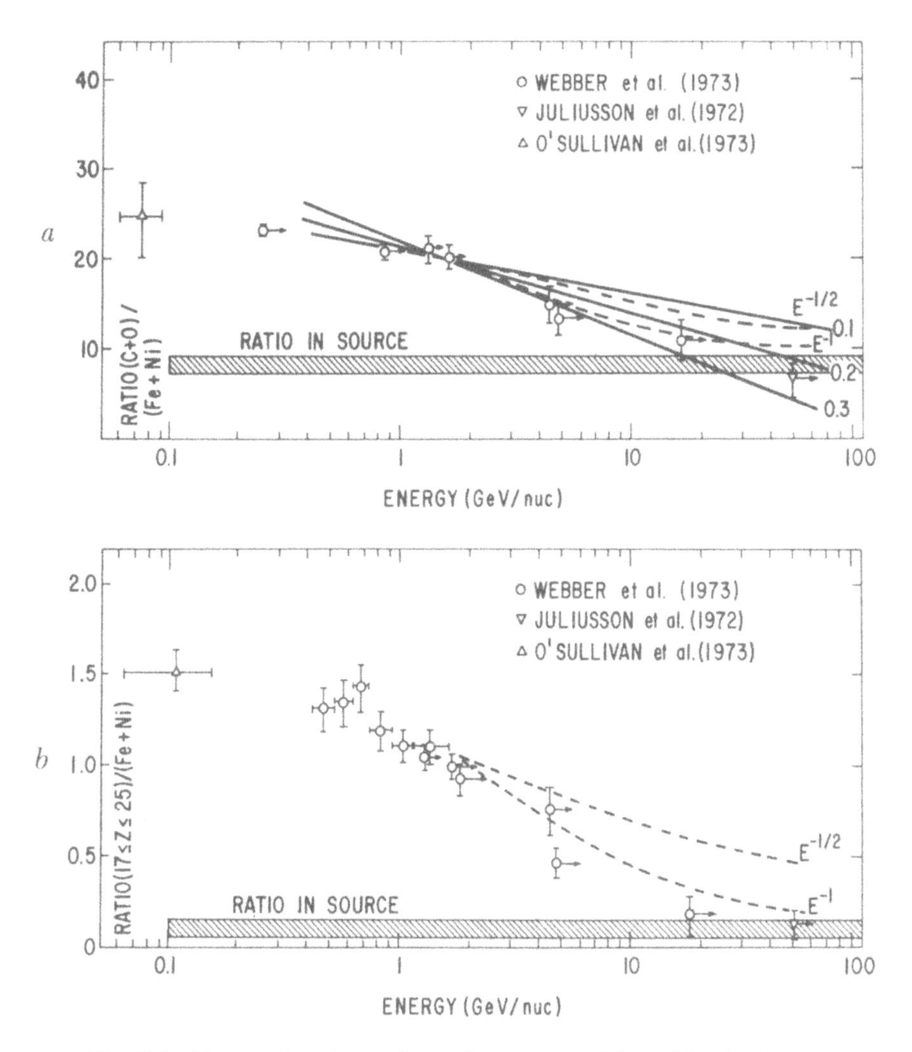

Fig. 5-9. *Energy dependence of cosmic ray composition:* (*a*) *abundance ratio* $[C + O]/[Fe + Ni]$; (*b*) *abundance ratio* $[17 \leq Z \leq 25]/[Fe + Ni]$. *The curves are from the model of Cowsik and Wilson* (*1973*).

with an energy-dependent mean pathlength and subsequently leak out of the Galaxy with an energy-independent mean pathlength. Thus low-energy heavy nuclei undergo considerable spallation while diffusing out of their source regions (e.g., in a supernova envelope), whereas high-energy heavy nuclei undergo spallation only after leaving the sources. The model appears to be consistent with the remarkable isotropy of cosmic rays, but remains to be tested against future measurements of the spectrum of high-energy positrons, which result from interactions of cosmic ray nuclei.

5.2.3. Transiron Cosmic Rays ($Z \gtrsim 30$)

It is natural to devote a separate discussion to the elements in the Periodic Table beyond iron. Because of their extreme rarity in nature (see Fig. 5-5), for many years there existed a psychological "Iron Curtain" that was only broken in 1965 with the observation of transiron tracks in meteorites (Walker et al., 1965; Fleischer et al., 1967). All subsequent studies have required special detectors with very large collecting power. Generally the dividing line between "subiron" and "transiron" is taken at $Z \approx 30$ rather than at $Z = 26$. The reason is that in nature the elements with $Z \gtrsim 30$ are synthesized by entirely different reactions involving neutron capture. We must review the various ways in which the elements in the Periodic Table are synthesized in stellar environments when we discuss models of the origin of cosmic rays. For the present we will discuss only experimental techniques and data.

Table 5-4 gives statistics and references to the various transiron cosmic ray experiments, most of which were cooperative efforts, often on an international scale, among several institutions. All but the first two entries and the Skylab and Electric Barndoor have involved both plastic detectors and nuclear emulsions. Texas I and II contained only nuclear emulsions; the Electric Barndoor was a combination of large ionization chambers and a Lucite Cerenkov counter; and the Skylab experiment consisted only of Lexan stacks.

Stimulated by the report of fossil tracks of transiron cosmic rays in meteorites, presented at the 1965 London Cosmic Ray Conference, Peter Fowler and his colleagues flew what was then considered a giant stack of nuclear emulsions—about 4.5 m^2 in area—and found two events to which they assigned a charge $Z = 90 \pm 4$. In hindsight, the observation of any such heavy nuclei from that experiment (Texas I) must be regarded as a fortunate statistical fluke which served to spur on both Fowler and ourselves to bigger and better experiments. Given his area-time factor of only 60 m^2h, the expected number of transbismuth nuclei, based on data from subsequent flights, is only $\sim$0.2!

The studies of transiron cosmic rays with large arrays of plastic sheets began with the simple idea of categorizing an event in a particular charge group, depending on which of several plastics of differing sensitivities recorded that event, but received considerable impetus with the development in 1967 of the etch rate method of particle identification (Price et al., 1967). In all current flights the etch rate in plastic appears to provide a better measure of charge than does the signal of a track in emulsion. Fig. 5-10 shows the track of the same ultraheavy nucleus in Lexan and in emulsion. The track of an Fe nucleus in emulsion is included for comparison.

In addition to the logistical and engineering problems of lifting a cumbersome platform as large as 40 m^2 in area to a float depth of 3 g/cm^2 and nondestructively recovering it hundreds of miles away from the launch site, there are other experi-

Table 5-4. Ultra-heavy Cosmic Ray Experiments

Experiment	Date	Mean Depth (g/cm^2)	Area X Time (m^2h)	Particles with $70 \leq Z \leq 83$	Particles with $Z > 83$	Reference
Texas I	10/6/66	4.5	60	0	2	Fowler et al. (1967).
Texas II	5/5/67	3.5	290	5	1	Fowler et al. (1970).
Barndoor I	9/23/67	3.7	120	0	0	Blanford et al. (1969).
Barndoor II	5/25/68	3.4	190	5	0	Blanford et al. (1969); revised in Blanford et al. (1973b).
Texas III a	9/18/68	3.0	600	6	2	Price et al. (1971b).
Barndoor III	9/19/68	~4.3	590	10	3	Blanford et al. (1973a).
Texas III b	9/22/68	3.5	300	7	1	Price et al. (1971b).
Texas III c	"	"	"	3	0	Blanford et al. (1973a).
Sioux Falls a	9/13/69	~3	75	2	1	O'Sullivan et al. (1971).
Sioux Falls b	"	"	"	0	0	Blanford (1971); Wefel (1971).
Texas IV a	9/27/69	2.9	240	6	1	Blanford et al. (1973b).
Texas IV b	"	"	700	8	1	see Figure 4 in Price (1972).
Minneapolis	9/4/70	3.7	1100	16	2	Shirk et al. (1973).
Sioux Falls II (2 flights)	5/(7,12)/71	4.5	610	9	1	Fowler et al. (1972).
Sioux Falls III (3 flights)	5/(7,14)/72	~3.8	680	20	6	Fowler et al. (1973).
Electric Barndoor (6 flights)	1/70 to 9/72	3 to 4.5	95	0	0	Binns et al. (1973).
Skylab	5/29/73	1.1	~8000	in progress		Price and Shirk (1973).
Sioux City	9/(18,25)/73	3	1950	in progress		Osborne et al. (1973a).

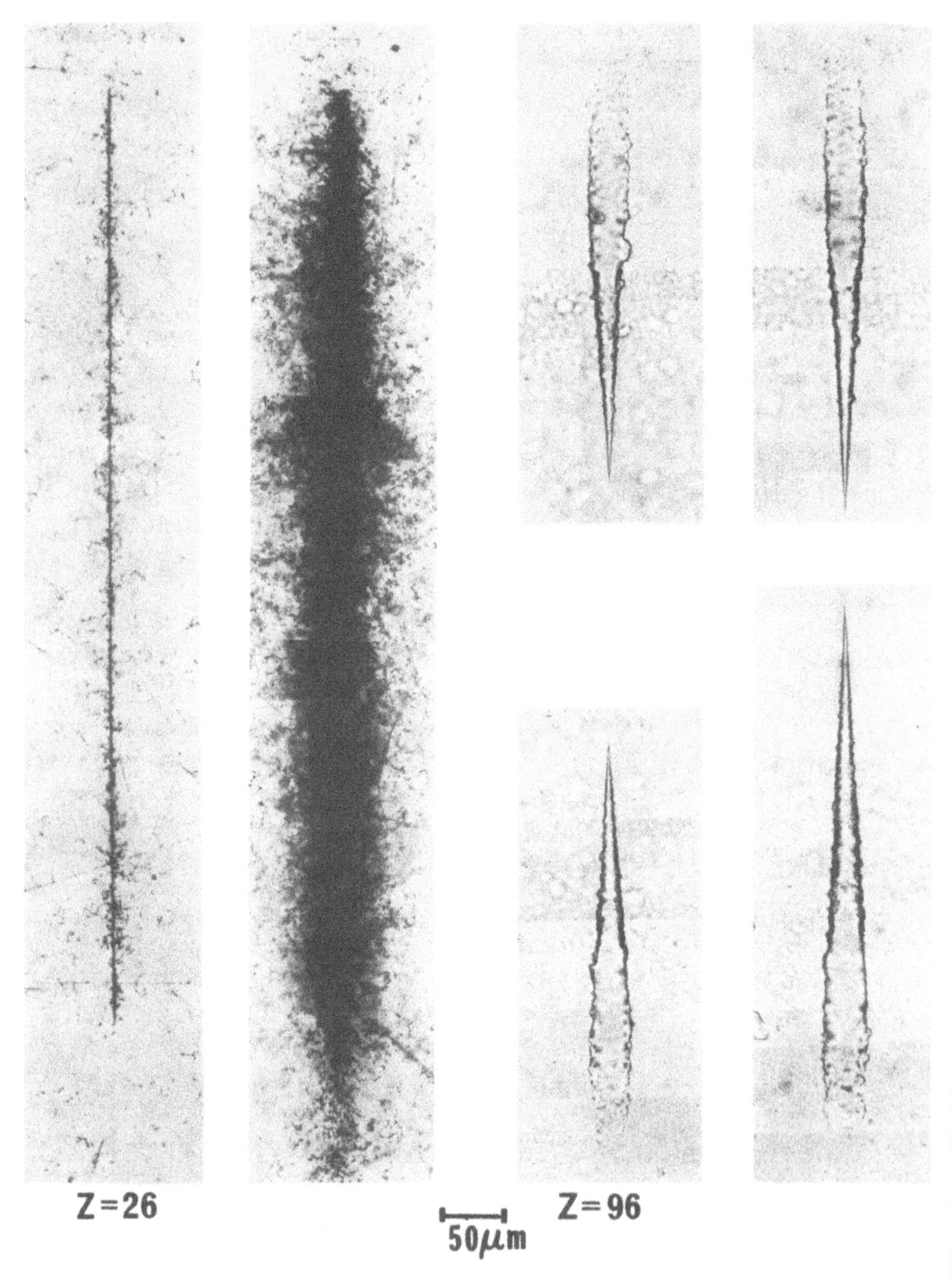

Fig. 5-10. *On the left is a track of an Fe nucleus in nuclear emulsion. The other three pictures show the track of an ultraheavy nucleus* ($Z \approx 96$) *that passed through a nuclear emulsion and two Lexan sheets, emerging from the bottom Lexan with a residual range of* 8 ± 2 *mm, during the Texas IV flight.* (*Courtesy of P. H. Fowler.*)

mental difficulties. The temperature of the plastics and emulsions must be kept down by means of a suitable sunshield. At least one layer in the stack must be shifted by several cm after the balloon reaches float level in order to allow the observer to reject tracks produced during ascent and descent, when the particles would have to pass through many g/cm^2 of air. Tracks of nuclei with $Z \gtrsim 30$, which are only $\sim 3 \times 10^{-4}$ times as abundant as Fe, must be located with high efficiency at a typical concentration of only several per m^2. For ionization rates high enough that etched cones from opposite sides of a sheet connect, both high-voltage spark techniques and ammonia-vapor techniques have been used for rapid scanning of large areas. (See Chapter 2 for details.) Assuming that sheets are etched to 50% of their original thickness, the approximate thresholds for detecting relativistic, minimum-ionizing particles at various zenith angles by a hole technique are given in Table 5-5 for several plastics. Commercially available cellulose nitrate is rather variable in response and has been used mainly to locate tracks of relativistic nuclei with $Z > 30$, which are then identified by their tracks in adjacent sheets of CTA or Lexan. Lexan is the most reproducible and thoroughly studied of the three plastics but cannot be used for rapid scanning of any but the heaviest events unless its sensitivity is increased by exposure to UV. A possible approach would be to leave a large array of Lexan sheets in bright sunlight to accumulate a suitable UV exposure. In several flights relativistic nuclei down to $Z \approx 50$ have been located with high efficiency by scanning a layer of nuclear emulsion in a stereomicroscope. Needless to say, the procedure is tedious. In the flights at Sioux Falls, Minneapolis, and Sioux City, where the geomagnetic cutoff is low enough to permit low energy particles to enter the earth's atmosphere, a technique that depends on holes being etched through the sheets can be used to find particles that increase in ionization rate as they slow down and come to rest in a Lexan stack. Shirk et al. (1973) found that particles with Z down to 30 that stop in a Lexan stack can be rapidly located by etching every fourth sheet and locating the holes with an ammonia technique. Fig. 5-11 shows the detection efficiency, which ranges from $\sim 10^{-4}$ for Fe to nearly unity for $Z \gtrsim 32$. The number of consecutive holes in the sequence of fourth sheets is a crude measure of Z. An accurate determination of range and charge results from processing portions of intervening sheets for times chosen such that the cones do not touch but are long enough to be

Table 5-5. Minimum Charges of Relativistic Particles Detectable by Hole Techniques

Detector	Zenith Angle		
	$\theta_z = 0^\circ$	$\theta_z = 45^\circ$	$\theta_z = 60^\circ$
Cellulose nitrate (Daicell)	~32	~34	~36
Cellulose triacetate (Kodacel, unplasticized)	46	51	57
Lexan polycarbonate	68	75	82

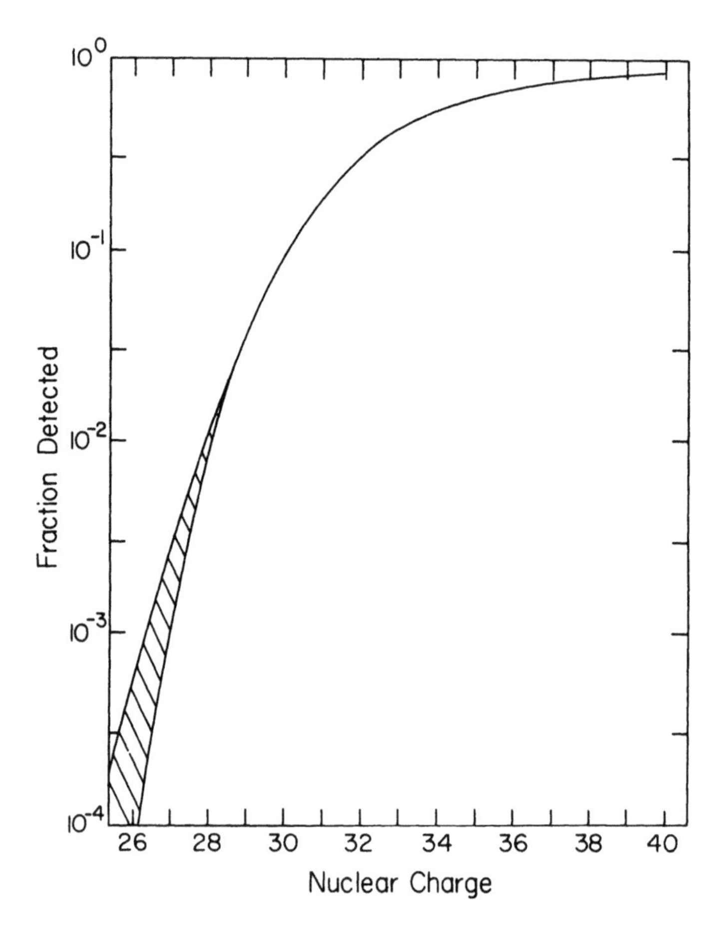

Fig. 5-11. *Scanning efficiency calculated for cosmic rays stopping in a Lexan stack in which every fourth sheet is etched 160h and treated with ammonia vapor to detect holes etched through the sheet. A Monte Carlo calculation took into account the angular distribution of incoming particles and variations of sheet thickness. The measured efficiency for Fe was $1.4 \pm 0.7 \times 10^{-4}$. (After Shirk et al., 1973.)*

easily measured. Tanks containing up to 1 m^3 of NaOH solution are necessary to process sheets from large payloads. For rough work such as locating the heavy events, the same tankful of solution can be used repeatedly, but for accurate charge determination small portions containing the latent tracks should be etched in a freshly precipitated solution (see Chapter 2 for recipe). Another experimental difficulty, that of determining velocity, has already been discussed in Chapter 3. Here we remind the reader of the methods that have been used: (1) The balloon or satellite can be situated such that the earth's field excludes particles with magnetic rigidity corresponding to velocity β less than $\sim$0.9, so that to within $\sim$5% the velocity of all particles can be assumed to be the same. This method may suffer from a background of slow, sub-iron secondaries (Blanford et al., 1972). (2) If the stack is sufficiently thick (usually 1 g/cm² of plastic is enough) adequate information about the particle velocity can be obtained from measurements of the change (or absence of change) of etching rate along the trajectory in the stack. (3) Fast film Cerenkov detectors (Pinsky, 1972) of thickness ~0.2 g/cm² (including packaging) yield a measure of velocity for nuclei with $Z \gtrsim 60$ and $\beta \gtrsim 0.7$.

Data for Transiron Nuclei at High Energy ($\beta \gtrsim 0.9$)

As a result of painstaking analyses by many individuals we now have a fairly good picture of the charge distribution of relativistic cosmic rays with $Z > 50$, with a resolution that varies from flight to flight but is typically known to within two to three units of charge. Fig. 5-12 is a collection of all data available as of June, 1973. For additional data and a recent discussion of ultraheavy cosmic rays the reader should see the rapporteur paper presented by Fowler (1973) at the Denver Cosmic Ray Conference. The abundances in Fig. 5-12 apply to a mean float depth of ~3.5 g/cm² and are uncorrected for breakup in the atmosphere, which will preferentially destroy the heaviest nuclei (about half the nuclei with $Z \gtrsim 80$ will have interacted) and obscure any valleys in the distribution. Even so, certain features are evident:

1. There appear to be abundance peaks at $Z \approx 52$ to 54 and at $Z \approx 76$, which imply synthesis by rapid-neutron capture, a process we will discuss in Section 5.2.5 on cosmic ray origin.

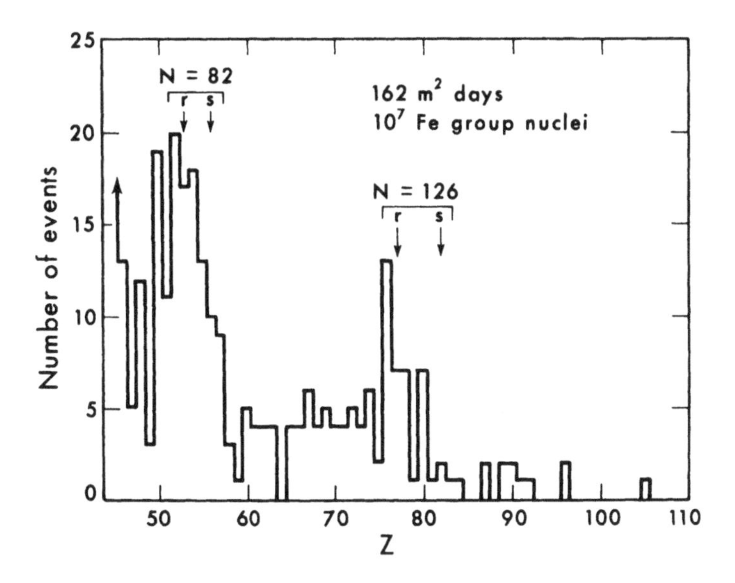

Fig. 5-12. *Charge distribution of ultraheavy cosmic rays with $\beta \gtrsim 0.9$ at an average atmospheric depth of ~3.5 g/cm². All data available as of July, 1973 are plotted. Arrows indicate expected positions of r- and s-process abundance peaks corresponding to the closed neutron shells $N = 82$ and 126.*

2. There appears to be a gap at $Z \approx 86$ due to the absence of short-lived radionuclides with $84 \leq Z \leq 89$ in the cosmic rays. A small population of such particles might be expected at balloon level due to fragmentation of heavier nuclei in the overlying atmosphere.
3. There is a finite number of nuclei with $Z \approx 90$ to 92.
4. There are at most three possible transuranic nuclei. One of them passed through and left tracks in both emulsion and plastic detectors (Price et al., 1971b), but the charge determined in emulsion ($Z \approx 104$) was quite different from the charge determined in plastic ($Z \approx 92$). The possibility that its true charge was 92 could not be ruled out. A second one, a fast nucleus that entered the Texas II emulsion stack (Fowler et al., 1970), was assigned a most probable charge $Z \approx 103$, with a very small probability that its velocity might have been low enough for its charge to be as low as 92. The third one, shown in Fig. 5-10, came to rest in the Texas IV plastics-emulsion stack, and its charge was estimated to be $Z \approx 96$.

In the charge interval $32 \lesssim Z \lesssim 50$ most of the experiments in Table 5-4 had detection efficiencies less than unity and decreasing with decreasing Z because of the proximity of the vastly more abundant Fe-peak elements. In the work of Blanford et al. (1969, 1973a) serious attempts were made, using Daicell cellulose nitrate, to locate events with Z as low as 35. The charge identifications were made with adjacent cellulose triacetate sheets or nuclear emulsions, depending on whether Z was greater than ~40 or not. The Electric Barndoor (Binns et al., 1973) has a high efficiency in this charge interval and appears to have a higher resolution than was achieved by Blanford et al., but in six flights during a two-year period has collected only 52 events with $32 \leq Z \leq 50$ and only one event with $Z \geq 44$. They have remarked on the low abundance of strontium ($Z = 38$) relative to its abundance in the solar system.

Data for Transiron Nuclei at Low Energy

There is a strong element of luck in balloon experiments. Shirk et al. (1973) launched a 20 m² array of plastics, emulsion, and Cerenkov film detector (Fig. 5-13) designed to study the energy spectrum of transiron nuclei down to energies of ~300 MeV/nucleon in what proved to be an exciting and nearly disastrous flight. It began at Minneapolis, which has a sufficiently low geomagnetic cutoff to permit studies of low-energy nuclei, but the balloon became derelict after 44 hours, as a result of failures in all of the three separate descent mechanisms. During the next 15 days it drifted 1000 miles west of the Oregon coast, then turned around and drifted eastward across North America until it finally descended enough of its own accord to have its payload ripped off by telephone lines in a flax field near Moose Jaw, Saskatchewan. The apparent carnage in Fig. 5-14 is somewhat misleading, since most of the modules of plastics suffered nothing worse than occa-

Fig. 5-13. *Array of Lexan sheets, nuclear emulsion and Cerenkov film being launched at Minneapolis on 4 September 1970.* (*Courtesy of W. Z. Osborne and L. S. Pinsky.*)

sional bullet holes from the rifles of local duck hunters (annoyed at the violation of Canadian air rights?).

Using the fourth sheet scanning technique and the graph shown in Fig. 5-11, Shirk et al. (1973) determined the relative abundances of nuclei with $26 \leq Z \leq 42$ at energies of ~300 to 400 MeV/nucleon at the top of the atmosphere. Because of the rapidly falling abundances with increasing Z from 26 to ~40, very little contamination in this charge region is to be expected from spallation in the atmosphere and in interstellar matter, since the still rarer, heavier nuclei will break up into lighter ones that will add little to the original numbers of lighter nuclei. They found that the cosmic ray abundance pattern for $26 \leq Z \leq 42$ is quite similar to that observed for the elements in our solar system. Like Binns et al. (1973), they also found a low abundance of strontium.

An unexpected result of the Minneapolis flight was the observation that the energy spectrum of the nuclei with $Z \geq 60$ appears to be steeper than the spectra of lighter cosmic rays such as Fe and He. Fig. 5-15 compares the spectrum of cosmic rays with $Z \geq 60$ with that of He at the same degree of solar modulation. The difference appears quite striking. By comparison, the differences among the

Fig. 5-14. *Nearly nondestructive landing of the same detector array in a flax field in Saskatchewan on 19 September 1970. (Courtesy of W. Z. Osborne and L. S. Pinsky.)*

energy spectra of various subiron cosmic rays that we spoke of in the section on Changing Composition at Energies above 3 GeV/nuc, p. 250, are much smaller and confined to energies higher than a few GeV/nucleon. In a detailed statistical analysis by the maximum likelihood method, which took into account adiabatic deceleration, nuclear interactions, and detector characteristics, Osborne et al. (1973b) found the most probable index of $\nu = 4.5^{+1.1}_{-1.0}$ for an assumed differential spectrum of the form $E^{-\nu}$ in interstellar space, as compared with an index $\nu = 2.5$ for the He spectrum. They found an overall confidence level of 99.99% for the existence of a real difference in the spectra of He and $Z \geq 60$. On the other hand, Fowler et al. (1973) have recently reported that the energy spectrum of nuclei with $Z > 60$ appears to be similar to the shape of the Fe energy spectrum. [Early results from Skylab (Price and Shirk, 1973) indicate no difference in the spectra of Fe and $Z \geq 70$.]

Results of Binns et al. (1973) for the energy spectrum of cosmic rays with $Z \geq 32$ are also displayed in Fig. 5-15. There is some indication that their spectrum may have a slope intermediate between the spectra of He and $Z \geq 60$.

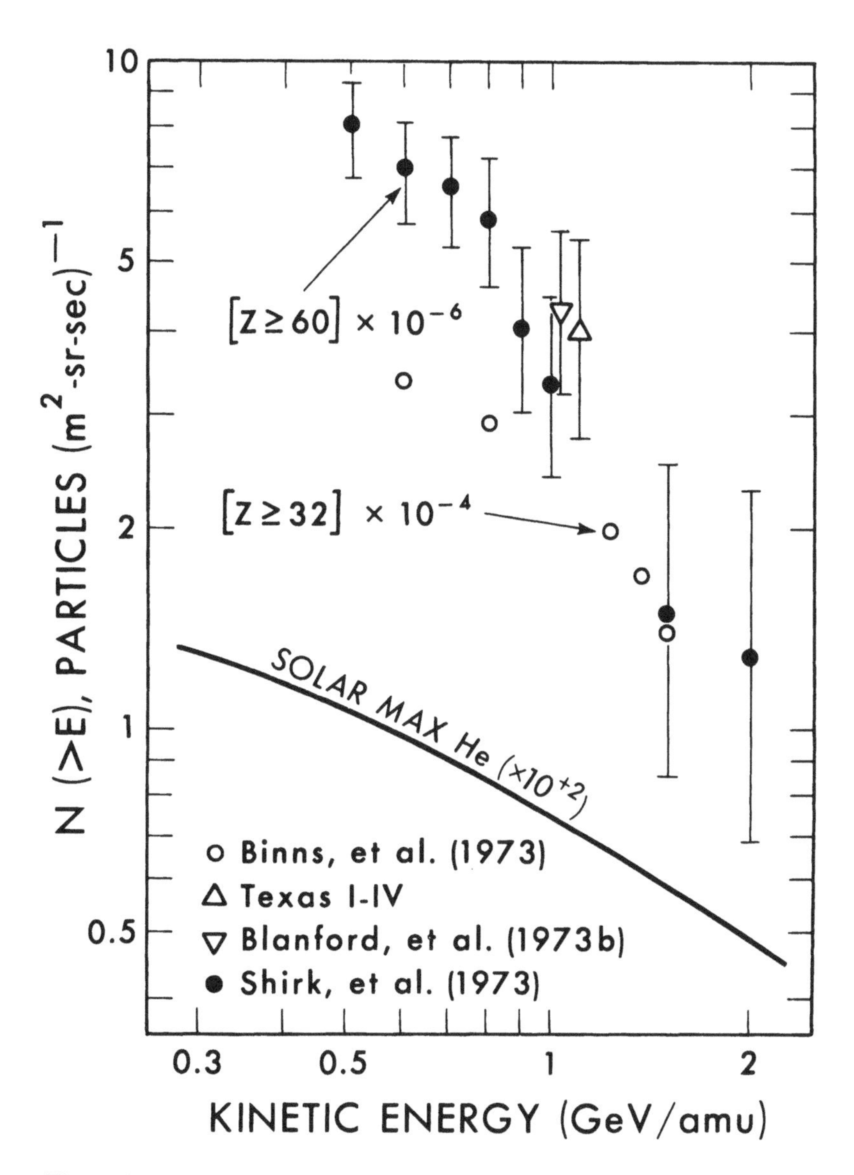

Fig. 5-15. *Integral energy spectrum of cosmic rays with $Z \geq 60$ at the top of the atmosphere at solar maximum compared with the helium spectrum. Data from Texas I to IV flights and from Blanford et al. (1973a,b) were corrected for differences in solar modulation relative to that for the flight of Shirk et al. (1973), which occurred at solar maximum. Also shown is the integral energy spectrum of cosmic rays with $Z \geq 32$ obtained by Binns et al. (1973).*

5.2.4. Search for Superheavy Elements in the Cosmic Rays

One of the most exciting developments in nuclear theory in the last decade is the prediction that nuclear shell effects cause the lifetimes of hypothetical nuclides with $Z \approx 114$ and $A \approx 184$ to be many orders of magnitude longer than those of neighboring nuclides (see Nix, 1972, for a very readable account of the theory). In Chapter 7 we will consider such superheavy elements from the viewpoint of their potential importance to nuclear physics. There we describe numerous searches for superheavy elements in nature and in laboratory reactions, all of which have given negative results.

Bearing in mind the early history of cosmic ray research, in which many of the elementary particles were first discovered in nature, it is perhaps not unreasonable to speculate that somewhere in the Galaxy conditions may be right for synthesizing superheavy elements and that some of them might reach the solar system in the form of cosmic rays. The most stable of the superheavy nuclides are expected to have lifetimes as long as 10^8 years, whereas the mean age of cosmic rays may be as little as 10^6 years if they spend most of their time in the galactic disc. In this case very few of the superheavies would have decayed in space.

The most likely means of synthesis in astrophysical processes would appear to be by a series of neutron captures at a rapid rate at high temperature and neutron density—the so-called *r*-process (see Fig. 5-16). Several authors have examined the question of whether superheavy nuclides can be synthesized by the *r*-process (Schramm and Fowler, 1971; Schramm and Fiset, 1973; Brueckner et al., 1973; Johns and Reeves, 1973; Howard and Nix, 1973). Necessary conditions for the production of nuclides with $A \gtrsim 290$ by neutron capture are that the value of the surface asymmetry parameter K in the surface energy term, $E_s = a_s A^{2/3} \cdot (1 - KI^2/A^2)$, of the semiempirical mass formula familiar to nuclear physicists must be less than $\sim$2.3, and the neutron densities during an appreciable fraction of the *r*-process must be less than about 10^{28} cm^{-3}. The coefficient a_s is an empirically determined constant and I is the difference between the number of neutrons and the number of protons in the nucleus. If K is too high, the heavy, neutron-rich nuclei that must be formed in order to reach the so-called "island" of superheavy nuclides shown in Fig. 5-16 will end the *r*-process by spontaneous fission. If the value of K turns out to be as low as $\sim$1.8, then a positive identification of superheavy elements in the cosmic rays would establish an association between cosmic ray production and *r*-process synthesis at neutron densities less than 10^{28} cm^{-3} and confirm that some superheavy nuclides have lifetimes greater than about 10^6 years.

The Skylab experiment listed at the bottom of Table 5-4 was specifically designed to search for transuranic and superheavy cosmic rays. Fig. 5-17, taken in the interior of the medical area of the Skylab, shows the cloth holder with 36 pockets containing Lexan stacks of total area $\sim$1.3 m^2. In addition to the large

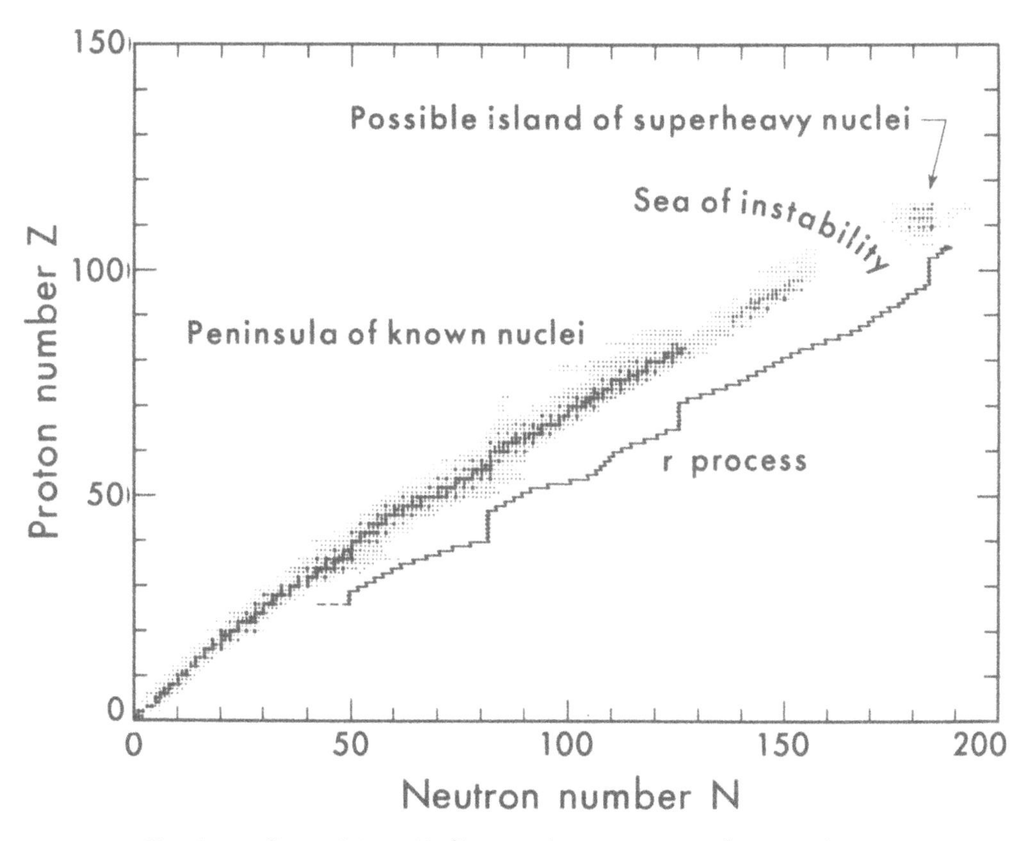

Fig. 5-16. *Chart of the stable (large dots) and radioactive (small dots) nuclides, including the hypothetical "island" of superheavy nuclides. In a sufficiently intense neutron flux, a given nucleus can increase its mass by a succession of neutron captures and its proton number by beta decay, following the zigzagged path in the figure. (After Nix, 1972.)*

area-time factor, having only $\sim$1 g/cm^2 of overlying aluminum meant that ultra-heavy cosmic rays with energies as low as $\sim$250 MeV/nucleon would reach the detectors and that only a very small fraction would undergo nuclear interactions. One of the stacks was brought to earth at the end of September, 1973, and the remaining 35 stacks were returned in February, 1974 after a total exposure of 253 days. Two tracks of relativistic ultraheavy nuclei were found in the first stack, one with $Z \approx 82$ and one with $Z \approx 90$ (Price and Shirk, 1973). The cone lengths for each event showed a very small spread from the top to the bottom of the stack, indicating that the uncertainty in charge should be less than ± 2 units. The manuscript for this book was completed just a few days after the successful completion of the Skylab mission. The 35 remaining stacks had been delivered to Berkeley and etching of every fourth sheet had just begun.

Fig. 5-17. *Experiment S-228 (Trans-uranic Cosmic Rays) deployed on the wall of Skylab behind and to the left of astronaut Bill Pogue. Each of the 36 pockets contains a stack of 32 Lexan sheets measuring 18 cm × 20 cm × 0.025 cm.*

After Skylab it may be quite a few years before electronic detectors with adequate collecting power to study transuranic cosmic rays will be carried on other satellites.

5.2.5. Origin of Cosmic Rays

A proper discussion of this old and unsolved problem would include arguments based on data from the electronic and electromagnetic components of cosmic rays as well as the nuclear components, and also from radio-astronomical and optical data. Here we shall content ourselves with an abbreviated discussion in which we derive elemental abundances at cosmic ray sources and compare that distribution with the abundances of the elements in our solar system and with theoretical abundances expected to arise in various nucleosynthetic processes. Then in Section 5.3 we shall look at the relationship between abundances of cosmic rays and solar flare particles.

Abundances of Elements in the Solar System

Because of extensive chemical fractionation early in their history, the crusts of the earth and moon are not as reliable a sample of solar system material as are the meteorites. Though there is still some difference of opinion over certain details, it now appears that the class of meteorites known as the Type I carbonaceous chondrites exemplifies the least fractionated, most primitive sample of solid matter in the solar system available for study. Elements with a low vapor pressure at a temperature of a few hundred degrees centigrade appear to have been quantitatively retained in these objects, and even some of the more volatile elements such as Hg, Br, and Tl are present to some degree. The abundances of elements that are gases or are present only as volatile compounds (e.g., H_2O) at ordinary temperatures must be studied by other means, usually involving a blend of solar spectroscopic observations and nucleosynthesis theory. Since the pioneering work of Suess and Urey (1956), Cameron (1968; 1973) has devoted considerable attention to the problem of the abundances of the elements in the solar system, and we use his latest compilation in Table 5-6 and in Fig. 5-5. It is important to point out that the compilation is not inviolate; in fact several entries have changed by large factors since the 1968 version, boron being a spectacular example (having increased by a factor of 56). We should thus not rest a case for a particular conclusion as to the origin of cosmic rays on one or two discrepancies between measured cosmic ray abundances and inferred solar system abundances.

Generally the agreement between meteoritic abundance data and astronomical data based on spectroscopic studies of the solar photosphere and corona is quite good, where there is overlap. Because of their high ionization potentials it is difficult to observe the noble gases in the sun, and their abundances are still quite poorly known. The solar data for very heavy elements are also quite limited. Of the actinides, U has not been measured, and only one line of Th has been observed. Its abundance, determined spectroscopically (Grevesse, 1969), appears to be a factor ten higher than the meteoritic abundance.

Nucleosynthesis of the Elements

In their "big bang" model of the origin of the universe, Gamow and his collaborators (Alpher et al., 1948; Alpher and Herman, 1950) attempted to account for the origin of the elements by a succession of neutron captures in the early stages of expansion, when the matter was still hot and dense. Because of the absence of stable nuclides with mass 5 or 8, they were unable to form elements heavier than helium from free nucleons, but they did contribute to current concepts by recognizing that the abundance peaks in the heavy-element region correspond to regions in which the neutron capture probabilities are small.

The recognition that most of the elements are synthesized in stellar interiors followed two discoveries in the 1950's: (1) The abundance of heavy elements relative to hydrogen is higher in recently formed stars than in the oldest stars in the Galaxy and is in fact an increasing function of the time of formation of the star. (2) Stars of certain types have an anomalous abundance of elements believed to be produced by a specific burning process. For example, the discovery of technetium in the atmospheres of red giant stars (Merrill, 1952) strongly suggests that element synthesis by neutron capture has occurred recently in those stars, since all the isotopes of technetium are radioactive with half-lives not exceeding two million years.

Beginning with the remarkable work of Burbidge et al. (1957), a number of theoretical nuclear astrophysicists have developed scenarios that account for the origin of the majority of the elements in terms of their synthesis in the hot interiors of stars, followed by mass loss into the interstellar medium. The increasing abundance of heavy elements with the time of formation of a star naturally follows from the enrichment of the raw material by ejection of matter into the interstellar medium. Matter is ejected continuously during certain phases of stellar evolution such as the red giant phase and can also be ejected with spectacular violence in supernova explosions. The book by Clayton (1968) gives a detailed account of the status of our understanding of nucleosynthesis in stars. The article by Cameron (1971) can be consulted for recent advances in the theory.

To prepare the reader for the discussion of cosmic ray origins we shall briefly discuss the main features of the solar system abundance distribution in Fig. 5-5 as they are presently understood. The general trend of decreasing abundance with increasing atomic number reflects the incompleteness of the ongoing process of fusion of hydrogen into successively heavier elements in stars. Many clues point to nuclear processes—the higher abundance of even-Z than odd-Z elements, the decline of abundances from H to Si because of the increasing Coulomb barrier for charged particle reactions, the abundance peak at the nuclide with the greatest binding energy per nucleon (Fe), and the abundance peaks near closed shells of the heavier nuclei. We list the main features under eight categories.

(1) *Nucleosynthesis in the "big bang."* Starting with an equilibrium between neutrons and protons at high temperature and density at the beginning of the universe, most of the helium in the universe today was produced in the very early stage of expansion of the universe. It is possible that most of the isotopes ^{2}H, ^{3}He, and ^{7}Li were also made in the big bang.

(2) *Hydrogen burning.* In general the history of a star is defined by a succession of stages of gravitational contraction and nuclear burning. During the contraction, the interior of the star is heated by the release of gravitational potential energy. When the temperature is increased to the

point at which the nuclear fuel can burn, contraction is temporarily halted. Thermonuclear reactions then provide the source of energy generation necessary to maintain the stellar luminosity. With the exhaustion of this nuclear fuel, gravitational contraction resumes. Hydrogen burning at temperatures of $\sim 10^7$K is responsible for the luminosity of most stars but is extremely slow (the process $4p + 2e \rightarrow {}^4He + 2\nu +$ 26.7 MeV is a weak interaction) and in $\sim 10^{10}$ years has increased the helium content of the Galaxy by only $\sim 10\%$. During its hydrogen-burning phase a star is situated on the so-called Main Sequence of the famous Hertzsprung-Russell diagram of luminosity vs. temperature.

(3) *Helium burning*. After the hydrogen in the center of a star is consumed, the core contracts, the temperature rises to $\sim 10^8$K, and helium begins to burn by means of the two reactions ${}^4He + {}^4He \rightarrow {}^8Be$ (lifetime only $\sim 2 \times 10^{-16}$ sec) and ${}^8Be + {}^4He \rightarrow {}^{12}C + \gamma$, the net result being $3\alpha \rightarrow {}^{12}C + 7$ MeV. The problem of bridging the gap at masses 5 and 8 is overcome because the density of ${}^4He$ is so high that ${}^8Be$ can react with ${}^4He$ before decaying. By the time helium is completely burned much of the ${}^{12}C$ has been converted to ${}^{16}O$ by alpha particle capture.

(4) *Carbon and oxygen burning*. The Coulomb barrier prevents carbon and oxygen from burning until the interior reaches a temperature of almost 10^9K. Then a variety of reactions can occur on a short time scale, leading to γ, p, α, n, and heavy species such as ${}^{23}Na$, ${}^{20}Ne$, ${}^{24}Mg$, and ${}^{28}Si$. At such high temperatures the intensity of γ rays is enormous (remember the Planck radiation law), the most weakly bound nuclei begin to photodisintegrate, and at the completion of carbon and oxygen burning the main product is ${}^{28}Si$. The γ rays also produce neutrinos which carry energy out of the core at a rate $\propto T^9$ and force evolution to proceed at an explosive rate.

(5) *The equilibrium process*. At a temperature of $\sim 2.5 \times 10^9$K, with the accompanying enormous γ ray intensities, there ensues a very complex pattern of (γ, α), (γ, p), and (γ, n) reactions and their inverses. Some of the ${}^{28}Si$ breaks up into ${}^{27}Al$, ${}^{24}Mg$ and other nuclides; some of the ${}^{28}Si$ is built up rapidly by (α, γ) and other reactions, and statistical equilibrium between capture and photodisintegration reactions soon prevails, Those nuclei are favored which have the largest binding energies—${}^{54}Fe$, ${}^{56}Fe$, ${}^{56}Ni$, and the nuclides that are multiples of alpha particles (e.g., ${}^{40}Ca$). The exact final nuclear abundances for the equilibrium process depend in detail on the density, temperature, and time scale, since nuclear beta decay can alter the ratio of both free and bound protons and neutrons $(\overline{Z}/\overline{N})$ and thus shift the equilibrium in favor of ${}^{56}Fe$, ${}^{56}Ni$ (which later decays to ${}^{56}Fe$), or ${}^{54}Fe + 2p$. The solar system material, with its high ${}^{56}Fe$ abundance, apparently went through the equilibrium

process at high density and with $\overline{Z}/\overline{N}$ high. The cosmic rays might have been synthesized under different conditions that could be inferred from the relative abundances of ^{54}Fe, ^{56}Fe, and other nuclides.

(6) *Neutron capture processes*. The Coulomb barriers for synthesis of elements with $Z \gtrsim 30$ by charged particle reactions are impossibly high and the only plausible mechanism is neutron capture. We may distinguish two qualitatively different kinds of neutron capture processes according to the time scale on which they occur in nature—the "*s*-process," which is *slow* compared with the time scale for β-decay (seconds to centuries), and the "*r*-process," which is rapid compared with β-decay rates. The *s*-process occurs in a weak neutron field (e.g., $\sim 10^5$ cm^{-3}) such as results from the reaction $\alpha + {}^{13}C \rightarrow {}^{16}O + n$ in stellar interiors, whereas the *r*-process occurs in a very intense neutron field ($\gtrsim 10^{24}$ cm^{-3}) such as accompanies a supernova explosion. The two processes follow different paths on a nuclidic chart of Z vs. N such as is shown in Fig. 5-16. The stable nuclides are denoted by large black dots and unstable nuclides by small dots. As a stable nuclide captures neutrons it moves to the right on the chart until it is converted into an unstable one. If the neutron flux is weak it moves diagonally up and to the left by beta decay before it can capture another neutron. The *s*-process thus fills out only the valley of β-stability along the center of the band of stable nuclides but does not populate the neutron-rich islands of stable nuclides on the right-hand fringe of the stability line or the proton-rich islands on the left-hand fringe. The *s*-process terminates at bismuth with the onset of alpha decay of the actinides. The relative abundances correlate inversely with neutron capture cross-sections. Since $\sigma(n, \gamma)$ is extremely small for nuclei with closed neutron shells, the *s*-process leads to abundance peaks at $N = 50$, 82, and 126. The corresponding mass numbers are $A = 90$, 140, and 208.

In the *r*-process, an intense neutron flux drives nuclides at constant Z to a closed neutron shell, whence they move toward the stability line by beta decay as the neutron flux subsides, thus populating the neutron-rich islands as well as some of the nuclides in the center of the valley of stability. The saw-toothed line to the neutron-rich side of the stable nuclides in Fig. 5-16 shows the path followed in the *r*-process, before beta decay. Because of spontaneous fission the saw-toothed path eventually terminates and fission fragments with $Z \approx 40$ to 50 become the seed nuclei for further neutron capture. The *r*-process leads to abundance peaks corresponding to the neutron shells $N = 50$, 82, and 126 but about 10 mass units lighter than in the *s*-process. Th and U, as well as a variety of transuranic nuclides, possibly even extending up to the superheavy region (shown by the island at $Z \approx 114$ and $N \approx 184$ in

Fig. 5-16), are synthesized by rapid neutron capture. The *r*-process is strongly nonequilibrium and is very difficult to treat quantitatively, whereas the *s*-process has been treated with considerable success.

(7) *The p-process.* To account for the rare, stable nuclides on the proton-rich fringes of the stability line apparently requires enormously intense proton fluxes. The details are not well understood, but one possibility is that nuclides previously synthesized by neutron capture undergo rapid proton capture at high temperature in a supernova shock wave.

(8) *Synthesis of Li, Be, and B.* The deep gorge at $Z = 3$ to 5 in Fig. 5-5 is a reflection of the fragility of these elements, which are rapidly disintegrated by (p, α) reactions in the interior of even the coldest stars. They can only be synthesized nonthermally, and it seems possible to account for their low abundances entirely by spallation reactions of low-energy cosmic rays with galactic gas and in shock waves in supernova envelopes.

The inevitable result of stellar nucleosynthesis is that, as a star evolves, the elements become heavier.

Abundances at Cosmic Ray Sources

Various corrections have to be applied to the abundances of cosmic rays within the solar system to determine their abundances at sources. After these corrections have been made, there is still no assurance that all nuclear species have been emitted and accelerated from the sources without bias, but we will postpone discussing that possibility until later. In the most successful model of cosmic ray transport (Cowsik et al., 1967), there is a steady-state distribution of the various nuclei in space, governed by production of primary nuclei in sources, leakage from the Galaxy, conversion to secondary nuclei by spallation and radioactive decay, loss of energy by ionization, and finally modulation within the heliosphere. The appropriate steady-state transport equation has been solved in various ways, the most elegant being the method of Cowsik and Wilson (1973), who convert a set of differential equations into simple algebraic matrix equations that do not need to be numerically integrated. Their calculations of source abundances, together with those of other workers, are compared with solar system abundances in Table 5-6. Laboratory cross section data are quite limited, and Silberberg and Tsao (1973) have developed empirical formulas for production of cosmic ray secondaries from interactions of primaries with interstellar hydrogen. Their formulas have been used in the computations. Cowsik and Wilson (1973) conclude that, within the uncertainties of the observational data (Webber et al., 1972) and of the cross section formulas, it is possible to account for the overabundance of the naturally rare elements (^{2}H, ^{3}He, Li, Be, B, Na, Al, P, F, $17 \leq Z \leq 25$, etc.) in the cosmic rays entirely by spallation of heavier elements during transport. This can be seen from the fact that the uncertainties are comparable to the calculated source abundances of these elements (column 2), which are much lower than the abundances

Table 5-6. Composition of Cosmic Rays at their Sources

Nuclide	Cosmic Ray Sources: Cowsik and Wilson (1973)	Cosmic Ray Sources: Shirk et al. (1973)	Cosmic Ray Sources: Cartwright et al. (1973)	Solar System Abundances Cameron (1973)
H	1.1×10^9	-	-	3.2×10^{10}
D	1.7×10^5 *	-	-	5.2×10^5
^3He	7.0×10^4 *	-	-	3.7×10^5
^4He	2.3×10^8	-	1.52×10^8	2.2×10^9
Li	9.0×10^0 *	-	-	5.0×10^1
Be	2.0×10^{-1}*	-	-	8.1×10^{-1}
B	2.2×10^0 *	-	-	3.5×10^2
C	6.3×10^6	-	5.4×10^6	1.2×10^7
N	1.6×10^6 *	-	6.5×10^5	3.7×10^6
O	6.0×10^6	-	5.9×10^6	2.2×10^7
F	1.9×10^3 *	-	-	2.5×10^3
Ne	1.8×10^6	-	5.6×10^5	3.4×10^6
Na	6.5×10^4 *	-	1.9×10^5	6.0×10^4
Mg	1.4×10^6	-	1.1×10^6	1.1×10^6
Aℓ	9.7×10^4 *	-	2.6×10^5	8.5×10^4
Si	$\equiv 1.0 \times 10^6$	-	$\equiv 1.0 \times 10^6$	$\equiv 1.0 \times 10^6$
P	8.8×10^3 *	-	5.4×10^4	9.6×10^3
S	1.9×10^5	-	1.7×10^5	5.0×10^5
Cℓ	4.9×10^2 *	-	-	5.7×10^3
Ar	2.5×10^4 *	-	2.7×10^4	1.2×10^5
K	3.7×10^3	-	-	4.2×10^3
Ca	1.6×10^5	-	9.8×10^4	7.2×10^4
Sc	6.7×10^1 *	-	-	3.5×10^1
Ti	4.4×10^3 *	-	-	2.8×10^3
V	1.6×10^3 *	-	-	2.6×10^2
Cr	2.1×10^4 *	-	-	1.3×10^4
Mn	1.4×10^4 *	-	-	9.3×10^3
Fe	1.3×10^6	$\equiv 1 \times 10^6$	1.3×10^6	8.3×10^5
Co + Ni	-	$3x8 \times 10^4$	1.1×10^5	5.0×10^4
Cu + Zn	-	$(5 \pm 1) \times 10^3$	-	1.8×10^3
Rb + Sr + Y	-	-	<23**	38
Ga to Mo	-	2.5×10^2	-	3.7×10^2
Hg to Bi	-	2.3 ± 1.0	-	4.7
Th + U	-	2.1 ± 1.3	-	0.08†
$Z \gtrsim 96$	-	$\lesssim 0.8$	-	0

* Not accurately determined because of large fraction of secondaries in observed cosmic rays.
† Value would be ~0.8 if we use solar Th value (Grevesse, 1969).
**Maehl et al. (1973).

of the neighboring, more abundant elements. Other workers (column 4) have reached a different conclusion with regard to the odd-Z elements Na, Al, and P. In particular, Cartwright (1971) and Simpson (1972) have concluded that these elements have to be present in sources at levels several times higher than solar system abundance levels, relative to neighboring, even-Z nuclei. The issue is an important one to resolve, as we shall see in the discussion to follow.

Do Cosmic Rays Come from Ordinary Galactic Gas?

It is, unfortunately, still not possible to answer this question definitively. One extreme viewpoint is to remark on the similarities between the composition of cosmic ray sources and of solar system material and to explain away the differences in terms of a process in which certain elements are preferentially accelerated at sources. At the other extreme might be the viewpoint that all of the cosmic rays are freshly synthesized and accelerated in a class of violent events such as supernova explosions, and that their composition reflects nucleosynthetic conditions during the explosion.

To argue the affirmative side of the question we must account for the discrepancies. Consider first the elements with $Z \leq 26$. Havnes (1971, 1973) and Cassé and Goret (1973) point out that there is a strong correlation between the ratio of cosmic ray source abundances to solar system abundances (CRS/SS) and the first ionization potential, as shown in Fig. 5-18. They can account for the relative abundances of all the elements up to Fe, with the exception of hydrogen, by assuming that cosmic rays come from the ionized fraction of ordinary interstellar gas that surrounds certain stars. To fit the data they assume a distribution of temperatures from a few thousand to a few $\times\ 10^4$ degrees. Havnes (1973) singles out hot, magnetic stars of the class known as peculiar A stars to be both the source of ionization and the driving force for acceleration of the interstellar gas to cosmic ray energies. The idea has attractive features, but he has difficulty in accounting

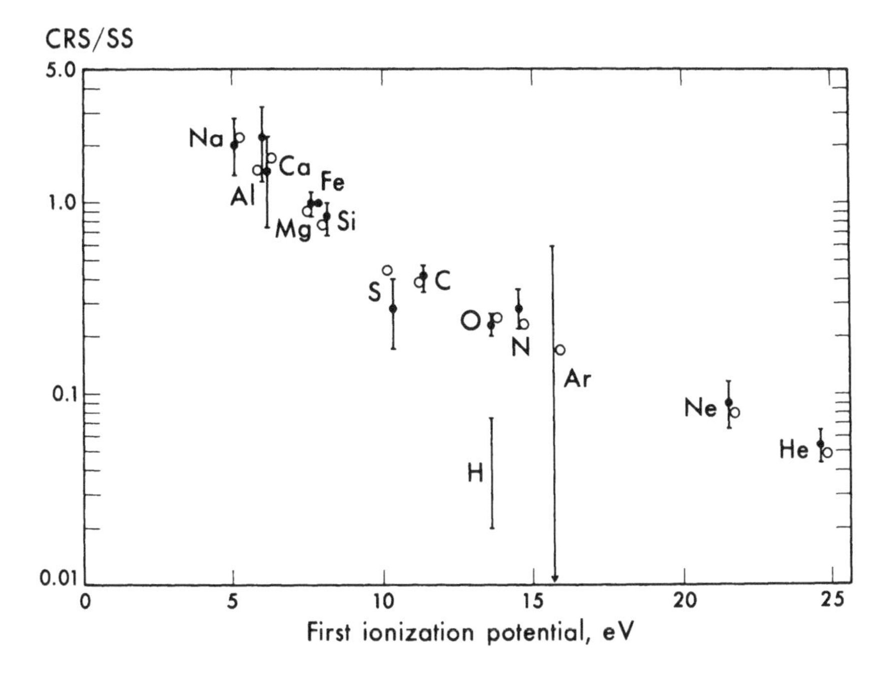

Fig. 5-18. *Correlation between the ratio of cosmic ray source abundance to solar system abundance (CRS/SS) and the first ionization potential. Solid points are deduced from observation; open circles are calculated for a dilute plasma. (After Cassé and Goret, 1973.)*

for even as much as 10% of the energy density in cosmic rays. Kristiansson (1972) believes that the ratio CRS/SS correlates better with cross sections for impact ionization by fast electrons than with first ionization potential. In neither case is the correlation perfect, and the apparently good agreement with calculation in Fig. 5-18 would be considerably worsened if somewhat different, probably equally valid, abundance data for the elements He, Ne, Na, and Al were used. Cowsik (1971) and coworkers (Cowsik and Price, 1971; Cowsik and Wilson, 1973) point out that CRS/SS increases rather smoothly with atomic number. (This is not surprising, since ionization potential and atomic number correlate reasonably well.) They suggest that there is a preferential acceleration of heavy nuclei in matter with solar system composition, the degree of enrichment increasing with Z as is observed in the case of solar flare particles (see Section 5.3). One set of candidates for the cosmic ray accelerators is white dwarf stars, which are extremely numerous, hot, rotating, magnetized objects that might be capable of accelerating ionized interstellar gas at their peripheries by electromagnetic interactions. Here again, getting adequate power to replace cosmic rays leaking out of the Galaxy is a problem.

Now consider the elements beyond Fe. In the region from Ni out to $Z \approx 70$ the ratio CRS/SS remains about the same as it is for Fe, and since most of these elements have ionization potentials similar to that of Fe, it is possible to retain the arguments used in support of cosmic ray sources drawing upon ordinary interstellar gas. For cosmic rays with $Z > 70$, difficulties are encountered. The prominent Pb peak characteristic of solar system material is absent and instead there appears to be an abundance peak centered at $Z \approx 76$ to 78, as seen in Fig. 5-12. The cosmic ray sources appear to be very greatly enriched in Th and U relative to the meteoritic abundances (Cameron, 1973), but would not present a major stumbling block to the models we have been discussing, provided we ignore the meteoritic data and adopt the solar spectroscopic value, which is an order of magnitude higher (Grevesse, 1969). Both values are listed in Table 5-6. The concept of cosmic rays originating entirely in ordinary galactic gas cannot survive if it turns out that the data for transuranic cosmic rays in Fig. 5-12 are correct. If the three events assigned charges $Z > 92$ were correctly identified, then the transuranic cosmic rays are about 30% ($\pm$10%) as abundant as are Th + U, in contrast to a relative abundance of 0 within our solar system (no replenishment of transuranics since the beginning of the solar system) and perhaps ~2% in interstellar gas that is frequently replenished with transuranics from *r*-process synthesis. But it is only fair to say that our confidence in the correctness of charge assignments for such heavy elements is less than complete, and we cannot rule out the possibility that no transuranic cosmic rays have been detected.

Measurements of abundances in both the cosmic rays and in the sun need to be improved, particularly for the ultraheavy elements. Future measurements of isotopic abundances in cosmic rays will provide strong constraints on models. For the present we must still consider seriously the possibility that the bulk of the cosmic rays may represent gas of average galactic composition that has been biased toward elements with a high Z or with a low ionization potential in the process of being raised to cosmic ray energies, the driving force perhaps coming from rotating, magnetized objects such as white dwarf stars, neutron stars, or peculiar A stars. Our discussion has been limited to the compositional evidence, but the problems of accounting for the power, energy spectrum, and isotropy of nuclear as well as electronic cosmic rays must also be satisfied by a model of their origin.

Are Cosmic Rays Synthesized in Explosive Events?

A traditional view with several attractive features is that both synthesis and acceleration of nuclei to cosmic ray energies are associated with the same event, usually thought to be a supernova explosion in which the *r*-process makes nuclei all the way up to and beyond uranium. The view goes back to Baade and Zwicky (1934) and has since been championed by a variety of authors including Ginzburg and Syrovatskii (1964), Hayakawa (1969), and Colgate and Johnson (1960). The idea is attractive because the power requirements seem to be met; the overabundance of heavy elements in cosmic rays seems consistent with the fact that a supernova is the last stage of evolution and the relative abundance of heavy elements increases as a star evolves; and synchrotron emission observed in supernova remnants is evidence for the acceleration of electrons to highly relativistic energies.

Acceleration might occur in the expanding shock wave (Colgate and Johnson, 1960), in the near field of the magnetized, rotating neutron star remnant (a pulsar) (Gunn and Ostriker, 1969; Goldreich and Julian, 1969), or in its far field (Kulsrud et al., 1972), where the nebula is pushing against the interstellar medium. It is not at all clear, however, whether nuclei raised to high energies can escape from the region of acceleration around a supernova without drastic decrease of energy. Moreover, no one has worked out all the details of synthesis in the path of a shock wave sweeping through material at a temperature of several billion degrees and at densities ranging up to at least 10^5 g/cm^3. Not the least of the problems is to follow the modifications of the newly synthesized material that occur during the rapid expansion and cooling to densities and temperatures characteristic of the interstellar medium (Clayton and Peters, 1970).

Arnett and Schramm (1973) have recently investigated the possibility of reproducing cosmic ray source abundances in supernova explosions. It is generally believed that stars with mass less than $\sim$4 $M_\odot$ (the subscript indicating solar) end life nonexplosively as planetary nebulae and white dwarfs. Stars in the mass range 4 to $\sim$8 $M_\odot$ turn into supernovas during the explosive carbon-burning phase but are rejected as candidates for cosmic ray sources because they convert all of their C and O into Fe. Arnett and Schramm propose that massive stars ($M > 8M_\odot$) are the cosmic ray sources since they seem to be the only candidates capable of producing elements all the way from He to Fe during the supernova explosion. They assume that cosmic rays come from both the envelopes and the remnants via a pulsar acceleration mechanism. Averaging over the death rates of stars in the mass range 8 to 70 $M_{\odot}$, they compute a composition of the major elements from He to Fe that is reasonably consistent with cosmic ray source composition. The hydrogen has to come from the envelope. The heavy elements are synthesized by the *r*-process. Though the model is *ad hoc* and very sketchy, it has interesting possibilities. It would seem to lead to spatial and time variations in composition, since the relative contributions of core and envelope might be quite variable. In Chapter 6 we will see that no variations have been detected, though the limits are not yet very restrictive.

Several examples of cosmic ray abundances might be singled out as evidence for a different nucleosynthetic history than for solar system material—the abundances of Na, Al, P, Sr, and the ultraheavy elements having already been mentioned. A famous example is the C/O ratio, which is ~ 0.9 in cosmic ray sources and ~ 0.5 in the sun. One can explain away most of these differences by invoking selective acceleration of elements with low ionization potential. Isotopic anomalies could not, however, be explained away in such fashion, and considerable effort is now going into the design of sophisticated instruments that may have sufficient resolution to determine isotopic abundances.

The strongest evidence favoring cosmic ray synthesis in explosive events is the distribution of ultraheavy nuclei shown in Fig. 5-12. Peaks in the distribution occur at $Z \approx 53$ and ≈ 76, indicative of *r*-process synthesis, rather than at $Z = 56$ and 82, which would be indicative of *s*-process synthesis. In the solar system, peaks occur at both places (Fig. 5-5) indicating a mixture of *r*- and *s*-process synthesis. Though there is still a question as to possible systematic errors in charge assignments, the relative abundances of the Pt group ($74 \leq Z \leq 79$), the Pb group ($80 \leq Z \leq 83$), the U group (Th + U), and the transuranic nuclei, taken at face value, support the view that heavy cosmic rays represent *r*-process material that has reached earth with a leakage lifetime of no more than a few million years. The use of radioactive elements to estimate the age of the cosmic rays has been discussed by most of the authors in Table 5-4 and especially by Mewaldt et al. (1970) and Schramm (1972). A leakage lifetime of a few million years, together with a mean leakage pathlength of ~ 5 g/cm^2 inferred from the observed abundances of secondary nuclei, implies a gas density close to one atom/cm^3 and favors the galactic disc rather than the much less dense halo as the containment volume for the cosmic rays.

Are There Several Sources of Cosmic Rays?

Neither of the two extreme models discussed above (pp. 270 and 273) seems capable of accounting for all of the features of cosmic rays. Nature may be more complicated in that a variety of sources may contribute. If such is the case one might expect the composition to change with energy. The gradually decreasing fraction of secondaries and increasing fraction of Fe relative to lighter primaries with increasing energy above a few GeV/nucleon seems to be a consequence of energy-dependent diffusion out of source regions, as mentioned earlier, rather than indicative of more than one kind of source operating at different energies. On the other hand, the steep energy spectrum of cosmic rays with $Z \geq 60$, shown in Fig. 5-15, does not fit into the picture of energy-dependent diffusion. Osborne et al. (1973b) have suggested that the excess of high-Z nuclei at energies below $\sim$2 GeV/nucleon comes from a special source whose output is concentrated at high Z and "low" energies (that is, below a few GeV/nucleon). In their discussion Osborne et al. pointed out that the sun accelerates particles of different Z differently, such that the relative abundance of heavy nuclei increases with decreasing energy (Fig. 5-24). The explanation for the different shapes of the energy spectra of solar particles of different Z is not known. The energy regime within which solar particles have been studied extends from $\sim$0.1 to $\sim$100 MeV/nucleon. Perhaps a similar mechanism, occurring at much higher energy in sources much more powerful than the sun (flare stars, for example), might be responsible for the excess high-Z cosmic rays. Another possible source might be peculiar A stars, whose surfaces are known to be extremely enriched in very heavy elements such as Hg, Au, Pb, U, and rare earths. These stars have surface magnetic fields of up to 10^4 G, and one of the explanations of their peculiar composition is that the heavy elements are synthesized in nuclear reactions produced during flares.

5.3. SOLAR FLARE PARTICLES

Being very close to one star—the sun—gives us an opportunity to study the processes of acceleration, storage, and propagation of energetic particles in great detail. In this section we will discuss the composition and energy spectra of particles emitted in solar flares and see if they have any bearing on the origin of cosmic rays.

Fig. 5-19 is a qualitative picture of some of the features of the quiet sun. It is taken from a book by one of the Skylab astronauts, Ed Gibson (1973), who has given an excellent, detailed description of the physics of the quiet sun. The exterior of the sun consists of the visible surface—called the photosphere—and the solar atmosphere, divided into a high-density, low-temperature chromosphere and a low-density, high-temperature corona. The sun is never completely quiet. Intense solar activity is usually associated with complex sunspot groups, which in turn represent regions of enhanced solar magnetic field strength, typically several thousand gauss. Within the sunspot group a solar flare such as the one shown in Fig. 5-20 may occur. The flare is characterized by bursts of x rays, optical line

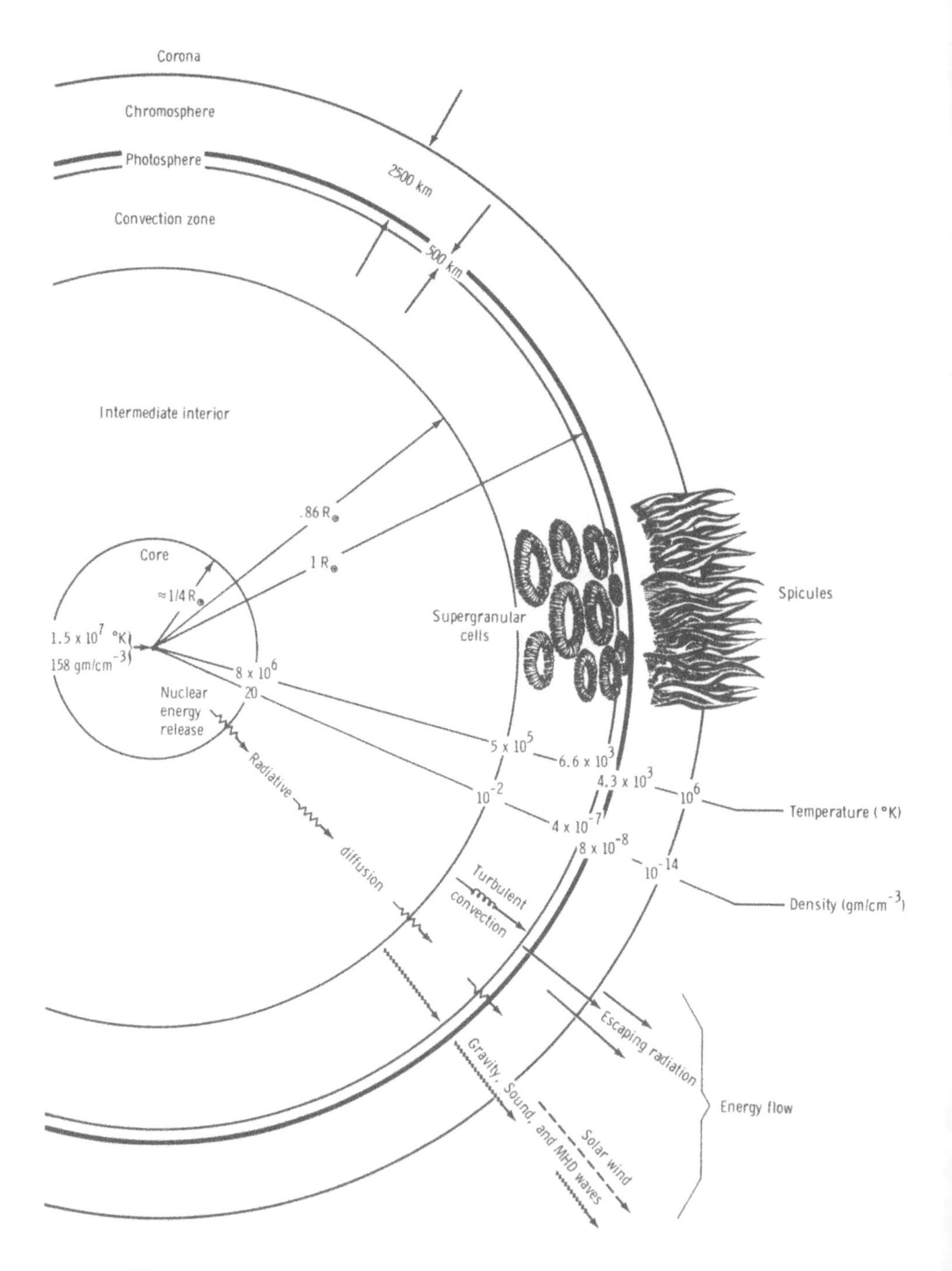

Fig. 5-19. *Idealized general solar properties, structure, and modes of outward energy flow.* (*After Gibson, 1973.*)

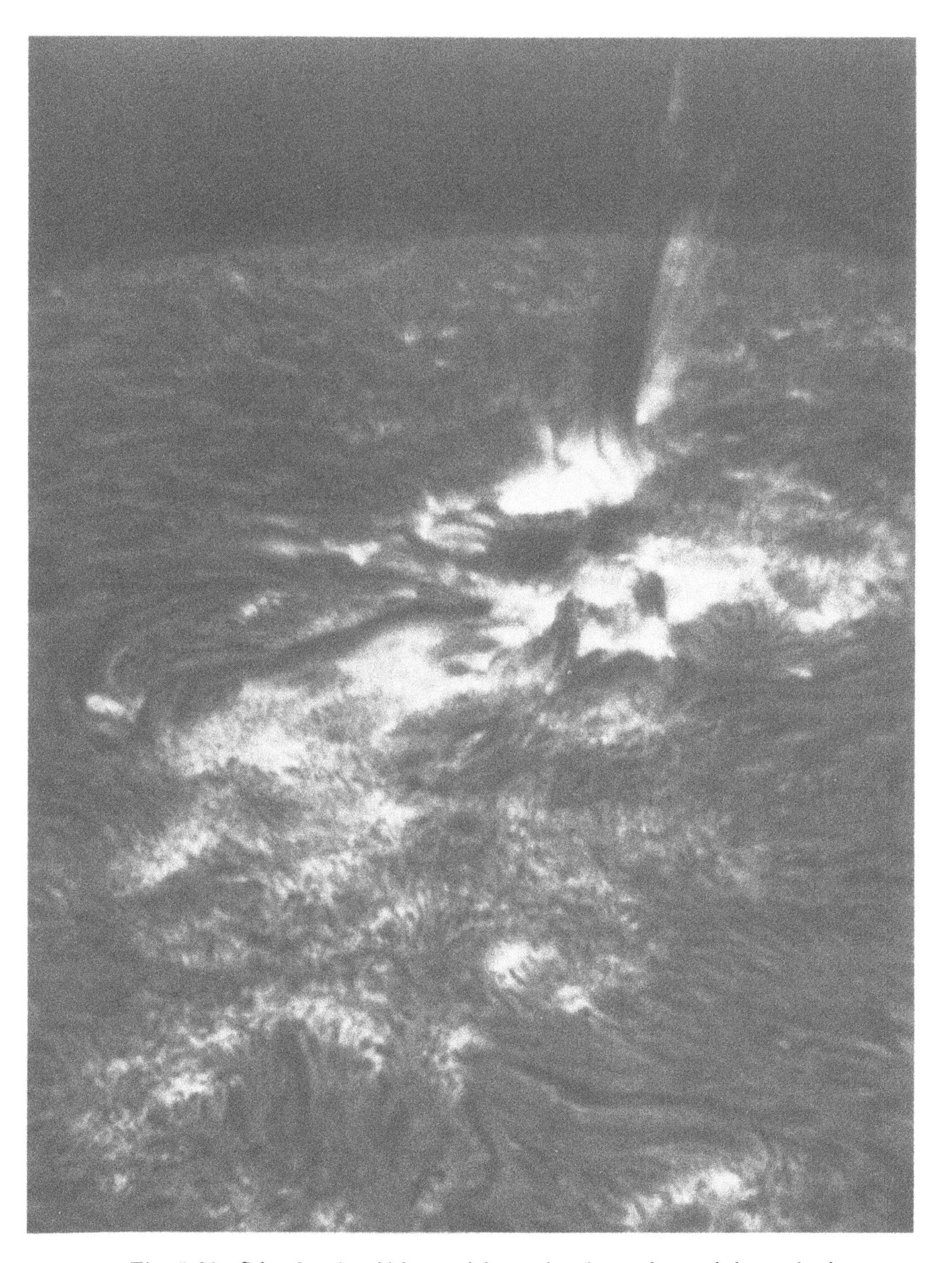

Fig. 5-20. *Solar flare in which material was ejected several tens of thousands of kilometers above the surface of the sun. (Courtesy of Big Bear Solar Observatory, Big Bear, California.)*

emission, and radio waves at frequencies of the order of hundreds of MHz. Strong solar flares may emit white light for a short period and may also accelerate charged electrons and nuclei to relativistic energies.

The nature of the process by which these solar flare particles are accelerated is not understood. Perhaps they are pre-accelerated in some large-scale magnetic trap by interaction with magnetic waves and only released during the flare. Alternatively the flare may reflect particle acceleration in a region where oppositely directed magnetic field lines annihilate each other and an intense electric field is set up. It is generally believed that solar particles originate in the chromosphere. Fichtel and McDonald (1967) have reviewed our knowledge of solar particles up to about 1966. Since then the picture has changed drastically. Recent studies of charge and energy spectra down to very low energies, using both solid track detectors and electronic detectors, have provided strong constraints on models of the acceleration process and have revealed possible similarities between solar particles and galactic cosmic rays.

5.3.1. Solar Flare Fe Tracks in Glass from the Surveyor Camera

The first opportunity to study present-day solar flare particles using track-recording solids came with the return of the Surveyor III television camera from the moon by the Apollo 12 astronauts. A flint glass filter over the camera lens, seen in Fig. 5-21, was incompletely shrouded by a mirror and camera housing, so that particles within a solid angle of $\sim$1.3 ster could directly impinge on the glass during a 2.6 year period in which there were seven major solar flares. Because particles at zenith angles less than $\sim 30°$ were excluded by the shroud from entering the glass, tracks which when etched on the exposed surface would have half cone angles greater than $\sim 60°$ were thus not revealed. We saw in Fig. 3-18 that at energies below $\sim$0.2 MeV/nucleon the half cone angle for Fe tracks in SiO_2 rapidly increases to 90° (and $v_t/v_g \rightarrow 1$) with decreasing energy. In addition, there was a 3/4 wavelength coating on the top surface. These two effects made the spectrum of Fe nuclei below $\sim$0.3 MeV/nucleon completely inaccessible to study.

Three groups received samples for track studies. In order to utilize the largest possible volume of glass, Price et al. (1971a) determined the track density gradient by sequentially etching and following all tracks to the end of their range. The statistics are good with this procedure, but one is prevented from observing tracks at nearly vertical incidence and there is in addition the danger of cumulative errors in determination of energy as more and more of the glass is etched away. Crozaz and Walker (1971) and Fleischer et al. (1971) ground their samples at a 60° angle so that the most probable angle of entry was normal to the ground surface and even tracks with low values of $(v_t/v_g - 1)$ could be revealed. This procedure avoids the problem of systematic errors in knowledge of residual range. In their

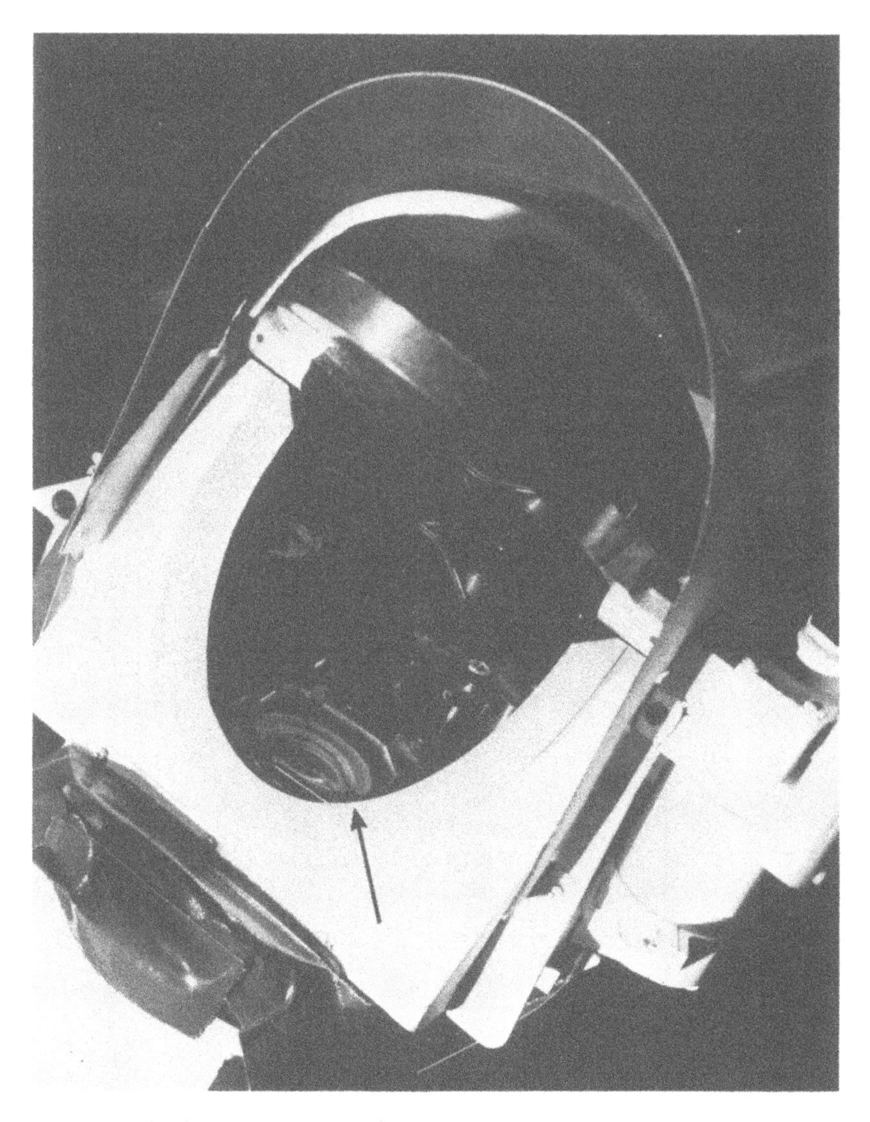

Fig. 5-21. *Portion of returned Surveyor III TV camera showing the flint glass filter located over the lens inside the shroud.*

reassessment of the Surveyor data Hutcheon et al. (1974) observed the growth of solar flare Fe etch pits during a sequence of etches of the Surveyor glass and showed that, at a residual range as great as ~100 μm, pits at a steep angle to the surface can still be etched. To extract the correct energy spectrum from the pit density profile obtained by Price et al. requires corrections that depend critically on a knowledge of the etch pit cone angle as a function of energy. Instead, Hutcheon et al. averaged the data of Crozaz and Walker and of Fleischer et al. and assumed

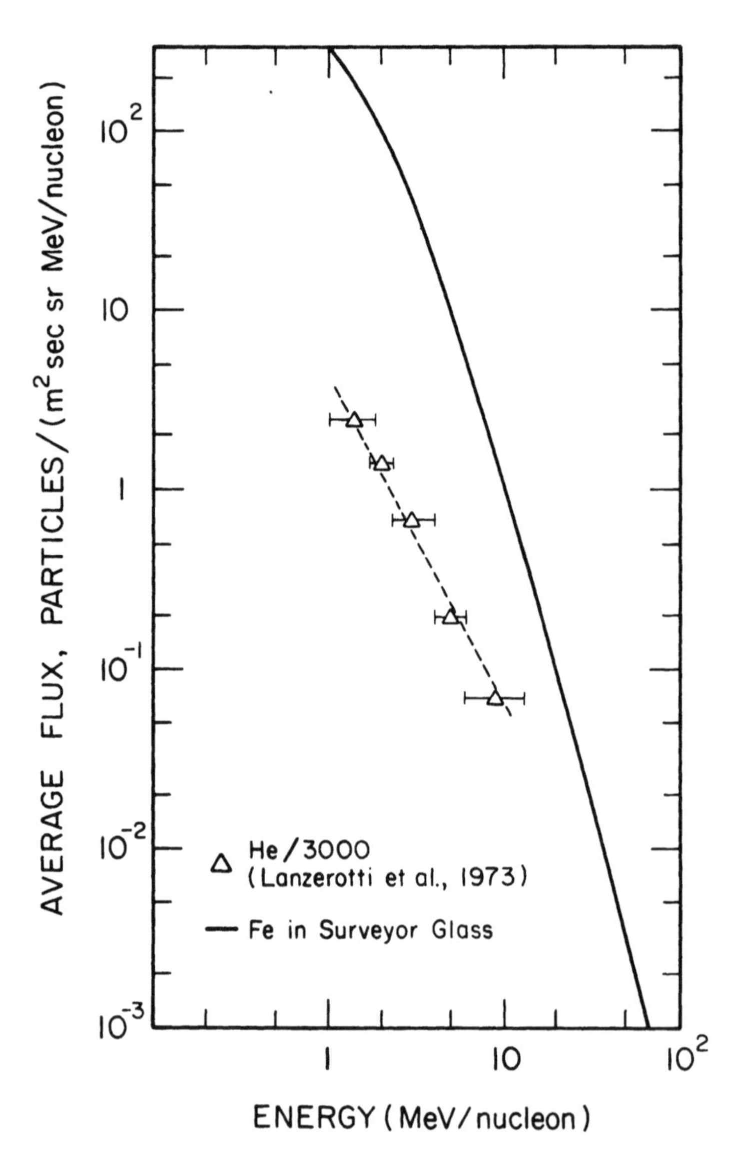

Fig. 5-22. *Average spectrum of Fe nuclei in space during 1967–69 (from tracks in Surveyor III glass) compared with He spectrum (from electronic detector on IMP-4). The Fe/He ratio is enhanced by about a factor 10^2 at the lowest energy. The enhancement decreases with energy.*

a 100 μm etchable range in order to derive the solar flare energy spectrum shown in Fig. 5-22. A reanalysis indicates that the maximum etchable range value of 100 μm should be reduced to ~60 μm when properly weighted for solid angle and etching efficiency. The ordinate gives Fe ions m^{-2} sec^{-1} $ster^{-1}$ $(MeV/nucleon)^{-1}$ assuming the 2.55 year exposure to be the same as the long-term average rate. In

Chapter 6, when we discuss ancient solar flares, we shall discuss the problem of properly normalizing the Surveyor data to a long-term average flare rate.

In their paper Price et al. (1971a) added up the daily He fluxes from Lanzerotti's alpha particle detector (Lanzerotti et al., 1973) for the interval in which Surveyor was exposed and concluded that the Fe/He ratio in low-energy solar flare particles is enhanced by a large factor, of order 20, relative to the ratio in the sun. To show the enhancement, in Fig. 5-22 we have plotted the He data, scaled down by the factor 3000 (believed to be the approximate solar ratio of He/Fe). Price et al. suggested that the high Fe/He ratio might result from the preferential leakage of incompletely ionized heavy nuclei from the accelerating region. Their reasoning was based on the fact that at a few MeV/nucleon the fractional charge of a heavy ion is much lower and the magnetic rigidity much higher than that of a light ion with the same velocity. It is perhaps worth pointing out here that they would not have discovered the phenomenon of heavy element enhancements if they had only looked at the Fe/H ratio as a function of energy. The reason is that Fe and H, which have quite different A/Z ratios, will have different energy spectra when compared on an energy/nucleon basis, because acceleration and propagation depend on magnetic rigidity, $R = (A/Z)m_pc^2\beta\gamma$, rather than on energy.

The discovery of a high Fe/He ratio in solar particles created considerable excitement for several reasons:

1. Observations at energies above $\sim$20 MeV/nucleon, using rocket-borne nuclear emulsions, had led to the appealing (though erroneous) result that solar particle abundances apparently are the same from event to event and the same as the abundances in the solar atmosphere (Biswas and Fichtel, 1965; Bertsch et al., 1972).
2. An analogous process, operating at much higher energy, might be responsible for converting a gas of roughly solar composition into galactic cosmic rays, which we saw to be enhanced in heavy elements.
3. The extremely high density of Fe tracks observed in the lunar soil at all depths down to at least 2.5 meters was difficult to understand if solar flare particles have normal solar composition but could be understood if there were a strong enhancement of heavy elements in solar flares (Barber et al., 1971).

Subsequently, Mogro-Campero and Simpson (1972a,b) reported evidence for enrichments of heavy nuclei by amounts that increased with Z, and Lanzerotti et al. (1972) used their alpha particle data in conjunction with measurements by Price and Sullivan (1971) to show that at low energies in the 25 Jan 1971 flare O, Si, and Fe were enhanced relative to He. In a comparison of measurements made at energies of $\sim$1 MeV/nucleon with electronic detectors and at energies above $\sim$20 MeV/nucleon with nuclear emulsions, Armstrong et al. (1972) suggested that the (C + N + O)/He ratio was energy dependent, being higher at low energies than at high ones.

5.3.2. Energy Dependence of Heavy Ion Enhancements in Flares

The systematics of heavy ion enhancements have been clarified as a result of a series of studies with plastic detectors exposed on rockets at appropriate times in flares during 1971 (Crawford et al., 1972) and 1972 (Price et al., 1973b; Sullivan et al., 1973; Crawford et al., 1975) and with glass, mica, and plastic detectors exposed on the moon during the Apollo 16 and 17 missions (Burnett et al., 1972; Fleischer and Hart, 1972, 1973; Price et al., 1972; Price and Chan, 1973; Shirk and Price, 1973; Woods et al., 1973). C. E. Fichtel and his colleagues at Goddard Space Flight Center generously made available space for plastic detectors on their recoverable nose cones and were responsible for launching and recovering the rockets as well as for analyzing nuclear emulsions. At energies accessible to both detectors, plastics and emulsions agree well in determinations of fluxes of the major charge groups. With plastics it is possible to say something about charge composition at energies down to ~0.2 MeV/nucleon and to determine energy spectra of elements from He on down through the Periodic Table, limited only by counting statistics. Fig. 5-23 shows etched tracks in the top sheet of a stack of Lexan detectors exposed for 4 minutes on a sounding rocket during the spectacular flare of 4 Aug 1972. In this section we discuss the way in which the composition changes with energy; in the next section we take a detailed look at the composition at energies above ~15 MeV/nucleon, where the composition in a particular flare does not appear to be changing.

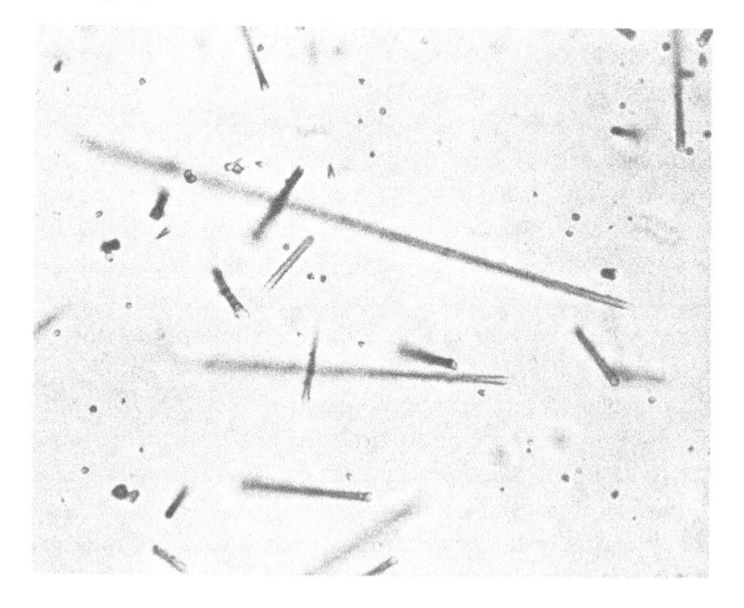

Fig. 5-23. *High density of tracks of nuclei with $Z \geq 6$ observed at a magnification of 1350× in the top sheet of a Lexan stack exposed 4 minutes in a rocket during the flare on 4 Aug 1972. (Price, unpublished.)*

Fig. 5-24 shows energy spectra for several elements from He to Fe in four of the six flares studied, one of which was the strongest in recorded history (4 Aug 1972). Referring back to the Apollo 17 results in Fig. 5-8, we see that during the period 11 to 13 Dec 1972 the energy spectra fell very steeply at energies below $\sim$1 MeV/nucleon. That the majority of the particles originated in the sun, even though there was no detectable flare at that time, was inferred from their measured O/(C + N) abundance ratio, which agrees with the solar value of $\sim$1.65 rather than the cosmic ray value of $\sim$0.74 (Chan et al., 1973).

The overall features of the data from the flares shown in Figs. 5-24 and 5-25 can be summarized as follows (Price et al., 1973b):

1. At sufficiently low energies heavy elements in solar flare particles are always enriched relative to their abundances at high energies.
2. The enrichment factor, defined as $Q(Z, E) \equiv (Z/\mathrm{He})_{SP}/(Z/\mathrm{He})_{\odot}$, is an increasing function of Z from He at least up to $Z \sim 44$ and possibly to the end of the Periodic Table (Shirk and Price, 1973; Mogro-Campero and Simpson, 1972b). Since only the elements He, O (or CNO), Si, Fe $[32 \leq Z \leq 38]$, and $[Z \geq 44]$ have thus far been studied, one cannot rule out possible fine-structure in the enrichments such as a correlation with first ionization potential.
3. The enrichment factor decreases with energy and appears to approach a constant value beyond some characteristic energy of the order of the mean energy of the flare particles. This constant value differs from flare to flare but is usually within a factor three of unity, indicating that large deviations of the high-energy particle population from average solar abundances do not occur.
4. The maximum value of the enrichment factor, observed at the lowest energy accessible with a track detector, $\sim$0.1 MeV/nucleon, is similar for all solar conditions ranging from nearly quiet sun to the most intense flare.
5. Even at energies as low as $\sim$2 MeV/nucleon, Fe seemed to be nearly completely ionized, even though the equilibrium charge in neutral matter at such energies would be only $\sim$13. The evidence, admittedly somewhat indirect, was based on the shapes of rigidity spectra in the 17 April 1972 flare (Braddy et al., 1973) and on the use of the geomagnetic field at Ft. Churchill as a magnetic analyzer in the 25 Jan 1971 flare (Sullivan and Price, 1973). This result implies that conditions at the accelerating region can be such as to remove even the inner electrons with ionization potentials up to several keV. A likely way of doing this is through impact ionization by free electrons, which have a high kinetic temperature in the region where the flare occurs. Once the stripped ions leave the flare region they will tend to reach charge equilibrium by capture of orbital electrons, but the cross section for this

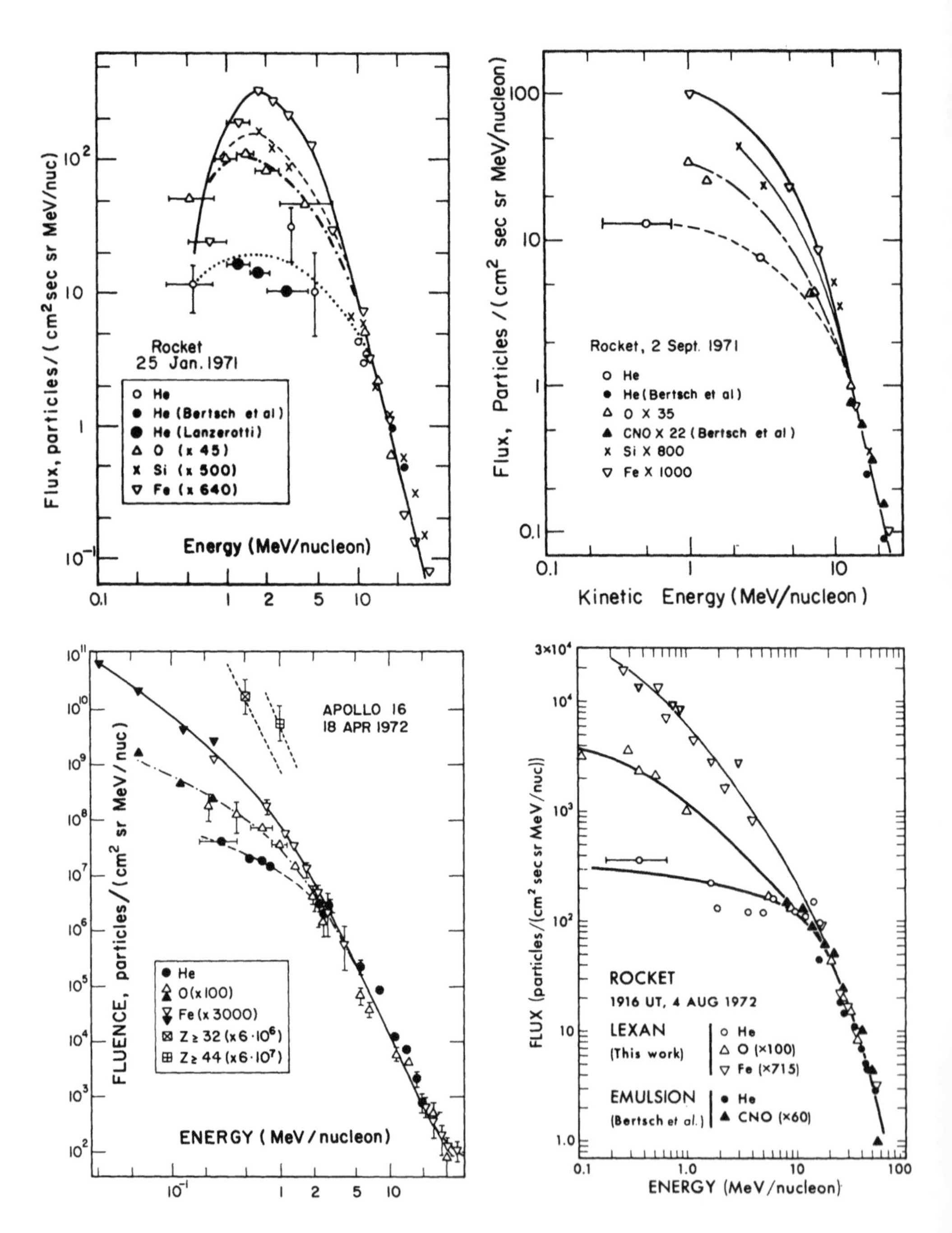

Fig. 5-24. *Energy-dependent composition of solar particles in events of different intensity. Spectra are scaled by constant factors so that they coincide at high energies. (After Price et al., 1973b,c; Shirk and Price, 1973.)*

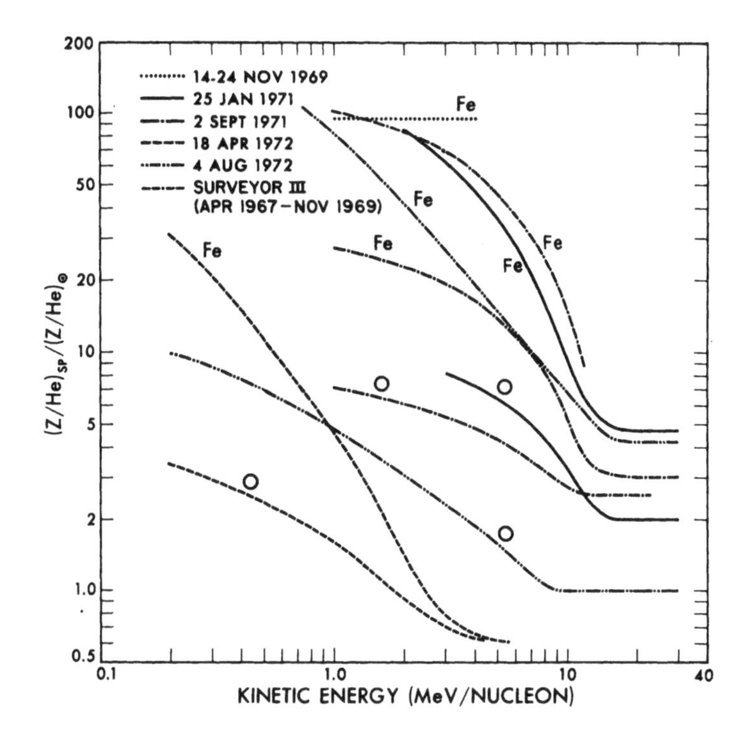

Fig. 5-25. *Energy-dependent enhancement of the abundance ratios O/He and Fe/He in various flares relative to their values in the sun. (After Crawford et al., 1975.)*

process is sufficiently low that they have a high probability of reaching the vicinity of earth without changing their charge if they leave the sun by a reasonably direct route.

Recent low-energy measurements in a weak flare on 30 October 1972, using electronic detectors on the IMP-7 spacecraft (Hovestadt et al., 1973; Gloeckler et al., 1973), complement the pattern of results obtained with solid track detectors. At an energy of 0.1 MeV/nucleon the ratios O/He, Si/He, and Fe/He were overabundant relative to the solar photospheric ratios by factors of 35, 70, and 100, respectively, and are similar to ratios observed in some of the flares studied with solid track detectors. Their results for ionization states supported the conclusions reached by means of plastic detectors. At an energy of 0.1 MeV/nucleon they inferred that the effective charges of C and O were 5.4 and 7.4, respectively—much higher than the values 2.5 and 2.9 expected if these ions were in charge equilibrium with neutral matter.

To the above list we add the observation of Bhandari et al. (1973), based on fossil track studies, that over the last few million years transiron elements in solar flares have been enriched relative to Fe, and the observations of Anglin et al. (1973) and Garrard et al. (1973) with electronic detectors, that the $^3He/^4He$ ratio in some flares is as high as $\sim 5\%$. The latter result implies the passage of low-energy solar particles through several g/cm^2 of solar material and the subsequent acceleration of the nuclear reaction products. We should also add the observation of Krimigis and Armstrong (1973) that the abundance ratios He/CNO and Fe/O at a given energy/nucleon vary by as much as a factor ten with time following the onset of a solar flare.

In a span of two years the picture of solar particle composition has changed from an appealingly simple one to an exceedingly complex one, but the truth about nature is always more interesting than the way we would have it be. At present there is no satisfactory explanation for the variations of composition with energy. Various ways have been proposed for selectively accelerating heavy elements (Korchak and Syrovatskii, 1958; Gurevich, 1960; Cowsik, 1971; Cartwright and Mogro-Campero, 1972; Ramadurai, 1973), but none can account for all of the observations.

Part of the compositional variation might result from temporal or spatial variations in the composition at the site of particle acceleration. For example, several models of the transition region between the chromosphere and the corona, where the temperature rises abruptly from $\lesssim 10^4$ to $\gtrsim 10^6$K, indicate that heavy ions preferentially diffuse outward in the strong temperature gradient (Jokipii, 1966; Nakada, 1969; Alloucherie, 1970). From solar spectroscopic observations it is not yet possible to rule out a strong, variable heavy element enhancement in a localized region of the solar atmosphere.

Most of the proposed explanations of heavy ion enrichments in flares incorporate the fact, first recognized by Niels Bohr, that the fractional charge of a heavy ion is lower than that of a light ion if both are traveling with the same velocity in neutral matter—see eq. (1-8). The heavy ion will thus have a greater magnetic rigidity and gyroradius than the light one. If, as was inferred to be the case with Fe ions, the ions escape from the sun fully stripped, then any preferential selection of heavy ions based on their incomplete ionization must occur in an early stage of the flare but then be followed by a stage in which electrons are removed, perhaps by impact with electrons at a high kinetic temperature as we have already suggested.

In order for sufficient ^{3}He to be made during the time of explosive release of energy in a flare ($\sim$100 sec), the ions that undergo nuclear reactions must pass through several g/cm^2 of material, which requires that they be accelerated within a region with a density of at least $\sim 10^{-12}$ g/cm^3, corresponding to the chromosphere (Fig. 5-19). Sullivan et al. (1973) have pointed out that one ought to be able to detect a finite flux of Li and Be in those flares for which the ^{3}He/^{4}He ratio is several percent. The reason is that ^{6}Li, ^{7}Li, and ^{7}Be can be produced at energies as low as 9 MeV/nucleon (by $\alpha + \alpha$ reactions) whereas the threshold for proton reactions leading to ^{3}He is at $\sim$24 MeV. The upper limit for the ratio $(\text{Li} + \text{Be} + \text{B})/(\text{C} + \text{N} + \text{O})$ in flares is $\sim 1\%$ (Teegarden et al., 1973; Sullivan et al., 1973), which is much lower than the ratio for galactic cosmic rays ($\sim 10\%$).

At a density of 10^{-12} g/cm^3 the rate of energy loss by ionization may be comparable in magnitude to the rate of energy gain by processes involving moving magnetic fields. Unfortunately the calculation of ionization loss in a plasma is complicated by the fact that ion velocities may be less than or close to thermal velocities, a regime for which the ionization loss rate is poorly known. As Ramadurai (1973) has pointed out, in a competition between energy gains due to collisions with moving magnetic fields (the famous Fermi acceleration process) and energy losses due to ionization, particles with large Z^2/A will have steeper spectra than particles with small Z^2/A, and indeed, such appears to be the case (Fig. 5-24). However, not only the slopes but also the magnitudes of the spectra must be explained. A difficulty with Ramadurai's model is that the energy spectra should cross, so that any enhancement at low energy is compensated by a depletion at high energy. This does not fit the observations. Further, he has used a simplified form of the ionization relation, valid only at ion velocities greater than thermal velocities. The problem of accounting both for the steeper spectra and the enhanced intensity level of heavy ions is still unsolved.

5.3.3. Charge Composition at High Energies in Flares

We have seen that the overabundance of heavy nuclei relative to light nuclei in individual flares tends to decrease with increasing energy. Plastic detectors have been particularly useful in their ability to record compositional information over a very wide energy interval. Nuclear emulsions cannot be used to identify particles with energies less than ~15 MeV/nucleon but at higher energies are valuable because they provide compositional data for the major elements in flares extending over a complete solar cycle (see Biswas and Fichtel (1965) for a review of the early measurements with emulsions). Neither the data from emulsions nor the data from plastic detectors show any evidence for a change of composition with energy above ~15 MeV/nucleon, though admittedly a subtle change might be overlooked because the fluxes are weak and statistics are poor at the higher energies.

With the acquisition of data from several flares in 1971 and 1972 it has become clear that there are real differences in composition from flare to flare, so that not even at high energies do solar particle abundances exactly mirror abundances in the sun or in the meteorites. Table 5-7 compares abundances of solar particles at energies greater than 15 MeV/nucleon in three recent flares with spectroscopically determined abundances in the solar photosphere and corona, with the solar system abundances compiled by Cameron (1973), and with cosmic ray source abundances.

Three points should be noted in connection with Table 5-7:

(1) Among the three flares the abundances of the heaviest (Fe) and lightest (He) elements definitely vary with respect to intermediate elements such as O or Si. Table 5-8 summarizes the available evidence for the variability of the abundances of He and Fe, drawn from flares on 2 Sept

Table 5-7. Relative Abundances in Solar Particles (E>15 MeV/nuc), Sun, Solar System and Cosmic Ray Sources

Ref. / Z	25 Jan 71 (a)	2 Sept 71 (b)	4 Aug 72 (c)	(g)	Photo-sphere (d)	Coronal (d)	(e)	Solar System (f)	Cosmic Ray Sources (Table 5-6)
He	$\sim9\text{x}10^4$	$\sim5\text{x}10^4$	-	$6.8\text{x}10^4$	-	$2\text{x}10^5$	-	*	2.0 to $2.3\text{x}10^4$
C	-	-	-	470±40	1050	1260	-	1180	520 to 630
O	1300±300	1500±250	-	1000±200	1900	1260	980	2150	550 to 600
Ne	-	180±60	-	170±30	-	79	250	*	70 to 180
Na	-	-	-	21±9	5	6	-	6	6 to 15
Mg	120±40	130±40	-	210±40	98	100	145	106	120 to 140
Aℓ	-	-	-	19±9	7	6	-	8.5	10 to 25
Si	≡100	≡100	≡100	≡100	≡100	≡100	≡100	≡100	≡100
S	15±6	11±5	13±5	↑	46	32	14	50	19
Ar	<4	<4	1.4^{+3}_{-1}	|	-	13	-	12	<3
Ca	8±5	5±3	8±3	70±30	6	4	-	7	16
Ti-Cr	<4	<4	(Ti<1)	|	1.6	1.6	-	1.5	<3
Fe	90±20	45±15	106±25	|	71	100	55	83	130
Ni-Zn	4±3	3±3	6±3	↓	5	8	3	5	4

(a) Crawford et al., 1972; (b) Sullivan et al., 1973; (c) Crawford et al., 1974; (d) Withbroe, 1971; (e) Malinovsky and Heroux, 1973; (f) Cameron, 1973; (g) Webber et al., 1973c.
* values not quoted since they were derived from solar particle data.

Table 5-8. Variability of Solar Particle Abundances at E≳15 MeV/nucleon

Event	Reference	He	CNO	O	Fe
2 Sept 1966	(Bertsch et al., 1973)	79±13	≡1.65	-	0.031±0.007
12 Apr 1969	(Bertsch et al., 1973)	91±13	≡1.65	-	<0.047
25 Jan 1971	(Bertsch et al., 1973)	46±10	≡1.65	-	0.059±0.014
"	(Sullivan et al., 1973)	70	-	≡1.0	0.067±0.013
6 Apr 1971	(Teegarden et al., 1973)	68±14	1.49±.13	≡1.0	0.17±0.08
2 Sept 1971	(Bertsch et al., 1973)	36±7	≡1.65	-	0.055±0.018
"	(Teegarden et al., 1973)	42±2.5	1.61±.04	≡1.0	0.028±0.005
"	(Sullivan et al., 1973)	32	-	≡1.0	0.030±0.01
4 Aug 1972	(Biswas et al., 1973)	66±12	≡1.65	-	-
"	(Webber et al., 1973c)	68±2	1.63	≡1.0	[Z≳16]=0.07±.03
"	(Crawford et al., 1974)	100	-	≡1.0	0.14
Solar abundances		100-160	1.65	≡1.0	0.04 to 0.07

1966, 12 April 1969, and 6 April 1971, as well as from the three flares referred to in Table 5-7. The He/O ratio ranges from 36 ± 7 to ~100, and the O/Fe ratio ranges from 0.03 ± 0.01 to 0.17 ± 0.08. The values of these ratios for the solar atmosphere are not at all well known, but estimates are included in Table 5-8 for comparison. The Fe/O ratio in most flares seems close to its solar value. The He/O ratio is lower than the coronal value of ~160 quoted by Withbroe (1971), but the He abundance adopted by him was extremely uncertain. Judging from recent measurements of the He/H ratio in the chromosphere (5 to 7%) and in the solar wind (~4 to 5%), we would favor lowering the coronal value of He/O from 160 to no more than 100, which is not far from the range of values observed in flares. There is thus no evidence for a systematic change of abundances with atomic number, if only measurements at energies above ~15 MeV/nucleon are considered. The variations from flare to flare may represent real fluctuations in the composition of the region in which the flare originated.

(2) The abundances of the elements other than He and Fe, relative to Si or O, are the same within statistical errors in the three flares referred to in Table 5-7, though for the least abundant elements the errors are so large that one cannot rule out variations as large as a factor two.

(3) The abundances of several elements in flares differ significantly from the coronal values adopted by Withbroe and the solar system values adopted by Cameron, and yet they agree rather well with the abundances in cosmic ray sources, as has been pointed out repeatedly (Crawford et al., 1972; Mogro-Campero and Simpson, 1972a,b; Price, 1973a). The abundances of S and Ar are low in flares whereas Ca, Na, and Al are possibly high. One could try to reconcile these "discrepancies" in terms of a preferential emission of elements with a low ionization potential, as we discussed earlier in the section on cosmic ray abundances. This suggestion encounters the difficulty that He/O, Fe/O, and C/O are not significantly different in flare particles and in the sun, and yet their ionization potentials range from 24.6 eV for He down to 7.9 eV for Fe.

Better solar particle abundance data, with smaller statistical errors, would be useful, but it already appears to us that the solar abundance values of certain elements should be modified, perhaps by a factor of as much as three. The solar Fe abundance is a famous example of a seriously erroneous photospheric value that was only recently revised upward by a factor of five to ten (Aller, 1972). The S/Si ratio should probably come down to ~0.15 (note the new, low value of Malinovsky and Heroux in column e of Table 5-7); and the Ar/Ca ratio should be less than ~0.5 rather than greater than ~2. This last result has important implications for theories of nucleosynthesis, which have difficulty in explaining the local peak at Ca observed in both solar particles and in cosmic rays.

5.4. THE SOLAR WIND

The atmospheres of ordinary stars are not static. In the case of our sun, ionized material in the outer corona continuously flows radially outward at supersonic speed, carrying with it magnetic field lines and interacting with the lunar surface, the magnetospheres of Earth and Jupiter, the cosmic rays, and the tails of comets. The distribution of ionization states in the solar wind is consistent with their origin in the corona at a temperature of ~1 to 2 × 10^6K.

Although our knowledge of the solar wind composition is still quite sketchy, direct measurements of a few of the lightest elements have been made with electrostatic detectors on satellites (Bame, 1972) and with mass spectroscopic analyses of the rare gases He, Ne, and Ar, which became embedded in metal foils while they were exposed for one to two days on the lunar surface during Apollo missions (Geiss et al., 1972). These studies have shown that the composition of the solar wind is quite variable and that the material appears to be fractionated with respect to solar surface material. For example, the He/H ratio in the solar wind varies from ~0.003 to ~0.25, with an average value of ~0.045 (Bame, 1972). This average value is probably a factor two to three lower than that in the outer convective zone of the sun. The He/O ratio varies from ~30 to >400, with an average value of ~100 (Bame, 1972). The ^{4}He/^{3}He abundance ratio is also variable, showing a positive correlation with the ^{4}He/^{20}Ne abundance ratio and with the level of disturbance in the solar wind as indicated by geomagnetic field changes (Geiss et al., 1972). Table 5-9 summarizes measurements of solar wind elemental abundances for one particularly favorable day, 6 July 1969, when the random motion in the solar wind was sufficiently small that good measurements could be made with an electrostatic instrument (Bame, 1972). The Ne and Ar values are average values relative to He (Geiss, 1972). This table should be used cautiously, with the realization that the abundances on 6 July 1969 are not necessarily good average values. Recently Geiss (1973) has summarized the results of more measurements and has concluded that the average abundances of the elements in the table agree to within a factor of two with abundances in the solar surface. Isotopic abundance ratios of the light rare gases (He, Ne, and Ar) have been determined (Geiss, 1972), but are not included in the table. Evidence for the existence of C, N, Mg,

Table 5-9. Corona and Solar Wind Relative Abundances (uncertainty at least ±50%)

	Coronal abundance (Withbroe, 1971)	Solar wind abundance	Reference
H	1780–2300	5000	Bame (1972)
He	360	150	Bame (1972)
O	≡1.00	≡1.00	
Ne	0.06	0.26	Geiss (1972)
Si	0.08	0.21	Bame (1972)
Ar	0.01	0.007	Geiss et al. (1972)
Fe	0.08	0.17	Bame (1972)

and S in the solar wind has been obtained with spacecraft detectors (Bame, 1972). Kr and Xe, presumably implanted from the solar wind, have been found embedded in grains in the lunar soil, in lunar breccias, and in gas-rich meteorites (Eberhardt et al., 1970).

Various physical processes are expected to deplete or enrich different ion species in quite different ways (Geiss, 1972). The charge/mass ratio of the ion would enter into electromagnetic separation processes. Gravitational fractionation would depend primarily on the mass. Diffusion processes at the photosphere-corona boundary region depend strongly on the degree of ionization. If dynamical friction (momentum transfer from protons to heavier ions) is significant in the solar wind acceleration region, a factor similar to Z^2/A is important. Fractionation effects can also result from wave-particle interactions. The relative importance of these various fractionation effects must be understood before one can draw conclusions about solar abundances from solar wind abundance measurements. To aid in the attempt to distinguish between the various processes one needs to study ion species over a very large range of values of Z and A and at various times during the solar cycle.

To this end, mica detectors were exposed to the solar wind on both the Apollo 16 and 17 missions with the expectation that the mass distribution of heavy ions in the solar wind could be estimated by measuring the diameters and depths of etch pits using the techniques described in Chapter 3. The solar flare that occurred during Apollo 16 produced such a high background of low-energy heavy ion tracks in the mica that it was difficult to see the shallow etch pits that would be expected from heavy solar wind ions. Burnett et al. (1972) divided the pits into three categories on the basis of diameter: > xenon, > lead, > thorium. The corresponding pit densities (for a seven-day exposure facing the sun) were 2×10^5, 4×10^4, and 4.5×10^3 cm^{-2}.

The contribution from solar flare particles in the mica exposed during the Apollo 17 mission (Walker and Zinner, 1973) was sufficiently small that reliable pit counts could be made. Fig. 5-26 compares pits in pieces of mica that were exposed facing toward and away from the sun on Apollo 17 and in a piece of mica irradiated with 1 keV/nucleon Xe ions. The mica that faced the sun has a high density of etch pits, some of which are similar in size to those from Xe ions. Walker and Zinner (1973) interpreted these and other micrographs of the mica as evidence that the solar wind contains ions much heavier than Fe. The recent, detailed calibration experiments of Borg et al. (1974) have shown that, because of inherent scatter in response, the etch pit size distribution must be treated on a statistical basis. From their analysis of the Apollo 17 mica they concluded that the Fe/CNO ratio in the solar wind is similar to that in the sun and that the very heavy element groups $Z > 45$ and $Z > 60$ are present in the solar wind and not overabundant relative to Fe by more than a factor of 7 and 2 respectively.

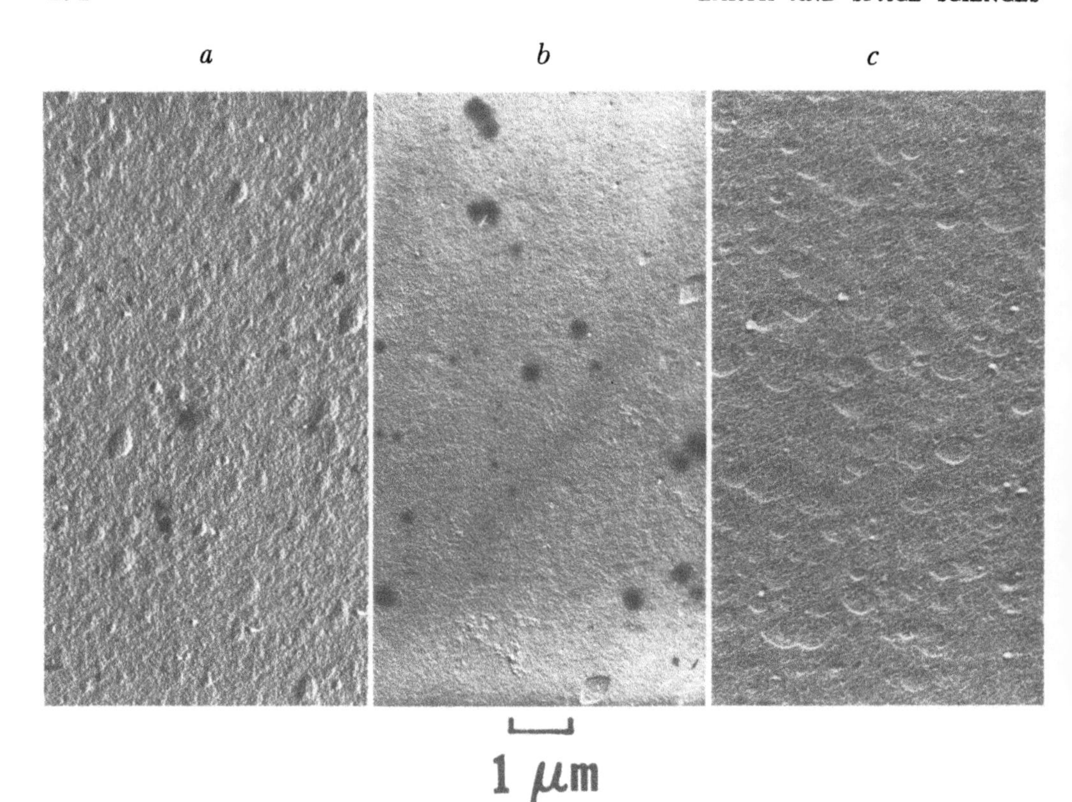

Fig. 5-26. *Etch pits in mica (a) exposed to solar wind on Apollo 17 experiment; (b) exposed in direction opposite to solar wind on Apollo 17 experiment; and (c) irradiated with 1 keV/nucleon Xe ions. (After Walker and Zinner, 1973; Burnett et al., 1972.)*

5.5. ENERGY DISTRIBUTION OF HEAVY PARTICLES IN INTERPLANETARY SPACE

Having discussed the three main categories of energetic particles—the galactic cosmic rays, the solar flare particles, and the solar wind—we are ready to examine the overall energy spectrum of Fe-group nuclei ($Z \geq 24$) shown in Fig. 5-3. The information is based partly on direct measurements and partly on plausible assumptions. Even though some of it is quite uncertain, it will be useful in our later discussion of fossil tracks in meteorites and lunar samples to be able to refer to these estimates of the energy distribution of Fe-group nuclei in the present-day radiation.

5.5.1. Galactic Cosmic Rays

At energies greater than a few GeV/nucleon, solar modulation has a negligible effect on the spectrum. The solid curve at highest energies in Fig. 5-3 is the average of recent data for the elements $Z \geq 24$ by Smith et al. (1973) and Balasubrahmanyan and Ormes (1973). The dashed curve at lower energies is our estimate of the flux of nuclei with $Z \geq 24$ averaged over a solar cycle. In the absence of adequate data for these nuclei over an entire solar cycle, we have supplemented the data with He data, scaled down by the relative abundance He/[$Z \geq 24$] = 294 (Webber et al., 1972), which seems justified since nuclei with similar Z/A should be modulated similarly. Helium data at solar minimum were taken from the review article by Meyer (1969) and data on VH nuclei (very heavy nuclei) at various times in the solar cycle from papers by Freier and Waddington (1968), Freier et al. (1971) and Fernandez (1971). The averaging was done by constructing correlation curves between the fluxes of He and nuclei with $Z \geq 24$ and neutron monitor counting rates. Neutron monitor data are available for the entire 11-year solar cycle.

5.5.2. Solar Flares

The Surveyor glass data in Fig. 5-22 are replotted in Fig. 5-3 as the curve labeled "average over 1967–69." As we discuss in the next chapter, the activity during this period is probably slightly higher but within a factor two of the long-term average level for solar flares.

5.5.3. The Solar Wind

The results from the mica samples on Apollo 17 (Borg et al., 1974), together with the conclusions of Geiss (1973), suggest that we can scale down the known flux distribution of protons by the expected ratio of Fe/H in the solar wind. We use the measured average He/H ratio of 0.045 in the solar wind and assume no fractionation of elements beyond He. With an assumed He/Fe ratio of 3,000, this gives $(Fe/H)_{SW} = 1.5 \times 10^{-5}$. The value, $(Fe/H)_{SW} = 3.4 \times 10^{-5}$, reported in Table 5-9 was measured with an electrostatic detector on a spacecraft on 6 July 1969. Because the composition is known to vary widely from day to day, we prefer to use the calculated value as representative of the average Fe/H ratio. Neglecting the contribution of the other elements with $Z \geq 24$ should introduce less than a 10% error since their abundance in the corona is quite low.

Gosling et al. (1971) have determined the distribution of solar wind flow speeds over the period of 1962 to 1970. We have averaged their speed distributions and have converted the average into a differential energy spectrum averaged over 4π steradians. This spectrum, scaled down by 1.5×10^{-5}, is plotted in Fig. 5-3.

5.5.4. Suprathermal Ions

At energies from a few keV/nucleon to ~50 keV/nucleon, the flux is normally so low as to be undetectable with electrostatic analyzers such as those used to study the solar wind. However, about ten to twenty times a year Frank (1970) has detected a large flux of suprathermal ions ($\sim$5 to $\sim$50 keV/nucleon) coming predominantly from the direction of the sun and lasting for a few hours to a day. The events are believed to be associated with small-scale solar activity rather than with the solar wind. Based on the solar flare data in Fig. 5-24, we have assumed an enrichment of the solar Fe/H ratio by a factor $\sim 10^2$ (giving H/Fe $\approx$ 300). Assuming 20 events per year, we have derived from Frank's data the average flux of suprathermal Fe ions shown in Fig. 5-3. This spectrum is more uncertain than any of the others in the figure and should not be taken seriously until direct measurements of Fe ions can be made. It will not be easy to do.

5.5.5. Choice of Spectrum at Inaccessible Energies

For all practical purposes we are completely ignorant of the behavior of the present-day average energy spectrum of nuclei with $Z \geq 24$ at energies below $\sim$1 MeV/nucleon. However, we shall see in Chapter 6 that fossil track densities in lunar and meteoritic crystals monotonically fall off in a smooth fashion with depth and there is no evidence for inflection points or minima in the distributions down to energies as low as $\sim$50 keV/nucleon. In a speculative vein, and for completeness, we have connected the regions labeled "suprathermal ions" and "solar flares" with a smooth curve (dotted line) obeying the power law $E^{-2.7}$.

At present there seems no justification for assuming a smooth connection between the solar wind flux and the suprathermal ion flux. The mechanisms are different; the sources are different (outer corona for the solar wind, chromosphere for flares); the compositions are different; and high-voltage electron microscope observations (Bibring et al., 1972) indicate a fairly well-defined range for solar wind damage in lunar grains, rather distinct from the deeper regions containing tracks of suprathermal ions and flare particles.

Chapter 5 References

L. H. Aller (1972), "The Abundances of the Elements," *Invited and Rapporteur Papers, 12th Inter. Conf. on Cosmic Rays*, Hobart, Australia, 39–52.

Y. Alloucherie (1970), "Diffusion of Heavy Ions in the Solar Corona," *J. Geophys. Res.* **75**, 6899–6914.

R. A. Alpher and R. C. Herman (1950), "Theory of the Origin and Relative Abundance Distribution of the Elements," *Rev. Mod. Phys.* **22**, 153–212.

R. A. Alpher, H. A. Bethe, and G. Gamow (1948), "The Origin of Chemical Elements," *Phys. Rev.* **73**, 803–804.

J. D. Anglin, W. F. Dietrich, and J. A. Simpson (1973), "Solar Flare Accelerated

Isotopes of Hydrogen and Helium," in *High Energy Phenomena on the Sun*, R. Ramaty and R. G. Stone (eds.), NASA Special Pub. 342, pp. 315–340.

T. P. Armstrong, S. M. Krimigis, D. V. Reames, and C. E. Fichtel (1972), "A Comparison of Measurements of the Charge Spectrum of Solar Cosmic Rays from Nuclear Emulsions and the Explorer 35 Solid-State Detector," *J. Geophys. Res.* **77,** 3607–3612.

W. D. Arnett and D. N. Schramm (1973), "Origin of Cosmic Rays, Atomic Nuclei and Pulsars in Explosions of Massive Stars," *Astrophys. J.* **184,** L47–L51.

W. Baade and F. Zwicky (1934), "Cosmic Rays from Supernovae," *Proc. Nat. Acad. Sci.* **20,** 259–263.

V. K. Balasubrahmanyan and J. F. Ormes (1973), "Results on the Energy Dependence of Cosmic Ray Composition," *Astrophys. J.* **186,** 109–122.

S. J. Bame (1972), "Spacecraft Observations of the Solar Wind Composition," in *Solar Wind*, C. P. Sonett, P. J. Coleman, and J. M. Wilcox (eds.), NASA Special Pub. 308, 535–558.

S. J. Bame, A. J. Hundhausen, J. R. Asbridge, and I. B. Strong (1968), "Solar Wind Ion Composition," *Phys. Rev. Letters* **20,** 393–395.

S. J. Bame, J. R. Asbridge, A. J. Hundhausen, and M. D. Montgomery (1970), "Solar Wind Ions: $^{56}Fe^{+8}$ to $^{56}Fe^{+12}$, $^{28}Si^{+7}$, $^{28}Si^{+8}$, $^{28}Si^{+9}$, and $^{16}O^{+6}$," *J. Geophys. Res.* **75,** 6360–6366.

D. J. Barber, R. Cowsik, I. D. Hutcheon, P. B. Price, and R. S. Rajan (1971), "Solar Flares, the Lunar Surface, and Gas-Rich Meteorites," *Proc. Second Lunar Sci. Conf.* **3,** 2705–2714. Cambridge: MIT Press.

K. P. Bartholomä, W. Enge, and K. Fukui (1973), "Chemische Zusammensetzung der Kosmischen Strahlung Niederenergetischer Teilchen von $Z = 12$ bis $Z = 28$, Gemessen in Plastik Detektoren," *Z. Physik* **259,** 75–91.

R. Beaujean and W. Enge (1971), "Study of the Charge and Isotope Composition of Medium Cosmic Ray Particles," *Proc. 12th Inter. Conf. on Cosmic Rays*, Hobart, Australia, **7,** 2583–2588.

R. Beaujean and W. Enge (1972), "Isotopenanalyse an Niederenergetischen Teilchen der Elemente Bor, Kohlenstoff, Stickstoff und Sauerstoff aus der Kosmischen Strahlung in Plastik-Detektoren," *Z. Physik* **256,** 416–440.

D. L. Bertsch, C. E. Fichtel, and D. V. Reames (1972), "Nuclear Composition and Energy Spectra in the 1969 April 12 Solar-Particle Event," *Astrophys. J.* **171,** 169–177.

D. L. Bertsch, S. Biswas, C. E. Fichtel, C. J. Pellerin, and D. V. Reames (1973), "Variations of the Relative Abundances of He, CNO and Fe-Group Nuclei in Solar Cosmic Rays and their Relationship to Solar Particle Acceleration," *Solar Physics* **31,** 247–257.

N. Bhandari, J. Goswami, D. Lal, and A. Tamhane (1973), "Charge Composition and Energy Spectrum of Solar Flare VVH and VH Nuclei," *Proc. 13th Inter. Cosmic Ray Conf.*, Univ. of Denver, **2,** 1464–1469.

J. P. Bibring, M. Maurette, and R. Meunier (1972), "Solar Wind Implantation Effects in the Lunar Regolith," *Lunar Science III*, C. Watkins (ed.), 71–73. Lunar Sci. Inst., Houston.

W. R. Binns, J. I. Fernandez, M. H. Israel, J. Klarmann, R. C. Maehl, and R. A. Mewaldt (1973), "Chemical Composition of Cosmic Rays with $Z \geq 30$ and $E \geq 325$ MeV/N," *Proc. 13th Inter. Cosmic Ray Conference*, Univ. of Denver, **1,** 260–264.

S. Biswas and C. E. Fichtel (1965), "Composition of Solar Cosmic Rays," *Space Sci. Rev.* **4,** 709–736.

S. Biswas, D. L. Bertsch, C. E. Fichtel, C. Pellerin, and D. V. Reames (1973), "Relative Abundances and Energy Spectra of Solar Cosmic Ray Nuclei in the August 4, 1972 Event," *Proc. 13th Inter. Cosmic Ray Conf.*, Univ. of Denver, **2,** 1543–1547.

G. E. Blanford (1971), *Calibration and Use of Plastic Track Detectors in the Study of Extremely Heavy Cosmic Rays*, Ph.D. Thesis, Washington University.

G. E. Blanford, M. W. Friedlander, J. Klarmann, R. M. Walker, J. P. Wefel, W. C. Wells, R. L. Fleischer, G. E. Nichols, and P. B. Price (1969), "Observation of Trans-Iron Nuclei in the Primary Cosmic Radiation," *Phys. Rev. Letters* **23,** 338–342.

G. E. Blanford, M. W. Friedlander, J. Klarmann, S. S. Pomeroy, R. M. Walker, and J. P. Wefel (1972), "Observation of Low-Charge Low-Energy Geomagnetically Forbidden Particles," *J. Geophys. Res.* **77,** 6037–6041.

G. E. Blanford, M. W. Friedlander, J. Klarmann, S. S. Pomeroy, R. M. Walker, J. P. Wefel, P. H. Fowler, J. M. Kidd, E. J. Kobetich, R. T. Moses, and R. T. Thorne (1973a), "Observation of Cosmic Ray Particles with $Z > 35$," *Phys. Rev.* **D8,** 1707–1722.

G. E. Blanford, M. W. Friedlander, J. Klarmann, R. M. Walker, and J. P. Wefel (1973b), "Observation of Cosmic Ray Particles with $Z \geq 50$ and Interpretation of the Charge Spectrum," *Phys. Rev.* **D8,** 1722–1729.

J. Borg, M. Maurette, R. M. Walker, and E. Zinner (1974), "Apollo 17 Lunar Surface Cosmic Ray Experiment—Measurement of Heavy Solar Wind Particles," *Lunar Science V*, 76–78, Lunar Sci. Inst., Houston.

D. Braddy, J. Chan, and P. B. Price (1973), "Charge States and Energy-Dependent Composition of Solar-Flare Particles," *Phys. Rev. Letters* **30,** 669–671.

H. L. Bradt and B. Peters (1950), "Abundance of Li, Be, B and Other Light Nuclei in the Primary Cosmic Radiation and the Problem of Cosmic Ray Origin," *Phys. Rev.* **80,** 943–953.

K. A. Brueckner, J. H. Chirico, S. Jorna, H. W. Meldner, D. N. Schramm, and P. A. Seeger (1973), "Superheavy Elements from *r*-Process Calculations with an Energy-Density Mass Formula," *Phys. Rev.* **C7,** 2123–2128.

E. M. Burbidge, G. R. Burbidge, W. A. Fowler, and F. Hoyle (1957), "Synthesis of the Elements in Stars," *Rev. Mod. Phys.* **29,** 547–650.

D. Burnett, C. Hohenberg, M. Maurette, M. Monnin, R. Walker, and D. Woolum (1972), "Solar Cosmic Ray, Solar Wind, Solar Flare, and Neutron Albedo Measurements," *Apollo 16 Preliminary Science Report*, NASA Special Pub. 315, Chap. 15.

A. G. W. Cameron (1968), "A New Table of Abundances of the Elements in the Solar System," in *Origin and Distribution of the Elements*, L. H. Ahrens (ed.), 125–143. London: Pergamon Press.

A. G. W. Cameron (1971), "Processes of Nucleosynthesis," *Comments Astrophys. Space Sci.* **2,** 153–160.

A. G. W. Cameron (1973), "Abundances of the Elements in the Solar System," *Space Sci. Rev.* **15,** 121–146.

B. G. Cartwright (1971), "The Origin of Fluorine, Sodium, and Aluminum in the Galactic Cosmic Radiation," *Astrophys. J.* **169,** 299–310.

B. G. Cartwright and A. Mogro-Campero (1972), "The Preferential Acceleration of Heavy Nuclei in Solar Flares," *Astrophys. J.* **177,** L43–L47.

B. G. Cartwright, M. Garcia-Munoz, and J. A. Simpson (1971), "Abundances of the Galactic Nuclei H through the Fe Group from Satellite Measurements," *Proc. 12th Inter. Conf. on Cosmic Rays*, Hobart, Australia, **1,** 215–220.

B. G. Cartwright, M. Garcia-Munoz, and J. A. Simpson (1973), "Abundances of Cosmic Ray Sources," unpublished report.

M. Cassé (1973a), "^{54}Mn and the Lifetime of Relativistic Cosmic Rays," *Astrophys. J.* **180,** 623–629.

M. Cassé (1973b), "^{56}Ni in Cosmic Rays?," *Proc. 13th Inter. Cosmic Ray Conference*, Univ. of Denver, **1,** 546–550.

M. Cassé and P. Goret (1973), "Atomic Properties of the Elements and Their Acceleration to Cosmic Ray Energies," *Proc. 13th Inter. Cosmic Ray Conf.*, Univ. of Denver, **1,** 584–589.

M. Cassé, L. Koch, N. Lund, J. P. Meyer, B. Peters, A. Soutoul, and S. N. Tandon (1971), "Chemical Composition of Heavy Cosmic Ray Nuclei Above 5 GV," *Proc. 12th Inter. Conf. on Cosmic Rays*, Hobart, Australia, **1,** 241–245.

J. H. Chan and P. B. Price (1974), "Anomalies in the Composition of Interplanetary Heavy Ions with $0.01 < E < 40$ MeV/amu," *Astrophys. J.* **190,** L39–L41.

J. H. Chan, P. B. Price, and E. K. Shirk (1973), "Charge Composition and Energy Spectrum of Suprathermal Solar Particles," *Proc. 13th Inter. Cosmic Ray Conf.*, Univ. of Denver, **2,** 1650–1654.

D. D. Clayton (1968), *Principles of Stellar Evolution and Nucleosynthesis*. New York: McGraw-Hill.

D. D. Clayton and J. G. Peters (1970), "A Comment on the Abundances of Heavy Cosmic Rays," *Acta Phys. Acad. Sci. Hung.* **29,** Suppl. 1, 381–384.

S. A. Colgate and M. H. Johnson (1960), "Hydrodynamic Origin of Cosmic Rays," *Phys. Rev. Letters* **5,** 235–238.

R. Cowsik (1971), "Particle Acceleration by White Dwarfs and the Charge

Spectrum of the Nuclear Component of Cosmic Rays," *Proc. 12th Inter. Conf. on Cosmic Rays*, Hobart, Australia, **1,** 329–333.

R. Cowsik and P. B. Price (1971), "Origins of Cosmic Rays," *Physics Today* **24,** No. 10, 30–38.

R. Cowsik and L. W. Wilson (1973), "Is the Residence Time of Cosmic Rays in the Galaxy Energy-Dependent?" *Proc. 13th Inter. Cosmic Ray Conf.*, Univ. of Denver, **1,** 500–505.

R. Cowsik, Y. Pal, S. N. Tandon, and R. P. Verma (1967), "Steady State of Cosmic-Ray Nuclei—Their Spectral Shape and Path Length at Low Energies," *Phys. Rev.* **158,** 1238–1242.

H. J. Crawford, P. B. Price, and J. D. Sullivan (1972), "Composition and Energy Spectra of Heavy Nuclei with $0.5 < E < 40$ MeV per Nucleon in the 1971 January 24 and September 1 Solar Flares," *Astrophys. J.* **175,** L149–L153.

H. J. Crawford, P. B. Price, J. D. Sullivan, and B. G. Cartwright (1975), "Solar Flare Particles: Energy-Dependent Composition and Relationship to Solar Composition," *Astrophys. J.* **195,** 213–221.

G. Crozaz and R. M. Walker (1971), "Solar Particle Tracks in Glass from the Surveyor 3 Spacecraft," *Science* **171,** 1237–1239.

P. M. Dauber (1971), editor of *Isotopic Composition of the Primary Cosmic Radiation*, Danish Space Research Institute, Lyngby, Denmark.

P. Eberhardt, J. Geiss, H. Graf, N. Grögler, V. Krähenbühl, H. Schwaller, J. Schwarzmüller, and A. Stettler (1970), "Trapped Solar Wind Noble Gases, Exposure Age and K/Ar-Age in Apollo 11 Lunar Fine Material," *Proc. Apollo 11 Lunar Sci. Conf.*, **2,** 1037–1070. London: Pergamon Press.

W. Enge, K. P. Bartholomä, and K. Fukui (1971), "Charge Composition of Low Energy Cosmic Ray Nuclei of Elements Neon to Nickel," *Proc. 12th Inter. Conf. on Cosmic Rays*, Hobart, Australia, **7,** 2589–2594.

J. I. Fernandez (1971), Ph.D. Thesis, Washington University.

C. E. Fichtel and F. B. McDonald (1967), "Energetic Particles from the Sun," *Ann. Rev. Astron. Astrophys.* **5,** 351–398.

L. A. Fisk, B. Kozlovsky, and R. Ramaty (1974), "An Interpretation of the Observed Oxygen and Nitrogen Enhancements in Low Energy Cosmic Rays," *Astrophys. J.* **190,** L35–L37.

R. L. Fleischer and H. R. Hart (1972), "Composition and Energy Spectra of Solar Cosmic Ray Nuclei," *Apollo 16 Preliminary Science Report*, NASA Special Pub. 315, Chap. 15, pp. 1–11.

R. L. Fleischer and H. R. Hart (1973), "Enrichment of Heavy Nuclei in the 17 April 1972 Solar Flare," *Phys. Rev. Letters* **30,** 31–34.

R. L. Fleischer, P. B. Price, R. M. Walker, M. Maurette, and G. Morgan (1967), "Tracks of Heavy Primary Cosmic Rays in Meteorites," *J. Geophys. Res.* **72,** 355–366.

R. L. Fleischer, H. R. Hart, and G. M. Comstock (1971), "Very Heavy Solar

Cosmic Rays: Energy Spectrum and Implications for Lunar Erosion," *Science* **171,** 1240–1242.
P. H. Fowler (1973), "Ultra Heavy Cosmic Rays," *Proc. 13th Inter. Cosmic Ray Conf.*, Univ. of Denver, **5,** 3627–3637.
P. H. Fowler, R. A. Adams, V. G. Cowen, and J. M. Kidd (1967), "The Charge Spectrum of Very Heavy Cosmic Ray Nuclei," *Proc. Roy. Soc.*, London, **A301,** 39–45.
P. H. Fowler, V. M. Clapham, V. G. Cowen, J. M. Kidd, and R. T. Moses (1970), "The Charge Spectrum of Very Heavy Cosmic Ray Nuclei," *Proc. Roy. Soc.*, London, **A318,** 1–43.
P. H. Fowler, A. Jurak, R. T. Thorne, C. O'Ceallaigh, D. O'Sullivan, Y. V. Rao, and A. Thompson (1972), "Preliminary Study of the Charge Spectrum of Ultra Heavy Cosmic Ray Primaries," *Proc. 8th Inter. Conf. Nuclear Photography and Solid State Track Detectors*, Bucharest, **1,** 476–484.
P. H. Fowler, R. T. Thorne, A. Mazumdar, C. O'Ceallaigh, D. O'Sullivan, Y. V. Rao, and A. Thompson (1973), "The Charge and Energy Spectrum of Ultra Heavy Cosmic Ray Primaries," post-deadline paper at 13th Inter. Cosmic Ray Conf., Univ. of Denver.
L. A. Frank (1970), "On the Presence of Low-Energy Protons ($5 \leq E \leq 50$ keV) in the Interplanetary Medium," *J. Geophys. Res.* **75,** 707–716.
P. S. Freier and C. J. Waddington (1968), "Very Heavy Nuclei in the Primary Cosmic Radiation. I. Observations on the Energy Spectrum," *Phys. Rev.* **175,** 1641–1648.
P. S. Freier, E. J. Lofgren, E. P. Ney, F. Oppenheimer, H. L. Bradt, and B. Peters (1948), "Evidence for Heavy Nuclei in the Primary Cosmic Radiation," *Phys. Rev.* **74,** 213–217.
P. S. Freier, C. E. Long, T. F. Cleghorn, and C. J. Waddington (1971), "The Charge and Energy Spectra of Heavy Cosmic Ray Nuclei," *Proc. 12th Inter. Conf. on Cosmic Rays*, Hobart, Australia, **1,** 252–257.
M. Garcia-Munoz and J. A. Simpson (1970), "Galactic Abundances and Spectra of Cosmic Rays Measured on the IMP-4 Satellite," *Acta Phys. Acad. Sci. Hung.* **29,** Suppl. 1, 317–323.
M. Garcia-Munoz, G. M. Mason, and J. A. Simpson (1973), "Isotopic Composition of Low Energy Beryllium in the Galactic Cosmic Rays," *Proc. 13th Inter. Cosmic Ray Conf.*, Univ. of Denver, **1,** 100–105.
T. L. Garrard, E. C. Stone, and R. E. Vogt (1973), "The Isotopes of H and He in Solar Cosmic Rays," in *High Energy Phenomena on the Sun*, NASA Special Pub. 342, pp. 341–354.
J. Geiss (1972), "Elemental and Isotopic Abundances in the Solar Wind," in *Solar Wind*, C. P. Sonett, P. J. Coleman, and J. M. Wilcox (eds.), NASA Special Pub. 308, pp. 559–581.
J. Geiss (1973), "Solar Wind Composition and Implications about the History of

the Solar System," *Proc. 13th Inter. Cosmic Ray Conf.*, Univ. of Denver,. **5,** 3375–3398.

J. Geiss, F. Buehler, H. Cerutti, P. Eberhardt, and Ch. Filleux (1972), "Solar Wind Composition Experiment," *Apollo 16 Preliminary Science Report*, NASA Special Pub. 315, Chap. 14.

E. G. Gibson (1973), *The Quiet Sun*, NASA Special Pub. 303.

V. L. Ginzburg and S. I. Syrovatskii (1964), *The Origin of Cosmic Rays*. New York: Macmillan.

G. Gloeckler, C. Y. Fan, and D. Hovestadt (1973), "Direct Observations of the Charge States of Low Energy Solar Particles," *Proc. 13th Inter. Cosmic Ray Conf.*, Univ. of Denver, **2,** 1492–1497.

P. Goldreich and W. H. Julian (1969), "Pulsar Electrodynamics," *Astrophys. J.* **157,** 869–880.

M. L. Goldstein, L. A. Fisk, and R. Ramaty (1970), "Energy Loss of Cosmic Rays in the Interplanetary Medium," *Phys. Rev. Letters* **25,** 832–835.

J. T. Gosling, R. T. Hansen, and S. J. Bame (1971), "Solar Wind Speed Distributions: 1962–1970," *J. Geophys. Res.* **76,** 1811–1815.

N. Grevesse (1969), "Abundances of Heavy Elements in the Sun," *Solar Physics* **6,** 381–398.

J. E. Gunn and J. P. Ostriker (1969), "Acceleration of High-Energy Cosmic Rays by Pulsars," *Phys. Rev. Letters* **22,** 728–731.

A. V. Gurevich (1960), "On the Amount of Accelerated Particles in an Ionized Gas under Various Accelerating Mechanisms," *Sov. Phys. JETP* **11,** 1150–1157.

O. Havnes (1971), "Abundances and Acceleration Mechanisms of Cosmic Rays," *Nature* **229,** 548–549.

O. Havnes (1973), "On Cosmic Rays and Magnetic Stars," *Astron. Astrophys.* **24,** 435–440.

S. Hayakawa (1969), *Cosmic Ray Physics*. New York: Wiley.

R. C. Haymes (1971), *Introduction to Space Science*. New York: Wiley.

D. Hovestadt, O. Vollmer, G. Gloeckler, and C. Y. Fan (1973), "Measurement of Elemental Abundance of Very Low Energy Solar Cosmic Rays," *Proc. 13th Inter. Cosmic Ray Conf.*, Univ. of Denver, **2,** 1498–1503.

W. M. Howard and J. R. Nix (1973), "Production of Superheavy Nuclei by the Multiple Capture of Neutrons," *Nature*, **247,** 17–20.

K. C. Hsieh and J. A. Simpson (1969), "The Isotopic Abundances and Energy Spectra of ^{2}H, ^{3}He, and ^{4}He of Cosmic-Ray Origin in the Energy Region $\sim$10-100 MeV/nucleon," *Astrophys. J.* **158,** L37–L41.

A. J. Hundhausen (1970), "Composition and Dynamics of the Solar Wind Plasma," *Rev. Geophys.* **8,** 729–811.

I. D. Hutcheon, D. Macdougall, and P. B. Price (1974), "Rock 72315: A New

Lunar Standard for Solar Flare and Micrometeorite Exposure," in *Lunar Science V*, 378–380, Lunar Sci. Inst., Houston.
M. H. Israel, J. Klarmann, R. C. Maehl, and W. R. Binns (1973), "Energy Spectra of Individual Cosmic-Ray Elements with $12 < Z < 28$," *Proc. 13th Inter. Cosmic Ray Conference*, Univ. of Denver, **1,** 255–259.
O. Johns and H. Reeves (1973), "Neutron Energy Distributions and Termination of the r-Process," *Astrophys. J.* **186,** 233–238.
J. R. Jokipii (1966), "Effects of Diffusion on the Composition of the Solar Corona and the Solar Wind," in *The Solar Wind*, R. J. Macklin and M. Neugebauer (eds.), 215. New York: Pergamon Press.
E. Juliusson, P. Meyer, and D. Muller (1972), "Composition of Cosmic-Ray Nuclei at High Energies," *Phys. Rev. Letters* **29,** 445–448.
A. A. Korchak and S. I. Syrovatskii (1958), "On the Possibility of a Preferential Acceleration of Heavy Elements in Cosmic Ray Sources," *Sov. Phys. Doklady* **3,** 983–985.
R. P. Kovar and A. J. Dessler (1967), "On the Anisotropy of Galactic Cosmic Rays," *Astrophys. Letters* **1,** 15–16.
S. M. Krimigis and T. P. Armstrong (1973), "Measurements of the Relative Abundances of Fe-Group, He, and M Nuclei During the October 29, 1972 Solar Particle Event," *Proc. 13th Inter. Cosmic Ray Conf.*, Univ. of Denver, **2,** 1510–1515.
S. M. Krimigis, P. Verzariu, J. A. Van Allen, T. P. Armstrong, T. A. Fritz, and B. A. Randall (1970), "Trapped Energetic Nuclei $Z \geq 3$ in the Earth's Outer Radiation Zone," *J. Geophys. Res.* **75,** 4210–4215.
K. Kristiansson (1972), "Further Evidence for a Cosmic Ray Selection Mechanism," *Astrophys. Space Sci.* **16,** 405–412.
K. Kristiansson, O. Mathiesen, and A. Stenman (1963), "The Relative Abundance of Nuclei Heavier than Lithium in Primary Cosmic Radiation," *Arkiv Fys.* **23,** 479–504.
R. M. Kulsrud, J. P. Ostriker, and J. E. Gunn (1972), "Acceleration of Cosmic Rays in Supernova Remnants," *Phys. Rev. Letters* **28,** 636–639.
L. J. Lanzerotti, C. G. Maclennan, and T. E. Graedel (1972), "Enhanced Abundances of Low-Energy Heavy Elements in Solar Cosmic Rays," *Astrophys. J.* **173,** L39–L43.
L. J. Lanzerotti, R. C. Reedy, and J. R. Arnold (1973), "Alpha Particles in Solar Cosmic Rays over the Last 80,000 Years," *Science* **179,** 1232–1234.
N. Lund, B. Peters, R. Cowsik, and Y. Pal (1970), "On the Isotopic Composition of Primary Cosmic Ray Nuclei," *Phys. Letters* **31B,** 553–556.
F. B. McDonald, B. J. Teegarden, J. H. Trainor, and W. R. Webber (1974), *Astrophys. J.* **187,** L105–L108.
M. Malinovsky and L. Heroux (1973), "An Analysis of the Solar Extreme-Ultraviolet Between 50 and 300 Å," *Astrophys. J.* **181,** 1009–1030.

M. Meneguzzi (1973), "The Variation of the High Energy Primary Nuclei Composition and the Confinement Region of Cosmic Rays," *Proc. 13th Inter. Cosmic Ray Conference*, Univ. of Denver, **1**, 378–383.

P. W. Merrill (1952), "Technetium in the Stars," *Science* **115**, 484.

R. A. Mewaldt, R. E. Turner, M. W. Friedlander, and M. H. Israel (1970), "The Propagation of Very Heavy Primary Cosmic Ray Particles," *Acta Phys. Acad. Sci. Hung.* **29**, Suppl. 1, 433–437.

P. Meyer (1969), "Cosmic Rays in the Galaxy," *Ann. Rev. Nuc. Sci.* **7**, 1–38.

A. Mogro-Campero and J. A. Simpson (1970), "Identification and Relative Abundances of C, N and O Nuclei Trapped in the Geomagnetic Field," *Phys. Rev. Letters* **25**, 1631–1634.

A. Mogro-Campero and J. A. Simpson (1972a), "Enrichment of Very Heavy Nuclei in the Composition of Solar Accelerated Particles," *Astrophys. J.* **171**, L5–L9.

A. Mogro-Campero and J. A. Simpson (1972b), "The Abundances of Solar Accelerated Nuclei from Carbon to Iron," *Astrophys. J.* **177**, L37–L41.

A. Mogro-Campero, N. Schofield, and J. A. Simpson (1973), "On the Origin of Low Energy Heavy Nuclei Below ~30 MeV/nucleon Observed in Interplanetary Space during Quiet Times, 1968–72," *Proc. 13th Inter. Cosmic Ray Conf.*, Univ. of Denver, **1**, 140–145.

M. P. Nakada (1969), "A Study of the Composition of the Lower Solar Corona," *Solar Physics* **7**, 302–320.

J. R. Nix (1972), "Predictions for Superheavy Nuclei," *Physics Today* **25**, No. 4, 30–38.

J. F. Ormes and V. K. Balasubrahmanyan (1973), "Charge Dependence of the Energy Spectra of Cosmic Rays," *Nature Phys. Sci.* **241**, 95–96.

W. Z. Osborne, L. S. Pinsky, P. B. Price, and E. K. Shirk (1973a), "Charge and Energy Spectra of $Z \geq 60$ Cosmic Rays," to appear in *Proc. 14th Inter. Cosmic Ray Conf.*, Munich, August, 1975.

W. Z. Osborne, L. S. Pinsky, E. K. Shirk, P. B. Price, E. J. Kobetich, and R. D. Eandi (1973b), "Energy Spectrum of Nuclei with $Z \geq 60$ as Evidence for a New Source of Cosmic Rays," *Phys. Rev. Letters* **31**, 127–130.

D. O'Sullivan, P. B. Price, E. K. Shirk, P. H. Fowler, J. M. Kidd, E. J. Kobetich, and R. Thorne (1971), "High Resolution Measurements of Slowing Cosmic Rays from Fe to U," *Phys. Rev. Letters* **26**, 463–466.

D. O'Sullivan, A. Thompson, and P. B. Price (1973), "Composition of Galactic Cosmic Rays with $30 < E < 130$ MeV/nucleon," *Nature Phys. Sci.* **243**, 8–9.

Y. Pal (1967), "Cosmic Rays and Their Interactions," in *Handbook of Physics*, 2nd ed., E. V. Condon and H. Odishaw (eds.), 9-272 to 9-328. New York: McGraw-Hill.

E. N. Parker (1965), "The Passage of Energetic Charged Particles Through Interplanetary Space," *Planet. Space Sci.* **13**, 9–49.

L. S. Pinsky (1972), "A Study of Heavy Trans-Iron Primary Cosmic Rays ($Z \geq 55$) with a Fast Film Cerenkov Detector," NASA Tech. Memo X-58102, Manned Spacecraft Center, Houston.
A. M. Poskanzer, G. W. Butler, and E. K. Hyde (1971), "Fragment Production in the Interaction of 5.5 GeV Protons with Uranium," *Phys. Rev.* **C3**, 882–904.
P. B. Price (1971), "The Study of Isotopes of Heavy Cosmic Ray Nuclei by Means of Tracks in Plastics," in *Isotopic Composition of the Primary Cosmic Radiation*, P. M. Dauber (ed.), Danish Space Research Institute, Lyngby, Denmark, 63–74.
P. B. Price (1972), "Ultra-Heavy Cosmic Rays," *Invited and Rapporteur Papers, 12th Inter. Conf. on Cosmic Rays*, Hobart, Australia, 453–473.
P. B. Price (1973a), "A Cosmochemical View of Cosmic Rays and Solar Particles," *Space Sci. Rev.* **15**, 69–88.
P. B. Price (1973b), "Composition and Energy Spectra of Solar Particles," in *High Energy Phenomena on the Sun*, R. Ramaty and R. G. Stone (eds.), NASA Special Pub. 342, 377–392.
P. B. Price and J. H. Chan (1973), "The Nature of Interplanetary Heavy Ions with $0.1 < E < 40$ MeV/nucleon," *Apollo 17 Preliminary Science Report*, NASA Special Pub. 330, pp. 19-15 to 19-20.
P. B. Price and E. K. Shirk (1973), "Search for Trans-uranic Cosmic Rays on Skylab," paper at Symposium on High-Energy Astrophysics, Tucson, Dec. 7.
P. B. Price and J. D. Sullivan (1971), "Composition and Spectrum of Particles with $0.2 \leq E \leq 30$ MeV/nuc in the 25 January 1971 Solar Flare," *Proc. 12th Inter. Conf. on Cosmic Rays*, Hobart, Australia, **7**, 2641–2646.
P. B. Price, R. L. Fleischer, D. D. Peterson, C. O'Ceallaigh, D. O'Sullivan, and A. Thompson (1967), "Identification of Isotopes of Energetic Particles with Dielectric Track Detectors," *Phys. Rev.* **164**, 1618–1620.
P. B. Price, R. L. Fleischer, D. D. Peterson, C. O'Ceallaigh, D. O'Sullivan, and A. Thompson (1968), "High Resolution Study of Low Energy Cosmic Rays with Lexan Track Detectors," *Phys. Rev. Letters* **21**, 630–633.
P. B. Price, D. D. Peterson, R. L. Fleischer, C. O'Ceallaigh, D. O'Sullivan, and A. Thompson (1970), "Composition of Cosmic Rays of Atomic Number 12 to 30," *Acta Phys. Acad. Sci. Hung.* **29**, Suppl. 1, 417–422.
P. B. Price, I. D. Hutcheon, R. Cowsik, and D. J. Barber (1971a), "Enhanced Emission of Fe Nuclei in Solar Flares," *Phys. Rev. Letters* **26**, 916–919.
P. B. Price, P. H. Fowler, J. M. Kidd, E. J. Kobetich, R. L. Fleischer, and G. E. Nichols (1971b) "Study of the Charge Spectrum of Extremely Heavy Cosmic Rays Using Combined Plastic Detectors and Nuclear Emulsions," *Phys. Rev.* **D3**, 815–823.
P. B. Price, D. Braddy, D. O'Sullivan, and J. D. Sullivan (1972), "Composition of Interplanetary Particles at Energies from 0.1 to 150 MeV/nucleon," *Apollo 16 Preliminary Science Report*, NASA Special Pub. 315, p. 15-11.

P. B. Price, J. H. Chan, D. O'Sullivan, and A. Thompson (1973a), "Galactic Heavy Cosmic Rays with $5 < E < 130$ MeV/nucleon," *Proc. 13th Inter. Cosmic Ray Conf.*, Univ. of Denver, **1,** 146–151.
P. B. Price, J. H. Chan, H. J. Crawford, and J. D. Sullivan (1973b), "Systematics of Heavy Ion Enhancements in Solar Flares," *Proc. 13th Inter. Cosmic Ray Conf.*, Univ. of Denver, **2,** 1479–1483.
P. B. Price, J. H. Chan, I. D. Hutcheon, D. Macdougall, R. S. Rajan, E. K. Shirk, and J. D. Sullivan (1973c), "Low-Energy Heavy Ions in the Solar System," *Proc. Fourth Lunar Sci. Conf.*, 2347–2361. Cambridge: MIT Press.
G. Raisbeck (1971), "^{7}Be and Other Electron Capture Isotopes in Cosmic Rays," in *Isotopic Composition of the Primary Cosmic Radiation*, P. M. Dauber (ed.), Danish Space Research Institute, Lyngby, Denmark, 139–146.
S. Ramadurai (1973), "Fermi Acceleration and the Energy Spectra of Heavy Nuclei," *Astrophys. Letters* **14,** 85–88.
B. Rossi (1964), *Cosmic Rays*. New York: McGraw Hill.
D. N. Schramm (1972), "The Propagation of Cosmic Rays with $Z \geq 74$," *Astrophys. J.* **177,** 325–339.
D. N. Schramm and E. O. Fiset (1973), "Superheavy Elements and the *r*-Process," *Astrophys. J.* **180,** 551–570.
D. N. Schramm and W. A. Fowler (1971), "Synthesis of Superheavy Elements in the *r*-Process," *Nature* **231,** 103–106.
P. A. Seeger, W. A. Fowler, and D. D. Clayton (1965), "Nucleosynthesis of Heavy Elements by Neutron Capture," *Astrophys. J. Suppl.* **11,** 121–166.
M. M. Shapiro (1971), "Composition of Cosmic Ray Nuclei, $Z < 26$," *Invited and Rapporteur Papers, Proc. 12th Inter. Conf. on Cosmic Rays*, Hobart, Australia, 422–452.
M. M. Shapiro and R. Silberberg (1970), "Heavy Cosmic Ray Nuclei," *Ann. Rev. Nuc. Sci.* **20,** 323–392.
E. K. Shirk and P. B. Price (1973), "Observation of Trans-Iron Solar Flare Nuclei in an Apollo 16 Command Module Window," *Proc. 13th Inter. Cosmic Ray Conf.*, Univ. of Denver, **2,** 1474–1478.
E. K. Shirk, P. B. Price, E. J. Kobetich, W. Z. Osborne, L. S. Pinsky, R. D. Eandi, and R. B. Rushing (1973), "Charge and Energy Spectra of Trans-Iron Cosmic Rays," *Phys. Rev.* **D7,** 3220–3232.
R. Silberberg and C. H. Tsao (1973), "Partial Cross-Sections in High-Energy Nuclear Reactions, and Astrophysical Applications. I. Targets with $Z \leq 28$. II. Targets Heavier than Nickel," *Astrophys. J. Suppl.* **25,** 315–367.
J. A. Simpson (1972), "Galactic Sources and the Propagation of Cosmic Rays," *Invited and Rapporteur Papers, 12th Inter. Conf. on Cosmic Rays*, Hobart, Australia, 324–356.
D. F. Smart and M. A. Shea (1972), "Daily Variation of Electron and Proton

Geomagnetic Cutoffs Calculated for Fort Churchill, Canada," *J. Geophys. Res.* **77**, 4595–4601.
L. H. Smith, A. Buffington, G. F. Smoot, L. W. Alvarez, and M. A. Wahlig (1973), "A Measurement of Cosmic-Ray Rigidity Spectra Above 5 GV/c of Elements from Hydrogen to Iron," *Astrophys. J.* **180**, 987–1010.
Space Science Board (1973), *International Magnetospheric Study: Guidelines for United States Participation*, National Academy of Sciences, Washington, D.C.
E. C. Stone (1973), "Cosmic Ray Isotopes," *Proc. 13th Inter. Cosmic Ray Conf.*, Univ. of Denver, **5**, 3615–3626.
O. Struve, B. Lynds, and H. Pillans (1959), *Elementary Astronomy*. New York: Oxford University Press.
H. E. Suess and H. C. Urey (1956), "Abundances of the Elements," *Rev. Mod. Phys.* **28**, 53–74.
J. D. Sullivan and P. B. Price (1973), "On the Charge State of Low Energy Fe Nuclei Accelerated by Solar Flares," *Proc. 13th Inter. Cosmic Ray Conf.*, Univ. of Denver, **2**, 1470–1473.
J. D. Sullivan, H. J. Crawford, and P. B. Price (1973), "Relative Abundances of Nuclei with $2 \leq Z \leq 36$ in Solar Flares," *Proc. 13th Inter. Cosmic Ray Conf.*, Univ. of Denver, **2**, 1522–1525.
B. J. Teegarden, T. T. von Rosenvinge, and F. B. McDonald (1973), "Satellite Measurements of the Charge Composition of Solar Cosmic Rays in the $6 \leq Z \leq 26$ Interval," *Astrophys. J.* **180**, 571–581.
I. H. Urch and L. J. Gleeson (1973), "Energy Losses of Galactic Cosmic Rays in the Interplanetary Medium," *Astrophys. Space Sci.* **20**, 177–185.
R. M. Walker and E. Zinner (1973), "Measurements of Heavy Solar Wind and Higher Energy Solar Particles during the Apollo 17 Mission," *Apollo 17 Preliminary Science Report*, NASA Special Pub. 330, pp. 19-2 to 19-11.
R. M. Walker, R. L. Fleischer and P. B. Price (1965), "Particle Tracks in Solids and the Isotopic Composition of Contemporary and Ancient Cosmic Rays," *Proc. 9th Inter. Cosmic Ray Conf.*, London, **2**, 1086.
W. R. Webber, S. V. Damle, and J. Kish (1972), "Studies of the Chemical Composition of Cosmic Rays with $Z = 3 - 30$ at High and Low Energies," *Astrophys. Space Sci.* **15**, 245–271.
W. R. Webber, J. A. Lezniak, and J. Kish (1973a), "Isotopic Composition Measurements of Cosmic-Ray Nuclei with $Z \geq 10$ Made Using a New Technique," *Astrophys. J.* **183**, L81–L86.
W. R. Webber, J. A. Lezniak, J. C. Kish, and S. V. Damle (1973b), "Evidence for Differences in the Energy Spectra of Cosmic Ray Nuclei," *Nature Phys. Sci.* **241**, 96–98.
W. R. Webber, E. C. Roelof, F. B. McDonald, B. J. Teegarden, and J. Trainor (1973c), "Pioneer 10 Measurements of the Charge and Isotopic Composition

of Solar Cosmic Rays during August 1972," *Proc. 13th Inter. Cosmic Ray Conf.*, Univ. of Denver, **2,** 1516–1521.

J. P. Wefel (1971), *Measurements of the Relative Abundances of Extremely Heavy Cosmic Rays*, Ph.D. Thesis, Washington University.

G. L. Withbroe (1971), "The Chemical Composition of the Photosphere and Corona," in *The Menzel Symposium on Solar Physics, Atomic Spectra and Gaseous Nebulae*, NBS Special Pub. 353, p. 127.

R. T. Woods, H. R. Hart, and R. L. Fleischer (1973), "Quiet Time Energy Spectra of Heavy Nuclei from 20 to 850 keV/amu," *Apollo 17 Preliminary Science Report*, NASA Special Pub. 330, chap. 19, pp. 11–15. Also to appear in *Astrophys. J.*, 15, 1975

Chapter 6

Ancient Energetic Particles in Space

6.1. INTRODUCTION TO TRACK STUDIES IN EXTRATERRESTRIAL MATERIALS

Although it was some time after the discovery of track etching that fossil tracks were first seen in meteorites (Maurette et al., 1964), the idea that extraterrestrial rocks might preserve a record of their particle bombardment in space provided the original, main incentive for our track studies and for the development of the chemical etching technique of track revelation (Price and Walker, 1962). This central theme has continued throughout the years, providing a constant well-spring of ideas and results that have interacted with our other track work, leading us into new domains of energy and mass in the study of energetic particles in space.

The two kinds of extraterrestrial material currently available to us—meteorites and lunar samples—contain generally similar types of silicate minerals and share other common features. However, the types of information that can be obtained from each are somewhat different. Examples of meteorites and lunar rocks are shown in Figs. 6-1 and 6-2.

Meteorites appear to be among the oldest and, in some cases, the least chemically differentiated objects in the solar system. For a good introduction into the field of meteoritics, we recommend the book by Wood (1968). Most of the meteorites were apparently formed in a rather short time interval of several tens of millions of years (Podosek, 1970a) at a time $\sim 4.6 \times 10^9$ yrs ago. Studies of the decay products of radioactive isotopes that are now extinct show that meteorites must have formed within $\sim 10^8$ yrs of the time when our solar system stopped acquiring newly formed radioelements—that is, at the time of separation and condensation of solar system material from the rest of the Galaxy. (See Section 6.5 for a fuller discussion.) For these reasons, and many others, meteorites appear to give us our best experimental handles on the processes operative in the early solar system.

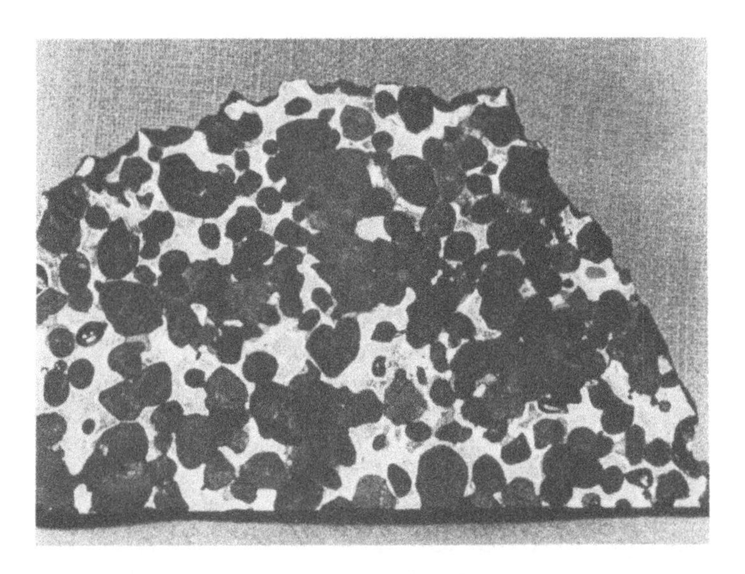

Fig. 6-1. *Polished section of the Springwater meteorite. This is a relatively rare type of stony-iron meteorite called a pallasite. The iron-nickel matrix material is studded with silicate inclusions that consist mostly of large single crystals of olivine. Fossil tracks in extraterrestrial materials were first found in a meteorite of this type. (Photo courtesy of F. Getsinger.)*

Although the materials that formed the moon apparently condensed nearly 4.6 billion years ago and remained as separate material in the solar system, crystalline lunar rocks are considerably younger than most meteorites, ranging in age from 3.2×10^9 to $\sim 4.0 \times 10^9$ yrs. Solidification ages as high as 4.2×10^9 yrs were reported for several small fragments at the Fourth Annual Lunar Science Conference held in March, 1973 (Schaeffer and Husain, 1973). Earlier work reported in the proceedings of the first three lunar science conferences (*Proc. Apollo 11 Lunar Science Conference*, Pergamon Press, 1970; *Proc. Second Lunar Science Conference*, MIT Press, 1971; *Proc. Third Lunar Science Conference*, MIT Press, 1972) shows that most samples lie in the 3.2 to 4.0×10^9 y range. (Although results on lunar samples are found in a number of journals, the conference proceedings represent the most complete compilation of lunar sample research.) The largely unexplored region from 4.0×10^9 yrs to 4.6×10^9 yrs presumably reflects a tumultuous period during the moon's history where the original record has been largely effaced.

Although both meteorites and lunar samples have been exposed to energetic particles in space for long times, there are some fundamental differences in the accessible record of such radiations in the two different types of materials. For one thing, the orbits are very different. Most meteorites probably spend the majority of time between 2 and 3 AU (1 AU is defined as the mean distance between the earth and the sun) and thus sample a different region of interplanetary space than do the lunar samples.

Fig. 6-2. *Lunar rock 15465 from the Apollo 15 mission. This rock is typical of a class of rocks called breccias that consist of complex assemblages of rock fragments. Some of the fragments are themselves breccias. In contrast to this rock, some breccias are very fine-grained, having apparently been formed by shock compression of lunar soil. Still other rocks are called "crystalline," meaning they formed by the cooling of a molten lava.*

The spectacular luminous display accompanying the arrival of a meteorite means that some of the outer regions of these objects have been heated. The direct thermal effects penetrate only a few millimeters at most and do not destroy the particle track record in the interior of the meteorite; however, the removal of material by ablation means that the record of low energy particles, such as those produced in solar flares, is destroyed. Lunar rocks carried to earth inside a spacecraft clearly do not suffer this limitation.

The great age of meteorites does not mean that they can be used to give a complete record of high energy radiations in space. Although of great antiquity, most meteorites have existed in space as meter-sized objects, capable of registering cosmic ray effects, for relatively short times. Stony meteorites, which are the most prevalent among those objects seen to arrive on earth, have typical cosmic ray exposure ages of $\sim 10^7$ yrs, and the oldest stone exposure age is only $\sim 10^8$ yrs (Herzog and Anders, 1971a). Metallic meteorites typically have much longer exposure ages but, at best, the record does not extend back more than $\sim 10^9$ yrs. The reasons why meteorites give us this biased sampling are not completely clear; one probable explanation is that the age distributions reflect the finite survival lifetimes of small objects in space set by collision with other interplanetary debris. For a good general review of meteorite cosmic ray exposure ages, see Anders (1962). A more recent discussion for stony meteorites is given by Herzog and Anders (1971b).

Lunar rocks also have relatively limited lifetimes for direct exposure to space. As we will discuss below in more detail, rocks picked from the present lunar surface have been there for relatively brief times of $\sim 10^6$ to $\sim 10^8$ yrs. Again the basic limitation on lifetime is set by the infall of meteoroids that either destroy or bury lunar surface rocks at a finite rate.

Despite these limitations, the energetic particle record can be extended back in time to an era close to the beginning of the solar system. The present lunar surface consists of a debris layer (commonly called the lunar regolith) that started to build up when the material forming the present crust first solidified. Radioisotope measurements of lunar rocks show that this began in different parts of the moon at least 3.2 to 4.0×10^9 yrs ago. At the beginning of that interval the sun was very young and, in analogy with other stars (Wilson and Woolley, 1970), was likely much more active than at present. The solar system was also in a different position with respect to the Galaxy and during the ensuing years may have experienced very different galactic bombardments as it wandered through and, perhaps, across the spiral arms of its local universe. The churning of the lunar surface by meteoroids was also certainly more intense in the early stages of its development. The complete history of the bombardments to which the moon (and Earth) have been subjected, as well as the dynamic processes that shaped the present lunar surface, are thus recorded in the individual fragments of the lunar regolith. As might be expected, this record is complicated and will take much time to decipher. In this chapter we will describe the progress as of 1973 on this most fascinating problem.

There are two special classes of meteoritic and lunar samples that give unique information on energetic particles emitted by the sun during the early history of the solar system. These special rocks—lunar gas-rich breccias and gas-rich meteorites—consist of mechanical assemblages of small particles that have been individually irradiated prior to being incorporated into the final larger object.

From the above brief description of the nature of extraterrestrial materials it should be clear that track studies in extraterrestrial materials can help us understand the processes operative in the early solar system as well as the nature of past radiations in space—both solar and galactic. The record of such radiations also allows us to measure dynamic processes such as erosion in space and the evolution of the lunar surface.

Although the unique characteristic of extraterrestrial materials is the ability to extend our knowledge back in time, fossil track studies in these materials have contributed substantially to defining the energies and masses of present-day energetic particles in space.

In this chapter we will first treat the general subject of sources of tracks in extraterrestrial materials from a theoretical point of view. Following this discussion we review the experimental status of track studies in meteorites and lunar samples. This chapter deals with a very active research area and, particularly in the case of lunar samples, is destined to be somewhat out of date by the time this book is read. Our goal is to give a general framework within which the interested reader will be able to follow the most recent work.

The experimental work on tracks in extraterrestrial materials has been done largely, though not exclusively, by scientists located in the following locations: The General Electric Laboratory, Schenectady, New York; the Laboratoire de Mineralogie, Paris; the Max Planck Institut für Kernphysik, Heidelberg, Germany; the Tata Institute, Bombay, India; the University of California, Berkeley; the University of California, San Diego; the University of Paris, Orsay, France; and Washington University, St. Louis. For convenience, in the text we will sometimes refer to these groups as the G.E. group, the Orsay group, etc. The work at the Tata Institute was directed by Prof. D. Lal, who has recently moved to the Physical Research Laboratory, Navrangpura, Ahmedabad; we designate work done under his direction simply as from the Indian group.

6.2. EXTERNAL SOURCES OF TRACKS IN EXTRATERRESTRIAL MATERIALS

6.2.1. General Considerations

Extraterrestrial materials differ in two fundamental ways from terrestrial ones. Because of the lack of shielding of the earth's atmosphere, they are exposed directly to energetic particles. These particles may either register directly or interact with the constituents of the material to produce secondary reaction products that may register as tracks. Another difference is their great ages; most meteorites and some lunar materials preserve a record of ancient isotopes that were once present in the early solar system but which had decayed to negligible proportions by the time most terrestrial samples were formed.

We first consider the production of tracks by radiations impinging on the materials from space. In Chapter 5 we have shown that modern energetic particles in space, be they from the sun or the Galaxy, consist of the complete range of elements represented in the Periodic Table. Fortunately, lighter components such as protons and helium nuclei never produce enough radiation damage to give etchable tracks in the common silicate minerals found in extraterrestrial materials; if this were not so, almost all crystals would be too densely radiation damaged to study.

As they slow down and their ionization rate increases, nuclei with $Z \geq 18$ will produce enough damage to register tracks and will leave a record of their presence that is permanent (neglecting heating and shock events). The abundance of elements in the charge range 18–30 is more than three orders of magnitude greater than the combined abundance of all elements with $Z > 30$, and, therefore, most of the tracks that register directly will be formed from those in the group $18 \leq Z \leq 30$; we will henceforth refer to these as VH (very heavy) ions in common with the terminology of cosmic ray scientists.

The qualitative features of VH tracks in extraterrestrial materials are as follows:

1) Because of the low sensitivity of silicate minerals, a track is formed only near the end of the range of a VH ion; typical etching conditions give track lengths of $\sim$10 μm to 20 μm.
2) Because of the high probability for nuclear interaction, the high rate of energy loss, and the form of the energy spectrum of VH ions in space, there is a rapid decrease of VH track density with depth. At very low energies profound changes occur on the scale of micrometers; at high energies on the scale of centimeters.
3) The rapid attenuation gives rise to characteristic angular anisotropies in the track distributions, with the resultant preferred orientation being toward the direction of the nearest surface.
4) Accompanying the short VH tracks are occasional much longer tracks produced by the lower, but finite, abundance of much heavier nuclei.

Because of their dominant role, we devote considerable space in this chapter to calculations of the detailed behavior of VH ions in extraterrestrial materials.

In principle, hypothetical elementary particles with magnetic charge—called magnetic monopoles—represent another possibility for impinging particles that might be capable of direct track registration. A single monopole would produce an ionization equivalent to a particle with $Z = 68.5n$ at relativistic energies, where n is an integer ≥ 1. In contrast to an electrically charged particle, a monopole loses *less* energy in its last few micrometers of range and hence can be detected only at higher energies. Our best estimate is that singly charged monopoles ($n = 1$) would not register in silicate minerals but that doubly charged monopoles ($n = 2$) would register in the more sensitive ones, namely, micas and some natural glasses. However, as discussed in Chapter 7, the flux of even singly charged monopoles is known to be extremely low and, although monopoles may one day be discovered, they do not constitute an important source of tracks in meteorites. The outstanding characteristic of a monopole track, if it existed, would be its extremely long length relative to that of any other known track-producing particle. Dyons—the particles with both magnetic and electric charge suggested by Schwinger (1969) to be the building blocks of elementary particles—would register in all known track detectors but, here too, their experimental existence has not been demonstrated.

The number of possibilities for track progenitors increases considerably when the production of tracks by nuclear interactions is taken into account. One of the earliest studies of tracks in solids (Price and Walker, 1962) showed that high energy protons, which constitute the majority of galactic cosmic rays, are capable of generating track-producing heavy recoils in high-energy spallation reactions. This is a significant part of the track record in extraterrestrial materials and is treated in a separate section of this chapter. The qualitative feature of spallation recoils is that they give unusually short tracks which, because of the lower cross section of protons for nuclear interaction and lower rate of slowing down by ionization, show much smaller spatial gradients than do tracks from VH particles.

High energy reactions also produce neutrons as secondaries. At high energies these may cause fission of a variety of heavy elements and, when thermalized, may cause fission of ^{235}U. These reactions are particularly important in connection with the study of anomalous fission effects in meteorites, and the neutron flux problem is treated in Section 6.2.6.

Another mechanism that was considered and then rejected by Fleischer et al. (1967a) was the production of "pseudo-particle tracks" by the collective action of high intensity meson jets produced in ultrahigh energy collisions of cosmic rays.

As shown explicitly by Crozaz et al. (1969), α-particles of several tens of MeV also produce short tracks ($\lesssim 0.5$ μm) by inelastic, compound nucleus reactions. Tracks of compound nuclei are not important for meteorites but possibly play a role in lunar samples, though this has not yet been demonstrated experimentally.

6.2.2. Form of the VH Track Density vs. Depth Profile at High Energies

Fig. 6-3 shows a schematic diagram of the sources of tracks in a spherical meteorite bombarded in space. A high energy particle of the VH group entering the meteorite cannot produce a track unless it slows down almost to the end of its range. The track density at a given depth thus depends on the number of incident particles that are capable of reaching that depth; that is, on the energy spectrum of the incident radiation. However, the number reaching a given depth is further attenuated by nuclear interactions of the incident nuclei with the target nuclei of the meteorite; at high energies (i.e., above ~1 GeV/nucleon) the attenuation by nuclear interaction dominates the depth dependence of track density. Conversely, at lower energies, where the range of the particles is small compared with an interaction mean free path, the depth dependence is determined almost completely by the energy spectrum of the incident radiation.

We now formulate the problem analytically for the case of high energy galactic particles. Consider first the case of a single incident charge species. In our previous work we have stressed the concept of a critical rate of ionization loss, J_c. Fast particles with less than this rate of ionization do not register; slower particles with $J > J_c$ do. A necessary concomitant of this concept is that there is a fixed, maxi-

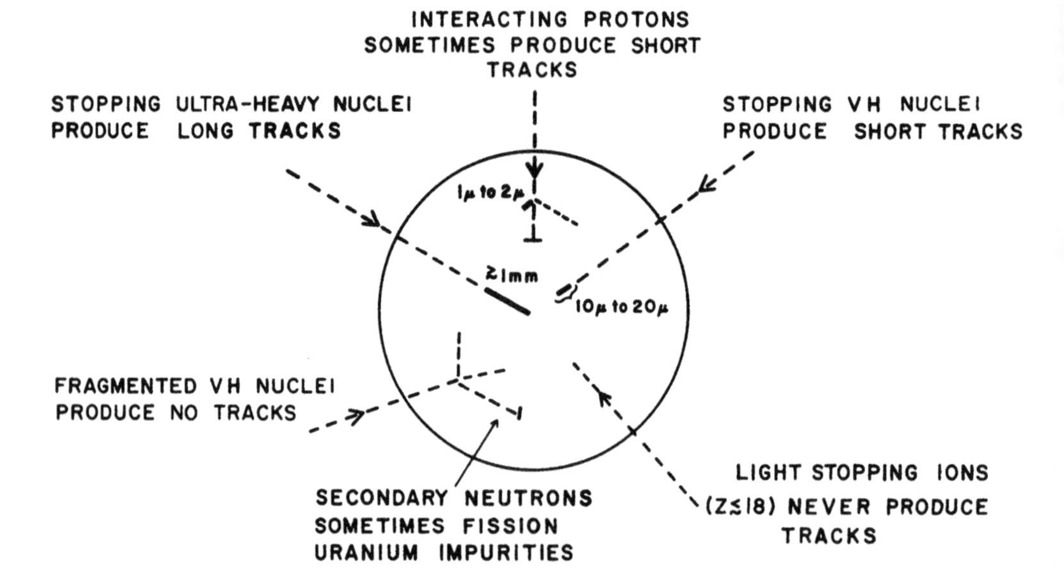

Fig. 6-3. *Schematic diagram illustrating production of tracks by external sources in meteorites. Not included are possible esoteric particles such as doubly-charged magnetic monopoles that have, as yet, not been observed in nature. The most abundant sources of tracks are stopping VH nuclei and spallation recoils from proton interactions. Some 25% of fragmented VH nuclei still produce tracks.*

mum etchable track length $\Delta R_c(Z)$ for any atomic species traversing a particular crystal. More recent measurements of track registration show that this concept must be modified; prolonged etching will sometimes continue to reveal tracks with lower and lower values of J; also, particles of the same energy have a distribution of track lengths after a given etching treatment. However, for a fixed etching time, the concept of a fixed track length for a given slowing species is still valid provided $\Delta R_c(Z)$ is taken as a properly averaged quantity. We thus continue to formulate the problem as previously.

The calculation of track density reduces to determining the number of particles that survive after traversing a given distance R and come to rest between R and $R + \Delta R_c(Z)$. For the case of a spherical meteorite, Fleischer et al. (1967a) showed that the rate of production of tracks on a polished or cleaved surface of a crystal at the center of such an object is given by

$$\frac{d\rho}{dt} = 2\pi\eta \left(\frac{dN_z}{dE}\right)\left(\frac{dE}{dR}\right)_{R_0} \Delta R_c(Z)\, S \exp\,(-\psi R_0). \qquad (6\text{-}1)$$

The various symbols are as follows: η is the etching efficiency (of the order unity for proper etching of most crystalline detectors); (dN_Z/dE) is the number of cosmic rays of charge Z between E_0 and $E_0 + dE$, where E_0 is the energy corresponding to a range of R_0; $\left(\frac{dE}{dR}\right)_{R_0}$ is the change in energy with range evaluated at E_0; $\Delta R_c(Z)$ is the averaged etchable track length of a slowing ion and ψ is the probability per unit path length for inelastic nuclear collisions. The term S given by the expression

$$S = [1 + \psi R_0 F_1 + \psi^2 R_0{}^2 F_1 F_2/2 + \cdots] \tag{6-2}$$

is a correction to the nuclear interaction term, taking into account those incoming nuclei that survive the collisions and are still capable of registering tracks. The fragmentation parameter F_1 represents the fraction that survive the first collision, F_2, those that have survived the second collision, etc. It should be noted that the fragmentation parameters as used here are not quite equivalent to those used in nuclear emulsion studies where the concept of fragmentation was first developed. In our case the fragmentation parameters are weighted by the effective etchable track length after collision. In practice only the first and second terms in S with values of F_0 between 0.15 and 0.25 have been used to calculate track densities.

The reader should be warned that a fully general solution to the problem would include an explicit energy dependence of the nuclear interaction probability and the fragmentation parameters. This has not been done, and the formula as given is strictly valid only for very high energy particles where ψ is constant to a good approximation or for very low energy particles where the nuclear interaction terms are unimportant.

The situation is more complicated when the test crystal lies at a point between the center and the surface of a spherically bombarded object. In this case the total density of tracks depends on the angle of inclination β (see Fig. 6-4) between the normal to the crystal surface on which the density is measured and the radius vector of the object. As shown by Fleischer et al. (1967a), the equation for the track production rate can be written as follows for the two special cases $\beta = 0$ and $\beta = \pi/2$:

$$\frac{d\rho}{dt} = 2\pi\eta \int_0^\pi \frac{dN_Z}{dE(r)} \cdot \frac{dE(r)}{dr} \Delta R_c(Z) \cdot [1 + \psi F_1 r(\theta) + \tfrac{1}{2}\psi^2 F_1 F_2 r^2(\theta)] \cdot \exp(-\psi r(\theta)) \sin\theta \,|\cos\theta|\, d\theta, \quad \text{if} \quad \beta = 0$$

$$\frac{d\rho}{dt} = 4\eta \int_0^\pi \frac{dN_Z}{dE(r)} \frac{dE(r)}{dr} \Delta R_c(Z) [1 + \psi F_1 r(\theta) + \tfrac{1}{2}\psi^2 F_1 F_2 r^2(\theta)] \cdot \exp(-\psi r(\theta)) \sin^2\theta \, d\theta, \quad \text{if} \quad \beta = \pi/2 \tag{6-3}$$

where $r(\theta) = -(R_0 - D)\cos\theta + (R_0{}^2\cos^2\theta + 2R_0 D\sin^2\theta - D^2\sin^2\theta)^{1/2}$ is the distance from a small crystal to the surface of the object at the time it is irradiated.

As shown by Maurette et al. (1969), the detailed angular distribution of tracks on a given crystal surface can be calculated if the track density is expressed in

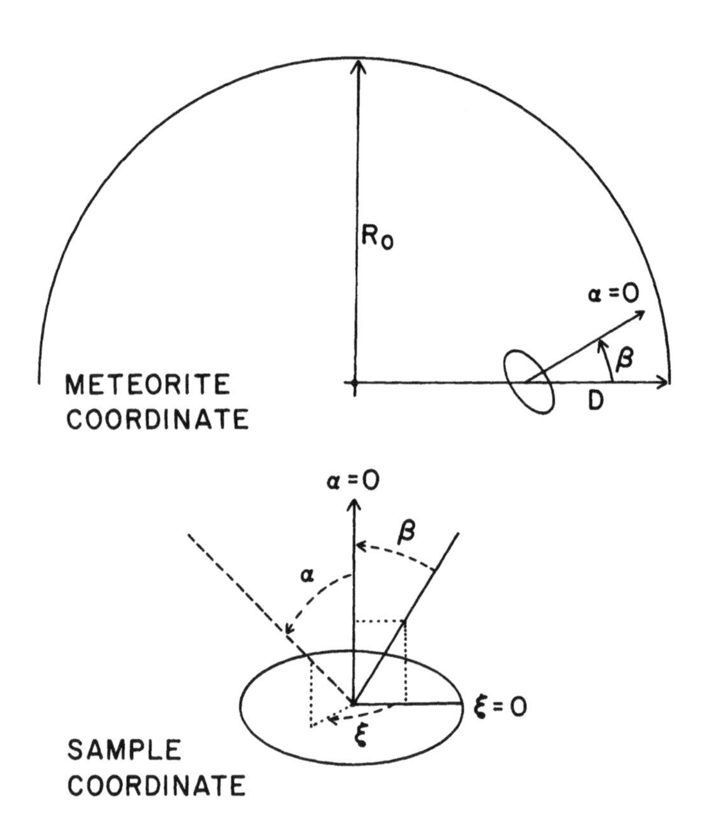

Fig. 6-4. *Coordinate systems. This figure shows the definition of the coordinates used in the text. R_0 is the radius of a spherical meteorite (or lunar rock) and D is the distance below the preatmospheric surface. The angle β is the angle between the radius vector and the normal to a polished or cleaved surface on which the measurements are made. The angles ξ and α are the azimuth and zenith angles, measured with respect to the normal to the surface of the sample.*

terms of the zenith angle α (the angle the track makes with respect to the normal to the viewing plane) and the azimuthal angle ξ (the angle made by the projection of a track on the viewing plane with respect to an axis in this plane) as defined in Fig. 6-4. In this case the track production rate is given by:

$$\frac{d^3\rho}{dt\,d\alpha\,d\xi} = \left\{ \left(\frac{dN_z}{dE(r)}\right)_{E[r(\alpha,\xi)]} \left(\frac{dE(r)}{dr}\right)_{r(\alpha,\xi)} [1 + \psi F_0 r(\alpha, \xi) \right. \tag{6-4}$$

$$+ \tfrac{1}{2}\psi^2 F_0 F_1 r^2(\alpha, \xi)] e^{-\psi r(\alpha,\xi)} + \left(\frac{dN_z}{dE(r)}\right)_{E[r(\pi-\alpha,\pi+\xi)]} \left(\frac{dE(r)}{dr}\right)_{r(\pi-\alpha,\pi+\xi)}$$

$$\cdot [1 + \psi F_0 r(\pi - \alpha, \pi + \xi) + \tfrac{1}{2}\psi^2 F_0 F_1 r^2(\pi - \alpha, \pi + \xi)]$$

$$\left. \cdot\, e^{-\psi r(\pi-\alpha,\pi+\xi)} \right\} \alpha \Delta R_c(Z) \sin\alpha \cos\alpha$$

where $r(\alpha, \xi) = -(R_0 - D)(\cos\beta\cos\alpha - \sin\beta\sin\alpha\cos\xi)$
$+ [(R_0 - D)^2 (\cos\beta\cos\alpha - \sin\beta\sin\alpha\cos\xi)^2 + (2\,R_0 - D)D]^{1/2}.$

In our original work on this problem (Fleischer et al., 1967a), we showed that the average energy spectrum of contemporary galactic cosmic rays could be approximated by the expression $\frac{dN}{dE} = C(.94 + E)^{-\gamma}$ where $C = 1.4$ nuclei/m² sec sr (GeV/nuc), $\gamma = 2.2$, and E is expressed in GeV/nuc. Assuming the constancy of cosmic rays, we then calculated curves of track density vs. depth. More recently, as described in Section 6.2.4, it has proven possible to measure directly the energy spectrum of track-producing particles averaged over a two million year span. In Figs. 6-5 and 6-6 we show calculated curves of the track production rate using this empirical energy spectrum. As discussed in Section 6.3, the high energy form of this average spectrum (corresponding to depths $\gtrsim 1$ cm) is identical with that of contemporary cosmic rays. The shapes of the track density profiles at great depths are thus similar to those previously given, although the absolute values are slightly different.

In Fig. 6-5 we show the case appropriate for a meteorite, namely, uniform irradiation from all sides (4π irradiation). In Fig. 6-6 we show track profiles for a spherical rock sitting on an infinite plane surface (2π irradiation); this figure is appropriate for a lunar rock with a simple irradiation history. For a discussion of track profiles for nonspherical geometries or for more complex irradiation geometries, the reader is referred to Bhattacharya et al. (1973).

As follows from eqs. (6-3) and (6-4), and as can be seen explicitly in Figs. 6-5

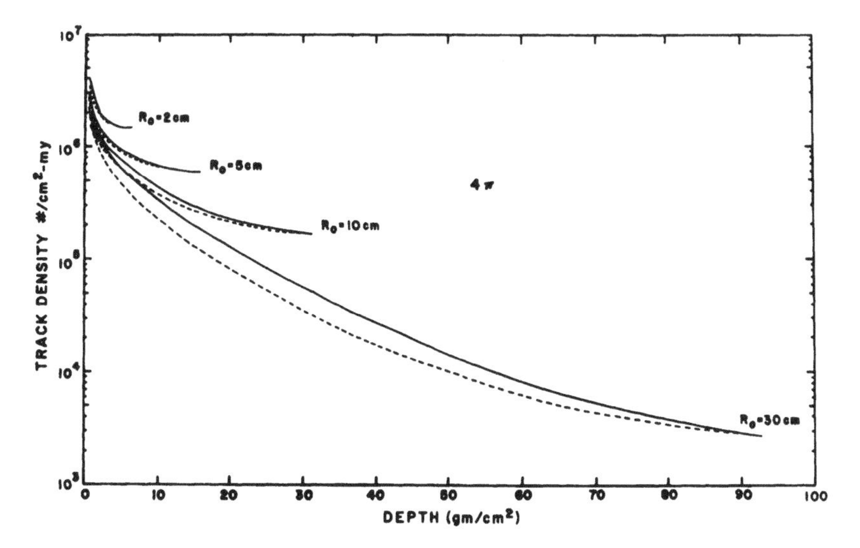

Fig. 6-5. *Track density vs. depth for (spherical) rocks of different size and with typical lunar compositions. The irradiation is assumed to be isotropic corresponding to a 4π irradiation in space or (with a decrease of a factor of two) to a uniform tumbling of a rock on a planetary surface. VH tracks are assumed to have a length of 10 μm; no correction for scanning efficiency has been included. Solid lines are for* $\beta = 0$*, dashed lines for* $\beta = \pi/2$.

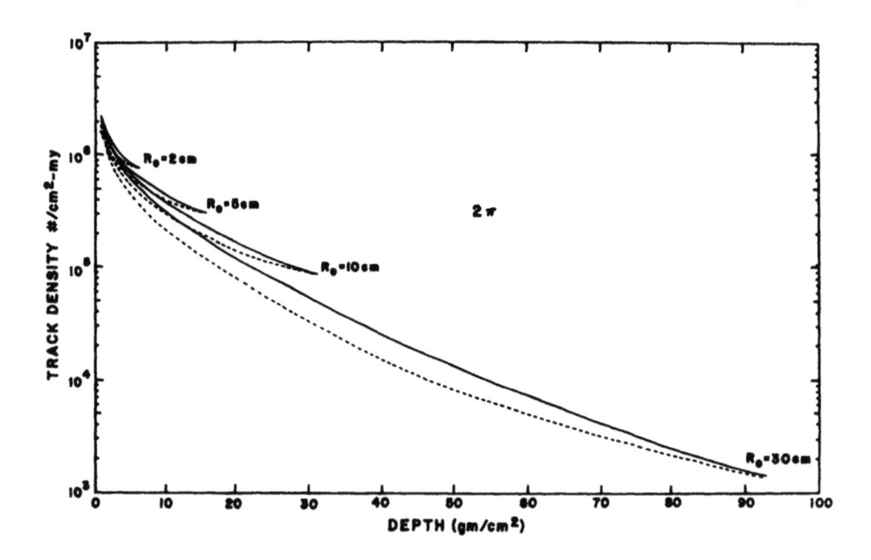

Fig. 6-6. *Same as Fig. 6-5 except that the solid angle for irradiation is 2π, corresponding to a rock sitting in a fixed position on a planetary surface.*

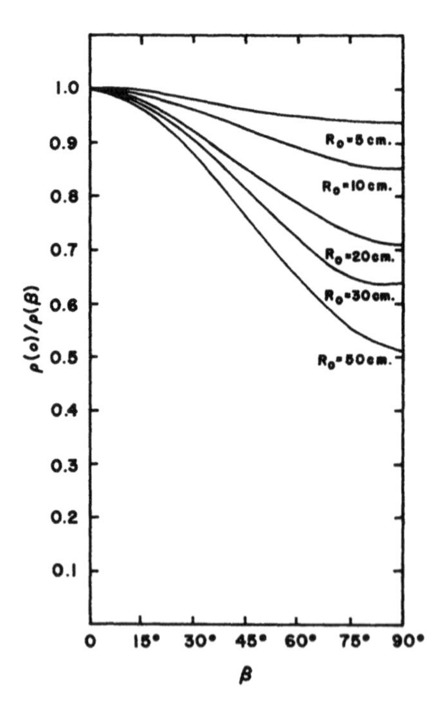

Fig. 6-7. *Track density as a function of sample inclination angle for a stony-iron meteorite. (After Maurette et al., 1969.) The total track density at an angle β relative to that at $\beta = 0°$ is shown for different values of R_0. These curves are good to about 5% for all values of $D/R_0 \lesssim 0.6$.*

and 6-6, the track density at any depth D, where D is $< R_0$, depends on the angle of inclination, β, with respect to the radius vector. Fig. 6-7, taken from Maurette et al. (1969), shows an explicit plot of this dependence for spheres of different size and with $D/R_0 \leq 0.6$. The values are good to $\sim 5\%$ for $D < 0.6\ R_0$. If the orientation of a rock is known, as is frequently the case in the later lunar missions, β can be directly measured. If, as in the early lunar missions, and with most meteorites, the orientation of the sample is *not* known, then measurements of a number of crystals from a given location can define an average track density that can be compared with the average of the curves for $\beta = 0$ and $\beta = \pi/2$.

The measurement of a large number of crystals with different orientations is not always convenient (for example, when rock sections are used) and Maurette et al. (1969) have described an alternate method for calculating β from measurements of the distribution of azimuthal angles in the plane of a test crystal. By explicit calculation these authors find that for $D < 0.5\ R_0$, the symmetry of the azimuthal angular distribution is determined primarily by β and is relatively independent of either D or R_0. Fig. 6-8 shows the nature of this dependence for one particular

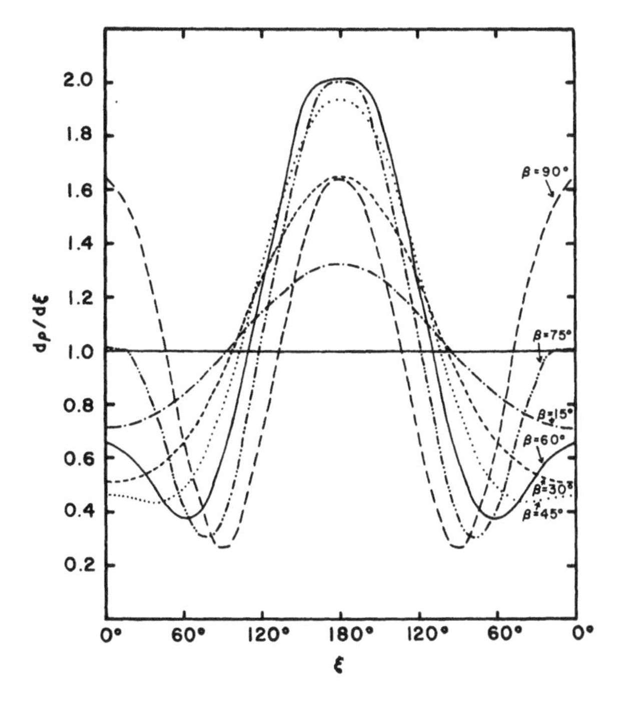

Fig. 6-8. *Calculated azimuthal angle distributions for a stony-iron meteorite. (After Maurette et al., 1969.) The azimuthal variation of track density is given for different choices of the sample inclination angle β. The radius was assumed to be 30 cm and the sample depth to be 5 cm. It can be seen that the distribution changes monotonically from an isotropic distribution to one that is strongly peaked when β goes from 0° to 45°. For $\beta \geq 60°$, a double maximum is apparent in the distribution.*

choice of D and R_0. It can be seen that when β goes from 0° to 45° the distribution changes monotonically from an isotropic one to one that is strongly peaked. For $\beta \geq 60°$, a double maximum is apparent in the distribution. As is the case with the dependence of track density on β (Fig. 6-7), the magnitude of the anisotropy depends primarily on R_0 and not on D when $D/R_0 \lesssim 0.5$.

6.2.3. Profile of the VH Track Density vs. Depth at Low Energies

At very low energies, the nuclear interaction terms can be ignored in the calculation of track density vs. depth. At low penetration distances, the dimensions of most samples of interest are large compared with the depths being measured. The usual calculation then reduces to finding the track density vs. depth below an infinite plane surface. The basic formulation remains the same and the proper calculation of track density vs. depth is obtained by using a computer program with tabulated range-energy values for the ion and absorber of interest. (See Chapter 3 for a discussion of the appropriate range-energy curve.)

Although we will shortly present our best estimate of the track density production curve for low energy particles based on our combined work on glass from the Surveyor III spacecraft, we would like to digress at this point to show that in some cases a particularly simple relationship exists between the energy spectrum of low energy particles and the depth dependence of track density.

Consider the case where the energy spectrum of particles can be represented by a simple power law in energy of the form $\frac{dN}{dE} = CE^{-\gamma}$. Consider further an energy domain where the range-energy relationship can also be approximated by a power law of the form $R = AE^{\beta}$. The track density at a given distance D below the surface is then given by the following relationship:

$$\rho = KD^{(1-\gamma-\beta)/\beta} \tag{6-5}$$

where K is a constant.

Thus a power law energy spectrum results in a power law in track density vs. depth. At ranges $\gtrsim 0.1$ mm, the exponent β is approximately 1, giving finally

$$\rho \approx KD^{-\gamma} \tag{6-6}$$

6.2.4. Absolute Values of VH Track Production Rates at the Orbit of Earth

Our procedure in the original papers on galactic track production rates was to take the best average energy spectrum for cosmic rays, as measured in balloons and satellites, and assume that the VH component had not changed in time. Here we take a different approach. The track production rates are derived empirically using track data on well-dated extraterrestrial materials. The question of how the empirical data compare with the best estimates of the average contemporary flux is treated separately in Section 6.3. We shall show that the two are closely similar and thus the two approaches give similar values for absolute track production rates. The values do not support the modified production rates given by Comstock (1972).

We saw in Fig. 5-3 that from 10 keV/nuc to ~100 MeV/nuc the average spectrum consists primarily of solar flare particles. For input data we used the measurements of particle tracks in the Surveyor III glass (Barber et al., 1971a; Crozaz and Walker, 1971; Fleischer et al., 1971a) that was exposed during a 2.55-year period near solar maximum (see Fig. 5-22).

When expressed as a power law, the average Fe flux during the Surveyor period has a spectral index that becomes more negative with increasing energy. At energies greater than ~5 MeV/nuc the average flux can be represented approximately by the expression

$$dN/dE = 6.2 \times 10^{6}\, E^{-3.3} \text{ particles/(cm}^2 \text{ sr y MeV/nuc)} \qquad (6\text{-}7)$$

The normalization of the Surveyor data to a long-term average is a difficult problem. The best evidence on the relative activity of the sun during the Surveyor exposure comes from a comparison of short-lived and long-lived radioisotopes in lunar rocks. The results and analysis of Finkel et al. (1971) show that the proton flux responsible for the production of ^{22}Na (half-life 2.6 yrs) is 1.8 ± 0.8 times the average flux of protons responsible for the production of ^{26}Al (half-life 7×10^5 yrs). Since the Surveyor exposure period is also ~2.6 yrs, this gives a crude first estimate of the ratio of the activity during the Surveyor period to the long-term average.

However, two additional points must be kept in mind. First, it is estimated (Arnold, private communication) that 25% of the ^{22}Na activity was induced in a giant flare in 1956. Inclusion of this factor reduces the activity ratio to ~1.4 ± 0.6. Second, as pointed out by Rancitelli et al. (1971), the measured ^{26}Al activity values (and hence inferred long-time average proton fluxes) can be lowered considerably by the inclusion of erosion. In particular, Rancitelli (private communication) estimates that the measured values can be reduced by a factor of 1.4 with an erosion rate of 10^{-7} cm/y. This supports the view that the Surveyor exposure period may have been representative of the long-term solar average. In previous papers we have assumed either that the Surveyor period was similar to the long-term average (this amounts to dividing the raw data by 2.6 yrs to obtain a production rate), or that the Surveyor period was more active by a factor of two than average. In this work we make the compromise assumption that the period was

50% more active than the long-term values. The final value for the time-averaged energy spectrum is thus

$$\overline{\frac{dN}{dE}} = 4.1 \times 10^6 E^{-3.3} \text{ particles/(cm}^2 \text{ sr y MeV/nuc).} \tag{6-8}$$

The corresponding track production rates, assuming $\overline{\Delta R_c} = 10\ \mu m$ for a typical mineral, and $\overline{\Delta R_c} = 100\ \mu m$ for the Surveyor glass, are shown in Fig. 6-11. As noted previously, reassessment of the value of $\overline{\Delta R_c}$ indicates that, after weighting for solid angle and efficiency, it should be replaced with an effective value of $\sim 60\ \mu m$. The reader is referred to the paper by Hutcheon et al. (1974) for a discussion of the dependence of the track density profile on the value of $\overline{\Delta R_c}$, which will vary from mineral to mineral. Because of the uncertainties discussed above, the accuracy of the absolute value is probably no better than a factor of two.

The track production rates at depths less than $\sim 10\ \mu m$ are very uncertain. The Surveyor III data show little increase in track density in the depth interval from 0 to 10 μm. In contrast, the Apollo 16 cosmic ray experiment (Burnett et al., 1972; Fleischer and Hart, 1972d; Price et al., 1972; Braddy et al., 1973; Fleischer and Hart, 1973b), which was exposed during a small solar flare, showed a monotonic increase in particles down to a depth only 0.5 μm below the surface.

At this writing, it is not known whether the Apollo 16 flare was anomalous or whether, for some reason (such as a dust cover or ultraviolet light fading), the Surveyor glass did not adequately register very low energy ions. As described below, the very high track densities found in micron-sized lunar grains suggest the latter alternative.

The values for track production rates at energies greater than 100 MeV/nuc, corresponding to galactic cosmic rays, are taken from the recent work of Walker and Yuhas (1973). The Apollo 16 landing site was dominated by two bright-rayed (and therefore fresh) craters. The ejecta blanket of the younger of these—South Ray Crater—was characterized by boulders that were sharp and angular in appearance with little evidence of rounding or spalling.

A rock chip from one of these boulders, 68815, had a spallation age, as determined by the Kr-Kr method, of $2.0 \pm 0.2 \times 10^6$ yrs (Behrmann et al., 1973a,c; Drozd et al., 1974). Based on previous work on lunar erosion rates, to be discussed later, a rock chip with the mechanical properties of 68815 would be expected to have suffered relatively little change in this period of time. To the extent that the measured spallation age is valid, the track profiles vs. depth in this rock can therefore be used to construct an empirical energy spectrum valid for calculating track production rates in rocks with different geometries.

In Fig. 6-9 we show the track density profile measured in the depth interval

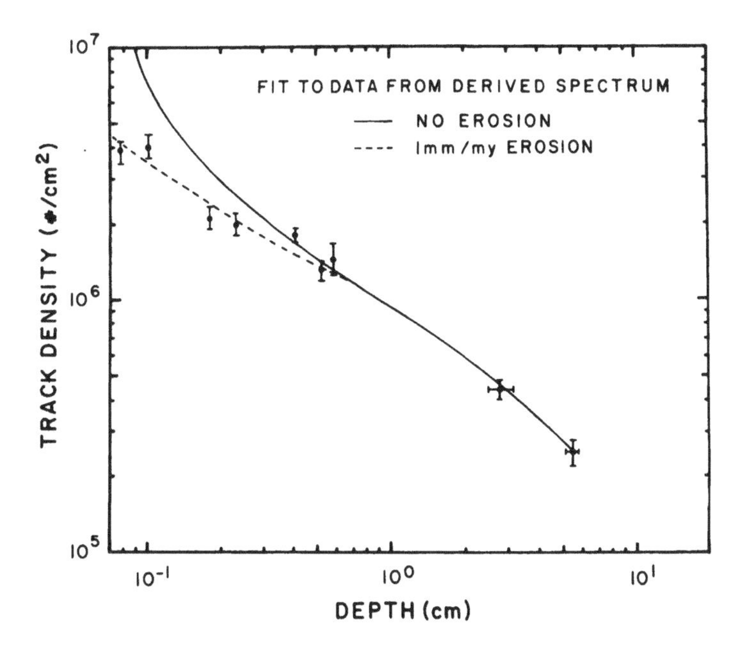

Fig. 6-9. *Track density vs. depth in lunar rock 68815. (After Walker and Yuhas, 1973.) This rock has a particularly young Kr-Kr exposure age of 2 × 10^6 yrs and, unlike most lunar rocks, appears to have had a simple one-stage exposure history. The solid line shows the calculated track density using the energy spectrum for galactic cosmic rays given in Fig. 6-10 coupled with the average solar flare contribution as discussed in the text. The dotted line below 0.6 cm includes the effects of an assumed microerosion rate of 10^{-7} cm/y.*

from <1 mm to 6 cm in feldspar crystals in a polished section. Since only long tracks were counted, these data refer only to VH ions. By empirical curve fitting using eq. (6-2) and the known geometry of the rock, the energy spectrum shown in Fig. 6-10 was found to give the best fit to the data at depths ≥ 1 cm. In these calculations, the interaction probability ψ was taken as 0.054 $(g/cm^2)^{-1}$ and the fragmentation parameter as 0.25. Shown for comparison are other spectra that have been used to calculate exposure ages.

The lower energy part of the spectrum shown in Fig. 6-10 is derived from the Surveyor data. Although the experimental points in rock 68815 do not fit this spectrum very well, essential agreement is restored if an erosion rate of 10^{-7} cm/y is assumed. This erosion rate has only a small effect on the calculated values at depths ≥ 5 mm. The low energy part of the spectrum is being explored in more detail using lunar rock 72315, whose outer surface was exposed for only $\sim 10^5$ years and has suffered negligible erosion (Hutcheon et al., 1974).

In Fig. 6-11 we give curves of track production rate vs. depth for an infinite slab with no erosion.

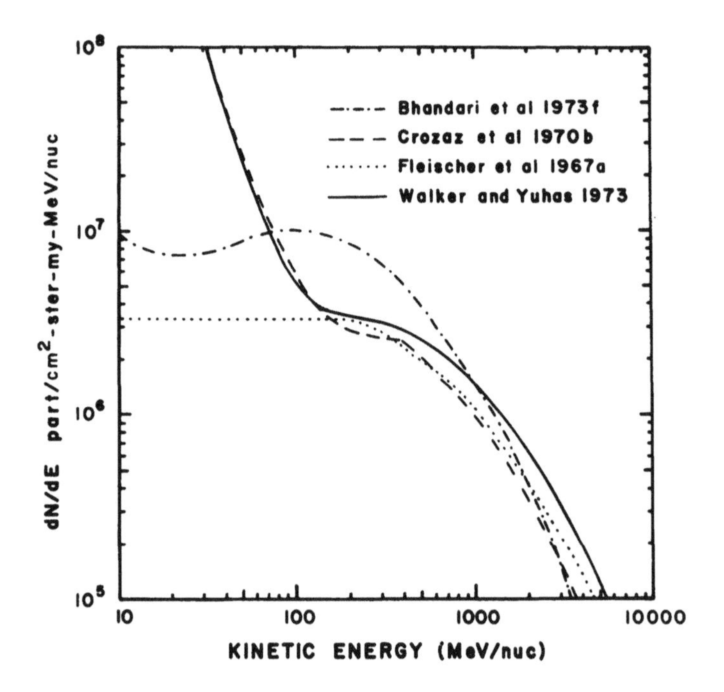

Fig. 6-10. *Average VH particle energy spectrum as derived from lunar rock 68815 and the Surveyor solar flare data (solid line). (After Walker and Yuhas, 1973.) Previously derived spectra are shown for comparison. As discussed in text, this curve must be coupled with track length measurements to calculate track density profiles in extraterrestrial samples. The dashed and dotted curves show several spectra previously used.*

We now consider the degree to which the curves in Figs. 6-10 and 6-11 can be used to calculate absolute exposure ages reliably. Above ~200 MeV/nuc corresponding to depths $\gtrsim 3$ g/cm², the solar contribution to the 68815 data becomes negligible ($<10\%$). The effect of the assumed erosion rate of 1 mm/10^6 y also becomes negligible ($<3\%$) at this depth. At great depths, therefore, the track production values are essentially determined by the track counts on rock 68815 and the measured Kr-Kr age of this rock.

To the extent that all lunar materials register and store tracks to the same degree as the feldspar crystals in 68815, the basic accuracy of the method at depths >3 g/cm² should be the combined error in the Kr-Kr age, the track counting statistics, the measurement of ΔR_c and the uncertainties in the geometry of 68815. These contribute a combined error with a one sigma deviation of $\pm 25\%$.

The empirical track production energy spectrum is strictly valid only over the energies corresponding to the depths measured in 68815—~50 MeV/nuc to ~650 MeV/nuc. This is the energy range of interest for most lunar samples. To

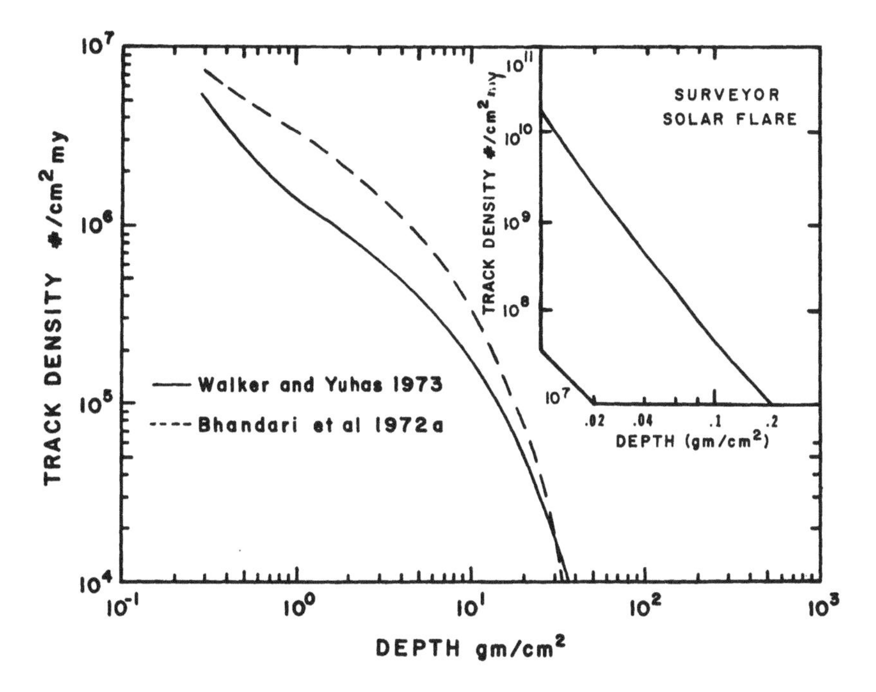

Fig. 6-11. *Minimum track density vs. depth below an infinite plane surface calculated from the energy spectrum of Fig. 6-10 and an assumed* $\overline{\Delta R_c}$ *of 10 μm. This calculation was done for a plane of observation perpendicular to the rock surface. Shown for comparison is the track density profile given by Bhandari et al.* (*1971b*) *for the* average *track density. Also shown in insert at right is the minimum track density vs. depth calculated from the compromise Surveyor spectrum* (*see text*). *In the case of a rock, the actual track densities at shallow depths depend critically on the rates of erosion.*

estimate rates at greater depths shown in Fig. 6-11, we have assumed an energy spectrum of the following form for $E > 0.65$ GeV/nuc:

$$\frac{dN}{dE} = \frac{6.43 \times 10^6}{(1 + E)^{2.2}} \text{ particles/cm}^2 \text{ sr my MeV/nuc} \qquad (6\text{-}9)$$

where E is expressed in GeV/nuc.

This form is chosen to give the best agreement with the most recent measurements of contemporary cosmic rays (see Chapter 5). This spectrum also gives agreement within experimental error with earlier experimental measurements on the St. Séverin meteorite (Cantelaube et al., 1967).

How realistic is it to assume that 68815 is representative of all lunar materials? Relative to fresh tracks, most fossil tracks are apparently somewhat shortened in lunar material; the total etchable ranges are also different from the normally obtained experimental lengths (Burnett et al., 1972; Plieninger and Krätschmer, 1972; Plieninger et al., 1972, 1973; Price et al., 1973a; Storzer et al., 1973a,b). Samples that have had different thermal or shock histories might, therefore, be expected to show different track length distributions. Measured track lengths also depend critically on the etching conditions used. These variations in track length can be taken into account, in a first approximation, by actually measuring the total track length distribution in any sample and substituting the experimental value $(\overline{\Delta R_c})_{\text{exp}}$ for ΔR_c in eq. (6-2).

In Fig. 6-11 we give the track production rate curve below a semi-infinite plane surface normalized to $\Delta R_c = 10$ μm. To get the appropriate value for a particular sample, this curve should be multiplied by $(\overline{\Delta R_c})_{\text{exp}}/10$ μm. The estimated track production rates should also be corrected for scanning efficiencies. Thus in the case of a measurement on a polished surface, where tracks <2 μm are excluded, the track production rates should be multiplied by $[(\overline{\Delta R_c})_{\text{exp}} - 2\ \mu\text{m}]/[\overline{\Delta R_c}]_{\text{exp}}$.

The fact that the lengths of fossil tracks in lunar and meteoritic samples are shorter than those of fresh tracks is potentially disturbing. Uniform $(\overline{\Delta R_c})_{\text{exp}}$ values would be obtained if the tracks were in thermal equilibrium, with as many fading away as being introduced. If this were the case, track exposure ages would be lower than the true values. However, as 68815 has a typical average track length and a short exposure age, this explanation would predict that apparent track ages much beyond 2.0 my could not be observed.

This is not the case. As shown in a later section, several track ages $\gtrsim 20 \times 10^6$ yrs have been measured. In one case studied in detail (Behrmann et al., 1973a,c) the difference between the Kr-Kr age of 50 my and the apparent track age of 29 my is best explained by assuming a finite rate of rock erosion; this explanation gives a much better fit to the depth profile than assuming that the difference is due to track fading.

An equally convincing argument that track fading does not introduce large errors in exposure ages based on VH tracks follows from the measurement of spallation track ages of up to several hundreds of millions of years (Fleischer et al., 1971b, see Section 6.6.3). Spallation tracks are even less thermally stable than those of VH ions (Maurette, 1970; also Fig. 6-12), and their demonstrated stability under lunar conditions assures that the VH tracks have not faded appreciably over long periods of time.

It should also be realized that track densities measured at the same place in a given rock vary from one type of mineral to the next. As first noted by Crozaz et al. (1970b) and Fleischer et al. (1970b), track densities in pyroxene crystals may be less than in adjoining feldspars by factors of 1.4 to 2. Olivine crystals give still lower densities (Bhandari et al., 1972c). Based on measurements of the registration of ^{252}Cf tracks, Fleischer et al. (1970b) view this effect as due to different etching

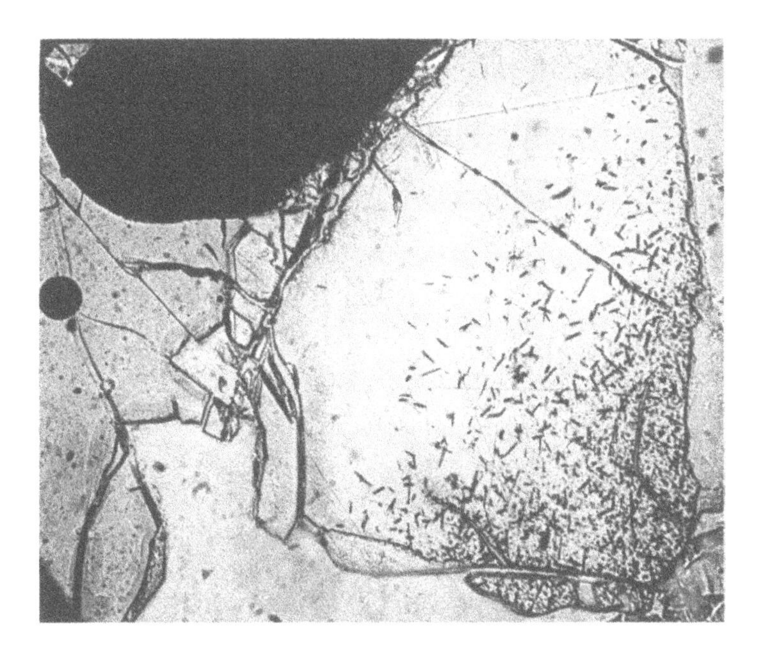

Fig. 6-12. *Tracks in a phosphate crystal adjacent to glassy zone in the Estherville meteorite. The (long) fission and cosmic ray VH tracks have been erased close to the glass boundary, showing that the glass was formed recently, probably during entry. The small dots that become more evident far from the boundary are due to spallation recoils. The preferential annealing of these smaller tracks is clearly visible. (Photograph courtsey of J. Shirck, unpublished work.)*

efficiencies. However, independent calibration work by Crozaz et al. (1970b) suggests that the difference is due to the higher sensitivity for registration of tracks in feldspar as opposed to the other minerals. As pointed out by Bhandari et al. (1972c), this latter explanation implies that the differences between various mineral phases will increase with increasing depth as the incoming VH nuclei become increasingly fragmented.

A different approach to defining the average rate of track production in meteorites and lunar materials has been taken by Bhandari et al. (1971b, 1973c). They found that at depths >1 mm the track profiles in rocks 12018, 12038, and 10017 were consistent with a single energy spectrum. Rock 12038 in particular seemed to have a simple irradiation history. From their data they constructed an empirical track production curve for an infinite plane surface of the following form:

$$\rho = K(A + BX)^{-\alpha}. \tag{6-10}$$

The absolute value of the track production rate was fixed by normalizing their data at 500 MeV/nuc to the data of von Rosenvinge et al. (1969) for contemporary

Table 6-1. Values of Various Parameters in Equation 6.10 as Given by Bhandari et al. (1972a) (units = tracks/cm^2my)

Constants	Depth		
	$2 \times 10^{-3} < x < .1$ cm	$0.1 < x < 1.$ cm	$1.0 < x < 25$ cm
K	1.2×10^6	1.4×10^6	6.36×10^{11}
α	0.75	0.7	6.15
A	0	0	7.5
B	1	1	1

cosmic rays. The values of the parameters in eq. 6-10, as given by Bhandari et al. (1972a,d), are shown in Table 6-1.

In Fig. 6-11 we show a comparison of the track production rates calculated as previously described by us (solid curve) with those obtained from eq. 6-10 (dashed curve). Because the solid curve is calculated for a crystal orientation giving the *minimum* track densities, while eq. 6-10 represents *average* track densities, the difference between the curves is exaggerated. From a comparison of the curves, we conclude that the Indian group, given the same track data, would calculate ages approximately 20% lower in the 1 to 2 mm depth interval, 20% to 50% lower in the 1 to 3 cm range, and essentially the same at depths between 3 and 8 cm.

Also shown for comparison in Fig. 6-10 are the energy spectra that have been previously used by various groups. At galactic cosmic ray energies ($E \geq 100$ MeV/nuc) where the comparison is meaningful (the earlier work took no account of solar flare particles), the present spectrum has the same shape as proposed originally by Fleischer et al. (1967a), which, in turn, is close to the values used by the Washington University group in their lunar work. However, the absolute value lies some 30% higher, and this has the effect of reducing previously cited exposure ages given by all save the Indian group.

The energy spectrum given by the Indian group, which is based on their track profile measurements and a normalization to the data of von Rosenvinge et al. (1969) for contemporary cosmic rays, is also shown. Although the absolute values differ by as much as a factor of three at one point, the curves cross each other and the calculated track densities vs. depth (Fig. 6-11) show considerably less spread. More work needs to be done to resolve the apparent differences in the spectra.

6.2.5. Production of Tracks in High-Energy Spallation Reactions

A high energy proton for which the deBroglie wavelength is much smaller than the geometric cross section of a nucleus can produce a "spallation" reaction that consists of three stages: (1) an initial nucleon-nucleon reaction, leading increasingly at higher energies to meson production; (2) the propagation of the incident nucleon with the accompanying initial reaction products through the nucleus (with possible subsequent secondary and higher order collisions); and (3) the de-excitation of the residual nucleus by nucleon emission, leaving a heavy recoiling nucleus. At low energies, the number of nucleons emitted is too small to give the residual nucleus much recoil energy, while at high energies the residual nucleus tends to have higher final energy but lower mass.

In practice it has been found that bombardment of silicate minerals with protons from 3 GeV to 30 GeV (typical energies of galactic cosmic ray protons) leads to the production of short tracks due to recoiling residual nuclei (Price and Walker, 1962; Maurette and Walker, 1964; Fleischer et al., 1967a; Crozaz et al., 1970b; Fleischer et al., 1970b; Fleischer et al., 1971b; Seitz, 1972; Crozaz et al., 1972c). In Fig. 6-12, we show a mixture of VH tracks and spallation recoils in a crystal located next to a rim of glass formed during entry. The "dotlike" appearance of spallation recoils is evident as well as the differential thermal stability of VH and spallation recoil tracks.

In practice, the spallation track production rates in most minerals are $\sim 10^{-9}$ track/proton, indicating that the recoils are due to trace elements rather than to the major constituents of the minerals. Whitlockite is, however, an exception. For this mineral the high rate of spallation track production (6% of all proton interactions according to Seitz, 1972) indicates that a major element, probably Ca, is producing the bulk of the recoils.

It is possible to obtain the spallation age of an extraterrestrial rock by first measuring the density of short tracks and then calibrating the response with a high energy bombardment. Both measurements should be made on the same crystal, with the same orientation, since trace impurities and angular dependence effects can influence spallation recoil production. There is, however, some ambiguity in the age so obtained because of the possible importance of secondary nuclear cascade products in producing recoils. Fig. 6-13, taken from Reedy and Arnold (1972), shows that the radioisotope production actually may *increase* as a function of depth due to these cascade effects. In general, the further the product isotope is in mass from the target nucleus, the less pronounced is the cascade rise. The production rate of spallation recoils thus probably rises less than the factor of two increase shown for ^{55}Fe.

With large amounts of shielding the effective proton flux drops off and conversion of spallation track count to a spallation age requires an independent estimate of the shielding; in this regard spallation track ages are no different from those rare gas spallation ages that are determined from measurements of the accumulated amount of a spallation isotope and an assumed production rate.

6.2.6. Cosmic Ray Induced Fission

If fission tracks in extraterrestrial materials are to be used either to date the samples from the accumulation of ^{238}U spontaneous fission tracks (Fleischer and

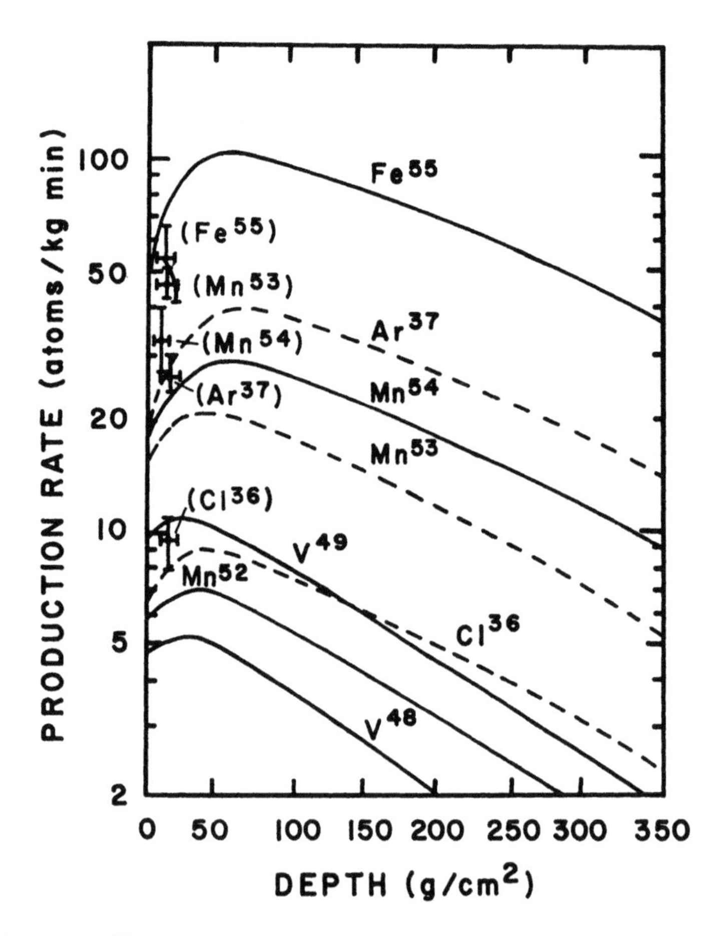

Fig. 6-13. *The calculated production rates of various radionuclides from galactic cosmic ray particles as a function of depth in the moon assuming the chemical composition of rock 12002. (After Reedy and Arnold, 1972.) The points for the radionuclides whose symbols are enclosed in parentheses are the experimental results of Finkel et al. (1971) and D'Amico et al. (1971) for this rock.*

Hart, 1972a) or to establish a fission "excess" due to the decay of an extinct isotope such as ^{244}Pu, then it is essential to estimate the amount of induced fission due to cosmic ray bombardment. The induced fission can either be produced directly by collisions of high-energy cosmic rays on heavy nuclei or indirectly, by the secondary neutrons generated in cosmic ray collisions with the nuclei of the medium. In the latter case the thermal neutron induced fission of ^{235}U is the dominant source of fission events. Because it is generally more important, we first discuss the problem of secondary neutron induced fission.

In our previous treatment of this problem (Fleischer et al., 1967a) we used theoretical estimates of the production of thermal neutrons as a function of depth.

Recently, however, the neutron flux at different depths below the lunar surface was directly measured by Woolum and Burnett (1974) on the Apollo 17 mission. Their Lunar Neutron Probe Experiment consisted of a long rod containing both ^{235}U and ^{10}B targets with mica and plastic detectors. As can be seen in Fig. 6-14, the rate of induced fission due to neutrons first rises to a maximum at a depth of $\sim$130 g/cm^2 and then decays away.

The data shown in Fig. 6-14 can be used directly to estimate the fission contribution for a sample exposed below the lunar surface for a given time. Because the concentration of ^{235}U was greater at an earlier time, the production rate is not constant at a given depth. Specifically, the total fission contribution per gram of ^{235}U for a sample exposed for a time ΔT_{exp} starting at time T_{exp} in the past is given by the following expression:

$$\frac{K}{\lambda_D} \exp(\lambda_D T_{exp})[1 - \exp(-\lambda_D \Delta T_{exp})] \tag{6-11}$$

where K is the experimentally measured present-day rate of neutron-induced fission per gram of ^{235}U.

A convenient way to estimate the potential importance of induced fission is to ask at what exposure time interval, ΔT_{exp}, the induced fission contribution will equal the total accumulated spontaneous fission contribution due to ^{238}U since the object was formed. Many lunar samples were formed $\sim 4.0 \times 10^9$ yrs ago. Assum-

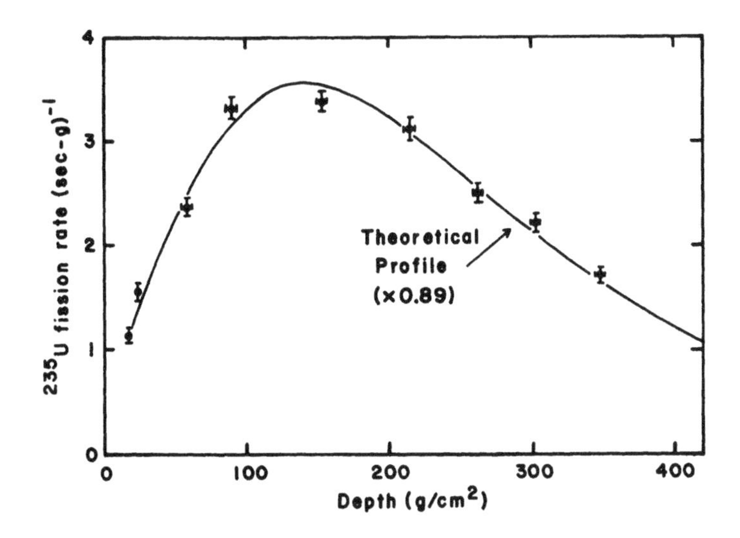

Fig. 6-14. *Measured variation in the rate of neutron-induced uranium fission as a function of depth below the lunar surface. This curve is from Woolum and Burnett (1974), who were responsible for the neutron probe experiment flown on the Apollo 17 mission.*

ing this total age, a sample residing at the position of maximum neutron flux would have equal induced and spontaneous fission contributions at an exposure interval of 8.8×10^8 yrs—provided the irradiation started at that time in the past and continued to the present.

However, if the exposure started 4.0×10^9 yrs ago and then stopped after a given time, the critical exposure interval is reduced to 2.9×10^7 yrs.

The data shown in Fig. 6-14 are taken for an infinite plane surface of fixed chemistry. Because neutrons will escape from objects of finite size, the rates can be used to calculate *maximum* induced fission contributions in smaller objects. The precise solution of the neutron flux problem for a finite object is complicated and will not be discussed here. The rates given in Fig. 6-14 should be approximately valid for stony meteorites; for iron meteorites the true rates would be lower because of the efficient capture of epithermal neutrons by the iron.

Direct collisions of high energy nucleons with heavy elements such as Pb, Bi, U, and Th will fission the elements with a reasonably high efficiency. However, fissions induced by particles with energies $\gtrsim 100$ MeV give characteristic, readily recognizable V-shaped tracks (Fleischer et al., 1967a). The characteristic shape stems from the fact that the fissioning nucleus is in effect an excited spallation recoil that fissions in flight, giving each half of the fission event a forward component of velocity in the laboratory system. The absence of V-tracks in calibrated tektite samples was used by Fleischer et al. (1965a) to set an upper limit of 300 yrs for the space exposure of tektites. In practice, high energy fission events have not yet been seen in extraterrestrial samples.

6.3. CONSTANCY OF HEAVY COSMIC RAYS IN TIME

Several studies (Arnold et al., 1961; Lal, 1972a) of radioisotopes produced by spallation have shown that the flux of galactic protons has been constant within at least a factor of two for the last few million years. In this section—following the discussion of Yuhas and Walker (1973)—we treat the independent question whether the flux of galactic VH nuclei ($20 \lesssim Z \lesssim 28$) has also been constant.

The depth dependence of fossil VH tracks in the meteorite St. Séverin (Cantelaube et al., 1967; Lal et al., 1969a; Maurette et al., 1969) was previously shown to be consistent with a long-term energy spectrum of the form $\frac{dN}{dE} = C(1 + E)^{-\gamma}$, $E \gtrsim 0.4$ GeV/nuc, and $\frac{dN}{dE} = C$, $E \lesssim 0.4$ GeV/nuc. The best fits to the constants C and γ were close to the best values then available for the galactic VH flux averaged over a solar cycle. However, in those analyses, the unknown amount of ablated material was treated as an adjustable parameter. The average distance of the meteorite from the sun was also larger than the distance of the earth from the sun where the contemporary cosmic ray measurements were made. Additional work by the Indian group on the meteorites Keyes and Patwar (Lal et al., 1969a; Tamhane, 1972) confirms the constancy of the VH flux to within a factor of two at meteoritic orbits.

Fig. 6-10 shows the empirical energy spectrum derived from rock 68815, which has an independently measured exposure age of 2.0 ± .2 my. We now inquire as to how this energy spectrum compares with the best estimate of the time-averaged spectrum determined by measurements of the contemporary flux.

To estimate the contemporary VH spectrum, the effect of solar modulation must be taken into account. This was done as follows:

A) The data of von Rosenvinge et al. (1969) were used to give an average value of the He to VH ratio of 200.

B) Measurements of α-particle fluxes taken at different times in the solar cycle were used to construct correlation curves between α-particle fluxes and neutron monitor data.

C) Neutron monitor data, which are continuous for solar cycle 19, were then used to construct the time-averaged galactic VH spectrum. As previously discussed, the low energy part of the spectrum was taken from the Surveyor III data. The resulting composite spectrum for solar cycle 19 is shown in Fig. 6-15.

The shape of the curve so obtained is indistinguishable from the empirical curve derived from the track counting data (Fig. 6-10). It is, therefore, apparent that the average shape of the heavy particle energy spectrum has not changed appreciably in the last 2.0 my.

In Fig. 6-16 we show a fit of our inferred track production rates to the experimental data in another rock, 67915, with a much longer exposure age of 50 my. In this case the energy spectrum derived from contemporary measurements does not give good agreement with the data. However, if an erosion rate of 1 mm/my is assumed, essential agreement is restored. There is thus no evidence that any changes in spectral shape have occurred in the last 50 my.

We turn now to the much more difficult question of the absolute value of the long-term flux. The contemporary spectrum reproduces the absolute value of the track densities in rock 68815, provided that $\overline{\Delta R}$ is taken equal to 9.6 ± 0.5 μm. This is close both to the experimental value of the maximum track length of 9 μm experimentally measured in 68815 and the value of 10 μm assumed in our original calculations (Fleischer et al., 1967a).

It might thus appear that it has been proven that the absolute value of the VH flux has also remained constant in time. However, recent information on the nature of track etching in crystals (Burnett et al., 1972; Plieninger et al., 1972, 1973; Price et al., 1973a; Storzer et al., 1973a,b) weakens this conclusion. It is known that prolonged etching increases the apparent length of tracks; thus the concept of a fixed $\overline{\Delta R}$ is not totally valid.

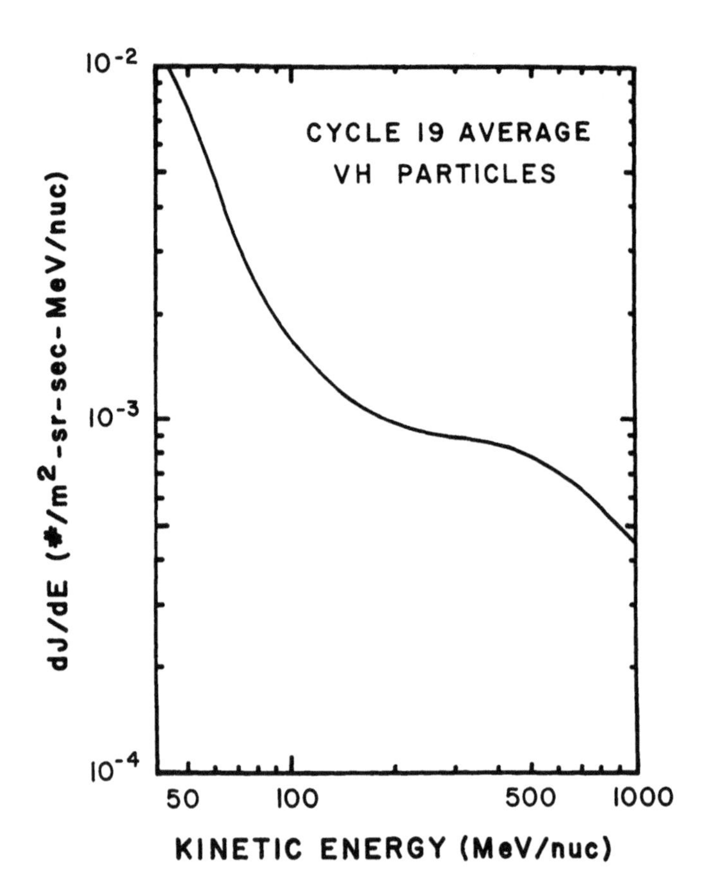

Fig. 6-15. *Solar cycle averaged VH differential energy spectrum for solar cycle 19 obtained as described in text. The low energy part (< 100 MeV/nuc) is taken from the compromise Surveyor III spectrum and does not represent solar cycle 19 values.* (*After Yuhas and Walker, 1973.*)

Strictly speaking, $\overline{\Delta R}$ is the range over which a slowing down particle will give a track long enough to be counted when etched under a standard set of conditions. As shown, for example, by Storzer et al. (1973a,b), prolonged etching of feldspars can give tracks ≥ 2 μm in length over the last 30 μm of a slowing iron particle. The observed length of any one track is, however, $\lesssim 15$ μm. Because the appropriate ion beams are just now becoming available, the complete set of appropriate calibration experiments has not yet been performed on the feldspars of the calibration rock 68815. A complete calibration should also include annealing experiments with extrapolation to lunar thermal conditions.

Although more work remains to be done, the data of Storzer et al. (1973a,b) for etching times used in the measurements of 68815 suggest that it is unlikely that the long-term absolute flux values will be appreciably lowered by the more complete analysis. Although the total flux of VH ions appears not to have changed

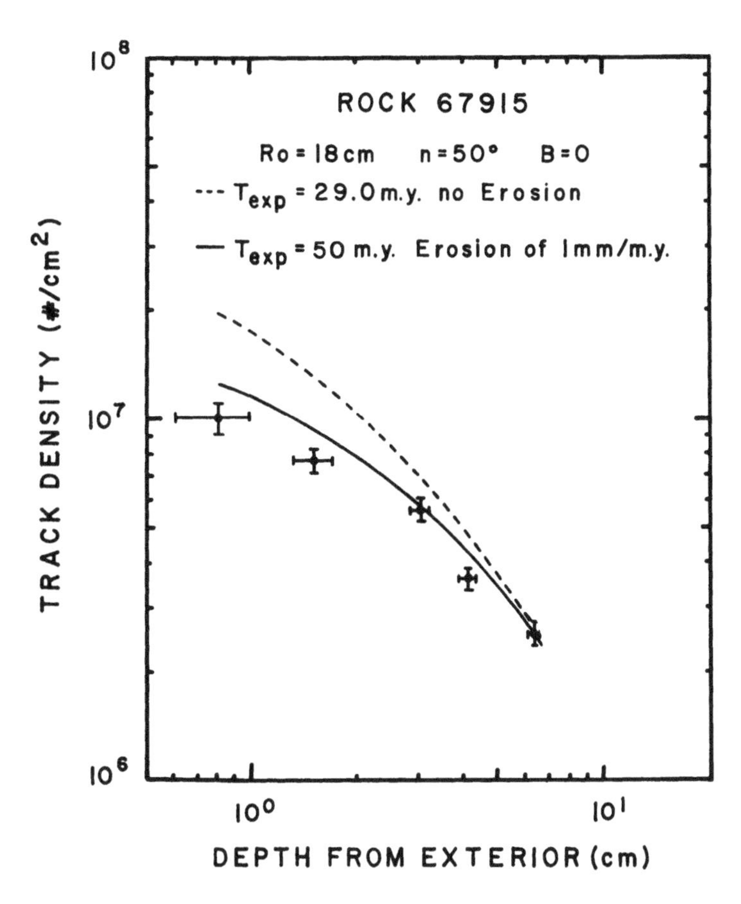

Fig. 6-16. *Track density vs. depth for rock 67915. (After Behrmann et al., 1973c.) This rock, in common with others removed from the same location, has a well-defined Kr-Kr age of 50 my apparently dating the age of formation of North Ray Crater on the Apollo 16 landing site. The depth dependence matches the predicted values from Fig. 6-10 fairly well, provided an erosion rate of* $\sim 10^{-7}$ *cm/y is assumed.*

appreciably, in the next section we will discuss evidence by Plieninger et al. (1973) indicating that changes in the average chemical composition may have occurred.

6.4. THE ELEMENTAL ABUNDANCES OF HEAVY PARTICLES IN SPACE

One of the outstanding early achievements of the study of fossil tracks in extraterrestrial materials was the measurement of the abundance of elements with $Z > 30$ in the cosmic radiation. Although emphasis quickly shifted to detailed measurements of these particles in the contemporary radiation, the future may well see a renewed emphasis on detailed fossil track measurements of the abundances of extremely heavy elements. In what follows below, we first discuss the measurements of the $Z > 30$ group and then turn to the composition cf particles within the VH group.

The initial analysis of chemical abundances was based on the idea of a fixed ionization threshold for track production. Since an element much heavier than iron would exceed this threshold over a longer portion of its path length, it would manifest itself as a very long track. Using short etching times that gave average VH track lengths of ~ 10 μm, Maurette (1965, 1966) and Fleischer et al. (1967b) noted that there were occasional tracks >50 μm in length in the meteorites Estherville and Eagle Station. Attributing all tracks ≥ 20 μm in length to elements with charges >30, Maurette (1966) and Fleischer et al. (1967b) found values of 1.6×10^{-4} and 3.7×10^{-4} for the ratio of the sum of all elements with charges >30 relative to the VH group (hereafter designated as the VVH/VH ratio). Although we now know that extended etching can give iron tracks >20 μm in length, these original estimates of the VVH/VH ratio are valid because of the short etching times used in the original experiments.

Subsequently Price et al. (1968b) gave an improved estimate of 4.4×10^{-4} for the VVH/VH ratio based on measurements in the Patwar meteorite where the shielding during irradiation was demonstrably less than in the first two meteorites studied. The authors also attempted to give a detailed abundance spectrum for atomic numbers >30 using improved track registration data obtained by Price et al. (1968a). However, in distinction to the early measurements of the contemporary flux, there appeared to be very few particles with $Z > 38$.

This apparent discrepancy was removed by the Washington University group (Maurette et al., 1969), who used a "coincidence" technique to measure the relative contributions of very long tracks in the meteorite Johnstown. In their work two large crystals removed from different shielding depths were sequentially etched and polished to follow long tracks through the bulk of the crystal. One track longer than 1 mm in length was found as well as two other tracks that were attributed to elements with $Z \gtrsim 70$. From these data the authors deduced a value of $\sim 6 \times 10^{-6}$ for the abundance ratio of $Z \geq 70$ ions relative to the VH group. This is close to the best value for contemporary cosmic rays and was the first indication that the flux of extremely heavy cosmic rays has not changed drastically with time.

In parallel work, the Indian group (Lal et al., 1969b; Lal, 1969) found a value of 3×10^{-4} for the VVH/VH ratio in Patwar using the TINT method (see Fig. 6-17). Although tracks up to 400 μm in length were also observed by etching large crystals, the longer tracks seen by Maurette et al. (1969) were not found.

Subsequently, Price et al. (1971b) reported measurements of long tracks in large pigeonite crystals from lunar rock 12021. Numerous tracks up to 500 μm

Fig. 6-17. *Extremely heavy cosmic ray tracks in the meteorite Patwar. Short tracks from the dominant VH group ions can be seen as "whiskers" on the very much longer tracks produced by the much less abundant extremely heavy ions. The tracks have been decorated with silver to enhance contrast.* (*Photograph courtesy of D. Lal.*)

were found and one track >1 mm in length was seen. Although there is some ambiguity regarding the precise exposure conditions of this rock, reasonable assumptions give abundance values for different charge groups that bracket the values for contemporary cosmic rays.

Recent work by the Indian group on particles with $Z > 30$ has focussed on the measurement of VVH/VH ratio at very low energies (Lal, 1972a; Bhandari et al., 1973d,e,f). Using olivine crystals, these authors find consistent values of $[Z > 30]/\text{Fe} \approx 1.2 \times 10^{-3}$ for particles with energies greater than 20 MeV/nuc in both lunar rocks and meteorites. To investigate lower energy values, characteristic of solar flares, they removed individual crystals from the lunar dust. Choosing crystals with a large gradient, indicative of a surface irradiation, they found that the VVH/Fe ratios increased by as much as an order of magnitude as the energy decreased to 5 MeV/nuc. They attribute this increase to the heavy element enhancement first found in contemporary solar flare studies by the Berkeley group (Price et al., 1971a; Chapter 5).

The long-term variation in the galactic VVH/Fe ratio was also investigated by the Indian group using crystals from different layers of the Apollo 15 drill stem.

Crystals with low track densities and no visible gradients were used to select crystals with tracks of a predominantly galactic origin. Values ranging from 0.9×10^{-3} to 2.0×10^{-3} were found for different layers indicating the possibility that small but measurable secular variations have occurred in the galactic VVH/VH ratio.

Measurements of the VVH/VH ratio have also been made in the meteorite Ilimaes by Kashkarov et al. (1971). In agreement with work in the other meteorites discussed, they found a value of 2×10^{-4}.

The study of VVH tracks deserves to be pursued further in both meteoritic and lunar samples. As shown by Fleischer et al. (1967a), the large nuclear interaction cross sections of the heavier elements make them most sensitive to changes in shielding, either in their propagation through the interstellar medium or on the local scale. Maurette et al. (1969) give explicit calculations of the changes in abundance ratios with shielding. The initial investigations of fossil cosmic ray tracks have shown no striking, order of magnitude, changes with time. Variations, if indeed they exist, will be revealed only by detailed, quantitative work.

As in all the work described above, one continuing problem will be the proper calibration of the track registration properties of different crystals. Progress in the production of energetic heavy particle beams can be expected to contribute greatly to our ability to read the fossil track record.

We now turn attention to measurements of the chemical abundances of elements within the VH group. Using their newly developed TINT method, the Indian group (Lal et al., 1969b; Lal, 1969) showed that the distribution of total track lengths for the abundant VH particles consisted of a series of separate peaks separated from one another by about 3 μm. By attributing the most abundant peak to iron, they set up a charge vs. track length calibration scale for different minerals. In pyroxenes, for example, tracks which fell in the observed peaks at 2.5 μm, 5.5 μm, 8.5 μm, 11.5 μm, 15 μm, and 18 μm were attributed, respectively, to atomic numbers 23 through 28. In a later paper Bhandari et al. (1972c) give track lengths of 16 μm and 9 μm, respectively, for the registration of iron ions in feldspar and olivine.

Acceptance of the charge scale as given above would lead to some unusual consequences. For example, as we shall discuss in more detail later, if the charge scale is used to separate cosmic ray tracks from fission tracks, Bhandari et al. (1971a,c,d) conclude that there is a large component of "superheavy element" fission in many extraterrestrial samples. The measurements reported by Lal (1969) also lead to the conclusion that there is almost as much Mn as Fe in the cosmic radiation. This is in contrast to what most observers have found in measurements of the contemporary radiation.

The charge scale adopted by the Indian group has been criticized directly by workers who have made measurements using calibration beams of heavy ions that have recently become available (Plieninger and Krätschmer, 1972; Plieninger et al., 1972; Plieninger et al., 1973; Price et al., 1973a; Storzer et al., 1973a,b). All these workers, as well as Burnett et al. (1972), have shown that prolonged etching can lead to much longer tracks. It has been further shown that a given etching may produce tracks of a certain length at different parts along the trajectory of a slowing particle. Thermal and shock effects may also affect length distributions. All these factors make it difficult to construct a reliable charge vs. track length calibration.

The most promising approach to this problem appears to be that taken by Plieninger et al. (1973). These authors bombard a test crystal from one side with a parallel beam of iron ions of 9.6 MeV/nuc. These calibration tracks are developed, as they slow down and reach the region where they can be etched, in the same way as the fossil tracks. Using the TINT method, Plieninger et al. (1973) found a broad peak for the stopping calibration iron ions at $\sim$18 μm in lunar pyroxenes from rock 10049. The fossil background showed a similar peak at 18 μm as well as other subsidiary peaks at $\sim$11 μm and 6 μm which these authors attribute to Cr and Ca respectively. The results are shown in Fig. 3-23.

Using their data, Plieninger et al. (1973) deduce a ratio of 0.7 for the ratio of (V + Cr + Mn)/Fe in rock 10049. This value is much higher than the value of 0.4 measured in contemporary cosmic rays (and in a crystal removed from the lunar fines) and the authors conclude that the chemical composition of VH cosmic rays has varied in the past. However, other explanations for the observed results are possible (more shielding than assumed, partial annealing of tracks, etc.) and their conclusion cannot be fully accepted at this time. The observations are nonetheless intriguing and additional work clearly needs to be done on this problem.

6.5. STUDY OF NUCLEOSYNTHESIS AND THE EARLY HISTORY OF THE SOLAR SYSTEM BY EXTINCT ISOTOPES

6.5.1. Extinct Isotopes in Meteorites

The presence of naturally radioactive elements such as uranium and thorium shows that element building must have occurred at a finite time in the past. When and how elements were formed are the central themes of the field of nucleosynthesis. (The reader is referred to the book by Clayton [1968] for a thorough treatment of this subject.) The combination of astrophysics with nuclear physics has led to the view that the elements were not formed in a single process but rather in a series of different processes occurring at different places and different times. It is surprising, but nonetheless true, that the elements that go to make up this page (and the reader himself) have been processed through stellar explosions that occurred in the far reaches of the Galaxy prior to the material being incorporated, and isolated, in the present solar system. Furthermore, we know that outside the solar system element building is still going on today in our Galaxy.

A brief discussion of nucleosynthesis was given in Chapter 5. Here we wish to treat only one aspect of this fascinating area—measurements of the decay products of the so-called extinct isotopes and, particularly, the contribution of track studies to the measurement of extinct ^{244}Pu. It will be seen that this area, which is still largely unexploited experimentally, has already resulted in important conclusions, both about the processes of nucleosynthesis and about the history of events in the early solar system.

Whatever nuclear processes formed the present abundances of isotopes in the solar system must also have produced a host of other radioactive isotopes. Because of their short half-lives, these have long since decayed to virtually undetectable levels. (Evidence for the positive identification of ^{244}Pu in terrestrial samples has recently been given by Hoffman et al. [1971]. However, the amounts are very small and we continue to classify ^{244}Pu as an extinct element. Attempts to find ^{244}Pu fission tracks in terrestrial samples have so far proven unsuccessful [Fleischer and Naeser, 1972; also see Chapter 7].) Harrison Brown (1947) suggested that some of these isotopes may have been present in sufficient concentrations at the time material became segregated in the early solar system to produce local isotopic anomalies. Specifically he suggested that local variations in the abundance of ^{129}Xe might be observed due to the decay of ^{129}I ($T_{1/2} = 17 \times 10^6$ yrs).

This prediction was beautifully verified by Reynolds (1960), who first observed a large ^{129}Xe anomaly in the xenon isotope pattern in the Richardton meteorite. The subsequent story of extinct isotopes is largely linked to measurements of the distribution of isotopes of xenon—an area of study that has come to be known as "Xenology" (Reynolds, 1963). Although originally there was some question about the association of the ^{129}Xe anomaly with the decay of ^{129}I, this was laid to rest when the ^{129}Xe was shown to be directly correlated with the presence of ^{127}I. The key experiment performed by Jeffery and Reynolds (1961) consisted of irradiating a meteorite with thermal neutrons to convert a fraction of any ^{127}I present to ^{128}Xe. A series of stepwise heating experiments then established a nearly perfect correlation between the high temperature release of excess ^{129}Xe and neutron-produced ^{128}Xe, demonstrating that the ^{129}Xe was clearly correlated with the element iodine.

The potential role of ^{244}Pu in producing isotopic anomalies in xenon was first discussed by Kuroda (1960). In this case the xenon is produced by the spontaneous fission of the isotope. Because of its relatively long half life of $\sim 8 \times 10^7$ yrs, the presence of ^{129}I ($T_{1/2} = 17 \times 10^6$ yrs) anomalies virtually demanded concomitant contributions from ^{244}Pu. Although anomalies in the heavy xenon isotopes, suggestive of a fission component, had been observed in several meteorites, the case for fission xenon did not become persuasive until the discovery of a very large anomaly in the meteorite Pasamonte by Rowe and Kuroda (1965).

In that same year Fleischer et al. (1965b) pointed out that the spontaneous fission of an extinct isotope should give rise to an excess of fission tracks in meteoritic crystals. Assuming that ^{244}Pu was the only spontaneously fissioning isotope of interest, they showed that the ratio of ^{244}Pu spontaneous fission tracks to those from ^{238}U is given by the following expression:

$$\frac{\rho_{\mathrm{Pu}}}{\rho_{\mathrm{U}}} = \frac{\lambda_{F\mathrm{Pu}}\lambda_{D\mathrm{U}}}{\lambda_{D\mathrm{Pu}}\lambda_{F\mathrm{U}}}\left(\frac{\mathrm{Pu}}{\mathrm{U}}\right)_0 \frac{[1 - \exp(-\lambda_{D\mathrm{Pu}}\Delta T_1)]}{[1 - \exp(-\lambda_{D\mathrm{U}}\Delta T_1)]} \exp[-(\lambda_{D\mathrm{Pu}} - \lambda_{D\mathrm{U}})\,\Delta T_0] \tag{6-12}$$

In this equation the λ_D and λ_F refer, respectively, to the total decay constants and to the decay constants for spontaneous fission. The time interval ΔT_0 represents the time from the initial collapse of the solar nebula (when newly synthesized materials stopped being added to the solar system) to the formation of cooled down crystals capable of retaining fission tracks. The time interval ΔT_1 is the total time from the end of the interval ΔT_0 to the present. The ratio $(\mathrm{Pu/U})_0$ is the ratio of ^{244}Pu to ^{238}U at the time solar system material became isolated from galactic inputs (beginning of ΔT_0). The process is illustrated schematically in Fig. 6-18.

In writing eq. (6-12) it is assumed that no geochemical segregation of Pu relative to U occurred in the process of crystal formation. As we will show shortly, this is probably not a good assumption.

Fleischer et al. (1965b) also gave experimental evidence to show that fission track excesses ranging from ~50 to ~200 were present in the meteorites Toluca and Moore County. Various sources of tracks were discussed and it was concluded that the spontaneous fission of extinct ^{244}Pu was the most likely source of the tracks.

The direct linkage between excess fission tracks and excess fission xenon gas was shown in an elegant experiment by Wasserburg et al. (1969a,b), who measured a xenon fission component in separated whitlockite crystals from the meteorite St. Séverin. Cantelaube et al. (1967) had previously showed that the whitlockite phase showed a considerable fission track excess. The correlation between the fission track excess and the fission xenon excess was not completely quantitative in the sense that some 25 times more fission gas was observed than would have been expected on the basis of the fission track excess. This difference was attributed to the difference in the retention temperatures of tracks and xenon gas.

Prior to the demonstration of the correlation between excess fission tracks and fission xenon in St. Séverin, the isotopic composition of the fission xenon component had been determined very closely in a milestone paper by Hohenberg et al. (1967a). Using temperature release patterns, these authors were able to separate and isotopically characterize trapped xenon, spallation produced xenon, and fission xenon in the meteorite Pasamonte. The fission isotope pattern found in

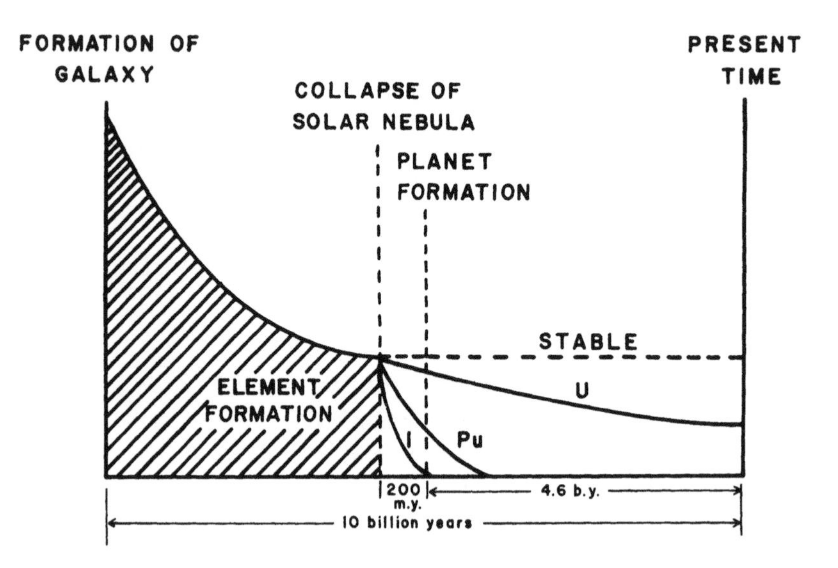

Fig. 6-18. *Schematic showing (on the left) the* rate *of element production, which declines with the age of the Galaxy, and (on the right) the* concentrations *of elements within the solar system. Isotopes such as* ^{129}I *($T_{1/2}$ = 17 my) and* ^{244}Pu *($T_{1/2}$ = 80 my) were present at the beginning of the collapse of the solar nebula. At this time newly formed elements stopped being added to the solar system from the Galaxy. Sufficient amounts of* ^{129}I *and* ^{244}Pu *remained after the formation of solid materials to produce characteristic isotope anomalies and fission track excesses in meteoritic and lunar materials.*

St. Séverin by Wasserburg et al. (1969a,b) was identical to the fission spectrum in Pasamonte and different from that produced by fission of any known nuclide.

Although by 1969 the fission nature of the excess xenon in certain meteorites had clearly been established and the heavy isotope pattern determined, there was no proof that the fission effects were, in fact, due to ^{244}Pu, as most workers assumed. Final proof of the origin of the fission xenon came in 1971 when Alexander et al. reported direct measurements of the spontaneous fission xenon isotopes produced in a sample of man-made ^{244}Pu. Table 6-2, taken from Alexander et al. (1971), shows that the fission spectrum measured in meteorites agreed precisely with the xenon yields from ^{244}Pu. With this work the prior existence of ^{244}Pu must be considered to be established beyond reasonable doubt. The agreement must also be considered a great triumph for the process of scientific deduction.

We now turn to the implications of the extinct isotope measurements for nucleosynthesis and the early chronology of the solar system.

The very existence of ^{129}I and ^{244}Pu in appreciable quantities at the beginning of the solar system establishes several qualitative facts. First, freshly produced elements must have been added to the solar nebula just prior to collapse. Thus, not all the elements were formed at the time of formation of the Galaxy; and some

Table 6-2. Comparison of Meteoritic Fission Yields with ^{244}Pu Fission Yields

Meteorite	Reference	Fission Yield			
		^{131}Xe	^{132}Xe	^{134}Xe	^{136}Xe
Pasamonte	Marti et al. (1966) from data of Rowe and Bogard (1966)	33 ± 3	93 ± 8	91 ± 2.5	≡ 100
Pasamonte	Hohenberg et al.(1967a)	25 ± 3	88.5 ± 3	94 ± 5	≡ 100
Whitlockite from St. Séverin	Wasserburg et al. (1969a,b)	31 ± 8	97 ± 8	93 ± 1	≡ 100
Kapoeta	Rowe (1970)	26 ± 3	88 ± 4	91 ± 5	≡ 100
^{244}Pu	Alexander et al.(1971)	25.1 ± 2.2	87.6 ± 3.1	92.1 ± 2.7	≡ 100

general, more or less continuous, element formation process must be present in the Galaxy. Second, the fact that the extinct isotope traces are found in meteorites means that the formation of solid objects, some of them highly differentiated, as for example the iron meteorites, must have occurred in a relatively short time following the collapse of the solar nebula.

A complete treatment of the constraints placed on nucleosynthesis, taking into account other information such as the ratio of ^{235}U to ^{238}U and of ^{238}U to ^{232}Th is beyond the scope of this book. For more detailed treatments, the reader is referred to papers by Hohenberg (1969), Wasserburg et al. (1969c), and Schramm and Wasserburg (1970).

One of the key questions that is still not settled is the ratio of Pu to U at the time of collapse, i.e., the quantity $(Pu/U)_0$ in eq. (6-12). If the Pu and U were formed continuously at a constant rate in the Galaxy prior to the collapse of the solar nebula, and if the mixing time for this continuously formed material were small compared to the half life of ^{244}Pu, then the ratio $(Pu/U)_0$ would be given by $\frac{P_{Pu}}{P_U}\frac{\lambda_{DU}}{\lambda_{DPu}}$ where P_{Pu} and P_U refer to the relative production rates in r-process nucleosynthesis events. Substitution of the parameters gives an equilibrium value of $(Pu/U)_0$ of $\sim$.017.

A limiting maximum value for $(Pu/U)_0$ is to assume that both are formed at the same time prior to collapse, i.e., $(Pu/U)_0 = \frac{P_{Pu}}{P_U}$. This (unrealistic) limit gives $(Pu/U)_0 = .65$.

Values of $(Pu/U)_0 < 0.017$ would indicate sporadic galactic production or incomplete mixing, while values >0.017 would indicate a "salting" of the solar nebula with recently formed elements just prior to its collapse.

From eq. (6-12) it can be seen that limiting values of $(Pu/U)_0$ can be measured by setting $\Delta T_0 = 0$. This sets the maximum track excess ρ_{Pu}/ρ_U (or fission xenon excess) that can be observed with a given value of $(Pu/U)_0$. With $(Pu/U)_0 = 0.017$ the maximum value of ρ_{Pu}/ρ_U is 92.

Results of the early track investigations in the meteorite El Taco indicated a rather higher value of ρ_{Pu}/ρ_U (Price et al., 1968c), suggesting a large, "sudden-synthesis" component. However, Shirck et al. (1969), in a more detailed analysis of the same meteorite, found a lower value consistent with a continuous synthesis model and a value of $\Delta T_0 \simeq 5 \times 10^7$ yrs.

The excess fission xenon observed in the separated crystals of whitlockite from St. Séverin gave a value for Pu/U at the time of xenon retention of $\sim$0.03, again suggestive of a large sudden-synthesis component. This high value has recently been confirmed by Lewis (1975). However, the measurement of Pu/U in a specific mineral phase (as would always be done in fission track measurements) suffers the possible complication that geochemical enrichment of Pu relative to U in the process of crystal formation could give misleading values of Pu/U.

That fractionation is a likely explanation of the high value measured by Wasserburg et al. (1969a,b) is suggested (but not yet proved) by two further observations on St. Séverin. Using a neutron irradiation method, Podosek (1970b) found a "whole rock" value of 0.0127 for the Pu/U value at the time of xenon retention. This number was later revised to 0.015 by Podosek (1972). Subsequently, Crozaz (1974) showed that the Th/U ratio in whitlockite crystals in St. Séverin was enriched over the whole rock value by a factor of ~3. Since Pu can follow Th in certain valence states, this observation suggests a similar enrichment of Pu in the whitlockite phase.

However, the complete story of extinct isotopes in meteorites is far from finished. For example, Podosek and Lewis (1972) measured an extraordinarily high ratio of 0.09 for white inclusions from the carbonaceous chondrite Allende. (The ^{129}I excess was, however, normal.) Also remarkable, and still puzzling, is the fact that similar ^{129}I excesses are observed in meteorites with quite different physical and chemical characteristics. As first noted by Hohenberg et al. (1967b) and studied in more detail by Podosek (1970a), various meteorites apparently started to retain xenon within a time span of several millions of years.

In Table 6-3 we give a summary of data for meteorites in which the ^{244}Pu excesses are well documented. Fleischer et al. (1967b), reporting the results of a preliminary survey of a number of meteorites, also listed the following meteorites as ones in which the fossil track background is predominantly due to fission: Linwood (diopside), Four Corners (diopside and enstatite), Odessa (diopside), and Moore County. However, additional work needs to be done to confirm these

Table 6-3. Fission Track Measurements in Meteorites

Meteorite	ρ_{Tot} (cm^{-2})	ρ_U *	ρ_{Pu}/ρ_U
St. Séverin [a] (whitlockite)	7.4×10^6	1.06×10^6	6 ± 1.2
Toluca [b] (diopside)	4.7×10^6	8×10^4	59 ± 10
" (albite)	7×10^4	1×10^4	6 ± 1.6
" (enstatite)	5×10^3	1.3×10^3	3 ± 1.2
El Taco [c] (diopside)	1.9×10^6	3.1×10^4	61 ± 14
Angra Dos Reis [d]	1.2×10^7	6.2×10^5	19

(a) Cantelaube et al. (1967); (b) Fleischer et al. (1968); (c) Shirck et al. (1969); (d) Bhandari et al. (1971d).

*This represents the calculated total fission contribution from the measured uranium concentration. ΔT_1 is assumed to be 4.5×10^9 yrs.

preliminary interpretations. The tracks in Kodaikanal were also listed by Fleischer et al. (1967b) as being "probably fission." However, subsequent work by Shirck et al. (1969) showed that the tracks were instead produced by heavy cosmic rays.

The fission track age of Serra de Magé has been measured by Carver and Anders (1970) to be 540 $\pm$ 90 my. Presumably this low age is caused by a shock-annealing event in the history of the meteorite.

All told, the amount of fission track data in meteorites is quite limited. This undoubtedly reflects the fact that most track groups have been primarily concerned with lunar samples in recent years. Much more work deserves to be done in meteorites in the future.

6.5.2. Extinct Isotopes in Lunar Samples

Extinct isotope effects in lunar samples have shown up much more prominently in rare gas measurements than in track studies. The first observation of a ^{244}Pu xenon fission component in lunar rocks was reported for Apollo 14 breccias at the Third Lunar Science Conference (Crozaz et al., 1972b). Subsequent work by Drozd et al. (1972), Behrmann et al. (1973b), Marti et al. (1973), and Reynolds et al. (1973) has confirmed and extended the original observations.

Three of the Apollo 14 breccias, 14301, 14313, and 14318, have large ^{244}Pu fission xenon contributions similar to those found in meteorites. The ratio of excess ^{244}Pu fission to ^{238}U fission is, respectively, 30, 125, and 45 for these three rocks. Since K-Ar and Rb-Sr measurements indicate that these Apollo 14 breccias were formed into rocks no earlier than 3.95×10^9 yrs ago, this large fission component is somewhat surprising.

All the breccias that show a large fission excess are gas-rich rocks with a large component of solar gas. This gas was presumably added to the individual grains that constitute the breccia at a time when the grains formed part of an ancient lunar soil. The solar gas may have been redistributed when the breccia was formed, but it has not been lost. The temperature release pattern of the ^{244}Pu xenon follows closely the temperature release pattern of this solar component. The two components thus appear to be closely associated, and it is likely that the ^{244}Pu gas being measured was once part of an early lunar atmosphere (Behrmann et al., 1973b).

The cosmic ray track background in lunar samples is generally considerably higher than in meteoritic samples. However, because the uranium concentrations are rather high and also because the uranium is always found in certain specific mineral phases (see Chapter 8), fission dating is possible, although difficult.

The Washington University group has reported fission track ages ranging from 1 to 4×10^9 yrs in a variety of crystalline and breccia rocks (Burnett et al., 1970, 1971a; Crozaz et al., 1972c; Graf et al., 1973). Most of the measurements were made in phosphate crystals, although several values were obtained by using buried U-rich inclusions in feldspar crystals. The major uncertainties in these measurements are the relatively large corrections necessary for spallation recoil tracks. No evidence of a ^{244}Pu excess was found in any of the samples. Specifically, Graf et al. (1973) found no evidence of a large ^{244}Pu *in situ* decay component in individual phosphate crystals of the rocks 14301 and 14318 that have the large fission xenon excesses. This work supports the idea that the xenon in these rocks may have been implanted from an early atmosphere rather than having been generated within the rocks themselves by ^{244}Pu decay.

In contrast, the Berkeley group reported measurements of a positive ^{244}Pu contribution amounting to about 50% in a particularly large and favorable whitlockite crystal in rock 14321 (Hutcheon and Price, 1972). This excess is consistent with the measured age of 3.95×10^9 yrs for this rock and an initial $(Pu/U)_0$ value of 0.013. Subsequent xenon gas work by Marti et al. (1973) confirmed the existence of a small but measurable fission xenon component in this rock. In contrast to the gas-rich Apollo 14 breccias, which show the large fission xenon excesses, rock 14321 has a very small solar component and was apparently effectively degassed at the time of its formation.

The Indian group has also reported fission track data in a number of samples. The major thrust of their work has been an attempted demonstration of the existence of a separate fission component due to the decay of superheavy elements. Because of this emphasis, we treat their work separately below.

The Question of Element X

In our discussion of excess fission tracks in extraterrestrial materials, we have so far treated only the contributions due to ^{244}Pu. This is based principally on the demonstrated identity of the isotope pattern of the excess fission xenon component with the isotope pattern measured in the laboratory for the spontaneous decay of ^{244}Pu.

The question remains, however, whether some fission tracks in some samples might have been produced by the decay of a still unidentified isotope, "element X." Based on the half-lives and fission yields of the known isotopes, element X, if it existed, would almost certainly have to be an as yet undiscovered "superheavy" element. (See Chapter 7 for a fuller discussion of this point.)

Several arguments, other than the xenon isotope patterns, can be advanced to show that the bulk of the excess fission tracks must be due to ^{244}Pu and not to the decay of superheavy elements. Fleischer et al. (1968) measured the track excesses in three mineral phases of the meteorite Toluca and found a decrease in the track excess with decreasing thermal stability of track retention. They attributed the observed differences in excess tracks to the slow cooling of the meteorite parent body and used measurements of track stability in the different phases to construct a cooling curve for the meteorite parent body assuming ^{244}Pu decay. Agreement was obtained between the cooling rate thus inferred and an independent determination based on the observed Ni profiles in Ni-Fe phase boundaries. This accord was taken to show that the half-life of the species responsible for fission was approximately that of ^{244}Pu.

A more direct and convincing argument was subsequently advanced by Price and Fleischer (1969) who measured the track length distributions of excess fission tracks in three meteorites. The lengths so found were close to those for uranium fission fragments. Since theoretical estimates had indicated that fission due to superheavy elements would give much longer tracks, they concluded that the tracks were probably produced by ^{244}Pu. They also failed to find any evidence for triple or quaternary fission—a decay mode again predicted for superheavy element decay.

In contrast to the above work, the Indian group (Bhandari et al., 1971a,c,d) has claimed to find large contributions from superheavy element fission in both lunar and meteoritic samples. The approach used by these workers was to measure the total track length distributions using either the TINT method (track in track) or TINCLE method (track detected in a cleavage boundary). Using calibration neutron irradiations, they first determined that fission tracks lie in the length range from 14.3 to 15.0 μm. As previously discussed, they used measurements of certain crystals with high cosmic ray backgrounds and low fission components to construct a track length vs. charge scale in which tracks of 2.5 μm, 5.5 μm, 8.5 μm, 11.5 μm, 15 μm, and 18 μm were attributed to V, Cr, Mn, Fe, Co, and Ni cosmic ray particles, respectively.

Using these calibrations, measurements of fossil track length distributions in other samples containing both fission and cosmic ray components were divided by the Indian group into the following three length categories: 10–13 μm, 13–16 μm, and 16–22 μm. The 10–13 μm category was assumed to give a measure of the background of iron nuclei from which the cosmic ray contribution to the longer length categories could be calculated (25% ρ_{10-13} for the 13–16 μm range and 18% ρ_{10-13} for the 16–22 μm interval). The tracks in the 13–16 μm range remaining after subtraction of the cosmic ray background were attributed to ^{238}U and ^{244}Pu fission; those remaining in the 16–22 μm range were attributed to the fission of a superheavy element. Table 6-4 summarizes the data of the Indian group.

In both lunar samples and meteorites, uranium generally tends to be concentrated in interstitial regions. Bhandari et al. (1971a,c,d) make a particular point of the fact that the proportion of tracks falling in the two fission categories *increases* as the total number of tracks in a cleavage *increases*. This is, of course, what would be expected from the known concentration of fissionable materials in grain boundaries.

If the interpretations of Bhandari et al. (1971a,c,d) are accepted, then the superheavy contributions in the meteorites Moore County, Norton County, Nakhla, and Steinbach are comparable to the ^{244}Pu contributions in these same meteorites. In contrast, the superheavy contribution is only 0.5% of the ^{244}Pu contribution in the meteorite Angra dos Reis.

Table 6-4. Relative Fission Contributions to Tracks in Meteorite and Lunar Samples (from Bhandari et al., 1971c)

Sample	Track type measured	$N_{c\ell}$	Fossil track density ρ (cm^{-2})	No. of tracks measured		Relative track abundances (Fe(10-13 μm) = 100 tracks)				Relative contributions to tracks	
				Total	Due to iron (10-13 μm)	Total in 13-16 μm interval	Estimated due to (U+Pu) fission (i.e. corrected for CR tracks)	Total in 16-22 μm interval	Estimated fission tracks from super-heavy nuclei	(U+Pu)/Fe	Super-heavy/U+Pu
Moore County	TINTS	-	3×10^{6}	144	29	210	185	101	83	1.8	0.45
Norton County	TINTS	-	1.7×10^{6}	162	50	126	101	54	36	1	0.36
Norton County	TINCLES	1-10	1.7×10^{6}	65	23	83	58	40	22	0.58	0.38
Norton County	TINCLES	10-25	1.7×10^{6}	250	36	280	255	178	160	2.5	0.63
Norton County	TINCLES	25-100	1.7×10^{6}	95	11	245	220	400	382	2.2	1.7
Norton County	TINCLES	1-5	1.7×10^{5}	40	13	77	52	92	74	0.52	1.4
Norton County	TINCLES	5-100	1.7×10^{5}	70	7	157	132	730	712	1.32	5.5
Steinbach	TINCLES	1-20	1.3×10^{6}	70	21	128	103	61	43	1.0	0.42
Nakhla	TINCLES	1-10	4.25×10^{5}	143	25	184	159	220	202	1.6	1.3
Apollo 11, 12 rocks	TINTS and TINCLES	1-5	7×10^{6}	323	123	37	12	2	~0	0.12	~0
Apollo 11, 12 fines	TINCLES	1-10	$\sim7\times10^{6}$	42	22	59	34	23	5	0.34	0.15
Apollo 11, 12 fines	TINCLES	10-20	$\sim7\times10^{6}$	44	13	108	83	77	59	0.85	0.69
Apollo 11, 12 fines	TINCLES	20-100	$\sim7\times10^{6}$	154	25	136	111	270	252	1.1	2.3

Applying their analysis to lunar samples, Bhandari et al. (1971a,c,d) found no appreciable excess fission tracks in crystalline rocks, but a large contribution in individual, low track density crystals removed from both Apollo 11 and Apollo 12 fines.

As of the moment of this writing, it is not clear whether the interpretation given by the Indian group can be accepted. For one thing, the data show some puzzling features. From other evidence, Angra dos Reis appears to have formed very early in the solar system; it has a large ^{244}Pu anomaly (Hohenberg, 1970) and a very primitive initial ratio of strontium isotopes (Papanastassiou, 1970). The very low apparent component of superheavy fission is thus surprising.

The lunar soil is composed mostly, although not entirely, of material derived from the lunar rocks characteristic of the area in which the soil is collected. The striking difference observed by Bhandari et al. (1971a) between the soil crystals and crystals from lunar rocks is, therefore, also puzzling.

The key question that must be answered in the Bhandari et al. treatment is whether the cosmic ray background has been adequately accounted for at the longer track lengths. As we have already discussed in Section 6.5, it is now known that iron nuclei can give tracks even longer than the 22 μm limit set by the Indian group for superheavy element fission events.

However, the actual track lengths observed in practice depend critically on the etching treatments. Bhandari et al. (1971a,c) internally calibrated their crystals and etching treatments using fossil tracks; the only assumption they made is that the most prominent peak in the fossil cosmic ray track distribution is produced by iron.

The only apparent flaw in their analysis is the fundamental assumption that the etching behavior of tracks is independent of the time the tracks have been stored. Although it is dangerous to compare etching treatments from one laboratory to the next (trace impurities can change etching characteristics), comparison of the results of Plieninger et al. (1972) with those of Bhandari et al. (1971a,c) suggest that the fossil tracks have been systematically shortened. As pointed out by Fleischer and Hart (1973f) in an explicit critique of the Bhandari et al. interpretations, such a shortening could result from shock events which are known to shorten tracks in lunar samples (Fleischer et al., 1972b; Fleischer and Hart, 1973e). Cosmic ray tracks added after the shocks would not be shortened and would thus be mistakenly assigned to the fission category.

Partial thermal annealing of tracks over time could also account for the results. Tracks added recently would be systematically longer than older tracks. This explanation could also account for the differences between lunar rocks and lunar soil samples. Because of the vastly different thermal conductivities of rock and soil, rocks get quite hot while soil grains below the surface tend to remain cold.

The progressive increase in the proportion of tracks in the 16 μm to 22 μm range with the total number of tracks in a cleavage observed by Bhandari et al. can also be explained as an etching effect; when the cleavage is highly eroded, the transfer of the etchant is more efficient. It is also worth noting that chemical gradients near crystal boundaries are common and can lead to accelerated chemical etching and therefore to longer etched tracks.

Clearly the full explanation of the intriguing observations of the Indian group will require additional work.

As a final note, we should mention that certain carbonaceous meteorites have xenon isotope patterns suggestive of an excess fission component different than ^{244}Pu and possibly due to the spontaneous decay of a volatile superheavy parent (Anders and Larimer, 1972). Unfortunately, these meteorites are very fine grained and fission track studies have not yet been possible in these objects. It is possible that in analogy with lunar breccias the unusual fission gas could have been added as a surface component to the constituents of the meteorite before it was assembled in its final form. Recent studies of the isotopic composition of oxygen in certain carbonaceous chondrites (Clayton et al., 1973) suggest that some grains may predate the solar system and have been formed from interstellar gas of somewhat different composition than solar system material. It thus seems possible that some of the xenon may have had a similar "exotic" origin without necessarily having resulted from fission of a superheavy element. After all, as we pointed out in Chapter 5, many theorists are skeptical that the so-called island of stability can even be reached by known processes of nucleosynthesis.

6.6. LUNAR SURFACE DYNAMICS

6.6.1. Introduction

In these next few sections we will be considering measurements and inferences that bear on the dynamic nature of the lunar surface. Although the moon is vastly less geologically active than the earth (at least for the last 3×10^9 years), it is far from being a dead planet; the surface of the moon is still developing and changing with time. Although, as we shall see, these changes are, on the average, slower than on earth, they are still far more rapid than we might have wished for the study of the ancient history of solar and galactic radiations!

Because much of our discussion will center on the interrelated questions of erosion and exposure ages, we begin with introductory remarks concerning these concepts. The simplest example of an exposure age is that period of time starting when a sample is removed from a position of complete shielding to a position of direct exposure to radiation and ending when it is removed from the radiation either by being reshielded or brought back to earth. But the same sample may have many different exposure ages depending on the nature of the radiation effect being studied. The solar wind penetrates only to depths of $\sim$1000 Å, which is much less than the depth of penetration of visible light in a reasonably transparent medium. Thus solar wind exposure ages are virtually synonymous with ages during which a sample was directly exposed to sunlight on the very surface of the moon—in the colorful terminology of our colleagues: the "sun-tan age" (Bhandari et al., 1971b).

Solar flare proton effects penetrate to a depth of several centimeters and solar flare VH ions to several millimeters. For fist-sized or larger rocks, exposure to solar flares also generally implies exposure directly to space. However, measurable effects may also be produced if the sample was covered by only a shallow layer of material. Galactic VH particles produce tracks at reasonable rates for the first few tens of centimeters (Figs. 6-5 and 6-6); galactic protons and their secondary nuclear cascade products, particularly neutrons (Fig. 6-13), penetrate to several meters.

Because of these different characteristic distances, exposure ages measured in different ways give different kinds of information about the formation and evolution of the lunar surface. It has also turned out that most lunar samples have not had simple, one-stage exposures. Instead most have had rather complex histories that require the application of a variety of techniques to unravel.

We will also discuss erosion rates, and it is important to understand the nature of the erosion concept as used here. The vast, crater-strewn fields seen through a simple pair of field glasses show immediately that the moon has been subjected to *large-scale erosion* in the past. One important result of the Apollo missions has been to measure the formation ages of specific large features such as Cone Crater (Apollo 14), South Ray Crater (Apollo 16), North Ray Crater (Apollo 16), and Copernicus (Apollo 12). Smaller impacting objects destroy large boulders as well as shatter even smaller rocks, finally producing the fine-grained dust and rubble layer that forms the lunar surface. None of these processes is included in what we call erosion in this book.

Instead, erosion as used here refers to two smaller scale processes—mass-wastage erosion and microerosion. Mass-wastage erosion refers to the removal of up to centimeters of material from the surface of a rock in steps up to several millimeters at a time. It is the process that gives rise to the characteristic rounding and "weathered" appearance of lunar boulders (see Fig. 6-19). Microerosion refers to the very fine-scale removal of material ranging from atomic sputtering processes to micrometeorite impacts that remove up to tens of microns of material at a time. It is important to realize that although microerosion sets a rigid lower limit on the mass-wastage erosion, the latter may be much larger than the former. Also, the erosion of a single grain, sitting on the surface of the moon, may proceed quite differently from that of a rock as a whole.

The subject of lunar surface dynamics is difficult to treat definitively at this time. Although many hard numbers exist, the experimental data are still being gathered at a rapid rate. Equally important, the conceptual models of how to analyze and interpret the data are still being developed. It is our hope that this section of the

Fig. 6-19. *Lunar rock erosion. The lower part of rock 14310 was buried in the lunar regolith and has retained a sharp, angular appearance. The upper exposed part has been eroded and rounded by repeated micrometeorite impacts.*

book will serve two purposes: Firstly, we hope the general reader will gain some sense of the current state of a field that we have personally found very exciting. Secondly, we hope that those who are attacking the problems of lunar surface dynamics from other points of view and, in particular, those who are using the track results in the interpretation of their own data will benefit from a review of the basic principles and a critical appraisal of existing data.

6.6.2. Heavy Particle Track Determinations of the Exposure History of Lunar Rocks

The early observations of Apollo 11 crystalline rocks showed clear evidence for the existence of galactic VH ion tracks at depths $\gtrsim 1$ cm (Crozaz et al., 1970a,b; Fleischer et al., 1970b,c). Using curves similar to those shown in Figs. 6-5 and 6-6, a number of authors have calculated galactic exposure ages of rocks from this and other missions (Crozaz et al., 1970b; Fleischer et al., 1970c; Lal et al., 1970; Price and O'Sullivan, 1970; Barber et al., 1971a; Bhandari et al., 1971b; Crozaz et al., 1971; Fleischer et al., 1971b; Behrmann et al., 1972; Berdot et al., 1972a,b,c; Bhandari et al., 1972a,b,d; Crozaz et al., 1972d,e; Fleischer and Hart, 1972b; Hart et al., 1972b; Lal, 1972a; Poupeau et al., 1972b; Yuhas et al., 1972; Behrmann et al., 1973c; Fleischer et al., 1973; Yuhas and Walker, 1973; Fleischer and Hart, 1974). The proton spallation ages of the same rocks, measured principally by rare gas techniques, are generally higher. This is usually attributed to the greater penetration of the protons compared to VH nuclei, which would permit the spallation products to accumulate when the rock was buried below the surface or was part of a larger boulder. For this reason the track ages are generally referred to as "surface exposure ages" or "sun-tan ages."

Unfortunately, converting track data to exposure ages always involves certain assumptions. These assumptions and the concomitant limitations on the meaning of the "surface exposure ages" have not always been noted by other investigators. For this reason, we wish to review the situation and summarize track exposure ages determined by different methods.

Before discussing these results, we would like to point out the inherent limitation on the accuracy of track exposure ages due to difficulties in obtaining accurate track counts. Because of the subjective factors involved, different groups counting crystals from the same rock, etched in exactly the same way, do not necessarily obtain the same track densities. The fact that different investigators also use different etching recipes increases the spread of track densities measured by different groups.

The best indication of the importance of these differences is a series of measurements made by five different groups (Yuhas et al., 1972) on the "track consortium lunar rock" 14310. Table 6-5 shows that the agreement is generally within 20% although occasional larger differences occur. Rock 14310 was neither the easiest nor the hardest rock in which to measure tracks and probably represents a fairly representative test of reproducibility when optical microscopy is possible.

Table 6-5. Interlaboratory Comparison of Track Densities in Feldspars: Track Counts by Different Groups on Samples from Identical Rock Locations

Sample Label	Laboratory Number	ρ_1 Density x $10^6/cm^2$	Laboratory Number	ρ_2 Density x $10^6/cm^2$	% Difference†
AB-24	1	3.5 ± 0.2*	5	3.0 ± 0.3*	15
AB-24	1	3.9 ± 0.2	5	4.0 ± 0.5	2.5
AB-24	1	3.4 ± 0.2	5	2.8 ± 0.4	19
AB-24	1	1.9 ± 0.1	5	1.7 ± 0.3	11
AB-19	2	1.6 ± 0.2	5	2.0 ± 0.2	22
AB-19	2	1.6 ± 0.2	5	1.3 ± 0.2	22
P-End	3	2.6 ± 0.3	4	2.4 ± 0.2	8
E	1	4.1 ± 0.1	4	2.7 ± 0.1	40
P-Top	1	2.3 ± 0.1	3	2.5 ± 0.2	8.5
P-Top	1	2.3 ± 0.1	2	2.5 ± 0.1	8.5
P-Top	2	2.5 ± 0.1	3	2.5 ± 0.2	0
A-3	1	1.5 ± 0.1	2	1.6 ± 0.1	6.5

*Errors here are only statistical.

†% difference defined as $\frac{2|\rho_1-\rho_2|}{\rho_1+\rho_2}$ x (100).

Crystalline rocks are generally straightforward to measure, though the quality and appearance of tracks may vary considerably from one rock to the next. Lunar breccias are another matter. Many crystals are intensely shocked and the etching of shock-produced features makes visibility so poor that they become nearly impossible to study. More important, crystals removed from the same location may have very different track densities. In these cases some of the crystals clearly have retained a "memory" of a prior irradiation that occurred before they were incorporated into coherent rocks (Crozaz et al., 1970b; Dran et al., 1972a,b; Hart et al., 1972a,b; Hutcheon et al., 1972a,b; Macdougall et al., 1973).

Crystals removed from a given location in a crystalline rock may also show small variations in track densities. Some of this variation probably arises from the differences in etching behavior from grain to grain due to chemical inhomogeneities; some of it also arises from shock effects (Fleischer et al., 1973).

We do not wish to exaggerate the difficulties of track counting; as can be seen from Table 6-5, satisfactory agreement is obtained by independent observers. We do wish to indicate that absolute values of track counts in general are probably no better than $\pm 20\%$ and that some rocks are much easier to work with than others.

The simplest procedure for obtaining an age is to make a single point determination at a depth where only galactic cosmic ray tracks are present and where surface erosion effects are minimal. This method, results for which are listed in column A in Table 6-6, proceeds as follows:

1) A track density is measured at a given position in the rock at a distance $\gtrsim 1$ cm from the surface. At this point the effects of solar flares are negligible and surface exposure ages are thus completely determined by galactic cosmic ray effects (neglecting the influence of erosion).
2) From all that is known about the rock, a decision is made as to how the rock was probably irradiated. Two of the simplest choices are to assume that the rock has always remained in the position in which it was found (2π irradiation) or that it was tumbled randomly to produce an isotropic irradiation on the surface (4π irradiation).
3) Figs. 6-5 to 6-6 are consulted to infer the age. In the cases where the rock cannot be reasonably represented by a single sphere, the computer program can be modified for appropriate ellipsoids or some average track density can be calculated from the limiting spheres that would define the rock.

Implicit in this procedure is the assumption that all the tracks were accumulated while the rock was exposed completely on the lunar surface. Because some of the tracks might, in fact, have been accumulated while the rock was buried at shallow depths or when it was part of a larger boulder, this method gives *maximum* surface exposure ages. This fact was clearly stated in the earlier papers but has tended to be overlooked by some investigators.

If the rock can be described reasonably well as a sphere, and if samples are ob-

Table 6-6: Surface Exposure Histories of Lunar Rocks

Rock	Rock Type	Weight (grams)	Surface Exposure Ages ($\times 10^6$ yr) A	B	C	D	E	F	Spallation Age ($\times 10^6$ yr)	Exposure† History SS-single stage MS-multistage	Comments
10003[a]	Crystalline	213.	10(4π)	-	-	-	-	-	129 ± 11[aa]	MS?	
10003[j]			-	-	-	-	5	-			
10017[b]	Crystalline	973.	11(4π)	-	-	-	-	-	450-510[aa,bb]	MS?	Gradient small but possible single stage exposure to VH nuclei.
10017[c]			6(4π)	-	-	-	-	-			
10017[d,g]			10(4π)	-	-	-	4	-			
10017[a]			10(4π)	-	-	-	-	-			
10044[c]	Crystalline	247.5	-	4	-	-	-	-	70 ± 17[cc]	MS?	
10047[b]	Crystalline	138.	16(4π)	-	-	-	-	-	86 ± 4[bb]	MS?	
10049[c,f]	Crystalline	193.	30(4π)	-	-	-	-	-	25[dd]	SS	Close agreement between tracks and spallation gives low limit on erosion.
10057[b]	Crystalline	919.	-	-	28	-	-	-	52 ± 2[aa] 47 ± 2[bb]	SS?	Erosion could account for difference in track and spallation ages.
10057[j]			-	-	-	-	8.5	0			
10057[c]			29(4π)	-	-	-	-	-			
10058[b]	Crystalline	282.	13(4π)	-	-	-	-	-	60[ee]	MS?	
10058[j]			-	-	-	-	4	-			

12002[f]	Crystalline	1529.5	-	24	-	-	-	-	94 ± 6[ff]	MS	Small gradient.
12002[g]			-	-	-	-	2	35			
12013[h]	Breccia	82.3	14(4π)	-	-	-	-	-	40-47[gg] 48[hh]	MS	
12017[f]	Crystalline	53.0	-	2	-	-	-	-		?	Possible single stage exposure to VH nuclei.
12017G[f]	-	-	.01(2π)	-	-	-	-	-		SS	This is a glass splash. Age assumes galactic production only at 1 mm depth.
12018[g]	Crystalline	787.0	-	-	-	-	1.7	0	195 ± 16[ff]	MS	Small grad. from 1 to 4 cm.
12020[g]	Crystalline	312.0	-	-	-	-	2.6	25	71[ii]	MS?	
12021[f]	Crystalline	1876.6	13(4π)	-	-	-	-	-	303 ± 18[ff]	MS?	
12022[i]	Crystalline	1864.3	10(2π)	-	-	-	-	-	220[dd]	MS?	Measurable gradient. Possible single stage exposure to VH nuclei.
12022[j]			-	-	-	-	4	0			
12038[g]	Crystalline	746.0	-	-	-	-	1.3	0	150-190[dd]	MS?	Measurable gradient. Possible single stage exposure to VH nuclei.
12040[h]	Crystalline	319.0	5(2π)	-	-	-	-	-	225 ± 11[aa]	?	
12063[h]	Crystalline	2426.0	5	-	<1.5 < .7	-	-	-	95 ± 5[ff]	MS	Rock pitted on all sides. Careful measurement shows small gradient.
12063[j]			-	-	-	-	2.8	0			
12064[h]	Crystalline	1214.3	1.5(2π)	-	-	-	-	-	190-220[ii]	SS?	Lowest track density observed at 1 cm depth. Possible single stage exposure to VH nuclei.

Rock	Rock Type	Weight (grams)	Surface Exposure Ages ($\times 10^6$ yr) A	B	C	D	E	F	Spallation Age ($\times 10^6$ yr)	Exposure[†] History SS-single stage MS-multistage	Comments
12065[f]	Crystalline	2109.0	14(2π)	-	-	-	-	-	180-200[ii]	MS?	
14047[k]	Breccia	242.01	-	-	-	3.4	-	-	<45[jj]	SS?	
14055[k]	Breccia	110.99	-	-	-	0.05	-	-		?	
14066[k]	Breccia	509.80	-	-	-	0.49	-	-	24 ± 2[kk]	SS?	
14068[l]	Breccia	35.47	-	-	-	-	-	15	(25)[ll]	SS?	
14270[k]	Breccia	25.59	-	-	-	1.4	-	-		?	
14301[l]	Breccia	1360.60	-	-	-	-	-	8	102[mm]	MS	Disagreement between groups on variability of ρ and absolute value of ρ min.
14301[k]			-	-	-	.34	-	-			
14303[l]	Breccia	898.40	-	-	-	-	2.5	0	29[nn]	MS	
14305[l]	Breccia	2497.50	-	-	-	-	-	35		SS?	Possible single stage exposure to VH nuclei.
14307[o]	Breccia	155.00	-	-	5	-	-	-	(125)[ll]	MS	
14310[m]			-	15	1-3	-	-	-			Most completely studied rock. Three orthogonal rock sections studied by five groups. Gradients generally very small. For details see Yuhas et al. (1972).
14310[n]	Crystalline	3439.00	10-30	-	-	-	-	-	265[oo]	MS	
14310[l]			-	-	-	-	1	-			

14311[k]	Breccia	3204.40	–	–	–	1-3	–	–	661[oo]	MS	
14311[l]			–	–	–	–	–	12			
14321[m]			–	25	–	–	–	–			
14321[k]	Breccia	8998.0	–	–	–	8	–	–	23-27[pp]	SS?	Basic disagreement on variation of ρ in large crystals. One group finds large variations, others do not.
14321[o]			–	23	–	–	–	–			
14321[l]			–	–	–	–	2-4	–			
15058[p]			25	–	–	–	–	–			Sun-tan age 1 to 2 my using solar flare tracks. Small gradient.
15058[q]	Crystalline	2672.5	–	–	–	–	2	10		MS	
15058[v]			10(2π)	–	–	–	–	–			Grain mount analysis. Small depth variation.
15058[w]			–	7	–	–	–	–			Time since last major shock.
15085[q]	Crystalline	458.9	–	–	–	–	<1	–		?	
15118[q]	Crystalline	27.6	–	–	–	–	1.3	–		?	
15233[k]	Breccia	3.8	–	–	–	0.3-7	–	–		?	
15265[q]	Breccia	314.1	–	–	–	–	<1	–		?	
15388[q]	Crystalline	9.0	–	–	–	–	<1	–		?	
15405[s]	Breccia	513.1	–	–	–	0.5	–	–		?	

Rock	Rock Type	Weight (grams)	Surface Exposure Ages (x 10^6 yr) A	B	C	D	E	F	Spallation Age (x 10^6 yr)	Exposure† History SS-single stage MS-multistage	Comments
15426[q]	Breccia	223.6	–	–	–	–	–	15		?	Soil clod.
15426[x]			0.5	–	–	–	–	–			
15475[v]	Crystalline	406.8	–	–	15.5	–	–	–	473 ± 20[oo]	MS	Peculiar depth variation implies at least two-stage surface exposure.
15505[w]	Crystalline	1147.4	0.6	–	–	–	–	–		MS	Age since last shock.
15535[q]	Crystalline	404.4	–	–	–	–	<1	<10		?	
15555[q]			–	–	–	–	1	26			
15555[r]	Crystalline	9613.7	34	–	15	–	–	–	81^{+17}_{-7} [qq]	MS	Although a multistage history, the measured gradient suggests that exposure history is still relatively simple.
15555[p]			26	–	–	–	–	–			
15557[q]	Crystalline	2518.0	–	–	–	–	<1	–		?	
61016[y]	Crystalline	11745.	20	–	–	–	–	–			Inclusion of erosion raises age to 40 my. } Plum Crater samples
61175[y]	Breccia	543.	–	20	–	–	–	–			} Plum Crater samples
62235[v]	Crystalline	320.	4	–	2	–	–	–	153.3 ± 6.5[oo]	SS?	Large grain-to-grain track density variations (30%-40%).
67915[v]	Breccia	255.	–	–	50	–	–	–	50.6 ± 3.0[oo]	SS	Boulder sample associated with North Ray Crater. 1 mm/my erosion rate gives agreement between track and rare gas age.

68415[v]	Crystalline	371.	4(2π)	-	-	-	-	-	92.5 ± 13.3[oo]	SS?	Age estimate is sensitive to topography and erosion.
68815[t]	Breccia	1789.	-	-	2.0	-	-	-	2.0[oo]	SS	Because of simple exposure history, this rock is used as cosmic ray track standard.
69935[v]	Breccia	128.	-	-	2.3	-	-	-	1.99 ± .37[oo]	SS	Boulder sample. Same track age as 68815.

Age A - Single point determination. Sample not at center so 2π or 4π irradiation must be assumed. This gives maximum surface age.
Age B - Single point determination at rock center. This age gives maximum time rock could have been exposed at surface in present configuration.
Age C - Determined from gradient measurement in rock. This is probably the best approximation to true "sun-tan" exposure.
Age D - Single point determination taking minimum track density measured in a number of crystals from same location. Applicable to breccias with pre-irradiation histories and to shocked rocks. This is also a maximum surface age.
Age E - Single point determination at a depth of 1 mm. This age is sensitive to surface topography and to rock erosion; it may be higher or lower than true surface exposure age.
Age F - Sub-decimeter age. See text for explanation.

† Single stage irradiation refers to track record only. Thus a rock which had been buried at 1 meter where the track production rate was very small would be considered to have a single stage exposure if it had recently been brought to the surface. The primary criterion for a single stage exposure is a steep track gradient or an agreement between track and spallation ages.

a) Price and O'Sullivan (1970)
b) Crozaz et al. (1970b)
c) Fleischer et al. (1970c)
d) Lal et al. (1970)
f) Fleischer et al. (1971b)
g) Bhandari et al. (1971b)
h) Crozaz et al. (1971)
i) Barber et al. (1971a)
j) Lal (1972a)
k) Hart et al. (1972b)
l) Bhandari et al. (1972a)
m) Crozaz et al. (1972e)
n) Berdot et al. (1972b)
o) Berdot et al. (1972a)
p) Poupeau et al. (1972b)
q) Bhandari et al. (1972b)
r) Behrmann et al. (1972)
s) Fleischer and Hart (1972b)
t) Yuhas and Walker (1973)
u) Behrmann et al. (1973c)
v) Yuhas (unpublished)
w) Fleischer et al. (1973)
x) Fleischer and Hart (1973c)
y) Fleischer and Hart (1974)
aa) Schwaller (1971)
bb) Marti et al. (1970a,b)
cc) Hohenberg et al. (1970)
dd) Bogard et al. (1971)
ee) Burnett et al. (1971b)
ff) Marti and Lugmair (1971)
gg) Kaiser (1971)
hh) O'Kelley et al. (1971)
ii) Hintenberger et al. (1971)
jj) Megrue and Steinbrunn (1972)
kk) Kaiser (1972b)
ll) Bogard and Nyquist (1972)
mm) Crozaz et al. (1972e)
nn) Kirsten et al. (1972a,b)
oo) Drozd et al. (1974)
pp) Lugmair and Marti (1972)

tained from the *center* of the rock, then it makes no difference whether the rock has remained in one position or has rolled about the lunar surface. Ages determined from center measurements are listed in column B in Table 6-6. For samples that are not located at the center, however, the calculated ages are sensitive to the choice of a 2π or 4π irradiation. Many lunar rocks are found to be cratered on all sides (a good example is lunar rock 12063) and/or to have track densities that increase in all directions from the center. For these reasons, the "rolling stone model" that takes a 4π irradiation has often been assumed. However, as we shall show later, a surface can become saturated with craters in a relatively short time, and the presence of craters on all sides is not a sufficient justification for assumption of a uniform 4π irradiation.

In Table 6-6 we list values given by various experimenters for maximum surface ages derived as described above. The values are taken directly from the literature and have not been corrected using the empirical spectrum given in Section 6.2.4. The corrected values would generally lie within 30% of these previously inferred ages.

A more sophisticated approach to the problem of determining surface exposure ages, results for which are listed in column C in Table 6-6, is to measure the track density-depth profile to see to what extent the assumed radiation condition (i.e., zero tracks accumulated prior to exposure of the rock in its present form on the lunar surface) is indeed fulfilled. In Figs. 6-20 and 6-21, taken from Crozaz et al. (1971) and Crozaz et al. (1972d,e), we show detailed measurements of track profiles compared to theoretical calculations. The predicted profiles do *not* match the observations and the implication is that both rocks have been exposed in their present form for very much shorter times than would be estimated from a single measure of the track density at the center of the rock. In this case a better estimate of the true surface exposure age can be obtained by subtracting an assumed constant track background from all points until the profile of the remaining tracks fits that expected for a simple exposure to galactic cosmic rays. Rocks 14310 and 12063, for example, which have maximum surface ages of 16×10^6 yrs and 5×10^6 yrs determined by points measured at a single depth, appear, in fact, to have been exposed in their current configuration for less than 3×10^6 yrs and less than 1.5×10^6 yrs when analyzed by the procedure just described.

Of course, this procedure assumes that the cosmic ray energy spectrum is constant in time. We have previously shown that this is a good assumption for lunar rocks with exposure ages of 2×10^6 yrs and 50×10^6 yrs (though to obtain detailed agreement for the longer age it was necessary to assume an erosion rate of $\sim 10^{-7}$ cm/y), as well as for the meteorite St. Séverin with an exposure age of 10×10^6 yrs. Nonetheless, we cannot completely rule out the possibility that deviations from the expected profiles found in many other lunar rocks may represent exposures during periods of time when the average cosmic ray spectrum was very

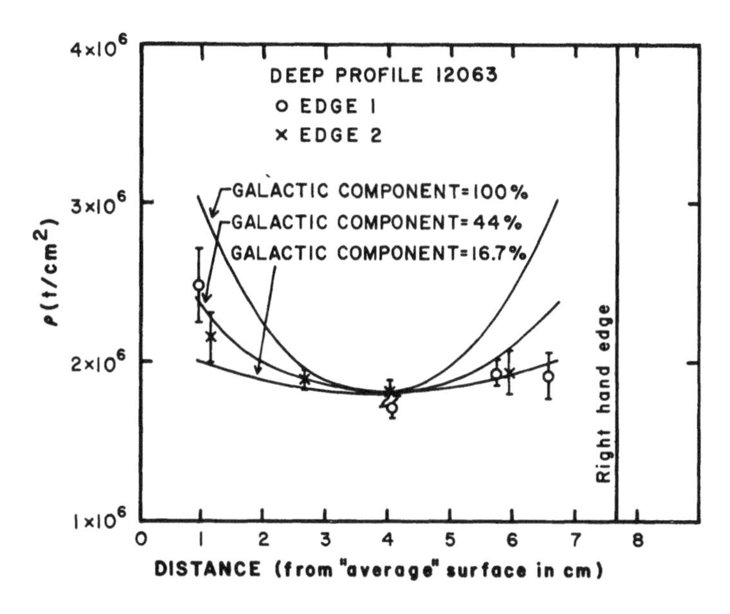

Fig. 6-20. *Track profile at depths well below the surface of rock 12063. (After Crozaz et al., 1971.) The presence of impact pits and solar flare tracks on all sides of this rock clearly show that it has been tumbled on the lunar surface. Even the tumbling, however, cannot account for the track profiles, which are very shallow. The curves indicate the profiles that would be expected if part of the track density was acquired when the rock was well below the surface in such a configuration as to show no gradient. The galactic contributions noted on the curves give the proportion of the total track density acquired while rolling freely on the surface.*

different (see, for example, Fig. 6-21). However, we consider this an unlikely possibility.

We have previously noted that crystals removed from a given location in a lunar breccia frequently show a wide range of track densities, indicating that some of the crystals have received a "preirradiation" dose prior to being incorporated in the rock. In such cases, Hart et al. (1972a,b) and Fleischer and Hart (1973a) have underlined the importance of measuring *minimum* track densities in order to get a *maximum* surface age for the breccia as a whole. Ages measured this way are listed in column D in Table 6-6.

Some care must be taken to recognize the possible meaning of such ages. For example, Hart et al. (1972b) give exposure ages of 0.49×10^6 years and 0.34×10^6 years respectively for rocks 14066 and 14301 using the minimum track density approach, while Keith et al. (1972) show that these rocks are saturated with ^{26}Al, indicating an exposure of at least a few million years. However, for 14301, Keith et al. note that the levels of ^{26}Al show that it was buried during most of its irradiation, so that the 0.34×10^6 years from the track work represents a minor portion

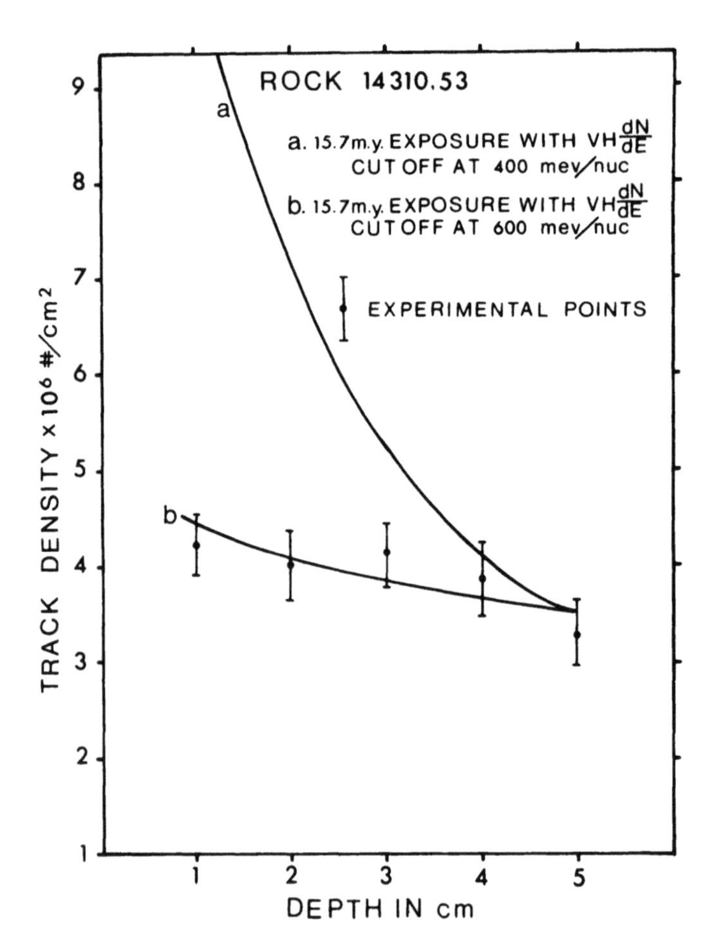

Fig. 6-21. *Track density vs. depth in rock 14310. (After Crozaz et al., 1972e.) The theoretical curves show that a severe modification of the galactic energy spectrum shown in Fig. 6-10 could account for the data. The theoretical curves assume that no galactic particles are found below the specified energies. However, the steeper profiles shown in Figs. 6-9 and 6-16 for both older and younger rocks indicate that the flat profile is probably due to a complex irradiation history. At this writing, however, large fluctuations in galactic energy spectra over times of the order of 10^7 yrs cannot be completely ruled out.*

of its near surface exposure following a longer subsurface exposure at a depth where it could be reached by protons. For 14066 the difference could be caused by a similar effect—with a near surface burial that was closer to the surface. Alternatively the track age could (1) represent the exposure of one side of the rock only (if the sample with the minimum track density did not come from the geometric center of the rock) or (2) give the time during a long surface residence of 14066 when an impact reset some of the crystals in the sub-sample used for the track work.

The Indian group has proposed two additional ways of defining track exposure ages (Bhandari et al., 1971b). Because the galactic track production rate rises rather sharply as the surface is approached, this group has proposed that the track densities in the near surface region of rocks are better indicators of true surface exposure ages than track densities measured at depths of several centimeters. The "sun-tan" age as defined by this group is based on track densities measured in the 1 mm to 2 mm depth range.

Ages determined in this way are listed in column E in Table 6-6. These ages are uniformly very young and again suffer the problem that they disagree with the conclusions reached by counting of induced radioactivities in certain rocks. For example, Bhandari et al. (1972b) give sun-tan ages of $<1 \times 10^6$ yrs for rocks 15085, 15265, and 15557 whereas the counting data of Keith et al. (1972) and Rancitelli et al. (1972) indicate that these rocks have been surface-exposed for several million years.

We have previously noted that the track production rates used by the Indian group are somewhat different from those given here. However, this is not the source of the discrepancy since the production rates in the 1 to 2 mm region are similar. The young sun-tan ages found by the Indian group are very likely related to the effects of rock erosion which limit the maximum track densities that can be obtained when very close to a surface.

The fact that Bhandari et al. (1971b) find many more tracks in the interiors of lunar rocks than would be predicted by their sun-tan ages is attributed by them to a partial burial history. In particular, they define a "subdecimeter" age by assuming that the excess tracks are accumulated while the rock is moving steadily up in the regolith from a position of complete shielding to its final position on the surface. Ages based on this model are given in column F of Table 6-6.

6.6.3. Track Spallation Ages

As previously noted, production rates for spallation recoil tracks vary widely from one type of mineral to the next and also from one grain to the next of the same mineral. In Table 6-7 we show spallation ages inferred by Fleischer et al. (1971b) using track production rates measured with a calibration irradiation of 3 GeV protons. Stored spallation tracks were distinguished from VH tracks by their shorter length. To obtain these ages a nominal flux of $3 \times 10^7/\text{cm}^2$ y of primary cosmic ray nucleons was assumed (Bazilevskaya et al., 1968).

Two ages, the surface age and the minimum spallation age, are given; the latter age corresponds to the assumption that the crystals were irradiated below the surface at the peak of the secondary nucleon cascade. In this work the ratio of the peak to the surface production rate was taken as 2.5 instead of the factor of 2 previously indicated in Section 6.2.5.

Table 6-7. Track Spallation Ages of Lunar Pyroxenes (from Fleischer et al., 1971b)

Sample Number	Production Rate (P)* (tracks/10^9 protons)	Observed Track Density (ρ_{sp}) (cm^{-2})	Proton Exposure (ρ_{sp}/P) ($protons/cm^2$)	Surface Age (10^6 yr)	Minimum Spallation Age (10^6 yr)	Radiometric Spallation Ages (10^6 yr)
10017	1.7†	2.1 (± 0.1) x 10^7	1.2 x 10^{16}	420	170	200-640 (a,e,f,g,i,j)
10044	1.0	8.2 (± 1.2) x 10^6	8.2 x 10^{15}	270	110	56-100 (f)
10049	2.39	1.49 (± 0.15) x 10^6	6.2 x 10^{14}	21	8.5	22.5-25 (g,ℓ)
12002	1.7†	2.66 (± 0.25) x 10^6	1.6 x 10^{15}	55	20	50-145 (b,d,h)
12017	1.45	4.57 (± 0.32) x 10^6	3.2 x 10^{15}	105	40	-
12021	2.31	5.1 (± 0.3) x 10^7	2.2 x 10^{16}	740	300	300 (h)
12065	1.35	6.81 (± 0.28) x 10^6	5.1 x 10^{15}	170	70	160-200 (c,k)

*Absolute values uncertain to ± 30%, but relative values are valid for 10049, 12002, 12017, 12021, and 12065.
†Average of other values.
a) Albee et al. (1970); b) Alexander (1971); c) Bloch et al. (1971); d) D'Amico et al. (1971); e) Eberhardt et al. (1970); f) Fireman et al. (1970); g) Hintenberger et al. (1970); h) Marti and Lugmair (1971); i) Marti et al. (1970b); j) O'Kelley et al. (1970); k) Stoenner et al. (1971); ℓ) From measurements by Funkhouser et al. (1970).

As can be seen from Table 6-7, the track spallation ages are in good agreement with spallation ages determined by independent, radiometric techniques.

A different approach to the problem of spallation has been taken by Crozaz et al. (1972c) in their study of lunar whitlockites. These minerals concentrate uranium, and the background of fossil tracks has a large contribution from the spontaneous fission of ^{238}U. The track densities are also so high that they require an electron microscope for measurement. Thus, it is not simple to separate the long (VH and fission) tracks from the shorter spallation tracks. In principle this can be done by length measurements on plastic replicas, but a clear-cut separation of the spallation recoils was not observed by Crozaz et al. (1972d,e) in their work.

The cosmic ray contribution was measured instead by making total pit counts on crystals with varying concentrations of uranium as shown in Fig. 6-22. The total cosmic ray background found by extrapolation of the data to zero uranium concentration contained only a small contribution of VH tracks (as measured on adjacent crystals of feldspar) and was due almost entirely to spallation recoils. Combination of the intercept value with the measured radiometric age of 210×10^6 yrs (Hintenberger et al., 1971) gave a spallation track production rate of $1 \pm .1/\text{cm}^2$ y for the whitlockite.

All spallation ages, whether determined by tracks or by isotopic measurements, have implicit assumptions built into their determination. They are strictly valid only for samples that have had a one-stage irradiation history in a fixed (and known) geometry. However, the isotopic measurements are capable of giving information about deviations from a simple history and are intrinsically superior to the track determinations of spallation ages. Since isotopic determinations of spallation

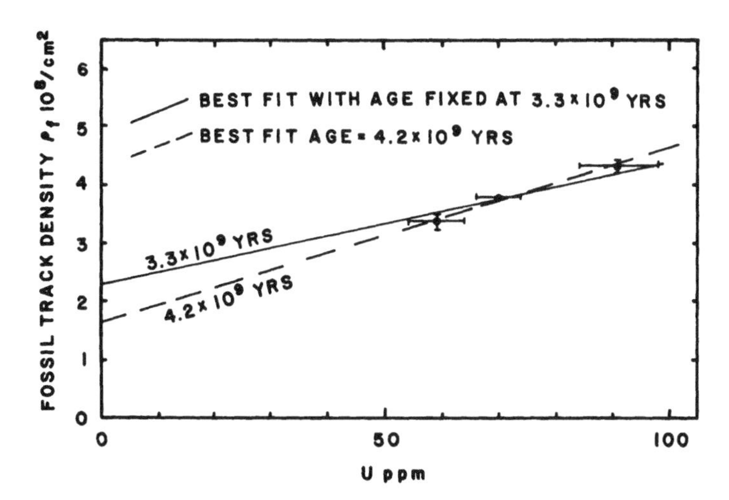

Fig. 6-22. *Total track densities vs. uranium concentrations for several phosphate grains in rock 12064. (After Crozaz et al., 1972c.) The intercept for a zero concentration of uranium gives the spallation track contribution in the phosphates.*

ages are possible in most samples of interest, the spallation track method has been little used. However, the track determinations may well find future extensive use in studying the spallation ages of individual small fragments that make up the lunar soil. For such fragments a major potential difficulty may be presented by the need to distinguish the recoil tracks from previously longer tracks that have been fragmented by shock-induced mechanical deformation (Fleischer et al., 1972a,b).

6.6.4. Measurement of Mass-Wastage Erosion from Comparison of Spallation and Heavy Particle Exposure Ages

The differences between the penetration of VH ions and that of high energy protons can be used to measure the rate of mass-wastage of lunar rocks and meteorites. If the dimensions of an irradiated object change by only a few centimeters during the time of irradiation, the apparent surface exposure age, as measured by VH tracks, will be lowered; however, the proton ages will be virtually unchanged. As we have seen in a previous section, discordancies between spallation and VH exposure ages are always in this sense. If the reverse were ever found for carefully measured ages, it would imply a changed proton to VH ratio in the past.

Price et al. (1967) were the first to develop the method and use it to derive an erosion rate for an extraterrestrial object. In a crystal removed at a given depth from the *present* surface of the meteorite Patwar they found a lower track density than would be predicted from the proton spallation age. Since some material must have been ablated away during atmospheric entry, the true depth of the crystal when the meteorite was in space must have been greater than the present depth. However, by assuming that no material was ablated, and attributing the difference completely to a mass-wastage erosion that occurred at a constant rate in space, Price et al. (1967) were able to set an *upper limit* of $<10^{-7}$cm/y for the erosion rate.

We have previously noted that stony meteorites have relatively low cosmic ray exposure ages of the order of 10^7 yrs. Rapid space erosion was one idea advanced to explain the low ages. If a particle has a characteristic absorption length L, a limiting apparent exposure age will be reached in a time T_{lim} given, approximately, by $T_{\text{lim}} \sim L/V_e$, where V_e is the erosion rate. Beyond this time the buildup of the spallation products in the meteorite is matched by the loss due to removal of material. Since L is $\sim$50 cm for high energy protons, the value of V_e of $<10^{-7}$ cm/y measured by Price et al. shows that $T_{\text{lim}} \approx 5 \times 10^8$ yrs. In short, small-scale erosion does *not* limit the lifetime of stony meteorites; the explanation of their short lifetimes must be sought elsewhere.

One of the most striking facts that emerged from the initial study of the first Apollo 11 samples was the frequent great disparity between proton and VH exposure ages. If the largest of these differences were attributed to erosion, the inferred rates would be higher by orders of magnitude than the limits set by Price et al. (1967) on the Patwar meteorite. However, in occasional samples, with mechanical properties similar to those with large discordancies, rather small differences in spallation and VH ages were found. Because a discordancy can be produced most simply by burial of a sample at shallow depths, it is natural to regard the large discordancies as arising from this cause. On the other hand, small discordancies can be used, as in the case of Patwar, to set upper limits on erosion rates. In short, the name of the game is to find samples whose spallation and VH ages are nearly the same.

Using the track density measured at the center of rock 10057 to derive a maximum surface age of 28×10^6 yrs, and combining this with measured spallation ages of 45 my, Crozaz et al. (1970b) set limits of 5×10^{-8} to 10^{-7} cm/y for the erosion rate of this rock. An even more stringent limitation was set by Fleischer et al. (1971b) who found a value $\lesssim 3 \times 10^{-8}$ cm/y for a relatively small rock, 10049, which had nearly identical spallation and track ages of 25 and 30 mv respectively.

The first results from the moon therefore showed that erosion processes were measurable and important, but vastly smaller (by a factor of $\geq 10^5$) than for common earth materials—a vivid demonstration of the effects of wind and water in shaping our own planet.

Two additional samples from the later Apollo missions have been analyzed in rather more detail for estimates of mass wastage. In Fig. 6-16 we showed the depth dependence of track densities for a rock chip removed from a large boulder on the Apollo 16 mission (67915). The inclusion of a mass-wastage term of 1.2×10^{-7} cm/y not only brings the track ages into concordance with the spallation ages, but significantly improves the agreement between theory and experiment for the entire depth profile. The same is true, but to a lesser extent, for one of the largest rocks yet brought back from the moon (15555—known to many as the "Great Scott" rock after Astronaut David Scott), for which a mass-wastage rate of 1.3×10^{-7} cm/y was obtained (Crozaz et al., 1972d,e).

Mass-wastage rates can also be determined by an independent method. Solar flare protons produce a characteristic gradient of induced radioactivities in the first few centimeters of a lunar rock. Though the analysis is somewhat complicated, the depth profiles of different isotopes can be used to determine an energy spectrum of the bombarding particles (Reedy and Arnold, 1972).

In several lunar rocks the energy spectra derived for the short-lived isotope ^{22}Na ($T_{1/2} = 2.6$ yrs) agree well with those for the longer lived isotope ^{26}Al ($T_{1/2} = 7 \times 10^5$ yrs) (Finkel et al., 1971; Wahlen et al., 1972). The agreement shows that the average flux of solar flare protons has not changed appreciably in the last million years.

The energy spectra derived from measurements of the still longer lived isotope ^{53}Mn ($T_{1/2} = 4.5 \times 10^6$ yrs) are, however, somewhat flatter. If the difference is attributed to erosion, mass wastages of $\lesssim 5 \times 10^{-8}$ cm/y to $\lesssim 2 \times 10^{-7}$ cm/y are obtained (Finkel et al., 1971; Wahlen et al., 1972), in substantial agreement with the track results.

As an aside, we note that comparison of spallation ages with galactic cosmic ray track densities can also be used to measure the atmospheric ablation of meteorites. Fleischer et al. (1967b, 1970a) (the latter paper discussing the wildly improbable Schenectady meteorite that fell within a few miles of a meteorite research group) have given original sizes for some 25 meteorites. Meteorites like Patwar and St. Séverin which have measured ablations as low as ~1 cm are rare; in most cases the present surface of a meteorite was greater than 5 to 15 cm below the true surface in space.

Bhandari et al. (1973g) made a systematic study of this problem in some 50 meteorites and came to the conclusion that the average mass loss was about 70%, independent of the size, class, or exposure age of the meteorite. From a theoretical analysis of the ablation process, these authors further concluded that most meteorites enter the earth's atmosphere at velocities around 12–15 km/sec.

6.6.5. Solar Flare Studies of Microerosion in Lunar Rocks

One of the primary interests of track studies in lunar rocks was the measurement of the long-term behavior of VH ions from solar flares. The first observation of increases of factors of 10 to 100 in track densities within the first millimeter of a rock surface was an exciting high spot of the early examinations of the first samples (Crozaz et al., 1970a,b; Fleischer et al., 1970b,c; Price and O'Sullivan, 1970). Fig. 6-23 shows an especially steep track gradient at the very surface of a rock. It soon became evident, however, that the analysis of the track gradients in terms of ancient energy spectra was a nontrivial matter. Inextricably intertwined in the analysis of the energy spectrum is the question of the microerosion of rock surfaces.

To study the track gradient with depth in rocks, sections are cut with their

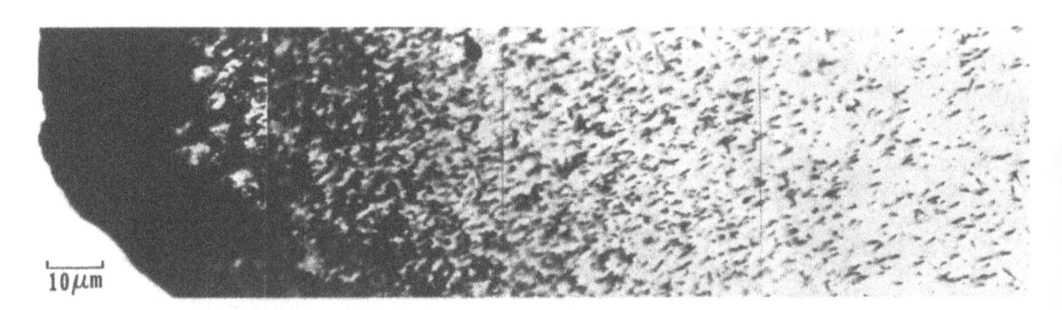

Fig. 6-23. *Steep track gradient due to solar flares at the surface of rock 72315, taken from a crevice in the top of an Apollo 17 boulder. In the outer ~50 μm where the track density exceeds $10^8/cm^2$, the gradient can be studied by scanning or transmission electron microscopy.*

planes perpendicular to the surfaces. These are then mounted in epoxy, polished, and etched. The surface boundary line is then searched to find the crystals with the highest track densities. The depth profiles are measured starting from these surface crystals and proceeding inward.

Singling out the crystals with the highest track densities has the advantage of minimizing the effects of a dust covering or of isolated recent impact pits, but has the added consequence that only crystals that have survived on the rock surface for some time are analyzed. The "average" crystal is not measured. Therefore, erosion rates are lower limits and apply to rates that represent *small* removals of material from the surface. We will return to this point later.

In Section 6.2.3 we noted that for low energy particles an energy spectrum of the form $dN/dE = CE^{-\gamma}$ results in a track density profile of the form $\rho = BD^{-\alpha}$. The exponents α and γ are related by the relationship $\alpha = (1 - \gamma - \beta)/\beta$ where β is the constant in the approximate range-energy relationship $R = AE^{-\beta}$. We now show that in the presence of rock erosion the track density profile is modified in such a way as to reduce the exponent of the power law by unity.

Let X represent the distance of a test crystal below the present surface of a rock. In the case of a constant erosion rate V_e, the distance below the surface at any time t in the past is simply $R(t) = X + V_e t$. The rate of track production as a function of time is $\dot{\rho}(t) = KR(t)^{-\gamma} = K[X + V_e t]^{-\gamma}$. Integrating this expression over the total exposure time of the rock gives $\rho(X) = [K/V_e(1 - \gamma)][X^{-\gamma+1} - (X + V_e T_{exp})^{-\gamma+1}]$. When $X \ll V_e T_{exp}$, that is, when we are considering a depth where erosion is dominant, this reduces simply to $\rho(X) = [K/V_e(1 - \gamma)][X^{-\gamma+1}]$.

This expression has several interesting consequences. Firstly, in erosion equilibrium the slope of the track density vs. depth is independent of erosion rate, provided the rate is constant. Secondly, the index describing the range spectrum of the solar flare particles producing the tracks is given simply by adding one to the slope on a log-log plot of track density vs. depth. Finally, the absolute value of the track density at any given depth depends directly on the solar activity and inversely on the erosion rate V_e.

In Fig. 6-24 we show data for a number of lunar rocks for which the solar flare track data are available. Similar data on other rocks are given by Lal (1972a). The spectra are reasonably well fitted by a power law but with significantly shallower slopes than those found in the Surveyor glass. Explicit evaluations of the coefficients in the power law depth dependence for several rocks are given in Table 6-8. Also shown are "corrected" values where dust layers of varying thickness have been assumed in order to give a best power law fit to the track data.

There are at least three possible explanations for the difference between the rock data and the Surveyor data: (1) In the lunar rocks we are not looking at the true surface. Either material was removed from the rocks between the time they were on the moon and the time they were studied by us, or they were covered with a dust layer on the moon. The effect of such a layer is to decrease the measured

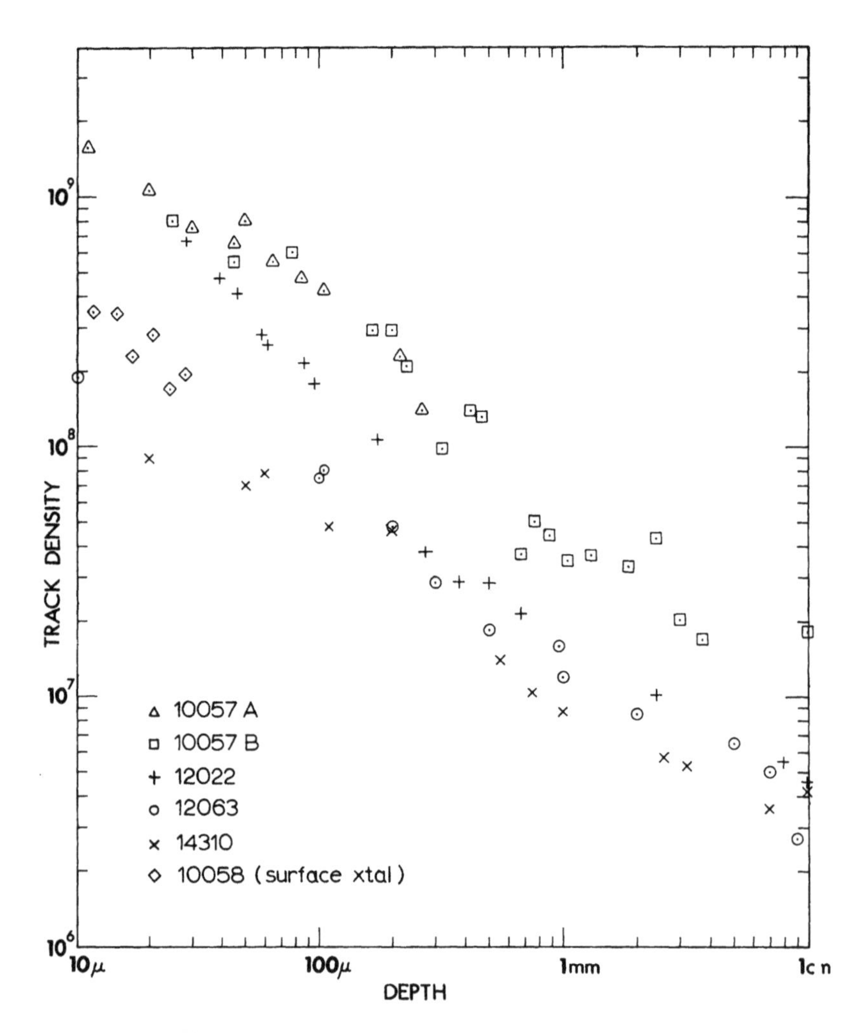

Fig. 6-24. *Typical data on the variation of solar-flare-produced track densities with depth in lunar rocks. (After Crozaz et al., 1972e.) When solar flare tracks are present (as evidenced by large surface gradients), the profiles reported by different groups generally fall within the limits of the data shown here. Occasional examples of steeper spectra are found in freshly exposed surfaces and in individual crystals removed from the fines. The data for rock 12022 are from Barber et al. (1971a), who showed that the track density continues to increase smoothly in this rock to a depth only 2 μm below the surface.*

slope relative to the true slope. (2) Erosion of the rock has modified the density vs. depth curve. (3) The properties of solar flares during the period of exposure of the Surveyor glass are not characteristic of the long-time average properties of solar flares.

Although photographs taken on the lunar surface show that lunar rocks are

Table 6-8. Summary of Depth Dependence of Track Density in Selected Lunar Rocks (100μm to 1000μm) (from Crozaz et al., 1972e)

Rock	α_1	α_2	Absolute ρ at 300μm
10057	1.04	1.14	1.1×10^8
12022*	1.10	1.21	4.4×10^7
12063	0.85	(0.91) 1.21†	3.3×10^7
14310	0.86	(1.10) 1.28†	2.5×10^7

α is the coefficient in the power law spectrum with depth, $\rho = \rho_o D^{-\alpha}$, α_1 is calculated from raw data, α_2 is corrected for galactic contribution.

*From Barber et al. (1971a).

†50μm dust layer assumed on surface, values without dust layer are shown in parentheses.

generally clean, dust is observed in pockets and depressions. It is also certain that some abrasion of the rocks occurs during their voyage from the moon to the laboratory. We know the only way to decide on the importance of (1) is to make a large number of measurements in different rocks. If each time a high density region is chosen the same gradient is found, then such regions are likely to be true surface crystals. Since numerous rocks so far studied appear to give similar slopes when regions of maximum density are measured, accidental abrasion does not appear to be an important effect.

If the difference is attributed to erosion and the rocks are assumed to be in erosion equilibrium ($D < V_e T$), then the slope of the uneroded spectrum can be determined. Table 6-9, taken from Crozaz et al. (1972d,e), shows that the final value of the spectral index so inferred for several rocks is 2.6 ± 0.1. This is close to the best value of 2.9 ± 0.2 derived from the Surveyor data for the same energy range and lends strong support to the rock erosion model.

Table 6-9. Comparison of Long and Short Term Energy Spectra (from Crozaz et al., 1972e)

	α_e	α_c	γ
Rocks (10μm to 1000μm)	1.20 ± 0.05	2.20 ± 0.05	2.6 ± 0.1
Surveyor (10μm to 300μm)	2.5 ± 0.2	2.5 ± 0.2	2.9 ± 0.2

Rocks 10057, 12022, 12063 and 14310 corrected for galactic contribution and for assumed 50μm dust covering on 12063 and 14310.

α is the coefficient in the power law spectrum with depth, $\rho = \rho_o D^{-\alpha}$, α_e is the experimental value, α_c has been corrected for rock erosion.

γ is the coefficient in the power law spectrum with energy, $\frac{dN}{dE} = AE^{-\gamma}$.

Assuming that solar flares have been constant in time, different authors have used the Surveyor III data to give absolute numbers for the erosion rates that range from $\sim 10^{-9}$ cm/y to $\gtrsim 10^{-7}$ cm/y (Barber et al., 1971a; Crozaz et al., 1971; Fleischer et al., 1971a). Some of the differences arise from the basic input data. For example, it can be seen from Fig. 6-24 that track densities at a given depth range over a factor of four in different rocks. Different treatments of the reference Surveyor data also caused some of the initial spread in erosion rate. Using the compromise flare spectrum given here, we now calculate track densities of $6.6 \times 10^8/\text{cm}^2$ and $3.6 \times 10^7/\text{cm}^2$ at depths of 100 μm and 1000 μm, respectively, for a rock in erosion equilibrium with an erosion rate of 10^{-8} cm/y. The results shown in Fig. 6-24 thus imply erosion values between 2×10^{-8} cm/y and 8×10^{-8} cm/y for different rocks.

It is important to note that limits on microerosion rates can be calculated from a combination of solar flare and galactic track data *independent of the Surveyor data.* The flattening of the Surveyor slope by one unit requires that the rock be in erosion equilibrium, i.e., that $D \ll V_e T$. If the true surface exposure age, T, is determined independently, then $V_e \geq D/T$ where D is the depth to which the erosion equilibrium occurs. Analysis of galactic track data on rocks 12063 (Crozaz et al., 1971) and 14310 (Crozaz et al., 1972d,e) leads to the conclusion that the rocks had been exposed in their present geometry for $<1.5 \times 10^6$ yrs and $<3 \times 10^6$ yrs, respectively. Since the erosion equilibrium condition applies to a depth of ~ 1 mm, this implies microerosion rates of $\gtrsim 6 \times 10^{-8}$ cm/y and $\geq 3 \times 10^{-8}$ cm/y for the two rocks, consistent with the values determined from the compromise Surveyor spectrum.

The microerosion rates are considerably higher than those that would be expected from atomic sputtering processes in the solar wind (McDonnell et al., 1972). Whether the derived rates apply to some unusual process such as microflaking due to surface stresses caused by the gradient in radiation damage at crystal surfaces (Seitz et al., 1970; Seitz and Wittels, 1971) or whether they can be accounted for by micrometeorite bombardment is not clear.

In this connection it is important to note that not *all* surfaces appear to be in erosion equilibrium. Hutcheon et al. (1972c) reported measurements of a steep, Surveyor-type spectrum in a crystal found at the bottom of a surface vug in a vesicular lunar basalt. They estimated an exposure time of $\sim 10^5$ yrs, which implies an erosion rate $<2 \times 10^{-8}$ cm/y. These authors suggest that the vug may have been oriented out of the plane of the ecliptic and hence subject to a smaller micrometeoroid flux. Fig. 6-25 shows their gradient for the vug crystal, together with more recent measurements by Hutcheon et al. (1974) of a steep gradient in a feldspar crystal that was freshly exposed at the surface of an Apollo 17 boulder about 10^5 years ago. Fig. 6-23 is an optical micrograph of a portion of this crystal. Being situated in a crevice in the boulder, the crystal was partly protected from erosion by micrometeorites. Hutcheon et al. concluded that the track profile

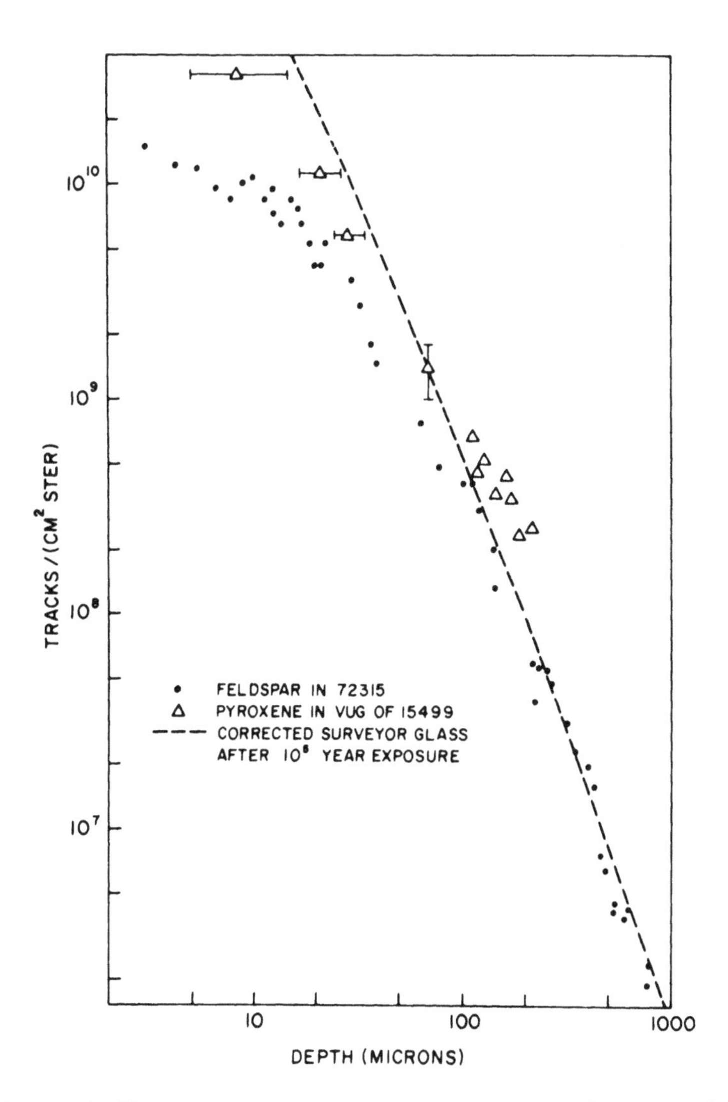

Fig. 6-25. *Track density gradients that nearly agree with that measured in the Surveyor glass. In both cases the crystals were partially shielded from full exposure to micrometeorites, and both crystals have a surface exposure of* $\sim 10^5$ *years to solar flare particles. (After Hutcheon et al., 1974.)*

would fit the Surveyor profile perfectly if the crystal had been covered by only 20 μm of matrix material, and that there was no measurable erosion during the 10^5 year exposure.

As shown in Fig. 6-26, steep gradients have also been observed in individual crystals removed from the lunar soil.

The critical assumption made in this section that the flux of flare-produced heavy particles has not changed in the past cannot be directly tested. However,

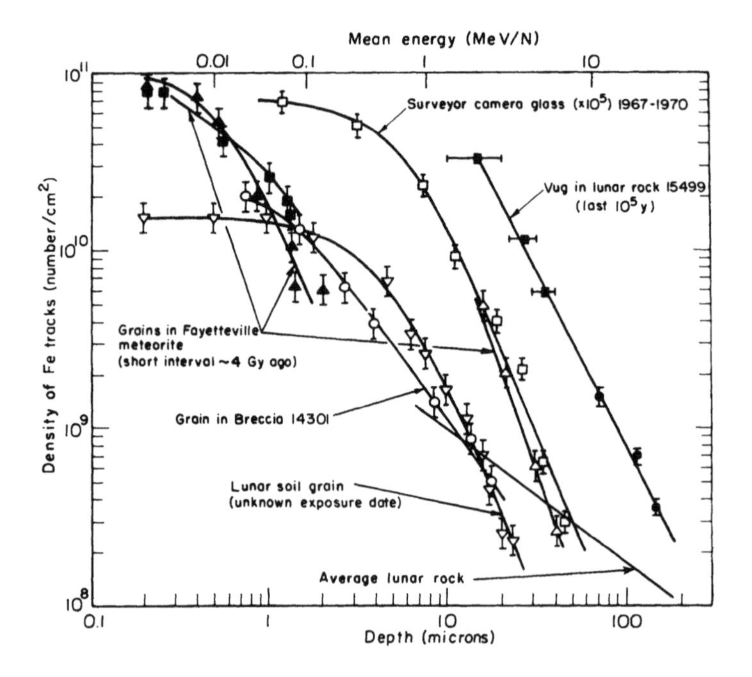

Fig. 6-26. *Track density gradients in various extraterrestrial objects. Erosional processes cause the gradient in a typical lunar rock to be shallow. Lunar soil grains, as well as grains inside certain lunar and meteoritic breccias, often have gradients as steep as that observed in Surveyor camera glass. Being protected from erosion, a crystal at the bottom of a vug in lunar rock 15499 also had a steep gradient.* (*After Rajan, 1973, and Hutcheon et al., 1972c.*)

studies of proton-induced radioactivities (Finkel et al., 1971; Wahlen et al., 1972) and proton-induced thermoluminescence in lunar rocks (Hoyt et al., 1973) have shown that the average flux of solar flare protons has remained constant within a factor of two for times ranging from $\sim 5 \times 10^3$ yrs to $\sim 2 \times 10^6$ yrs. The assumption that the flux of heavy flare particles has also remained constant is thus plausible but not proven.

We have also made the assumption that only erosion affects the track profiles. However, in some cases, particularly the rocks from Apollo 17, "patinated" surfaces are observed, indicating that *accretional* processes may also be important. Morrison et al. (1973) indicate that in certain rocks the accretion rate may exceed the rate of microerosion. Additional work on patinated rocks will be necessary to sort out these two effects.

6.6.6. General Aspects of Track Studies in Lunar Soils

Lunar soil, typically greyish in hue, dirties an astronaut's space suit at least as effectively as most terrestrial soils. Seen with the naked eye, there is little to distinguish it from familiar earth materials. But even a cursory glance with an ordinary magnifying glass reveals an astonishing variety of shapes and forms, totally unlike those seen in terrestrial samples. When lunar soils were first received in various laboratories in September of 1969, many of us, particularly those with scanning electron microscopes, went on an orgy of picture taking. It was hard in those early days to stop looking and to get on with the business of making measurements. Even now, when much of the early excitement has slackened, we find ourselves lacking in sufficient moral fiber to resist including photographs of individual lunar dust grains (see Figs. 6-27 and 6-28).

The most striking aspect of lunar soil is the presence of a large amount of glass; glass in every shape ranging from nearly perfect spherules to complicated agglutinates. It is these glassy objects, some produced originally in large impacts, others by micrometeoroids, that are principally responsible for the visual diversity of the soil.

But a more profound examination of all components of a soil, ranging from individual mineral grains and small rock fragments to the very finest particles that are less than a micron in size, reveal still deeper diversities. For one thing,

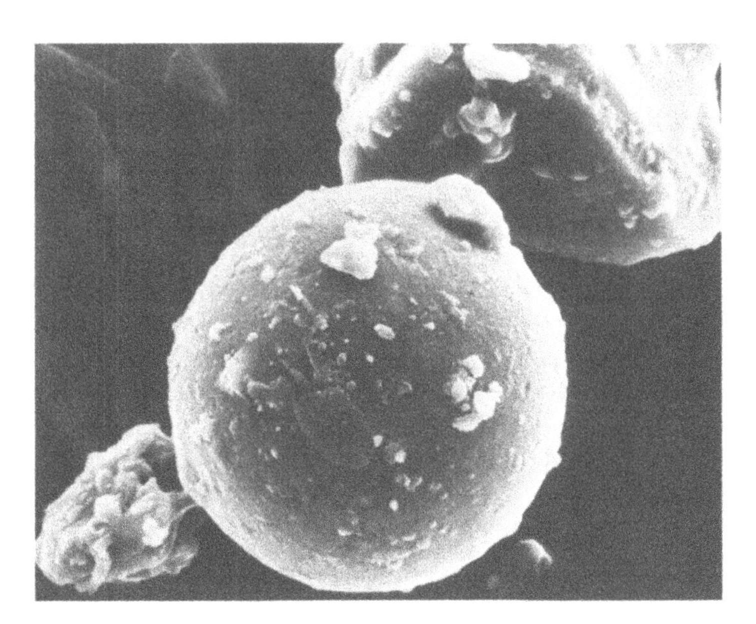

Fig. 6-27. *Glass spherules up to several millimeters in size are a common constituent of most lunar soil samples. The colors are quite variable. One soil from the Apollo 15 mission is composed primarily of bright green spherules; another from Apollo 17 is composed almost entirely of bright orange spherules. Although many spherules are quite round and very homogeneous, some are inhomogeneous and deformed, indicating incomplete melting. This one, from the Apollo 11 mission, is 25 μm in diameter.* (*Courtesy of E. Lifshin.*)

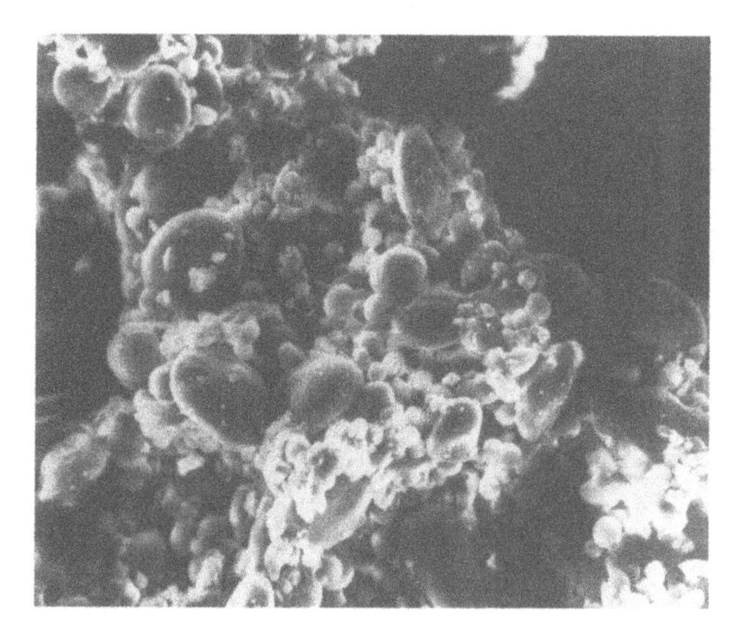

Fig. 6-28. *Cluster of small glass spherules.*

a sample of soil from any one location on the moon contains material originally thrown out from many parts of the moon. (For an excellent popular discussion of the nature of the lunar soil, see Wood, 1970.) One might almost say that with sufficient intelligence it should be possible to decipher the history of the entire moon from a single scoop of the appropriate lunar dust!

The diversity within a given sample of lunar soil is matched by remarkable differences between samples taken from different locations—sometimes even on the same mission. Many readers will remember the stir created during the Apollo 17 mission by the discovery of an "orange" soil. On close examination this soil turned out to be composed almost entirely of reddish brown glass spherules. On the Apollo 15 mission, many of the soil samples contained large numbers of very pretty, bright green spherules.

Quite diverse sampling techniques were used in collecting lunar soils. The most abundant type of sample was obtained with a simple scoop in which 100–200 g of material were typically collected. However, even these simple scoop samples were sometimes taken with an elaborate protocol. For example, a sequence of samples might be taken in the general area surrounding a boulder, from a "fillet" in contact with the boulder itself, and finally from underneath the boulder after it had been rolled aside by the astronauts. Many trenches were also dug on the various missions, with scoop samples being removed from different levels on the walls of the trench.

In the early Apollo missions the vertical stratigraphy of the lunar regolith was sampled with either single or double drive tubes giving a total penetration of up to 80 cm. On Apollo missions 15–17, an electrically driven drill provided vertical sampling up to 3 m in depth. On one mission, samples of the very surface itself were collected on special cloth materials placed in contact with the surface.

In this section we will not give a complete treatment of track studies in all these types of samples. Rather we shall attempt to give the reader a feel for the nature of the problems being studied and to summarize the major results so far obtained.

Every gram of lunar dust contains many millions of crystals and glass fragments that can be individually studied for their track records. Typically, crystals of the same type and size removed from a given soil sample show large variations in track densities. In Fig. 6-29 we show the results of track measurements in feldspar

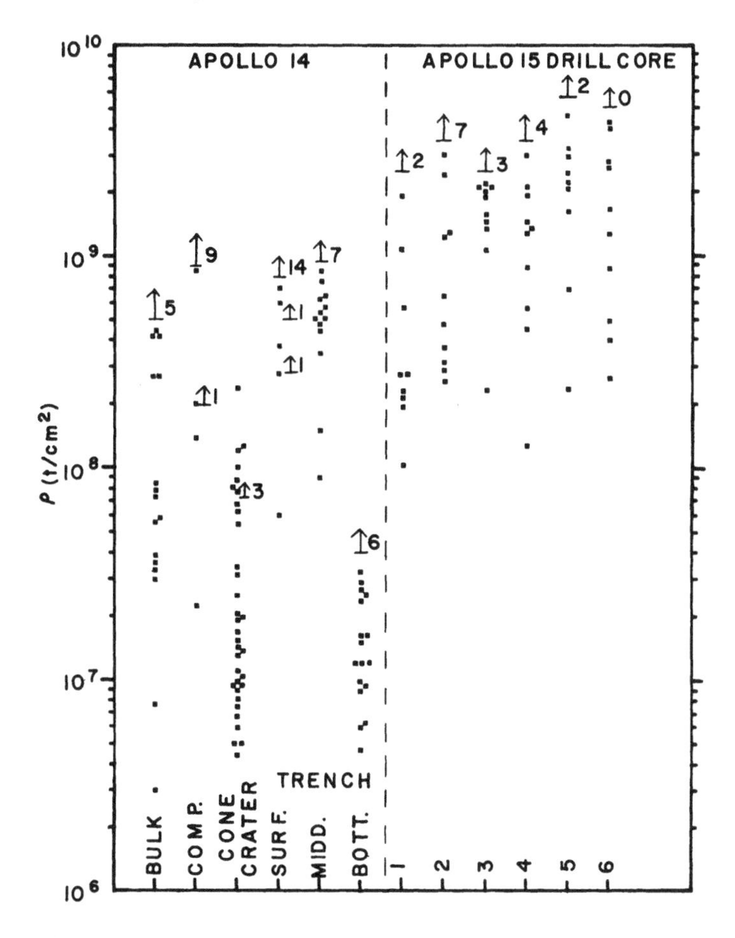

Fig. 6-29. *Representative distribution of track densities measured by the Washington University group in several Apollo 14 and 15 samples. (After Crozaz et al., 1972e.) The crystals are all feldspars and are greater than 150 μm in size. The samples labeled "Cone crater" and "Trench bottom" are clearly very much less irradiated than the other (more typical) samples, many of which have no measured crystals with* $<10^8$ *tracks/cm²*.

crystals of $>150\ \mu m$ in size from different soil samples. It can be seen that the range of track densities even in this selected fraction is frequently at least three orders of magnitude in a given soil sample. The statistical distribution of densities obviously reflects the time-depth history of different grains and gives information about the dynamic processes that shape the regolith.

The lunar soils are also characterized by extremely high track densities; in many samples virtually all the crystals have densities in excess of 10^8 t/cm^2. This is true even for samples removed from the deepest portions of the drill stems. It is easy to show that track production by galactic cosmic rays alone is simply not sufficient to give the high track densities observed at great depths. The high track densities clearly imply that the crystals were once at, or very near, the surface where they were exposed to solar flare particles. This is also shown by the fact that many grains (typically 10–20%) have observable gradients (see Fig. 6-30), implying irradiation by particles with a steep energy spectrum characteristic of solar flares.

The fact that many of the crystals have apparently been directly exposed to the sun at one time in their history is not surprising. Rare gas measurements by a

Fig. 6-30. *Replicated tracks in a feldspar crystal from the lunar fines as seen in a transmission electron microscope. Each track shows up as a black dot with a light flag off to its side, produced by shadowing the replica at a small angle with a Pt film. Track densities up to almost $10^{11}/cm^2$ can be resolved. A definite track gradient is evident in the crystal, indicating a solar flare irradiation. This crystal was once on the very surface of the moon.* (*Photo courtesy of R. S. Rajan.*)

number of groups have shown that most soil samples are heavily charged with gas of solar composition implanted in the surface layers of the grains by solar wind bombardment. What is remarkable is the efficiency of the exposure process that results in the high proportion of surface exposed grains.

Another characteristic and remarkable feature of the soil is the marked dependence of track density on grain size. The Orsay group under the direction of M. Maurette was the first to show (Borg et al., 1970) that the micron-sized fraction of the soil typically contains track densities of the order of 10^{11} t/cm^2 (see Fig. 6-31). These observations and others on the very fine grained fraction have a number of important implications, and we discuss them in a separate section.

It can be seen from Fig. 6-29 that different soil samples have very different radiation histories. Some, such as the sample labeled "Cone crater," have a large proportion of grains with rather low track densities. Others, such as most of the core samples from Apollo 15 that are shown, have crystals virtually all of which contain track densities $\geq 10^8$ t/cm^2. The distribution of track densities also varies from sample to sample.

The track record of a soil can be precisely defined only by giving the complete distribution function of track densities. However, several track indices can be

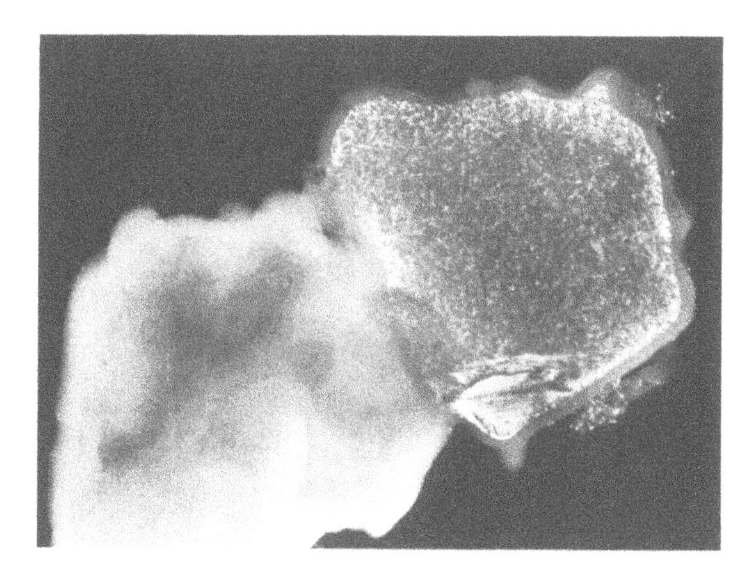

Fig. 6-31. *Ultrahigh resolution transmission electron microscope photographs of micron-sized lunar dust grains. Grains of this size typically have internal track densities of 10^{10} to 10^{12}/cm^2, indicating a high abundance of very low energy particles in space. Many grains are also coated with an amorphous layer probably produced by solar-wind-induced radiation damage. Rounding, probably due to solar wind erosion, is another characteristic feature. Sticking of the grains due to interaction of the amorphous layers can also be seen. (Photo courtesy of M. Maurette.)*

defined that have proven useful in characterizing different samples. Among the indices that have been used are the following:

a) the *minimum track density*,
b) the *quartile track density*, defined as the density above which 75% of the grains are found,
c) the *median track density*, and
d) the *percentage of track-rich grains* (TR%), defined as the percentage of grains with densities $>10^8$ t/cm².

As should be expected, these indices are roughly correlated and serve to define the general level of radiation that a soil sample has received. In Fig. 6-32, taken from Fleischer et al. (1974), we show measurements of median, quartile, and minimum track densities for a number of samples.

The radiation level as monitored by tracks is also correlated with other parameters that are indicative of the length of time a soil has been near the surface (Crozaz et al., 1972d,e; Poupeau et al., 1972a). This is shown explicitly in Tables 6-10 and 6-11 where we give the levels of CH_4 and ^{36}Ar (indicative of the solar

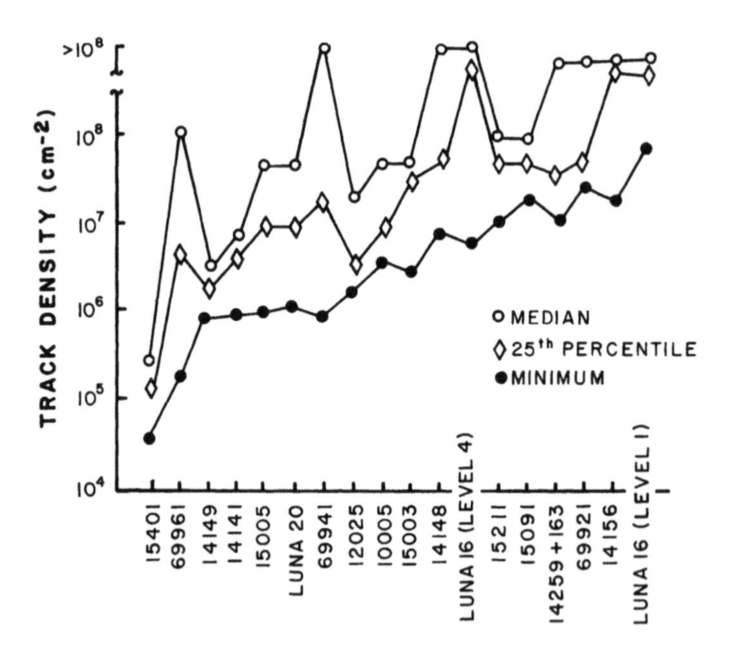

Fig. 6-32. *Graphical listing in an arbitrary sequence of median, 25th percentile, and minimum track densities for a series of lunar soils investigated by the General Electric group. (After Fleischer et al., 1974.) Median and minimum values are not well correlated (reflecting differences in the number and type of various irradiated components in different soils) and both vary by more than three orders of magnitude.*

Table 6-10. Correlation of Track Densities with Other Soil Parameters: Heavily Irradiated Samples (from Crozaz et al., 1972e)

Sample	Description	$>10^8/cm^2$ (%)	Average ρ*	Median Grain Size (mm)	% Aggregates[e] 0.25 to 1 mm	% Aggregates[e] 90μm–150μm	CH_4 μg/g[f]	^{36}Ar ccSTP x 10^{-8}
12042	20 m N.W. of Halo	100	1.2×10^9	0.094[a]	-	-	-	-
12044	S. Rim Surveyor	100	1.2×10^9	-	26	-	-	-
12025 and 12028 (except unit VI)	Double core	78	8.6×10^8	0.050-0.070[b]	2.4-35	-	0.9-2.5	2.2 to 2.3[f] x 10^4
14148	Surface of trench	95	1.5×10^9	0.087[a,c]	32	50	3.0	5.5×10^5[f]
14156	Middle of trench	90	1.0×10^9	0.068[c]	25	44	4.0	-
14163	Bulk, near LEM	48	3×10^8	0.065[c]	20	-	2.6	3.0×10^5[f]
14259	Comp., near Doublet	85	1.5×10^9	0.050[c]	30	52	-	-
Apollo 15	Deep drill	99	2.1×10^9	-	-	-	-	-
15231	Sta. 2, bott. of boulder	100	4.7×10^8	-	-	-	5.1	-
15471	Sta. 4, near Dune Crater	100	9.7×10^8	-	-	-	3.2	-
L16A14	6-8 cm Luna 16	100	1.71×10^9	0.070[d]	67[b]		-	8.6×10^4[g]
L16G14	29-31 cm	100	5.1×10^8	0.090				

*Calculated assuming all crystals with unresolvably high track densities have $\rho = 2 \times 10^9$ tracks/cm^2.
a) King et al. (1971); b) Quaide et al. (1971); c) LSPET (1971); d) A. P. Vinogradov (1971); e) McKay et al. (1971) Apollo 14 (private communication); f) Cadogan et al. (1972); g) W. A. Kaiser (1972a).

Table 6-11. Correlation of Track Densities with Other Soil Parameters: Light to Medium Irradiated Samples (from Crozaz et al., 1972e)

Sample	Description	$>10^8/cm^2$ (%)	Average ρ	Median Grain Size (mm)	% Aggregates[d] 0.25 to 1 mm	% Aggregates[d] 90μm-150μm	CH_4 μg/g[e]	^{36}Ar[e] ccSTP x 10^{-8}
12028,61,67,69	Coarse layer double core	0	4.2×10^7	0.570[a]	0	-	-	-
12030	Between LEM and Head Crater	0	4.4×10^7	0.105[b]	0.5	-	-	-
12033	Trench sample near Head Crater	0	1.35×10^7	0.097[a]	2	-	0.1	2.4×10^3
14141	Cone Crater	11	5.2×10^7	0.735[c]	1	5-12	0.5	-
14149	Bottom of trench	26	6.4×10^7	0.410[c]	16	26.4	3.0	2.6×10^4

a) Quaide et al. (1971); b) King et al. (1971); c) LSPET (1971); d) McKay et al. (1971); e) Cadogan et al. (1972).

wind content of the soils), the fraction of glassy agglutinates (indicative of the micrometeoroid bombardment), and the median grain size, for soils with high and low TR values, respectively (Crozaz et al., 1972d,e).

In finishing this brief survey of the general track characteristics of soils, it should be pointed out that differences in experimental techniques make it difficult to compare results from one laboratory to the next by better than about a factor of two. The Indian group, for example, sometimes estimates whether a crystal has $>10^8$ t/cm^2 by its appearance in an optical microscope. Several other groups use scanning electron microscopes to make total pit counts. In typical cases where only "well-formed, deep pits" are counted, the total pit counts are approximately twice as high as the track counts that would be made using optical microscopy. Another standard technique is to make surface replicas of the etched crystals and count these either with a scanning electron microscope or with a transmission electron microscope. If everything is counted on the replica, then this is equivalent to the total pit counting in the scanning electron microscope. However, in the case of replicas, a length criterion can be employed. In several samples where an overlap between the various techniques was studied, Crozaz et al. (1972d,e) found that agreement with optical microscope measurements was found for feldspar crystals if only tracks longer than one-sixth of the maximum track length were counted.

It is generally assumed that the tracks that are counted in the soil grains are produced by VH ions, either from solar flares or galactic cosmic rays. This is not necessarily always the case; at high densities ($\gtrsim 2 \times 10^7$ t/cm^2) where accurate optical measurements become impossible, all the counting techniques count short tracks such as are also produced by inelastic proton and alpha particle collisions. However, this fact is not critical for most of the discussion that follows. Track exposure ages for soils are generally determined by counting crystals with low track densities where the optical methods distinguish clearly the tracks produced by VH ions.

6.6.7. The Unique Radiation Record of the Micron-Sized Fraction of the Lunar Soil

The smallest size fraction of the lunar soil, consisting of particles in the range of one micron in size, has a distinct and unique radiation history. In a series of remarkable papers, the Orsay group delineated the radiation record of this fine-grained material and showed by means of calibration irradiations how the observed features could be caused by a combination of solar irradiations. They have further discussed the implications of their results for a number of important lunar and astrophysical problems. The Berkeley group has also contributed important information to the study of the fine-grained fraction.

The small size of the particles makes it possible to study them directly using high voltage transmission electron microscopy. The Orsay group has used the

1.5 MeV microscope at Toulouse, France, while the Berkeley results were obtained with a 650 keV microscope located on the Berkeley campus.

Fig. 6-31 illustrates the four remarkable features of the fine-grained material first reported by Borg et al. (1970) and Dran et al. (1970). The interiors of the grains are characterized by very high track densities with typical values ranging from 10^{10} t/cm^2 to $>3 \times 10^{11}$ t/cm^2. The grains themselves are noticeably rounded, and a positive correlation exists between the degree of rounding and the presence of high track densities. A faint, but distinct, thin coating is observed to rim the grains; the lack of electron diffraction shows that these coatings are amorphous. Finally, it can be seen that grains can be fused or stuck together when the coatings are in contact.

Dran et al. (1970) originally suggested that the amorphous layer surrounding the individual grains was due to the accumulative effect of radiation by an intense solar wind bombardment of the grains while on the lunar surface. Subsequent experiments using laboratory irradiations with low energy ions have given strong support to this interpretation (Bibring and Maurette, 1972; Bibring et al., 1972a,b; 1973a). Amorphous coatings, as well as the characteristic rounding, are produced in angular crystals originally removed from the interiors of lunar rocks. Further, the thickness of the coatings so produced is found to reach an equilibrium value compatible with that found on the lunar dust grains. The coatings on both the natural and irradiated samples are also found to anneal at the same temperatures as tracks. In contrast, condensation films produced on grains that have been exposed to vaporized lunar material tend to form layers of small crystallites. The thicknesses of such vapor-deposited films also increase with time.

The thickness of an amorphous layer produced by ion bombardment is found to vary as $E^{0.5}$. If the amorphous films in the lunar dust are indeed produced by solar wind ion bombardment, then measurements of the coating thicknesses in different core tube layers give information on the past variations of solar wind energies. We shall return to this point in our discussion of solar particle paleontology.

The Orsay group has emphasized the potential importance of the "sticking" phenomenon observed in the fine-grained material. They propose that the radiation damage in the amorphous layer is accompanied by a large amount of stored energy (this is a well-known effect whose importance for nuclear reactors was first noted by E. Wigner). When two grains collide, raising the temperature of contact, some of this energy is released, providing a still further increase in temperature. In this way the fusing of two grains is facilitated. Quite apart from the mechanism involved, the Orsay group has demonstrated that the presence of the amorphous layers affects the formation of some lunar breccias. Bibring et al. (1972a,b) showed that a sample of lunar dust that had been leached to remove the amorphous coatings was very much more resistant to heat sintering than a sample of unleached material. Bibring et al. (1972a) also found small particles of lunar dust welded to the sides of Surveyor III materials returned on the Apollo 12 mission. Since the dust particles were probably low-velocity ejecta blown out by the Apollo 12 descent engine, this points again to an efficient low-velocity sticking mechanism.

Bibring et al. (1972b, 1973b) have emphasized the important point that the sticking process could give a mechanism for building up aggregates of matter in a dust cloud surrounding an active star—perhaps the nucleation mechanism responsible for our own planet.

The origin of the very high track densities seen in the micron-sized grains is still not clear. The Berkeley group (Barber et al., 1971b) originally proposed that the micron-sized grains were not indigenous to the moon but had received their intense irradiations in space while floating as free particles. This argument was mostly based on the view that the track densities were simply too high to have been produced *in situ* on the moon. Borg et al. (1971), however, took a different point of view and attributed the high track densities to bursts of "suprathermal" ions released sporadically by the sun. To bolster this view, Borg et al. (1972) report evidence that the particles producing the high track densities have more of a "pulse spectrum" than the inverse power law spectrum characteristic of solar flares, i.e., the high track densities are concentrated in the outer micrometer or two and then drop precipitously to track densities more characteristic of solar flare irradiations seen in larger grains.

However, the observation of enhanced emission of heavy particles in solar flares (Price et al., 1971a; also see Chapter 5) has relaxed many of the problems connected with the high track densities. Although it is likely that tracks were produced by an *in situ* irradiation of the grains while they were on the moon, our state of knowledge of emissions of particles from the sun in the range 10 keV/nuc to 100 keV/nuc is too meager to decide whether the tracks are produced by pulses of "suprathermal" ions or simply by low-energy solar flare particles.

The fact that the micron-size grains are *uniformly* coated with amorphous coatings suggests that these fine particles are more easily stirred on the lunar surface than the coarser grained fragments which frequently show a pattern of anisotropic irradiation. This tendency of larger grains to be less uniformly irradiated has been shown to persist to even larger fragments by Macdougall et al. (1972). Gold (1971) and Criswell (1972) have both proposed that electrostatic effects may be important in moving small particles in the dust. In particular, Criswell has given an analysis of a limb glow observed by the Surveyor VII spacecraft which implies an annual churning rate of 10^{-3} g/cm^2 for particles of $\lesssim 6$ μm in size.

Durrieu et al. (1971) have pointed out that heating of heavily irradiated grains results in the formation of a characteristic pattern of tiny microcrystallites. The presence of these microcrystallites in grains collected on earth might therefore be used to identify grains that had once been irradiated in space and then heated as they slowed down in the earth's atmosphere. Durrieu et al. (1971) show a picture of a dust grain removed from a sample of antarctic ice that is suggestive of such a history.

6.6.8. Calculation of Model Ages for Soil Samples

Consider a single impact event that deposits a layer of fresh, unirradiated material at least several centimeters thick on the lunar surface. If no stirring of the layer takes place, then only the uppermost millimeter or so of material will be affected by solar flares. The tracks in the bulk of the layer will be produced by galactic cosmic rays. Because of the nature of the track production rate curve, most of the crystals will have a rather low track production rate. If the sample is then scooped up and thoroughly mixed, the track distribution function (neglecting solar flare effects) for crystals chosen at random will resemble that shown in Fig. 6-33, the lower cut-off corresponding to those crystals that were at the point of deepest penetration of the scoop.

If stirring of the layer occurs, more and more particles will be brought to the surface and exposed to solar flare particles. The net result will be an increase in the high track density tail. If the stirring continues for a long time, then a large number of the crystals will be exposed to solar flares; the average track density will increase, as will the spread of track densities.

Of course, more complicated stirring can occur, with lightly irradiated material being brought up from depth at infrequent intervals and fresh or partly irradiated material being added from nearby impacts. Fig. 6-33 indicates schematically the way the track density distributions change with time. These diagrams are based on a specific Monte Carlo approach to the problem of mixing given in an early paper by Comstock et al. (1971).

In the case where a well-defined layer of material is found that exhibits the simple track distribution function of Fig. 6-33 (bottom), then the exposure age of a soil sample has a clear and precise meaning. If either the maximum or the average depth of the sample can be independently determined, the track exposure age for the soil is as valid as the track exposure age of a (rare) surface rock with a measured track gradient consistent with a single exposure history.

Although, as we shall show, some simply irradiated soil layers exist, by far the largest majority of soil samples have a complex pattern of track densities indicating a complicated irradiation history. However, by searching through a large number of crystals, picking out those with the minimum track densities, and assuming that the crystal resided at the bottom of the layer, it is possible to define a "minimum ρ" model age.

Strictly speaking, this age defines the time when the *freshest* crystal was last added to the total assemblage of grains in the sample. Because even this crystal

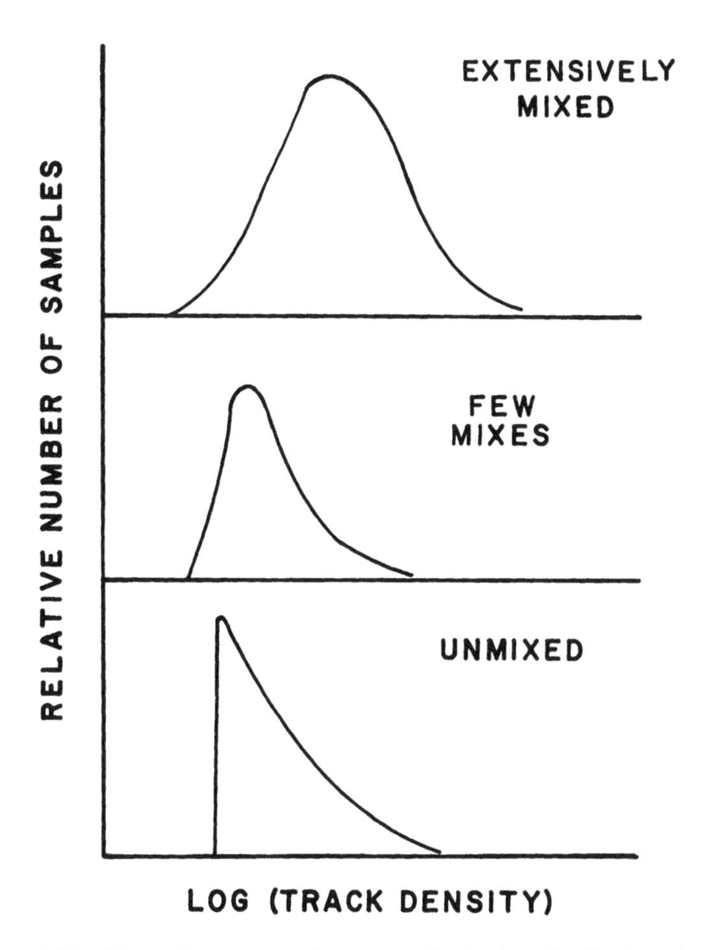

Fig. 6-33. *Schematic representation of track density distributions in upper portion of a soil layer that is given (top) many, (middle) few, or (bottom) no mixing events. (After Fleischer and Hart, 1973c.) The characteristic increase in track densities as a lower cutoff is approached in the unmixed soil is simply a reflection of the form of the track density vs. depth profile for galactic cosmic rays. The upper two curves are based on the mixing model of Comstock et al. (1971).*

may have had an inherited track density accumulated prior to final deposition, the actual time when it was added may be shorter than the model age.

Another useful model age was first defined by the Indian and San Diego groups (Arrhenius et al., 1971). Taking the expression given by Bhandari et al. (1971b) for the track production rate below an infinite plane surface (eq. 6-10), these authors showed that the age T(my) for a simply exposed layer could be expressed as follows:

$$T = \rho_{.25}/K(A + BX)^{-\alpha} \qquad (6\text{-}13)$$

where $\rho_{.25}$ is the quartile track density, and $X = .75\ X_m$ for a linear scoop with a maximum sampling depth of X_m.

The rationale for using the quartile track density is that it includes a representative sampling of crystals while still being relatively insensitive to the presence of a substantial fraction of preirradiated material with a high track density.

As might be anticipated, ages based on average track densities are systematically higher by a factor of two to five than those based on the quartile track densities; these, in turn, are systematically higher than model ages based on minimum densities (Fleischer and Hart, 1973a,d; Fleischer et al., 1973).

Still another model age has been given by Macdougall (1972). This method uses measured azimuthal angular distributions to define a burial depth following the theory previously outlined (see Figs. 6-7 and 6-8).

Model ages can be particularly useful for two different problems. If a distinct cratering event sprays unirradiated or reset crystals (whose tracks were removed by heat or mechanical shock in the cratering event [Fleischer and Hart, 1973e]) onto a mature, well-irradiated regolith, then a scoop of regolith material may contain a sizeable population of distinctly young crystals that can be used to date the cratering event. The potentialities and limitations for crater dating will become apparent in the next section when we discuss certain specific soil samples from the Apollo 16 mission.

Another application of model ages is the construction of a detailed chronology for the deposition of individual layers seen in core tubes. This important subject is treated in a separate section.

6.6.9. Track Results in Some Lunar Soil Samples of Special Interest

There are at least six soil samples that appear to be distinct, relatively unmixed, single units. In sequence of their discovery, these include the following:

a) A very distinct coarse-grained layer ~4 cm thick found in an Apollo 12 double core (12028) at a (corrected) depth of ~18 cm.
b) Another coarse-grained sample 12033 removed from a trench.
c) Sample 14141 removed from the edge of Cone Crater.
d) From the same Apollo 14 mission, sample 14149 removed from the bottom of a trench.
e) A scoop sample containing an abundance of green glass spherules 15401, collected on the Apollo 15 mission [Fleischer and Hart, 1973c].
f) A scoop sample from Apollo 17, 74220 (Fleischer et al., 1974) containing reddish brown spherules.

All of these samples are visually distinctive (primarily by virtue of their coarse grain size) from the majority of lunar samples. In all these samples the track densities are low and have the characteristic distribution shown in Fig. 6-33 (bottom). In addition to these simply irradiated soils, there exist a number of other samples with a distinctive low track density component coupled with a higher track density fraction. These soils can also be used to estimate surface dwell times and the formation ages of nearby craters.

The first three soils listed above can be used immediately to set constraints on the rate of deep mixing of the lunar soil. Using the minimum ρ method, Crozaz et al. (1971) conclude that the coarse-grained layer of the Apollo 12 core remained on the surface for no more than $\sim 15 \times 10^6$ yrs (as shown later, the stratigraphic analysis of Arrhenius et al. gives a much lower age of ~ 1 my). The corresponding age for the Apollo 12 trench sample is $< 40 \pm 5 \times 10^6$ yrs by the minimum ρ method (Crozaz et al., 1971) and ≤ 80 my by the quartile method (Arrhenius et al., 1971). Three groups (Bhandari et al., 1972a,d; Hart et al., 1972a,b; Phakey et al., 1972a,b) gave ages for the Cone Crater sample of 6 to 8 my. Crozaz et al. (1972d,e) originally estimated a much higher age of 15 to 25 my but now believe that a better estimate is 10 to 15 my. Interpretation of the Apollo 14 trench sample 14149 is somewhat ambiguous; although it appears to have a distinctive track distribution, independent evidence exists that some material from the top has spilled in.

The Apollo 16 soils give an interesting illustration of the potentialities and limitations of model soil ages as indicators of formation ages. The landing site was dominated by two craters, South Ray and North Ray. The ray ejecta pattern from the former, as seen on orbital photographs, is very distinct and fresh looking; it also appears to overlap that of the older looking, more subdued North Ray Crater. Rare gas measurements on North Ray boulders (Behrmann et al., 1973c; Marti et al., 1973; Drozd et al., 1974) have clearly established the age of North Ray Crater as 50 ± 1 my. Although the picture is somewhat more complex, similar measurements in the boulder-strewn field of South Ray establish its age as 2 ± 0.1 my (Behrmann et al., 1973c; Drozd et al., 1974).

Because of the obvious overlap of the ray ejecta (as seen from orbit) on a number of the stations that were sampled, it was originally thought that many of the soil samples would be dominated by South Ray material. However, crystals with low track densities corresponding to a 2 my exposure age are notably lacking in these samples. For example, Behrmann et al. (1973c) found not one crystal with track density less than $10^7/cm^2$ out of 274 studied from different soil samples supposedly dominated by South Ray material. Similarly, Bhandari et al. (1973b) report a model soil age of 17 my that they attribute to the formation of South Ray Crater. The measurements of Fleischer et al. (1974), with one exception, also give high model ages for material supposedly from South Ray Crater. They report values of 27 my, 6 my, and 36 my, respectively, for trench samples taken at different levels. The spread on the youngest of these ages would almost overlap the rare gas age of 2.0 my, and this sample possibly contains genuine South Ray material.

Independent measurements of albedo and glass agglutinate content confirm that the so-called South Ray samples are relatively mature soils. The missing young material remains something of an enigma but is possibly associated with the patchy distribution of the South Ray ejecta. For a fuller discussion, see Behrmann et al. (1973c), McKay and Heiken (1973), and Drozd et al. (1974).

Although model ages for North Ray soil samples give somewhat better agreement with the rare gas ages of 50 my, the spread of values shows the inherent difficulty in precise age determinations by this method. For example, Bhandari et al. (1973b) give model ages of 24 my and >100 my, while Behrmann et al. (1973c) find a model age of ~100 my for a North Ray sample.

The samples returned from the Russian Luna 16 and 20 missions have proven extremely interesting from the track standpoint (as well as from many other points of view). Track measurements in the larger grains (>50 μm) from different levels in the Luna 16 core (Berdot et al., 1972a; Comstock et al., 1972a,b; Phakey and Price, 1972; Walker and Zimmerman, 1972; Bhandari et al., 1973a; Comstock et al., 1973; Poupeau et al., 1973) show that the material at all levels is highly irradiated (TR% from 64% to 100%), though not outside the limits that have been measured for some of the more heavily irradiated samples returned on the Apollo missions. However, measurements of the finer-grained fraction of the Luna 16 material reported by Phakey and Price (1972) give a more extreme view of the intensity of radiation to which samples have been exposed. Only 7 to 9% of the micron-sized grains show strong diffraction patterns, contrasted to 30 to 60% for a wide variety of Apollo specimens. The track densities in these crystals, while resolvable, are much higher on the average than in the comparable fractions of the Apollo soils. The very high proportion of grains showing weak diffraction spots or diffuse scattering is attributed by the authors to an extreme state of radiation damage.

The Orsay group (Durrieu et al., 1973), who studied the very fine-grained fraction of a different level of the Luna 16 core, found track densities that were comparable to the most heavily irradiated fraction of the Apollo samples that they had previously studied. In common with Phakey and Price, they also found a high proportion of grains with weak or diffuse diffraction spots. High voltage electron microscopy showed, further, that many of these grains contained small crystallites similar to those observed in calibration crystals that had been heated following a heavy radiation dose. Durrieu et al. thus propose that the unusual properties of the Luna 16 material stem both from heavy irradiation and heat-induced metamorphism.

The high state of radiation damage of the Luna 16 material is also confirmed by the relatively high spallation exposure age of 930 my measured by Kaiser (1972a) and the very large Sm and Gd neutron-capture isotope anomalies observed by Russ (1972).

The reason for the relatively high state of radiation damage is unclear. Phakey and Price (1972) proposed that the soil was relatively unstirred compared with typical Apollo soils. They argued that, if stirring of very fine grains takes place by an electrostatic mechanism, as originally suggested by Gold (1971) (see also Criswell, 1972), the lowered efficiency could be explained by the proximity of the Luna 16 landing site to the limb, with a consequent lower exposure of the material to energetic electrons associated with the earth's magnetospheric tail. However, Bhandari et al. (1972a,d) take the opposite point of view that rapid stirring would give a longer average surface exposure and that, in any event, the Luna 20 samples taken at about the same longitude do not show the heavy irradiation effects. Whether the geographical location of the sample is important, or whether the effect is connected to the local topography, or whether, indeed, Luna 16 simply reflects a statistical fluctuation, may only be settled when we revisit the moon.

A number of authors (Vinogradov, 1972; Bhandari et al., 1973a; Comstock et al., 1973; Crozaz et al., 1973; Phakey and Price, 1973) found that the Luna 20 sample contained a relatively large fraction of crystals with low track densities. However, one group (Berdot et al., 1972b) reported a very high irradiation level in the sample that they studied. Based on the assumption that the material was laid down as a single layer, Crozaz et al. (1973) calculate a model age of 270 my, possibly associated with the formation of the crater Apollonius C.

6.6.10. The Layering Chronology of Lunar Cores

Examination of the lunar cores and drill stems has shown that the lunar regolith is more or less stratified on a local scale. In some cases, for example, the coarse-grained layer of the Apollo 12 double core or the even more impressive, extended, coarse-grained region in the Apollo 17 drill stem, the stratification is striking. In still other cases the strata are very subtle and their identification as distinct layers may be open to debate. The Apollo 15 drill stem has some 58 separate identified units typically 1 to 5 cm in thickness. The horizontal extent of the layers is not known, and they may be quite local discontinuities.

The importance of a gradual build up of the regolith, either in layers or in a particle-by-particle deposition, was emphasized by Phakey et al. (1972a,b) in their discussion of the Apollo 15 drill stem. They pointed out that the purely mixing aspects of the model of Comstock et al. (1971) simply could not account for the (relatively) uniform high irradiation levels found at all depths in the 3 meter core. Comstock (1972), however, has described how his model is consistent with discrete layers and the observed high track densities.

The possibility of obtaining a layering chronology for core tubes through track studies was first pointed out in the pioneering paper of Arrhenius et al. (1971). They first showed that the layers that had been visually identified differed from one another in their track record (see Fig. 6-34). They then proceeded to build up a layering chronology based on the calculation of quartile model ages for each of the layers. The surface exposure ages of the layers varied from <1 my to a maximum of 60 my. The overall deposition rate (total core length/total build-up

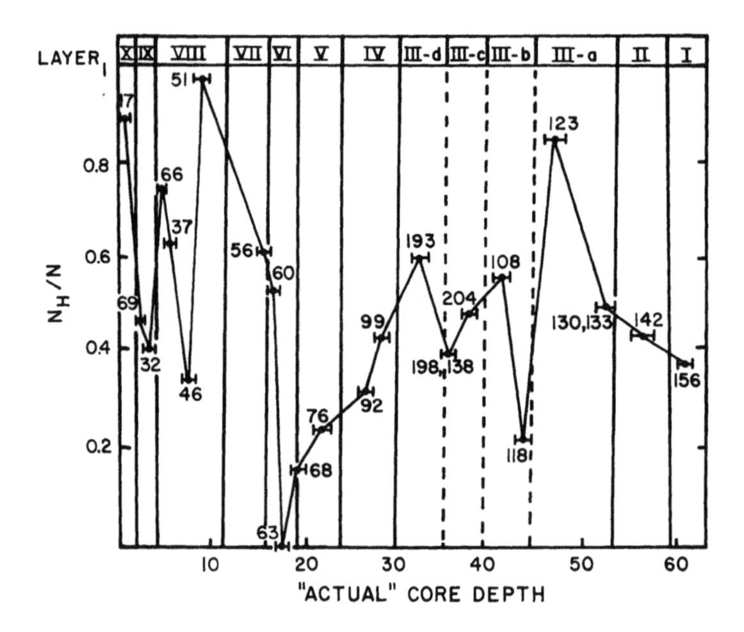

Fig. 6-34. *The fraction (number of grains with* $\rho > 10^8/cm^2$ *plus grains with gradients)/(all grains counted), or* N_H/N*, is plotted against location of sample in the double core of Apollo 12. (After Arrhenius et al., 1971.)* N_H *can also be thought of as the number of grains irradiated without shielding. The numbers refer to the number of grains counted at each position.*

time) was found to be ~0.2 cm/my. Substantial agreement was found by Macdougall (1972) in his work on angular distributions in core crystals. However, Fleischer and Hart (1973e) inferred a higher deposition rate of ≥0.35 cm/my. Subsequent work on layering chronologies has been done principally by the General Electric and Indian groups.

In their analysis of one of six stems comprising the Apollo 15 deep drill core, Bhandari et al. (1972b, 1973b) found a dispersion of layer ages similar to that found in their analysis of the Apollo 12 core and a deposition rate of 0.15 cm/my. In a complementary analysis of two other sections of the Apollo 15 deep drill stem, Fleischer and Hart (1972c) and Fleischer et al. (1974) found somewhat shorter exposure ages of layers and larger deposition rates of ~0.4 cm/my. Fleischer et al. (1974) also found a deposition rate of 0.22 cm/my in a limited sampling of a section of the Apollo 16 drill stem. As pointed out by Fleischer and Hart (1972c), the inferred deposition rate is independent, to a first approximation, of whether layers have been correctly identified and separated. This follows from the fact that the track production rate is nearly inversely proportional to depth in the range from 1 to 10 cm.

The apparent agreement between the Indian and General Electric analyses is to some extent fortuitous. Track production rates used by the two groups differed by a factor of 2.5 and one used the quartile density method and the other the minimum track method. The track production curve used here would raise the ages computed by the Indian group and lower those of the General Electric group, thereby increasing the differences in deposition rates. The General Electric group interpret the difference as due to the effects of predepositional irradiation, which would raise the ages inferred from quartile track densities but would not alter the minimum values.

The two groups also differ in their view of the layering process. The General Electric group stresses the fact that the track distribution functions measured in many individual layers are qualitatively similar to those in Fig. 6-33 (top), indicating that most of the crystals have a complicated preirradiation history before being finally deposited as part of a distinct layer.

The qualitative picture that emerges from the core studies is as follows: Cratering events, both large and small, throw out material onto the surrounding terrain. If the material that is thrown out comes from the surface layer of an old terrain, it will already possess all the features of what we have described as a "mature" soil. If, on the other hand, the material comes from considerable depth, it will have an initial low track density and start to record tracks rapidly at the instant of deposition. Because small impacts are more frequent than large impacts, the surficial layer of a fresh deposit will be rapidly stirred, exposing a large proportion of the surficial grains to the sun. As time goes on the layer will become more deeply stirred and will eventually mix with the underlying material, blurring the original boundary marking the layer. Eventually, an impact in an adjoining area will either throw a new layer on top of the previous one, or, if it hits in the exact spot that we are considering, will destroy the layer structure completely to a depth that depends on the size of the projectile.

Apart from the obvious information obtained about lunar surface dynamics, the great importance of these core studies is the ability to identify samples that were irradiated for various, relatively brief times, at specified times in the past. Detailed studies of solar wind and solar flare effects in dated layers could then give a detailed record of the past variations in solar activity.

In the case of the Apollo 15 deep drill core, it appears that the total interval of time that can be studied is $\sim 5 \times 10^8$ yrs to 10^9 yrs. Clearly, additional work on cores, both by track techniques and by independent chronological methods (see, for example, Russ et al., 1972) should have a high priority for future lunar sample investigations.

In our discussion we have so far assumed that layers build up in a simple depositional pattern with the oldest layers at the bottom and the youngest layers at the top. Unfortunately, this is not necessarily the case. Layered material near the lip of a crater can be folded back on itself, preserving the original layering structure but reversing the time vs. depth sequencing (Shoemaker, 1960). It is for this reason that measurements by a variety of techniques are necessary to establish a true chronological stratigraphy.

6.6.11. Absolute Rates of Microcratering on Lunar Rocks

Impact craters are a ubiquitous lunar phenomenon. They can be seen using a pair of field glasses from earth, they are obvious on any hand-held lunar rock (see Figs. 6-35 and 6-36), and still more show up when an electron microscope is used to examine lunar samples at high magnification (see Fig. 6-37). In this section we treat the problem of determining the absolute rates of infall of extralunar material based on direct measurements of dated lunar samples. In the next section we consider the effects of these infall rates on the various dynamic processes treated in previous sections.

Gault et al. (1972) give a critical discussion of the best current values for the rate of infall of extralunar material. Their values, shown in Fig. 6-38, fall into two separate ranges. The fluxes of particles in the range 10^{-12} g to 10^{-6} g are taken from instruments on space probes; those in the range 10^{-6} g to 1 g are inferred from meteor data. Strict criteria were employed in the selection of the spacecraft data and a number of data points included in other recent reviews (Kerridge, 1970; McDonnell, 1971; Soberman, 1971) were rejected. The reader is referred to the original article for a fuller discussion of the input data used for Fig. 6-38.

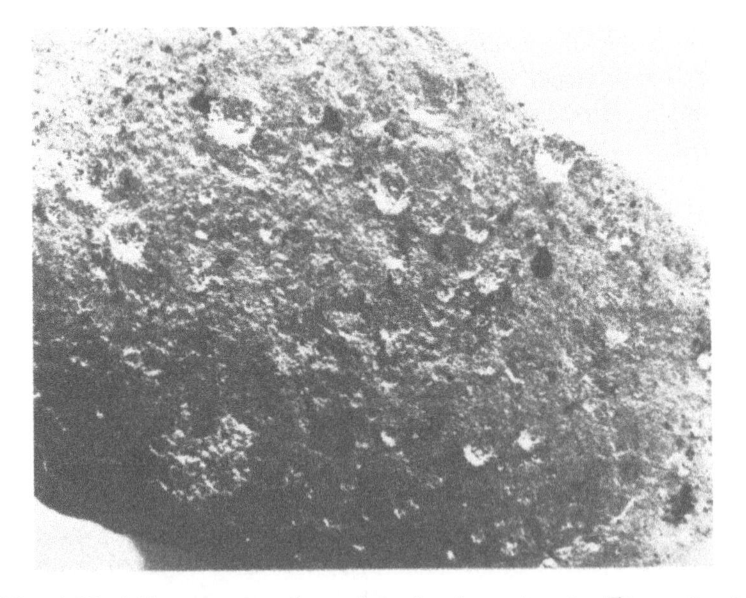

Fig. 6-35. *Microcraters on the surface of a lunar breccia. The easily visible microcratering gives a characteristic, unmistakable appearance to a lunar rock.*

Fig. 6-36. *Surface of rock 14305 showing a deep crack running through a large impact crater (dark, glassy region 6 mm in diameter). A slightly larger micrometeoroid would have shattered the rock. (Photo courtesy of F. Hörz.)*

Fig. 6-37. *Impact pit on a lunar glass spherule as seen in a scanning electron microscope. (After Crozaz et al., 1970b.) The characteristic spall zone around the center of the pit shows that it was produced by the hypervelocity impact of an interplanetary dust grain.*

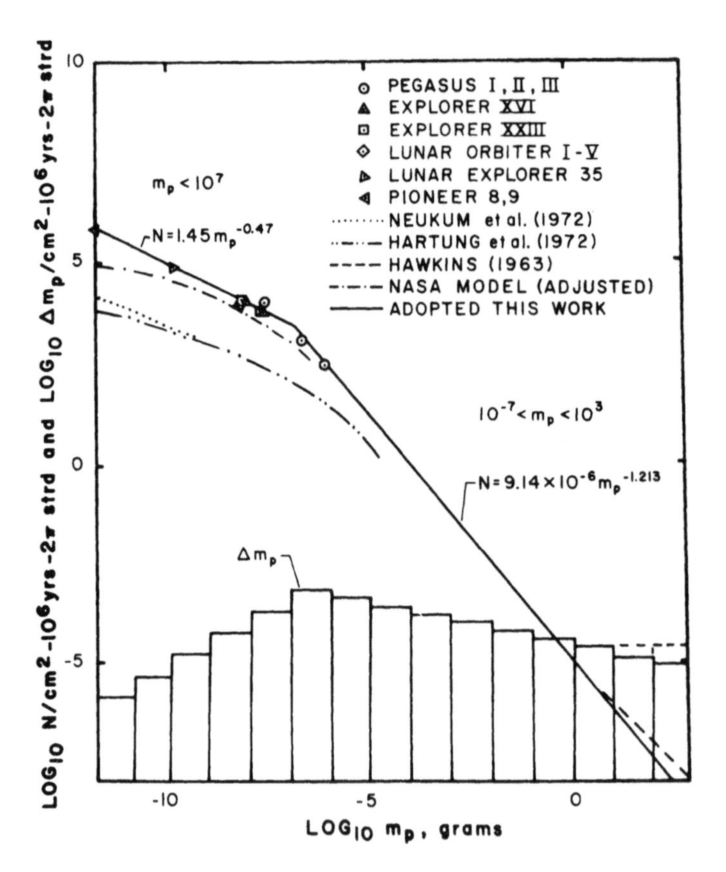

Fig. 6-38. *Mass-flux distribution from Gault et al. (1972). This curve shows a selection of the best data for micrometeoroid flux from spacecraft experiments. The solid line shows the distribution used by Gault et al. to predict the rate of impact processes with contemporary fluxes. Also shown is the incremental mass flux contained within each decade of M_p.*

Also shown in Fig. 6-38 is the incremental mass contributed by meteoroids in different mass ranges. The objects making up most of the total bombarding mass lie in the range 10^{-9} g to 1 g, with relatively little being contributed at lower and at higher mass values. However, Gault et al. (1972) point out that *very* large objects with masses $\gtrsim 10^{10}$ g may also contribute a significant fraction.

As discussed by Hartung et al. (1972), the relationship between the mass of an impacting projectile and the size of the impact crater it produces is somewhat uncertain. In their "Calibration I" model, which has been used to derive erosion values and absolute pitting rates cited in this chapter, Hartung et al. (1972) have relied on the experimental work of Bloch et al. (1971), Vedder (1971), and Mandeville and Vedder (1971), who showed that for particles from 10^{-14} g to 10^{-9} g the ratio of the diameter of a hypervelocity crater to the diameter of the impacting projectile is independent of mass. This independence is assumed to hold for projectiles of higher mass as well—short of those that produce catastrophic rupture of the rocks. The ratio of the crater diameter to the projectile diameter, however, depends on the velocity, and this must be estimated from meteor photography. Although some evidence exists for complex velocity distributions, the value of 20 km/sec is taken by most workers as representative of the average impact velocity. At this velocity, the crater/projectile diameter ratio measured for both iron and polystyrene projectiles is $\sim$2.

Assuming that the incident meteoroids are spherical and have a density of $\sim$3 g/cm^3 (characteristic of silicate minerals), craters that range from 1 mm to 100 μm in diameter correspond to the mass range from 10^{-3} g to 10^{-6} g, whereas craters in the size range from 100 μm to 0.1 μm correspond to the mass range from 10^{-6} g to 10^{-15} g. Craters larger than 100 μm correspond to a transition region above which data on present fluxes are derived primarily from observations of meteors. Only craters in the region 1 μm to 100 μm overlap with measurements in space probes.

As a surface is exposed longer, the number of pits increases until eventually they start overlapping and erasing one another. For any given size pit, there is a maximum exposure age beyond which the number density of that size crater (and smaller ones) does not change. The surface is then said to be "equilibrated." Contrary to the intuitive reaction of many people, a surface that is equilibrated for a certain size crater is not completely covered with overlapping craters of this size. Instead the surface merely has a characteristic distribution of craters. The reason that the craters do not overlap is that occasional impacts of still larger particles continually "reset" different parts of the surface.

When the crater population is sufficiently small that the interaction between craters may be ignored, the surface is said to be in a "production" state. In this state the slope of the logarithmic pit distribution curve is ≈ -3 for pits greater than several hundred microns in diameter. For older surfaces the crater distribution function flattens, reaching a limiting slope of about -2. Morrison et al. (1972) give examples of the progressive changes in crater distribution functions with age.

Clearly, if a production surface can be identified and then independently dated, the absolute production rate of impact craters can be measured. If the diameters can, in turn, be related to the masses of the bombarding particles, then the lunar data can be compared directly to the contemporary measurements shown in Fig. 6-38. Alternatively, once the rate of cratering is established, crater counts can be used to date other samples (assuming a constant flux of micrometeoroids) also found to be in a production state.

Unfortunately, attempts to obtain absolute pitting rates have led to a certain amount of confusion and controversy. For one thing, it appears (Neukum et al., 1973) that different *types* of rocks may have different crater populations. In Fig. 6-39 we show the areal density coefficients for a number of rocks. These coeffi-

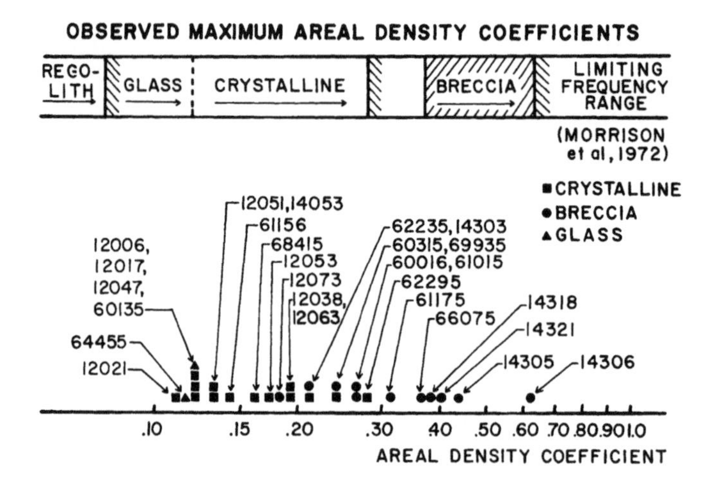

Fig. 6-39. *Comparison of highest areal density coefficients observed to date on various lunar surface materials from Neukum et al. (1973). The coefficient is defined as the extrapolated number density for craters of 1 cm based on measurements of the number vs. size for smaller craters. Note the systematic difference between various lunar rocks. The differences may be attributed to exposure age, rock properties, or a combination of both. The areal density coefficients for glass do not necessarily mean the highest approachable densities because only production populations have been examined. For comparison, coefficients for loose regolith are also included.*

cients are obtained by fitting a portion of a curve of pit frequency vs. size distribution with a production slope of -3 and then extrapolating to a crater size of 1 cm. It can be seen that there is a clear progression in pit populations from glassy surfaces on one extreme to breccias on the other with crystalline rocks in between. These differences cannot be attributed solely to differences in surface exposure ages and most probably reflect differences in the mechanical properties of the rocks. It is thus difficult to extrapolate data from one rock to another or to convert from crater diameters to impacting masses without considerably more calibration work than has yet been done.

There is also the outside chance that localized showers of micrometeorites may cause fluctuations in the crater production rates from one site to another or even from one sample to the next.

The most recent summary of absolute crater production rates is given by Hörz et al. (1973), who give a list of crater counts, determined by specialists in this field, for craters with central pits $\geq 500\ \mu m$ in size for a number of lunar samples, along with exposure ages determined by several different methods. A modified listing using the best available current data is given here as Table 6-12. We conclude from this table that the absolute production rate of craters $>500\ \mu m$ in size lies in the range of $1/cm^2$ my to $3/cm^2$ my for most rocks. This is somewhat lower than the value given by Hörz et al. (1973). The data for rocks 12017, 12054, and 14301

Table 6-12. Selected Surface Exposure Ages and Crater Counts for Craters > 0.05 cm in Diameter

Rock	No. of Craters Counted	ρ_c (cm^{-2})	Surface Exposure Age (my)	Cratering Rates cm^{-2} my^{-1}
12017	12[a]	2.3	$\lesssim$ 0.7[f]	$\gtrsim$ 3.3
12038	30[a]	3.6	$\lesssim$ 1.3[g]	$\gtrsim$ 2.8
12054	4[b]	0.4	0.05 to 0.5[h]	$\lesssim$ 0.8 to 8
14301	54[c]	2.5	> 1.5[i]	$\lesssim$ 1.7
14303	10[d]	2.3	$\lesssim$ 2.5[j]	$\gtrsim$ 0.9
14310	30[b]	2.0	1.5 to 3[k]	$\gtrsim$ 0.7 to 1.3
60315	14[e]	3.4	> 1.5[ℓ]	$\lesssim$ 2.3
61175	196[e]	13.2	> 1.5[m]	$\lesssim$ 8.8
62295	34[e]	4.0	$\lesssim$ 2.7[n]	$\lesssim$ 1.5
68415	57[e]	2.3	> 1.5[o]	$\lesssim$ 1.5

a) Hörz et al. (1971). b) Hartung et al. (1972). c) Morrison et al. (1972). d) Hartung et al. (1973). e) Neukum et al. (1973). f) Fleischer et al. (1971b). g) Bhandari et al. (1971b). h) Schönfeld (1971). i) Keith et al. (1972). j) Bhandari et al. (1972a). k) Crozaz et al. (1972e). ℓ) Clark and Keith (1973). m) Eldridge et al. (1973). n) Bhandari et al. (1973b). o) Rancitelli et al. (1973).

are particularly interesting because these surfaces appear to be in a production state.

Independent estimates of cratering rates have been given by the Washington University group for rocks 12063 and 68815. The surface exposure age of 12063 was determined by a detailed gradient measurement and that for 68815 by a Kr-Kr measurement. The rates are, respectively, $2/cm^2$ my for pits ≥ 1 mm in 12063 and $\geq 40/cm^2$ my for pits ≥ 30 μm in size for rock 68815. Extrapolation of these data to crater sizes of 500 μm using typical crater distribution curves (Morrison et al., 1972) give rates of $\sim 6/cm^2$ my and $\sim 3/cm^2$ my, compatible with those shown in Table 6-12.

As discussed in the next section, all these rates may be somewhat low due to the neglect of crater erosion.

Both Neukum (1973) and Morrison et al. (1972, 1973) have estimated exposure ages for Apollo 16 rocks based on assumed absolute cratering rates. Although the ages for some of the South Ray rocks agreed well with independently determined rare gas exposure ages (Behrmann et al., 1973c; Drozd et al., 1974), it appears that the agreement may be fortuitous. The somewhat different approaches used by Neukum (1973) and Morrison et al. (1973) have both been strongly criticized by Hartung et al. (1973), who maintain that these techniques are premature at this time. Neukum et al. (1973) also indicate that the South Ray rocks that were previously thought to be in production states may, in fact, be in transitional states where the crater counts would not be linearly proportional to age.

Extensive measurements of small pits ≥ 0.8 μm have also been reported by Neukum et al. (1972) and Schneider et al. (1972, 1973) on glass surfaces found either on lunar surface rocks or in fragments removed from the soil. In all these surfaces the track profiles closely matched that measured in the Surveyor III glass and the exposure ages were determined by these solar flare tracks. In making the age determinations, correction had to be made for thermal fading of the tracks in the glass; this was done either using the track diameter method of Storzer and Wagner (1969) or by normalizing to track counts in pyroxene crystals trapped in the glass. In their original paper these authors used the direct track counts measured by Barber et al. (1971a), Crozaz and Walker (1971), and Fleischer et al. (1971a) in the Surveyor III glass. In the later summary paper by Hörz et al. (1973), the ages have been raised by a factor of three based on the measurements of Storzer et al. (1973b) of the etchable range of iron ions in a glass similar to the Surveyor III glass. This adjustment in ages turns out not to be warranted, if one uses the revised Surveyor glass spectrum shown in Fig. 5-22 and discussed by Hutcheon et al. (1974).

Conversion of these crater data to fluxes of micrometeoroids leads to a large discrepancy with the contemporary data shown in Fig. 6-38. The lunar values averaged over the last 10^5 yrs appear to be systematically lower by one to two orders of magnitude than the satellite measurements in the mass range 10^{-9} to 10^{-11} g. The possibility that the present flux of micrometeoroids might be higher than the long-term average had previously been raised by Gault et al. (1972) and Morrison et al. (1972) in their discussion of rock exposure ages and erosion rates.

Perhaps the strongest evidence for possible changes of the meteoroid flux with time comes from the recent work of Storzer and Hartung (1973), who measured flare track exposure ages of 56 individual craters larger than 20 μm on rock 15205. The results indicate that the formation ages were not uniformly distributed in time; approximately ten times as many craters per unit time were produced during the last 10,000 years as the rate averaged over the total exposure of 10^5 yrs. The values obtained for the past 3,000 yrs are in good agreement with the satellite data. However, many of the lunar impact glasses have imperfect track retention, which may increase the apparent number of recent impact events.

In spite of the uncertainties in the absolute cratering rates, it should be noted that the detailed study of the characteristics of the craters has given a wealth of information on the micrometeoroids responsible for their production. For example, Hörz et al. (1973) conclude that over 90% of the pits were produced by micrometeoroids that were roughly spherical in shape, had densities that probably lie in the range 2 to 4 g/cm^3, and impacted the moon at velocities in excess of 5 km/sec. The existence of a finite flux of particles below 10^{-14} g has also been clearly established and negates the existence of a radiation pressure cut-off as had been suggested by a number of authors.

Many intriguing problems remain to be resolved. Different surfaces have crater distributions that are distinct from others and, in particular, a bimodal distribution of craters is clearly apparent on certain surfaces. The origin of the particles is also not certain, but a cometary source is compatible with the existing data. For a fuller discussion of these subjects, the reader is referred to the review article by Hörz et al. (1973) and the references contained therein.

6.6.12. Meteoroid Impacts as the Major Driving Mechanism for Dynamic Lunar Processes

There can be no doubt that the processes that form craters also stir the regolith, break up boulders, and contribute to mass wastage of lunar rocks. The key question is to what extent the infall of extralunar matter can quantitatively explain all these processes. Although this problem is far from completely solved, *all* the rates for different dynamic processes so far determined appear to fit comfortably into the meteoroid impact model.

The discussion in this section is based primarily on the fundamental article by Gault et al. (1972). However, the conclusions differ in detail since the selection of data is not entirely the same.

Consider first the question of the true surface exposure ages of lunar rocks. Using the satellite data of Fig. 6-38, Gault et al. (1972) estimate that lunar rocks in the 1 kg to 10 kg range should stay on the lunar surface from 2 to 6 my on the average before being catastrophically ruptured by impact of particles in the 10 mg to 100 mg range (see Figs. 6-36 and 6-40). The agreement between theory and experiment regarding the maximum size craters that are found on whole rocks given by Hartung et al. (1973) provides support for this model with a compressive strength of $\sim$3 kb for lunar rocks (1 kb = 10^9 dyne/cm^2).

Based on a comparison of maximum surface exposure ages with these estimates, Gault et al. (1972) advanced the idea that the long-term average flux of micrometeoroids was much lower than given by the satellite data. However, the track exposure age data in Table 6-12 provide no compelling evidence that this is the case. Almost all rocks have multistage exposure histories and the true surface exposure ages, where they have been reliably measured, are generally found to be much lower and to lie in the 1 my to 10 my range. However, one glaring exception to this statement is a chip from a much larger North Ray boulder (see Section 6.3) that appears to have remained relatively undisturbed for 30 to 50 million years. In addition, the number of rocks found to be undersaturated in ^{26}Al (and hence with an exposure of $<1.5 \times 10^6$ yrs) appears to be smaller than would be predicted by Gault et al. (1972).

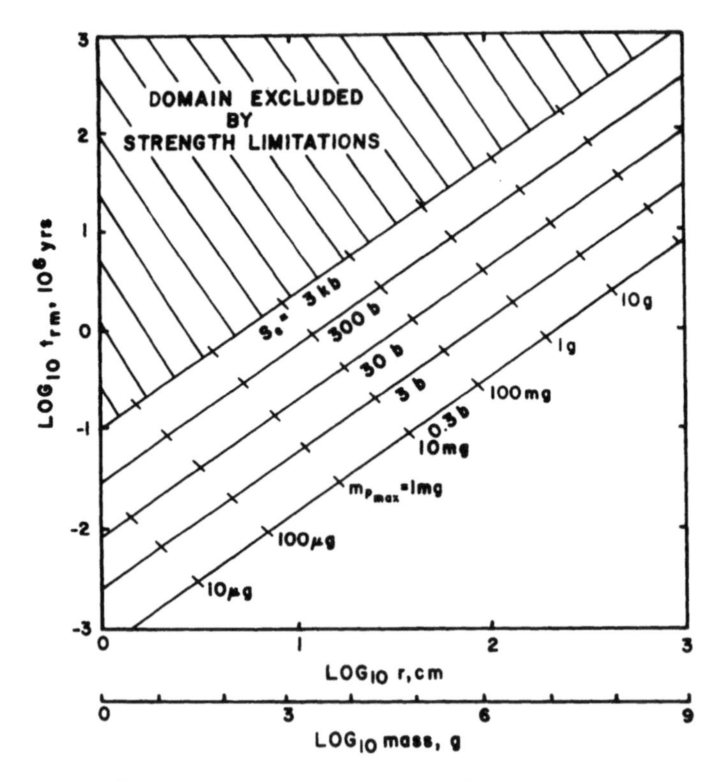

Fig. 6-40. *Calculated mean residence time before destruction by catastrophic rupture* t_{rm} *for spherical rocks of radius r and with compressive strength* S_e *that are exposed on the lunar surface to the meteoritic environment. (After Gault et al., 1972.) Masses of the largest particles,* $M_{p_{max}}$*, that are considered to impact the rocks and contribute to the destruction process are indicated.*

Erosion estimates based on the satellite data range from 0.5×10^{-8} cm/y to 2×10^{-7} cm/y (Gault et al., 1972; Hörz et al., 1973). These are in excellent accord with the measured mass-wastage erosion rates and provide no evidence for a decreased micrometeoroid flux.

The core layering chronology developed by the General Electric and Indian groups, as well as the work by various groups on special soil samples, indicates that layers in the thickness range of 0.5 cm to 5 cm maintain their distinctive character for times of the order of 10^6 to 10^7 yrs. Gault et al. (1972) originally estimated that 17 g/cm^2 of the regolith are stirred every million years. By far the largest amount of stirring—13 g/cm^2—takes place in the first 3 mm of depth. From 3 mm to 6 cm depth the turnover rate was estimated at 10^2 to 10^0 times per million years. The original predicted stirring rates are thus considerably higher than the measured rates. However, Gault (personal communication, Jan., 1974) has recently lowered the calculated stirring rates considerably. Using current flux values, he would now estimate that a layer 1 cm thick would have only a 50% chance of being completely stirred in 1 million years; the corresponding time for a 5 cm layer is 100 my. These new rates seem generally compatible with the experimental values.

Gault et al. also originally estimated a production rate of 8 to 10 craters/cm^{2}my $>$ 500 μm in size. Although this is somewhat higher than the value of 1 to 3 craters/cm^{2}my given in the previous section, it should be remembered that the experimental value is a lower limit since it neglects any loss of craters due to erosional processes. Typical depth to diameter ratios vary from 0.3 to 1.3 (Brownlee et al., 1973). Taking 0.8 as an average value, it can be seen that pits of 500 μm in diameter will be eroded in times on the order of 10^6 yrs at the microerosional rates of 2 to 8 $\times$ 10^{-8} cm/y discussed in Section 6.6.5. Many of the rocks shown in Table 6-12 have been exposed $\geq$2 my and it is thus not unreasonable that the true rate of microcratering may be higher than the 1 to 3 craters/cm^{2}my arrived at in the last section.

It thus appears that the current flux of micrometeoroids is capable of explaining almost all of the dynamic effects that have been measured. Two exceptions are the variable rates of crater production measured by Hartung and Storzer (1974) and the low average rates of crater production given by Neukum et al. (1972) and Schneider et al. (1972, 1973). However, both measurements refer to much smaller particles than the $\sim 10^{-6}$g projectiles that are responsible for the effects discussed in this section. Another possible exception is the unique radiation record stored in the micron-size fraction of the soil. The efficient, isotropic solar bombardment seen in these grains may require additional transport mechanisms such as those suggested by Gold (1971) and Criswell (1972).

6.7. SOLAR WIND AND SOLAR FLARE PALEONTOLOGY: STUDIES OF ANCIENT SOLAR RADIATIONS IN LUNAR SOILS, LUNAR BRECCIAS, AND GAS-RICH METEORITES

6.7.1. Introduction

In our discussion of solar flare effects in lunar rocks, we have shown that the depth dependence of solar flare tracks is similar to that expected from measurements of modern-day solar flares (provided erosion is taken into account). However, because the rocks have been exposed only relatively recently on the lunar surface, and also because the depth dependence is dominated by erosion, the conclusions on the constancy of solar flares are roughly limited to the last million years. To go back further in time, particularly toward the beginning of the solar system, it is necessary to examine the track record in lunar soils, breccias, and gas-rich meteorites.

6.7.2. Lunar Soils

The Orsay group has calibrated the thickness of amorphous coatings produced by ion bombardment and shown that it varies as $E^{0.5}$ (Bibring et al., 1972a,b, 1973a). In Fig. 6-41 we show the results of Bibring et al. (1972a) on the measurement of coating thicknesses in 130 grains chosen at random from lunar soil samples. Also shown is an energy scale for the bombarding particles presumed to have produced the coatings. Based on their calibrations, they estimate that the time to establish the equilibrium thickness by natural solar wind bombardment is on the order of 10^2 to 10^3 years. Thus, the distribution of coating thicknesses measures the variation in average solar wind energies over this time interval.

The figure shows a high frequency of periods of rather low solar wind energy (0.2 keV/nuc $< E <$ 1.0 keV/nuc) and a steep decrease in periods where the average solar wind energy is $>$1.0 keV/nuc. The present-day average energy of $\sim$0.8 keV/nuc falls in the peak of their distribution, indicating that we are now in a typical period of solar activity.

The Orsay group has also extended its measurements to individual layers of the Apollo 15 drill core where the track and gadolinium isotope measurements previously discussed indicate a total deposition time on the order of 5×10^8 to 10^9 yrs. Although somewhat different distributions were observed at different depths, they did not change markedly with depth and, in particular, no secular trend was ob-

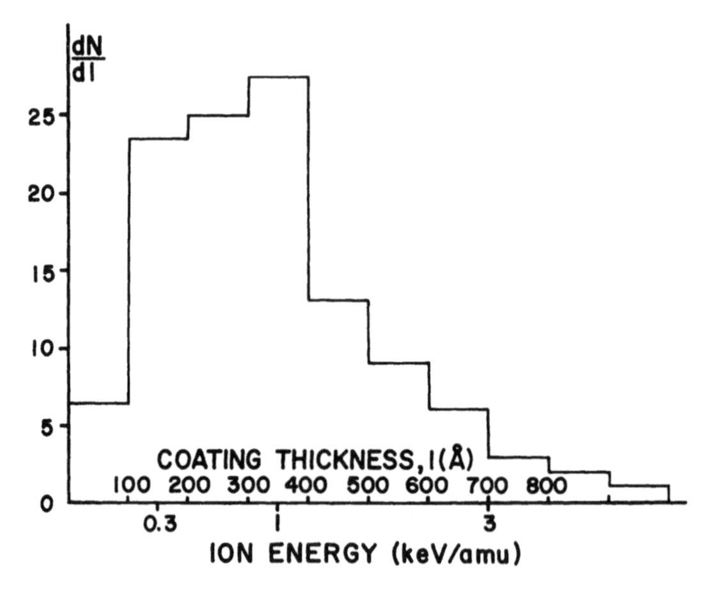

Fig. 6-41. *Distribution of the thickness of amorphous coatings on lunar dust grains. (After Bibring et al., 1972a.) Also shown is an energy scale relating the coating thickness to the energy of bombarding particles. The results show that periods of high solar wind energies lasting for times of 10^2 to 10^3 yrs are relatively rare.*

served (Bibring et al., 1973a). A typical distribution was observed at the deepest portion of the core, showing that the solar wind, pretty much in its present form, has been bombarding the moon for the last 10^9 yrs.

Bibring et al. (1972b) also point out that the existence of solar wind effects in the past sets important constraints on both the lunar atmosphere and any general dipole magnetic field. In particular, they estimate that the lunar atmosphere must have been $<10^{-9}$ atm for the solar wind effects to be observed.

No detailed study has yet been made of solar flare tracks in different soils or in different layers of the various cores (the reader will remember that the track chronology of the cores is derived from low track density crystals assumed to have been bombarded only with galactic cosmic rays). Certain heavily irradiated layers have been observed, as, for example, in the Apollo 15 deep drill (Crozaz et al., 1972d,e), indicating a possible period of enhanced solar activity. However, more work remains to be done on this problem.

The Paris group (Poupeau et al., 1973) has made an extremely interesting observation of the Luna 16 material that may shed important light on the past activity of the sun. These authors showed that feldspar crystals of different types fell into different irradiation categories. Anorthositic feldspars, characteristic of highland rocks, were systematically more irradiated by a factor of about five than feldspars of the type found in mare basalts. Poupeau et al. (1973) suggest that the results indicate a more active sun early in the history of the solar system. The anorthositic crystals, having been added (statistically) to the regolith at an earlier stage than the mare crystals, would be more influenced by this early solar activity. Although other explanations for the observed effect can be advanced, this is an important observation that deserves to be further explored.

Apart from the above observations, there is little qualitative difference observed in the track record in different soils or core layers; the exceptions previously noted are the coarse-grained layers that have clearly spent only a limited time on the surface, and the Luna 16 sample. It thus appears that any long-term changes in solar activity have been relatively subtle and will require painstaking work to establish with certainty.

6.7.3. Lunar Breccias

Lunar breccias range from extremely soft rocks that are best described as loose soil clods to extremely hard rocks with large clasts and generally complex textures. Although almost all track groups have reported measurements showing that some breccias are largely made up of individual particles which have retained a preirradiation history prior to being incorporated into the final rock form, the most detailed studies have been made by the Berkeley and Orsay groups.

In their paper at the Third Lunar Science Conference, Dran et al. (1972a,b) show that the Apollo 14 breccias can be divided into two categories—those with generally high track densities $\geq 10^9$ t/cm^2 and those with low track densities $\leq 10^7$ t/cm^2. The former had a wide range of track densities in crystals removed from the same (interior) regions, clearly giving evidence for a preirradiation history. The individual grains also occasionally showed large track gradients, giving further evidence that the tracks were accumulated while the crystals were exposed directly on the surface. In at least one breccia the micron-sized grains were found to contain very high track densities; they were also coated with amorphous layers and appeared in every way similar to the fine particles of the lunar dust.

Not surprisingly, the presence of preirradiation track records was found to be correlated with the degree of metamorphism in the breccias as determined by petrographic examination. The scratch hardness was also found to be inversely correlated with the retention of tracks predating the breccia formation. Based on laboratory heating experiments, Dran et al. (1972b) conclude that the temperatures of individual grains never exceeded ~600°C in the breccias with preirradiation histories. Certain Apollo 11 breccias were found to contain some crystals with high track densities and still others with the characteristic microcrystalline patterns produced by heating crystals with high track densities. In these they estimate that the temperatures during brecciation were in the range of ~600°C to ~700°C.

At the same conference Hutcheon et al. (1972a,b) reported complementary measurements on a variety of breccias. They also found a general correlation with metamorphism, but were able to grade the breccias even more finely by finding rather subtle track effects; in 14321, for example, they were able to resolve very faint diffraction contrast track images in crystals in which the tracks could no longer be chemically etched.

In all cases studied there is a good correlation between track and rare gas data. All breccias having an abundant number of crystals with high track densities also have a large amount of trapped rare gases with solar composition.

Extended results on lunar breccias have recently been reported by the Berkeley group in two papers, Macdougall et al. (1973) and Price et al. (1973b). In the first of these, the observations are extended to other breccias, and formation temperatures ranging from $\lesssim$300°C to >700°C are given for a number of breccias. In the second paper the authors give measurements of track gradients in different grains and show, in particular, that several grains in rock 14301 have track gradients similar to those observed in the Surveyor III glass.

The observations in rock 14301 are particularly significant. As we discussed in Section 6.5, the Washington University group has shown that this rock contains a large amount of xenon produced by the decay of now-extinct ^{244}Pu. This fact would seem to imply that the rock was formed very early in the history of the solar system and Price et al. (1973b) conclude that the sun has been emitting solar flare particles with a characteristic Surveyor-type spectrum for at least 4.0×10^9 yrs (and, from observations on gas-rich meteorites described below, for even longer times).

However, it is not absolutely certain that breccia 14301 was *assembled* 4×10^9 yrs ago. Although the ^{244}Pu xenon must have been added to the constituents of the rock very early in the history of the solar system, it is possible that the constituents could have been assembled into a rock fairly recently without a substantial gas loss. Since the bulk of the ^{244}Pu gas is found in the fine-grained matrix and not in the coarse grains where the large gradients were measured, it is conceivable that the latter were irradiated recently. However, all arguments considered, the bulk of the evidence points to a great antiquity for rock 14301 and a remarkable constancy in the properties of the sun over most of the lifetime of the solar system.

A rather different point of view has been stressed by the Paris group in their recent discussion of the preirradiation history of the Apollo 14 breccia 14307 (Berdot et al., 1973). In this rock they find that the steepest gradient is very much less than either the Surveyor III spectrum or even the lower slope observed in the majority of lunar dust grains showing gradients. They also point out that practically all the crystals measured have track densities greater than 10^9 t/cm^2 and, thus, much higher than in most lunar soils. These authors interpret these differences as pointing to an early very active sun which emitted solar flare particles with a much harder energy spectrum than at present.

The resolution of these differences in interpretation will clearly require additional detailed work on the suite of Apollo samples.

6.7.4. Gas-Rich Meteorites

The study of tracks (and rare gases) in lunar soils and breccias is inextricably intertwined with the study of a fascinating class of meteorites known as the gas-rich meteorites. For some time (Signer and Suess, 1963; Suess et al., 1964; Wänke, 1965) it had been suggested that the large amounts of gas of solar composition found in these objects resulted from the implantation of solar wind gases in the individual grains of the meteorite while the grains had been exposed to the sun. This interpretation was greatly bolstered by Eberhardt et al. (1966) when they showed by means of etching experiments that the gas was contained in the outermost layers of the grains. Final proof of the correctness of the hypothesis was obtained by Pellas et al. (1969) and Lal and Rajan (1969) when they showed that individual grains of several gas-rich meteorites showed very high track densities and sharp track gradients indicating direct irradiation by the sun (see Fig. 6-42).

Since the initial observations, a number of authors have made track measurements on gas-rich meteorites and have commented on the comparison of their measurements in the gas-rich meteorites with those in lunar breccias (Dran et al., 1970; Arrhenius et al., 1971; Barber et al., 1971a; Schultz et al., 1971; Wilkening,

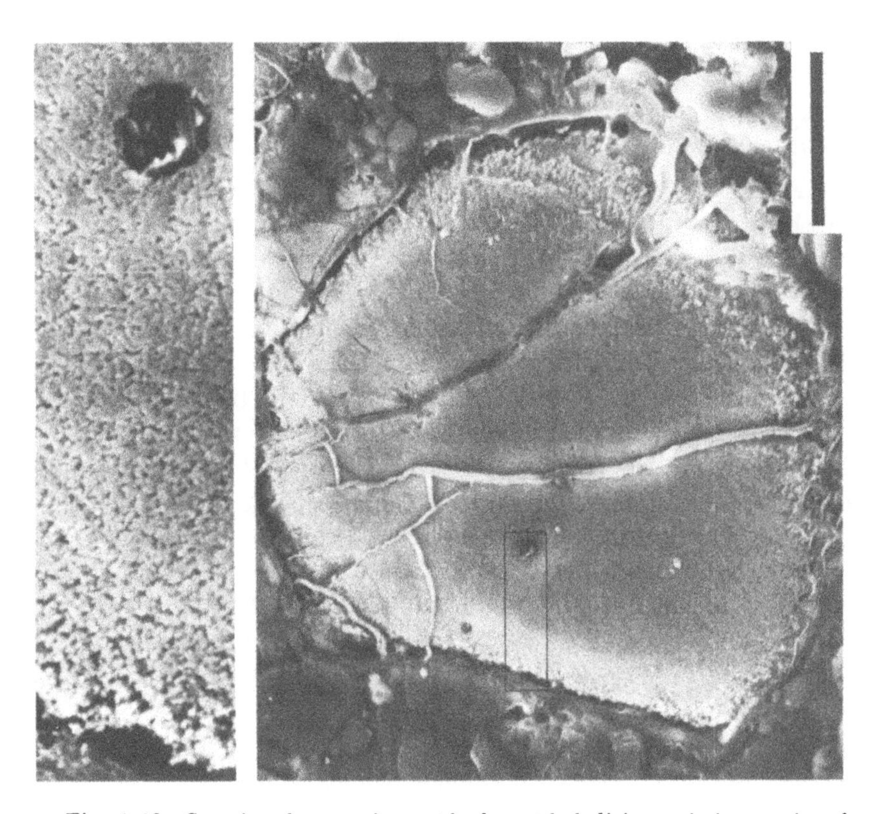

Fig. 6-42. *Scanning electron micrograph of an etched olivine grain in a section of the gas-rich meteorite Fayetteville. This broken olivine chondrule was irradiated after fracturing, as inferred from the track gradient at the broken edge. The area in the box is shown at left at 5.1x greater magnification; the scale bar is 20 μm for the main micrograph. Track density at the broken edge is* $\sim 10^{10}$ *t/cm*^{2}*. In SEM photos like this, etched tracks show up as tiny pits.* (*After Macdougall et al., 1974.*)

1971; Wilkening et al., 1971; Berdot et al., 1972a; Dran et al., 1972a,b; Lal, 1972b; Pellas, 1972; Poupeau and Berdot, 1972; Schultz et al., 1972; Brownlee and Rajan, 1973; Macdougall et al., 1973; Macdougall et al., 1974). At first the differences between the two types of breccias were emphasized, but as time has gone on the striking similarities have become more apparent. Fig. 6-43, from the paper by Barber and Price (1971), is a transmission electron micrograph of a grain from within the Fayetteville gas-rich meteorite containing a track density as high as those observed in typical, micron-size lunar soil grains (Fig. 6-31).

One point stressed by Pellas et al. (1969) and Lal and Rajan (1969) was the apparent *isotropy* of the irradiation for different grains. This suggested to these authors that the grains had been irradiated in space where they were free to rotate and not on a planetary surface. However, a recent careful study by Macdougall et al. (1974) has shown that anisotropically irradiated grains *are* found in gas-rich

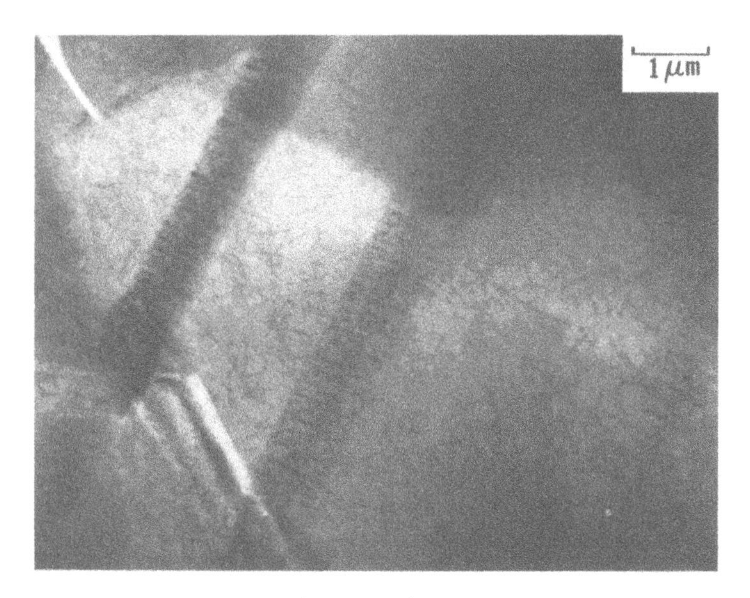

Fig. 6-43. *Solar flare tracks ($\rho \approx 10^{11}/cm^2$) in a crystal deep inside Fayetteville, observed by transmission electron microscopy of a sputter-thinned section. (After Barber and Price, 1971.)*

meteorites. The proportion of such grains is less than in a typical lunar breccia, suggesting a more efficient turning process for the meteoritic grains. Macdougall et al. (1974) suggest that this may have resulted from the lower gravity field of an asteroid-sized meteorite parent body. In any event, the Berkeley group (Phakey et al., 1972a,b) had previously shown that the proportion of isotropically irradiated grains in a lunar soil sample depends on the surface topography of the sample collection site. Samples collected on steep slopes, where the grains would tend to roll more easily, contained a higher percentage of uniformly irradiated grains.

One of the striking differences between a typical lunar breccia and a typical gas-rich meteorite has been the notable lack of glass spherules in the meteorite and their abundance in the lunar rock. However, the recent observation of chondrulelike objects and glass spherules with impact pits in the meteorite Kapoeta by Brownlee and Rajan (1973) has removed this difference. The relative paucity of the spherules in the meteorite can again be explained by a low-gravity field which permits crater ejecta to escape into space more readily than on the moon.

Significant differences between the two types of objects remain. Gas-rich meteorites are characterized by a light-dark structure with the gas concentrated in the dark regions; the same phenomenon has not been observed in gas-rich lunar breccias. The fraction of irradiated grains is much smaller in gas-rich meteorites than in gas-rich lunar breccias. The correlation between scratch hardness and tracks due to prior irradiation (Dran et al., 1972b) also does not appear to hold. No intermediate stages of metamorphism such as found in rock 14321 have yet been observed in the meteorites (Macdougall et al., 1973). It is likely, however, that these differences arise because of differences in the details by which the lunar and meteoritic breccias have formed and not on the basic nature of the processes.

To complete the parallel between meteoritic and lunar gas-rich breccias, Price et al. (1973b) have also found steep Surveyor-type gradients in the Fayetteville meteorite (see Fig. 6-26). Although it cannot be proved, it is likely that the meteorite was assembled at a time very close to the beginning of the solar system 4.6×10^9 yrs ago, again indicating the constancy of solar flare activity.

Chapter 6 References

A. L. Albee, D. S. Burnett, A. A. Chodos, O. J. Eugster, J. C. Huneke, D. A. Papanastassiou, F. A. Podosek, G. P. Russ III, H. G. Sanz, F. Tera, and G. J. Wasserburg (1970), "Ages, Irradiation History, and Chemical Composition of Lunar Rocks from the Sea of Tranquillity," *Science* **167,** 463–466.

E. C. Alexander, Jr. (1971), "Spallogenic Ne, Kr, and Xe from a Depth Study of 12002," *Proc. Second Lunar Sci. Conf.*, **2,** 1643–1650. Cambridge: MIT Press.

E. C. Alexander, Jr., R. S. Lewis, J. H. Reynolds, and M. C. Michel (1971), "Plutonium-244: Confirmation as an Extinct Radioactivity," *Science* **172,** 837–840.

E. Anders (1962), "Meteorite Ages," *Rev. Mod. Phys.* **34,** 287–325.

E. Anders and J. W. Larimer (1972), "Extinct Superheavy Elements in Meteorites: Attempted Characterization," *Science* **175,** 981–983.

J. R. Arnold, M. Honda, and D. Lal (1961), "Record of Cosmic Ray Intensity in the Meteorites," *J. Geophys. Res.* **66,** 3519–3531.

G. Arrhenius, S. Liang, D. Macdougall, L. Wilkening, N. Bhandari, S. Bhat, D. Lal, G. Rajagopalan, A. S. Tamhane, and V. S. Venkatavaradan (1971), "The Exposure History of the Apollo 12 Lunar Regolith," *Proc. Second Lunar Sci. Conf.* **3,** 2583–2598. Cambridge: MIT Press.

D. J. Barber and P. B. Price (1971), "Solar Flare Particle Tracks in Lunar and Meteoritic Minerals," *Proc. 25th Anniversary Meeting of EMAG*, British Institute of Physics.

D. J. Barber, R. Cowsik, I. D. Hutcheon, P. B. Price, and R. S. Rajan (1971a), "Solar Flares, the Lunar Surface, and Gas-Rich Meteorites," *Proc. Second Lunar Sci. Conf.* **3,** 2705–2714. Cambridge: MIT Press.

D. J. Barber, I. Hutcheon, and P. B. Price (1971b), "Extralunar Dust in Apollo Cores?," *Science* **171,** 372–374.

G. A. Bazilevskaya, A. N. Charakchyan, T. N. Charakchyan, A. N. Kvashmin, A. K. Pankratov, and A. A. Stepanyan (1968), "The Energy Spectrum of Primary Cosmic Rays and the Secondary Radiation Background in the Vicinity of Earth," *Can. J. Phys.* **46,** S515–S517.

C. Behrmann, G. Crozaz, R. Drozd, C. M. Hohenberg, C. Ralston, R. M. Walker, and D. Yuhas (1972), "Rare Gas and Particle Track Studies of Apollo 15 Samples," *The Apollo 15 Lunar Samples*, J. W. Chamberlain and C. Watkins (eds.), 329–332, Lunar Sci. Inst., Houston.

C. Behrmann, G. Crozaz, R. Drozd, C. Hohenberg, C. Ralston, R. Walker, and D. Yuhas (1973a), "Radiation History of the Apollo 16 Site," *Lunar Science IV*, J. W. Chamberlain and C. Watkins (eds.), 54–56, Lunar Sci. Inst., Houston.

C. J. Behrmann, R. J. Drozd, and C. M. Hohenberg (1973b), "Extinct Lunar Radioactivities: Xenon from ^{244}Pu and ^{129}I in Apollo 14 Breccias," *Earth Planet. Sci. Lett.* **17**, 446–455.

C. Behrmann, G. Crozaz, R. Drozd, C. Hohenberg, C. Ralston, R. Walker, and D. Yuhas (1973c), "Cosmic-Ray Exposure History of North Ray and South Ray Material," *Proc. Fourth Lunar Sci. Conf.* **2**, 1957–1974. New York: Pergamon Press.

J. L. Berdot, G. C. Chetrit, J. C. Lorin, P. Pellas, and G. Poupeau (1972a), "Track Studies of Apollo 14 Rocks, and Apollo 14, Apollo 15 and Luna 16 Soils," *Proc. Third Lunar Sci. Conf.* **3**, 2867–2881. Cambridge: MIT Press.

J. L. Berdot, G. C. Chetrit, J. C. Lorin, P. Pellas, and G. Poupeau (1972b), "Irradiation Studies of Lunar Soils: 15100, Luna 20, and Compacted Soil from Breccia 14307," *The Apollo 15 Lunar Samples*, J. W. Chamberlain and C. Watkins (eds.), 333–335, Lunar Sci. Inst., Houston.

J. L. Berdot, G. C. Chetrit, J. C. Lorin, P. Pellas, G. Poupeau, and H. Reeves (1972c), "Preliminary Track Data on Some Rocks from Apollo 14," *Lunar Science-III*, C. Watkins (ed.), 62–64, Lunar Sci. Inst., Houston.

J. L. Berdot, G. C. Chetrit, J. C. Lorin, P. Pellas, and G. Poupeau (1973), "Irradiation Records in a Compacted Soil: Breccia 14307," *Lunar Science IV*, J. W. Chamberlain and C. Watkins (eds.), 63–65, Lunar Sci. Inst., Houston.

N. Bhandari, S. Bhat, D. Lal, G. Rajagopalan, A. S. Tamhane, and V. S. Venkatavaradan (1971a), "Spontaneous Fission Record of Uranium and Extinct Transuranic Elements in Apollo Samples," *Proc. Second Lunar Sci. Conf.* **3**, 2599–2609. Cambridge: MIT Press.

N. Bhandari, S. Bhat, D. Lal, G. Rajagopalan, A. S. Tamhane, and V. S. Venkatavaradan (1971b), "High Resolution Time Averaged (Millions of Years) Energy Spectrum and Chemical Composition of Iron-Group Cosmic Ray Nuclei at 1 A.U. Based on Fossil Tracks in Apollo Samples," *Proc. Second Lunar Sci. Conf.* **3**, 2611–2619. Cambridge: MIT Press.

N. Bhandari, S. G. Bhat, D. Lal, G. Rajagopalan, A. S. Tamhane, and V. S. Venkatavaradan (1971c), "Super-Heavy Elements in Extraterrestrial Samples," *Nature* **230**, 219–224.

N. Bhandari, S. Bhat, D. Lal, G. Rajagopalan, A. S. Tamhane, and V. S. Venkatavaradan (1971d), "Fossil Tracks in the Meteorite Angra dos Reis: A Predominantly Fission Origin," *Nature* **234**, 540–543.

N. Bhandari, J. N. Goswami, S. K. Gupta, D. Lal, A. S. Tamhane, and V. S. Venkatavaradan (1972a), "Collision Controlled Radiation History of the Lunar Regolith," *Proc. Third Lunar Sci. Conf.* **3,** 2811–2829. Cambridge: MIT Press.

N. Bhandari, J. N. Goswami, and D. Lal (1972b), "Apollo 15 Regolith: A Predominantly Accretion or Mixing Model?," *The Apollo 15 Lunar Samples*, J. W. Chamberlain and C. Watkins (eds.), 336–341, Lunar Sci. Inst., Houston.

N. Bhandari, J. N. Goswami, D. Lal, D. Macdougall, and A. S. Tamhane (1972c), "A Study of the Vestigial Records of Cosmic Rays in Lunar Rocks Using a Thick Section Technique," *Proc. Ind. Acad. Sci.* **LXXVI, 1,** Sec. A, 27.

N. Bhandari, S. G. Bhat, J. N. Goswami, D. Lal, A. S. Tamhane, and V. S. Venkatavaradan (1972d), "Study of Heavy Cosmic Rays in Lunar Silicates," *Lunar Science-III*, C. Watkins (ed.), 65–67, Lunar Sci. Inst., Houston.

N. Bhandari, J. Goswami, and D. Lal (1973a), "Cosmic Ray Irradiation Patterns of Luna 16 and 20 Soils: Implications to Lunar Surface Dynamic Processes," *Earth Planet. Sci. Lett.* **20,** 372–380.

N. Bhandari, J. Goswami and D. Lal (1973b), "Surface Irradiation and Evolution of the Lunar Regolith," *Proc. Fourth Lunar Sci. Conf.* **3,** 2275–2290, New York: Pergamon Press.

N. Bhandari, J. N. Goswami, D. Lal, and A. S. Tamhane (1973c), "Long-Term Cosmic Ray Heavy Nuclei Fluxes Based on Observations in Meteorites and Lunar Samples," *Astrophys. J.* **185,** 975–983.

N. Bhandari, J. Goswami, D. Lal, and A. Tamhane (1973d), "Time Averaged Flux of Very Very Heavy Nuclei in Solar and Galactic Cosmic Rays," *Lunar Science IV*, J. W. Chamberlain and C. Watkins (eds.), 69–71, Lunar Sci. Inst., Houston.

N. Bhandari, D. Lal, and A. S. Tamhane (1973e), "Temporal Variations in the Relative Abundances of Galactic VH and VVH Nuclei During Last 0.1 B.Y.," *Proc. 13th Inter. Cosmic Ray Conference*, Univ. of Denver, **1,** 287–292.

N. Bhandari, J. N. Goswami, D. Lal, and A. S. Tamhane (1973f), "Charge Composition and Energy Spectrum of Solar Flare VVH and VH Nuclei," *Proc. 13th Inter. Cosmic Ray Conference*, Univ. of Denver, **2,** 1464–1469.

N. Bhandari, D. Lal, R. S. Rajan, J. R. Arnold, and K. Marti (1973g), unpublished report.

S. K. Bhattacharya, J. N. Goswami, S. K. Gupta, and D. Lal (1973), "Cosmic Ray Effects Induced in a Rock Exposed on the Moon or in Free Space: Contrast in Patterns for 'Tracks' and 'Isotopes'," *The Moon* **8,** 253–286.

J. P. Bibring and M. Maurette (1972), "Stellar Wind Radiation Damage in Cosmic Dust Grains: Implications for the History of Early Accretion in the Solar Nebula," *On the Origin of the Solar System*, H. Reeves (ed.), 284–292, Centre National de la Recherche Scientifique, Paris.

J. P. Bibring, M. Maurette, R. Meunier, L. Durrieu, C. Jouret, and O. Eugster (1972a), "Solar Wind Implantation Effects in the Lunar Regolith," *Lunar Science-III*, C. Watkins (ed.), 71–73, Lunar Sci. Inst., Houston.

J. P. Bibring, J. P. Duraud, L. Durrieu, C. Jouret, M. Maurette, and R. Meunier (1972b), "Ultrathin Amorphous Coatings on Lunar Dust Grains," *Science* **175,** 753–755.

J. P. Bibring, J. Chaumont, G. Comstock, M. Maurette, R. Meunier, and R. Hernandez (1973a), "Solar Wind and Lunar Wind Microscopic Effects in the Lunar Regolith," *Lunar Science IV*, J. W. Chamberlain and C. Watkins (eds.), 72–74, Lunar Sci. Inst., Houston.

J. P. Bibring, J. Chaumont, B. Jouffret, M. Maurette, and R. Meunier (1973b), "The Sticking Process in the Early Solar Nebula," unpublished report.

M. R. Bloch, H. Fechtig, W. Gentner, G. Neukum, and E. Schneider (1971), "Meteorite Impact Craters, Crater Simulations, and the Meteoroid Flux in the Early Solar System," *Proc. Second Lunar Sci. Conf.* **3,** 2639–2652. Cambridge: MIT Press.

D. D. Bogard and L. E. Nyquist (1972), "Noble Gas Studies on Regolith Materials from Apollo 14 and 15," *Proc. Third Lunar Sci. Conf.* **2,** 1797–1819. Cambridge: MIT Press.

D. D. Bogard, J. G. Funkhouser, O. A. Schaeffer, and J. Zahringer (1971), "Noble Gas Abundances in Lunar Material—Cosmic Ray Spallation Products and Radiation Ages from the Sea of Tranquility and Ocean of Storms," *J. Geophys. Res.* **76,** 2757–2779.

J. Borg, J. C. Dran, L. Durrieu, C. Jouret, and M. Maurette (1970), "High Voltage Electron Microscope Studies of Fossil Nuclear Particle Tracks in Extraterrestrial Matter," *Earth Planet. Sci. Lett.* **8,** 379–386.

J. Borg, M. Maurette, L. Durrieu, and C. Jouret (1971), "Ultramicroscopic Features in Micron-Sized Lunar Dust Grains and Cosmophysics," *Proc. Second Lunar Sci. Conf.* **3,** 2027–2040. Cambridge: MIT Press.

J. Borg, M. Maurette, L. Durrieu, C. Jouret, J. Lacaze, and P. Peter (1972), "Search for Low Energy ($10 \lesssim E \lesssim 300$ keV/amu) Nuclei in Space: Evidence from Track and Electron Diffraction Studies in Lunar Dust Grains and in Surveyor III Material," *Lunar Science-III*, C. Watkins (ed.), 92–94, Lunar Sci. Inst., Houston.

D. Braddy, J. Chan, and P. B. Price (1973), "Charge States and Energy-Dependent Composition of Solar Flare Particles," *Phys. Rev. Letters* **30,** 669–671.

H. Brown (1947), "An Experimental Method for the Estimation of the Age of the Elements," *Phys. Rev.* **72,** 348.

D. E. Brownlee and R. S. Rajan (1973), "Micrometeorite Craters Discovered on Chondrule-Like Objects from Kapoeta Meteorite," *Science* **182,** 1341–1344.

D. E. Brownlee, F. Hörz, J. F. Vedder, D. E. Gault, and J. B. Hartung (1973),

"Some Physical Parameters of Micrometeoroids," *Proc. Fourth Lunar Sci. Conf.* **3,** 3197–3212. New York: Pergamon Press.

D. S. Burnett, M. Monnin, M. Seitz, R. Walker, D. Woolum, and D. Yuhas (1970), "Charged Particle Tracks in Lunar Rock 12013," *Earth Planet. Sci. Lett.* **9,** 127–136.

D. Burnett, M. Monnin, M. Seitz, R. Walker, and D. Yuhas (1971a), "Lunar Astrology—U-Th Distributions and Fission-Track Dating of Lunar Samples," *Proc. Second Lunar Sci. Conf.* **2,** 1503–1519. Cambridge: MIT Press.

D. S. Burnett, J. C. Huneke, F. A. Podosek, G. P. Russ III, and G. J. Wasserburg (1971b), "The Irradiation History of Lunar Samples," *Proc. Second Lunar Sci. Conf.* **2,** 1671–1679. Cambridge: MIT Press.

D. Burnett, C. Hohenberg, M. Maurette, M. Monnin, R. Walker, and D. Woolum (1972), "Solar Cosmic Ray, Solar Wind, Solar Flare, and Neutron Albedo Measurements," *Apollo 16 Preliminary Science Report*, NASA Special Pub. 315, Chap. 15, pp. 19–32.

P. H. Cadogan, G. Eglinton, J. N. M. Firth, R. Maxwell, Jr., B. J. Mays, and C. T. Pillinger (1972), "Survey of Lunar Carbon Compounds, II: The Carbon Chemistry of Apollo 11, 12, 14, and 15 Samples," *Lunar Science-III*, C. Watkins (ed.), 113–115, Lunar Sci. Inst., Houston.

Y. Cantelaube, M. Maurette, and P. Pellas (1967), "Traces d'ions Lourds dans les Mineraux de la Chondrite de Saint Séverin," Intl. Atomic Energy Agency Symposium, Monaco, *Radioactive Dating and Methods of Low-Level Counting*, 215–229, IAEA, Vienna, Austria.

E. A. Carver and E. Anders (1970), "Serra de Magé: A Meteorite with an Unusual History," *Earth Planet. Sci. Lett.* **8,** 214–220.

R. S. Clark and J. E. Keith (1973), "Determination of Natural and Cosmic Ray Induced Radionuclides in Apollo 16 Lunar Samples," *Proc. Fourth Lunar Sci. Conf.*, **2,** 2105–2113. New York: Pergamon Press.

D. D. Clayton (1968), *Principles of Stellar Evolution and Nucleosynthesis*. New York: McGraw-Hill.

R. N. Clayton, L. Grossman, and T. K. Mayeda (1973), "A Component of Primitive Nuclear Composition in Carbonaceous Meteorites," *Science* **182,** 485–488.

G. M. Comstock (1972), "The Particle Track Record of the Lunar Surface," Proc. IAU Symposium 47, *The Moon*, H. C. Urey and K. Runcorn (eds.), 226–230, Newcastle.

G. M. Comstock, A. O. Evwaraye, R. L. Fleischer, and H. R. Hart, Jr. (1971), "The Particle Track Record of Lunar Soil," *Proc. Second Lunar Sci. Conf.* **3,** 2564–2582. Cambridge: MIT Press.

G. M. Comstock, R. L. Fleischer, and H. R. Hart, Jr. (1972a), "Particle Track Record of the Sea of Plenty," *Lunar Science-III*, C. Watkins (ed.), 154–156, Lunar Sci. Inst., Houston.

G. M. Comstock, R. L. Fleischer, and H. R. Hart, Jr. (1972b), "Particle Track Record of the Sea of Plenty," *Earth Planet. Sci. Lett.* **13,** 407–409.

G. M. Comstock, R. L. Fleischer, and H. R. Hart, Jr. (1973), "Particle Track Record of the Luna Missions," *The Moon* **7,** 76–83.

D. R. Criswell (1972), "Lunar Dust Motion," *Proc. Third Lunar Sci. Conf.* **3,** 2671–2680. Cambridge: MIT Press.

G. Crozaz (1974), "U, Th, and Extinct ^{244}Pu in the Phosphates of the St. Séverin Meteorite," *Earth Planet. Sci. Lett.* **23,** 164–169.

G. Crozaz and R. M. Walker (1971), "Solar Particle Tracks in Glass from the Surveyor 3 Spacecraft," *Science* **171,** 1237–1239.

G. Crozaz, M. Hair, M. Maurette, and R. Walker (1969), "Nuclear Interaction Tracks in Minerals and Their Implications for Extraterrestrial Materials," *Proc. Intl. Topical Conf. on Nuclear Track Registration in Insulating Solids and Applications*, Clermont-Ferrand **2,** Part VII, pp. 41–54.

G. Crozaz, U. Haack, M. Hair, H. Hoyt, J. Kardos, M. Maurette, M. Miyajima, M. Seitz, S. Sun, R. Walker, M. Wittels, and D. Woolum (1970a), "Solid State Studies of the Radiation History of Lunar Samples," *Science* **167,** 563–566.

G. Crozaz, U. Haack, M. Hair, M. Maurette, R. Walker, and D. Woolum (1970b), "Nuclear Track Studies of Ancient Solar Radiations and Dynamic Lunar Surface Processes," *Proc. Apollo 11 Lunar Sci. Conf.* **3,** 2051–2080. New York: Pergamon Press.

G. Crozaz, R. Walker, and D. Woolum (1971), "Nuclear Track Studies of Dynamic Surface Processes on the Moon and the Constancy of Solar Activity," *Proc. Second Lunar Sci. Conf.*, **3,** 2543–2558. Cambridge: MIT Press.

G. Crozaz, D. Burnett, and R. Walker (1972a), "Uranium and Thorium Microdistributions in Meteorites," Symposium on Cosmochemistry, Cambridge, Mass., August 14–16, 1972.

G. Crozaz, R. Drozd, H. Graf, C. M. Hohenberg, M. Monnin, D. Ragan, C. Ralston, M. Seitz, J. Shirck, R. M. Walker, and J. Zimmerman (1972b), "Evidence for Extinct Pu^{244}: Implications for the Age of the Preimbrium Crust," *Lunar Science-III*, C. Watkins (ed.), 164–166, Lunar Sci. Inst., Houston.

G. Crozaz, R. Drozd, H. Graf, C. M. Hohenberg, M. Monnin, D. Ragan, C. Ralston, M. Seitz, J. Shirck, R. M. Walker, and J. Zimmerman (1972c), "Uranium and Extinct Pu^{244} Effects in Apollo 14 Materials," *Proc. Third Lunar Sci. Conf.* **2,** 1623–1636. Cambridge: MIT Press.

G. Crozaz, R. Drozd, C. M. Hohenberg, H. P. Hoyt, Jr., D. Ragan, R. M. Walker, and D. Yuhas (1972d), "Solar Flare and Galactic Cosmic Ray Studies of Apollo 14 Samples," *Lunar Science-III*, C. Watkins (ed.), 167–169, Lunar Sci. Inst., Houston.

G. Crozaz, R. Drozd, C. M. Hohenberg, H. P. Hoyt, Jr., D. Ragan, R. M.

Walker, and D. Yuhas (1972e), "Solar Flare and Galactic Cosmic Ray Studies of Apollo 14 and 15 Samples," *Proc. Third Lunar Sci. Conf.*, **3,** 2917–2931. Cambridge: MIT Press.

G. Crozaz, R. Walker, and D. Zimmerman (1973), "Fossil Track and Thermoluminescence Studies of Luna 20 Material," *Geochim. Cosmochim. Acta* **37,** 825–830.

J. D'Amico, J. DeFelice, E. L. Fireman, C. Jones, and G. Spannagel (1971), "Tritium and Argon Radioactivities and Their Depth Variations in Apollo 12 Samples," *Proc. Second Lunar Sci. Conf.* **2,** 1825–1839. Cambridge: MIT Press.

J. C. Dran, L. Durrieu, C. Jouret, and M. Maurette (1970), "Habit and Texture Studies of Lunar and Meteoritic Materials with a 1 MeV Electron Microscope," *Earth Planet. Sci. Lett.* **9,** 391–400.

J. C. Dran, J. P. Duraud, M. Maurette, L. Durrieu, C. Jouret, and C. Legressus (1972a), "The High Resolution Track and Texture Record of Lunar Breccias and Gas-Rich Meteorites," *Lunar Science-III*, C. Watkins (ed.), 183–185, Lunar Sci. Inst., Houston.

J. C. Dran, J. P. Duraud, M. Maurette, L. Durrieu, C. Jouret, and C. Legressus (1972b), "Track Metamorphism in Extraterrestrial Breccias," *Proc. Third Lunar Sci. Conf.* **3,** 2883–2903. Cambridge: MIT Press.

R. Drozd, C. M. Hohenberg, and D. Ragan (1972), "Fission Xenon from Extinct ^{244}Pu in 14301," *Earth Planet. Sci. Lett.* **15,** 338–346.

R. J. Drozd, C. M. Hohenberg, C. J. Morgan, and C. E. Ralston (1974), "Cosmic Ray Exposure History at the Apollo 16 Site," *Geochim. Cosmochim. Acta* **38,** 1625–1642.

L. Durrieu, C. Jouret, J. C. Leroulley, and M. Maurette (1971), "Applications of High Voltage Electron Microscopy to Cosmophysics," *Jernkont. Ann.* **155,** 535–540.

L. Durrieu, C. Jouret, L. Kashkarov, M. Maurette, and B. Vassent (1973), "Radiation Damage Metamorphism in the Luna 16-19 Sample," unpublished report.

P. Eberhardt, J. Geiss, and N. Groegler (1966), "Distribution of Rare Gases in the Pyroxene and Feldspar of the Khor Temiki Meteorite," *Earth Planet. Sci. Lett.* **1,** 7–12.

P. Eberhardt, J. Geiss, H. Graf, N. Groegler, U. Kraehenbuehl, H. Schwaller, H. Schwarzmueller, and A. Stettler (1970), "Trapped Solar Wind Noble Gases, Radiation Ages and K/Ar in Lunar Material," *Science* **167,** 558–560.

J. S. Eldridge, G. D. O'Kelley, and K. J. Northcutt (1973), "Radionuclide Concentrations in Apollo 16 Lunar Samples by Non-Destructive Gamma-Ray Spectrometry," *Proc. Fourth Lunar Sci. Conf.* **2,** 2115–2122. New York: Pergamon Press.

R. C. Finkel, J. R. Arnold, M. Imamura, R. C. Reedy, J. S. Fruchter, H. H. Loosli, J. C. Evans, A. C. Delany, and J. P. Shedlovsky (1971), "Depth

Variations of Cosmogenic Nuclides in a Lunar Surface Rock and Lunar Soil," *Proc. Second Lunar Sci. Conf.*, **2,** 1773–1789. Cambridge: MIT Press.
E. L. Fireman, J. D'Amico, and J. DeFelice (1970), "Tritium and Argon Radioactivities in Lunar Material," *Science* **167,** 566–568.
R. L. Fleischer and H. R. Hart, Jr. (1972a), "Fission Track Dating: Techniques and Problems," *Calibration of Hominoid Evolution*, W. W. Bishop, J. A. Miller, and S. Cole (eds.), 135–170, Edinburgh: Scottish Academic Press.
R. L. Fleischer and H. R. Hart, Jr. (1972b), "Particle Track Record of Apollo 15 Green Soil and Rock," *The Apollo 15 Lunar Samples*, J. W. Chamberlain and C. Watkins (eds.), 368–370, Lunar Sci. Inst., Houston.
R. L. Fleischer and H. R. Hart, Jr. (1972c), "Particle Track Record in Apollo 15 Deep Core from 54 to 80 Cm Depths," *The Apollo 15 Lunar Samples*, J. W. Chamberlain and C. Watkins (eds.), 371–373, Lunar Sci. Inst., Houston.
R. L. Fleischer and H. R. Hart, Jr. (1972d), "Composition and Energy Spectra of Solar Cosmic Ray Nuclei," *Apollo 16 Preliminary Science Report*, NASA Special Pub. 315, Chap. 15, pp. 2–11.
R. L. Fleischer and H. R. Hart, Jr. (1973a), "Surface History of Lunar Soil and Soil Columns," *Lunar Science IV*, J. W. Chamberlain and C. Watkins (eds.), 251–253, Lunar Sci. Inst., Houston.
R. L. Fleischer and H. R. Hart, Jr. (1973b), "Enrichment of Heavy Nuclei in the 17 April 1972 Solar Flare," *Phys. Rev. Letters* **30,** 31–34.
R. L. Fleischer and H. R. Hart, Jr. (1973c), "Particle Track Record of Apollo 15 Green Soil and Rock," *Earth Planet. Sci. Lett.* **18,** 357–364.
R. L. Fleischer and H. R. Hart, Jr. (1973d), "Particle Track Record in Apollo 15 Deep Core from 54 to 80 Cm Depths," *Earth Planet. Sci. Lett.* **18,** 420–426.
R. L. Fleischer and H. R. Hart, Jr. (1973e), "Mechanical Erasure of Particle Tracks, A Tool for Lunar Microstratigraphic Chronology," *J. Geophys. Res.* **78,** 4841–4851.
R. L. Fleischer and H. R. Hart, Jr. (1973f), "Tracks from Extinct Radioactivity, Ancient Cosmic Rays, and Calibration Ions," *Nature* **242,** 104–105.
R. L. Fleischer and H. R. Hart, Jr. (1974), "Particle Track Record of Apollo 16 Rocks from Plum Crater," *J. Geophys. Res.* **79,** 766–768.
R. L. Fleischer and C. W. Naeser (1972), "Search for Plutonium-244 Tracks in Mountain Pass Bastnaesite," *Nature* **240,** 465.
R. L. Fleischer, C. W. Naeser, P. B. Price, R. M. Walker, and M. Maurette (1965a), "Cosmic Ray Exposure Ages of Tektites by the Fission Track Technique," *J. Geophys. Res.* **70,** 1491–1496.
R. L. Fleischer, P. B. Price, and R. M. Walker (1965b), "Spontaneous Fission Tracks from Extinct Pu^{244} in Meteorites and the Early History of the Solar System," *J. Geophys. Res.* **70,** 2703–2707.
R. L. Fleischer, P. B. Price, R. M. Walker, and M. Maurette (1967a), "Origins of Fossil Charged Particle Tracks in Meteorites," *J. Geophys. Res.* **72,** 333–353.

R. L. Fleischer, P. B. Price, R. M. Walker, M. Maurette, and G. Morgan (1967b), "Tracks of Heavy Primary Cosmic Rays in Meteorites," *J. Geophys. Res.* **72,** 355–366.

R. L. Fleischer, P. B. Price, and R. M. Walker (1968), "Identification of Pu^{244} Fission Tracks and the Cooling of the Parent Body of the Toluca Meteorite," *Geochim. Cosmochim. Acta* **32,** 21–31.

R. L. Fleischer, E. Lifshin, P. B. Price, R. T. Woods, R. W. Carter, and E. L. Fireman (1970a), "Schenectady Meteorite," *Icarus* **12,** 402–406.

R. L. Fleischer, E. L. Haines, R. E. Hanneman, H. R. Hart, Jr., J. S. Kasper, E. Lifshin, R. T. Woods, and P. B. Price (1970b), "Particle Track, X-Ray, Thermal, and Mass Spectrometric Studies of Lunar Material from Apollo 11," *Science* **167,** 568–571.

R. L. Fleischer, E. L. Haines, H. R. Hart, Jr., R. T. Woods, and G. M. Comstock (1970c), "The Particle Track Record of the Sea of Tranquillity," *Proc. Apollo 11 Lunar Sci. Conf.*, **3,** 2103–2120. New York: Pergamon Press.

R. L. Fleischer, H. R. Hart, Jr., and G. M. Comstock (1971a), "Very Heavy Solar Cosmic Rays: Energy Spectrum and Implications for Lunar Erosion," *Science* **171,** 1240–1242.

R. L. Fleischer, H. R. Hart, Jr., G. M. Comstock, and A. O. Evwaraye (1971b), "The Particle Track Record of the Ocean of Storms," *Proc. Second Lunar Sci. Conf.*, **3,** 2559–2568. Cambridge: MIT Press.

R. L. Fleischer, H. R. Hart, Jr., and G. M. Comstock (1972a), "Particle Track Dating of Mechanical Events," *Lunar Science-III*, C. Watkins (ed.), 265–267, Lunar Sci. Inst., Houston.

R. L. Fleischer, G. M. Comstock, and H. R. Hart, Jr. (1972b), "Dating of Mechanical Events by Deformation-Induced Erasure of Particle Tracks," *J. Geophys. Res.* **77,** 5050–5053.

R. L. Fleischer, H. R. Hart, Jr., and W. R. Giard (1973), "Particle Track Record of Apollo 15 Shocked Crystalline Rocks," *Proc. Fourth Lunar Sci. Conf.*, **3,** 2307–2317. New York: Pergamon Press.

R. L. Fleischer, H. R. Hart, Jr., and W. R. Giard (1974), "Surface History of Lunar Soil and Soil Columns," *Geochim. Cosmochim. Acta* **38,** 365–380.

J. Funkhouser, O. Schaeffer, D. Bogard, and J. Zähringer (1970), "Gas Analysis of the Lunar Surface," *Science* **167,** 561–563.

D. E. Gault, F. Hörz, and J. B. Hartung (1972), "Effects of Microcratering on the Lunar Surface," *Proc. Third Lunar Sci. Conf.*, **3,** 2713–2734. Cambridge: MIT Press.

T. Gold (1971), "Evolution of Mare Surfaces," *Proc. Second Lunar Sci. Conf.*, **3,** 2675–2680. Cambridge: MIT Press.

H. Graf, J. Shirck, S. Sun, and R. Walker (1973), "Fission Track Astrology of Three Apollo 14 Gas-Rich Breccias," *Proc. Fourth Lunar Sci. Conf.*, **2,** 2145–2155, New York: Pergamon Press.

H. R. Hart, Jr., G. M. Comstock, and R. L. Fleischer (1972a), "The Particle Track Record of Fra Mauro," *Lunar Science-III*, C. Watkins (ed.), 360–362, Lunar Sci. Inst., Houston.

H. R. Hart, Jr., G. M. Comstock, and R. L. Fleischer (1972b), "The Particle Track Record of Fra Mauro," *Proc. Third Lunar Sci. Conf.*, **3,** 2831–2844. Cambridge: MIT Press.

J. B. Hartung and D. Storzer (1974), "Meteoroid Mass Distributions and Fluxes from Microcraters on Lunar Sample 15205," *Space Research XIV*, M. J. Rycroft and R. D. Reasenberg (eds.), 719–721. Berlin: Akademie Verlag.

J. B. Hartung, F. Hörz, and D. E. Gault (1972), "Lunar Microcraters and Interplanetary Dust," *Proc. Third Lunar Sci. Conf.*, **3,** 2735–2753. Cambridge: MIT Press.

J. B. Hartung, F. Hörz, F. K. Aitken, D. E. Gault, and D. E. Brownlee (1973), "The Development of Microcrater Populations on Lunar Rocks," *Proc. Fourth Lunar Sci. Conf.*, **3,** 3213–3234. New York: Pergamon Press.

G. F. Herzog and E. Anders (1971a), "Radiation Age of the Norton County Meteorite," *Geochim. Cosmochim. Acta* **35,** 239–244.

G. F. Herzog and E. Anders (1971b), "Absolute Scale for Radiation Ages of Stony Meteorites," *Geochim. Cosmochim. Acta* **35,** 605–611.

H. Hintenberger, H. W. Weber, H. Voshage, H. Wanke, F. Begemann, E. Vilscek, and F. Wlotzka (1970), "Concentrations and Isotopic Compositions of Rare Gases, Hydrogen, and Nitrogen in Lunar Dust and Rocks," *Science* **167,** 543–545.

H. Hintenberger, H. W. Weber, and N. Takaoka (1971), "Concentrations and Isotopic Abundances of the Rare Gases in Lunar Matter," *Proc. Second Lunar Sci. Conf.*, **2,** 1607–1625. Cambridge: MIT Press.

D. C. Hoffman, F. O. Lawrence, J. L. Mewherter, and F. M. Rourke (1971), "Detection of Plutonium-244 in Nature," *Nature* **234,** 132–134.

C. M. Hohenberg (1969), "Radioisotopes and the History of Nucleosynthesis in the Galaxy," *Science* **166,** 212–215.

C. M. Hohenberg (1970), "Xenon from the Angra dos Reis Meteorite," *Geochim. Cosmochim. Acta* **34,** 185–191.

C. M. Hohenberg, M. N. Munk, and J. H. Reynolds (1967a), "Spallation and Fissiogenic Xenon and Krypton from Stepwise Heating of the Pasamonte Achondrite. The Case for Extinct ^{244}Pu in Meteorites. Relative Ages of Chondrites and Achondrites," *J. Geophys. Res.* **72,** 3139–3177.

C. M. Hohenberg, F. A. Podosek, and J. H. Reynolds (1967b), "Xenon-Iodine Dating: Sharp Isochronism in Chondrites," *Science* **156,** 202–206.

C. M. Hohenberg, P. K. Davis, W. A. Kaiser, R. S. Lewis, and J. H. Reynolds (1970), "Trapped and Cosmogenic Rare Gases from Stepwise Heating of Apollo 11 Samples," *Proc. Apollo 11 Lunar Sci. Conf.*, **2,** 1283–1309. New York: Pergamon Press.

F. Hörz, J. B. Hartung, and D. E. Gault (1971), "Micrometeorite Craters on Lunar Rocks," *J. Geophys. Res.* **76,** 5770–5798.

F. Hörz, D. E. Brownlee, H. Fechtig, J. B. Hartung, D. A. Morrison, G. Neukum, E. Schneider, and J. F. Vedder (1973), "Lunar Microcraters: Implications for the Micrometeoroid Complex," summary of papers presented at the COSPAR meeting in 1973, Konstanz, Germany; Planetary and Space Sciences, in press.

H. P. Hoyt, Jr., R. M. Walker, and D. W. Zimmerman (1973), "Solar Flare Proton Spectrum Averaged Over the Last 5×10^3 Years," *Proc. Fourth Lunar Sci. Conf.*, **3,** 2489–2502. New York: Pergamon Press.

I. D. Hutcheon and P. B. Price (1972), "Plutonium-244 Fission Tracks: Evidence in a Lunar Rock 3.95 Billion Years Old," *Science* **176,** 909–911.

I. D. Hutcheon, P. P. Phakey, P. B. Price, and R. S. Rajan (1972a), "History of Lunar Breccias," *Lunar Science III*, C. Watkins (ed.), 415–417, Lunar Sci. Inst., Houston.

I. D. Hutcheon, P. P. Phakey, and P. B. Price (1972b), "Studies Bearing on the History of Lunar Breccias," *Proc. Third Lunar Sci. Conf.*, **3,** 2845–2865. Cambridge: MIT Press.

I. D. Hutcheon, D. Braddy, P. P. Phakey, and P. B. Price (1972c), "Study of Solar Flares, Cosmic Dust and Lunar Erosion with Vesicular Basalts," *The Apollo 15 Lunar Samples*, J. W. Chamberlain and C. Watkins (eds.), 412–414, Lunar Sci. Inst., Houston.

I. D. Hutcheon, D. Macdougall, and P. B. Price (1974), "Rock 72315: a New Lunar Standard for Solar Flare and Micrometeorite Exposure," *Lunar Science V*, 378–380, Lunar Sci. Inst., Houston.

P. M. Jeffery and J. H. Reynolds (1961), "Origin of Excess Xe^{129} in Stone Meteorites," *J. Geophys. Res.* **66,** 3582–3583.

W. A. Kaiser (1971), "Rare Gas Measurements in Three Mineral Separates of Rock 12013,10,31," *Proc. Second Lunar Sci. Conf.*, **2,** 1627–1641. Cambridge: MIT Press.

W. A. Kaiser (1972a), "Rare Gas Studies in Luna-16-G-7 Fines by Stepwise Heating Technique. A Low Fission Solar Wind Xe," *Earth Planet. Sci. Lett.* **13,** 387–399.

W. A. Kaiser (1972b), "Rare Gas Measurements in Three Apollo 14 Samples," *Lunar Science III*, C. Watkins (ed.), 442, Lunar Sci. Inst., Houston.

L. L. Kashkarov, L. I. Genaeva, and A. K. Lavrukhina (1971), "Composition of Heavy Cosmic Rays in the $Z \gtrsim 23$ Region," *Proc. 12th Inter. Conf. on Cosmic Rays*, Hobart, Australia, **5,** 1782–1791.

J. E. Keith, R. S. Clark, and K. A. Richardson (1972), "Gamma-Ray Measurements of Apollo 12, 14, and 15 Lunar Samples," *Proc. Third Lunar Sci. Conf.*, **2,** 1671–1680. Cambridge: MIT Press.

J. F. Kerridge (1970), "Micrometeorite Environment at the Earth's Orbit," *Nature* **228,** 616–619.

E. A. King, J. C. Butler, and M. F. Corman (1971), "The Lunar Regolith as Sampled by Apollo 11 and 12: Grain Size Analyses, and Origins of Particles," *Proc. Second Lunar Sci. Conf.*, **1,** 737–746. Cambridge: MIT Press.

T. Kirsten, J. Deubner, H. Ducati, W. Gentner, P. Horn, E. Jessberger, S. Kalbitzer, I. Kaneoka, J. Kiko, W. Krätschmer, H. W. Müller, T. Plieninger, and S. K. Thio (1972a), "Rare Gases and Ion Tracks in Individual Components and Bulk Samples of Apollo 14 and 15 Fines and Fragmental Rocks," *Lunar Science III*, C. Watkins (ed.), 452–454, Lunar Sci. Inst., Houston.

T. Kirsten, J. Deubner, P. Horn, I. Kaneoka, J. Kiko, O. A. Schaeffer, and S. K. Thio (1972b), "The Rare Gas Record of Apollo 14 and 15 Samples," *Proc. Third Lunar Sci. Conf.*, **2,** 1865–1889. Cambridge: MIT Press.

P. K. Kuroda (1960), "Nuclear Fission in the Early History of the Earth," *Nature* **187,** 36–38.

D. Lal (1969), "Recent Advances in the Study of Fossil Tracks in Meteorites Due to Heavy Nuclei of the Cosmic Radiation," *Space Sci. Rev.* **9,** 623–650.

D. Lal (1972a), "Hard Rock Cosmic Ray Archeology," *Space Sci. Rev.* **14,** 3–102.

D. Lal (1972b), "Accretion Processes Leading to Formation of Meteorite Parent Bodies," *Proc. Nobel Symposium #21*, "*From Plasma to Planet*," Saltsjobaden, Sweden, Sept. 6–10, 1971, 49–64. Stockholm: Almqvist and Wiksell (Wiley-Interscience).

D. Lal and R. S. Rajan (1969), "Observations on Space Irradiation of Individual Crystals of Gas-Rich Meteorites," *Nature* **223,** 269–271.

D. Lal, J. C. Lorin, P. Pellas, R. S. Rajan, and A. S. Tamhane (1969a), "On the Energy Spectrum of Iron-Group Nuclei as Deduced from Fossil-Track Studies in Meteoritic Minerals," Intl. Atomic Energy Agency Symposium on Meteorite Research, Vienna, *Meteorite Res.*, P. Millman (ed.), 275–285. Dordrecht, Holland: D. Reidel.

D. Lal, R. S. Rajan, and A. S. Tamhane (1969b), "Chemical Composition of Nuclei of $Z \geq 22$ in Cosmic Rays Using Meteoritic Minerals as Detectors," *Nature* **221,** 33–37.

D. Lal, D. Macdougall, L. Wilkening, and G. Arrhenius (1970), "Mixing of the Lunar Regolith and Cosmic Ray Spectra: Evidence from Particle-Track Studies," *Proc. Apollo 11 Lunar Sci. Conf.*, **3,** 2295–2303. New York: Pergamon Press.

R. S. Lewis (1975), "Rare Gases in Separated Whitlockite from the St. Séverin Chondrite: Xenon and Krypton from Fission of Extinct ^{244}Pu," *Geochim. Cosmochim. Acta*, to be published.

LSPET (Lunar Sample Preliminary Examination Team) (1971), "Preliminary Examination of Lunar Samples from Apollo 14," *Science* **173,** 681–693.

G. W. Lugmair and K. Marti (1972), "Exposure Ages and Neutron Capture Record in Lunar Samples from Fra Mauro," *Proc. Third Lunar Sci. Conf.*, **2,** 1891–1897. Cambridge: MIT Press.

J. A. M. McDonnell (1971), "Review of In Situ Measurements of Cosmic Dust Particles in Space," *Space Res.* **XI,** 415–435.

D. Macdougall, B. Martinek, and G. Arrhenius (1972), "Regolith Dynamics," *Lunar Science III*, C. Watkins (ed.), 498–500, Lunar Sci. Inst., Houston.

D. Macdougall, R. S. Rajan, I. D. Hutcheon, and P. B. Price (1973), "Irradiation History and Accretionary Processes in Lunar and Meteoritic Breccias," *Proc. Fourth Lunar Sci. Conf.*, **3,** 2319–2336. New York: Pergamon Press.

D. Macdougall, R. S. Rajan, and P. B. Price (1974), "Gas-Rich Meteorites: Possible Evidence for Origin on a Regolith," *Science* **183,** 73–74.

J. A. M. McDonnell, D. G. Ashworth, R. P. Flavill, and R. C. Jennison (1972), "Simulated Microscale Erosion on the Lunar Surface by Hypervelocity Impact, Solar Wind Sputtering, and Thermal Cycling," *Proc. Third Lunar Sci. Conf.*, **3,** 2755–2765. Cambridge: MIT Press.

J. D. Macdougall (1972), "Particle Track Records in Natural Solids from Oceans on Earth and Moon," Ph.D. Thesis, Univ. of California, San Diego.

D. S. McKay and G. H. Heiken (1973), "The South Ray Crater Age Paradox," *Proc. Fourth Lunar Sci. Conf.*, **1,** 41–47. New York: Pergamon Press.

D. S. McKay, D. A. Morrison, U. S. Clanton, G. H. Ladle, and J. F. Lindsay (1971), "Apollo 12 Soil and Breccia," *Proc. Second Lunar Sci. Conf.*, **1,** 755–773. Cambridge: MIT Press.

J. C. Mandeville and J. F. Vedder (1971), "Microcraters Formed in Glass by Low Density Projectiles," *Earth Planet. Sci. Lett.* **11,** 297–306.

K. Marti and G. W. Lugmair (1971), "Kr^{81}-Kr and K-Ar^{40} Ages, Cosmic-Ray Spallation Products, and Neutron Effects in Lunar Samples from Oceanus Procellarum," *Proc. Second Lunar Sci. Conf.*, **2,** 1591–1605. Cambridge: MIT Press.

K. Marti, P. Eberhardt, and J. Geiss (1966), "Spallation, Fission, and Neutron Capture Anomalies in Meteoritic Krypton and Xenon," *Z. Naturforsch.* **A21,** 398–413.

K. Marti, G. W. Lugmair, and H. C. Urey (1970a), "Solar Wind Gases, Cosmic-Ray Effects and the Irradiation History," *Science* **167,** 548–550.

K. Marti, G. W. Lugmair, and H. C. Urey (1970b), "Solar Wind Gases, Cosmic-Ray Spallation Products and Irradiation History of Apollo 11 Samples," *Proc. Apollo 11 Lunar Sci. Conf.*, **2,** 1357–1367. New York: Pergamon Press.

K. Marti, B. D. Lightner, and T. W. Osborn (1973), "Krypton and Xenon in Some Lunar Samples and the Age of North Ray Crater," *Proc. Fourth Lunar Sci. Conf.*, **2,** 2037–2048. New York: Pergamon Press.

M. Maurette (1965), *Étude des Traces d'ions Lourds dans les Minéraux Naturels d'Origine Terrestre et Extra-Terrestre*, Ph.D. Thesis, University of Paris.

M. Maurette (1966), "Étude des Traces d'Ions Lourds dans les Minéraux Naturels d'Origine Terrestre et Extra-Terrestre," *Bull. Soc. Franc. Min. Crist.* **89,** 41–79.

M. Maurette (1970), "On Some Annealing Characteristics of Heavy Ion Tracks in Silicate Minerals," *Rad. Effects* **5,** 15–19.

M. Maurette and R. M. Walker (1964), "Étude des Traces de Particules Induites par les Interactions de Protons de 3 GeV dans Différents Minéraux," *J. de Physique* **25,** 661–666.

M. Maurette, P. Pellas, and R. M. Walker (1964), "Cosmic Ray Induced Particle Tracks in a Meteorite," *Nature* **204,** 821–823.

M. Maurette, P. Thro, R. Walker, and R. Webbink (1969), "Fossil Tracks in Meteorites and the Chemical Abundance and Energy Spectrum of Extremely Heavy Cosmic Rays," Intl. Atomic Energy Agency Symposium on Meteorite Research, Vienna, *Meteorite Res.*, P. Millman (ed.), 286–315. Dordrecht, Holland: D. Reidel.

G. H. Megrue and F. Steinbrunn (1972), "Classification and Source of Lunar Soils; Clastic Rocks; and Individual Mineral, Rock, and Glass Fragments from Apollo 12 and 14 Samples as Determined by the Concentration Gradients of the Helium, Neon, and Argon Isotopes," *Proc. Third Lunar Sci. Conf.*, **2,** 1899–1916. Cambridge: MIT Press.

D. A. Morrison, D. S. McKay, G. H. Heiken, and H. J. Moore (1972), "Microcraters on Lunar Rocks," *Proc. Third Lunar Sci. Conf.*, **3,** 2767–2791. Cambridge: MIT Press.

D. A. Morrison, D. S. McKay, R. M. Fruland, and H. J. Moore (1973), "Microcraters on Apollo 15 and 16 Rocks," *Proc. Fourth Lunar Sci. Conf.*, **3,** 3235–3253. New York: Pergamon Press.

G. Neukum (1973), "Micrometeoroid Flux, Microcrater Population Development and Erosion Rates on Lunar Rocks, and Exposure Ages of Apollo 16 Rocks Derived from Crater Statistics," *Lunar Science IV*, J. W. Chamberlain and C. Watkins (eds.), 558–560, Lunar Sci. Inst., Houston.

G. Neukum, E. Schneider, A. Mehl, D. Storzer, G. A. Wagner, H. Fechtig, and M. R. Bloch (1972), "Lunar Craters and Exposure Ages Derived from Crater Statistics and Solar Flare Tracks," *Proc. Third Lunar Sci. Conf.*, **3,** 2793–2810. Cambridge: MIT Press.

G. Neukum, F. Hörz, D. A. Morrison, and J. B. Hartung (1973), "Crater Populations on Lunar Rocks," *Proc. Fourth Lunar Sci. Conf.*, **3,** 3255–3276. New York: Pergamon Press.

G. D. O'Kelley, J. S. Eldridge, E. Schönfeld, and P. R. Bell (1970), "Elemental Compositions and Ages of Apollo 11 Lunar Samples by Nondestructive Gamma-Ray Spectrometry," *Science* **167,** 580–582.

G. D. O'Kelley, J. S. Eldridge, E. Schönfeld, and P. R. Bell (1971), "Cosmogenic Radionuclide Concentrations and Exposure Ages of Lunar Samples from

Apollo 12," *Proc. Second Lunar Sci. Conf.*, **2,** 1747–1755. Cambridge: MIT Press.

D. A. Papanastassiou (1970), "The Determination of Small Time Differences in the Formation of Planetary Objects," Ph.D. Thesis, California Institute of Technology.

P. Pellas (1972), "Irradiation History of Grain Aggregates in Ordinary Chondrites. Possible Clues to the Advanced Stages," *Proc. Nobel Symposium #21, "From Plasma to Planet,"* Saltsjobaden, Sweden, Sept. 6–10, 1971, 65–93. Stockholm: Almqvist and Wiksell (Wiley-Interscience).

P. Pellas, G. Poupeau, J. C. Lorin, H. Reeves, and J. Audouze (1969), "Primitive Low-Energy Particle Irradiation of Meteoritic Crystals," *Nature* **223,** 272–274.

P. P. Phakey and P. B. Price (1972), "Extreme Radiation Damage in Soil from Mare Fecunditatis," *Earth Planet. Sci. Lett.* **13,** 410–418.

P. P. Phakey and P. B. Price (1973), "Radiation Damage in Luna 20 Soil," *Geochim. Cosmochim. Acta* **37,** 975–977.

P. P. Phakey, I. D. Hutcheon, R. S. Rajan, and P. B. Price (1972a), "Radiation Damage in Soils from Five Lunar Missions," *Lunar Science III*, C. Watkins (ed.), 608–610, Lunar Sci. Inst., Houston.

P. P. Phakey, I. D. Hutcheon, R. S. Rajan, and P. B. Price (1972b), "Radiation Effects in Soils from Five Lunar Missions," *Proc. Third Lunar Sci. Conf.*, **3,** 2905–2915. Cambridge: MIT Press.

T. Plieninger and W. Krätschmer (1972), "Registration Properties of Pyroxenes for Various Heavy Ions and Consequences to the Determination of the Composition of the Cosmic Radiation," *Proc. 8th Inter. Conf. on Nuc. Photog. and Solid State Track Detectors*, Bucharest, 537–544.

T. Plieninger, W. Krätschmer and W. Gentner (1972), "Charge Assignment to Cosmic Ray Heavy Ion Tracks in Lunar Pyroxenes," *Proc. Third Lunar Sci. Conf.*, **3,** 2933–2939. Cambridge: MIT Press.

T. Plieninger, W. Krätschmer, and W. Gentner (1973), "Indications for Time Variations in the Galactic Cosmic Ray Composition Derived from Track Studies on Lunar Samples," *Proc. Fourth Lunar Sci. Conf.*, **3,** 2337–2346. New York: Pergamon Press.

F. A. Podosek (1970a), "Dating of Meteorites by the High-Temperature Release of Iodine-Correlated Xe^{129}," *Geochim. Cosmochim. Acta* **34,** 341–365.

F. A. Podosek (1970b), "The Abundance of Pu^{244} in the Early Solar System," *Earth Planet. Sci. Lett.* **8,** 183–187.

F. Podosek (1972), "Gas Retention Chronology of Petersburg and Other Meteorites," *Geochim. Cosmochim. Acta* **35,** 755–772.

F. A. Podosek and R. S. Lewis (1972), "^{129}I and ^{244}Pu Abundances in White Inclusions of the Allende Meteorite," *Earth Planet. Sci. Lett.* **15,** 101–109.

G. Poupeau and J. L. Berdot (1972), "Irradiations Ancienne et Recente des Aubrites," *Earth Planet. Sci. Lett.* **14,** 381–396.

G. Poupeau, J. L. Berdot, G. C. Chetrit, and P. Pellas (1972a), "Predominant Trapping of Solar-Flare Gases in Lunar Soils," *Lunar Science III*, C. Watkins (ed.), 613–615, Lunar Sci. Inst., Houston.
G. Poupeau, P. Pellas, J. C. Lorin, G. C. Chetrit, and J. L. Berdot (1972b), "Track Analysis of Rocks 15 058, 15 555, 15 641 and 14 307," *The Apollo 15 Lunar Samples*, J. W. Chamberlain and C. Watkins (eds.), 385–387, Lunar Sci. Inst., Houston.
G. Poupeau, G. C. Chetrit, J. L. Berdot, and P. Pellas (1973), "Etude par la Methode des Traces Nucleaires du Sol de La Mer de La Fecondite (Luna 16)," *Geochim. Cosmochim. Acta* **37,** 2005–2016.
P. B. Price and R. L. Fleischer (1969), "Are Fission Tracks in Meteorites from Super-Heavy Elements?" *Phys. Letters* **30B,** 246–248.
P. B. Price and D. O'Sullivan (1970), "Lunar Erosion Rate and Solar Flare Paleontology," *Proc. Apollo 11 Lunar Sci. Conf.*, **3,** 2351–2359. New York: Pergamon Press.
P. B. Price and R. M. Walker (1962), "Electron Microscope Observation of Etched Tracks in Mica," *Phys. Rev. Letters* **8,** 217–219.
P. B. Price, R. S. Rajan, and A. Tamhane (1967), "On the Pre-Atmospheric Size and Maximum Space Erosion Rate of the Patwar Stony-Iron Meteorite," *J. Geophys. Res.* **72,** 1377–1388.
P. B. Price, R. L. Fleischer, and C. D. Moak (1968a), "On the Identification of Very Heavy Cosmic Ray Tracks in Meteorites," *Phys. Rev.* **167,** 277–282.
P. B. Price, R. S. Rajan, and A. Tamhane (1968b), "The Abundance of Nuclei Heavier than Iron in the Cosmic Radiation in the Geological Past," *Astrophys. J.* **151,** L109–116.
P. B. Price, R. L. Fleischer, and R. M. Walker (1968c), "The Utilization of Nuclear Particle Tracks in Solids to Study the Distribution of Certain Elements in Nature and in the Cosmic Radiation," *Origin and Distribution of the Elements*, L. H. Ahrens (ed.), Intl. Series of Monographs in Earth Sciences, **30,** 91–100, U. of Capetown, South Africa.
P. B. Price, I. Hutcheon, R. Cowsik, and D. J. Barber (1971a), "Enhanced Emission of Iron Nuclei in Solar Flares," *Phys. Rev. Letters* **26,** 916–919.
P. B. Price, R. S. Rajan, and E. K. Shirk (1971b), "Ultra-Heavy Cosmic Rays in the Moon," *Proc. Second Lunar Sci. Conf.*, **3,** 2621–2627. Cambridge: MIT Press.
P. B. Price, D. Braddy, D. O'Sullivan, and J. D. Sullivan (1972), "Composition of Interplanetary Particles at Energies from 0.1 to 150 MeV/nucleon," *Apollo 16 Preliminary Science Report*, NASA Special Pub. 315, chap. 15, pp. 11-19.
P. B. Price, D. Lal, A. S. Tamhane, and V. P. Perelygin (1973a), "Characteristics of Tracks of Ions of $14 \leq Z \leq 36$ in Common Rock Silicates," *Earth Planet. Sci. Lett.* **19,** 377–395.
P. B. Price, R. S. Rajan, I. D. Hutcheon, D. Macdougall, and E. K. Shirk (1973b),

"Solar Flares, Past and Present," *Lunar Science IV*, J. W. Chamberlain and C. Watkins (eds.), 600–602, Lunar Sci. Inst., Houston.

W. Quaide, V. Oberbeck, T. Bunch, and G. Polkowski (1971), "Investigations of the Natural History of the Regolith at the Apollo 12 Site," *Proc. Second Lunar Sci. Conf.*, **1,** 701–718. Cambridge: MIT Press.

R. S. Rajan (1973), "On the Irradiation History and Origin of Gas-Rich Meteorites," Ph.D. thesis, University of California, Berkeley.

L. A. Rancitelli, R. W. Perkins, W. D. Felix, and N. A. Wogman (1971), "Erosion and Mixing of the Lunar Surface from Cosmogenic and Primordial Radionuclide Measurements in Apollo 12 Lunar Samples," *Proc. Second Lunar Sci. Conf.*, **2,** 1757–1772. Cambridge: MIT Press.

L. A. Rancitelli, R. W. Perkins, W. D. Felix, and N. A. Wogman (1972), "Lunar Surface Processes and Cosmic Ray Characterization from Apollo 12–15 Lunar Sample Analyses," *Proc. Third Lunar Sci. Conf.*, **2,** 1681–1691. Cambridge: MIT Press.

L. A. Rancitelli, R. W. Perkins, W. D. Felix, and N. A. Wogman (1973), "Lunar Surface and Solar Process Analyses from Cosmogenic Radionuclide Measurements at the Apollo 16 Site," *Lunar Science IV*, J. W. Chamberlain and C. Watkins (eds.), 609–611, Lunar Sci. Inst., Houston.

R. C. Reedy and J. R. Arnold (1972), "Interaction of Solar and Galactic Cosmic-Ray Particles with the Moon," *J. Geophys. Res.* **77,** 537–555.

J. H. Reynolds (1960), "Determination of the Age of the Elements," *Phys. Rev. Letters* **4,** 8–10.

J. H. Reynolds (1963), "Xenology," *J. Geophys. Res.* **68,** 2939–2955.

J. H. Reynolds, E. C. Alexander, Jr., P. K. Davis, and B. Srinivasan (1973), "Studies of K-Ar Dating and Xenon from Extinct Radioactivities in Breccia 14318; Implications for Early Lunar History," *Geochim. Cosmochim. Acta* **38,** 401–418.

M. W. Rowe (1970), "Evidence for Decay of Extinct Pu^{244} and I^{129} in the Kapoeta Meteorite," *Geochim. Coschim. Acta* **34,** 1019–1028.

M. W. Rowe and D. D. Bogard (1966), "Isotopic Composition of Xenon from Ca-Poor Achondrites," *J. Geophys. Res.* **71,** 4183–4188.

M. W. Rowe and P. K. Kuroda (1965), "Fissiogenic Xenon from the Pasamonte Meteorite," *J. Geophys. Res.* **70,** 709–714.

G. P. Russ (1972), "Neutron Capture on Gd and Sm in the Luna 16 G-2 Soil," *Earth Planet. Sci. Lett.* **13,** 384–386.

G. P. Russ, III, D. S. Burnett, and G. J. Wasserburg (1972), "Lunar Neutron Stratigraphy," *Earth Planet. Sci. Lett.* **15,** 172–186.

O. A. Schaeffer and L. Husain (1973), "Early Lunar History: Ages of 2 to 4 mm Soil Fragments from the Lunar Highlands," *Proc. Fourth Lunar Sci. Conf.*, **2,** 1847–1863. New York: Pergamon Press.

E. Schneider, D. Storzer, and H. Fechtig (1972), "Exposure Ages of Apollo 15 Samples by Means of Microcrater Statistics and Solar Flare Particle Tracks," *The Apollo 15 Lunar Samples*, J. W. Chamberlain and C. Watkins (eds.), 415–419, Lunar Sci. Inst., Houston.

E. Schneider, D. Storzer, J. B. Hartung, H. Fechtig, and W. Gentner (1973), "Microcraters on Apollo 15 and 16 Samples and Corresponding Cosmic Dust Fluxes," *Proc. Fourth Lunar Sci. Conf.*, **3**, 3277–3290. New York: Pergamon Press.

E. Schönfeld (1971), personal communication.

D. N. Schramm and G. J. Wasserburg (1970), "Nucleochronologies and the Mean Age of the Elements," *Astrophys. J.* **162,** 57–69.

L. Schultz, P. Signer, P. Pellas, and G. Poupeau (1971), "Assam: A Gas Rich Hypersthene Chondrite," *Earth Planet. Sci. Lett.* **12,** 119–123.

L. Schultz, P. Signer, J. C. Lorin, and P. Pellas (1972), "Complex Irradiation History of the Weston Chondrite," *Earth Planet. Sci. Lett.* **15,** 403–410.

H. Schwaller (1971), "Isotopenanalyse von Krypton und Xenon in Apollo 11 und 12 Gesteinen," Ph.D. Thesis, Universität Bern.

J. Schwinger (1969), "A Magnetic Model of Matter," *Science* **165,** 757–761.

M. G. Seitz (1972), "Heavy Ion Radiation Studies in Terrestrial and Extraterrestrial Materials," Ph.D. Thesis, Washington University.

M. G. Seitz and M. C. Wittels (1971), "A Limit on the Radiation Erosion in Lunar Surface Material," *Earth Planet. Sci. Lett.* **10,** 268–270.

M. Seitz, M. C. Wittels, M. Maurette, R. M. Walker, and H. Heckman (1970), "Accelerator Irradiations of Minerals: Implications for Track Formation Mechanisms and for Studies of Lunar and Meteoritic Materials," *Rad. Effects* **5,** 143–148.

J. Shirck, M. Hoppe, M. Maurette, and R. Walker (1969), "Recent Fossil Track Studies Bearing on Extinct Pu^{244} in Meteorites," Intl. Atomic Energy Agency Symposium on Meteorite Research, Vienna, *Meteorite Res.*, P. Millman (ed.), 41–50. Dordrecht, Holland: D. Reidel.

E. M. Shoemaker (1960), "Penetration Mechanics of High Velocity Meteorites, Illustrated by Meteor Crater, Arizona," Intl. Geol. Cong., 21st, Copenhagen, Rept., Pt. 18, pp. 418–434.

P. Signer and H. E. Suess (1963), "Rare Gases in the Sun, in the Atmosphere and in Meteorites," *Earth Science and Meteoritics*, J. Geiss and E. D. Goldberg (eds.), 241–272, Amsterdam: North-Holland Publishing Co.

R. K. Soberman (1971), "The Terrestrial Influx of Small Meteoritic Particles," *Rev. Geophys. Space Phys.* **9,** 239–258.

R. W. Stoenner, W. Lyman, and R. Davis, Jr. (1971), "Radioactive Rare Gases and Tritium in Lunar Rocks and in the Sample Return Container," *Proc. Second Lunar Sci. Conf.*, **2,** 1813–1823. Cambridge: MIT Press.

D. Storzer and G. A. Wagner (1969), "Correction of Thermally Lowered Fission Track Ages of Tektites," *Earth Planet. Sci. Lett.* **5,** 463–468.

D. Storzer, G. Poupeau, W. Krätschmer, and T. Plieninger (1973a), "The Track Record of Apollo 15, 16, and Luna 16, 20 Samples and the Charge Assignment of Cosmic Ray Tracks," *Lunar Science IV*, J. W. Chamberlain and C. Watkins (eds.), 694–696, Lunar Sci. Inst., Houston.

D. Storzer, G. Poupeau, and W. Krätschmer (1973b), "Track-Exposure and Formation Ages of Some Lunar Samples," *Proc. Fourth Lunar Sci. Conf.*, **3,** 2363–2377, New York: Pergamon Press.

H. E. Suess, H. Wänke, and F. Wlotzka (1964), "On the Origin of Gas-Rich Meteorites," *Geochim. Cosmochim. Acta* **28,** 595–607.

A. S. Tamhane (1972), "Abundance of Heavy Cosmic Ray Nuclei from the Induced Micrometamorphism in Meteoritic Minerals," Ph.D. Thesis, University of Bombay.

J. F. Vedder (1971), "Microcraters in Glass and Minerals," *Earth Planet. Sci. Lett.* **11,** 291–296.

A. P. Vinogradov (1971), "Preliminary Data on Lunar Ground Brought to Earth by Automatic Probe 'Luna-16,'" *Proc. Second Lunar Sci. Conf.*, **1,** 1–16. Cambridge: MIT Press.

A. P. Vinogradov (1972), "Preliminary Data on the Lunar Samples Returned by the Automatic Station Luna 20," NASA TT F14, 615—Translation of Article in *Geokhimiya* **7,** 763–774.

T. T. von Rosenvinge, W. R. Webber, and J. F. Ormes (1969), "A Comparison of the Energy Spectra of Cosmic Ray Helium and Heavy Nuclei," *Astrophys. Space Sci.* **5,** 342–359.

M. Wahlen, M. Honda, M. Imamura, J. S. Fruchter, R. C. Finkel, C. P. Kohl, J. R. Arnold, and R. C. Reedy (1972), "Cosmogenic Nuclides in Football-Sized Rocks," *Proc. Third Lunar Sci. Conf.*, **2,** 1719–1732. Cambridge: MIT Press.

R. Walker and D. Yuhas (1973), "Cosmic Ray Track Production Rates in Lunar Materials," *Proc. Fourth Lunar Sci. Conf.*, **3,** 2379–2389. New York: Pergamon Press.

R. Walker and D. Zimmerman (1972), "Fossil Track and Thermoluminescence Studies of Luna 16 Material," *Earth Planet. Sci. Lett.* **13,** 419–422.

H. Wänke (1965), "Der Sonnenwind als Quelle der Uredelgase in Steinmeteoriten," *Z. Naturforsch.* **20A,** 946–949.

G. J. Wasserburg, J. C. Huneke, and D. S. Burnett (1969a), "Correlation Between Fission Tracks and Fission Type Xenon in Meteoritic Whitlockite," *J. Geophys. Res.* **74,** 4221–4232.

G. J. Wasserburg, J. C. Huneke, and D. S. Burnett (1969b), "Correlation Between Fission Tracks and Fission-Type Xenon from an Extinct Radioactivity," *Phys. Rev. Letters* **22,** 1198–1201.

G. J. Wasserburg, D. N. Schramm, and J. C. Huneke (1969c), "Nuclear Chronologies for the Galaxy," *Astrophys. J.* **157**, L91–L96.

L. L. Wilkening (1971), "Particle Track Studies and the Origin of Gas-Rich Meteorites," Nininger Meteorite Award Paper, Arizona State Univ., Tempe, Arizona.

L. Wilkening, D. Lal, and A. M. Reid (1971), "The Evolution of the Kapoeta Howardite Based on Fossil Track Studies," *Earth Planet. Sci. Lett.* **10,** 334–340.

O. Wilson and R. Woolley (1970), "Calcium Emission Intensities as Indicators of Stellar Age," *Mon. Not. Roy. Astr. Soc.* **148,** 463–475.

J. A. Wood (1968), *Meteorites and the Origin of Planets.* New York: McGraw-Hill.

J. A. Wood (1970), "The Lunar Soil," *Sci. Amer.* **223,** 14–23.

D. S. Woolum and D. S. Burnett (1974), "In-Situ Measurement of the Rate of ^{235}U Fission Induced by Lunar Neutrons," *Earth Planet. Sci. Lett.* **21,** 153–163.

D. Yuhas and R. Walker (1973), "Long Term Behavior of VH Cosmic Rays as Observed in Lunar Rocks," *Proc. 13th Inter. Cosmic Ray Conf.*, Univ. of Denver, **2,** 1116–1121.

D. E. Yuhas, R. M. Walker, H. Reeves, G. Poupeau, P. Pellas, J. C. Lorin, G. C. Chetrit, J. L. Berdot, P. B. Price, I. D. Hutcheon, H. R. Hart, Jr., R. L. Fleischer, G. M. Comstock, D. Lal, J. N. Goswami, and N. Bhandari (1972), "Track Consortium Report on Rock 14310," *Proc. Third Lunar Sci. Conf.*, **3,** 2941–2947, Cambridge: MIT Press.

PART III

Nuclear Science and Technology

Chapter 7

Nuclear Physics at High and Low Energies

7.1. INTRODUCTION

Nuclear physicists and chemists are among the most active users of track-etching techniques. Solid state track detectors are now being used in over fifty nuclear laboratories, most of which have their own accelerator. About one-third of the work is done in North America, one-third in the Soviet Union, and one-third in Europe. Many nuclear research groups who once used nuclear emulsions have found it easy to switch over to track-etch techniques since appropriate microscope equipment was already on hand.

It is interesting to compare the history of the development of nuclear emulsions and of dielectric track detectors.

Alpha particle tracks from a radioactive source were observed as early as 1910 (Kinoshita) in ordinary commercial photographic emulsion, but because of a general pessimism about the potentialities of the technique, no serious attempts were made to improve the sensitivity and stability of emulsions for the next twenty years. It was not until 1946 that Demers (1946), and, independently, Dodd and Waller, of Ilford Limited (cited by Powell et al., 1946), produced emulsions with a high concentration of silver halide grains.

Only then did real progress in the application of nuclear emulsions to nuclear physics begin, and, with the discovery of the pion by Lattes et al. (1947) and the discovery of heavy primary cosmic rays by Freier et al. (1948), the importance of nuclear emulsions was established. A year later Kodak developed the first emulsions that were sensitive to singly charged, minimum-ionizing particles, and in 1952 thick stacks of emulsion layers made by Ilford Ltd. began to be used to study interactions of fundamental particles in the cosmic rays.

In contrast, important nuclear physics experiments began to be done very quickly after the discovery of tracks in dielectric solids, and relatively little effort has gone into modifying the structure or composition of dielectric detectors. Today the favorite detectors in most nuclear laboratories are ordinary glass, muscovite mica, and commercially mass-produced plastics.

Admittedly, track-etching was discovered as a by-product of our attempts to improve the sensitivity of mica. After the initial observation of spallation recoil tracks (Price and Walker, 1962a) we spent considerable effort trying to increase the sensitivity so that we could see tracks of lighter particles. Gradually we realized that as they stood, mica and other dielectrics were already unique in being able to discriminate against light particles and to see the heavy ones. Of course, with subsequent improvements in our ability to control sensitivity, new applications have continued to appear.

G. N. Flerov, at the Joint Institute for Nuclear Research in Dubna, was quick to capitalize on the potentialities of the discovery. Since about 1962 he had been looking for an insensitive detector for his studies of transuranic elements and, in particular, to intensify work on spontaneously fissioning isomers, which his group had just discovered. When Zvara and Polikanov attended a Heavy Ion Conference at Asilomar in 1963 and heard of our ternary fission studies with mica (Price et al., 1963), they immediately recognized the potential utility of dielectric detectors.

Basically the only technological development needed to bring dielectric detectors into general use was to learn to magnify the tracks to optically visible size by a prolonged chemical etching. Before that discovery (Price and Walker, 1962a), we viewed the high spatial resolution achievable with the electron microscope as a major advantage of the mica detector and we spent considerable time searching for tails on V-shaped tracks of recoiling, fissioning compound nuclei in a fruitless attempt to measure lifetimes of $\sim 10^{-15}$ sec or greater. It was not until 1968, after channeling and blocking effects of fast ions in crystals were well understood, that dielectric detectors were successfully used in an indirect way to study compound nucleus lifetimes (Brown et al., 1968). The lifetimes amenable to study, $\sim 10^{-16}$ to $\sim 10^{-19}$ sec, are orders of magnitude shorter than what we had originally hoped to measure directly by electron microscopy.

Because dielectric detectors filled a long-standing need, interesting discoveries began to be made within a year or so after we learned to make the tracks optically visible and several years before we learned how to measure atomic numbers of particles by relating etching rate to ionization rate. Ninety percent of the nuclear physics studies with etched tracks have involved fission, with no attempt made to identify the particles other than to classify them as fission fragments. Most of the remaining ten percent have also used them as high-threshold detectors, relying on a measurement of track length to exclude unwanted events. Only recently have a few groups begun to exploit the particle-identifying capabilities of the track-etch technique, so that each event is characterized by orientation, energy, and charge. As familiarity and experience with particle identification techniques increase, we can expect a growing number of discoveries that depend on knowledge of atomic number, Z, as well as energy, E. Of course, because of its simplicity the most popular use of dielectric detectors may continue to be as a threshold detector. Table 7-1 lists some of the highlights in the one-decade history of the applications of dielectric detectors to nuclear physics and chemistry.

It is not our intention to discuss in great detail all of the contributions that have been made to nuclear science. It is not even feasible, as the reader can appreciate by turning to the bibliography at the end of this chapter. Instead, we shall discuss only a few experiments which are noteworthy either because they illustrate one of the very useful features of dielectric detectors or because of their likely future importance or impact.

Tables 7-2 to 7-4 cite some of the ways in which the important features of track-recording solids have been exploited. The one feature that sets them apart from all other detectors is their ability to record tracks of heavily ionizing particles under a variety of adverse conditions such as intense background radiation, high temperature, or low event rate (Table 7-2). Sandwiches consisting of thin, suitably chosen solid detectors in contact with thin target foils can be placed in the direct beam of an accelerator and subjected to large doses of particles such as 10 MeV/nucleon argon ions (Fleischer et al., 1966) or relativistic nitrogen ions (Katcoff and Hudis, 1972). Because of the absence of background counts, extremely low fission rates can be detected. Cross sections as low as 10^{-36} cm^2 have been measured (Burnett et al., 1964; Khodai-Joopari, 1966). One fission per day was recorded in experi-

Table 7-1. Landmarks in the History of Nuclear Applications of Solid State Track Detectors

1961	Attempts with electron microscopy to measure short lifetimes
1962	Observation of a complete-fusion reaction, $^{16}O + ^{27}A\ell$
1963	Observation of ternary fission
1964	Determination of fission barrier and saddle-point mass of Tl^{201}
1964	Synthesis of element 104 demonstrated
1965	Lifetimes of spontaneously fissioning isomers measured.
1968	Compound nucleus lifetimes measured by "blocking" in crystals
1969	Intensive searches made for superheavy elements and magnetic monopoles
1969	Excitation functions measured for (p,α) reactions
1972	Determination of Z and E of fragments in relativistic heavy ion reactions
1973	Electrofission of ^{24}Mg with picobarn cross-sections

Table 7-2. Exploiting the Insensitivity of Dielectric Detectors to Adverse Conditions Such as High Background Radiation

Detector	Background to Be Discriminated Against	Particles Studied or Sought	Reference
mica	100μamp-hr beam of ~30 MeV alphas in scattering chamber	Fissions with σ_f as low as $10^{-36} cm^2$	Burnett et al. (1964)
glass and mica	radiation from cyclotron and ambient temperature ~400°C	~1 fission/day of an isotope of element 104	Zvara et al. (1966; 1970)
mica	10^{13} alphas/cm^2 from decay of ^{226}Ra target in contact with detector	photofission and neutron-induced fission	Zhagrov et al. (1968); Babenko et al. (1968)
minerals	radioactivity, cosmic rays and solar flare particles summed over $>10^7$ years	magnetic monopoles; fissioning superheavy elements	see refs. in Appendix
Makrofol polycarbonate	p, α, fission fragments, ^{16}O and other reaction products from 10μamp electron beam	0.3 to 1.3 MeV/amu ^{12}C ions from $^{24}Mg(\gamma, ^{12}C)^{12}C$ with σ_f as low as $10^{-36} cm^2$	Chung (1973)

Table 7-3. Exploiting the Ability of Dielectric Detectors to Record Spatial Relationships

Detector	Geometry	Study	Reference
Lexan	Curved sheets enclose target (4π)	Structure of fission transition nuclei	Behkami et al. (1968a)
mica	Sandwich with inner film target	High-energy fission	Brandt et al. (1967); Hudis and Katcoff (1969)
$ThSiO_4$	Crystal serves as target and detector; all 3 fragments are made visible.	Ternary fission	Fleischer et al. (1966)
Makrofol	Cone or strips pointed downstream	Shape isomers	Lark et al. (1969)
glass	Planar sheet maps intensity distribution of fragments emitted along <111> planes of single crystal target	Lifetimes of 10^{-16}–10^{-19} sec	See entry 3 in Appendix.

Table 7-4. Exploiting the Capability of Particle Identification Under Adverse Conditions

Detector	Reaction	Study	Reference
cellulose nitrate	$^{10}B(p,\alpha)^{7}Be$	excitation function of the emitted alphas down to $\sigma \sim 10^{-30} cm^2$	Szabo et al. (1972)
Lexan	34 GeV $^{16}O + Au \rightarrow$ fragment	energy spectra down to 0.8 MeV/amu; cross-sections of fragments with $2 \lesssim Z \lesssim 16$ in a feeble beam	Sullivan et al. (1973)
Makrofol	$^{24}Mg(\gamma,^{12}C)^{12}C$	identify 0.3 to 1.3 MeV/amu ^{12}C ions with σ_f as low as $10^{-36} cm^2$	Chung (1973)

ments leading to the chemical and physical identification of element 104 (Zvara et al., 1966, 1971; Flerov et al., 1971a). Different plastic detectors, each 1 m^2 in area, were exposed in contact with ores of heavy elements for several months in a search for spontaneously fissioning superheavy elements (Flerov and Perelygin, 1969; Geisler et al., 1973), and 18 m^2 of plastic have been exposed for one year in a search for magnetic monopoles (Fleischer et al., 1971). If we include natural glasses and minerals, then even lower event rates can be studied. Tracks of magnetic monopoles and superheavy elements have been unsuccessfully sought in crystals that have resided on the lunar surface for more than 10^7 years and in crystals contained in meteorites that solidified 4.6 billion years ago. In both cases tracks accumulated without detectable fading over the entire time span.

The ability to record spatial relationships (Table 7-3) is of course shared by nuclear emulsions and only becomes unique to dielectric solids when coupled with one of the other distinguishing features such as insensitivity to background radiation.

The ability to determine the atomic number and energy of desired particles in a high background of undesired, lightly ionizing particles or at extremely low event rates (Table 7-4) uniquely distinguishes dielectric solids from detectors such as semiconductor particle identifiers or other electronic devices.

We turn now to a more detailed discussion of some of the scientific problems that illustrate the unique features of dielectric track detectors.

7.2. HALF-LIVES FOR SPONTANEOUS FISSION AND ALPHA DECAY

Table 7-5 summarizes the measurements of partial half-lives for spontaneous fission that have relied on mica, glass, or plastic detectors. The half-lives span an enormous range, from $\sim 3 \times 10^{18}$ y for ^{237}Np to $\sim 3.8 \times 10^{-4}$ sec for ^{258}Fm. The largest number of independent measurements has been made on ^{238}U, whose spontaneous fission decay forms the basis of the fission track dating method (Chapter 4). In addition to the standard procedure of placing a solid track detector against a foil of the isotope and counting the rate of accumulation of tracks, it is possible to make an independent check of the fission decay constant of ^{238}U by comparing the fission track age of a natural substance with its age determined by decay of another long-lived nuclide such as ^{40}K or ^{87}Rb. We point out in Chapter 4 that independent fission track age determinations of 27×10^6 year-old crystals of apatite and zircon in several fission-track laboratories agree remarkably well provided the value $\lambda_{SF} = 7 \times 10^{-17}$ y^{-1} is used by each laboratory.

Techniques for measuring very short half-lives involve great difficulties and demand ingenuity in coping with the problems of a high fission background, competing decay modes, and the rapid rate of disappearance of the sample itself. It is interesting to follow the competition between the transuranium research groups led by Flerov at Dubna and by Ghiorso at Berkeley. Track detectors have played

Table 7-5. Determination of Partial Half-lives for Spontaneous Fission

Nuclide	τ_{SF} (or λ_{SF})	Reference
^{233}U	$(1.2 \pm 0.3) \times 10^{17} y$	Aleksandrov et al. (1967)
^{235}U	$(3.5 \pm 0.9) \times 10^{17} y$	Aleksandrov et al. (1967)
^{238}U	$(6.85 \pm 0.20) \times 10^{-17} y^{-1}$	Fleischer and Price (1964)
^{238}U	$(7.03 \pm 0.11) \times 10^{-17} y^{-1}$	Roberts et al. (1968)
^{238}U	$(6.8 \pm 0.6) \times 10^{-17} y^{-1}$	Kleeman and Lovering (1971)
^{238}U	$(7.30 \pm 0.16) \times 10^{-17} y^{-1}$	Leme et al. (1971)
^{238}U	$(8.4) \times 10^{-17} y^{-1}$	Gentner et al. (1972)
^{238}U	$(6.82 \pm 0.55) \times 10^{-17} y^{-1}$	Khan and Durrani (1973)
^{238}U	$7.5 \times 10^{-17} y^{-1}$	Suzuki (1973)
^{237}Np	$3 \times 10^{18} y$	Perelygin et al. (1964)
^{241}Am	$(1.147 \pm 0.024) \times 10^{14} y$	Gold et al. (1970)
^{243}Am	$(2 \pm 0.5) \times 10^{14} y$	Gvozdev et al. (1966)
^{243}Am	$(3.3 \pm 0.3) \times 10^{13} y$	Aleksandrov et al. (1967)
^{244}Cm	$10^{7} y$	Perelygin et al. (1964)
^{244}Fm	(3.3 ± 0.5) msec	Nurmia et al. (1967)
^{246}Fm	~20 sec	Druin et al. (1971)
^{246}Fm	15 ± 5 sec	Nurmia et al. (1967)
^{248}Fm	~60 h	Druin et al. (1971)
^{248}Fm	10 ± 5 h	Nurmia et al. (1967)
^{248}Cf	$3 \times 10^{4} y$	Perelygin et al. (1964)
^{249}Bk	$(1.65 \pm 0.17) \times 10^{9} y$	Vorotnikov et al. (1970c)
^{250}Fm	~10y	Druin et al. (1971)
^{252}No	~7 sec	Ghiorso et al. (1967)
^{256}No	~20 min	Ghiorso et al. (1967)
^{256}No	1500 sec	Druin et al. (1964)
^{258}Fm	$380 \pm 60 \mu sec$	Hulet et al. (1971)
$^{260}104$	~0.1 sec *	Flerov et al. (1964; 1971a)
$^{258}104$	11 ± 2 msec	Ghiorso et al. (1969)
$^{259}104$	3.2 ± 0.8 sec total half-life ($SF/\alpha = 7\%$)	Flerov et al. (1971a); Ghiorso et al. (1969); Druin et al. (1973)
$^{261}105$	1.8 ± 0.6 sec	Flerov et al. (1971a,b)

* disputed

an important role in attempts at both laboratories to provide proof of the synthesis of isotopes of the elements beyond 100. Because spontaneous fission is not as specific a fingerprint as is alpha decay, the former usually has provided preliminary or supporting evidence rather than a definitive proof that a new element has been made.

Fig. 7-1, from the paper by Zvara et al. (1971), illustrates the unique capabilities of dielectric detectors. In order to show that element 104 has chemical properties more like those of Hf than of the actinides, they used a thermatographic column lined with mica strips at a temperature decreasing from ~400°C to room temperature over a distance of ~1 m. They bombarded a thin ^{242}Pu target with ^{22}Ne ions, chlorinated the recoil atoms, and passed the resulting chlorides along the column. After about one month of irradiation they measured the fission track distribution along the mica strips. The fission events indicated by the black dots in Fig. 7-1 were attributed to decay of the chloride of element 104, which condensed on the wall where the temperature was ~200°C. The fission events indicated by the circles were attributed to decay of actinide chlorides which condensed at the beginning of the column where the temperature was much higher. In a control experi-

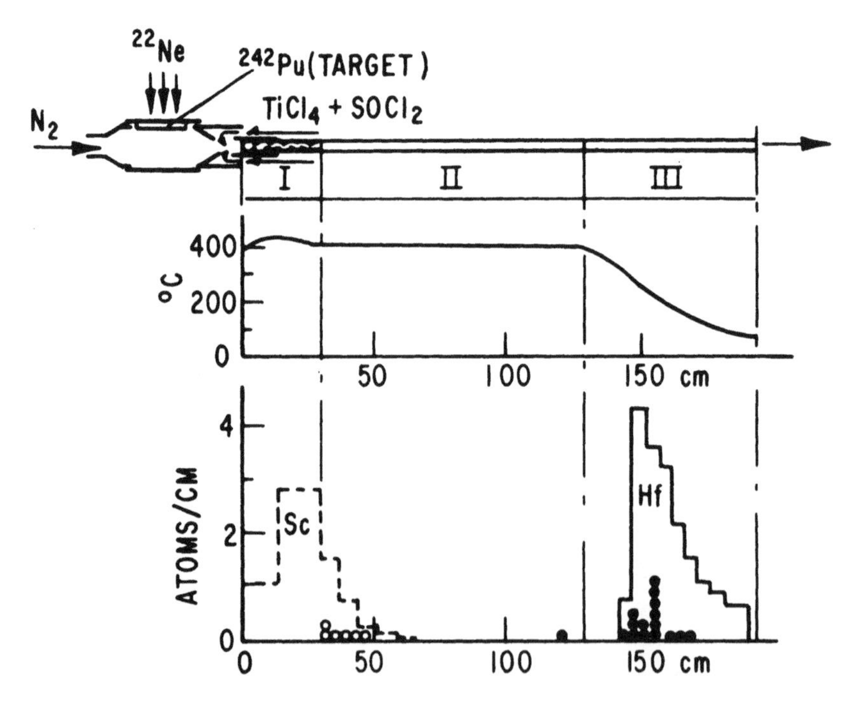

Fig. 7-1. *Thermo-chromatographic method of separating element 104 from actinides. Spontaneous fission tracks from element 104 are found (closed circles) at the cool end of the column at the same position as radioactivity from* ^{170,171}*Hf (histogram), whereas fission tracks from actinides are found (open circles) near the beginning of the hot column at the same position as radioactivity from* ^{44m}*Sc (dashed histogram). (After Zvara et al., 1971.)*

ment with counters, the chloride of ^{170}Hf and ^{171}Hf was found to decay near the end of the column where the temperature was ~200°C, whereas the chloride of ^{44m}Sc, whose chemistry is similar to that of the actinide chlorides, decayed in the high-temperature part of the column.

In their initial experiments with glass detectors on a moving belt, the Dubna group (Flerov et al., 1964) reported a 0.3 sec half-life for the spontaneously fissioning isotope of element 104 made in their ^{22}Ne irradiations of ^{242}Pu, whereas they later reported that there were, instead, two fissioning isotopes, one with a 0.1 sec half-life and the other with a half-life of ~4 sec (Flerov et al., 1971a). Though the conclusions of Zvara et al. (1971) have been contested (Ghiorso et al., 1971; see rebuttal and more recent evidence by Druin et al., 1973), the experiment provides an impressive example of the ability of dielectric detectors to perform under adverse conditions.

New information on the systematics of spontaneous fission has come from measurements of half-lives with dielectric detectors, shown in Fig. 7-2. The measure-

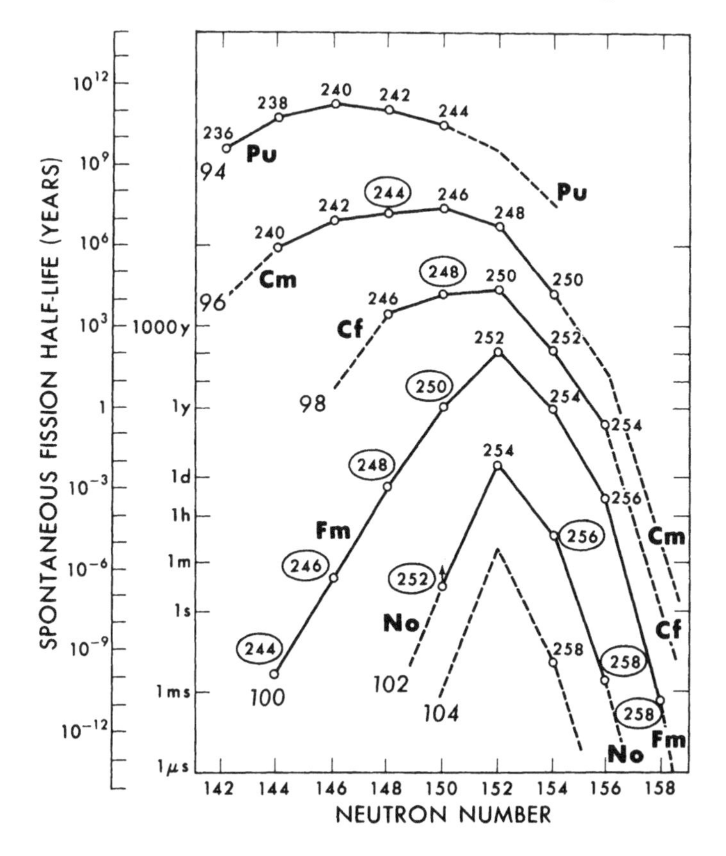

Fig. 7-2. *Systematics of the spontaneous-fission half-lives of even-even isotopes of transuranic elements. Measurements made with dielectric detectors are circled. Beyond the closed subshell at $N = 152$ the half-lives decrease precipitously.*

ment on ^{258}Fm, which employed mica strips on a rapidly rotating drum, is a particularly dramatic example (Hulet et al., 1971). The observed 380 μsec fission half-life is more than 10^8 times shorter than had been predicted independently by three groups and showed that even-even nuclei with more than 158 neutrons are catastrophically unstable toward fission. This result has forced a re-evaluation of the present techniques of calculating fission lifetimes of superheavy nuclei by superimposing single-particle effects upon the liquid-drop model.

Although no measurements of half-lives for alpha decay utilizing dielectric detectors have been made, they appear to be feasible. Perhaps the best approach would be to use cellulosic polymers or UV-sensitized polycarbonate film (Stern and Price, 1972) in conjunction with a high-frequency spark scanning device (Geisler and Philips, 1972) to locate coincident holes in a pair of very thin (~4 μm) Makrofol sheets that had been in contact with a large foil of an alpha emitter for a long time. The Makrofol would have to be UV-sensitized, so that an alpha particle would produce an etchable track through both of the 4 μm Makrofol sheets. One could, in principle, determine half-lives for alpha particle decay up to $\sim 10^{25}$ years by this method.

7.3. LIFETIMES OF COMPOUND NUCLEI BY THE BLOCKING EFFECT

To understand nuclear reactions it is important to know the reaction time, which relates to the number of degrees of freedom excited. In contrast to so-called *direct* interaction processes, which occur in times of $\sim 10^{-22}$ sec, approximately the time of passage of a nucleon through a nucleus, reactions involving a compound nucleus have much longer lifetimes of $\sim 10^{-15}$ to $\sim 10^{-20}$ sec. Because distances traveled by compound nuclei moving with typical speeds of $\sim 10^6$ to 10^9 cm/sec are usually less than one Ångstrom, direct observation of their tracks is not possible. However, an extremely ingenious way of using dielectric detectors to visualize such short recoil distances has been developed, and several groups in the Soviet Union and in the United States have succeeded in measuring lifetimes ranging from $\sim 10^{-18}$ to $\sim 10^{-16}$ sec.

The method, which was suggested independently by Tulinov (1965) and by Gemmell and Holland (1965), depends on the fact that a charged particle originating on a lattice site is prohibited from leaving the crystal along a row of atoms because of Rutherford scattering by those atoms. The angular distribution of such particles shows a characteristic blocking pattern of dips or shadows corresponding to the major crystal axes and planes. Fig. 7-3 is a beautiful example of a blocking pattern showing the intensity distribution of alpha particles scattered off of a single crystal of ThO_2. Now if, because of recoil, the charged particle originates at a point removed from the lattice site by a distance greater than the Thomas-Fermi screen-

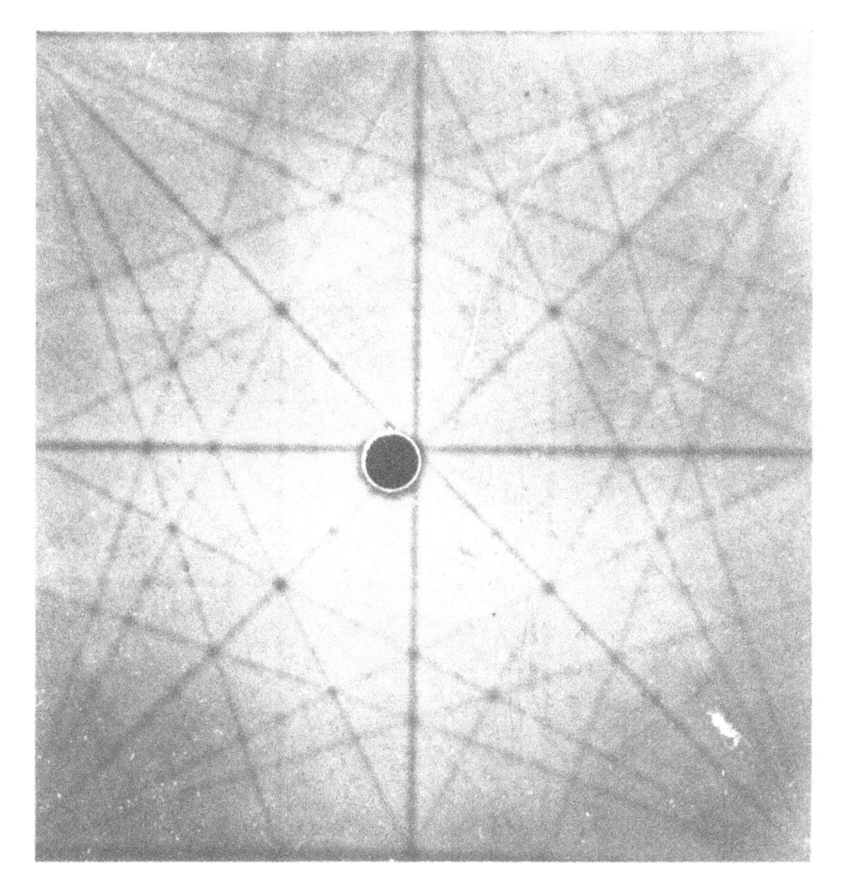

Fig. 7-3. *Blocking pattern around ⟨100⟩ axis of a thorium oxide single crystal (courtesy of G. R. Bellavance).*

ing distance (~0.1 Å), the blocking will be eliminated and no characteristic pattern should be detected. In an actual experiment, because of scattering from lattice vibrations and crystal defects the intensity along atomic planes would never drop entirely to zero, even if the charged particle were emitted from a lattice site. One therefore needs to determine the difference between two blocking patterns in which recoils do and do not occur.

In the first successful measurement of a compound nucleus lifetime, Gibson and Nielsen (1970) bombarded a single crystal of UO_2 with protons of two different energies and compared the blocking patterns of fission fragments emitted from the crystal and imaged in a Makrofol detector. The dashed curve in Fig. 7-4 shows the fission-fragment intensity distribution for fission induced by 12 MeV protons, which deposit so much excitation energy in the compound nucleus that it fissions extremely rapidly, before any significant recoil can occur. The minimum in the dashed curve is thus a measure of scattering effects. The data points and the solid

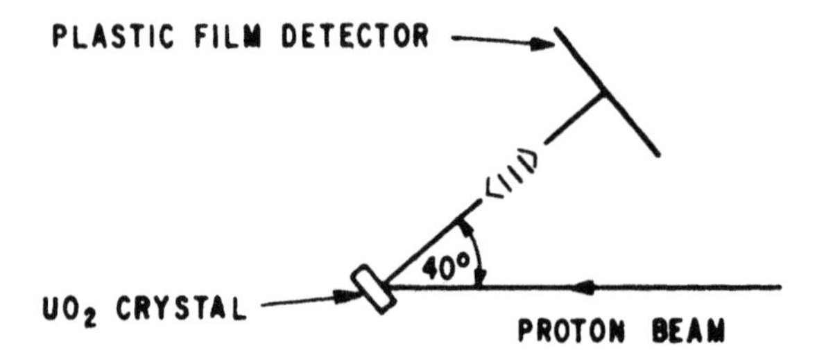

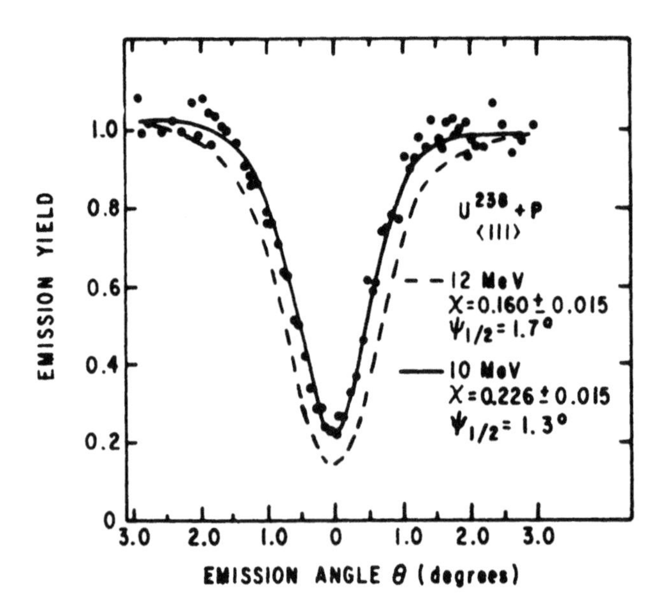

Fig. 7-4. *Measured angular distributions, normalized at large emission angles, of fission fragments emitted parallel to a ⟨111⟩ axis for bombardment of UO_2 crystals with 10-MeV protons (solid line and closed circles) and with 12- and 9-MeV protons (dashed line). The values indicated for the minimum yield for the 12-MeV curve are the averages, weighted by their respective statistical uncertainty, of three runs at 12 MeV and one at 9 MeV. For the 10-MeV case, the minimum yield is the result of two runs at 10 MeV. (After Gibson and Nielson, 1970.)*

curve in Fig. 7-4 were obtained with 10 MeV protons which deposit just enough excitation energy that the compound nucleus, after emission of a neutron, has a finite lifetime and travels an appreciable fraction of 1 Å. From the difference in valley depths in Fig. 7-4 and from a knowledge of the recoil velocity, Gibson and Nielsen computed a lifetime of $(1.4 \pm 0.6) \times 10^{-16}$ sec for decay of the ^{238}Np compound nucleus.

By using a plastic that is sensitive to alpha particles, one could also measure lifetimes of compound nuclei that decay by alpha emission.

7.4. FISSION BARRIERS

Here and in Section 7.5, on isomers, we discuss some of the important experiments with dielectric track detectors that have guided theorists in the development of a two-component theory of the shape dependence of the nuclear Hamiltonian. Fission, nuclear shapes and masses, heavy-ion reactions, and the question of superheavy nuclei are now being treated in a unified way by a combined macroscopic-microscopic approach first suggested by W. J. Swiatecki (1963) and later developed quantitatively by V. M. Strutinsky (1966). The macroscopic or liquid-drop contribution to the nuclear Hamiltonian successfully describes the smooth trends throughout the periodic table of such quantities as nuclear masses, fission-barrier heights, and fission-fragment kinetic energies, but not the local fluctuations. The microscopic or single-particle effects cause systematic deviations of the nuclear potential energy from the liquid-drop energy, and these effects can be taken into account by appropriate shell and pairing corrections. Single-particle effects are responsible for such important phenomena as the occurrence of deformed rather than spherical ground-state shapes for midshell nuclei and the occurrence of secondary minima in the fission barriers of most actinide nuclei (Fig. 7-5), which in turn accounts for the existence of short-lived, spontaneously fissioning isomers. The extra stability arising from single-particle shell closures leads to the possibility of an island of relatively long-lived superheavy nuclei near 114 protons and 184 neutrons. Nix (1972) gives a very readable account of the macroscopic-microscopic theory and its application to nuclear fission and superheavy nuclei. The recent text by Vandenbosch and Huizinga (1973) is an excellent treatise on all aspects of nuclear fission.

In 1963 S. G. Thompson and his colleagues at the Lawrence Berkeley Laboratory made the first accurate determination of the fission barrier and saddle-point mass of a nucleus (Burnett et al., 1964). With mica detectors and the variable energy, 88-inch cyclotron it was possible to determine the fission cross section for the reaction $^{4}He + {}^{197}Au \rightarrow {}^{201}Tl \rightarrow$ fission, as a function of energy, down to excitation energies below the fission barrier. The insensitivity of mica to the helium ions made it possible to measure cross sections from $\sim 10^{-24}$ to $\sim 10^{-35}$ cm^2, the lower limit being set by the presence of about 1 part in 10^{10} of uranium in the gold target. From the detailed shape of their excitation function, shown in Fig. 7-6, they were able to arrive at an accurate value for the fission barrier and saddle-point mass of ^{201}Tl. Since then the Berkeley group has been very active in determining fission barrier heights (Khodai-Joopari, 1966; Thompson, 1967; Moretto et al., 1968),

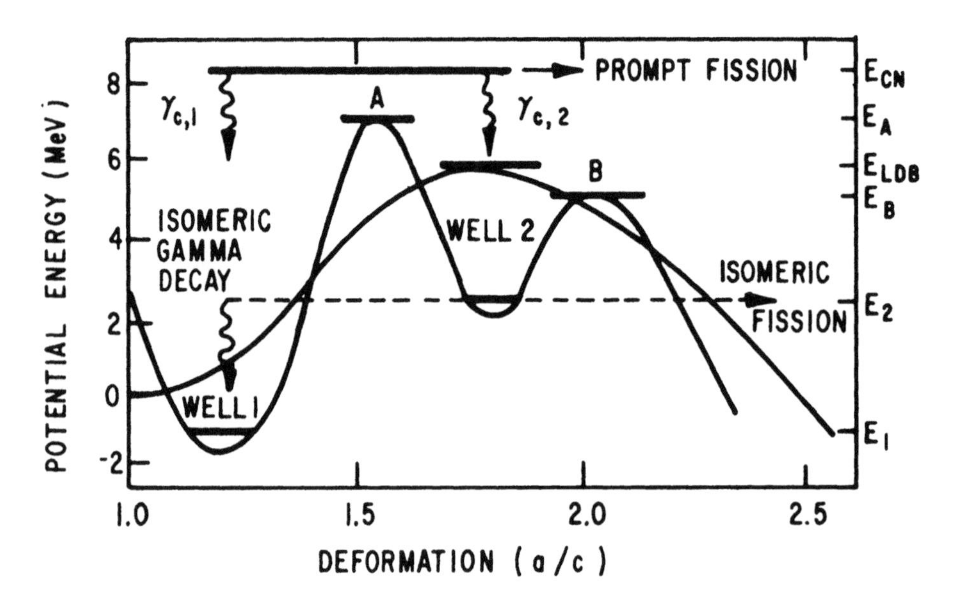

Fig. 7-5. *Potential energy of deformation varies with ratio a/c of the major and minor semiaxes of a prolate nucleus. Wiggly curve is the double-hump fission barrier; smooth curve is the fission barrier for the traditional liquid-drop model. A nucleus in well 2 with energy* E_2 *is isomeric (metastable) because of the potential barrier on either side. Decay to the right is by fission, to the left by gamma emission. Energy* E_1 *is the ground state of well 1;* E_2 *is the ground state of well 2;* E_A, E_B *and* E_{LDB} *are the heights of the inner barrier A, the outer barrier B and the liquid drop barrier; and* E_{CN} *is the excitation energy of the compound nucleus. Competing de-excitation processes (top) for the compound nucleus are gamma decay into wells 1 and 2* ($\gamma_{c,1}$ *and* $\gamma_{c,2}$) *and prompt fission. (After Clark, 1971.)*

measuring angular distributions that provide information on saddle-point pairing effects (Moretto et al., 1969a; 1972), and accounting for the systematics of fission probabilities in terms of the macroscopic-microscopic theory (Moretto et al., 1972).

The bibliography on induced fission in the appendix at the end of this chapter includes important experiments by other groups who have used beams of gamma rays, electrons, neutrons, and various charged particles and detectors of mica, glass, or plastic.

7.5. SPONTANEOUSLY FISSIONING ISOMERS

After Polikanov et al. (1962) discovered an isomeric state of ^{242}Am that spontaneously fissioned with a half-life of 14 msec, it was impossible for several years to understand the delicate balance that prevented the isomer from decaying to the ground state by gamma emission but allowed it to fission with a lifetime $\sim 10^{19}$

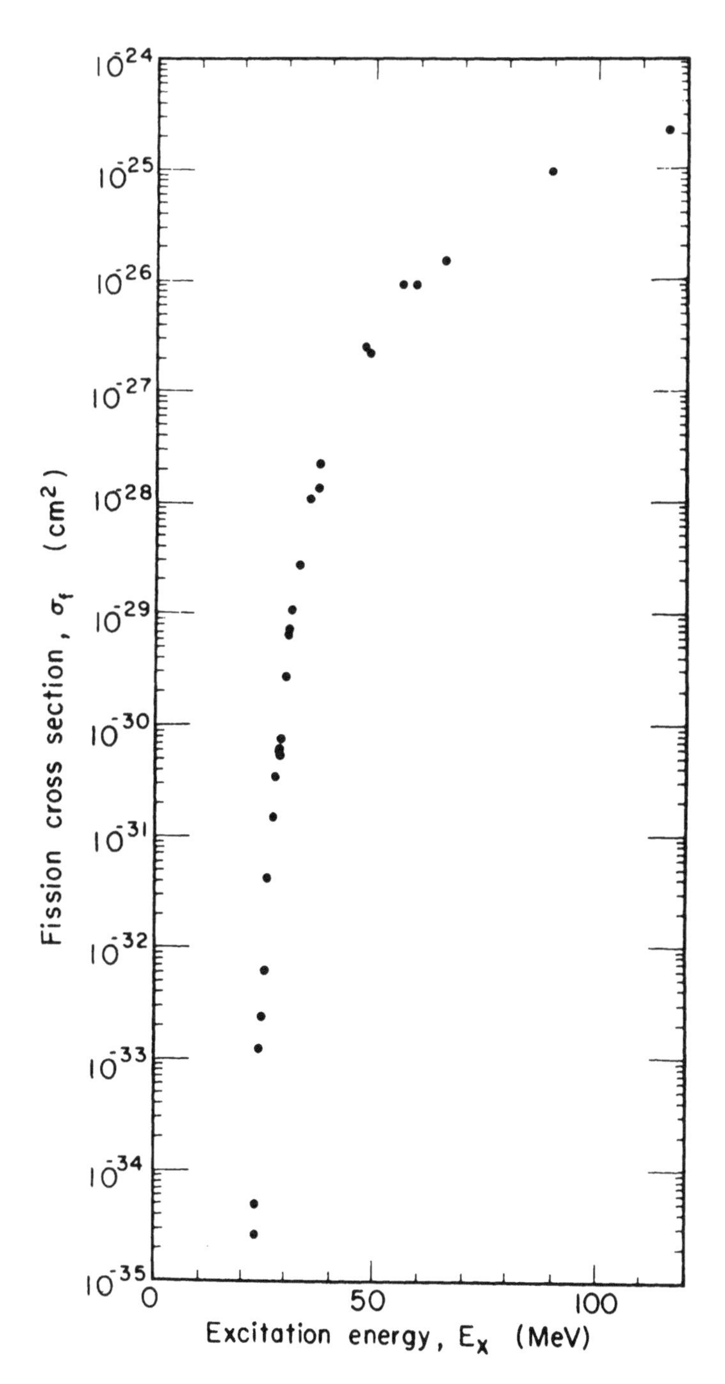

Fig. 7-6. *Measured fission cross section versus excitation energy for the reaction* $^{197}Au + {}^{4}He \rightarrow {}^{201}Tl$. (*After Burnett et al., 1964.*)

times shorter than the spontaneous fission lifetime of the ground state. In a lucid article in *Physics Today*, Clark (1971) tells the story of the exciting interplay of experiments and theory that led to the discovery of the double-hump fission barrier that characterizes the actinide elements. We use his sketch in Fig. 7-5 to show how the concept of a double-hump barrier, which comes directly from the combined (liquid-drop) + (single-particle) theory, accounts for spontaneously fissioning isomers (sometimes called shape isomers because they are stuck in a metastable configuration that is far from spherical). One can see that the liquid-drop model by itself predicts a single-hump barrier and a spherical configuration in the ground state. The inclusion of single-particle effects leads to a slightly deformed ground-state shape (roughly a prolate spheroid). The relative heights of barriers A and B determine whether the isomer tunnels toward the ground state via gamma decay or toward larger deformations that lead to fission. The height of the secondary minimum (well 2) above the ground state (well 1) accounts for the short half-life: for ^{242m}Am an elevation of ~2.5 MeV shortens the half-life by a factor of 10^{19}.

Calculations and experiments agree that spontaneously fissioning isomers are found only among the actinides. For a nucleus that contains more protons than berkelium the outer portion of the barrier (B in Fig. 7-5) is so low that the isomer decays too rapidly by spontaneous fission to be observed experimentally. For a nucleus that contains fewer protons than uranium the first peak (at A) is so low that the isomer decays primarily back to the first well by the emission of gamma rays. Fig. 7-7 summarizes the measurements of spontaneous fission lifetimes and indicates which nuclides were identified by track detectors. Half-lives range from 14 msec for ^{242m}Am down to about 0.5 nsec, which is the shortest lifetime that has been measured for these isomers.

Table 7-6 lists the four methods that have been used to measure lifetimes. With a suitable combination of excitation energy, projectile, and target, the isomer is produced and recoils out of the target. If its lifetime exceeds ~10^{-3} sec it can be caught on a rotating wheel or moving belt and transported past a series of solid track detectors. By adjusting the speed of the belt or wheel until a conveniently measurable distribution of fission tracks is observed along the detector strips, the lifetime can be determined from the slope of the distribution and a knowledge of the recoil velocity. For lifetimes between ~10^{-9} and 10^{-6} sec the recoils can be allowed to fly downstream and fission in flight. Practical flight paths range from

Table 7-6. Methods of Measuring Lifetimes of Spontaneously Fissioning Isomers

Lifetime	Method
≳ 1 msec	rotating wheel or moving belt + track detector
~ 1 μsec to ~1 msec	pulsed beam + catcher + semiconductor detector
~ 1 nsec to ~1 μsec	recoiling isomer fissions in flight onto track detector
~10 nsec to ~50 psec	fission in flight with geometric amplification of flight path (Limkilde and Sletten, 1973)

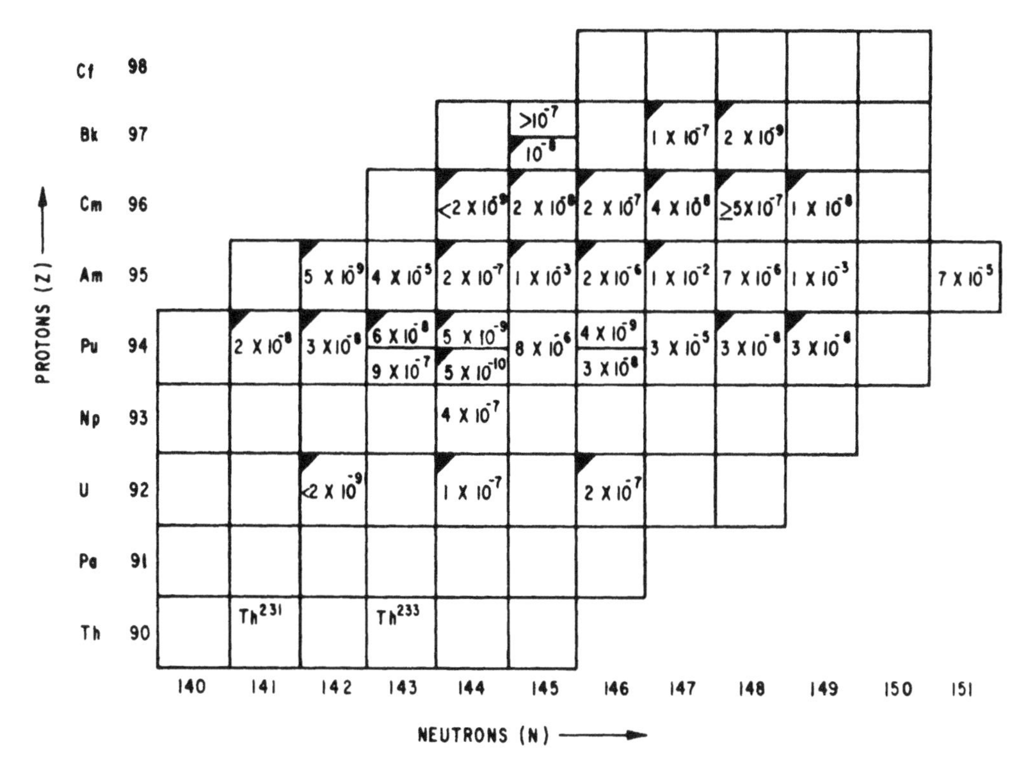

Fig. 7-7. *Actinide region of the nuclide chart, summarizing experimental evidence for the double-hump barrier. The entries indicate half-lives, in seconds, for spontaneous fission from the second potential well rather than from the inner potential well (see Fig. 7-5). A shaded corner indicates that the half-life was determined with a track technique. For two Pu isotopes the half-lives were determined in both an excited state and in the ground state of the second potential well.*

$\sim$1 mm to $\sim$100 cm. By an ingenious geometric amplification technique, Limkilde and Sletten (1973) have been able to work with flight paths as short as $\sim$30 microns, corresponding to lifetimes as short as 5×10^{-11} sec. The few isomers in Fig. 7-7 that were not studied with dielectric detectors have lifetimes in the inconvenient range $\sim 10^{-6}$ to $\sim 10^{-3}$ sec where neither method works. One then can pulse the accelerator beam and use fast electronics to time the fission counts in a nearby semiconductor detector.

Experiments that measure properties other than fission lifetimes have strengthened the case for a double-hump barrier. Some of these, such as the measurements of resonances in the fission cross section due to vibrational levels associated with the secondary minimum, are discussed in the papers by Clark (1971) and Nix (1972). Some of the track papers listed under *Induced Fission* at the end of this chapter describe determinations of fission cross sections and angular distributions of actinides that support the double-hump model.

7.6. THE SEARCH FOR SUPERHEAVY NUCLEI

By itself, the liquid-drop model predicts that with increasing Z^2/A the disruptive Coulomb forces grow faster than the cohesive nuclear forces until, at $Z^2/A \gtrsim 50$, the fission barrier disappears and nuclei are unstable. In addition, the large Coulomb forces cause heavy nuclei to decay rapidly by the emission of alpha particles. The inclusion of single-particle effects drastically changes the picture, such that a nucleus with a completely filled shell of neutrons or protons is bound extra tightly and has greater stability than its neighbors in the nuclidic chart.

The combined (liquid-drop) + (shell) approach led Myers and Swiatecki (1966) to predict that the fission barrier of a nucleus with $Z \approx 114$ should be several MeV high rather than vanishingly small. Two years later lifetimes up to $\sim 10^9$ years were calculated for certain nuclides near the center of the hypothetical island of stability centered on the closed shells $Z = 114$, $N = 184$ (Nilsson et al., 1969). It was also shown that such nuclides, or their daughters, would eventually undergo spontaneous fission. Fig. 7-8 shows schematically a recent version of the calculations (Nix, 1972). Of the beta-stable nuclides, shown in closed circles, $^{294}110$ is believed to have the longest half-life, $\sim 10^9$ years. Of course, a small uncertainty in calculated barrier heights leads to an uncertainty of several orders of magnitude in lifetime, so the magnitudes must not be taken too seriously, especially when we recall the overestimate of the spontaneous fission half-life of ^{258}Fm by a factor of $\sim 10^8$ (Section 7.2).

These theoretical developments triggered a number of searches for fossil spontaneous fission tracks in various substances where superheavies might have become geochemically concentrated. Assuming that long-lived superheavy nuclides were produced by galactic nucleosynthesis and have been decaying since the formation of the solar system some 4.5 billion years ago, one sees that if their half-life is shorter than $\sim 2 \times 10^8$ years there will not be enough superheavy atoms left on earth today to detect with conventional counters. But by studying fossil tracks acquired either when the solar system was very young or over a very extended period of time, one gains a tremendous advantage. For example, in meteorites, which solidified soon after the condensation of the nebula which is thought to have formed our sun, one could look for evidence for the former existence of now-extinct superheavies with lifetimes as short as $\sim 10^7$ years, and in old minerals to which heavy elements have segregated, one could look for tracks of surviving superheavies with lifetimes greater than $\sim 2 \times 10^8$ years with much higher sensitivity than attainable with counters.

Table 7-7 summarizes the results of track searches for superheavy nuclei, categorized according to the method used. Among the seven groups who looked for fossil track evidence, only Bhandari et al. (1971a,b) claim to have found conclusive evidence for superheavies. In both meteorites and lunar soil they have observed

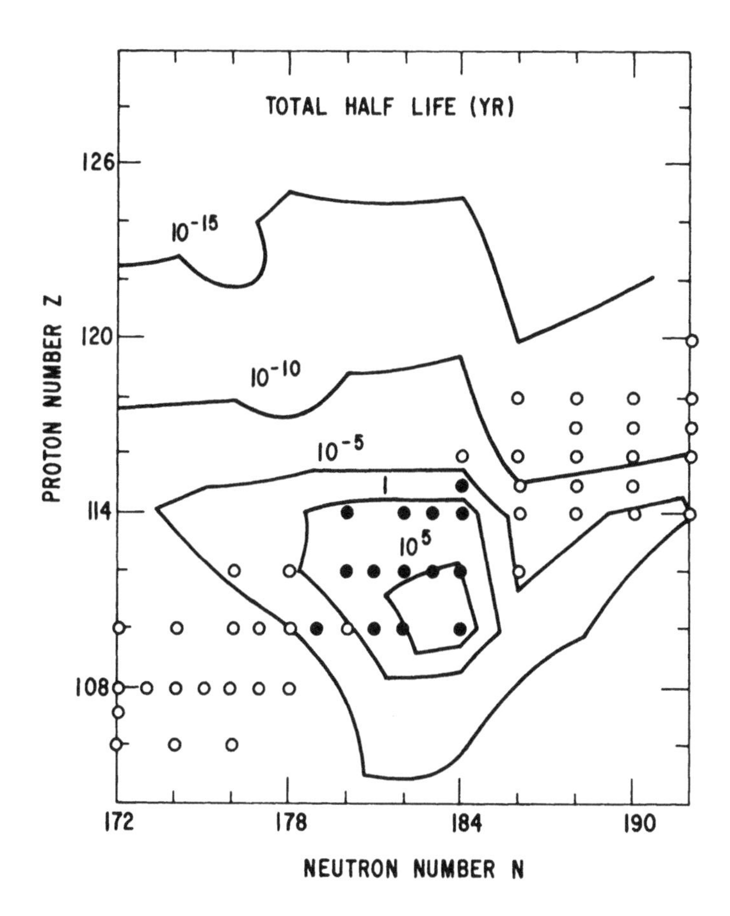

Fig. 7-8. *Total half-lives for even-Z superheavy nuclei. The contours are labeled by logarithms (to base 10) of the half-life in years. Solid points are used for beta-stable nuclei whose total half-life is longer than 1 year and open circles for beta-stable nuclei whose total half-life is shorter than 1 year. (After Nix, 1972).*

tracks too long to have come from spontaneous fission of ^{238}U or ^{244}Pu and not consistent in number or length with the heavy cosmic ray tracks commonly observed in extraterrestrial samples. On the basis of their length Bhandari et al. attribute the long tracks to spontaneous fission of now-extinct superheavy nuclei. We must point out, however, that theoretical opinion is divided on the question of whether superheavy fission fragments will have a greater kinetic energy and range than ^{238}U fission fragments (Nix, 1969; Schmitt and Mosel, 1972). Furthermore, recent calibration experiments with heavy ion beams and solar flare iron tracks show that fresh iron tracks are as long as those attributed to superheavies. The results point up the uncertainties in our knowledge of etching of ancient tracks and indicate that it may be premature to rule out a cosmic ray origin

Table 7-7. Searches for Superheavy Nuclei with Track Techniques

Method	Limitation or Result	Reference
Ancient tracks		
1) old glasses rich in heavy elements	Events probably due to cosmic-ray-induced fission	Cieslak (1969); Flerov et al. (1972)
2) meteoritic minerals	No ternary fission or long-range fragments	Price and Fleischer (1969)
3) Pb-rich, Au-rich minerals	$<10^{-25}$/y per Au atom and $<3 \times 10^{-24}$/y per Pb atom	Price et al. (1970)
4) feldspar inclusions in Mn nodules	Events might be due to spontaneous fission of ^{238}U	Otgonsuren et al. (1970)
5) meteorites and lunar soil	Long tracks attributed to superheavies; likely cosmic ray interference	Bhandari et al. (1971a, b)
6) Ta-rich, Bi-rich, Pb-rich minerals	$<10^{-25}$/y per Bi atom, $<10^{-24}$y per Pb, $<3 \times 10^{-27}$/y per Ta	Haack (1973)
7) veins in mica; quartz with pyrite inclusions	No long-range fragments	Otgonsuren et al. (1972)
Plastic film		
1) Mylar on Pb	Events probably due to cosmic-ray-induced fission	Flerov and Perelygin (1969); Flerov et al. (1972)
2) Makrofol on ores of many heavy elements	$<2 \times 10^{-23}$/y per Hg; $<4 \times 10^{-21}$/y per Th; $<6 \times 10^{-13}$g/g of superheavy with half-life of 10^9y.	Geisler (1972); Geisler et al. (1973)
Cosmic Rays		
1) Stacks of nuclear emulsion + Lexan	2 events with Z~104 to ~107; large errors do not rule out Cm, Pu or U	Fowler et al. (1970); Price et al. (1971a); Blanford et al. (1971); Shirk et al. (1973); Price (1972)
2) Minerals at lunar surface	2 long tracks; cannot rule out Pb or U cosmic rays	Price et al. (1971b)
Accelerator reactions		
1) $^{40}Ar + ^{248}Cm$; fission tracks in mica on-line	$\sigma < 5 \times 10^{-32}$ cm^2 for $\tau > 10^{-9}$ sec	Thompson (1968)
2) $^{136}Xe + ^{238}U$; chemical separations; next to plastic	$\sigma \lesssim 10^{-31}$ cm^2 for 1 hr $< \tau <$ 10 d	Flerov (1972)
3) 24 GeV p + W; chemical separation; next to plastic	excess fissions, possibly due to ^{252}Cf	Marinov et al. (1971a,b)
4) 28 GeV p + W, Au, U + glass detectors on line	$\sigma < 10^{-32}$ cm^2 for long-range recoils casts doubt on Marinov et al.	Katcoff and Perlman (1971)

Table 7-7. (Continued)

Method	Limitation or Result	Reference
5) 28 GeV p + W, U; chemical separations next to Makofol	$\sigma < 10^{-39}$ to 10^{-40} cm^2 for elements 107 to 116; conflicts with Marinov et al.	Unik et al. (1972)
6) 29 GeV p + Pb	Negative results with 10X higher sensitivity than Marinov et al. (1971a)	Geisler et al. (1973)
7) Mass-separation from ores + neutron-induced fission + tracks on quartz detector	possible interference from UCO_3 at mass 300	Sowinski et al. (1971)
8) 24 GeV p + U	$\sigma \lesssim 2 \times 10^{-40}$ cm^2 for volatile superheavies; conflicts with Marinov et al.	Westgaard et al. (1972)

(Plieninger et al., 1972; Burnett et al., 1972; Price et al., 1973). Thermal and shock effects (Fleischer and Hart, 1973) can shorten pre-existing tracks, making more recently produced tracks look anomalously long.

Plastic film detectors have been used both in the Soviet Union (Flerov and Perelygin, 1969; Flerov et al., 1972) and in the United States (Geisler 1972; Geisler et al., 1973) to look for possible contemporary superheavy nuclei. In the latter study, carried out at a depth of $\sim$600 m in a salt mine, 1 m^2 sheets of Makrofol were exposed for 0.3 yr in contact with ores of various heavy elements and afterwards etched and spark-scanned without detecting any anomalous fissions.

Fowler et al. (1970) generated considerable excitement when they announced the discovery of an extremely heavy cosmic ray track in a nuclear emulsion that had been exposed at the top of the atmosphere. Their best estimate of its charge was $Z \approx 105$ with a small probability that it might be as light as $Z \approx 92$. In a subsequent exposure of superimposed Lexan, cellulose triacetate, and nuclear emulsion Price et al. (1971a) found another track that was estimated by the emulsion technique to have a charge of $\sim$104. However, in both kinds of plastic the same event was estimated to have a charge of only $\sim$90 to 92. Since then one other flight has yielded a possible trans-uranic cosmic ray that passed through both emulsion and Lexan (see Fig. 5-10); its charge was estimated to be $\sim$96. The uncertainties in response of emulsions and plastics to ultraheavy nuclei are sufficiently great, and knowledge of velocity sufficiently uncertain (Blanford et al., 1971; Shirk et al., 1973), that it is unwarranted to claim with assurance that transuranic cosmic rays have been discovered. It is certainly unrealistic to regard the three events just discussed as evidence for superheavy elements. However, balloon flights and other searches in nature are continuing. Just as this manuscript was completed, a stack of Lexan 1.3 m^2 in area was successfully brought back to earth by astronauts Carr, Gibson, and Pogue following a 253 day exposure in the Skylab orbiting workshop (see Chapter 5). During that time enough events were recorded that it should be possible to use abundance peaks at Pb and U, together with a gap at the short-lived nuclei $84 \leq Z \leq 89$, as an internal calibration. Searches of this kind are particularly important because cosmic rays may send nuclei into the solar system within $\sim 10^6$ years after they have been explosively synthesized.

Accelerator searches have yielded negative or conflicting results. Marinov et al. (1971a,b) initially claimed to have made eka-mercury ($Z = 112$) by bombarding a thick tungsten target with 24 GeV protons, but subsequent experiments failed to substantiate their claim (Unik et al., 1972; Westgaard et al., 1972; Geisler et al., 1973; Batty et al., 1973) or showed that their proposed mechanism would be extremely improbable (Katcoff and Perlman, 1971). Their mechanism required that a heavy fragment of a tungsten nucleus be given enough energy to initiate a nuclear reaction with another tungsten nucleus.

Low-energy heavy ion reactions seem more promising than high-energy proton reactions. Even if half-life calculations are off by many orders of magnitude, as the theorists warn us may be true, it may still be possible to make small quantities of superheavies in the laboratory and to study their physical and chemical properties. We believe there is a good chance that dielectric track detectors will play an important role in their eventual discovery.

7.7. LOW-ENERGY CHARGED PARTICLE REACTIONS RELEVANT TO ASTROPHYSICS

A virtually untapped application of alpha-sensitive plastics is to determine excitation functions for reactions such as (p, α) and (d, α) that take place at extremely low energies and with extremely low cross sections in stellar interiors. Much of the theory of stellar evolution and of nucleosynthesis in stars relies upon extrapolations of excitation functions to cross sections many orders of magnitude below what has been measured in the laboratory. The mean energy of a thermonuclear reaction in a star is governed by a competition between a cross section that rises extremely rapidly with energy and a Maxwellian energy distribution of particles that falls rapidly with energy. In most common stars (those on the so-called "main sequence") the mean energy in reactions between light elements is only $\sim$20 keV, corresponding to a cross section of $\sim 10^{-36}$ cm^2. This is lower by many orders of magnitude than has ever been measured with conventional detectors, but we believe it to be within a factor of 10 to 100 of that attainable with good plastic detectors.

At Debrecen, Hungary, cellulose acetate detectors are being used to determine excitation functions and angular distributions of alpha particles emitted in (p, α) and (d, α) reactions (Mesko et al., 1969; Somogyi and Schlenk, 1970; Hunyadi et al., 1971; Szabo et al., 1972). Fig. 7-9 shows the excitation function for the reaction $^{10}B(p, \alpha)^{7}Be$ measured at bombarding energies down to 60 keV (Szabo et al., 1972). This reaction is an important destructive process and may bear on the extremely low boron abundance in the universe. The dashed curve, calculated from the formula

$$\sigma(E) = (S(E)/E) \exp(-W_0),$$

diverges from the data at low energy if one assumes the cross section factor $S(E)$ to be independent of energy. W_0, the penetration factor for $l = 0$ (s-wave), depends on Z and A and the Coulomb barrier of the interacting p and ^{10}B. The reliability of the extrapolation of the cross section to $\sim$20 keV depends critically on how well the variation of $S(E)$ with energy is known and can be extrapolated to low energy.

We believe it should be possible to extend measurements such as these down to

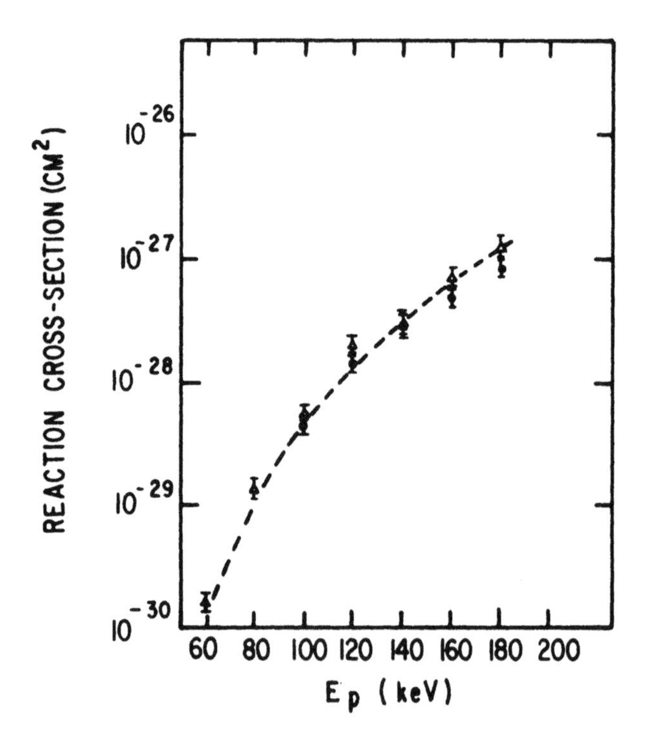

Fig. 7-9. *Total cross section values for* $^{10}B(p, \alpha)^7Be$ *reaction versus bombarding energy. The dashed line was calculated using constants* $S(E)$ *and was normalized to the experimental points at* $E_p = 140$ *keV. The cross section values are marked with symbols* • *and* Δ *measured by the activation method and by imaging of alpha tracks, respectively.* (*After Szabo et al., 1972.*)

cross sections as low as $\sim 10^{-34}$ cm^2. Lexan that has been UV-sensitized (Stern and Price, 1972) is being used routinely in experiments at Berkeley to measure alpha particle track densities as low as 10/cm^2 (cf. discussion of Fig. 5-24). The main limitation of etched cellulose nitrate and cellulose acetate is their rough surface, which makes it difficult to count low track densities. The alpha tracks in Lexan are very easy to recognize and the main problem is to eliminate background tracks from the starting material. Such a background can arise during manufacture, if a Po anti-static brush is used, or in storage because of decay of alpha emitters such as ^{222}Rn and its daughter nuclides in the atmosphere.

With ingenuity, one should be able to use plastic detectors to study nucleosynthesis of elements produced in the advanced stages of stellar evolution after the completion of hydrogen burning. Important examples are the carbon-burning reactions $^{12}C + ^{12}C \rightarrow$ products such as ^{20}Ne, ^{23}Na, ^{24}Mg, and ^{16}O. These reactions take place in nature at such high temperatures ($\sim 2 \times 10^9$ K, equivalent to energies ~ 0.2 MeV) that the important cross sections can, in principle, be measured in the laboratory. Experimentally, what is required is to distinguish low-energy reaction products like ^{20}Ne from scattered ^{12}C ions. One thus has to use the particle identification techniques discussed in Chapter 3.

7.8. ENERGETIC FRAGMENTS FROM RELATIVISTIC HEAVY-ION REACTIONS

The study of fragments emitted in relativistic heavy-ion reactions with heavy nuclei exploits several of the features of dielectric detectors—their insensitivity to background radiation, their ability to record spatial relationships of reaction products, and their ability to identify particles down to very low energies. These features have made it possible, with Lexan detectors, to determine angular distributions, energy spectra, and charge distributions of low-energy fragments emitted from heavy nuclei struck by 2.1 GeV/nuc ^{12}C and ^{16}O ions at the Berkeley Bevatron. In this particular experiment Lexan has two advantages over semiconductor particle identifiers: (1) The intensity of high-energy heavy-ion beams at the Bevatron, though gradually increasing, is still only $\sim 10^7$ ions per minute, so that semiconductor telescopes, which have a small angular aperture, have too small an event rate. (2) The most probable energy of fragments emitted from heavy nuclei turns out to be considerably lower in heavy-ion reactions than in proton reactions. Lexan can be used to estimate charges of particles with energies down to ~ 0.2 MeV/nucleon, whereas typical semiconductor detectors are limited to energies above ~ 3 MeV/nucleon and miss some of the new phenomena that we are going to discuss.

The motivation for the initial experiments at Berkeley (Sullivan et al., 1973) was to see if qualitatively new phenomena are initiated when a heavy nucleus is struck by an ensemble of relativistic nucleons all having the same momentum vector. One exciting possibility, suggested years ago by Glassgold et al. (1959), was that a hydrodynamic wave might be propagated through the nucleus, with particles emerging from the nucleus preferentially at the Mach angle given by the ratio of sound speed in the nucleus to the speed of the incoming projectile.

In bombardments of an Au target with 34 GeV ^{16}O nuclei, Sullivan et al. saw no evidence in the angular distribution of emitted fragments for hydrodynamic wave effects. What they did see were clear differences in the energy distributions and cross sections of charged particles emitted when ^{16}O ion beams were used compared with proton beams of the same energy/nucleon (2.1 GeV/nuc).

Fig. 7-10 allows one to compare their results for fragments with $Z > 6$ emitted at 45° to the beam in (a) the 2.1 GeV proton bombardment and (b) the 2.1 GeV/nuc ^{16}O bombardment. Etch rate is plotted as a function of range for a sample of ~ 300 events in each irradiation. The three-stage scheme of etch + UV + re-etch described in Chapter 3 was used for track identification. Superimposed on each plot is a grid showing energies and atomic numbers.

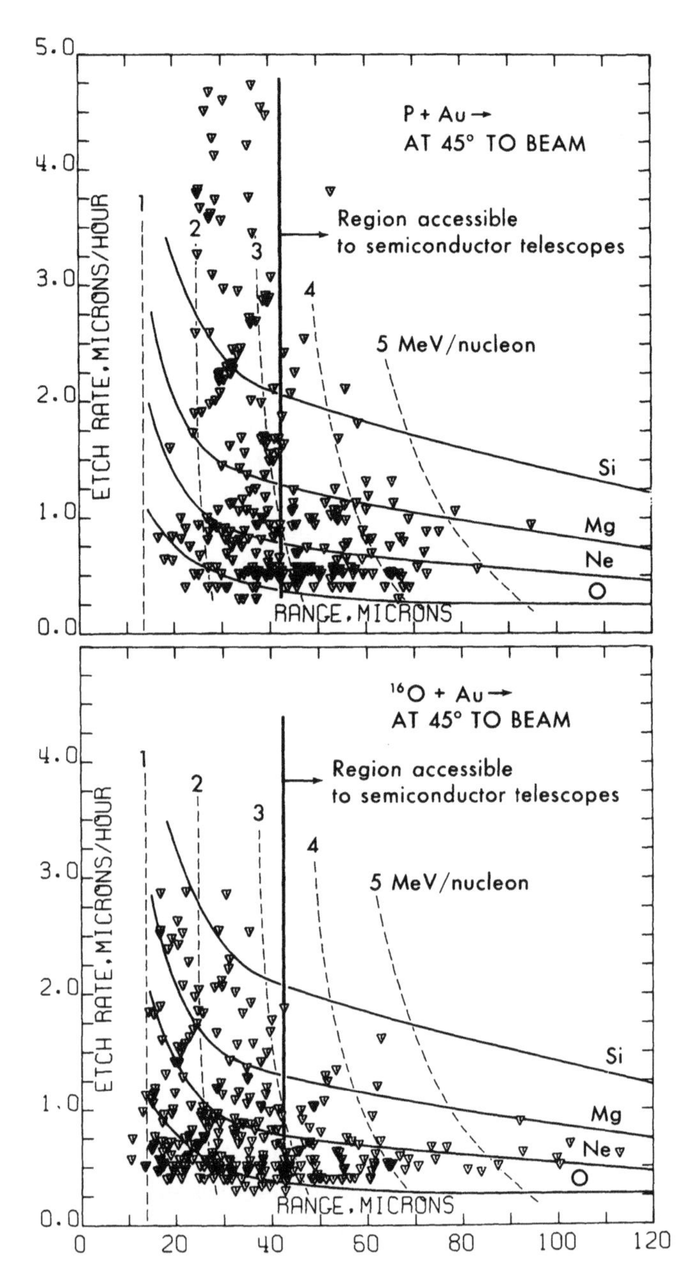

Fig. 7-10. *Track etching rate versus particle range for fragments at 45° to the beam from (a)* $^{16}O + Au$ *and (b)* $p + Au$. *Fission fragments have been excluded. The charge grid is based on a semiempirical fit to calibrations with beams of low energy* ^{16}O, ^{28}Si, ^{40}Ar, *and* ^{56}Fe *ions. Though in each case 300 events were plotted, the event rate per unit fluence was an order of magnitude greater for* $^{16}O + Au$ *than for* $p + Au$. *(After Sullivan et al., 1973.)*

Even in the small sample of data shown, it is clear that there are strong differences in the distribution of charges and energies of emitted fragments. In their detailed analysis Sullivan et al. (1973) found that for each charge the most probable energy is lower, the distribution is broader, and the minimum energy is lower, for the ^{16}O bombardment than for the proton bombardment. The surprising result is that in the ^{16}O bombardments fragments were detected with kinetic energies as low as $\sim$0.2 times that permitted by the Coulomb barrier for a spherical target nucleus. The cross section for producing fragments with $Z \geq 6$ is about 12 times higher in the ^{16}O bombardment than in the proton bombardment. When integrated over all angles, the cross section for fragments with $Z \geq 6$ produced by ^{16}O ions is higher than the geometric cross section for ^{16}O + Au and implies multiple fragment emission. Finally, the ratio of cross sections for producing specific charges decreases from $\sigma(^{16}\text{O})/\sigma(\text{p}) \approx 25$ for $Z \approx 8$ to $\sigma(^{16}\text{O})/\sigma(\text{p}) \approx 3$ for $Z \geq 12$. The latter ratio is consistent with that reported by Katcoff and Hudis (1972), who studied fission with mica detectors.

Sullivan et al. (1973) interpreted their results as evidence that heavy ions can deposit more energy in a nucleus than can protons, resulting in higher nuclear

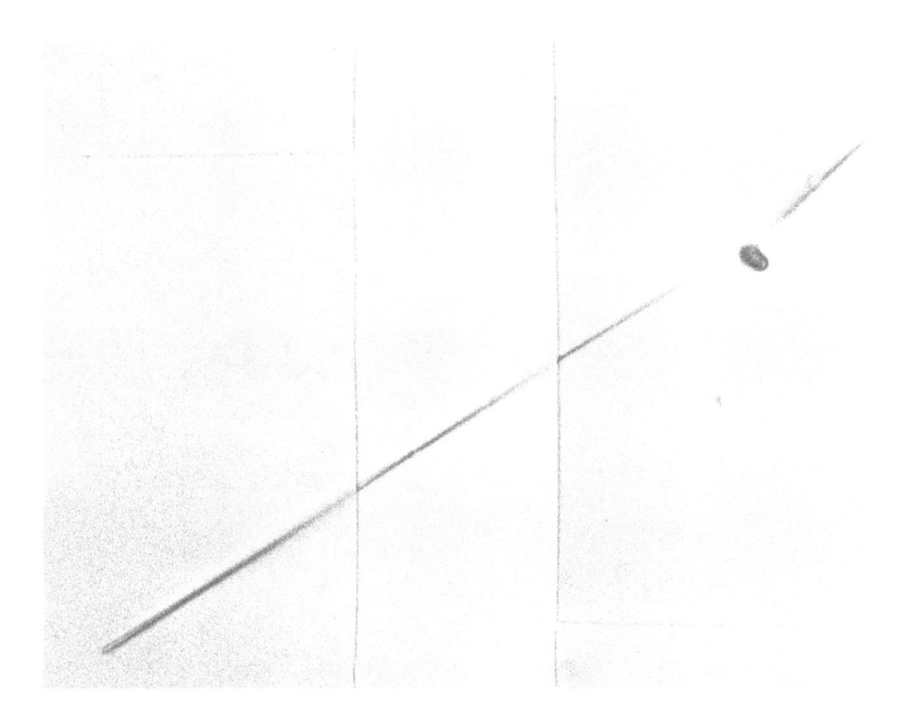

Fig. 7-11. *Tracks of three fragments emitted from a Au nucleus struck by a 34 GeV ^{16}O nucleus and recorded in a Lexan sheet in contact with the downstream side of the Au foil. To conserve momentum one or more fragments must at the same time have been emitted in the backward direction. (Crawford et al., 1974, unpublished).*

temperatures and shape elongations. Considered as a liquid drop, the nucleus has a surface tension that decreases with increasing temperatures and may vanish at a critical point. With reduced surface tension, the fissility increases and multiple fragment emission at low kinetic energies becomes possible. In a recent bombardment of a thin gold target sandwiched between two closely adhering Lexan detectors, Crawford et al. (1974) established that three or more fragments with $Z > 6$ are commonly emitted from the same nucleus when struck by a 34 GeV ^{16}O ion. Fig. 7-11 shows tracks of three particles emitted into the forward Lexan sheet from the same Au nucleus. In multiple fragment emission it is easy to see how, because of recoil of the emitting nucleus, one or more of the fragments may have a very low kinetic energy in the laboratory system.

In a continuation of their search for evidence of nuclear shock waves, Crawford et al. (1975) used very thick Lexan stacks to detect fragments with velocities comparable to or greater than the expected speed of sound in nuclear matter, ~0.1 to ~0.2 c. Fig. 7-12 shows angular and energy distributions of fragments with $5 \leq Z \leq 9$ emitted from a 0.4 g/cm² Au target bombarded with 10^{12} carbon ions of kinetic energy 25 GeV.

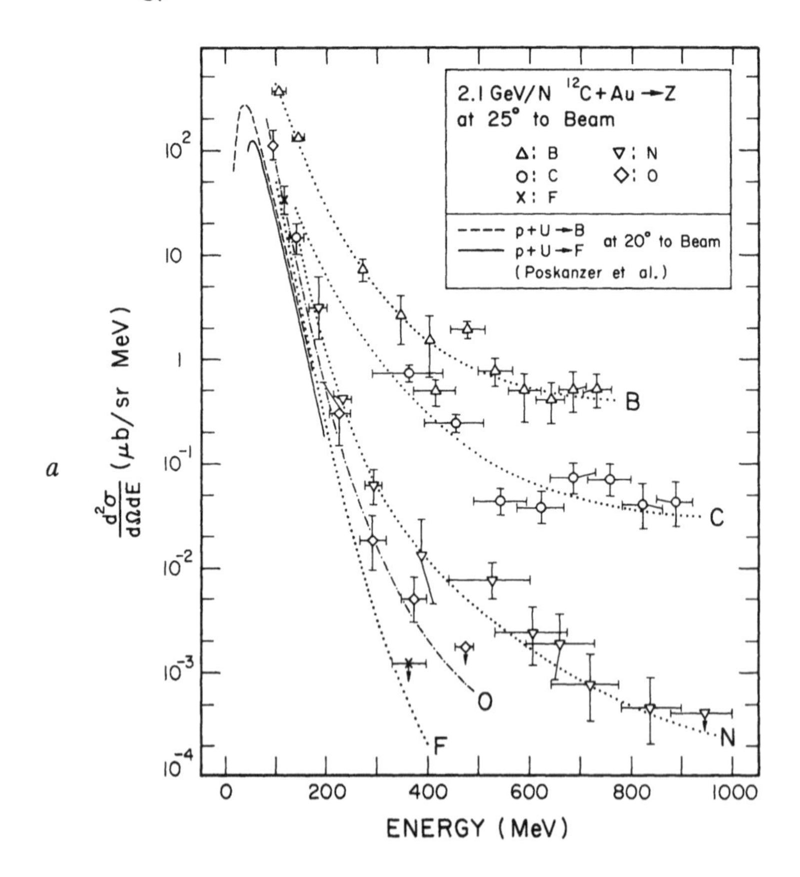

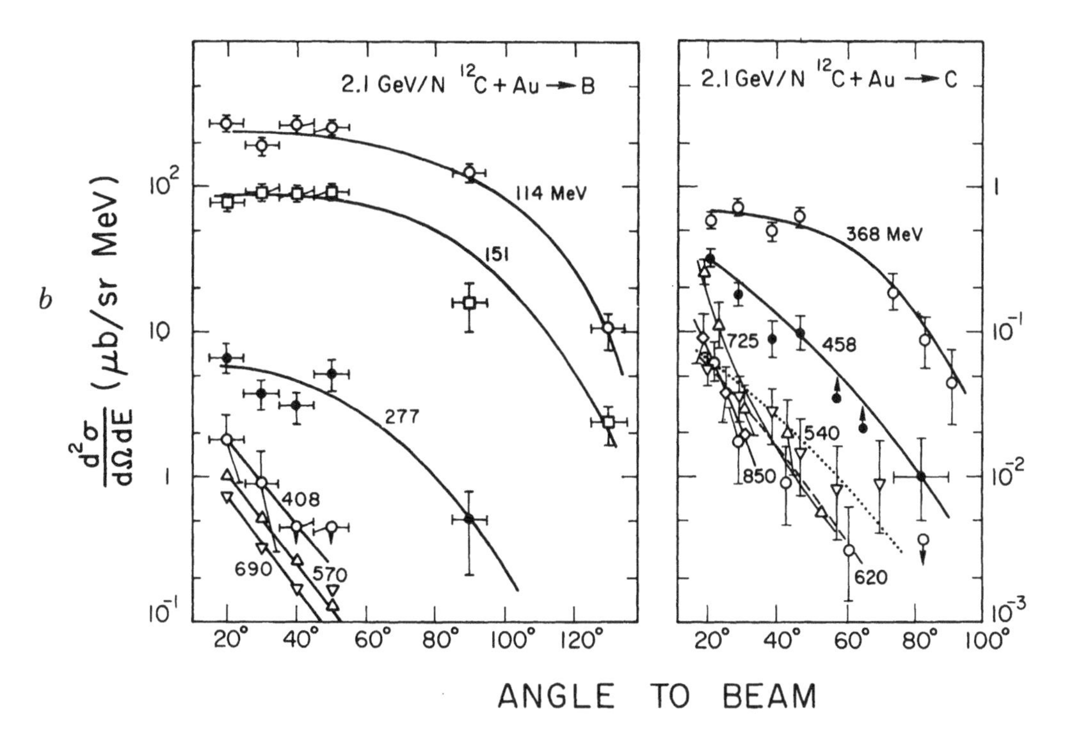

Fig. 7-12. (*a*) *Differential cross sections for production of fragments with Z = 5, 6, 7, 8 and 9 at 25° to the beam in reactions of 2.1 GeV/nucleon ^{12}C ions with a Au target. Where no events were seen (arrows), points indicate upper limits with confidence coefficient 0.84. Dashed curve and solid curve from Poskanzer et al. (1971) are for reactions of 5.5 GeV protons with a U target.* (*b*) *Angular distribution of boron and carbon fragments with various lab energies.* (*After Crawford et al., 1975.*)

The data exhibit several striking features:

(1) At energies above ~150 MeV the differential cross sections deviate strongly from the roughly exponential behavior characteristic of evaporation and fall off much more slowly, roughly as power laws, E^{-n}. The value of n increases from ~2.7 for boron to ~5 for oxygen.

(2) In the power law region the cross sections fall off very rapidly with fragment charge, with a possible even-odd effect enhancing the yields of carbon and oxygen. In contrast, differential cross sections for all fragments are comparable in the exponential part of the evaporation spectrum (not shown in the figure) and differ mainly in their peak values and in their cutoff energies.

(3) In the power law region the angular distributions are very anisotropic, being increasingly forward-peaked with increasing energy.

It is impossible to fit the fragment distributions in the power law region with a "thermal" model. Assumptions of such a model—that the excited residual nuclei have some reasonable distribution of lab velocities and that fragments are evaporated isotropically in the rest frame of each moving nucleus—lead to energy distributions that decrease much too rapidly and angular distributions that are much too flat to reproduce the data. Nor can the kinematics of quasi-elastic processes fit the data. A new, non-thermal process is required, but it is too early yet to say whether a hydrodynamic process will account for the power law spectra and the strong forward peaking.

7.9. ADDITIONAL EXAMPLES

7.9.1. Ternary Fission

The contributions that we have discussed in this chapter were chosen to illustrate some of the unusual capabilities of dielectric detectors. Many important topics were not even mentioned. For example, the first nuclear application of dielectric track detectors was the conclusive observation of ternary fission (Price et al., 1963), some twenty years after it was predicted. Compound nuclei with $Z^2/A \approx 43.3$ and large excitation energy will fission into three roughly equal fragments with a frequency $\sim 3\%$ that of binary fission (Fleischer et al., 1966). Fig. 7-13 shows photomicrographs of a number of three-pronged tracks resulting from ternary fission of the compound nucleus Ar + Th. Such studies (reviewed by Brandt, 1971) have a bearing on the search for superheavy nuclei, which should undergo ternary fission with a higher probability than do nuclei with a smaller Z^2/A.

7.9.2. Complete Fusion Reactions

Another problem bearing on the search for superheavy nuclei is to understand how the cross section for complete fusion of two nuclei depends on their masses and relative energies. The question is relevant because of the attempts being made at heavy-ion accelerators to synthesize superheavies in compound nucleus reactions. The experiments of Kowalski et al. (1968) with mica detectors showed that the cross sections for forming the compound nuclei $^{20}Ne + ^{27}Al$, $^{16}O + ^{27}Al$, and $^{16}O + ^{59}Co$ fall off rapidly above ~ 100 MeV. Their work, and subsequent experiments by Natowitz (1970a) and Zebelman and Miller (1973), indicate that there is a mass-dependent upper limit to the angular momentum that a compound nucleus can have. On the liquid-drop model, a high-energy projectile striking a nucleus with large impact parameter would produce a rotating liquid drop that would immediately fly apart. Thus, the higher the bombarding energy, the smaller the range of impact parameters over which stable compound nuclei can be formed.

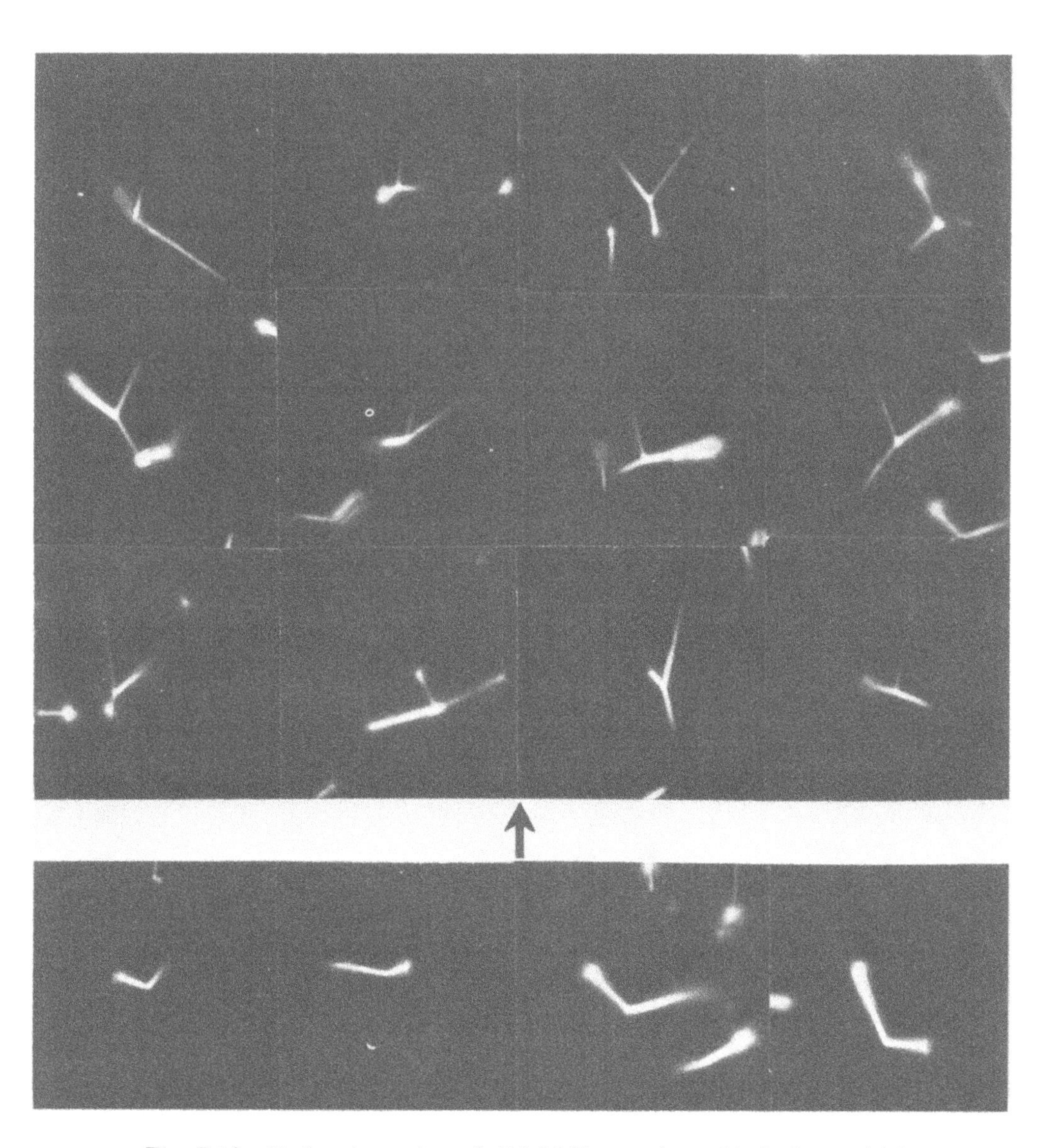

Fig. 7-13. *Nuclear interactions of 400-MeV argon ions with thorium nuclei in a crystal of thorite* ($ThSiO_4$)*, photographed with dark-field illumination after etching for 1 minute in boiling* H_3PO_4 *at 250°C. The argon ions, which do not register tracks, entered the crystal at a grazing angle of 15 degrees along the arrow. The V-tracks along the bottom row are binary fissions, whereas the three-pronged tracks along the top three rows are ternary fissions in which each fragment is sufficiently massive to register a track.* (*After Fleischer et al., 1965.*)

The complete-fusion cross section is given semiclassically (Natowitz, 1970b) by

$$\sigma_{CF} = (\pi\hbar^2/2\mu E_{cm})(J_{crit})^2$$

where μ is the reduced mass, E_{cm} is the energy in the center of mass, and the limiting angular momentum that is consistent with the experimental data is approximated by the simple function

$$J_{crit} = 100\hbar[1 - \exp(-0.00654A_{CF})],$$

where A_{CF} is the sum of the target and projectile masses (Natowitz, 1970b). In support of these ideas, experiments with the new ^{84}Kr beam at Orsay (Bimbot et al., 1971) show that the complete-fusion cross section with heavy nuclei is extremely small.

7.9.3. The Search for Magnetic Monopoles

Some of the most exhaustive searches for magnetic monopoles have been made with dielectric detectors, using a variety of ingenious collectors including ferromagnetic manganese deposits from the mid-Atlantic floor, ancient obsidian and subsurface mica, large arrays of plastic in balloons and at sea level, and crystals in lunar rocks (see papers on monopoles at end of chapter). Fig. 7-14 illustrates the principles on which several of these searches were based. The stringent limits set in these experiments have served to close several loopholes left in the early searches: in the new studies monopoles with a large multiple of the Dirac charge, $g = \hbar c/2e$, and a wide range of masses and energies should have been detected. Of course, all of these searches presupposed that the signature of a monopole is a distinctive, highly ionizing track like that of a relativistic nucleus with $Z = 137/2$ or some integral multiple thereof (Dirac, 1931; 1948). Magnetically charged particles with considerably lower ionization rate would have escaped detection, but their existence would then pose an even deeper mystery for theorists than the present situation of a universe seemingly without symmetry between electric and magnetic charge.

7.9.4. Photofission of Very Heavy and Very Light Elements

Great strides have been made in understanding photofission of very heavy elements, both by gamma rays and through electromagnetic interaction of high-energy electrons, thanks to a variety of measurements of angular distributions and fission cross sections using dielectric detectors. Forkman and Schroder (1972) have reviewed the experiments and mechanisms at low energy (in the giant resonance region), at intermediate energy (by interaction with quasi-deuterons in a nucleus),

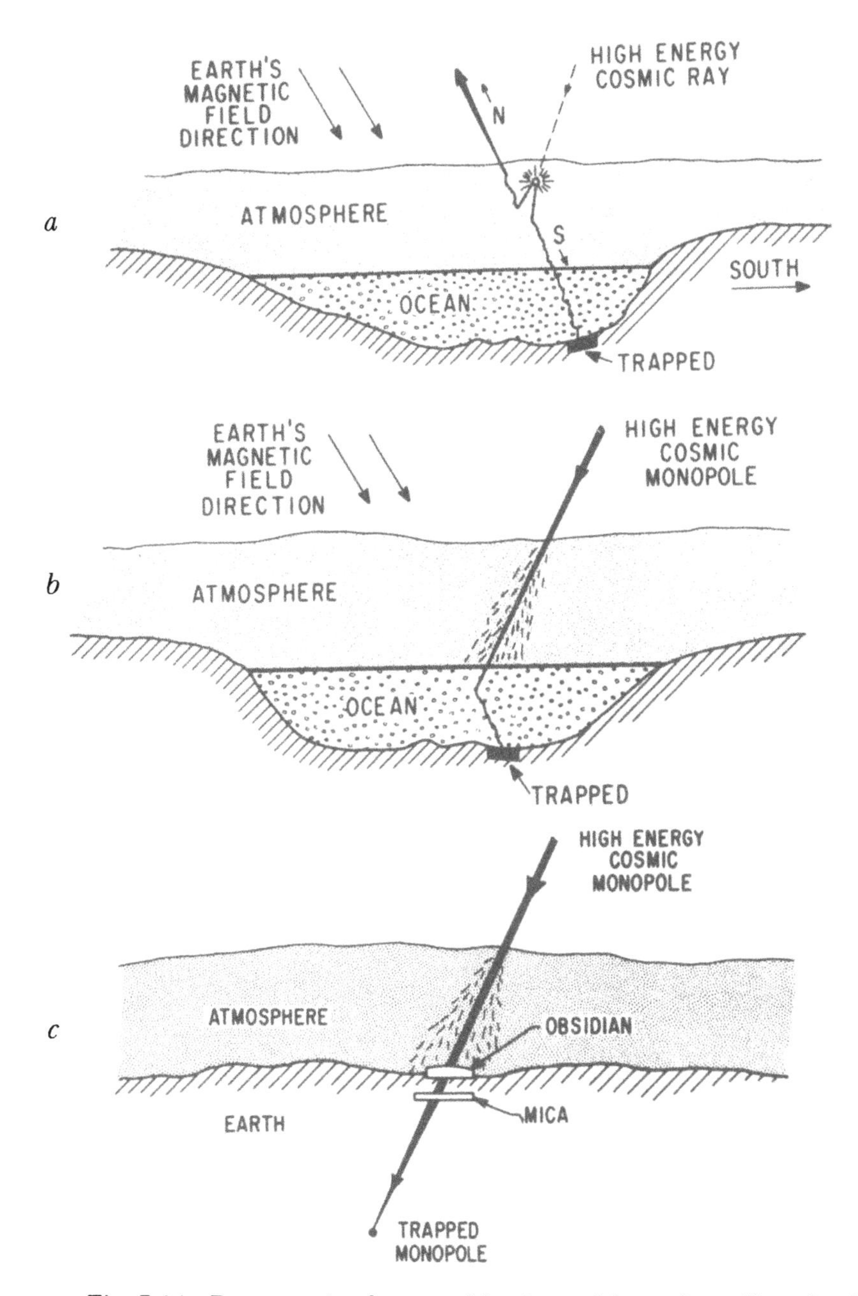

Fig. 7-14. *Free magnetic poles are envisioned as arriving at the earth's surface by either of two processes. In (a) a high-energy cosmic ray particle collides with a nucleus in the upper atmosphere and creates a pair of magnetic monopoles, which are slowed down in the atmosphere and then drift along geomagnetic field lines until either trapped at the ocean bottom or lost into space. In (b) and (c) the highest energy cosmic rays themselves are monopoles which are either stopped and trapped or penetrate and leave tracks in natural detectors. (After Fleischer et al., 1970b.)*

and at high energy (by photomeson production in a nucleus). One of the interesting disagreements among experimenters is whether the fissility as a function of Z^2/A goes through a minimum at some intermediate value ($Z^2/A \approx 20$) as predicted by the liquid-drop model or rises monotonically. The disagreement occurs at small Z^2/A where the fragments are light and close to the threshold for being recorded in a dielectric detector. We see this as a problem that could be solved by using Lexan as a particle identifier, where the atomic numbers of the light fragments are to be measured.

The fission of a light element such as magnesium is qualitatively different from the fission of a heavy element such as uranium. The Q-value is negative instead of positive, the cross section is fantastically small, and shell effects play a much more important role relative to liquid-drop effects than they do with heavy nuclei. Using 6 μm Makrofol detectors Chung (1973) has convincingly detected for the first time the fission of a light element, using an intense electron beam as a source of virtual photons with which to induce photofission. For the reaction $^{24}Mg(\gamma, ^{12}C)^{12}C$ he measured a maximum fission cross section of $\sim$40 picobarns (1 pb = 10^{-36} cm^2) at an electron energy of $\sim$24 MeV and decreasing to less than 5 pb at lower and higher energies. Fig. 7-15 shows his result. In order to discriminate against back-

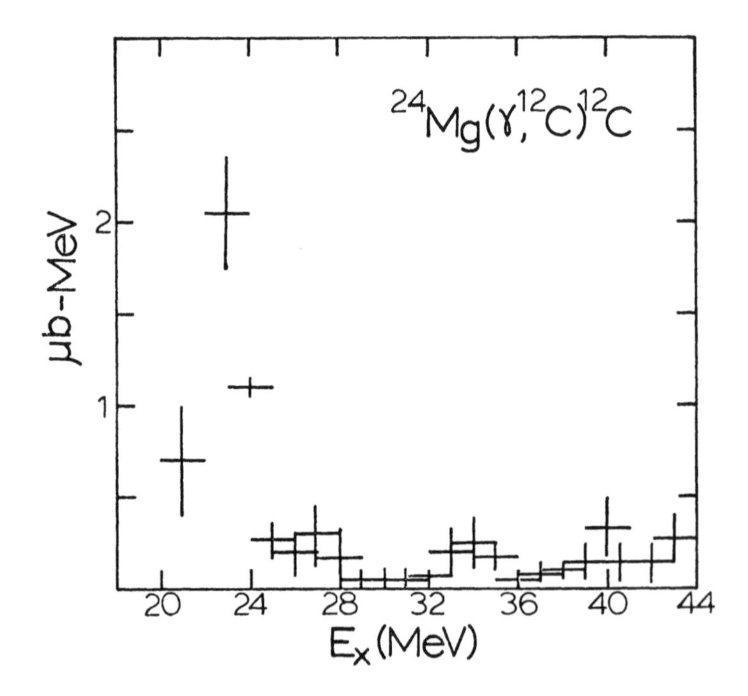

Fig. 7-15. *The equivalent photon-induced fission cross section as a function of excitation energy in* ^{24}*Mg. In terms of the incident electron beam the maximum in the curve corresponds to* $\sim$*40 picobarns. The* ^{12}*C ions, of energy 4 to 15 MeV, were identified as described in Fig. 3-13. (After Chung, 1973.)*

ground such as fission fragments from U and Th impurities in the target and low-energy charged particles such as ^{16}O and ^{20}Ne ions, he used the arrangement discussed in Fig. 3-13, together with absorber foils of various thicknesses, to pick out the ^{12}C ions, which were emitted with energies from 4 to 15 MeV. He was able to isolate tracks of ^{12}C nuclei emitted in their ground states. The resonance at ~24 MeV shown in Fig. 7-15 suggests that the photofission is initiated by a giant quadrupole resonance, probably associated with shell effects in ^{24}Mg. We hope the reader appreciates the technological achievement of identifying extremely low-energy carbon ions (0.3 to 1.25 MeV/nucleon) produced with picobarn cross sections amid a sea of other charged particles. Further studies along these lines should give a clearer picture of shell effects in fissioning nuclei.

7.9.5. Decay of High-Spin Nuclei

As a final example, we want to mention an interesting experiment still in progress. Because some nuclei are highly deformed in the ground state, the probability of radioactive decay by barrier penetration will depend on the direction of emission with respect to the symmetry axis of the nucleus. Nuclei of trivalent transuranium elements such as ^{253}Es can be oriented in a rare-earth ethyl sulfate crystal at very low temperature, by means of the coupling between the nuclear magnetic moment of the Es and the spins of unpaired valence electrons in the rare-earth element. The orientation can be maintained if the crystal is kept at ~0.01°K, which is possible if the energy deposited by the radioactive ^{253}Es is sufficiently low. Using a Ge detector, Shirley (1966) and his coworkers found that alpha particles were preferentially emitted along the long axis of the ^{253}Es, which has a prolate spheroid shape. With mica detectors and using the spontaneously fissioning isotope ^{254}Es, he and his colleagues are now attempting to determine the angular distribution of fission fragments with respect to the long axis. Because of the small branching ratio for spontaneous fission and the necessity to avoid overheating, the count rate is extremely low. Mica (or Lexan) has the advantage that the entire angular distribution can be obtained at one time.

APPENDIX. EXPERIMENTS WITH DIELECTRIC TRACK DETECTORS

1. *Spontaneous fission; trans-uranic nuclides.*— Fleischer and Price (1964); Flerov et al. (1964; 1966a; 1970a,b; 1971a,b); Zvara et al. (1966; 1970); Gvozdev et al. (1966); Aleksandrov et al. (1967); Nurmia et al. (1967); Ghiorso et al. (1967); Kuznetsov et al. (1966); Perelygin et al. (1964); Druin et al. (1964; 1971); Natowitz et al. (1968); Roberts et al. (1968); Ghiorso et al. (1969); Gold et al. (1970); Hulet et al. (1971); Leme et al. (1971); Kleeman and Lovering (1971); Flerov

(1972); Ruddy and Alexander (1972); Khan (1971); Gentner et al. (1972); Khan and Durrani (1973); Druin et al. (1973); Suzuki (1973).

2. *Spontaneously fissioning isomers.*— Flerov et al. (1965); Kuznetsov et al. (1965; 1967); Lobanov et al. (1965); Linev et al. (1965); Gangrskii et al. (1966; 1967; 1968; 1972); Markov et al. (1966); Bjornholm et al. (1967); Flerov et al. (1967a, b); Lark et al. (1969); Metag et al. (1969a,b); Natowitz and Archer (1969); Montgomery and Brandt (1970); Peter (1970); Repnow et al. (1970; 1971); Polikanov and Sletten (1970); Ruddy and Alexander (1969); Clark (1971); Ruddy et al. (1971); Alexander and Bimbot (1972); Limkilde and Sletten (1973); Kuks et al. (1973); Vandenbosch et al. (1973).

3. *Lifetimes of compound nuclei by the blocking effect.*— Brown et al. (1968); Melikov et al. (1969; 1970; 1972); Gibson and Nielsen (1970); Karamyan et al. (1970; 1971); Kamanin et al. (1972a,b; 1973); Skakun and Dikii (1972); Noelpp et al. (1972).

4. *Fission induced by charged particles.*— Burnett et al. (1964); Maurette and Stephan (1965); Nemilov et al. (1965; 1966); Khodai-Joopari (1966); Kon'shin et al. (1966a,b; 1967); Perfilov et al. (1966); Raisbeck (1966); Natanson and Dakowski (1967); Raisbeck and Cobble (1967); Thompson (1967); Smirenkin et al. (1968); Brodzinski and Cobble (1968); Kuvatov et al. (1968; 1970; 1972); Moretto et al. (1968; 1969a; 1972); Ralarosy et al. (1969; 1973); Kowalski et al. (1971); Natowitz and Chulick (1971); Pate and Peter (1971); Nakanishi and Sakanoue (1972); Itkis et al. (1973).

5. *Neutron-induced fission.*— Csikai and Nagy (1967); Ermagambetov et al. (1967); Gönnenwein et al. (1967); Loveland et al. (1967); Nesterov and Smirenkin (1968); Vorotnikov et al. (1967a,b; 1970a,b,c; 1971; 1972); de Mevergnies and del Marmol (1968); Ait-Salem et al. (1968); Androsenko and Smirenkin (1968; 1971); Behkami et al. (1968a,b); Shpak and Smirenkin (1968; 1969); Babenko et al. (1968; 1969); Huizenga et al. (1969); Fomushkin and Gutnikova (1970); Barutcugil et al. (1971); Shpak et al. (1971a,b; 1972); Yuen et al. (1971); Fomushkin et al. (1972); Emma et al. (1973).

6. *Photofission and electrofission.*— Soldatov et al. (1965a,b); Ranyuk and Sorokin (1967); Bowman et al. (1968); Mitrofanova et al. (1968); Rabotnov et al. (1968; 1970); Zhagrov et al. (1968; 1971); Kapitza et al. (1969); Moretto et al. (1969b); Methasiri (1970); Medveczky and Somogyi (1970); Medveczky et al. (1970); Dowdy and Krysinski (1971); Maly (1971); Manfredini et al. (1971); Methasiri and Johansson (1971); Anderson et al. (1972); Charlesworth et al. (1972); Chung et al. (1972); David et al. (1972); Forkman and Schroder (1972); Ignatyuk et al.

(1972); Kasilov et al. (1972); Kroon and Forkman (1972); Vartapetyan et al. (1972); Chung (1973).

7. *Ternary fission.*— Price et al. (1963); Fleischer et al. (1966); Flerov et al. (1966b); Brandt et al. (1966); Perelygin et al. (1969); Boos and Brandt (1971); Brandt (1971); Juric et al. (1970); Medveczky and Somogyi (1970); Medveczky et al. (1970).

8. *High-energy reactions.*— Price and Walker (1962a); Debeauvais and Cüer (1965); Brandt et al. (1967); Debeauvais et al. (1967; 1968); Matusevich and Regushevskii (1968); Hudis and Katcoff (1969); Remy et al. (1970; 1971); Aiguabella et al. (1970); Husain and Katcoff (1971); Katcoff and Hudis (1972); Sullivan et al. (1973); Rahimi et al. (1973); Crawford et al. (1974; 1975).

9. *Low-energy reactions other than fission.*— Price and Walker (1962b); Kowalski et al. (1968; 1970); Mesko et al. (1969); Natowitz (1970a); Somogyi and Schlenk (1970); Zoran and Popescu (1971); Hunyadi et al. (1971); Szabo et al. (1972); Zebelman and Miller (1973).

10. *Superheavy elements.*— Price and Fleischer (1969); Cieslak (1969); Flerov and Perelygin (1969); Price et al. (1970; 1971a,b); Otgonsuren et al. (1970; 1972); Bhandari et al. (1971a,b); Marinov et al. (1971a,b); Katcoff and Perlman (1971); Brandt et al. (1972); Flerov et al. (1972); Unik et al. (1972); Westgaard et al. (1972); Haack (1973); Sowinski et al. (1971); Geisler (1972); Fowler et al. (1970); Blanford et al. (1970); Shirk et al. (1973); Batty et al. (1973); Geisler et al. (1973). See also references in Table 5-3.

11. *Magnetic monopoles.*— Fleischer et al. (1969a,b,c; 1970a,b,c; 1971).

Chapter 7 References

R. Aiguabella, R. Pfohl, and R. Schmitt (1970), "Study of Heavy Ions Ejected in Interactions of 19 GeV/c Protons on Au Target, Deflected by Magnetic Field and Registered in Cellulose Nitrate," *Nuclear Reactions Induced by Heavy Ions*, 696–700. Amsterdam: North-Holland.

M. Ait-Salem, H. Gerhardt, F. Gonnenwein, H. Hipp, and H. Paap (1968), "Mass and Energy Distribution of Fission Fragments from Their Ranges in Solid State Detectors," *Nucl. Instr. Methods* **60,** 45–51.

B. M. Aleksandrov, L. S. Krivokhatskii, L. E. Malkin, and K. A. Petrzhak (1967), "Determination of the Probabilities of Spontaneous Fission of ^{233}U, ^{235}U and ^{243}Am," *J. Nucl. Energy* **21,** 193–195.

J. M. Alexander and R. Bimbot (1972), "Search for Spontaneous-Fission Isomerism in Nuclei of Medium Mass," *Phys. Rev.* **C5,** 799–804.

G. Anderson, I. Blomqvist, B. Forkman, G. G. Jonsson, A. Jarund, I. Kroon, K. Lindgren, B. Schroder, and K. Tesch (1972), "Photon-Induced Nuclear Reactions Above 1 GeV (I). Experimental," *Nucl. Phys.* **A197,** 44–70.

Kh. D. Androsenko and G. N. Smirenkin (1968), "Angular Anisotropy of Fission of U^{238} by Neutrons Near Threshold," *JETP Letters* **8,** 108–110.

Kh. D. Androsenko and G. N. Smirenkin (1971), "Fission of Pu^{240} by Neutrons Near Threshold," *Sov. J. Nucl. Phys.* **12,** 142–146.

Yu. A. Babenko, Yu. A. Nemilov, Yu. A. Selitskii, and V. B. Funshtein (1968), "Neutron-Induced Fission Cross-section of Ra^{226} and Angular Distribution of Fragments," *Sov. J. Nucl. Phys.* **7,** 186–188.

Yu. A. Babenko, V. T. Ippolitov, Yu. A. Nemilov, Yu. A. Selitskii, and V. B. Funshtein (1969), "Neutron Fission of Ra^{226} Near Threshold," *Sov. J. Nucl. Phys.* **10,** 133–136.

E. Barutcugil, S. Juhasz, M. Varnagy, S. Nagy, and J. Csikai (1971), "Angular Distributions of Fragments in the Fission of Uranium and Thorium Near the (n, 2n′f) Threshold," *Nucl. Phys.* **A173,** 571–576.

C. J. Batty, A. I. Kilvington, J. L. Weil, G. W. A. Newton, M. Skarestad, and J. D. Hemingway (1973), "Search for Superheavy Elements and Actinides Produced by Secondary Reactions in a Tungsten Target," *Nature* **244,** 429–430.

A. Behkami, J. H. Roberts, W. Loveland, and J. R. Huizenga (1968a), "Fragment Angular Distributions for Neutron-Induced Fission of U^{234}," *Phys. Rev.* **171,** 1267–1277.

A. Behkami, J. R. Huizenga, and J. H. Roberts (1968b), "Angular Distributions and Cross Sections of Fragments from Neutron-Induced Fission of ^{232}Th," *Nucl. Phys.* **A118,** 65–67.

N. Bhandari, S. G. Bhat, D. Lal, G. Rajagopalan, A. S. Tamhane, and V. S. Venkatavaradan (1971a), "Super-heavy Elements in Extraterrestrial Samples," *Nature* **230,** 219–224.

N. Bhandari, S. Bhat, D. Lal, G. Rajagopalan, A. S. Tamhane, and V. S. Venkatavaradan (1971b), "Spontaneous Fission Record of Uranium and Extinct Transuranic Elements in Apollo Samples," *Proc. Second Lunar Science Conf.*, **3,** 2599–2609. Cambridge: MIT Press.

R. Bimbot, C. Deprun, D. Gardes, H. Gauvin, Y. LeBeyec, M. Lefort, J. Peter, and B. Tamain (1971), "Complete Fusion Induced by Krypton Ions: Indications for Synthesis of Superheavy Nuclei," *Nature* **234,** 215–216.

S. Bjornholm, J. Borggreen, L. Westgaard, and V. A. Karnaukhov (1967), "Excitation Energy of the Spontaneously Fissioning Isomeric State in ^{240}Am," *Nucl. Phys.* **A95,** 513–525.

G. E. Blanford, P. H. Fowler, M. W. Friedlander, J. Klarmann, J. M. Kidd,

R. M. Walker, J. P. Wefel, and W. C. Wells (1970), "Primary Cosmic Ray Particles with $Z > 40$," *Acta Phys. Acad. Sci. Hung.* **29,** Suppl. 1, 423–427.

G. E. Blanford, M. W. Friedlander, J. Klarmann, S. S. Pomeroy, R. M. Walker, J. P. Wefel, and W. C. Wells (1971), "Comments on the Observation of Transuranic Nuclei in the Primary Cosmic Radiation," *Proc. 12th Inter. Cosmic Ray Conf.*, Hobart, Australia, **1,** 269–272.

A. H. Boos and R. Brandt (1971), "Ternary Fission Induced by Ar and Ca Ions in $ThSiO_4$ Crystals," *Proc. Seventh Inter. Colloq. of Corpuscular Photography and Solid Detectors*, Barcelona, 511–520.

H. R. Bowman, M. Croissiaux, R. C. Gatti, J. H. Heisenberg, R. Hofstadter, R. C. Jared, G. Kilian, L. M. Middleman, L. G. Moretto, S. G. Thompson, and M. R. Yearian (1968), "Electron-Induced Fission in ^{238}U, ^{209}Bi and ^{181}Ta," *Phys. Rev.* **168,** 1396–1398.

R. Brandt (1971), "Ternary Fission," *Angew. Chem. Internat. Edit.* **10,** 890–900.

R. Brandt, A. Kapustsik, V. P. Perelygin, S. P. Tretiakova, and N. Kh. Shadieva (1966), "Search for Ternary Fission in Ar Interaction with Heavy Element Using Mica Detectors," *Proc. Int. Conf. on Heavy Ion Phys.*, Dubna, **3,** 39–44.

R. Brandt, F. Carbonara, E. Cieslak, M. Dakowski, Ch. Gfeller, H. Piekarz, J. Piekarz, W. Riezler, R. Rinzivillo, E. Sassi, M. Sowinski, and J. Zakrzewski (1967), "Studies of High-Energy Nuclear Fission by Means of Mica Detectors," *Nucl. Phys.* **A90,** 177–185.

R. Brandt, D. Molzahn, and P. Patzelt (1972), "Search for Superheavy Elements in Uranium Irradiated with High-Energy Protons," *Radiochim. Acta* **18,** 157–158.

R. L. Brodzinksi and J. W. Cobble (1968), "Fission of Iridium at Intermediate Excitation Energies," *Phys. Rev.* **172,** 1194–1198.

F. Brown, D. A. Marsden, and R. D. Werner (1968), "Use of Blocking in Crystals to Study the Lifetime for Fission of U^{238} by 12-MeV Protons," *Phys. Rev. Letters* **20,** 1449–1451.

D. S. Burnett, R. C. Gatti, F. Plasil, P. B. Price, W. J. Swiatecki, and S. G. Thompson (1964), "Fission Barrier of Thallium-201," *Phys. Rev.* **134B,** 952–963.

D. Burnett, C. Hohenberg, M. Maurette, M. Monnin, R. Walker, and D. Woolum (1972), "Preliminary Report on Panel 4 of the Apollo 16 Cosmic Ray Experiment," *Apollo 16 Preliminary Science Report*, NASA Special Pub. 315, Chap. 15, pp. 19–32.

A. M. Charlesworth, J. Goldemberg, H. L. Pai, and B. B. P. Sinha (1972), "Electron and Positron Induced Fission in ^{238}U," *Bull. Am. Phys. Soc.* **17,** 440.

A. H. Chung (1973), *Electrofission of Mg-24*, Ph.D. Thesis, University of Toronto.

A. H. Chung, A. E. Litherland, J. Goldemberg, and H. L. Pai (1972), "Electrofission of ^{24}Mg," *Bull. Am. Phys. Soc.* **17,** 440.

E. Cieslak (1969), "Search for Long-Lived Spontaneously Fissioning Emitter in the Samples with Bi, Pb, Hg, W," Dubna Preprint P15-4738.

D. D. Clark (1971), "Shape Isomers and the Double-Humped Barrier," *Physics Today*, **24** (12), 23–31.

H. J. Crawford, J. D. Sullivan, and P. B. Price (1974), unpublished results.

H. J. Crawford, P. B. Price, J. Stevenson, and L. W. Wilson (1975), "Very Energetic Heavy Fragments from Relativistic Heavy Ion Reactions," *Phys. Rev. Letters* **34,** February 10.

J. Csikai and S. Nagy (1967), "Investigations on the Angular Distribution of Neutron-Induced ^{238}U Fission by Means of a Solid-State Track Detector," *J. Nucl. Energy* **21,** 375–376.

P. David, J. Debrus, U. Kim, G. Kumbartzki, H. Mommsen, W. Soyez, K. H. Speidel, and G. Stein (1972), "High-Energy Photofission of Gold and Uranium," *Nucl. Phys.* **A197,** 163–176.

M. Debeauvais and P. Cüer (1965), "Quelques resultats preliminaires sur la detection des produits de fission et de fragmentation de differents metaux moyens et lourds sous l'effect de 3 GeV par enregistrement dans le nouveaux detecteurs solides," *Compt. rend.*, Paris **261,** 2633–2636.

M. Debeauvais, R. Stein, J. Ralarosy, and P. Cüer (1967), "Spallation and Fission Fragments of Heavy Nuclei Induced by 18 GeV Protons Registered by Means of Solid Plastic Detectors," *Nucl. Phys.* **A90,** 186–198.

M. Debeauvais, R. Stein, G. Remy, and J. Ralarosy (1968), "Results on Fission and Fragmentation of U and Pb by 3, 18, and 24 GeV Protons Using Solid State Track Detectors," *J. de Physique* **29,** C1, 127–129.

P. Demers (1946), "New Photographic Emulsions Showing Improved Tracks of Ionizing Particles," *Phys. Rev.* **70,** 86.

M. de Mevergnies and P. del Marmol (1968), "Fission Cross-section of ^{232}Th for Thermal Neutrons," National Bureau of Standards Special Pub. 299.

P. A. M. Dirac (1931), "Quantized Singularities in the Electromagnetic Field," *Proc. Roy. Soc.*, London **A133,** 60–72.

P. A. M. Dirac (1948), "The Theory of Magnetic Poles," *Phys. Rev.* **74,** 817–830.

E. J. Dowdy and T. L. Krysinski (1971), "Angular Distributions of ^{238}U Photofission Fragments," *Nucl. Phys.* **A175,** 501–512.

V. A. Druin, N. K. Skobelev, B. V. Fefilov, and G. N. Flerov (1964), Dubna Preprint P-1580.

V. A. Druin, N. K. Skobelev, and V. I. Rud (1971), "Spontaneous Fission of Some Fm Isotopes," *Sov. J. Nucl. Phys.* **12,** 24–26.

V. A. Druin, Yu. S. Korotkin, Yu. P. Kharitonov, V. I. Krashonkin, Yu. V. Lobanov, D. M. Nadkarni, and S. P. Tretyakova (1973), "On Non-Observation of the Spontaneously Fissioning Activity of Kurchatovium-259 by the Berkeley Group," Dubna Preprint E7-7023.

V. Emma, S. LoNigro, and C. Milone (1970), "Fission Measurements by Means of Glass Detectors," *Lettere al Nuovo Cimento* **3,** 542–544.
V. Emma, S. LoNigro, and C. Milone (1971), "Photofission Cross-Section of Bi and Pb between 300 and 1000 MeV," *Lettere al Nuovo Cimento* **2,** 117–120.
V. Emma, S. LoNigro, and C. Milone (1973), "Angular Distributions of Fragments in the Fission of Thorium Induced by Neutrons in the Range 12–18 MeV," *Nucl. Phys.* **A199,** 186–192.
S. B. Ermagambetov, L. D. Smirenkina, G. N. Smirenkin, and A. S. Tishin (1967), "Fragment Angular Distribution in the Fission of Th^{232} by 1.6 MeV Neutrons," *JETP Letters* **5,** 30–33.
R. L. Fleischer and H. R. Hart (1973), "Tracks from Extinct Radioactivity, Ancient Cosmic Rays and Calibration Ions," *Nature* **242,** 104–105.
R. L. Fleischer and P. B. Price (1964), "Decay Constant for Spontaneous Fission of U^{238}," *Phys. Rev.* **133B,** 63–64.
R. L. Fleischer, P. B. Price, and R. M. Walker (1965), "Tracks of Charged Particles in Solids," *Science* **149,** 383–393.
R. L. Fleischer, P. B. Price, R. M. Walker, and E. L. Hubbard (1966), "Ternary Fission of Heavy Compound Nuclei in Thorite Track Detectors," *Phys. Rev.* **143,** 943–946.
R. L. Fleischer, I. S. Jacobs, W. M. Schwarz, P. B. Price, and H. G. Goodell (1969a), "Search for Multiply-Charged Dirac Magnetic Poles," *Phys. Rev.* **177,** 2029–2035.
R. L. Fleischer, H. R. Hart, I. S. Jacobs, P. B. Price, W. M. Schwarz, and F. Aumento (1969b), "Search for Magnetic Monopoles in Deep Ocean Deposits," *Phys. Rev.* **184,** 1393–1397.
R. L. Fleischer, P. B. Price, and R. T. Woods (1969c), "Search for Tracks of Massive, Multiply Charged Magnetic Poles," *Phys. Rev.* **184,** 1398–1401.
R. L. Fleischer, H. R. Hart, I. S. Jacobs, P. B. Price, W. M. Schwarz, H. G. Goodell, and F. Aumento (1970a), "Deep Sea Search for Multiply-Charged Magnetic Poles," *Rad. Effects* **3,** 131–138.
R. L. Fleischer, H. R. Hart, I. S. Jacobs, P. B. Price, W. M. Schwarz, and R. T. Woods (1970b), "Magnetic Monopoles: Where Are They and Where Aren't They," *J. Appl. Phys.* **41,** 958–965.
R. L. Fleischer, H. R. Hart, I. S. Jacobs, P. B. Price, W. M. Schwarz, R. T. Woods, F. Aumento, and H. G. Goodell (1970c), "Search for Cosmic Magnetic Monopoles," *Acta Phys. Acad. Sci. Hung.* **29,** Suppl. 3, 27–30.
R. L. Fleischer, H. R. Hart, G. E. Nichols, and P. B. Price (1971), "Sea Level Search for Cosmic Monopoles," *Phys. Rev.* **D4,** 24–27.
G. N. Flerov (1972), "Heavy Ion Research at Dubna," *IEEE Trans. Nucl. Sci.* **NS-19** (2), 9–15.
G. N. Flerov and V. P. Perelygin (1969), "On Spontaneous Fission of Lead—Search for Far Transuranic Elements," *At. Energ.* **26,** 520.

G. N. Flerov, Yu. Ts. Oganesyan, Yu. V. Lobanov, V. I. Kuznetsov, V. A. Druin, V. P. Perelygin, K. A. Gavrilov, S. P. Tretiakova, and V. M. Plotko (1964), "Synthesis and Physical Identification of the Isotope of Element 104 with Mass Number 260," *Phys. Letters* **13,** 73–75.

G. N. Flerov, A. A. Pleve, S. M. Polikanov, E. Ivanov, N. Martalogu, and I. Vilcov (1965), "The Excitation Function and the Isomeric Yield Ratio for the 14 msec Fissioning Isomer from Deuteron Irradiation of Plutonium," *Rev. Roum. Phys.* **10,** 217–222.

G. N. Flerov, V. I. Kuznetsov, and N. K. Skobelev (1966a), "On Spontaneous Fission of Isotope 102^{254}," Dubna Preprint E7-3043.

G. N. Flerov, S. A. Karamian, Z. V. Kuznetsov, Yu. Ts. Oganesyan, and Yu. E. Penionshkevitch (1966b), "Ternary Fission of U^{238} Induced by Ne Ions," Dubna Preprint E7-2924.

G. N. Flerov, A. A. Pleve, S. M. Polikanov, P. Tretiakova, M. Martalogu, D. Poenaru, M. Sezon, I. Vilcov, and N. Vilcov (1967a), "Excitation Energy of Spontaneously Fissioning Isomer ^{242m}Am," *Nucl. Phys.* **A97,** 444–448.

G. N. Flerov, A. A. Pleve, S. M. Polikanov, S. P. Tretiakova, I. Boca, M. Sezon, I. Vilcov, and N. Vilcov (1967b), "A Study of the Spontaneously-Fissioning Isomer of ^{242}Am Through the ^{241}Am (n, γ) Reaction," *Nucl. Phys.* **A102,** 443–448.

G. N. Flerov, Yu. Ts. Oganesyan, Yu. V. Lobanov, Yu. A. Lazarev, V. I. Kuznetsov, and S. P. Tretiakova (1970a), "Investigation of the Regularities of the Production of Spontaneously Fissionable Isotope of Element 105," Dubna Preprint P7-5108.

G. N. Flerov, V. A. Druin, and A. A. Pleve (1970b), "The Stability of Heavy Nuclei and the Limit of the Periodic System of Elements," *Sov. Phys. Uspekhi* **13,** 24–50.

G. N. Flerov, Yu. A. Lazarev, Yu. V. Lobanov, Yu. Ts. Oganesyan, and S. P. Tretiakova (1971a), "Spontaneously Fissioning Isotopes of Kurchatovium," *Proc. Int. Conf. Heavy Ion Physics*, Dubna Report D7-5769, p. 125.

G. N. Flerov, Yu. Ts. Oganesyan, Yu. V. Lobanov, Yu. A. Lasarev, S. P. Tretiakova, I. V. Kolesov, and V. M. Plotko (1971b), "On the Synthesis of Element 105," *Nucl. Phys.* **A160,** 181–192.

G. N. Flerov, V. P. Perelygin, and O. Otgonsuren (1972), "On the Origin of Tracks of Fission Fragments in Lead Glasses," *At. Energ.* **33,** 979–984.

E. F. Fomushkin and E. K. Gutnikova (1970), "Cross Sections and Angular Distributions of Fragments in the Fission of Pu^{238}, Pu^{242}, and Am^{241} by Neutrons of Energy 0.45–3.6 MeV," *Sov. J. Nucl. Phys.* **10,** 529–532.

E. F. Fomushkin, E. K. Gutnikova, A. N. Maslov, G. V. Novoselov, and V. I. Panin (1972), "Neutron-Fission Cross Sections of Bk^{249} and Cf^{249}," *Sov. J. Nucl. Phys.* **14,** 41–44.

B. Forkman and B. Schroder (1972), "A Review of Intermediate Energy Photo-fission," *Physica Scripta* **5,** 105–115.

P. H. Fowler, V. M. Clapham, V. G. Cowen, J. M. Kidd, and R. T. Moses (1970), "The Charge Spectrum of Very Heavy Cosmic Ray Nuclei," *Proc. Roy. Soc.*, London **318A,** 1–43.

P. Freier, E. J. Lofgren, E. P. Ney, F. Oppenheimer, H. L. Bradt, and B. Peters (1948), "Evidence for Heavy Nuclei in the Primary Cosmic Radiation," *Phys. Rev.* **74,** 213–217.

Yu. P. Gangrskii, B. N. Markov, S. M. Polikanov, and H. Jungclaussen (1966), "Spontaneously Fissioning Isomer with Half-Life 10^{-7} Sec," *JETP Letters* **4,** 289–290.

Yu. P. Gangrskii, B. N. Markov, S. M. Polikanov and H. Jungclaussen (1967), "Investigation of the Reaction $U^{238} + B^{11}$ Leading to a Spontaneously Fissionable Isomer Am^{242}," *Sov. J. Nucl. Phys.* **5,** 22–25.

Yu. P. Gangrskii, B. N. Markov, S. M. Polikanov, I. F. Kharisov, and H. Jungclaussen (1968), "Investigation of an Isomer of Cf-246 that Undergoes Spontaneous Fission," *Bull. Acad. Sci. USSR* **32,** 1525.

Yu. P. Gangrskii, Nguyen Cong Khanh, and D. D. Pulatov (1972), "Production of Spontaneously Fissioning Isomers, with Nanosecond Half-Life, in the α-Particle Reactions," Dubna Preprint P7-6286.

F. H. Geisler (1972), "Search for Superheavy Elements in Terrestrial Minerals and Cosmic Ray Induced Fission of Heavy Elements," Ph.D. Thesis, Washington University.

F. H. Geisler and P. R. Phillips (1972), "An Improved Method for Locating Charged Particle Tracks in Thin Plastic Sheets," *Rev. Sci. Instr.* **43,** 283–284.

F. H. Geisler, P. R. Phillips, and R. M. Walker (1973), "Search for Superheavy Elements in Natural and Proton-Irradiated Materials," *Nature* **244,** 428–429.

D. S. Gemmell and R. E. Holland (1965), "Blocking Effects in the Emergence of Charged Particles from Single Crystals," *Phys. Rev. Letters* **14,** 945–948.

W. Gentner, D. Storzer, R. Gijbels, and R. Van det Linden (1972), "Calibration of the Decay Constant of ^{238}U Spontaneous Fission," *Amer. Nucl. Soc. Trans.* **15,** 125.

A. Ghiorso, T. Sikkeland, and M. J. Nurmia (1967), "Isotopes of Element 102 with Mass 251 to 258," *Phys. Rev. Letters* **18,** 401–404.

A. Ghiorso, M. Nurmia, J. Harris, K. Eskola, and P. Eskola (1969), "Positive Identification of Two Alpha-Particle-Emitting Isotopes of Element 104," *Phys. Rev. Letters* **22,** 1317–1320.

A. Ghiorso, M. Nurmia, E. Eskola, and P. Eskola (1971), "Comments on 'Chemical Separation of Kurchatovium'," *Inorg. Nucl. Chem. Letters* **7,** 1117–1119.

W. M. Gibson and K. O. Nielsen (1970), "Direct Determination of the Lifetime of Excited Compound Nuclei by Angular Distribution Measurements of Fission Fragments Emitted from Single Crystals," *Phys. Rev. Letters* **24,** 114–117.

A. E. Glassgold, W. Heckrotte, and K. M. Watson (1959), "Collective Excitations of Nuclear Matter," *Ann. Phys.*, New York **6,** 1–36.

R. Gold, R. J. Armani, and J. H. Roberts (1970), "Spontaneous-Fission Decay Constant of ^{241}Am," *Phys. Rev.* **C1,** 738–740.

F. Gönnenwein, H. Heinrich, and H. Hipp (1967), "On the Measurement of Fission Angular Distribution in Solid State Detectors," *Z. für Naturf.* **22a,** 1133–1134.

B. A. Gvozdev, B. B. Zakhvataev, V. I. Kuznetsov, V. P. Perelygin, S. V. Pirozhkov, E. G. Chudinov, and I. K. Shvetsov (1966), "Period of Spontaneous Fission of ^{243}Am," *Radiokhimiya* **8,** 493–494.

U. Haack (1973), "Suche nach Überschweren Transuranelementen," *Naturwiss.* **60,** 65–70.

J. Hudis and S. Katcoff (1969), "High Energy Fission Cross Sections of U, Bi, Au and Ag Measured with Mica Track Detectors," *Phys. Rev.* **180,** 1122–1130.

J. R. Huizenga, A. N. Behkami, and J. H. Roberts (1969), "Channel Analysis of Neutron-Induced Fission of ^{236}U," *Second IAEA Symp. Phys. Chem. Fission*, IAEA, Vienna, 403–418.

E. K. Hulet, J. F. Wild, R. W. Lougheed, J. E. Evans, B. J. Qualheim, M. Nurmia, and A. Ghiorso (1971), "Spontaneous-Fission Half-Life of ^{258}Fm and Nuclear Instability," *Phys. Rev. Letters* **26,** 523–526.

I. Hunyadi, B. Schlenk, G. Somogyi, and D. S. Srivastava (1971), "Investigation of ^{27}Al(d, α) ^{25}Mg Nuclear Reaction in the Energy Range E_d = 650 – 540 keV Using Plastic Track Detector," *Acta Phys. Acad. Sci. Hung.* **30,** 73–79.

L. Husain and S. Katcoff (1971), "Antiproton- and Pion-Induced Fission at 2.5 GeV/c," *Phys. Rev.* **C4,** 263–267.

A. V. Ignatyuk, N. S. Rabotnov, G. N. Smirenkin, A. S. Soldatov, and Yu. M. Tsipenyuk (1972), "Sub-Barrier Photofission of Even-Even Nuclei," *Sov. Phys. JETP*, **34,** 684–693.

M. G. Itkis, K. G. Kuvatov, V. N. Okolovich, G. Ya. Ruskina, G. N. Smirenkin, and A. S. Tishin (1973), "Po^{210} Fission in Pb^{206} (α, f) Reaction," *Sov. J. Nucl. Phys.* **16,** 144–150.

M. Juric, R. Antanasijevic, and J. Vukovic (1970), "On the Yield of Ternary Fission of ^{236}U Induced by Thermal Neutrons," *Proc. Seventh Inter. Colloq. on Corpuscular Photography and Solid Detectors*, Barcelona, 481–487.

V. V. Kamanin, S. A. Karamian, A. M. Kucher, F. Normuratov, and Yu. Ts. Oganesyan (1972a), "On Compound Nucleus Life-Time Measurements Using the Blocking Effect," Dubna Preprint P7-6291.

V. V. Kamanin, S. A. Karamian, F. Normuratov, and S. P. Tretiakova (1972b), "Life-Times of the Excited Compound Nuclei in the Range $79 < Z < 89$ Measured Using the Shadow Effect," Dubna Preprint P7-6302.

V. V. Kaminin, S. A. Karamian, A. M. Kucher, F. Normuratov, and Yu. Ts.

Oganesyan (1973), "Measurement of Compound Nucleus Lifetime by Means of the Shadow Effect," *Sov. J. Nucl. Phys.* **16,** 140–143.
S. P. Kapitza, N. S. Rabotnov, G. N. Smirenkin, A. S. Soldatov, L. N. Usachev, and Yu. M. Tsipenyuk (1969), "Photofission of Even-Even Nuclei and Structure of the Fission Barrier," *JETP Letters* **9,** 73–76.
S. Karamian, Yu. Ts. Oganesyan, and F. Normuratov (1970), "Experiments on Measurement of the Compound Nucleus Lifetime in the W (^{22}Ne, f) Reaction," Dubna Preprint P7-5512.
S. Karamian, Yu. Melikov, F. Normuratov, O. Otgonsuren, and G. Solovyeva (1971), "Estimate of the Life-Time of Excited Compound Nucleus in Reaction W (Ne^{22}, f) (Experiment)," *Sov. J. Nucl. Phys.* **13,** 543–546.
V. I. Kasilov, A. V. Mitrofanova, Yu. N. Ranyuk, and P. V. Sorokin (1972), "Photofission and Photofragmentation of Tantalum," *Sov. J. Nucl. Phys.* **15,** 228–231.
S. Katcoff and J. Hudis (1972), "Fission of U, Bi, Au and Ag Induced by 29 GeV ^{14}N Ions," *Phys. Rev. Letters* **28,** 1066–1068.
S. Katcoff and M. L. Perlman (1971), "Experiments Related to Possible Production of Superheavy Elements by Proton Irradiation," *Nature* **231,** 522–524.
H. A. Khan (1971), "Determination of Fission Rate Decay Constant Profiles, and Distributions of ^{252}Cf, ^{239}Pu and ^{235}U Fission Sources by Solid State Track Detectors," *Nucleus* (Karachi) **8,** 63.
H. A. Khan and S. A. Durrani (1973), "Measurement of Spontaneous-Fission Decay Constant of ^{238}U with a Mica Solid State Track Detector," *Rad. Effects* **17,** 133–135.
A. Khodai-Joopari (1966), "Fission Properties of Some Elements Below Radium," Ph.D. Thesis, University of California, UCRL 16489.
S. Kinoshita (1910), "The Photographic Action of the α-Particles Emitted from Radioactive Substances," *Proc. Roy. Soc.*, London **83,** 432–453.
J. D. Kleeman and J. F. Lovering (1971), "A Determination of the Decay Constant for Spontaneous Fission of Natural Uranium Using Fission Track Accumulation," *Geochim. Cosmochim. Acta* **35,** 637–640.
V. A. Kon'shin, E. S. Matusevich, and V. I. Regushevskii (1966a), "Cross Section for Fission Fragment Production by 660 MeV Protons in Yb, Er, and Tb Nuclei," *Sov. J. Nucl. Phys.* **3,** 569.
V. A. Kon'shin, E. S. Matusevich, and V. I. Regushevskii (1966b), "Cross-Sections for Fission of Ta^{181}, Re, Pt, Au^{197}, Pb, Bi^{209}, Th^{232}, U^{235}, and U^{238} by 150–660 MeV Protons," *Sov. J. Nucl. Phys.* **2,** 489–492.
V. A. Kon'shin, E. S. Matusevich, and V. I. Regushevskii (1967), "Fissility of Nuclei in the Region $Z^2/A > 35$, in Fission by High Energy Protons," *Sov. J. Nucl. Phys.* **4,** 69.
L. Kowalski, J. C. Jodogne, and J. M. Miller (1968), "Complete-Fusion Collisions in Heavy-Ion Reactions," *Phys. Rev.* **169,** 894–898.

L. Kowalski, J. M. Miller, J. D'Auria, and J. C. Jodogne (1970), "Measurements of the Cross Sections for Formation of Compound Nuclei in Heavy-Ion Reactions," *Rad. Effects* **5,** 277–278.

L. Kowalski, A. Zebelman, A. Kandel, and J. M. Miller (1971), "Angular Distribution of Fission Fragments from the Yb^{170} Compound Nucleus Excited to 107 MeV," *Phys. Rev.* **C3,** 1370–1372.

I. Kroon and B. Forkman (1972), "Photon-Induced Nuclear Reactions Above 1 GeV (III). Fission in Gold and Lead," *Nucl. Phys.* **A197,** 81–87.

I. M. Kuks, V. I. Matvienko, Yu. A. Nemilov, Yu. A. Selitskii, and V. B. Funshtein (1973), "Search for Fission Isomers in the Radium Region," *Sov. J. Nucl. Phys.* **16,** 244.

K. G. Kuvatov, V. N. Okolovich, and G. N. Smirenkin (1968), "Angular Anisotropy of Fission of Nuclei Lighter than Gold by 40 MeV α-Particles," *JETP Letters* **8,** 171.

K. G. Kuvatov, V. N. Okolovich, and G. N. Smirenkin (1970), "Angular Anisotropy of the Fission of Pb^{204} and Pb^{208} by Alpha Particles Near the Threshold," *JETP Letters* **11,** 26–28.

K. G. Kuvatov, V. N. Okolovich, L. A. Smirina, G. N. Smirenkin, V. P. Bochin, and V. S. Romanov (1972), "Angular Anisotropy and Cross Section for Fission of Nuclei in the Region $Z = 73 - 83$ by 38-MeV α Particles," *Sov. J. Nucl. Phys.* **14,** 45–49.

V. I. Kuznetsov, N. K. Skobelev, and G. N. Flerov (1965), "Spontaneous Fission from an Isomeric State of One of Light Isotopes of Np with Half-Life of 60 Seconds," Dubna Preprint R-2435.

V. Kuznetsov, Yu. V. Lobanov, and V. P. Perelygin (1966), "The Decay Half-Time of the Isotope of the Element 102 with the Mass Number 256," *Sov. J. Nucl. Phys.* **4,** 332–334.

V. I. Kuznetsov, N. K. Skobelev, and G. N. Flerov (1967), "Detection of a Spontaneously Fissile Isomer with 2.6 Min. Half-Life in the Nuclear Reactions $U^{233} + B^{11}$ and $U^{233} + B^{10}$," *Sov. J. Nucl. Phys.* **4,** 70–71.

N. L. Lark, G. Sletten, J. Pedersen, and S. Bjornholm (1969), "Spontaneously Fissioning Isomers in U, Np, Pu and Am Isotopes," *Nucl. Phys.* **A139,** 481–500.

C. M. G. Lattes, H. Muirhead, G. P. S. Occhialini, and C. F. Powell (1947), "Processes Involving Charged Mesons," *Nature* **159,** 694–697.

M. P. T. Leme, C. Renner, and M. Cattani (1971), "Determination of the Decay Constant for Spontaneous Fission of ^{238}U," *Nucl. Instr. Methods* **91,** 577–579.

P. Limkilde and G. Sletten (1973), "A Subnanosecond and a Nanosecond Fission Isomer in ^{238}Pu," *Nucl. Phys.* **A199,** 504–512.

A. F. Linev, B. N. Markov, A. A. Pleve, and S. M. Polikanov (1965), "The Formation of a Spontaneously Fissioning Isomer in the Capture of Neutrons by Am," *Nucl. Phys.* **63,** 173–176.

Yu. V. Lobanov, V. I. Kuznetsov, V. P. Perelygin, S. M. Polikanov, Yu. Ts. Oganesyan, and G. N. Flerov (1965), "Spontaneously Fissile Isomer with 0.9 msec Half-Life," *Sov. J. Nucl. Phys.* **1,** 45–47.

W. Loveland, J. R. Huizenga, A. Behkami, and J. H. Roberts (1967), "Identification of Single-Particle States in the Fission Transition Nucleus ^{235}U," *Phys. Letters* **24B,** 666–668.

J. Maly (1971), "Evidence for Long-Range Fission Fragments Induced by Irradiation with Electrons," *Phys. Letters* **35B,** 148–150.

A. Manfredini, L. Fiore, C. Ramorino, and W. Wolfli (1971), "Photofission of ^{232}Th near Threshold: Angular Distribution," *Il Nuovo Cimento* **4A,** 421–430.

A. Marinov, C. J. Batty, A. I. Kilvington, G. W. A. Newton, V. J. Robinson, and J. D. Hemingway (1971a), "Evidence for the Possible Existence of a Superheavy Element with Atomic Number 112," *Nature* **229,** 464–467.

A. Marinov, C. J. Batty, A. I. Kilvington, J. L. Weil, A. M. Friedman, G. W. A. Newton, V. J. Robinson, J. D. Hemingway, and D. S. Mather (1971b), "Spontaneous Fission Previously Observed in a Mercury Source," *Nature* **234,** 212–215.

B. N. Markov, A. A. Pleve, S. M. Polikanov, and G. N. Flerov (1966), "Experiments on Synthesis of a Spontaneously Fissile Isomer in the Reaction Am^{241} (n, γ) Am^{242}," *Sov. J. Nucl. Phys.* **3,** 329–330.

E. S. Matusevich and V. I. Regushevskii (1968), "Cross Sections for Fission of Bi^{209}, U^{235}, U^{238}, Np^{237}, Pu^{239} by 1–9 GeV Protons," *Sov. J. Nucl. Phys.* **7,** 708–709.

M. Maurette and C. Stephan (1965), "Measurement of Absolute Fission Cross-Sections Induced by 156-MeV Protons, Using Mica as a Fission Fragment Detector," *Proc. IAEA Symp. Phys. Chem. Fission*, Salzburg **2,** 307–313.

L. Medveczky and G. Somogyi (1970), "Investigations on Ternary Photofission with Track Recorders of Different Sensitivity," *Rad. Effects* **5,** 51–59.

L. Medveczky, G. Somogyi, and G. Gotz (1970), "Investigations on Ternary Photofission by Means of Plastic and Mica Track Detectors," *Acta Phys. Acad. Sci. Hung.* **28,** 169–184.

Yu. V. Melikov, Yu. D. Otstavnov, and A. F. Tulinov (1969), "The Blocking Effect in Uranium Fission," *Sov. Phys. JETP* **29,** 968–969.

Yu. V. Melikov, Yu. D. Otstavnov, and A. F. Tulinov (1970), "Measurement of the Time of Fission of U^{238} by Neutrons Having an Energy of Approximately 3 MeV," *Sov. J. Nucl. Phys.* **12,** 27–28.

Yu. V. Melikov, Yu. D. Otstavnov, A. F. Tulinov, and N. G. Chetchenin (1972), "Determination of the Lifetime of Excited Compound Nuclei in Fission Using the Shadow Effect," *Nucl. Phys.* **A180,** 241–253.

L. Mesko, B. Schlenk, G. Somogyi, and A. Valek (1969), "^{19}F(d, α)^{17}O Angular Distributions at $E_d = 300 - 650$ KeV," *Nucl. Phys.* **A130,** 449–455.

V. Metag, R. Repnow, P. von Brentano, and J. D. Fox (1969a), "Fission Isomerism Induced by Helium Ions," *Z. Physik* **226,** 1–6.

V. Metag, R. Repnow, P. von Brentano, and J. D. Fox (1969b), "Charged-Particle Studies of Isomeric Fission," *Physics and Chemistry of Fission*, IAEA, Vienna, 449–456.

T. Methasiri (1970), "High-Energy Photofission Cross Sections of Uranium and Thorium," *Nucl. Phys.* **A158,** 433–439.

T. Methasiri and S. A. E. Johansson (1971), "High-Energy Photofission of Heavy and Medium-Heavy Elements," *Nucl. Phys.* **A167,** 97–107.

A. V. Mitrofanova, Yu. N. Ranyuk, and P. V. Sorokin (1968), "Photofission of Bi, Pb, Tl, Au, Pt, Os, Re, Ta and Hf at Energies up to 1600 MeV," *Sov. J. Nucl. Phys.* **6,** 512–515.

D. Montgomery and R. Brandt (1970), "Search for Spontaneously Fissioning Isomers with Nanosecond Half-Lives Produced by 600 MeV Protons," *Proc. Seventh Inter. Colloq. Corpuscular Photography and Solid Detectors*, Barcelona, 503–509.

L. G. Moretto, R. C. Gatti, and S. G. Thompson (1968), *Nuclear Chemistry Division Annual Rept.*, Lawrence Rad. Lab. Rept. 17989.

L. G. Moretto, R. C. Gatti, S. G. Thompson, J. R. Huizenga, and J. O. Rasmussen (1969a), "Pairing Effects at the Fission Saddle Point of ^{210}Po and ^{211}Po," *Phys. Rev.* **178,** 1845–1854.

L. G. Moretto, R. C. Gatti, S. G. Thompson, and J. T. Routti (1969b), "Electron- and Bremsstrahlung-Induced Fission of Heavy and Medium-Heavy Nuclei," *Phys. Rev.* **179,** 1176–1187.

L. G. Moretto, S. G. Thompson, J. Routti, and R. C. Gatti (1972), "Influence of Shells and Pairing on the Fission Probabilities of Nuclei Below Radium," *Phys. Letters* **38B,** 471–474.

W. D. Myers and W. J. Swiatecki (1966), "Nuclear Masses and Deformations," *Nucl. Phys.* **81,** 1–60.

T. Nakanishi and M. Sakanoue (1972), "Low-Energy-Deuteron Fission Cross Section of ^{231}Pa Measured with Mica Track Detector," *Radiochim. Acta* **16,** 24–27.

L. Natanson and P. Dakowski (1967), "The Fission of ^{197}Au by 13.4 MeV Deuterons," *Acta Phys. Polonica* **31,** 599–600.

J. B. Natowitz (1970a), "Complete-Fusion Cross Sections for the Reactions of Heavy Ions with Cu, Ag, Au, and Bi," *Phys. Rev.* **C1,** 623–630.

J. B. Natowitz (1970b), "Calculated Complete-Fusion Cross Sections for 5 to 10.5 MeV/amu ^{64}Ni, ^{86}Kr and ^{136}Xe Projectiles," *Phys. Rev.* **C1,** 2157–2159.

J. B. Natowitz and J. K. Archer (1969), "Production of Spontaneous Fission Isomers in the Reactions of ^{237}Np with 30 to 70 MeV ^{4}He," *Phys. Letters* **30B,** 463–464.

J. B. Natowitz and E. T. Chulick (1971), "Angular Distributions of Fission Fragments in the First Chance Fission of ^{198}Hg," *Nucl. Phys.* **A172,** 185–192.

J. B. Natowitz, A. Khodai-Joopari, J. M. Alexander, and T. D. Thomas (1968), "Detection of Long-Range Fragments from Decay of Cf^{252}," *Phys. Rev.* **169,** 993–999.

Y. A. Nemilov, V. V. Pavlov, Y. A. Selitskii, S. M. Solov'ev, and V. P. Eismont (1965), "Total and Differential Cross-Sections of Uranium and Thorium Fission by Low Energy Deuterons," *At. Energ.* **18,** 456–459.

Y. A. Nemilov, Y. A. Selitskii, S. M. Solov'ev, and V. P. Eismont (1966), "Angular Anisotropy of Fission by Sub-barrier Deuterons," *Sov. J. Nucl. Phys.* **2,** 330–334.

V. G. Nesterov and G. N. Smirenkin (1968), "Cross Section Ratios for Fission of U^{233}, U^{235}, and Pu^{239} by Fast Neutrons," *Sov. J. of Atom. Energy* **24,** 224–226.

S. G. Nilsson, C. F. Tsang, A. Sobiczewski, Z. Sgymanski, S. Wycech, C. Gustafson, I. L. Lamm, P. Möller, and B. Nilsson (1969), "On the Nuclear Structure and Stability of Heavy and Superheavy Elements," *Nucl. Phys.* **A131,** 1–66.

J. R. Nix (1969), "Predicted Properties of the Fission of Super-Heavy Nuclei," *Phys. Letters* **30B,** 1–4.

J. R. Nix (1972), "Calculation of Fission Barriers for Heavy and Superheavy Nuclei," *Ann. Rev. Nucl. Sci.* **22,** 65–120.

V. Noelpp, R. Abegg, J. Schacher, and R. Wagner (1972), "Blocking-Effekt bei der Uranspaltung mit 14 MeV-Neutronen," *Helv. Phys. Acta* **45,** 55.

M. Nurmia, T. Sikkeland, R. Silva, and A. Ghiorso (1967), "Spontaneous Fission of Light Fermium Isotopes; New Nuclides ^{244}Fm and ^{245}Fm," *Phys. Letters* **26B,** 78–80.

O. Otgonsuren, V. P. Perelygin, and G. N. Flerov (1970), "The Search for Far Transuranium Elements in Ferromanganese Nodules," *Sov. Phys. Doklady* **189,** 1200–1203.

O. Otgonsuren, V. P. Perelygin, S. P. Tretiakova, and Yu. A. Vinogradov (1972), "On Search for Tracks of the Spontaneous Fission Fragments of Far Transuranium Elements in Natural Minerals," *At. Energ.* **32,** 344–347.

K. Otozai, J. W. Meadows, A. N. Behkami, and J. R. Huizenga (1970), "Fragment Angular Distributions from Neutron-Induced Fission of ^{242}Pu," *Nucl. Phys.* **A144,** 502–512.

B. D. Pate and J. Peter (1971), "Fission of Medium Mass Nuclei Induced by 167 MeV Alpha Particles," *Nucl. Phys.* **A173,** 520–536.

V. Perelygin, S. P. Tretiakova, and G. I. Khlubnikov (1964), "Decay Times of Np^{237}, Cm^{244}, and Cf^{248}," Dubna Preprint 1635.

V. P. Perelygin, N. H. Shadieva, S. P. Tretiakova, A. H. Boos, and R. Brandt (1969), "Ternary Fission Produced in Au, Bi, Th, and U with Ar Ions," *Nucl. Phys.* **A127,** 577–585.

N. A. Perfilov, A. I. Obukhov, and O. E. Shigaev (1966), "Cross-Section for Fission Induced by Irradiation of Tin by Oxygen Ions," *Sov. J. Nucl. Phys.* **3,** 715.

J. Peter (1970), "Disintegration par Fission Spontanee de Tres Courte Demi-Vie des Noyaux Obtenus par Reaction Nucleaire de ^{14}N sur ^{169}Tm," *Phys. Letters* **31B,** 124–125.

T. Plieninger, W. Krätschmer, and W. Gentner (1972), "Charge Assignment to Cosmic Ray Heavy Ion Tracks in Lunar Pyroxenes," *Proc. Third Lunar Sci. Conf.*, **3,** 2933–2939. Cambridge: MIT Press.

S. M. Polikanov and G. Sletten (1970), "Spontaneously Fissioning Isomers in U, Pu, Am and Cm Isotopes," *Nucl. Phys.* **A151,** 656–672.

S. M. Polikanov, V. A. Druin, V. A. Karnaukhov, V. L. Mikheev, A. A. Pleve, N. K. Skobelev, V. G. Subbotin, G. M. Ter-akopyan, and V. A. Fomichev (1962), "Spontaneous Fission with an Anomalously Short Period. I." *Sov. Phys. JETP* **15,** 1016–1021.

A. M. Poskanzer, G. W. Butler, and E. K. Hyde (1971), "Fragment Production in the Interaction of 5.5 GeV Protons with Uranium," *Phys. Rev.* **C3,** 882–904.

C. F. Powell, G. P. S. Occhialini, D. L. Livesey, and L. V. Chilton (1946), "A New Photographic Emulsion for the Detection of Fast Charged Particles," *J. Sci. Instr.* **23,** 102.

P. B. Price (1972), "Ultra-heavy Cosmic Rays," *Invited and Rapporteur Papers, Twelfth Inter. Conf. on Cosmic Rays*, Hobart, Australia, 453–473.

P. B. Price and R. L. Fleischer (1969), "Are Fission Tracks in Meteorites from Super-heavy Elements?," *Phys. Letters* **30B,** 246–248.

P. B. Price and R. M. Walker (1962a), "Electron Microscope Observation of Etched Tracks from Spallation Recoils in Mica," *Phys. Rev. Letters* **8,** 217–219.

P. B. Price and R. M. Walker (1962b), "A New Track Detector for Heavy Particle Studies," *Phys. Letters* **3,** 113–115.

P. B. Price, R. L. Fleischer, R. M. Walker, and E. L. Hubbard (1963), "Ternary Fission of Heavy Compound Nuclei," in *Reactions between Complex Nuclei*, A. Ghiorso, R. M. Diamond and H. E. Conzett (eds.), 332–337. Berkeley: U. C. Press.

P. B. Price, R. L. Fleischer, and R. T. Woods (1970), "Search for Spontaneously Fissioning Trans-Uranic Elements in Nature," *Phys. Rev.* **C1,** 1819–1821.

P. B. Price, P. H. Fowler, J. M. Kidd, E. J. Kobetich, R. L. Fleischer, and G. E. Nichols (1971a), "Study of the Charge Spectrum of Extremely Heavy Cosmic Rays Using Combined Plastic Detectors and Nuclear Emulsions," *Phys. Rev.* **D3,** 815–823.

P. B. Price, R. S. Rajan, and E. K. Shirk (1971b), "Ultra-Heavy Cosmic Rays in the Moon," *Proc. Second Lunar Sci. Conf.* **3,** 2621–2627, Cambridge: MIT Press.

P. B. Price, D. Lal, A. S. Tamhane, and V. P. Perelygin (1973), "Characteristics

of Tracks of Ions with $14 \leq Z \leq 36$ in Common Rock Silicates," *Earth Planet. Sci. Lett.* **19,** 377–395.
N. S. Rabotnov, G. N. Smirenkin, A. S. Soldatov, L. N. Vsachev, S. P. Kapitza, and Yu. M. Tsipeniuk (1968), "Angular Distribution of Photofission Fragments near Threshold," *Phys. Letters* **26B,** 218–219.
N. S. Rabotnov, G. N. Smirenkin, A. S. Soldatov, L. N. Vsachev, S. P. Kapitza, and Yu. M. Tsipeniuk (1970), "Photofission of Th^{23}, U^{238}, Pu^{238}, Pu^{240}, Pu^{242} and the Structure of the Fission Barrier," *Sov. J. Nucl. Phys.* **11,** 285–294.
F. Rahimi, D. Gheysari, G. Remy, J. Tripier, J. Ralarosy, R. Stein, and M. Debeauvais (1973), "Fission of U, Th, Bi, Pb and Au Induced by 2.1 GeV ^{2}H Ions," *Phys. Rev.* **C8,** 1500–1503.
Grant M. Raisbeck (1966), "Excitation Functions for the Helium-Ion-Induced Fission of Re, Lu, Tm," Ph.D. Thesis, Purdue University.
G. M. Raisbeck and J. W. Cobble (1967), "Excitation Functions for the Helium-Ion-Induced Fission of Rhenium, Lutetium, and Thulium," *Phys. Rev.* **153,** 1270–1282.
J. Ralarosy, M. Debeauvais, R. Stein, G. Remy, and J. Tripier (1969), "Utilisation des detecteurs visuels plastiques pour l'etude de la fission induite par les ions lourds," *Compt. rend.*, Paris **269,** 593–596.
J. Ralarosy, M. Debeauvais, G. Remy, J. Tripier, R. Stein, and D. Huss (1973), Fission Cross Sections of Uranium, Thorium, Bismuth, Lead and Gold Induced by 58- to 100-MeV Alpha Particles," *Phys. Rev.* **C8,** 2372–2378.
Yu. N. Ranyuk and P. V. Sorokin (1967), "Fission of Uranium by 35–260 MeV Electrons," *Sov. J. Nucl. Phys.* **5,** 377–379.
G. Remy, J. Ralarosy, R. Stein, M. Debeauvais, and J. Tripier (1970), "Heavy Fragment Emission in High Energy Reactions of Heavy Nuclei," *J. de Physique* **31,** 27–34.
G. Remy, J. Ralarosy, R. Stein, M. Debeauvais, and J. Tripier (1971), "Cross-Sections for Binary and Ternary Fission Induced by High-Energy Protons in Uranium and Lead," *Nucl. Phys.* **A163,** 583–591.
R. Repnow, V. Metag, J. D. Fox, and P. VonBrentano (1970), "Evidence for a Direct Reaction Mechanism in the Production of Fission Isomers," *Nucl. Phys.* **A147,** 183–192.
R. Repnow, V. Metag, and P. VonBrentano (1971), "Fission Isomers in Cm and Bk Isotopes," *Z. Physik* **243,** 418–430.
J. H. Roberts, R. Gold, and R. J. Armani (1968), "Spontaneous-Fission Decay Constant of U^{238}," *Phys. Rev.* **174,** 1482–1484.
F. H. Ruddy and J. M. Alexander (1969), "Fissile Nuclei of Medium Mass with Nanosecond Lifetimes," *Phys. Rev.* **187,** 1672–1679.
F. H. Ruddy and J. M. Alexander (1972), "Long-Range Fragments from Fission of U^{236}," *Phys. Rev.* **C5,** 549–551.
F. H. Ruddy, M. N. Namboodiri, and J. M. Alexander (1971), "Limiting Cross

Sections for Fission Isomers with Nanosecond Lifetimes Produced in Heavy-Ion Reactions," *Phys. Rev.* **C3,** 972–974.

H. W. Schmitt and U. Mosel (1972), "Fission Properties of Heavy and Superheavy Nuclei," *Nucl. Phys.* **A186,** 1–14.

E. K. Shirk, P. B. Price, E. J. Kobetich, W. Z. Osborne, L. S. Pinsky, R. D. Eandi, and R. B. Rushing (1973), "Charge and Energy Spectra of Trans-Iron Cosmic Rays," *Phys. Rev.* **D7,** 3220–3233.

D. Shirley (1966), "Thermal Equilibrium Nuclear Orientation," *Ann. Rev. Nucl. Sci.* **16,** 89–118.

D. L. Shpak and G. N. Smirenkin (1968), "Angular Anisotropy of Fission of Pu^{238} by Neutrons," *JETP Letters* **8,** 332–334.

D. L. Shpak and G. N. Smirenkin (1969), "Angular Anisotropy of Am^{241} Neutron-Fission Fragments," *JETP Letters* **9,** 114–116.

D. L. Shpak, B. I. Fursov, and G. N. Smirenkin (1971a), "Angular Anisotropy of Neutron-Induced Np^{237} and Am^{241} Fission," *Sov. J. Nucl. Phys.* **12,** 19–21.

D. L. Shpak, Yu. B. Ostapenko, and G. N. Smirenkin (1971b), "Angular Anisotropy of Nucleon Pairing Effects in the Fission of Pu^{239}," *Sov. J. Nucl. Phys.* **13,** 547–553.

D. L. Shpak, A. I. Blokhin, Yu. B. Ostapenko, and G. N. Smirenkin (1972), "Angular Anistropy of the Fragments from Fission of Th^{232} by 13.40–14.80 MeV Neutrons," *JETP Letters* **15,** 228–229.

N. A. Skakun and N. P. Dikii (1972), "Observation of Shadows in the Angular Distributions of Fragments from Photofission of Tantalum, and Evaluation of the Fission Lifetime," *Sov. J. Nucl. Phys.* **15,** 341–342.

G. N. Smirenkin, V. G. Nesterov, and A. S. Tishin (1968), "Angular Anistropy and Nucleon Pair Correlation Effects in Nuclear Fission," *Sov. J. Nucl. Phys.* **6,** 671–676.

A. S. Soldatov, G. N. Smirenkin, S. P. Kapitza, and Yu. M. Tsipeniuk (1965a), "Quadrupole Fission of ^{238}U," *Phys. Letters* **14,** 217–218.

A. S. Soldatov, Z. A. Aleksandrova, L. D. Gordeeva, and G. N. Smirenkin (1965b), "Angular Distribution of Fragments in Photofission of ^{238}U and ^{232}Th by γ Rays from the Reaction $^{19}F(p, \alpha\gamma)^{16}O$," *Sov. J. Nucl. Phys.* **1,** 335.

G. Somogyi and B. Schlenk (1970), "The Application of Solid-State Track Detectors for Measuring Alpha-Particle Angular Distributions in Nuclear Reactions," *Rad. Effects* **5,** 61–68.

M. Sowinski, C. Stephan, T. Szyzewski, and J. Tys (1971), "Search for Superheavy Elements in Nature by Neutron-Induced Fission of Mass-Separated Nuclei," *Proc. Inter. Conf. Heavy Ion Physics*, Yu. P. Gangrskii (ed.), 79–84, Dubna.

R. A. Stern and P. B. Price (1972), "Charge and Energy Information from Heavy Ion Tracks in 'Lexan'," *Nature Phys. Sci.* **240,** 83–85.

V. M. Strutinsky (1966), "Influence of Nucleon Shells on the Energy of a Nucleus," *Sov. J. Nucl. Phys.* **3,** 449–457.

J. D. Sullivan, P. B. Price, H. J. Crawford, and M. Whitehead (1973), "Fragments with $6 \leq Z \leq 15$ Emitted from Au Bombarded with 2.1 GeV/Nucleon ^{16}O Ions and 2.1 GeV Protons," *Phys. Rev. Letters* **30,** 136–138.
M. Suzuki (1973), "Chronology of Prehistoric Human Activity in Kanto, Japan, Part I—Framework for Reconstructing Prehistoric Human Activity in Obsidian," *J. Fac. Sci.*, Univ. of Tokyo, Sec. V, **3,** 241–318.
W. J. Swiatecki (1963), *Proc. Second Inter. Conf. on Nuclidic Masses*, Springer-Verlag, Vienna, 58.
J. Szabo, J. Csikai, and M. Varnagy (1972), "Low-Energy Cross Sections for $^{10}B(p, \alpha)^{7}Be$," *Nucl. Phys.* **A195,** 527–533.
S. G. Thompson (1967), "Some Aspects of the Study of Fission," *Arkiv for Fysik* **36,** 267.
S. G. Thompson (1968), "The Search for Element 114," *Bull. Am. Phys. Soc.* **13,** 1442.
A. F. Tulinov (1965), "On an Effect Accompanying Nuclear Reactions in Single Crystals and Its Use in Various Physical Investigations," *Sov. Phys. Doklady* **10,** 463–465.
J. P. Unik, E. P. Horwitz, K. L. Wolf, I. Ahmad, S. Fried, D. Cohen, P. R. Fields, C. A. A. Bloomquist, and D. J. Henderson (1972), "Production of Actinides and the Search for Super-Heavy Elements Using Secondary Reactions Induced by GeV Protons," *Nucl. Phys.* **A191,** 233–244.
R. Vandenbosch and J. R. Huizenga (1973), *Nuclear Fission.* New York: Academic Press.
R. Vandenbosch, P. A. Russo, G. Sletten, and M. Mehta (1973), "Relative Excitations of the ^{237}Pu Shape Isomers," *Phys. Rev.* **C8,** 1080–1083.
G. A. Vartapetyan, N. A. Demekhina, V. I. Kasilov, Yu. N. Ranyuk, P. V. Sorokin, and A. G. Khudaverdyan (1972), "Photofission Cross Sections Up to 5 BeV. Super-giant Resonance in Photonuclear Reactions," *Sov. J. Nucl. Phys.* **14,** 37–40.
P. E. Vorotnikov, S. M. Dubrovina, G. A. Otroshchenko, and V. A. Shigin (1967a), "Channel Analysis of the Fission of Th^{230} by Neutrons," *Sov. J. Nucl. Phys.* **5,** 207–216.
P. E. Vorotnikov, S. M. Dubrovina, G. A. Otroshchenko, and V. A. Shigin (1967b), "Angular Distribution of Fragments from Fission of Am^{241} by Neutrons," *Sov. Phys. Doklady* **11,** 591–592.
P. E. Vorotnikov, S. M. Dubrovina, V. N. Kosyakov, L. V. Chistyakov, V. A. Shigin, and V. M. Shubko (1970a), "Angular Distributions of Fission Fragments and the Cross Section for Neutron-Induced ^{249}Bk Fission," *Nucl. Phys.* **A150,** 56–60.
P. E. Vorotnikov, S. M. Dubrovina, G. A. Otroshchenko, and V. A. Shigin (1970b), "Angular Distributions of Fission Fragments in the Reaction Pa^{231} (n, f)," *Sov. J. Nucl. Phys.* **10,** 280–281.

P. E. Vorotnikov, S. M. Dubrovina, G. A. Otroshchenko, L. V. Chistyakov, V. A. Shigin, and V. M. Shubko (1970c), "Cross-Section for Neutron Fission of Bk^{249}," *Sov. J. Nucl. Phys.* **10,** 419–421.

P. E. Vorotnikov, S. M. Dubrovina, G. A. Otroshchenko, V. A. Shigin, A. V. Davydov, and E. S. Pal'shin (1971), "Fission of U^{232} by 0.1–1.5 MeV Neutrons," *Sov. J. Nucl. Phys.* **12,** 259–260.

P. E. Vorotnikov, B. M. Gokhberg, S. M. Dubrovina, V. N. Kosyakov, G. A. Otroshchenko, L. V. Chistyakov, V. A. Shigin, and V. M. Shubko (1972), "Cf^{249} Fission Cross Section in the Neutron Energy Range 0.16–1.6 MeV," *Sov. J. Nucl. Phys.* **15,** 20–21.

L. Westgaard, B. R. Erdal, P. G. Hansen, E. Kugler, G. Sletten, and S. Sundell (1972), "Search for Super-Heavy Elements Produced by Secondary Reactions in Uranium," *Nucl. Phys.* **A192,** 517–523.

G. Yuen, G. T. Rizzo, A. N. Behkami, and J. R. Huizenga (1971), "Fragment Angular Distributions of the Neutron-Induced Fission of ^{230}Th," *Nucl. Phys.* **A171,** 614–624.

A. M. Zebelman and J. M. Miller (1973), "Role of Angular Momentum in Complete Fusion Reactions $B^{11} + Tb^{159}$, $C^{12} + Gd^{158}$, $O^{16} + Sm^{154}$," *Phys. Rev. Letters* **30,** 27–30.

E. A. Zhagrov, Yu. A. Nemilov, and Yu. A. Selitskii (1968), "Yield and Angular Anistropy of Ra^{226} Photofission Fragments," *Sov. J. Nucl. Phys.* **7,** 183–185.

E. A. Zhagrov, Yu. A. Nemilov, N. V. Nikitina, and Yu. A. Selitskii (1971), "Photofission Cross Section of Ra^{226}," *Sov. J. Nucl. Phys.* **13,** 537–539.

V. Zoran and Gh. Popescu (1971), "On the Use of Solid State Track Detectors in Nuclear Reaction Studies," *Stud. Cercet. Fiz.*, Rumania, **23,** 1127–1130.

I. Zvara, Yu. T. Chuburkov, R. Tsaletka, T. S. Zvarova, M. R. Shalayevsky, and B. V. Shilov (1966), "Chemical Properties of Element 104," *Sov. J. At. Energy* **21,** 709.

I. Zvara, Yu. T. Chuburkov, V. Z. Belov, G. V. Buklanov, B. B. Zakhvatoyev, T. S. Zvarova, O. D. Maslov, R. Tsaletka, and M. R. Shalayevsky (1970), "System for Chemical Separation of Khurchatovium," *J. Inorg. Nucl. Chem.* **32,** 1885.

I. Zvara, V. Z. Belov, L. P. Chelnokov, V. P. Domanov, M. Hussonois, Yu. S. Korotkin, V. A. Schegolev, and M. R. Shalayevsky (1971), "Chemical Separation of Khurchatovium," *Inorg. Nucl. Chem. Letters* **7,** 1109–1116.

Chapter 8

Element Mapping and Isotopic Analysis

8.1. INTRODUCTION

Solid state track detectors have unique capabilities for measuring the concentration and spatial distribution of certain elements. In principle any isotope is capable of being studied if it emits heavy nuclear particles, either directly because of its natural radioactivity, or, more important, as a result of specific nuclear reactions when bombarded in an accelerator or nuclear reactor. Certain isotopes with large cross sections for specific reactions are, of course, more suitable than others. This area of research naturally began with the measurement of uranium via the detection of fission fragments in samples irradiated with thermal neutrons (Price and Walker, 1963). As time has gone on, however, more and more nuclides have been studied and the number of diverse problems that can be approached by track methods is enlarging rapidly.

In this chapter we present the basic techniques of element mapping and give a summary of various applications. Here we deal primarily with natural systems; in Chapter 10 we give additional results on problems of a more technological nature. The literature review for this chapter was completed in early 1973; however, references to several important articles that appeared during the latter part of 1973 and early in 1974 have also been included.

We begin with a discussion of the measurement of uranium in different types of samples. The problem of bulk measurements is treated first, following which we illustrate the methods used for determining the spatial distribution of uranium with a resolution of several microns. This ability to "micromap" isotopes is one of the most unusual and important features of track detectors. Following a brief discussion of other possibilities involving fission fragments we turn to the problem of nuclides that emit lighter fragments such as α-particles and protons. Although relatively little work has been done in this area, it appears to be one of the most interesting for future studies, particularly for biological applications.

8.2. BULK DETERMINATIONS OF URANIUM

8.2.1. External Detector Method

This method, which is suitable for homogeneous solids with a concentration of uranium $\gtrsim 50$ ng/g, consists of placing a sample whose uranium concentration is to be measured in intimate contact with a uranium-poor track detector. A standard glass sample of known uranium concentration is also included. To avoid problems due to any spatial inhomogeneities of the neutron flux, the standard glass is usually placed on the opposite side of the same track detector used to measure the unknown. Following neutron irradiation, the detector is etched and scanned. In the event that the unknown has nearly the same elemental composition as the standard, the uranium concentration expressed in weight fraction in the unknown is given by

$$C_x(\mathrm{U}) = \left(\frac{I_s}{I_x}\right) \frac{\rho_x}{\rho_s} C_s(\mathrm{U}) \tag{8-1}$$

where the subscripts s and x refer to the standard and the unknown, respectively, I is the ratio of ^{235}U to ^{238}U, and ρ is the density of induced fission tracks.

An important point that is sometimes ignored arises when the unknown has an elemental composition that is appreciably different from that of the standard glass. In this case, the concentration expressed in weight fraction is given by

$$C_x(\mathrm{U}) = C_s(\mathrm{U}) \frac{I_s}{I_x} \frac{\rho_x}{\rho_s} \frac{R_s}{R_x} \tag{8-2}$$

where R represents the effective range of fission fragments in mg/cm^2. This quantity increases with increasing Z because of the tighter bonding of atomic electrons; neglect of this term can cause gross errors in the uranium determination. In Fig. 8-1 we show values of R_s/R_z as a function of Z using the experimental data of Abdullaev et al. (1967) and Mory et al. (1970). Below $Z \sim 50$ the two sets of measurements show a monotonic increase of R with Z (with the exception of one point for Cu) and agree with each other to about 5%, which is the stated precision of the measurements. Above $Z \sim 50$ the results of Mory et al. show a great deal of scatter and diverge from those of Abdullaev et al. In this region the absolute determination of U is probably no better than $\pm 20\%$ at the present time. Standards of high Z composition would improve the ability to measure absolute values. Relative values can of course be measured to any desired precision.

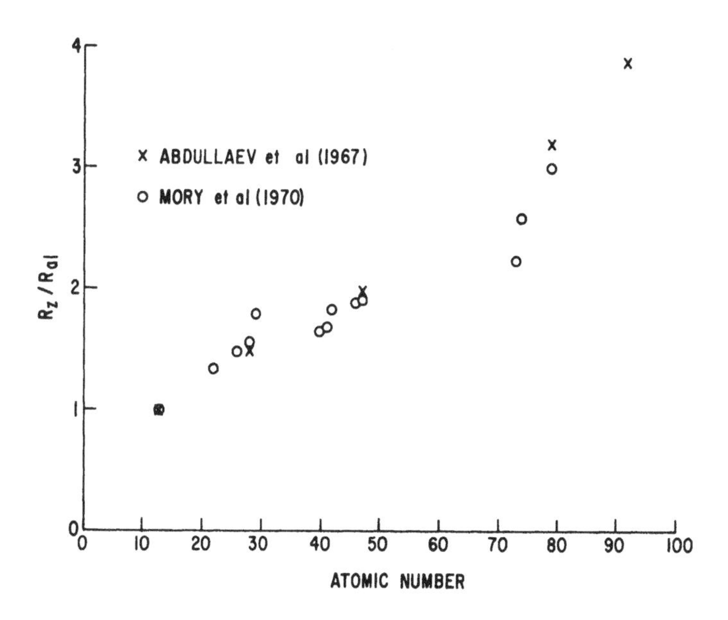

Fig. 8-1. *Relative ranges (expressed in gm/cm²) for average fission fragments in absorbers of different atomic number.*

An additional complication can arise if the unknown is crystalline rather than amorphous. In this case it is possible to get anomalously high transmission along certain crystal axes (Mory, 1969; also see Fig. 10-17) or along grain boundaries (see Fig. 10-16). Although these effects have not been studied in most materials, they are expected to contribute $<10\%$ error to the absolute uranium determination. The foils used by Mory et al. (1970) were cold-worked to minimize these anomalous transmission effects, but it is possible that some of the scatter seen in Fig. 8-1 is caused this way. Shell effects such as described by Chesire and Poate (1970) may also play a role.

The detector should be thin to minimize the effects of the spatial variation of neutrons in the reactor and must have a uranium concentration much lower than that of the standard glass or the sample. Both sample and standard should be thicker than the range of fission fragments in the unknown but should otherwise be as thin as possible to avoid self-absorption effects.

High-grade muscovite mica, fused or crystalline quartz, and Lexan plastics are all suitable detectors—the final choice being somewhat a matter of personal taste. Although, of this group, the concentration of uranium is the highest in muscovite (typically $\sim$5 ng/g), this material has the advantage that the tracks can be easily and reliably counted both at very high and very low densities. The best technique is to etch the mica for a time appropriate to the density to be measured ($\lesssim$10 min for $\gtrsim 10^6$ t/cm² and up to 2 hours for very low densities) and then to coat the sur-

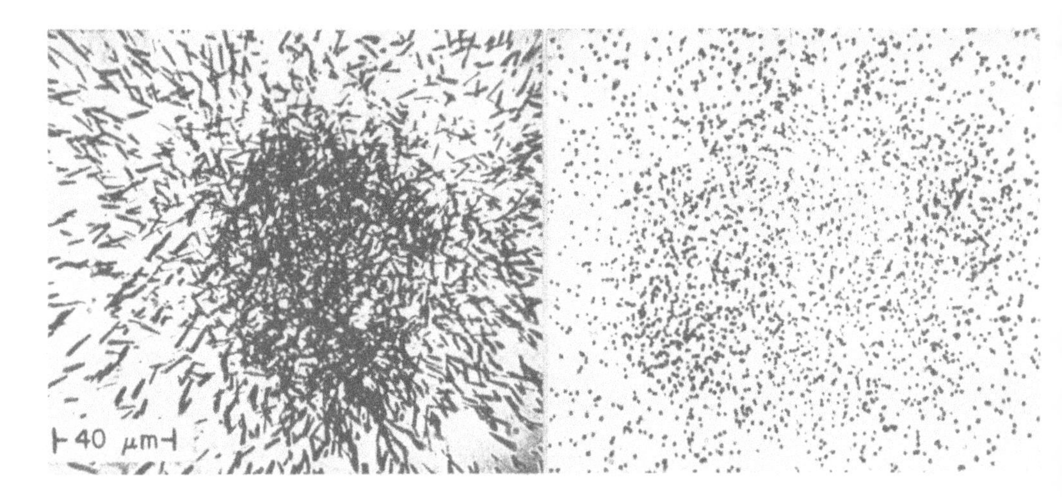

Fig. 8-2. *Optical counting of fission tracks. The photograph on the left shows a fission star in a mica detector placed in contact with a zircon grain. Counting of individual tracks is clearly very difficult. The right-hand photo shows a silvered surface of the same detector in reflected light. Counting of individual tracks is clearly simplified.* (*Photograph courtesy of D. Zimmerman.*)

face with a reflecting material such as Al or Ag. Track counts made with reflected light will give highly reproducible results with $>95\%$ of all tracks present counted (Seitz et al., 1973) (see Fig. 8-2). The mica should be annealed to remove fossil tracks and then be freshly cleaved just prior to mounting to minimize contamination.

Lexan has a much lower concentration of uranium ($<10^{-10}$ atom fraction) and is also suitable. However, special care must be taken to avoid overheating of the sample during irradiation, or track fading may lead to spurious results. The plastic also is affected by irradiation and at a dose much beyond 2×10^{17} n/cm^2 may become discolored and brittle. This apparently is a function of the temperature and, possibly, the proportion of fast neutrons. With care, 10^{18} n/cm^2 is a feasible dose. Special care also must be taken to clean the Lexan properly, as it has a tendency to gather dust electrostatically. Plastics have the great advantage that they can be counted automatically by the sparking method (see Chapter 2). In this case two detector sheets, one for the sample and one for the standard glass, must be used.

Fused quartz and many crystalline detectors (e.g., quartz, feldspar, olivine) generally contain very little uranium and are also satisfactory detectors. However, fused quartz detectors do not cleave and must be carefully cut and polished. They are also expensive. Commercial, optically polished, fused quartz plates frequently show extensive microcracks after etching and must be repolished before being used. Because contamination problems generally limit the applicability of the external detector method to samples with a uranium concentration >50 ng/g, the use of quartz detectors is seldom warranted.

Whatever the detector, it is practically mandatory that all mountings be made in a dust-free environment such as a clean room or a laminar flow clean hood. The number of uranium-rich particles in ordinary dust is quite astonishing. Dusting of surfaces can be done conveniently using commercial pressurized freon. With reasonable care it is possible to obtain clean detectors that correspond to uranium concentrations of $\lesssim 5$ ng/g. Fused quartz is particularly useful in establishing the blank levels. At the 5 ng/g level we have found that ordinary de-ionized water, typical laboratory detergents, and reagent grade alcohol and acetone are all satisfactory cleaning agents. *Decontaminant soaps, however, contain large quantities of uranium and should never be used.*

So far we have discussed the problem of uranium determination in a homogeneous solid material. In a variety of cases, ranging from man-made pottery to lunar rocks, uranium is *not* distributed uniformly. Although the distribution patterns may themselves be of interest (this is the subject of Section 8.3), this heterogeneity adds special complications in determining a bulk, average uranium concentration.

There are essentially three approaches to measuring the average uranium concentration of a heterogeneous sample. The first two consist of removing the heterogeneities either by dissolving the sample and measuring the uranium concentration in the dried solute product (Fleischer and Lovett, 1968) or crushing the sample to a scale that is much finer than the scale of the heterogeneities (Fisher, 1970b). A third approach (Geisler et al., 1974a) is shown schematically in Fig. 8-3. In this method the detector is moved away from the sample by a fixed distance and the irradiations are performed in vacuum. The separation has the effect of fanning out the fission "stars" into a uniform pattern on the detector. The detailed discussion of this method given by Geisler et al. shows that it is possible to obtain 5% accuracy in highly heterogeneous lunar samples as small as 15 mg in mass. Note again, however, that a large enough area must be sampled that a representative number of high-uranium regions will be exposed.

If the sample to be studied contains the uranium in solution or in suspension, as for example in blood or sea water, an aliquot can be dried on a slide or other suitable sample holder to give a solid target. In this case care must be taken to see that the uranium distribution is uniform and not clumped, either in the form of discrete particles or around the edges of the original drop. A wetting agent can be useful in producing a uniform sample. The thickness of the film, T, must also be monitored to see that it falls either in the "thin" or "thick" category, $T \ll R_x$ or $T \gg R_x$. The intermediate case is to be avoided since this introduces additional corrections in arriving at a final uranium value. Direct measurement of the uranium in a liquid using a sealed glass, quartz, or plastic vial would also be feasible but, to our knowledge, has not been tried.

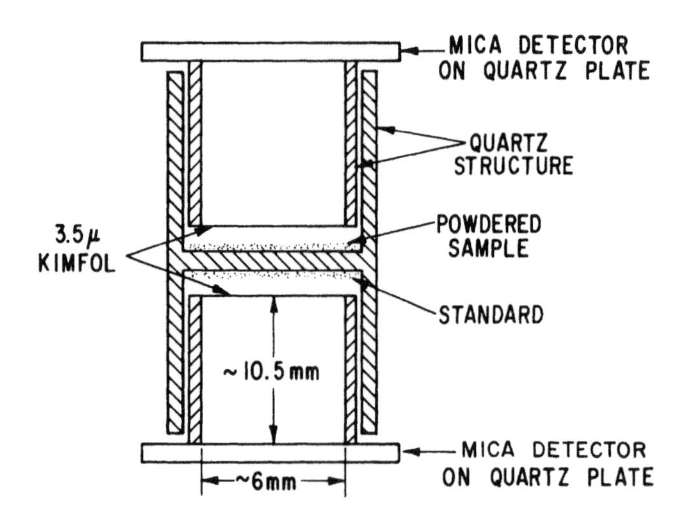

Fig. 8-3. *Schematic diagram of cell used for determination of average uranium concentration in samples with a highly heterogeneous distribution of uranium. Typically a number of such cells are mounted in an evacuated quartz cylinder for neutron irradiation.* (*After Geisler et al., 1974a.*)

In cases where the sample exists in powdered or liquid form, a completely different approach to the U comparison standard is possible. Murali et al. (1970), Yabuki (1971), and Carpenter (1972a) have used the method of standard addition. In this technique known amounts of different uranium standards are added to aliquots of the sample. The track densities are measured for each case and the results extrapolated to zero addition to get the uranium content of the sample. Although this method requires more measurements, it avoids the problem of glass standards and also gives a good measure of the reproducibility of the sample preparation procedures.

In addition to the measurements of uranium made on samples that have been dated by the fission track technique (listed in Chapter 4), track techniques have been used to measure bulk uranium concentrations in a wide variety of substances. These include air contaminants (Hamilton, 1970a), human blood (Carpenter and Cheek, 1970; Hamilton, 1970b; Wiechen, 1971), seaweed and other plants (Abdullaev et al., 1968; Su, 1972), archaeological and recent glasses (Fleischer and Price, 1964; Schöner and Herr, 1969; Fleischer et al., 1971; Carpenter, 1972b), Baltic amber (Uzgiris and Fleischer, 1971), petroleum (Berzina et al., 1969), manganese nodules from oceans (Yabuki and Shima, 1971), sea water and fresh water (Fleischer and Lovett, 1968; Bertine et al., 1970; Dörschel and Stolz, 1970; Hashimoto, 1971b), ancient and modern bones (Fleischer et al., 1965b; Otgonsuren et al., 1970), tektites (Durrani and Khan, 1971), meteorites (Fisher 1969a,b; 1972a; 1973), lunar samples (Graf et al., 1973), ocean sediments (Lahoud et al.,

1966; Fisher and Bostrom, 1969; Haglund et al., 1969; Bertine et al., 1970; Bonatti et al., 1971; Bostrom and Fisher, 1971; Rydell and Fisher, 1971), and terrestrial rocks (Lobanov et al., 1967; Wakita et al., 1967; Akimov et al., 1968; Komarov and Skovorodkin, 1969; Chalov et al., 1970; Fisher, 1970a; 1972b; Hashimoto et al., 1970; Murali et al., 1970; Nishimura, 1970; Aumento 1971; Aumento and Hyndman, 1971).

The above list includes only measurements of *average* uranium and not measurements made of either the *distribution* of uranium in complex samples or in separated phases from complex systems. These will be discussed in the next section.

8.2.2. Internal Detector Method

Much lower levels of uranium concentration can be measured, with no problem whatsoever of contamination, in any substance that itself registers tracks. In the internal detector method the unknown may have to be first heated to remove fossil tracks (because of the possibility of redistributing uranium this step is undesirable if the fossil density is low compared with the expected induced track density) and then irradiated along with a standard glass sample. After the sample is returned it is either cleaved or polished to expose an internal surface that was at a distance greater than R_x from any external surface during irradiation. The track density in the sample is then given by $\rho = N_U \sigma I \phi \eta R_x$ where ϕ is the neutron flux determined from the measurements on the glass standard, N_U is the number of atoms of uranium per cm^3, and η is the etching efficiency. Expressed this way it is important to note that the constant relating track density to neutron flux ($N_U I \sigma \eta R_x$) must be independently determined for the standard glass. As discussed in Chapter 4, this has been done for at least three sets of glasses that are generally available to investigators.

The absolute accuracy of this method of uranium determination is proportional to the degree to which the value of $R_x \eta$ is established for the unknown. This can be determined in a separate experiment using a thin, calibrated ^{252}Cf source to determine track densities, track lengths, cone angles, etc.

In practice it is possible to obtain total neutron doses $>10^{20}/cm^2$. The sensitivity for uranium detection at this dose is truly extraordinary. Setting a limit of 10^2 t/cm^2 corresponds to a detectable U concentration in track-recording materials of 5×10^{-15} atom fraction. At all doses great care should be taken to avoid materials that become too radioactive (graphite and quartz are convenient structural materials). Although epoxies used to hold polished rock sections do not become excessively radioactive, they physically deteriorate and may buckle or break at doses much greater than $\sim 10^{18}/cm^2$. Chlorine-containing materials such as polyvinyl chloride (PVC) and Saran wrap should be avoided; they become very radioactive and also release corrosive gases.

8.3. MICROMAPPING OF FISSIONABLE MATERIALS WITH THERMAL NEUTRONS

Although it may be useful to measure the average uranium content of a sample, it is frequently much more important to measure the detailed spatial distribution of uranium. For example, rocks of both terrestrial and extraterrestrial origin are generally extremely heterogeneous on the microscale, with the uranium being concentrated in small, accessory or trace-mineral phases. Using dielectric track detectors it is easy to measure the uranium distribution on a micron scale. In this section we describe the techniques that have been employed and some of the more interesting results that have been obtained.

8.3.1. Techniques for Uranium Mapping

All the techniques for micromapping start with a track detector pressed tightly against a flat surface that is to be studied. In the *in situ* technique developed by Fleischer (1966) the detector is a sheet of thin plastic whose thickness is less than the range of fission fragments. Provided the edges of the sandwich are properly sealed, the plastic can be etched directly in place to reveal tracks. If the specimen is transparent the sandwich is transferred directly to a microscope where the uranium-rich regions can be revealed by focussing on the plastic, and the regions themselves studied on the polished section by focussing down to the plane of the section.

The thin plastic covering is fixed in place using a combination of thermal-pressure casting and solvent casting. In the thermal-pressure casting step, Makrofol polycarbonate film 8×10^{-4} cm thick is pressed against the surface of interest and the assembly is heated to a temperature of 190°C, where the film becomes softened. The pressure produces intimate contact, and upon cooling a tight bond is produced. A nonsticking material (such as H film) is used as the surface of the pressure transmitting medium. The solvent casting stage involves coating around the thermal-pressure cast area so as to prevent the removal of the thin film during the chemical treatment. For small samples the entire surface area not already covered with the thin film may be coated with the plastic track detector by merely brushing on a solution of plastic in a volatile solvent and then allowing the solvent to evaporate. For a large flat surface it may merely be necessary to coat an area around the thin film and then expose only the thin film area to the chemical treatment—that is, allow no edges of the coated area to be given the chemical treatment. When practical, the coating of the entire sample is the more reliable method. Lexan polycarbonate, dissolved in methylene chloride, can be used for solvent casting.

Although this method is conceptually attractive, practical problems have limited its application. The bonding of the thin plastic to the section can be difficult, and etchant itself tends to leak onto the material being studied and may destroy or obscure the features that one wants to study. The plastic also forms a barrier between the observer and the section, making it difficult to apply optical petrographic techniques or to make microprobe measurements of the uranium-rich regions. (The technique does have the virtue, however, of being suitable for tracing back individual induced tracks to their origins.)

For these reasons most micromapping has been done by separating the detector from the material after irradiation and using it as a map to guide the observer to interesting regions on the unknown. These can then be studied by either optical or microprobe techniques. Used this way, the utility of the fission map depends on the accuracy and ease with which regions on the section being studied can be referenced to the fission map.

As shown first by Kleeman and Lovering (1967a), the simplest method of referencing is to *look* at the detector. If Lexan is used as the detecting material, many rock sections will give a "print" that closely resembles a photograph of the rock section. Such a print is shown in Fig. 8-4, taken from Kleeman and Lovering

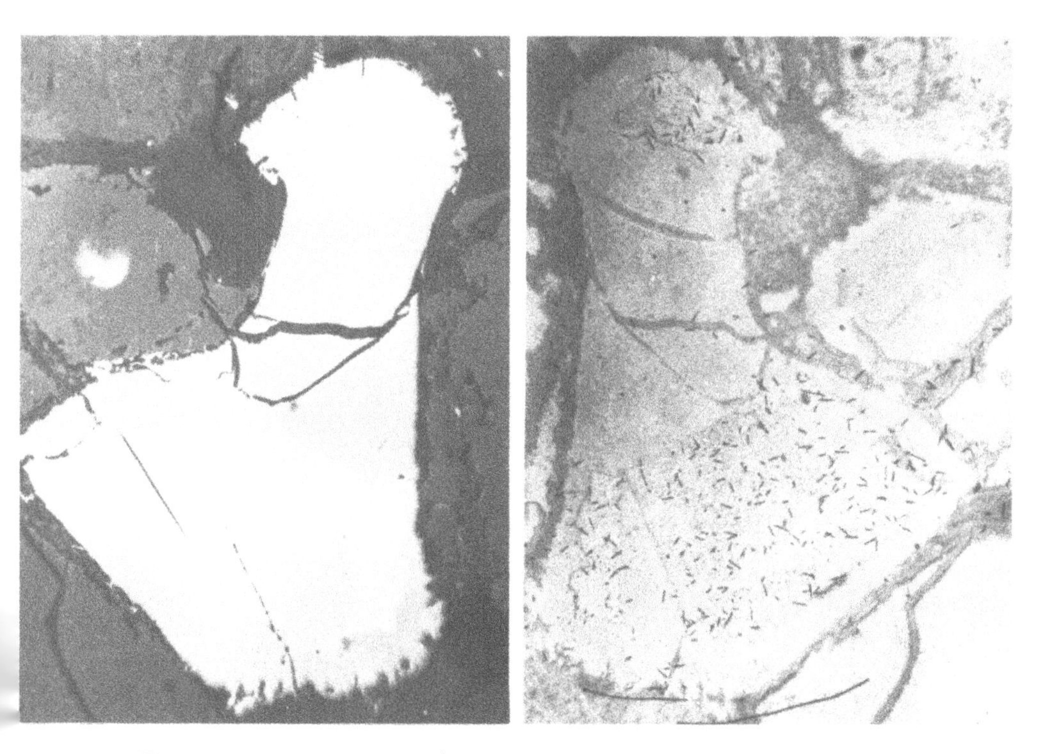

Fig. 8-4. *Optical photograph (left) and Lexan print (right) of a terrestrial rock. Variations in the unresolvably high density of induced alpha tracks in different minerals and in grain boundaries provide an easily recognizable image of the rock. The fission track distribution shows that the light phase, rutile, contains 1.1 parts per million of uranium. (After Kleeman and Lovering, 1967b.)*

(1967b). The remarkable (and unexpected) fact that an actual image of the rock section is produced on the plastic is due to the contrast provided by myriads of very tiny pits that are produced in differing amounts in the vicinity of grain boundaries and in the interior regions of different mineral phases. Although the details of the production of track contrast images probably differ from one rock to another, the dominant source appears to be the production of α-particle pits from (n, α) reactions on B and/or Li (Kleeman and Lovering, 1969). Both these elements have very high neutron cross sections and apparently tend to segregate differently among the various mineral constituents of rocks. Although B and Li play a role in giving prints in terrestrial rocks, good prints may also be obtained from fission tracks alone as shown in Fig. 8-5.

The Lexan print method, while elegant, is not always satisfactory. If there is a wide range of uranium concentrations, fission track counting becomes very difficult in high-density regions; there is the additional constraint of insuring that the temperature during irradiation does not exceed the track-fading temperature. It is also true that not all materials that one wishes to study give good Lexan prints. Finally, as we have pointed out in the discussion of fission track dating of inclusions, it may be necessary to have an external detector of the same composition as the minerals being dated.

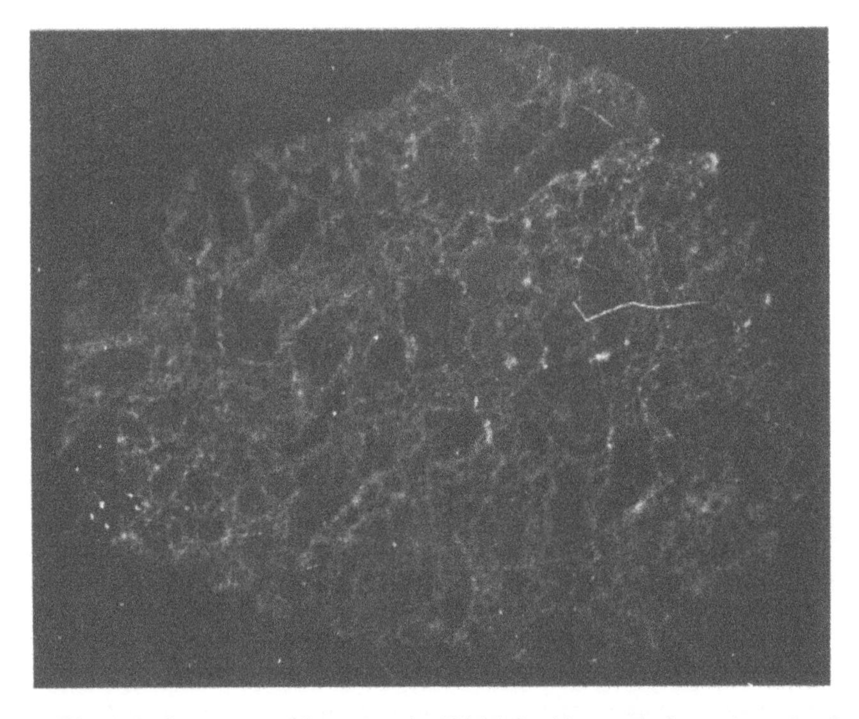

Fig. 8-5. *Lexan map of lunar breccia 66055. In this case the image is produced by the varying concentrations of fission tracks arising from the heterogeneous distribution of uranium in the sample.* (*After Macdougall et al., 1973.*)

Several techniques have been found satisfactory for referencing a fission map to a rock section with a precision of ± 10 μm. In one variant (Burnett et al., 1970), a mica sheet is firmly clamped onto a rock section using either an aluminum or graphite ring. The flatness of the mica against the section can be judged by looking at interference rings that show up under normal illumination. The surface of the mica is then scratched in an irregular grid pattern with a needle. Following this, the rock section with its overlay of gridded mica is photographed at low magnification. When fission stars are found at particular locations on the mica, this composite photograph can be used to find the appropriate region on the rock section (see Fig. 8-6). A minor variant of this basic method is to lay a photograph of the mica fission map directly on a photograph of the rock section and mark the position of interesting locations on the rock section by inserting pins at the appropriate locations. This general mapping method has also been used with feldspar detectors. An advantage of this method is the ability to measure a wide variety of uranium concentrations in a single irradiation. Fig. 8-7, for example, shows a densely populated fission pattern that is still resolvable in the mica using a scanning electron microscope.

In another application of this same basic technique, Haines (1972) has outlined a method for establishing congruent coordinate systems respectively on the rock section and on the mirror-image fission map. Once the x, y coordinates of an interesting point have been measured on the fission map, a simple transformation

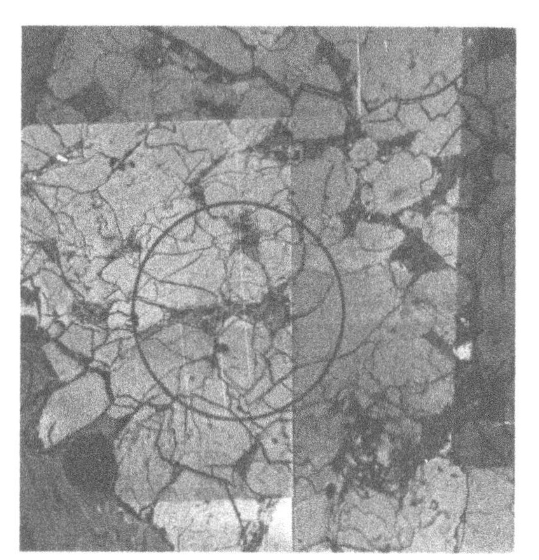

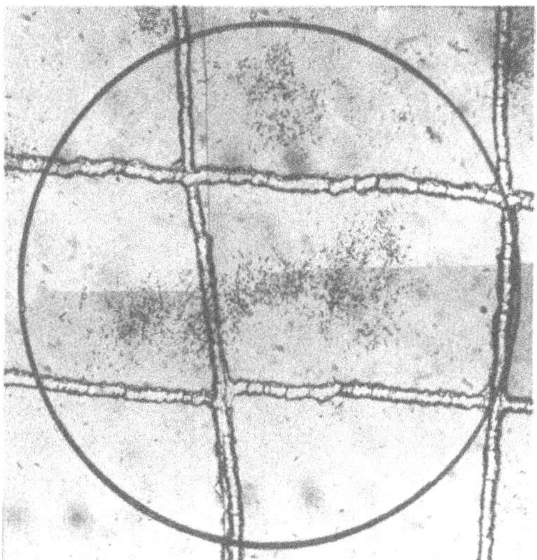

Fig. 8-6. *Mica fission map of the meteorite Nakhla. In this technique, an additional photograph (not shown) is taken of the rock section with the mica detector in place. The (nonuniform) fiducial grid system on the mica makes it simple to locate equivalent regions on the rock section following demounting and etching. (After Crozaz et al., 1974.)*

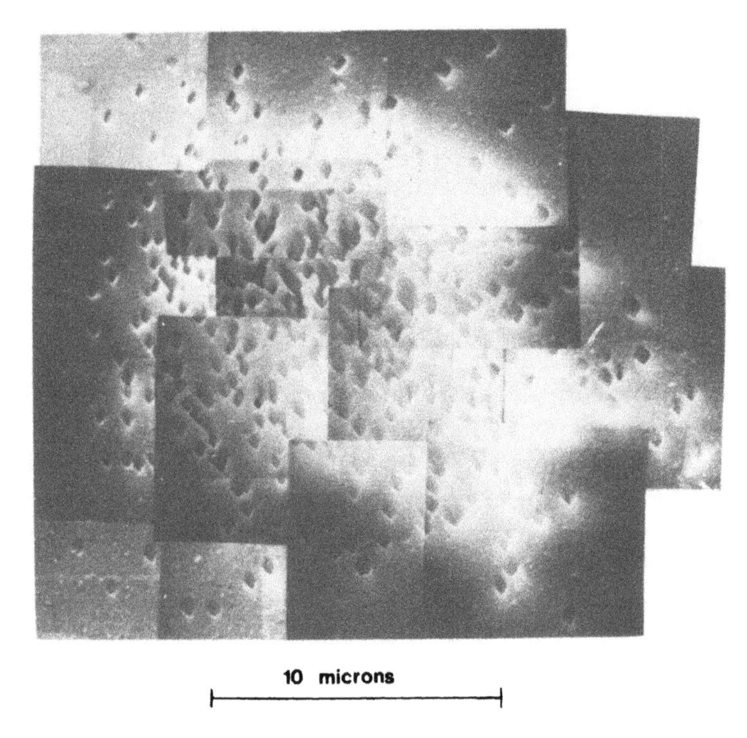

Fig. 8-7. *A densely populated fission star in mica as seen in a scanning electron microscope.*

gives the equivalent x, y coordinates of the equivalent point on the rock section. This procedure is particularly useful if one has a computer-controlled mechanical stage for the rock section. Given a sequence of x, y map points, the controller can be programmed to automatically put the equivalent points of the rock section in the field of view of a viewing instrument (in Haines' case an electron microprobe). Because only a few such computer-controlled systems exist, this method has not seen general use.

Still another system, appropriate for optical microscope work, has been described by Dooley et al. (1970). These workers put the section to be studied and the fission map on a single mechanical stage. Separate microscopes are used to view each. Once the map and the section have been put into the proper relative orientations, fission-rich areas found in the mapping microscope will automatically appear on the rock section in the second microscope.

8.3.2. Applications of Uranium Mapping

TERRESTRIAL MATERIALS

Uranium mapping has been applied to a number of natural terrestrial samples. Much of the work has been done on igneous rocks or minerals (Hamilton, 1966;

Kleppe and Roger, 1966; Kleeman and Lovering, 1967a,b; Komarov et al., 1967; Komarov, 1968; Wollenberg and Smith, 1968; Berzina et al., 1969; 1970; 1971b,c; Condie et al., 1969; Kleeman et al., 1969; Komarov and Skovorodkin, 1969; Berzina and Kravchenko, 1970; 1971; Kleeman and Cooper, 1970; Aumento and Hyndman, 1971; Nishimura, 1972; Tezuka, 1972; Haines and Zartman, 1973; Macdougall, 1973; Seitz and Hart, 1973; Grauert et al., 1974). The basic goal of most of this work is to understand the geochemistry of uranium—its partitioning in various parent magma systems and its microscopic association with geochemically related elements, particularly potassium.

In general the uranium concentrations in the parent igneous minerals are found to be extremely low, posing a problem for the rather large heat flow characteristic of oceanic basins. Most of the uranium in the rocks appears to be in interstitial weathering products and thus has probably been taken up from the environment.

A systematic laboratory study of the partitioning of U and Th in high-pressure silicate melts, as well as the diffusion of these elements, has recently been undertaken by Seitz and his colleagues (Seitz and Shimizu, 1972; Seitz, 1973a,b; 1974b). This work has greatly aided the interpretation of results found on natural systems. For example, the partitioning of uranium between diopside and a melt does not seem to be appreciably dependent on the composition of the melt, the temperature or pressure of the system, or the concentration of uranium. Thus, conclusions concerning fractionation processes based on U and Th measurements (e.g., Seitz and Hart, 1973) are independent of the temperature and pressure at which fractionation occurred.

Uranium distributions have also been measured in a number of other natural systems including modern bones (Becker and Johnson, 1970; Schlenker and Oltman, 1973), ancient bones (Seitz, 1972; Seitz and Taylor, 1974), skeletal carbonates (Schroeder et al., 1970; Szabo et al., 1970; Amiel et al., 1972), skarn deposits (Dikov et al., 1971), micas (Hamilton, 1966; Komarov and Shukolyukov, 1966; Sandru and Danis, 1971), cave concretions (Bigazzi and Rinaldi, 1968), as well as other types of rocks (Berzina and Demidova, 1966; Berzina and Dolomanova, 1967; Berzina et al., 1967a,b,c; 1968; 1972; Alekseyev et al., 1969; Fielding, 1970; Mel'gunov and Varvarina, 1970; Pleskova and Zheleznova, 1970; Gavshin, 1972). Many of these latter studies have been concerned with the *exchange* of uranium between the environment and the substance being examined, e.g., the increase in uranium content of bones with time. A specific study of the exchange of uranium with Ca ions was reported by Thompson (1970), who studied the distribution of uranium in hydroxyapatite and showed that the uranium exchange was strongly favored at crystal end faces.

One interesting, recent application of uranium mapping has led to the suggestion of a new type of thermoluminescent dating of archaeological ceramics (Zimmerman, 1971). Thermoluminescent dating is based on the gradual filling of electron traps by internal ionizing radiation. When a test sample is heated, the traps are emptied and the electrons recombine, giving off visible light. The total light output, coupled with measurements of the natural dose rate and the radiation sensitivity, can be used to measure the last time a ceramic object was brought to high temperature. For most materials this represents the time that the ceramic material was fired.

Because the *average* radioactivity of most ceramics is not high, corrections for environmental radiation levels due to cosmic rays and differing burial conditions limit the accuracy of the method (Aitken et al., 1968). However, in a study of archaeological ceramics from a wide variety of sources, Zimmerman (1971) showed that the uranium in these materials is concentrated in small, trace-mineral phases such as zircon and apatite—much as in the case of igneous rocks (however, many of the brightest phosphors in a typical ceramic are notably deficient in uranium). Because of the high specific activity of the zircon and apatite inclusions, environmental effects become negligible and, in principle, thermoluminescence dates can be measured with higher accuracy on these crystals. The first successful application of this inclusion method of dating was recently reported by Zimmerman et al. (1974) in their study of the authenticity of a well-known bronze horse in the collection of the Metropolitan Museum of Art in New York (see Fig. 8-8). The thermoluminescence measurements on zircon crystals removed from the ceramic core proved that the horse, which some experts had claimed was a modern forgery, was indeed made in ancient times. Because the horse had been extensively x-rayed, the normal methods of thermoluminescence authentication were not applicable, and only the presence of uranium-rich inclusions, revealed by uranium mapping techniques, made the authentication possible.

Using a spark imaging technique, Malik et al. (1973) have further studied the correlation between thermoluminescence and the distribution of uranium in archaeological materials. Their work demonstrates again the heterogeneity of the uranium distribution in most materials and the general lack of correlation (with the exception of zircon and apatite) between thermoluminescence and uranium concentration.

In one of the earliest papers on fission track mapping, Malmon (1964) explored some of the possibilities for localization of uranium in biological material. Although he used a thin film rather than an etched dielectric as a fission fragment detector, the principles are the same. The ability to map an increasing number of isotopes, described later in this chapter, should lead to increased work on the localization of specific atoms or atom clusters in biological systems.

Extraterrestrial Materials

Uranium distributions have also been measured in extraterrestrial objects, first in meteorites and then in lunar samples. Initial studies of separated minerals from the Vaca Muerta meteorite by Fleischer et al. (1965a) showed a total variation in uranium concentration of 4×10^6 for different phases. Although additional

Fig. 8-8. *An antique bronze horse. The antiquity of this object, which had been the subject of considerable debate, was established by the inclusion method of thermoluminescence dating (see text). This method was established in part through the use of fission maps. (Photograph courtesy of the Metropolitan Museum of Art, New York.)*

work has since been done in separated mineral phases in connection with fission-track dating work and the measurement of ^{244}Pu (see Chapter 6), systematic uranium mapping in meteorites has been the subject of only a few papers. In one of the early ones, Hamilton (1966), using mica maps, showed that the uranium distribution in the St. Marks meteorite was highly heterogeneous but made no attempt to identify the phases. Using the *in situ* method of track mapping, Fleischer (1968) extended this type of work to a number of stone meteorites, again demonstrating the basic heterogeneity in all classes of meteorites studied. Most recently, Crozaz et al. (1974) used uranium micromaps in conjunction with an electron microprobe to identify the uranium-rich phases in some 15 stony and iron meteorites. With few exceptions the uranium was found to be localized in phosphate inclusions ranging in size from ~10 μm to ~300 μm and with uranium concentrations ranging from 0.2 μg/g to 10 μg/g.

Recently Seitz (1974d) and Seitz et al. (1974) have studied the partitioning of U, Th, and Pu in crystal melt systems, the crystals representing the major host minerals for actinide elements in meteorites. This work lends support to the idea that ^{244}Pu can be preferentially segregated in certain meteoritic crystals (see Chapter 6 for a fuller discussion of this point).

Even a cursory look at the first returned lunar samples showed that the uranium in all types of samples—crystalline rocks, breccias, and soils—was also concentrated in trace, accessory mineral phases. Since 1969, samples from all lunar missions have been extensively micromapped. The distributions have been studied partly for geochemical reasons, for example, to better understand Pb-U-Th ages of lunar materials, and partly to identify uranium-rich grains that could be "microdated" either with fission tracks or with electron and ion probes. The track-dating aspect is discussed more fully in the chapter on extraterrestrial materials.

Crozaz et al. (1970), Fleischer et al. (1970), and Lovering and Kleeman (1970) demonstrated the fundamental heterogeneity of uranium distribution but they did not fully characterize the mineral phases. Burnett et al. (1970) showed that phosphates were important in rock 12013, while Haines et al. (1971) emphasized the importance of Zr-Ti phases in this same rock. Subsequent work by a variety of investigators (Burnett et al., 1971; Lovering and Wark, 1971; Lovering et al., 1971; Rice and Bowie, 1971; Brown et al., 1972; Busche et al., 1972; Crozaz et al., 1972; Durrani and Khan, 1972; Fleischer and Hart, 1972; Glass et al., 1972; Haines et al., 1972; Hutcheon et al., 1972; Lovering et al., 1972; Peckett et al., 1972; Thiel et al., 1972; Graf et al., 1973; Macdougall et al., 1973) has now established the pattern of uranium distributions in lunar material.

In rocks and soils uranium is found concentrated in familiar phases, including phosphates (whitlockite and apatite), zircon, baddeleyite, and glasses. The concentrations are variable even within a given phase in a given rock (smaller grains tend to have higher uranium concentrations) but typically run from 10 μg/g to 1000 μg/g for phosphates, 100 μg/g to 2000 μg/g for zircons, 50 μg/g to 1000 μg/g for ZrO_2, and up to 15 μg/g in lunar glasses.

Although most of the uranium-rich minerals are similar to those found in terrestrial igneous rocks, there is a new class of minerals rich in both Zr and Ti that are unique to lunar samples. In some of these grains the uranium concentration has been found to be as high as 3% (Haines et al., 1971), making it possible to obtain approximate U-Pb ages directly by electron microprobe and ion microprobe analysis. Although various compositions have been reported and delineated by *ad hoc* names [Phase A (Ramdohr and El Goresy, 1970), Phase β (Haines et al., 1971), Phase B (Lovering and Wark, 1971), Phase X and Phase Y (Brown et al.,

1972; Peckett et al., 1972), "Zr-Ti-Fe" phases (Burnett et al., 1970; Rice and Bowie, 1971)], at least two well-defined minerals have been established. The first, having the composition $Fe_8^{2+}(Zr + Y)_2 \cdot Ti_3Si_3O_{24}$, is unique to lunar samples. Lovering et al. (1971) have proposed the name "tranquillityite," derived from the Sea of Tranquillity from which the first rocks found to contain the mineral were collected during the Apollo 11 mission. The name has been approved by the Commission on New Minerals and Mineral Names of the International Mineralogical Association. Another mineral of composition ($CaZrTiO_5$), analogous to the previously defined terrestrial mineral, zirkelite, has also been identified (Busche et al., 1972).

The mass of uranium accounted for by different phases is more difficult to measure and is variable from rock to rock. For example, in the crystalline rock 12040 from Apollo 12, Burnett et al. (1971) showed that 78% of the uranium resides in phosphate phases, while in the brecciated rock 12013 the contribution of the zirconium phases clearly dominates (Burnett et al., 1971; Haines et al., 1971).

The problem of determining the concentration of uranium in individual grains, and hence the mass balance of uranium, becomes difficult when the dimensions of the grains are comparable to the etchable range of fission fragments ($\sim$15 μm). This problem has been discussed by Crozaz et al. (1972).

8.3.3. Neutron Detection of Other Elements

Thermal neutron-induced fission can also be used to detect the presence of artificially produced heavy isotopes in addition to naturally occurring ^{235}U. In particular, several papers on the detection of Pu, mostly related to health, safety or biological applications, have appeared in recent years (Andersen, 1965; Brackenbush and Baumgartner, 1965; Bleany, 1969; Sakanoue et al., 1971).

8.4. FAST PARTICLE FISSION MAPPING OF HEAVY ELEMENTS

We pointed out in an early review paper (Fleischer et al., 1965c) that both bulk concentration measurements and micromapping techniques could be extended to heavy elements other than uranium by bombarding with energetic particles.

8.4.1. Thorium Mapping

Fast particle micromapping has so far found application principally in the measurement of thorium, which usually is more abundant than uranium by a factor of $\sim$4 in natural samples. Because of the intimate association of thorium and uranium, it is generally necessary to perform two irradiations—one with thermal neutrons to determine the concentration of ^{235}U (and, by inference, the concentration of ^{238}U), the other with high-energy particles that fission both uranium and thorium. The concentration of thorium is then determined by subtraction.

The first results of the double irradiation technique were reported by Bimbot et al. (1967) using 85 MeV protons; a somewhat simpler technique was subsequently described by Hair et al. (1971) using 30 MeV α-particles. In the α-particle method an unknown and a standard sample of known uranium concentration are first irradiated in a flux of thermal neutrons. The two are next irradiated with 30 MeV α-particles in a cyclotron. Fission events are recorded in auxiliary mica detectors placed in contact with the samples.

As shown by Hair et al. (1971), the Th/U ratio of an unknown is given by the following expression:

$$\frac{(C_{\mathrm{Th}})_x}{(C_{\mathrm{U}})_x} = \left(\frac{\rho_x(\alpha)}{\rho_s(\alpha)}\right)\left(\frac{\rho_s(\mathrm{n})}{\rho_x(\mathrm{n})}\right) - 1 \tag{8-3}$$

where the ρ's are the track densities measured in the standard (s) and the unknown (x) for irradiations with the α-particles (α) and neutrons (n) respectively. This expression follows from the fact that the uranium and thorium fission cross sections are equal for 30 MeV α-particles.

Table 8-1, taken from Hair et al. (1971), is a comparison of this method of thorium determination with independent methods applied to the same samples. In two samples of obsidian the values are about 10% lower than the values determined by an independent (n,γ) analysis. In contrast, the values for several USGS rock standards are about 10% higher than the average of many other determinations using a variety of analytic methods. However, the independent determinations bracket the track values and it is difficult to assign any meaning to the observed difference. In particular, the results agree best (±5%) with a delayed neutron activation analysis done by N. H. Gale of Oxford.

Table 8-1. Comparison of Different Methods of Thorium Determination

	Th Conc. (μg/g)		
	Track	Other	Difference
Obsidian			
#84	32.6 ± 3	35.3 ± 0.7 (n,γ)	-7.6%
#23	17.7 ± 1.3	20.3 ± 2.2 (n,γ)	-12.8%
USGS Standards			
Basalt	7.71 ± 0.6	6.81 (USGS Average)	+12.2%
Andesite	7.74 ± 0.6	6.96 (USGS Average)	+11.2%
Tl Glass	40.6 ± 0.8	41.6 ± 4 (Wt Added)	-2.4%

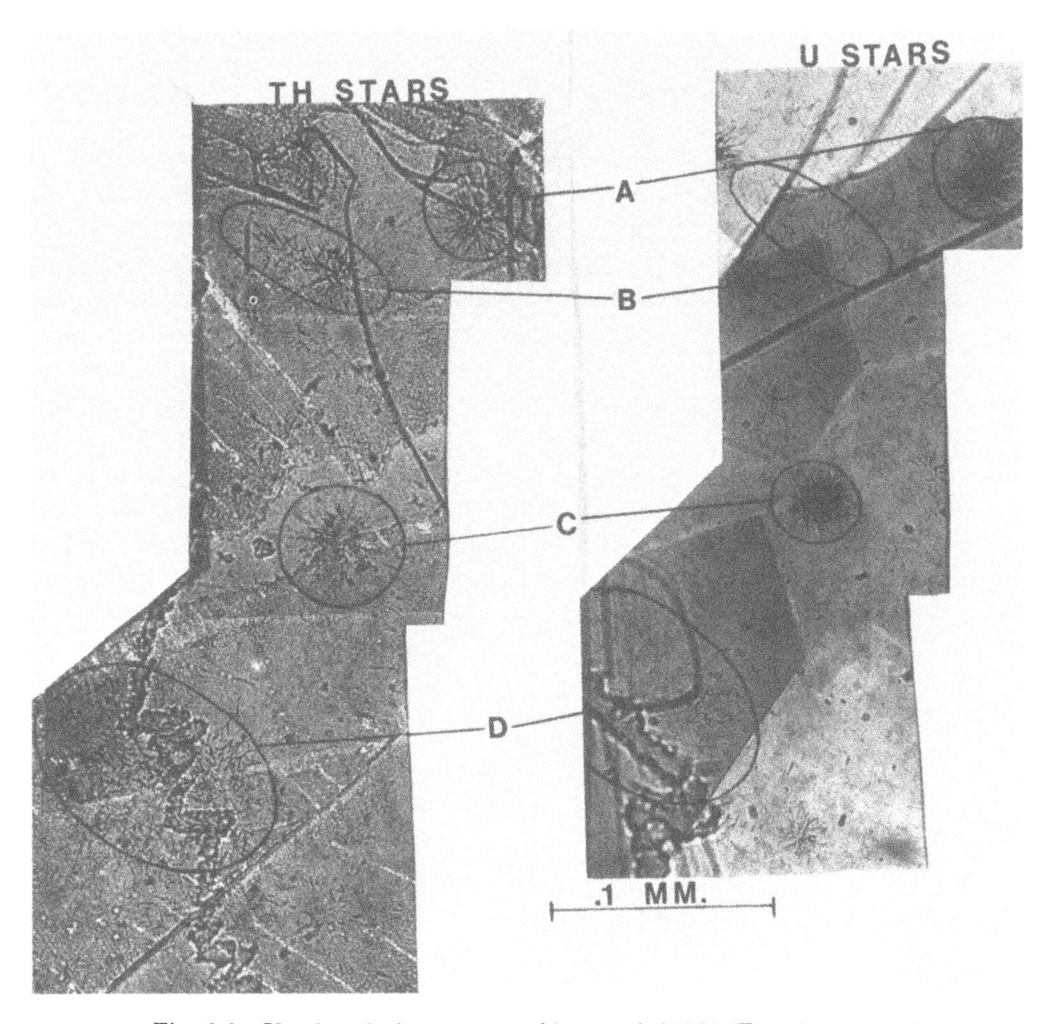

Fig. 8-9. *Uranium-thorium star map of lunar rock 12013. The mica map on the left was obtained by bombarding with 30 MeV α-particles which fission both U and Th nuclei. The map on the right was obtained by bombarding with thermal neutrons which fission only uranium. Comparison of the maps shows striking differences in the Th/U ratios for different grains. The grains corresponding to stars A and C clearly have lower Th/U ratios than those corresponding to stars B and D. (Photograph courtesy of D. Yuhas.)*

In Fig. 8-9 we show the results of Th-U micromapping of a lunar rock sample performed by Burnett et al. (1971). Although whole rock values for the Th/U ratios are quite constant, it can easily be seen that the Th/U ratios for different uranium-rich inclusions vary greatly. The observed ratio varies from $\sim$0.2 for some zircons to $>$20 for some phosphates. As discussed in Chapter 6, this result is important in connection with the study of extinct ^{244}Pu in extraterrestrial materials.

Thorium and uranium correlations determined by track techniques were used by Fleischer et al. (1971) to verify the identifications of obsidian artifacts used to establish trade routes.

Determining the Th/U ratios in trace meteorite minerals is much more difficult than in lunar samples and points up the limitations of the α-particle technique. For small inclusions $\gtrsim 30$ μm in size, the method is applicable at the 10% level for Th concentrations $\gtrsim 10^2$ μg/g. Below this level, the time for the cyclotron irradiations starts to become prohibitive (because of heating problems, increasing the beam current is self-defeating). There is the further difficulty that small tracks, which are produced by the interaction of the α-particles with the constituents of the mica (Crozaz et al., 1969), start to produce an unacceptable background for accurate counting of the fission tracks.

To measure Th/U ratios in samples with low uranium concentration such as meteorites (uranium $\lesssim 250$ ng/g), it is necessary to choose other particles. As shown by Berzina et al. (1968; 1971c), Reimer and Carpenter (1973), and Crozaz et al. (1974), fast neutrons are a feasible alternative. The method is somewhat more cumbersome since $\sigma_f(\text{U}) \neq \sigma_f(\text{Th})$ and a thorium standard such as that described by Schreurs et al. (1971) must be included and separately counted.

Important considerations in the neutron method are the control of the neutron energy spectrum and (for low levels of thorium) the problem of obtaining a sufficient neutron flux. As first calculated by Huang et al. (1967) and demonstrated explicitly by Crozaz et al. (1974), a pure fission spectrum of neutrons from ^{235}U fission is satisfactory when the Th/U ratio is $\sim$4. In this case the number of thorium fissions is about equal to the number of uranium fissions (mostly ^{238}U) and a statistically meaningful separation of the Th contribution can be made. Moderation and thermalization of the neutron flux, leading to enhanced ^{235}U fission, obviously degrades the thorium determination. Crozaz et al. (1974) used a special facility developed by T. P. Blewitt of the Argonne National Laboratory for their experiments. This facility, which uses a uranium converter and a low-temperature sample container, is not generally available, and other sources of fast neutrons would have to be found for similar work. The use of 14 MeV neutrons generated by the D, T reactions has been discussed by Nishiwaki et al. (1971); these are suitable only for large Th concentrations.

8.4.2. Mapping Other Elements

One promising facility for future thorium measurements is the LAMPF accelerator or "meson factory" that is being developed at Los Alamos. When fully operative, this may give an induced neutron flux as high as 10^{13} n/cm^2/sec > 10 MeV with an energy spectrum that would extend to $\sim$800 MeV. As originally estimated by Burnett and Woolum (private communication), the LAMPF facility will also make it possible to extend the fission mapping to many more elements. If we assume that the element is uniformly distributed throughout a clean target containing no other heavy element impurities, and if we further take 100 t/cm^2 as a counting limit, we estimate that Pb and Bi could be measured at the ng/g level and Au at about four times as high a concentration.

If the heavy elements were located in clusters 100 μm on a side, and at least ten fission tracks were demanded to locate the cluster, then clusters with concentrations in the μg/g range would be measurable.

Still another way to assess the future uses of heavy element mapping is to calculate the size of pure metallic clusters that could be detected with a one-day irradiation—again assuming 10 fission tracks/cluster. For Pb and Bi we obtain $\sim 2 \times 10^{-14}$ g and for Au $\sim$4 times as much. This corresponds to submicron particle sizes.

The use of complex bombarding particles such as ^{16}O ions affords still another approach to the problem of fission mapping of heavy elements. As given in Table 2.16 of Hyde (1964), even elements as light as Ho can have fission cross sections approaching 1 barn at high energies (for comparison, natural U has an effective cross section of $\sim$4 barns for thermal neutrons). Since the fissionability is highest for the heaviest nuclei, going down the Periodic Table to lighter and lighter elements will ultimately be stopped by the interference from fission reactions in the heavier elements that will inevitably be present. Although the effect of heavy elements could be assessed by measuring the energy dependence of track production, it is likely that, in practice, only the heaviest elements such as Pb, Bi, Tl, Hg, and Au can be mapped—and those under special conditions of noninterference from uranium and thorium. The use of gamma rays, as well as neutrons, to map elements in geological materials has been discussed by Flerov et al. (1971). Preliminary results on the photofission of uranium, thorium, and bismuth have also been reported by Carpenter (1969).

The presence of $^{239}PuO_2$ in the biosphere is a problem of increasing seriousness. For the normal relative abundances of uranium and thorium ($\sim$1:4), Pu may be distinguished from U plus Th by separate irradiations with thermal neutrons and with 30 MeV alpha particles. For equal amounts of Pu and U the neutron irradiation gives 180 fissions of ^{239}Pu per U plus Th fission; the alpha particle irradiation gives 20 U plus Th fissions per ^{239}Pu event.

8.5. MAPPING OF ALPHA-PARTICLE AND PROTON EMITTERS

The number of possibilities for element mapping increased sharply once detectors capable of registering α-particles became available. In analogy with uranium-fission mapping, the detector may be placed next to the substance to be studied during irradiation and the α-particles from various specific nuclear reactions such as $^{10}B(n,\alpha)^{7}Li$ may then be studied. Alternatively the α-particles may be produced by natural emitters or by isotopes such as ^{239}Pu and ^{210}Po that have been produced by *prior* bombardment. When the sample is counted after irradiation, the process is usually referred to as alpha autoradiography.

8.5.1. Induced Alpha Emission

Because of its large cross section of 760 barns due to the isotope ^{10}B (σ = 3840 barns), natural boron is particularly easy to map and a number of papers have appeared on this subject. If an α-sensitive detector is placed next to a flat sample, the density of tracks recorded by a given neutron dose ϕ is given by

$$\rho = C(\mathrm{B}) \frac{N_v \sigma_\mathrm{B} \phi}{4} (R_\alpha \cos^2 \theta_\alpha + R_\mathrm{Li} \cos^2 \theta_\mathrm{Li}) \qquad (8\text{-}4)$$

where $C(\mathrm{B})$ is the atomic concentration of boron, N_v is the number of atoms per unit volume, and θ_α and θ_Li are the critical angles for recording tracks of α-particles and Li ions in the detector. R_α and R_Li refer to the ranges of the α-particles (1.47 MeV) and Li (0.84 MeV) fragments respectively. Several authors, including Armijo and Rosenbaum (1967), Chrenko (1971), and Garnish and Hughes (1972), have discussed the problem of calculating R_α and R_Li for different substances in order to obtain absolute values for $C(\mathrm{B})$ for different substances. However, as in the case of uranium, it is generally simpler and more reliable to include boron-doped standards to measure concentrations. This eliminates problems of counting criteria, variable behavior of different plastics, and the necessity of making absolute neutron flux measurements. Standards that have been used include B compounds (Armijo and Rosenbaum, 1967), glasses (Carpenter, 1972b), and B-doped Si (Chrenko, 1971).

Boron can have a profound effect on the metallurgical properties of various alloys and is particularly important in nuclear materials. The same (n,α) reaction that makes B easy to detect can produce bubbles of helium gas that may act as nuclei for crack formation (Harries, 1966) and for swelling. It is therefore not surprising that most of the reported work of B mapping has been in metals (Bean et al., 1966; Armijo and Rosenbaum, 1967; Hughes and Rogers, 1967; Loveridge and McInnes, 1968; Elen and Glas, 1970; Kawasaki et al., 1971; Garnish and Hughes, 1972).

Boron is a fairly ubiquitous element and has also been measured in a variety of other substances including water (Fleischer and Lovett, 1968), minerals and rocks (Berzina and Malinko, 1969; Seitz and Hart, 1973), natural and man-made diamonds (Chrenko, 1971), glass (Carpenter, 1972a), and a variety of biological materials (Carpenter, 1972b). The partitioning of boron in silicate melts as a function of temperature and pressure has also recently been studied by Seitz (1973c, 1974a).

The practical sensitivity for the detection of boron so far achieved is ~1 μg/g for a homogeneous sample. This limit has arisen from the presence of other competing (n,α) reactions, either in the material being studied or in the detector itself. An important (n,α) reaction contributing background in the plastic detectors is $^{17}O(n,\alpha)^{14}C$, which gives an irreducible limit of 0.2 μg/g for the detectors used (Armijo and Rosenbaum, 1967). However, the use of an oxygen-free alpha detector such as polypropylene would obviously improve the lower limit for detection. For example, Profax 500 polypropylene contains as little as 750 ppm of oxygen, which permits boron to be measured at the ppb level.

The very short range of the α-particles and Li fragments makes the spatial resolution for mapping extremely good. Using an electron microscope, Hughes et al. (1969) claim a resolution of 0.3 μm; these authors also give the minimum detectable particle size as 10^{-16} g.

Other isotopes also have large (n,α) cross sections and are suitable for α-mapping. The most notable is $^{6}Li(n,\alpha)^{3}H$ which, in its natural abundance, has an effective cross section of 67 barns. This large cross section, while implying a large sensitivity for lithium, also means that lithium may cause an undesirable background in measurements of other elements such as boron. One way to distinguish between the two is to choose a plastic that will be sensitive to the ^{7}Li fragment from the ^{10}B reaction but not the α-particles and tritons from the ^{6}Li reaction (Fleischer and Lovett, 1968). One paper devoted to the distribution of Li in lithium silicate glasses has been published (Wood and Robinson, 1970) as well as one on the spatial distribution of Li in minerals and rocks (Berzina et al., 1971a). Carpenter (1974) has also discussed the problem of Li determination, particularly in biological materials.

Other α-emitting reactions suitable for the study of specific isotopes also suggest themselves. Particularly attractive are those reactions that involve isotopes of elements that could be useful in biological work, namely, H, C, N, and O. In an early paper Malmon (1965) pointed out the potential importance of the $^{17}O(n,\alpha)^{14}C$, $^{18}O(p,\alpha)^{15}N$, and $^{15}N(p,\alpha)^{12}C$ reactions. To illustrate potential applications of the technique, imagine a system into which ^{17}O has been added as a tracer. The normal abundance of this isotope relative to ^{16}O is 4×10^{-4}, so that any enrichment of the isotope above this value is potentially detectable. If the replacement were 100%, we estimate that $\sim 10^{11}$ t/cm^2 would be produced by a (feasible) thermal neutron radiation of 10^{18} n/cm^2 in a detector placed next to a typical organic substance. This high track density implies that if the ^{17}O replacement were localized, say in a particular region of a cell, then this region would be revealed with a high spatial resolution. The observations would, of course, have to be done with an electron microscope, where mapping of ^{17}O-enriched regions ~0.1 μm on a side is theoretically possible.

The first experimental work on the ^{17}O problem was reported in 1973 by Car-

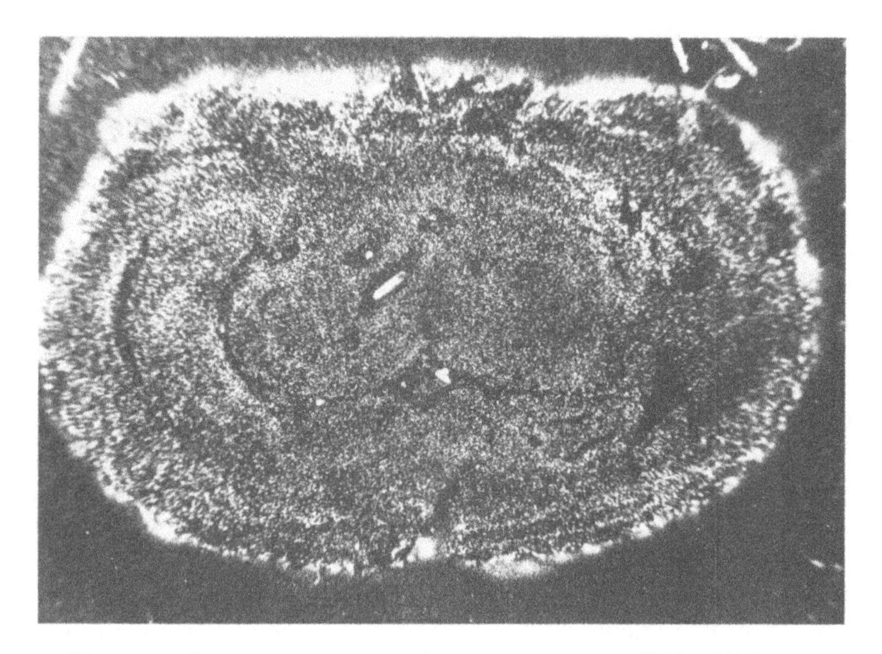

Fig. 8-10. *Oxygen mapping in rat brain tissue using the $^{17}O(n,\alpha)^{14}C$ reaction and the dyed Kodak detector CN(LR-118). Following inhalation of air enriched in ^{17}O, the animal was sacrificed and the brain was sectioned.* (*Photograph from unpublished work of B. S. Carpenter.*)

penter et al. who prepared and measured enriched ^{17}O standards of Al_2O_3 and citric acid. Subsequent, as yet unpublished, work by this group has demonstrated the utility of ^{17}O as a tracer in biological systems (see Fig. 8-10).

The localization of deuterium and tritium via the D(T,n) reaction has been demonstrated by Geisler and his colleagues (Geisler et al., 1973, 1974b; Slatkin et al., 1973), who have emphasized the potential biomedical applications of the method. In particular they have measured lymphocytes labeled with deuterated thymidine and have suggested the use of this method in human cancer detection.

8.5.2. Alpha Autoradiography

Another interesting area of research is the use of plastic α-autoradiographs analogous to those that have been made for many years using photographic emulsions. When α-emitters are involved, emulsions can clearly be replaced with plastics, giving a gain in simplicity accompanied by a suppression of background. Fading problems are also less severe.

Alpha radiography has been applied to a number of problems including the distribution of Pu and U in bones (Becker and Johnson, 1970; Cole et al., 1970; Simmons and Fitzgerald, 1970; Schlenker and Oltman, 1973), the electromigration of radioactive species (Hashimoto and Iwata, 1969a,b; Shinagawa et al.,

1969), the determination of U/Th ratios (Hashimoto, 1970), the uniformity of electrodeposited sources (Hashimoto, 1971a), and the degree of enrichment of uranium used in reactors (Chapuis et al., 1970).

An ingenious method of mapping Pb based on α-radiography was recently demonstrated by Hamilton (1971). The method consists of bombarding a lead-bearing sample with high energy α-particles to produce the isotope ^{210}Po from the reaction $^{208}Pb(\alpha,2n)Po$. After irradiation the sample is covered with plastic and the α-particles from the decay of ^{210}Po (half-life 138 days) are detected. Since the energy of the ^{210}Po α-particles is 5.3 MeV, which is above the threshold for detection in most α-sensitive plastics, only tracks from somewhat below the surface are counted. The α-emitter ^{208}Po is also produced by the α-particle bombardment of Pb but in smaller amounts than the ^{210}Po. The cross section for $^{208}Pb(\alpha,2n)^{210}Po$ is 1.016 b at 30 MeV. This implies a relatively high sensitivity and measurements in the 0.1 μg/g level should be possible.

Because of lead isotope production in the radioactive decay of parent uranium and thorium, isotope ratios of leads from different sources are not constant. This adds a complication in the determination of total lead from the ^{210}Po measurement but, conversely, implies that the track measurements combined with an independent measurement of total lead can be used to obtain lead isotopic information.

Table 8-2, taken from Hamilton (1971), shows that the track method gives satisfactory agreement with independent radioactivation measurements of lead at the 10 μg/g level.

Seitz (1974c) has also used α-radiography to study the partitioning of Sm in rock melts at high pressure.

8.5.3. Induced Proton Emission

Still another important reaction is $^{14}N(n,p)^{14}C$, which has the relatively high thermal neutron cross section of 1.8 barns. Using sensitive cellulose nitrate as the detecting material, Carpenter and LaFleur (1972a,b) have reported measure-

Table 8-2. A Comparison of the Concentration of Lead by the Track and Radioactivation Techniques for Samples of Known Isotopic Composition (from Hamilton, 1971)

	Lead Concentration (μg/g)	
Material	Track Method	Radioactivation Technique
Sectioned obsidian glass	7.3	8.4
Powdered granite	15.0	16.2
Animal bone	12.2	10.8

Precision ± 10%

ments of nitrogen at the 3% level. Obviously the use of low-nitrogen detectors such as Lexan treated with an intense dose of UV would give the ability to measure very much lower concentrations and, in fact, to map nitrogen at the microscopic level. The short recordable range of a proton in plastic is an advantage in terms of spatial resolution.

Chapter 8 References

Kh. Abdullaev, A. Kapustsik, O. Otgonsuren, V. P. Perelygin, and D. Chultém (1967), "Determining the Concentration of Fissionable Materials in Solid Bodies," Joint Inst. for Nucl. Research, Dubna. Also see *Instr. and Expt. Tech.* (1968), 319–321.

Kh. Abdullaev, B. B. Zakhvataev, and V. P. Perelygin (1968), "Determination of Uranium Concentration in Plants from Tracks of Uranium Fission Fragments," *Radiobiologiya* **8,** 765–766.

M. J. Aitken, D. W. Zimmerman, and S. J. Fleming (1968), "Thermoluminescent Dating of Ancient Pottery," *Nature* **219,** 442–445.

A. P. Akimov, I. Berzina, M. Yu. Gurvich, and B. G. Lutts (1968), "Uranium Content in Eclogitic Inclusions from Kimberlite Pipes," *Dokl. Akad. Nauk SSSR* **181,** 1245–1248; (1968) *Chem. Abs.* **69,** 8321.

F. A. Alekseyev, R. P. Gottikh, V. Ya. Vorob'yeva, and L. V. Murav'yeva (1969), "Uranium Distribution in Sedimentary Rocks in the Western Part of the Amu Darya Petroleum Basin," *Geochemistry International* **6,** 963–970. (Translation from *Geokhimiya* **10,** 1238–1247.)

A. J. Amiel, D. S. Miller, and G. M. Friedman (1972), "Uranium Distribution in Carbonate Sediments of a Hypersaline Pool, Gulf of Elat, Red Sea," *Israel J. Earth Sci.* **21,** 187–191.

B. V. Andersen (1965), "New Technique for Plutonium Particle Size Analysis," AEC Accession No. 27189, Rept. No. BNWL-SA-26, pp. 1–14.

J. S. Armijo and H. S. Rosenbaum (1967), "Boron Detection in Metals by Alpha-Particle Tracking," *J. Appl. Phys.* **38,** 2064–2069.

F. Aumento (1971), "Uranium Content of Mid-Oceanic Basalts," *Earth Planet. Sci. Lett.* **11,** 90–94.

F. Aumento and R. D. Hyndman (1971), "Uranium Content of the Oceanic Upper Mantle," *Earth Planet. Sci. Lett.* **12,** 373–384.

C. P. Bean, R. L. Fleischer, P. S. Swartz, and H. R. Hart, Jr. (1966), "Effect of Thermal-Neutron Irradiation on the Superconducting Properties of Nb_3Al and V_3Si Doped with Fissionable Impurities," *J. Appl. Phys.* **37,** 2218–2224.

K. Becker and D. R. Johnson (1970), "Nonphotographic Alpha Autoradiography and Neutron-Induced Autoradiography," *Science* **167,** 1370–1372.

K. K. Bertine, L. H. Chan, and K. K. Turekian (1970), "Uranium Determinations in Deep-Sea Sediments and Natural Waters Using Fission Tracks," *Geochim. Cosmochim. Acta* **34,** 641–648.

I. G. Berzina and P. G. Demidova (1966), "Determination of the Growth of Minerals by the Tracks from Fission Fragments from Uranium," *Atomnaya Energiya* **21,** 304–306.

I. G. Berzina and E. I. Dolomanova (1967), "Uranium Content in Cassiterites Determined from Tracks of Fragments Caused by Uranium Fission," *Dokl. Akad. Nauk SSSR* **175,** 171–174.

I. G. Berzina and S. M. Kravchenko (1970), "Fission-Track Radiographic Study of the Spatial Distribution of Uranium in Basalts, Alkaline-Trachytic Ignimbrites, and Syenites," *Dokl. Akad. Nauk SSSR* **193,** 181–183.

I. G. Berzina and S. M. Kravchenko (1971), "The Behavior of Uranium During Crystallization of Tholeitic and Alkaline Continental Basalts from F-Radiography Data," *International Geochemical Congress*, A. I. Tubarinov (ed.), 134. Moscow: Acad. Sci. USSR.

I. G. Berzina and S. V. Malinko (1969), "Determination of Boron Spatial Distribution and Concentration in Minerals and Rocks," *Dokl. Akad. Nauk SSSR* **189,** 849–851.

I. G. Berzina, I. B. Berman, and M. Yu. Gurvich (1967a), "Determination of Uranium Clarkeite Concentrations in Ionic Crystals," *Atomnaya Energiya* **22,** 504–505.

I. G. Berzina, I. B. Berman, M. Yu. Gurvich, G. N. Flerov, and Yu. S. Shimelevich (1967b), "Determining the Concentration of Uranium and Its Manner of Distribution in Minerals and Rocks," *Atomnaya Energiya* **23,** 520–527.

I. G. Berzina, A. N. Stolyarova, G. N. Flerov, and Yu. S. Shimelevich (1967c), "Possibility of Demonstrating the Migration of Uranium and Its Decay Products in Minerals," *Dokl. Sov. Phys.* **12,** 1155–1157 (1968).

I. G. Berzina, M. Yu. Gurvich, and G. I. Khlebnikov (1968), "Determination of the Concentration and Spatial Distribution of Thorium in Minerals and Rocks According to the Tracks from Fission Fragments," Inst. At. Energii, Rept. IAE-1519; *Nucl. Sci. Abs.* **23,** 3625 (1969).

I. G. Berzina, D. P. Popenko, and Yu. S. Shimelevich (1969), "Determination of Trace Amounts of Uranium in Petroleums from the Fission-Fragment Tracks," *Geokhimiya* **8,** 1024–1027.

I. G. Berzina, S. M. Kravchenko, M. Yu. Gurvich, and B. P. Zolotarev (1970), "Determination of Concentrations of Uranium and Its Spatial Distribution in Cenozoic Basaltoids from Fission-Fragment Tracks," *International Geology Review* **12,** 493–502.

I. G. Berzina, I. B. Berman, and A. S. Nazarova (1971a), "Detection of the Spatial Distribution and Determination of Lithium Concentration in Minerals and Rocks," *Dokl. Akad. Nauk SSSR* **201,** 686–689.

I. G. Berzina, O. P. Eliseeva, and A. N. Stolyarova (1971b), "Uranium in Accessory Apatites," *Izv. Akad. Nauk SSSR*, Ser. Geol. **7,** 79–86.

I. G. Berzina, M. Yu. Gurvich, and G. I. Khlebnikov (1971c), "Determination of Thorium Concentration and Spatial Distribution in Minerals and Rocks from Tracks of Fission Fragments," Tr., *Vses. Nauch.—Issled. Inst. Yad. Geofiz. Geokhim.* No. 9, 158–161.

I. G. Berzina, I. V. Mel'nikov, and D. P. Popenko (1972), "Determination of the Quantity and Spatial Distribution of Uranium in Fluorites from the Tracks of Uranium Fission Fragments," *Atomnaya Energiya* **32,** 211–215.

C. Bigazzi and G. F. Rinaldi (1968), "Variazione del Rapporto U/$CaCo_3$ Nelle Concrezioni di Grotta," *Atti Soc. Tosc. Sc. Nat. Mem.*, Serie A, **LXXV,** 647–653.

R. Bimbot, M. Maurette, and P. Pellas (1967), "A New Method for Measuring the Ratio of the Atomic Concentrations of Thorium and Uranium in Minerals and Natural Glasses, Preliminary Application to Tektites," *Geochim. Cosmochim. Acta* **31,** 263–274.

B. Bleaney (1969), "The Radiation Dose-Rates Near Bone Surfaces in Rabbits after Intravenous or Intramuscular Injection of ^{239}Pu," *Brit. J. Radiol.* **42,** 51–56.

E. Bonatti, D. E. Fisher, O. Joensuu, and H. S. Rydell (1971), "Postdepositional Mobility of Some Transition Elements, Phosphorus, Uranium and Thorium in Deep Sea Sediments," *Geochim. Cosmochim. Acta* **35,** 189–201.

K. Bostrom and D. Fisher (1971), "Volcanogenic Uranium, Vanadium and Iron in Indian Ocean Sediments," *Earth Planet. Sci. Lett.* **11,** 95–98.

L. W. Brackenbush and W. V. Baumgartner (1965), "Detection of Plutonium in Bioassay Programs Using Nonelectrical Fission Fragment Track Detectors," AEC Accession No. 26742, Rept. No. BNWL-SA-58, pp. 1–11.

G. M. Brown, C. H. Emeleus, J. G. Holland, A. Peckett, and R. Phillips (1972), "Mineral-Chemical Variations in Apollo 14 and Apollo 15 Basalts and Granitic Fractions," *Proc. Third Lunar Sci. Conf.*, **1,** 141–157. Cambridge: MIT Press.

D. S. Burnett, M. Monnin, M. Seitz, R. M. Walker, D. Woolum, and D. Yuhas (1970), "Charged Particle Track Studies in Lunar Rock 12013," *Earth Planet. Sci. Lett.* **9,** 127–136.

D. Burnett, M. Monnin, M. Seitz, R. Walker, and D. Yuhas (1971), "Lunar Astrology—U-Th Distributions and Fission-Track Dating of Lunar Samples," *Proc. Second Lunar Sci. Conf.*, **2,** 1503–1519. Cambridge: MIT Press.

F. D. Busche, M. Prinz, K. Keil, and G. Kurat (1972), "Lunar Zirkelite: A Uranium-Bearing Phase," *Earth Planet. Sci. Lett.* **14,** 313–321.

B. S. Carpenter (1969), "Applications of the Nuclear Track Technique to Trace Analysis," National Bureau of Standards Technical Note 505, pp. 110–114.

B. S. Carpenter (1972a), "Quantitative Applications of the Nuclear Track Technique," *Microscope* **20,** 175–182.

B. S. Carpenter (1972b), "Determination of Trace Concentration of Boron and Uranium in Glass by the Nuclear Track Technique," *Analyt. Chem.* **44,** 600–602.

B. S. Carpenter (1974), "Lithium Determination by the Nuclear Track Technique," *J. Radioanal. Chem.* **19,** 233–234.

B. S. Carpenter and C. H. Cheek (1970), "Trace Determination of Uranium in Biological Material by Fission Track Counting," *Analyt. Chem.* **42,** 121–123.

B. S. Carpenter and P. D. LaFleur (1972a), "Observing Proton Tracks in Cellulose Nitrate," *Int. J. Appl. Rad. Isotopes* **23,** 157–159.

B. S. Carpenter and P. D. LaFleur (1972b), "Nitrogen Determination in Biological Samples Using the Nuclear Track Technique," *Am. Nucl. Soc. Trans.* **15,** 118.

B. S. Carpenter, D. Samuel, and I. Wasserman (1973), "Quantitative Applications of ^{17}O Tracer," *Rad. Effects* **19,** 59.

P. I. Chalov, U. Mamyrov, and Ya. A. Musin (1970), "Comparative Study of Some Methods for Measuring the Relative Uranium-235 Content in Natural Uranium Samples," *Izv. Akad. Nauk Kirg. SSR* **5,** 13–19.

A. M. Chapuis, H. Francois, and N. Gerard-Nicodeme (1970), "Controle de l'Enrichissement de l'Uranium Metallique a l'Aide du Nitrate de Cellulose," *Rad. Effects* **5,** 91–97.

I. M. Chesire and J. M. Poate (1970), "Shell Effects in Low-Energy Atomic Collisions," *Atomic Collision Phenomena in Solids*, D. W. Palmer, M. W. Thompson, and P. D. Townsend (eds.), 351–360. Amsterdam: North-Holland.

R. M. Chrenko (1971), "Boron Content and Profiles in Large Laboratory Diamonds," *Nature Phys. Sci.* **229,** 165–167.

A. Cole, D. J. Simmons, H. Cummins, F. J. Congel, and J. Kastner (1970), "Application of Cellulose Nitrate Films for Alpha Autoradiography of Bone," *Health Phys.* **19,** 55–56.

K. C. Condie, C. S. Kuo, R. M. Walker, and V. R. Murthy (1969), "Uranium Distribution in Separated Clinopyroxenes from Four Eclogites," *Science* **165,** 57–59.

G. Crozaz, M. Hair, M. Maurette, and R. M. Walker (1969), "Nuclear Interaction Tracks in Minerals and Their Implications for Extraterrestrial Materials," *Proc. Intl. Topical Conf. on Nuclear Track Registration in Insulating Solids and Applications*, Clermont-Ferrand, **2,** Section VII, pp. 41–54.

G. Crozaz, U. Haack, M. Hair, M. Maurette, R. Walker, and D. Woolum (1970), "Nuclear Track Studies of Ancient Solar Radiations and Dynamic Lunar Surface Processes," *Proc. Apollo 11 Lunar Sci. Conf.*, **3,** 2051–2080. New York: Pergamon Press.

G. Crozaz, R. Drozd, H. Graf, C. M. Hohenberg, M. Monnin, D. Ragan, C.

Ralston, M. Seitz, J. Shirck, R. M. Walker, and J. Zimmerman (1972), "Uranium and Extinct Pu^{244} Effects in Apollo 14 Materials," *Proc. Third Lunar Sci. Conf.*, **2,** 1623–1636. Cambridge: MIT Press.

G. C. Crozaz, D. Burnett, and R. M. Walker (1974), "Uranium and Thorium Distributions in Meteorites," *Meteoritics*, in press.

Yu. P. Dikov, T. M. Kaikova, and V. P. Perelygin (1971), "Geochemistry of Uranium in Scarn Deposits," *International Geochemical Congress* (book II), A. I. Tubarinov (ed.), 532. Moscow: Acad. Sci. USSR.

J. R. Dooley, Jr., R. B. Taylor, and F. J. Jurceka (1970), "Dual Microscope Comparator for Fission Track Studies," *Rev. Sci. Instr.* **41,** 887–888.

E. Dörschel and W. Stolz (1970), "Urangehaltsbestimmung von Wasser mit Hilfe der Festkorperspurmethode," *Radiochem. Radioanal. Letters* **4,** 277–283.

S. A. Durrani and H. A. Khan (1971), "Ivory Coast Microtektites: Corrected Values of Uranium Content," *Nature Phys. Sci.* **232,** 175.

S. A. Durrani and H. A. Khan (1972), "Charged-Particle Track Parameters of Apollo 15 Lunar Glasses," *The Apollo 15 Lunar Samples*, J. W. Chamberlain and C. Watkins (eds.), 352–356, Lunar Sci. Inst., Houston.

J. D. Elen and A. Glas (1970), "Precipitation of Trace Amounts of Boron in AISI 304L and AISI 316L," *J. Nucl. Materials* **34,** 182–188.

P. E. Fielding (1970), "The Distribution of Uranium, Rare Earths, and Color Centers in a Crystal of Natural Zircon," *Am. Mineralogist* **55,** 428–440.

D. E. Fisher (1969a), "Uranium Measurements in Hypersthene Chondrites and Their Relation to the 600–700 Million Year 'Event'," *Earth Planet. Sci. Lett.* **7,** 278–280.

D. E. Fisher (1969b), "Uranium Content of Some Stone Meteorites and Their Pu-Xe Decay Interval," *Nature* **222,** 1156.

D. E. Fisher (1970a), "Homogenized Fission Track Analysis of Uranium in Some Ultramafic Rocks of Known Potassium Content," *Geochim. Cosmochim. Acta* **34,** 630–634.

D. E. Fisher (1970b), "Homogenized Fission Track Determination of Uranium in Whole Rock Geologic Samples," *Analyt. Chem.* **42,** 414–416.

D. E. Fisher (1972a), "Uranium Content and Radiogenic Ages of Hypersthene, Bronzite, Amphoterite and Carbonaceous Chondrites," *Geochim. Cosmochim. Acta* **36,** 15–33.

D. E. Fisher (1972b), "U/He Ages as Indicators of Excess Argon in Deep Sea Basalts," *Earth Planet. Sci. Lett.* **14,** 255–258.

D. E. Fisher (1973), "Achondritic Uranium," *Earth Planet. Sci. Lett.* **20,** 151–156.

D. E. Fisher and K. Bostrom (1969), "Uranium Rich Sediments on the East Pacific Rise," *Nature* **224,** 64–65.

R. L. Fleischer (1966), "Uranium Micromaps: Technique for In Situ Mapping of Distributions of Fissionable Impurities," *Rev. Sci. Instr.* **37,** 1738–1739.

R. L. Fleischer (1968), "Uranium Distribution in Stone Meteorites by the Fission Track Technique," *Geochim. Cosmochim. Acta* **32,** 989–998.

R. L. Fleischer and H. R. Hart, Jr. (1972), "Particle Track Record of Apollo 15 Green Soil and Rock," *The Apollo 15 Lunar Samples,* J. W. Chamberlain and C. Watkins (eds.), 368–370, Lunar Sci. Inst., Houston; *Earth Planet. Sci. Lett.* **18,** 357–364 (1973).

R. L. Fleischer and D. B. Lovett (1968), "Uranium and Boron Content of Water by Particle Track Etching," *Geochim. Cosmochim. Acta* **32,** 1126–1128.

R. L. Fleischer and P. B. Price (1964), "Uranium Content of Ancient Man-Made Glass," *Science* **144,** 841–842.

R. L. Fleischer, C. W. Naeser, P. B. Price, R. M. Walker, and U. B. Marvin (1965a), "Fossil Particle Tracks and Uranium Distributions in the Minerals of the Vaca Muerta Meteorite," *Science* **148,** 629–632.

R. L. Fleischer, P. B. Price, and R. M. Walker (1965b), "Applications of Fission Tracks and Fission Track Dating to Anthropology," *7th Intl. Conf. on Glass,* Brussels, **II,** paper 224, pp. 1–8. New York: Gordon and Breach.

R. L. Fleischer, P. B. Price, and R. M. Walker (1965c), "Solid State Track Detectors: Applications to Nuclear Science and Geophysics," *Ann. Rev. Nuc. Sci.* **15,** 1–28.

R. L. Fleischer, E. L. Haines, H. R. Hart, Jr., R. T. Woods, and G. M. Comstock (1970), "The Particle Track Record of the Sea of Tranquillity," *Proc. Apollo 11 Lunar Sci. Conf.,* **3,** 2103–2120. New York: Pergamon Press.

R. L. Fleischer, M. Maurette, P. B. Price, and R. M. Walker (1971), "Application of Solid-State Nuclear Track Detectors to Archaeology," *Science and Archaeology,* R. H. Brill (ed.), 279–283. Cambridge: MIT Press.

G. N. Flerov, I. G. Berzina, F. I. Wolfson, M. Yu. Gurvich, and I. V. Mel'nikov (1971), "On the Possible Application of (n,f) and (γ,f) Reactions for Certain Geological Problems," *International Geochemical Congress,* A. I. Tubarinov (ed.), 129. Moscow: Acad. Sci. USSR.

J. D. Garnish and J. D. H. Hughes (1972), "Quantitative Analysis of Boron in Solids by Autoradiography," *J. Mat. Sci.* **7,** 7–13.

V. M. Gavshin (1972), "Uranium Concentration in Natural Stratified Alumosilicates," *Dokl. Akad. Nauk SSSR* **205,** 956–959.

F. H. Geisler, K. W. Jones, H. W. Kraner, D. N. Slatkin, A. P. Wolf, J. S. Fowler, and E. P. Cronkite (1973), "Sensitive Radiography of Deuterium and Tritium Using the D(t,n)α Reaction and Plastic Track Detectors," *Bull. Am. Phys. Soc.* **18,** 580.

F. H. Geisler, J. Shirck, and R. Walker (1974a), "A New Method of Average U-Determination in Heterogeneous Samples," unpublished report.

F. H. Geisler, K. W. Jones, J. S. Fowler, H. W. Kraner, and D. N. Slatkin (1974b), "In Vitro Labeling of Lymphocytes with Deuterated Thymidine," *Bull. Am. Phys. Soc.* **19,** 373.

B. P. Glass, D. Storzer, and G. A. Wagner (1972), "Chemistry and Particle Track Studies of Apollo 14 Glasses," *Proc. Third Lunar Sci. Conf.*, **1,** 927–937. Cambridge: MIT Press.

H. Graf, C. Hohenberg, J. Shirck, S. Sun, and R. Walker (1973), "Astrology of Apollo 14 Extinct Isotope Breccias," *Lunar Science IV*, J. W. Chamberlain and C. Watkins (eds.), 312–314, Lunar Sci. Inst., Houston.

B. Grauert, M. G. Seitz, and G. Soptrajanova (1974), "Uranium and Lead Gain of Detrital Zircons Studied by Isotopic Analyses and Fission-Track Mapping," *Earth Planet. Sci. Lett.* **21,** 389–399.

D. S. Haglund, G. M. Friedman, and D. S. Miller (1969), "The Effect of Fresh Water on the Redistribution of Uranium in Carbonate Sediments," *J. Sed. Petrol.* **39,** 1283–1296.

E. L. Haines (1972), "Precise Coordinate Control in Fission Track Uranium Mapping," *Nucl. Instr. Methods* **98,** 183–184.

E. L. Haines and R. E. Zartman (1973), "Uranium Concentration and Distribution in Six Peridotite Inclusions of Probable Mantle Origin," *Earth Planet. Sci. Lett.* **20,** 45–53.

E. L. Haines, A. L. Albee, A. A. Chodos, and G. J. Wasserburg (1971), "Uranium-Bearing Minerals of Lunar Rock 12013," *Earth Planet. Sci. Lett.* **12,** 145–154.

E. L. Haines, A. J. Gancarz, A. L. Albee, and G. J. Wasserburg (1972), "The Uranium Distribution in Lunar Soils and Rocks 12013 and 14310," *Lunar Science-III*, C. Watkins (ed.), 350–352, Lunar Sci. Inst., Houston.

M. W. Hair, J. Kaufhold, M. Maurette, and R. M. Walker (1971), "Th Microanalysis Using Fission Tracks," *Rad. Effects* **7,** 285–287.

E. I. Hamilton (1966), "Distribution of Uranium in Some Natural Minerals," *Science* **151,** 570–572.

E. I. Hamilton (1970a), "The Concentration of Uranium in Air from Contrasted Natural Environments," *Health Phys.* **19,** 511–520.

E. I. Hamilton (1970b), "Uranium Content of Normal Blood," *Nature* **227,** 501–502.

E. I. Hamilton (1971), "New Technique for Determining the Concentration and Distribution of Lead in Materials," *Nature* **231,** 524–525.

D. R. Harries (1966), "Neutron Irradiation Embrittlement of Austenitic Stainless Steels and Nickel-Base Alloys," *J. Brit. Nucl. Energy Soc.* **5,** 74–87.

T. Hashimoto (1970), "Comparison of Fission-Track Method and α-Particle Method for Detection of Actinide Nuclides," *Japan Analyst* **19,** 508–513.

T. Hashimoto (1971a), "Electrodeposition of Americium and Observation of the Surface by Means of α-Particle Tracks on a Cellulose Nitrate Film," *J. Radioanal. Chem.* **9,** 251–258.

T. Hashimoto (1971b), "Determination of the Uranium Content in Sea Water by a Fission Track Method with Condensed Aqueous Solution," *Analyt. Chim. Acta* **56,** 347–354.

T. Hashimoto and S. Iwata (1969a), "A New Method Based on Alpha Particle Track for Detection of Radionuclides on Electromigrated Paper," *Japan Analyst* **18,** 527–528.

T. Hashimoto and S. Iwata (1969b), "Rapid Separation of Natural Radionuclides by Paper Electromigration and a New Detection Method of These Alpha Emitters Based on Alpha Particle Track," *Japan Analyst* **18,** 1382–1388.

T. Hashimoto, S. Hayashito, and S. Iwata (1970), "Determination of Uranium Contents in Some Rocks Based on Fission Track Method," *Japan Analyst* **19,** 1538–1543.

W. H. Huang, M. Maurette, and R. M. Walker (1967), "Observation of Fossil α-Particle Recoil Tracks and Their Implications for Dating Measurements," Intl. Atomic Energy Agency Symposium, Monaco, *Radioactive Dating and Methods of Low-Level Counting*, 415–429, IAEA, Vienna.

J. D. H. Hughes and G. T. Rogers (1967), "High-Resolution Autoradiography of Trace Boron in Metals and Solids," *J. Inst. Metals* **95,** 299–302.

J. D. H. Hughes, M. A. P. Dewey, and G. W. Briers (1969), "Boron Autoradiography with the Electron Microscope," *Nature* **223,** 498–499.

I. D. Hutcheon, P. P. Phakey, and P. B. Price (1972), "Studies Bearing on the History of Lunar Breccias," *Proc. Third Lunar Sci. Conf.*, **3,** 2845–2865. Cambridge: MIT Press.

E. K. Hyde (1964), *The Nuclear Properties of the Heavy Elements III: Fission Phenomena.* Englewood Cliffs: Prentice-Hall.

S. Kawasaki, A. Hishinuma, and R. Nagasaki (1971), "Behavior of Boron in Stainless Steel Detected by Fission Track Etching Method and Effect of Radiation on Tensile Properties," *J. Nucl. Materials* **39,** 166–174.

J. D. Kleeman and J. A. Cooper (1970), "Geochemical Evidence for the Origin of Some Ultramafic Inclusions from Victorian Basanites," *Phys. Earth Planet. Interiors* **3,** 302–308.

J. D. Kleeman and J. F. Lovering (1967a), "Uranium Distribution in Rocks by Fission-Track Registration in Lexan Plastic," *Science* **156,** 512–513.

J. D. Kleeman and J. F. Lovering (1967b), "Uranium Distribution Studies by Fission Track Registration in Lexan Plastic Prints," *At. Energy in Australia* **10,** 3–8.

J. D. Kleeman and J. F. Lovering (1969), "Lexan Plastic Prints: How Are They Formed?," *Proc. Intl. Topical Conf. on Nuclear Track Registration in Insulating Solids and Applications*, Clermont-Ferrand, **2,** Section VI, 41–52.

J. D. Kleeman, D. H. Green, and J. F. Lovering (1969), "Uranium Distribution in Ultramafic Inclusions from Victorian Basalts," *Earth Planet. Sci. Lett.* **5,** 449–458.

L. Kleppe and Sister M. Roger (1966), *Procedures for Finding Fission Tracks in Mica and Mylar*, Clearinghouse for Federal and Scientific Information, Springfield, Va.

A. N. Komarov (1968), "Radiographic Methods Applied in Mineralogical and Geochemical Experimental Practice," *Izv. Akad. Nauk SSSR*, Ser. Geol. **1,** 50–61.

A. N. Komarov and Yu. A. Shukolyukov (1966), "The Form of Occurrence of Uranium in Micas," *Geokhimiya* **11,** 1322–1330.

A. N. Komarov and N. V. Skovorodkin (1969), "Investigation of the Abundance and Distribution of Uranium in Ultra-Mafic and Mafic Rocks by Fission Track Methods," *Geochemistry International* **6,** 127–133. (Translation from *Geokhimiya* **2,** 170–176.)

A. N. Komarov, Yu. A. Shukolyukov, and N. V. Skovorodkin (1967), "Investigation of the Content and Distribution of Uranium in Some Rocks and Minerals by Neutron Activation Analysis by Counting Fission Fragment Tracks," *Geochemistry International* **4,** 647–659. (Translation from *Geokhimiya* **7,** 763–776.)

J. A. Lahoud, D. S. Miller, and G. M. Friedman (1966), "Relationship Between Depositional Environment and Uranium Concentrations of Molluskan Shells," *J. Sed. Petrology* **36,** 541–547.

E. M. Lobanov, E. A. Isabaev, A. V. Yankovskii, A. Kh. Abil'daev, and V. S. Vasil'ev (1967), "Determination of Contents of Uranium and Thorium in Samples of Rocks and Ores from Fission Fragments," *Atomnaya Energiya* **23,** 555–556.

B. A. Loveridge and C. A. J. McInnes (1968), "Microanalytical Estimation of Boron in Steel by Using the $^{10}B(n,\alpha)^{7}Li$ Reaction," *Microscope* **16,** 105–114.

J. F. Lovering and J. D. Kleeman (1970), "Fission Track Uranium Distribution Studies on Apollo 11 Lunar Samples," *Proc. Apollo 11 Lunar Sci. Conf.*, **1,** 627–631. New York: Pergamon Press.

J. F. Lovering and D. A. Wark (1971), "Uranium-Enriched Phases in Apollo 11 and Apollo 12 Basaltic Rocks," *Proc. Second Lunar Sci. Conf.*, **1,** 151–158. Cambridge: MIT Press.

J. F. Lovering, D. A. Wark, A. F. Reid, N. G. Ware, K. Keil, M. Prinz, T. E. Bunch, A. El Goresy, P. Ramdohr, G. M. Brown, A. Peckett, R. Phillips, E. N. Cameron, J. A. V. Douglas, and A. G. Plant (1971), "Tranquillityite: A New Silicate Mineral from Apollo 11 and Apollo 12 Basaltic Rocks," *Proc. Second Lunar Sci. Conf.*, **1,** 39–45. Cambridge: MIT Press.

J. F. Lovering, D. A. Wark, A. J. W. Gleadow, and D. K. B. Sewell (1972), "Uranium and Potassium Fractionation in Pre-Imbrian Lunar Crustal Rocks," *Proc. Third Lunar Sci. Conf.*, **1,** 281–294. Cambridge: MIT Press.

D. Macdougall (1973), "Uranium Distribution in Weathered Marine Basalts," *Trans. Amer. Geophys. Un.* **54,** 988.

D. Macdougall, R. S. Rajan, I. D. Hutcheon, and P. B. Price (1973), "Irradiation History and Accretionary Processes in Lunar and Meteoritic Breccias," *Proc. Fourth Lunar Sci. Conf.*, **3,** 2319–2336. New York: Pergamon Press.

S. R. Malik, S. A. Durrani, and J. H. Fremlin (1973), "A Comparative Study of the Spatial Distribution of Uranium and of TL-Producing Minerals in Archaeological Materials," *Archaeometry* **15,** 249–253.
A. G. Malmon (1964), "Fission-Fragment Tracks as Uranium Tracers in Biological Electron Microscopy," *Biophys. J.* **4,** 1–10.
A. G. Malmon (1965), "High-Resolution Isotope Tracing in Electron Microscopy Using Induced Nuclear Reactions," *J. Theoret. Biol.* **9,** 77–92.
S. V. Mel'gunov and E. K. Varvarina (1970), "Use of Neutron-Fission Track Radiography in Studying Uranium Distribution in Some Metamorphic Formations," *Geol. Geofiz.* **10,** 38–44.
J. Mory (1969), "Charged Particle Tracks in Insulating Solids," *Solid State Dosimetry*, S. Amelinckx et al. (eds.), 393–424. New York: Gordon and Breach.
J. Mory, D. DeGuillebon, and G. Delsarte (1970), "Mesure du Parcours Moyen des Fragments de Fission avec le Mica comme Detecteur—Influence de la Texture Cristalline," *Rad. Effects* **5,** 37–40.
A. V. Murali, P. P. Parekh, and M. S. Das (1970), "Fission Track Method for the Determination of the Uranium Content of Whole Rock Samples," *Analyt. Chim. Acta* **50,** 71–77.
S. Nishimura (1970), "Determination of Uranium Contents of Standard Rocks by Fission-Track Registration in Muscovite," *Radioisotopes* **19,** 194–196.
S. Nishimura (1972), "Partition of Uranium Between Peridotite Nodules and Host Basalts," *Chem. Geol.* **10,** 211–221.
Y. Nishiwaki, H. Kawai, H. Morishima, S. Iwata, and T. Tsuruta (1971), "Rapid Method of Confirmation of Fissionable Materials with Etchpit Counting Method," Inter. Symp. on Rapid Methods for Measurement of Radioactivity in the Environment, München.
O. Otgonsuren, V. P. Perelygin, and D. Chultem (1970), "Build-Up of Uranium in Animal Bones," *Atomnaya Energiya* **29,** 301–302.
A. Peckett, R. Phillips, and G. M. Brown (1972), "New Zirconium-Rich Minerals from Apollo 14 and 15 Lunar Rocks," *Nature* **236,** 215–217.
M. A. Pleskova and E. I. Zheleznova (1970), "Content of Uranium in Fluorite Determined from Fission Tracks of Uranium," *Sb. Kratk. Soobshch. Mineral. Geokhim.* 1968 god 1, 23–27.
P. B. Price and R. M. Walker (1963), "A Simple Method of Measuring Low Uranium Concentrations in Natural Crystals," *Appl. Phys. Letters* **2,** 23–25.
P. Ramdohr and A. El Goresy (1970), "Opaque Minerals of the Lunar Rocks and Dust from Mare Tranquillitatis," *Science* **167,** 615–618.
G. M. Reimer and B. S. Carpenter (1973), "Fission Track Analysis for Thorium in Glasses and Minerals," Geochronology Conference, "ECOG II," Oxford, 3–8 Sept. 1973.
C. M. Rice and S. H. U. Bowie (1971), "Distribution of Uranium in Apollo 11

Rock 10017," *Proc. Second Lunar Sci. Conf.*, **1,** 159–166. Cambridge: MIT Press.

H. Rydell and D. E. Fisher (1971), "Uranium Content of Caribbean Core P6304-9," *Bull. of Marine Sci.* **21,** 787–789.

M. Sakanoue, M. Nakaura, and T. Imai (1971), "The Determination of Plutonium in Environmental Samples," Inter. Symp. on Rapid Methods for Measurement of Radioactivity in the Environment, München.

P. Sandru and A. Danis (1971), "Determination of Uranium and Thorium Impurities in Mica by the Fission Track Method in Order to Use This Mineral as a Detector of Heavy Particles," *Rev. Roum. Phys.* **16,** 95–103.

R. A. Schlenker and B. G. Oltman (1973), "Fission Track Autoradiographs," ANL-8060, Radiological and Environmental Research Division Annual Report, Part II, Center for Human Radiobiology, 163–168.

P. Schöner and W. Herr (1969), "Uranium Analysis by the Fission Track Method and Applications to Dating Natural and Man-Made Glasses," *Proc. Intl. Topical Conf. on Nuclear Track Registration in Insulating Solids and Applications*, Clermont-Ferrand, **2,** Section VI, 12–13.

J. W. H. Schreurs, A. M. Friedman, D. J. Rokop, M. W. Hair, and R. M. Walker (1971), "Calibrated U-Th Glasses for Neutron Dosimetry and Determination of Uranium and Thorium Concentration by the Fission Track Method," *Rad. Effects* **7,** 231–233.

J. H. Schroeder, D. S. Miller, and G. M. Friedman (1970), "Uranium Distributions in Recent Skeletal Carbonates," *J. Sed. Petrol.* **40,** 672–681.

M. G. Seitz (1972), "Uranium Variation in Bovid Teeth of the Olduvai Gorge," *Carnegie Institution Year Book* **71,** 557–559.

M. G. Seitz (1973a), "Uranium and Thorium Partitioning in Diopside-Melt and Whitlockite-Melt Systems," *Carnegie Institution Year Book* **72,** 581–586.

M. G. Seitz (1973b), "Uranium and Thorium Diffusion in Diopside and Apatite," *Carnegie Institution Year Book* **72,** 586–588.

M. G. Seitz (1973c), "Boron Partitioning Between Minerals and Melt and Microdistribution in a Garnet Lherzolite and Two Chondrites by Particle Track Mapping," *Carnegie Institution Year Book* **72,** 588–593.

M. G. Seitz (1974a), "Boron Partitioning in Synthetic Crystal-Melt Assemblages and Its Microscopic Distribution in a Garnet Lherzolite," *Chem. Geol.*, in press.

M. G. Seitz (1974b), "Uranium and Thorium Partitioning in a Simple Basalt at High Temperatures and Pressures from 0 to 25 kbar," *Geochim. Cosmochim. Acta*, in press.

M. G. Seitz (1974c), "Partitioning of Samarium Between Diopside and Liquid at High Pressure," *Chem. Geol.*, in press.

M. G. Seitz (1974d), "U, Th, Pu Fractionation in Geologic Systems: Early Pu/U Abundance in Meteorites," *Trans. Amer. Nucl. Soc.* **18,** 87–88.

M. G. Seitz and S. R. Hart (1973), "Uranium and Boron Distributions in Some Oceanic Ultramafic Rocks," *Earth Planet. Sci. Lett.* **21,** 97–107.
M. G. Seitz and N. Shimizu (1972), "Partitioning of Uranium in the Diopside-Albite-Anorthite System and a Spinel Lherzolite System by Fission Track Mapping," *Carnegie Institution Year Book* **71,** 548–553.
M. G. Seitz and R. E. Taylor (1974), "Uranium Variations in a Dated Fossil Bone Series from Olduvai Gorge, Tanzania," *Archaeometry* **16,** 129–135.
M. G. Seitz, R. M. Walker, and B. S. Carpenter (1973), "Improved Methods for Measurement of Thermal Neutron Dose by the Fission Track Technique," *J. Appl. Phys.* **44,** 510–512.
M. G. Seitz, D. S. Burnett, and P. M. Bell (1974), "U, Th and Pu Partitioning Between Major Host Minerals in Equilibrated Chondrites," to be published.
M. Shinagawa, H. Takemi, Y. Kinaga, K. Nishizawa, and K. Narisada (1969), "Application of Solid-State Track Detector for RdTh Daughter Nuclides and Their Alpha Particles," *Proc. Jap. Conf. Radioisotop.*, 9th, Tokyo, 1969, 226–228 (Japanese).
D. J. Simmons and K. T. Fitzgerald (1970), "Application of Cellulose Nitrate Films for Alpha Autoradiography of Bone," Argonne Nat. Lab. Rept. ANL-7760, Pt. II, 208.
D. N. Slatkin, K. W. Jones, F. H. Geisler, A. P. Wolf, J. S. Fowler, H. W. Kraner, and E. P. Cronkite (1973), "Medical Autoradiography with Stable Isotope Thymidine: Theory and Preliminary Experiments," USAEC Conf.—730525, *Proc. 1st Intl. Conf. on Stable Isotopes in Chemistry, Biology, and Medicine*, May 9–11, 1972, Argonne, Ill., 410–420.
C. S. Su (1972), "The Determination of Uranium Concentrations in Seaweeds by Nuclear Track Detectors," *Rad. Effects* **14,** 109–112.
B. Szabo, J. R. Dooley, Jr., R. B. Taylor, and J. N. Rosholt (1970), "Distribution of Uranium in Uranium-Series Dated Fossil Shells and Bones Shown by Fission Tracks," U.S. Geol. Surv., Prof. Pap. 700-B, pp. B90–B92.
A. Tezuka (1972), "Uranium and Thorium Contents and Uranium Distribution in Igneous Apatite," *Chikyukagaku* (Geochemistry) **5,** 28–37.
K. Thiel, W. Herr, and J. Becker (1972), "Uranium Distribution in Basalt Fragments of Five Lunar Samples," *Earth Planet. Sci. Lett.* **16,** 31–44.
R. C. Thompson (1970), "Uranium Localization on Hydroxyapatite by Analysis of Fission Fragment Tracks," *Science* **167,** 1494–1497.
E. Uzgiris and R. L. Fleischer (1971), "Charged Particle Track Registration in Amber," *Nature* **234,** 28–30.
H. Wakita, H. Nagasawa, S. Uyeda, and H. Kuno (1967), "Uranium, Thorium and Potassium Contents of Possible Mantle Materials," *Geochem. J.* **1,** 183–198.
A. Wiechen (1971), "Die Uranbestimmung in Biologischem Material durch die

Fission Track-Methods," Inter. Symp. on Rapid Methods for Measurement of Radioactivity in the Environment, München.

H. A. Wollenberg and A. R. Smith (1968), "Radiogeologic Studies in the Central Part of the Sierra Nevada Batholith, California," *J. Geophys. Res.* **73,** 1481–1495.

F. W. Wood and A. H. Robinson (1970), "Neutron-Induced Autoradiography of Lithium Silicate Glasses," *J. Mat. Sci.* **5,** 425–433.

H. Yabuki (1971), "Fission Track Analysis of Uranium in Manganese Nodules Collected from Pacific Ocean Floor," *Sci. Papers Inst. Phys. Chem. Res.* **65,** 100–104.

H. Yabuki and M. Shima (1971), "Measurement of Uranium, Thorium in Manganese Nodules Using Fission Tracks and Alpha Particle Tracks," *Rept. Inst. Phys. & Chem. Res.* **47,** 27–32.

D. W. Zimmerman (1971), "Uranium Distribution in Archeologic Ceramics: Dating of Radioactive Inclusions," *Science* **174,** 818–819.

D. W. Zimmerman, M. P. Yuhas, and P. Meyers (1974), "Thermoluminescence Authenticity Measurements on Core Material from the Bronze Horse of the New York Metropolitan Museum of Art," *Archaeometry* **16,** 19–30.

Chapter 9

Radiation Dosimetry

9.1. INTRODUCTION

We can divide dosimeters into two categories—those that measure fluence, or particles/cm^2, and those that measure absorbed energy expressed in rads (1 rad = 100 erg/g), or dose equivalent expressed in rems (Roentgen equivalents for man). Rems = rads × *QF*, where *QF* (quality factor) depends upon the type of particle and may depend on its energy. For γ rays and high-energy protons, $QF \approx 1$; for thermal neutrons, $QF \approx 3$; for fast neutrons, *QF* ranges up to 10; and for alpha particles and heavy ions, *QF* ranges up to $\sim$20.

Fluence monitors are useful in a stable environment such as an accelerator or a research reactor where the energy spectrum and types of particles are known and one is concerned only with the number of particles/cm^2.

For purposes of radiation biology an ideal dose monitor should measure a quantity related to biological effect. With penetrating γ rays, electrons, and protons, this quantity is simply the total ionization per gram. With heavy ions the radiation damage is extremely concentrated along their trajectories, and their biological effects, though being actively investigated, are still poorly understood. With neutrons the absorbed dose varies strongly with energy as shown in Fig. 9-1. Neutrons produce biological effects in two ways: fast neutrons can collide several times with hydrogen nuclei, which produce damage through ionization loss; when the energy of the neutron is sufficiently low it can be captured through the ^{1}H(n,γ)^{2}D reaction that releases a 2.2 MeV γ ray, which also produces biological damage. The capture reaction ^{14}N(n,p)^{14}C has a lower cross section and is of secondary importance. The energy deposited by recoil protons and capture gammas is distributed nonuniformly through the body. The dose is a maximum at a depth of $\sim$4 cm, but the biological effects are more harmful to the organism at a depth of $\sim$10 cm, where the vital organs are situated. Curves for both depths are shown in Fig. 9-1.

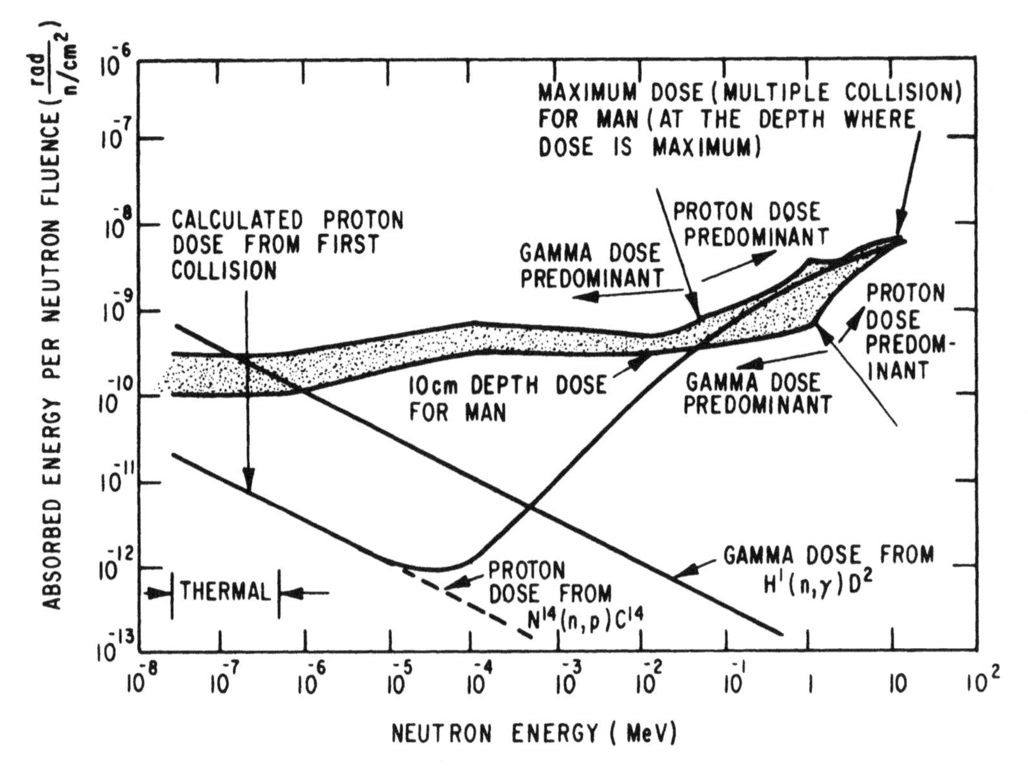

Fig. 9-1. *Absorbed dose in tissue irradiated by neutrons. The depth at which dose is a maximum may be considered to be 4 cm except in the energy region between 0.1 to 2 MeV. In this region the 4 cm depth dose will be approximately 2/3 the maximum dose. The first collision dose curve is shown for comparison.* (*After Pretre et al., 1968.*)

Much of this chapter will be concerned with ways in which dielectric detectors are being used to measure fluences and biological doses of neutrons. There is an outstanding need for sensitive dosimeters capable of monitoring routine doses encountered by persons in radiation environments under normal conditions, as well as for less sensitive dosimeters that can be left unprocessed for long periods until an accident occurs and a high dose is experienced. We will also discuss the use of dielectric detectors in accelerator dosimetry, in alpha particle dosimetry, and in heavy particle dosimetry.

9.2. NEUTRON DETECTION BY TRACKS FROM (n, FISSION) REACTIONS

In the discussion on element mapping in the preceding chapter we saw how the concentration of a fissionable element could be determined by bombarding the material with a known fluence of neutrons and counting the fission tracks. In 1963 we suggested that the procedure could be turned around and that the fluence of neutrons could be measured by counting the fission tracks in a thin, track-recording sheet in contact with a foil of fissionable material (Walker et al., 1963). To simplify readout we suggested several analog methods, such as using the etched sheet as a membrane separating two halves of an electrolytic cell and monitoring the current through the sheet.

Such a detector has a number of advantages over the method of counting recoil protons in nuclear emulsion and the method of measuring radioactivity in a fissile foil:

1. No need for specialized electronic counting equipment; ease of processing and evaluation.
2. No need for immediate retrieval.
3. No need for gram quantities of fissile materials.
4. Insensitivity to β, X, and γ radiation (which makes track counting in nuclear emulsion impossible at high doses).
5. No fogging, fading, or other storage problems at ambient temperatures.
6. Huge range of doses amenable to study.
7. Ease of activation or inactivation by separating fissile and detector foils.

Engineers within several components of the General Electric Company became convinced of the merit of such a concept, and work on practical dosimeters was initiated at Hanford (Prevo et al., 1964; Baumgartner et al., 1966; Baumgartner and Brackenbush, 1966), Knolls Atomic Power Laboratory (Schultz, 1966; 1970), and Vallecitos Atomic Laboratory (Davies and Darmitzel, 1965; and unpublished work). At about the same time work along similar lines began in France (Debeauvais et al., 1964; Mory and Walker, 1964); the U.S. Naval Radiological Defense Laboratory (Collver et al., 1965; Becker, 1966; Pretre et al., 1968); Oak Ridge National Laboratory (Kerr and Strickler, 1966; Becker, 1968); Canada (Bhatt, 1966); and Hungary (Medveczky and Somogyi, 1966). Since about 1967 a growing number of groups at centers for health physics and radiological protection throughout the world have contributed to the development of working dosimeters that can be read automatically or with a minimum of labor. Many of the problems and contributions are discussed in a recent review by Becker (1972), who has played a major role himself.

To within about 1%, at neutron energies encountered in reactor environments, the sensitivity of a plastic detector in contact with a thick fission foil depends only on the fission cross section and is given by the relation

$$\rho \text{ (tracks/cm}^2) = 1.6 \times 10^{-5} \text{ n (neutrons/cm}^2) \ \sigma \text{ (barns)} \qquad (9\text{-}1)$$

(Becker, 1966; Pretre et al., 1968). The reason for the constant factor is that all fissionable isotopes emit fission fragments with roughly the same range and with the same detection efficiency in plastics and other solids with a small value of θ_{crit}. For solids such as silicate glass, which do not record tracks at shallow angles to the surface, the numerical factor in eq. (9-1) should be multiplied by $\cos^2 \theta_{crit}$. For high-energy neutrons in accelerator environments the factor is likely to be somewhat different, but uncertainties in cross section are more serious.

9.2.1. Thermal Neutron Fluences

For a thick foil ($\gtrsim 8$ mg/cm^2) of pure ^{235}U, which has a 582 barn cross section for fission by thermal neutrons, eq. (9-1) becomes $\rho = 6.75 \times 10^{-3}$ tracks/neutron, whereas for a thick foil of natural U, the sensitivity is lower: $\rho = 4.87 \times 10^{-5}$ tracks/neutron. If one portion of the foil and detector is wrapped in a thick Cd foil so that only epithermal and fast neutrons can penetrate, the flux of thermal neutrons in a spectrum that includes neutrons of higher energy can be determined from the difference between track densities in the unshielded and shielded portions.

The first scientific application of fission track dosimeters was the use of ordinary microscope slide glass to monitor thermal neutron fluences in fission track data (Fleischer et al., 1965c). We found that the uranium was distributed quite uniformly at a level of about 0.4 ppm by weight in a typical group of slides. Samples of such glass are routinely used now in several fission track dating laboratories (see discussion in Chapter 4). In addition to being easy to read, they are a more reliable measure of neutron fluence than are the older monitors based on counting of radioisotropes in metal foils. Maurette and Mory (1968) have fabricated a variety of glasses with various levels of uranium so that, either by etching tracks in the glass or in an external detector, thermal neutron fluxes from $\sim 10^6$ to $\sim 10^{21}$/cm^2 can be determined. Two separate sets of special glass standards with various levels of uranium have recently been prepared in conjunction with the Corning Glass Works (Schreurs et al., 1971) and the Bureau of Standards (Seitz et al., 1973). The latter is designated as standard reference material, Certificates #610 to 617. By varying the concentration of ^{235}U in foils or glasses, and by using a variety of readout techniques ranging from the automatic spark-count method (Cross and Tommasino, 1970), which can rapidly determine track densities as low as ~ 1/cm^2, to electron microscopy, which allows track densities at least as high as 10^9/cm^2 to be measured, it is possible to monitor thermal neutron fluences covering the enormous range from $\sim 10^3$ to $\sim 10^{22}$/cm^2. At the highest fluences one must correct for the depletion by fission during irradiation of uranium in the glass and use a dosimeter such as quartz, which is very low in uranium (Kosanke, 1972).

Becker (1966) and Yokota et al. (1968) have shown that both gamma rays and thermal neutrons can be monitored with silver-activated phosphate glass that has either been coated with ^{235}U or doped with ^{235}U. The gamma ray dose is monitored by radiophotoluminescence in the glass, and the neutron dose is determined by etching tracks in the same material. It is also possible to measure the fast neutron dose if the glass is doped or coated with ^{232}Th, ^{237}Np, or natural U.

9.2.2. Fast Neutron Fluences

Fast neutron dosimetry is much more difficult than thermal neutron dosimetry, because the exact energy spectrum is seldom known, and because one may have to work with a complex spectrum of fast, epithermal, and thermal neutrons. The thermal neutrons are almost completely eliminated when the detector is covered with cadmium or boron, but the problem of determining the distribution of higher energy neutrons remains.

In the simple situation where the energy spectrum does not vary with time and can once be determined over a convenient time interval with more elaborate means, it is practical to monitor fluences in that same environment with a combination of track-recording solid and a material such as ^{232}Th or ^{237}Np that has a threshold for fission at an energy greater than ~0.1 MeV, so that it does not respond to thermal or epithermal neutrons. Fig. 9-2 gives some idea of the range of fast neutron fluence that can be measured using various concentrations of

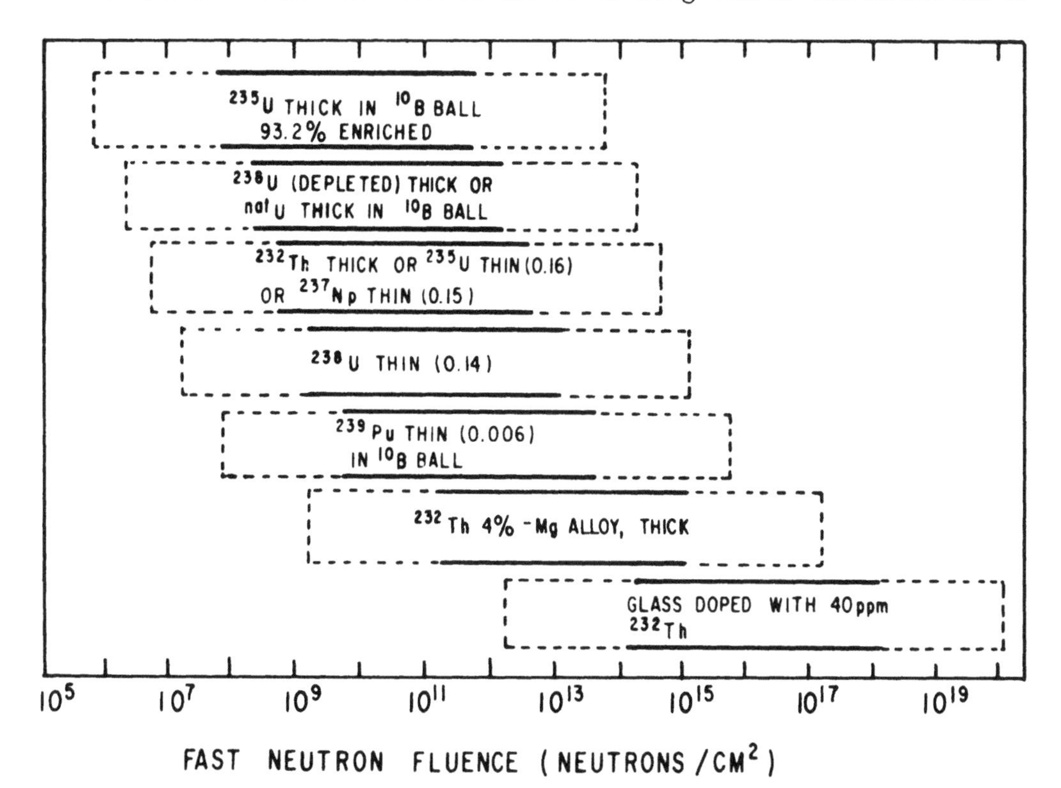

Fig. 9-2. *Range of fast neutron fluence that can be measured with fission foils.*

fissionable isotopes. The solid bars indicate the optimum range of fluences if optical microscopic counting is done; the corresponding range of track densities is $\sim 10^3$ to $\sim 10^7$ tracks/cm^2. The hatched region at lower fluences can easily be studied with the automatic spark-counting technique. The hatched region at higher fluences is less convenient but can be studied by scanning or transmission electron microscopy (Schreurs et al., 1971; Besant, 1970). The number given in parentheses for thin foils is the counting sensitivity for a conveniently producible foil relative to that for a thick foil (Pretre et al., 1968). To monitor fast fluences above about 10^{15}/cm^2 a low-uranium glass can be doped with a few ppm of ^{232}Th (Schreurs et al., 1971), whose thermal neutron fission cross section is only ~ 40 μb (de Mevergnies and del Marmol, 1968).

One way of getting information about the neutron energy distribution is to use several fissile materials with different thresholds. Fig. 9-3 shows the fission cross sections of various isotopes for neutron energies up to 10 MeV. The earliest and most thorough study of a practical personnel dosimeter was made by Baumgartner and Brackenbush (1966) and Baumgartner et al. (1966), who used the following criteria in choosing their threshold foils:

1. The dosimeter should have a fast neutron response that is an improvement over the response of nuclear film emulsion, which has a low-energy cutoff at about 0.7 to 0.8 MeV when the direction of the neutron is not known.
2. The energy response of the dosimeter should be proportional to the biological dose response for neutrons of different energies.
3. The dosimeter must have the capability for distinguishing doses from fast and thermal neutrons and should give some indication of the dose from intermediate neutrons.

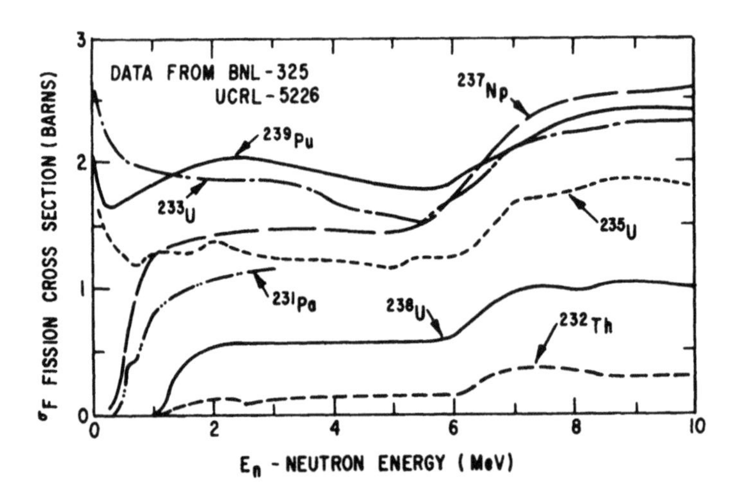

Fig. 9-3. *Energy-dependence of fission cross sections for various nuclides.*

It is obvious from Fig. 9-3 that no one nuclide by itself will satisfy all three of the criteria. ^{238}U has a threshold of 1 MeV and will not be able to detect the intermediate neutrons. Furthermore, it is impossible commercially to obtain ^{238}U completely free of ^{235}U, but only depleted. ^{232}Th has a threshold of 1.2 MeV and its cross section is even smaller than is that in the case of ^{238}U. ^{235}U and ^{239}Pu are very sensitive to thermal and epithermal neutrons. Furthermore, ^{239}Pu must be handled in gloveboxes.

^{237}Np obviously has the best characteristics for neutron dosimetry. Its response falls sharply below $\sim$0.4 MeV but has a finite cross section even for 0.1 MeV neutrons. This is an important advantage for personnel who will be exposed to degraded fission-spectrum neutrons around a reactor. Because of its low threshold, high cross section, and absence of high-energy gamma sensitivity, ^{237}Np is far superior to ^{238}U and ^{232}Th. The main drawback is its tendency to oxidize and flake in air. Baumgartner and Brackenbush embedded their ^{237}Np in resin and made it the basis of a compact dosimeter whose only drawbacks were the high background gamma activity, 4mR/h at the surface, and the inherent hazards of the fissionable material. The radioactivity could easily be decreased by a large factor by using thinner ^{237}Np. Fig. 9-4 shows the construction of their dosimeter, which contains three fissionable isotopes—^{237}Np, ^{235}U, and ^{238}U.

Calibrations with a variety of neutron sources having energies up to 14 MeV

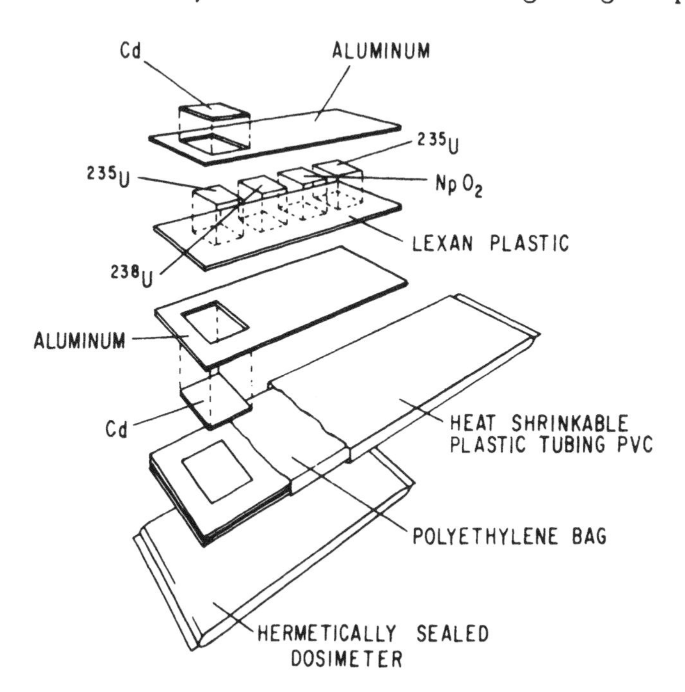

Fig. 9-4. *The personnel dosimeter of Baumgartner and Brackenbush (1966). The NpO_2 is incorporated in a polyester resin.*

Table 9-1

Energy	Comparative Track Counts
<0.5 eV	bare ^{235}U and Cd-covered ^{235}U
0.5 eV < E < 0.4 MeV	Cd-covered ^{235}U and ^{237}Np
0.4 MeV < E < 1.3 MeV	^{237}Np and ^{238}U
E > 1.3 MeV	^{238}U and bare ^{235}U
E > 0.4 MeV	^{237}Np

showed that their dosimeter could be used to measure fluences in five energy ranges shown in Table 9-1.

Sohrabi and Becker (1972) have discussed the advantages and difficulties associated with ^{237}Np fission foil dosimeters. They support the conclusion that it is the single most useful isotope for fast neutron dosimetry. For a 2.5 cm^2 active area and a thin Kimfol foil sandwiched between two 2.8 mg/cm^2 layers of ^{237}Np, the investment in ^{237}Np would be about $2 per badge; the sensitivity would be $\sim 6 \times 10^{-4}$ tracks per neutron/cm^2 for energies above 0.6 MeV; the dose rate outside the badge would be $\lesssim$0.05 mR/day; and the background track accumulation rate due to spontaneous fission would be <0.1 track/month. Fig. 9-5 shows the response as a function of neutron dose. To avoid the apparent saturation in spark counts at doses above a few rads, one can abandon the spark-counting technique in favor of track-counting by optical microscopy. The background spark count is zero, and 1 mR of fission neutrons would result in ~6 spark counts, permitting a dose measurement at this level with about ±40% statistical accuracy. If the Kimfol is sandwiched between two ^{237}Np foils the count rate in the graph should be doubled. To reduce the hazard, each ^{237}Np foil is sealed in a thin (~2 μ) plastic bag, which reduces the spark count rate by only about 20%.

Sohrabi and Becker have studied the response of both ^{237}Np and ^{232}Th dosimeters to 14 MeV neutrons and to reactor fission neutrons as a function of direction and distance of the dosimeter from a water-filled plastic phantom used to simulate the thorax of a human body. They found that the ^{237}Np badge was about ten times more sensitive than the ^{232}Th badge to fission neutrons and about three times more sensitive than the ^{232}Th badge to 14 MeV neutrons. Both showed rather strong directional variations, with about a threefold higher response if worn on the part of the body facing the neutron source than if worn on the opposite side. The separation between body and badge was more critical for the ^{237}Np, whose response decreased by a factor three if the separation was increased from 0 to 2 cm. There is less than a 15% variation of response with the orientation of the plane of the detector.

Of the many methods that have been tried for producing a reasonably stable thin layer of ^{237}Np on an inert carrier, the best seems to be the "burning-in" of a viscous solution of neptunium nitrate in acetone and alcohol at ~500°C to the

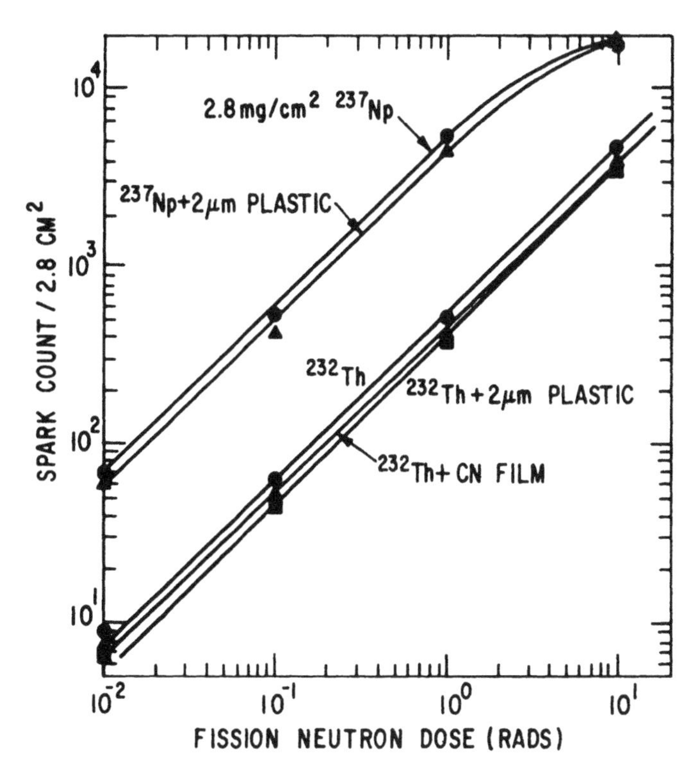

Fig. 9-5. *Spark count after optimized etching of a thin Makrofol film as a function of fast fission-neutron dose at the Oak Ridge Health Physics Research Center. Two of the curves are for a fission foil sealed in a 2μ plastic bag (Sohrabi and Becker, 1972). Note that for fast neutrons 1 rad = 10 rem.*

carrier after "painting" it or dipping it into the solution. The method has been used successfully at Argonne and in the Soviet Union for preparing tenacious, uniform layers of the desired thickness.

9.2.3. Determination of Biological Dose

Several attempts have been made to design detectors that match the biological dose curve in Fig. 9-1 over various energy intervals. Above ~2 MeV the fission cross section of natural uranium has nearly the same energy dependence as does the biological dose/(n/cm^2), but huge distortions arise at lower energy. To reduce their oversensitivity to thermal neutrons, Rago et al. (1970) have encapsulated detectors containing ^{235}U and ^{239}Pu in a 1/E thermal neutron absorber. A shield of 1.65 g/cm^2 ^{10}B results in the energy dependence shown in Fig. 9-6.

Another approach is to utilize the back-scattering of moderated neutrons from the human body (Tatsuta and Bingo, 1970). Using two detectors, one with natural

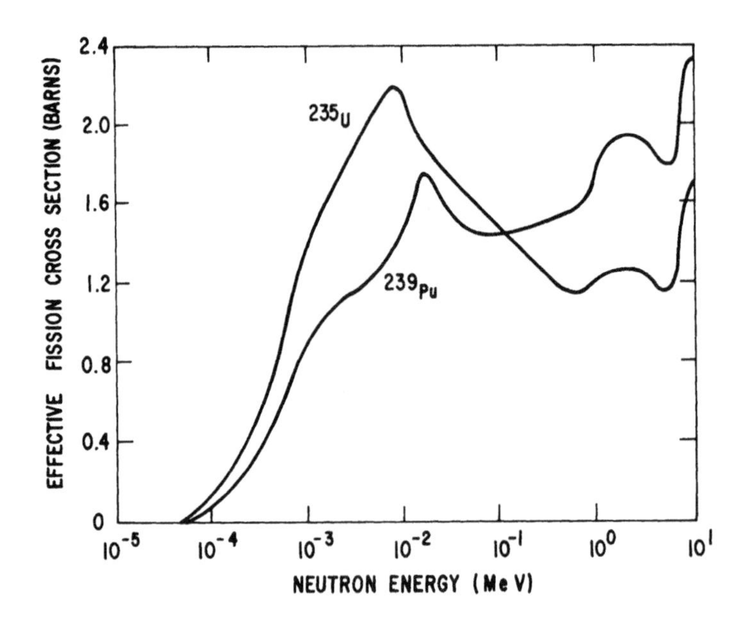

Fig. 9-6. *Effective fission cross section of* ^{235}U *and* ^{239}Pu*, encapsulated in 1.65 g/cm* ^{10}B*, as a function of energy. (After Rago et al., 1970.)*

and the other with slightly enriched UO_2, and a thermal neutron shield (Cd) only on the exterior side (no thermal neutron absorber between body and detector), it is possible to match the dose curve for neutron energies below 10 eV and above 1 MeV, as shown in Fig. 9-7.

At the Nuclear Research Center, Würenlingen, Switzerland, track-etch dosimeters that allow the measurement of either routine or accidental doses are being

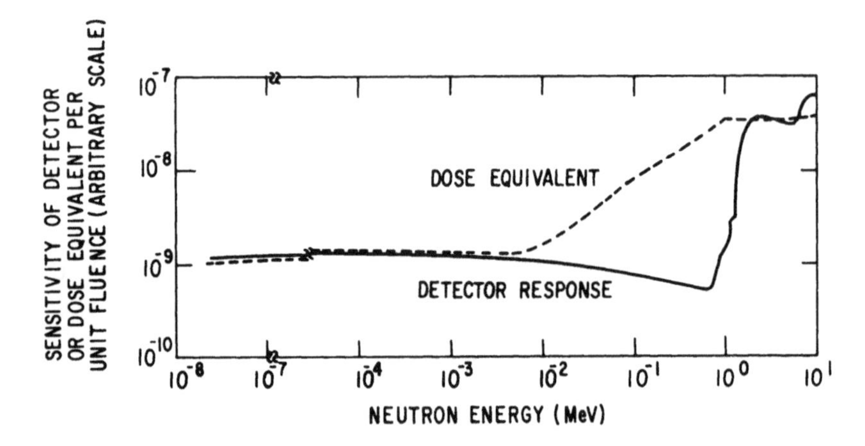

Fig. 9-7. *Sensitivity of a special neutron personnel dosimeter for perpendicularly incident neutrons and dose equivalent per unit fluence, as a function of neutron energy. (After Tatsuta and Bingo, 1970.)*

worn (Pretre, 1970). Each dosimeter contains a disc of ^{232}Th for fast neutron detection and a disc made of an alloy of 99% Al and 0.8% U (90% ^{235}U) for thermal neutron detection. The detector is a thin Makrofol sheet that can be etched and spark-counted. Fig. 9-8 shows the relationship between fission cross section and absorbed dose per neutron as a function of energy for the ^{232}Th foil and the Al-^{235}U foil. At energies above ~1.3 MeV and below ~1 eV the energy dependence of the cross section and of the absorbed dose per neutron are reasonably similar. In the important intermediate region the dose would be seriously underestimated. ^{237}Np, with its much lower fission threshold and higher sensitivity, would have attractive advantages over ^{232}Th.

Both Burger et al. (1970) and Remy et al. (1970) have pointed out that one can extend the approach of Baumgartner and Brackenbush (1966) by using a large number of foils with different energy dependence, thus arriving at an estimate of the neutron energy spectrum by an appropriate linear combination of responses. Remy et al. suggest not only fissile foils such as ^{234}U, ^{235}U, ^{236}U, ^{238}U, ^{237}Np, and ^{232}Th, with appropriate absorbers, but also foils containing Be and C in which (n,α) reactions are induced. The alpha tracks are then detected in cellulose nitrate. Their approach is quite comprehensive and includes all the reactions to be discussed in Section 9.3. If one has some idea of the distribution of neutron energies,

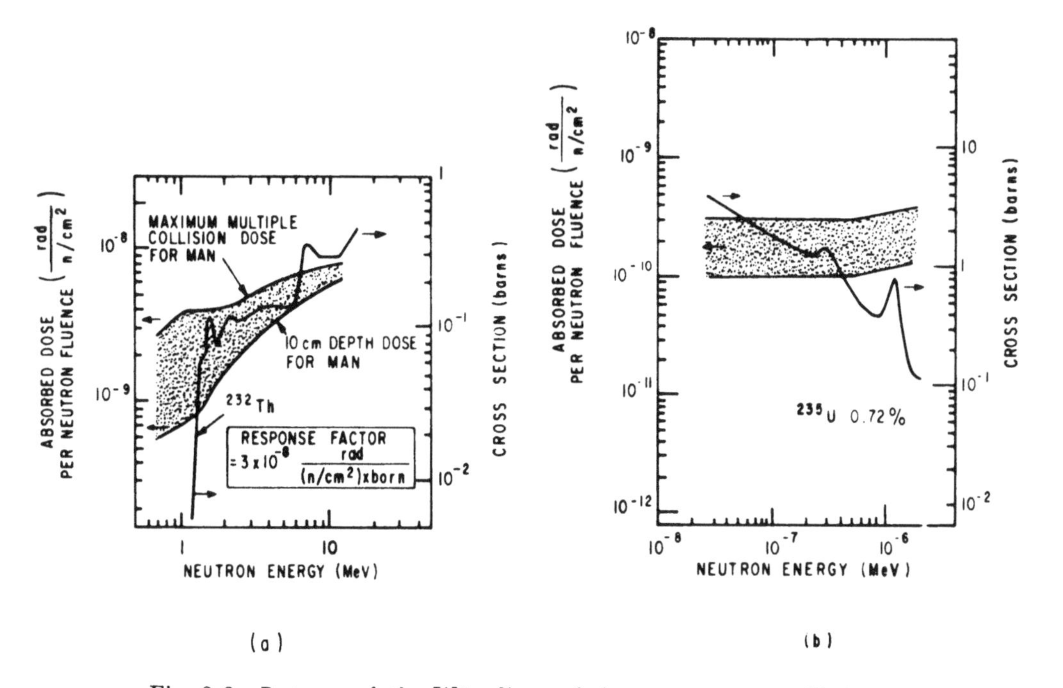

Fig. 9-8. *Response of the Würenlingen dosimeter to neutrons. Fission cross section and absorbed dose per neutron are plotted as a function of neutron energy for a* ^{232}*Th foil and for an Al foil containing 0.72%* ^{235}*U. (After Pretre, 1970.)*

the analysis is greatly simplified. For example, one can represent a spectrum of fission neutrons in equilibrium in air by the following functions:

$$\begin{aligned} \phi(E) &= \alpha\sqrt{E}\exp(-E/kT), && (\text{thermal} \leq E \leq 0.45 \text{ eV}) \\ \phi(E) &= \beta/E, && (0.45 \text{ eV} \leq E \leq 1.5 \text{ MeV}) \\ \phi(E) &= \gamma \exp(-0.75\, E), && (E \geq 1.5 \text{ MeV}) \end{aligned}$$

The interested reader should refer to the paper by Remy et al. (1970) for the details of the analysis.

Fission track dosimeters are being used at Oak Ridge National Laboratory (Becker, 1972) and at Karlsruhe and Leopoldshafen (Buijs et al., 1972) to monitor the hands of persons working with ^{238}Pu and other transuranium isotopes. A finger ring dosimeter contains a thin ^{232}Th foil attached to the top and a 10 μ Kimfol foil to be evaluated by spark counting. The Kimfol is glued to a retainer ring and pressed gently against the Th by a lead disc that reduces the gamma dose to the skin.

Agard et al. (1971) have shown that the high sensitivity of an (n,f) dosimeter that can be spark-counted is useful for monitoring the doses received by humans subjected to $\sim 10^7/\text{cm}^2$ fast neutrons from a Pu-Be source during *in vivo* neutron activation analysis for calcium in bones. Fast neutrons are used to penetrate to the bones, where thermalized neutrons produce the ^{48}Ca(n,γ)^{49}Ca reaction. The gamma rays provide a measure of calcium content in the bones. Their tests showed that the necessary dose information could be obtained from a single ^{232}Th foil.

Though the number of fission track dosimeters in use is still small (in addition to previous references, see papers by Běhounek et al., 1968; Gomaa et al., 1972; Preston and Peabody, 1972; Trousil et al., 1972), in many applications such dosimeters offer advantages over the method of proton-recoil tracks in nuclear emulsion, radioactive counting methods, and thermoluminescent dosimeters. The only serious drawback to their use is that they contain fissionable material, which might accidentally be exposed by a child or an animal if it were inadvertently brought home by the employee. It is certainly sufficiently safe that it could be given to carefully instructed personnel in high-risk environments.

9.2.4. Applications to Reactors and Neutron Generators

Information on fast neutron spectra in different reactors has been obtained with dielectric detectors in contact with various fissile isotopes inside a ^{10}B absorber sphere whose function is to reduce the oversensitivity of ^{235}U and ^{239}Pu to slow neutrons (Becker, 1966; Kerr and Strickler, 1966). Such an arrangement is now widely employed as a standard method for neutron flux determinations.

Dielectric detectors in a variety of configurations are being used inside reactors to map out fission density distribution inside fuel elements, near interfaces, and along reactor holes (Fleischer et al., 1965c; Tuyn, 1967; 1969; 1970; Sakanoue and Nakanishi, 1969a; Debrue et al., 1968; de Coster and Langela, 1970; Popa et al., 1970a); and to determine absolute fission rates (Gold et al., 1968; Roberts et al., 1968), fractional burn-up (Popa et al., 1970b), $^{238}U/^{235}U$ fission rates (Balducci et al., 1969; Besant and Ipson, 1970; Jowitt, 1971; Malykhin et al., 1972), fission product yields (Armani et al., 1970), precise fission cross sections (Rago and Goldstein, 1967; de Meverngnies and del Marmol, 1968; Sakanoue and Nakanishi, 1969b; Fabry et al., 1970; Benjamin et al., 1972), neutron energy spectra (Köhler, 1970), and effective thermal neutron temperatures (Liu and Su, 1971). Nakanishi and Sakanoue (1969) have mapped the neutron flux density in the vicinity of the tritium target of a 14 MeV neutron generator. In applications to flux mapping, high spatial resolution has been achieved by using electron microscopy to look at tracks in plastic films (Besant, 1970). Kosanke (1972) has found that fluence monitors made of natural crystalline quartz in contact with an extremely low concentration of fissile material will record tracks for periods exceeding one year at temperatures up to ~600°C without fading. This kind of monitor can thus be left in a high-temperature fast reactor to monitor high fractional burnup. In practice, of course, the thermal retentivity of a quartz detector should be checked before use, because Kosanke found that a sample of synthetic quartz was much less retentive of tracks than his natural quartz.

The General Electric Company has built and tested a tamper-proof nuclear safeguards device, shown in Fig. 9-9, that can be used for reactor surveillance (Weidenbaum et al., 1970). Information on the neutron energy spectrum is recorded in the sealed unit as a function of time by fissions induced in ^{235}U, ^{238}U, ^{237}Np, and ^{232}Th and recorded as tracks on a moving Makrofol tape. A ^{252}Cf fission source provides a known flux of particles to permit the tape speed to be determined. Its gradual decay of intensity (2.2 year half-life) provides an absolute time calibration. Once the flux level and neutron energy spectrum are known, the uses to which a reactor is being put can be recognized. The Arms Control and Disarmament Agency has indicated its approval of the device, but it is not yet clear whether the International Atomic Energy Association will have a large enough budget in the future to use the device to enforce the nonproliferation treaty. Commercial models are intended for use in Canadian reactors to count fuel-loading cycles.

9.2.5. High-Energy Radiation Near Accelerators

Threshold fission foil techniques are being used to monitor fluxes of high-energy nucleons around accelerators. At energies above a few hundred MeV, neutrons and protons induce fission with about the same cross sections and represent nearly

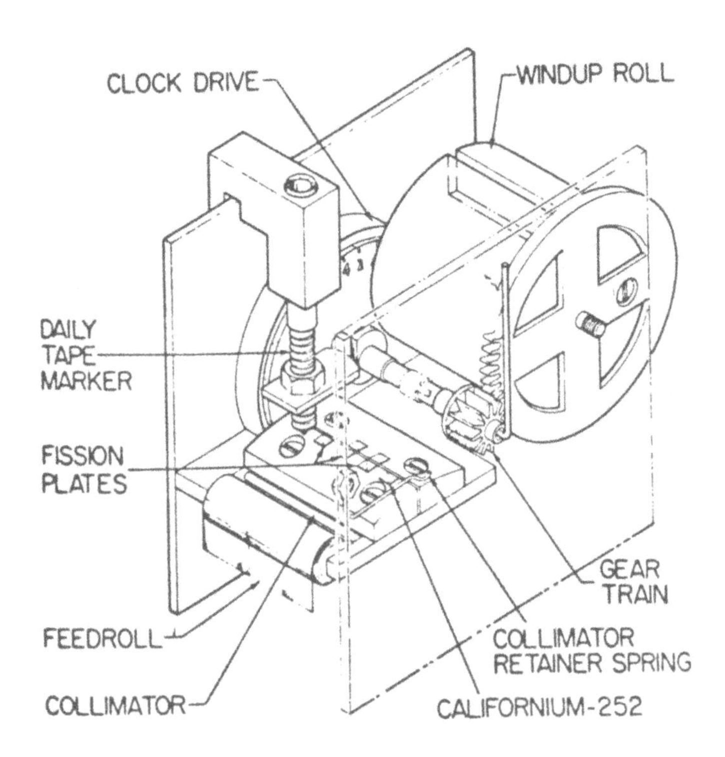

Fig. 9-9. *Nuclear safeguards device to monitor use of nuclear reactors. The neutron energy spectrum is recorded as a function of time by fission tracks induced in* ^{235}U, ^{238}U, ^{237}Np, *and* ^{232}Th *and recorded on a moving polycarbonate tape. A* ^{252}Cf *fission source provides a known flux of particles to permit calibration of the tape speed. (After Weidenbaum et al., 1970.)*

equal hazards since ionization damage is less than that from nuclear collisions. Elements somewhat lighter than Th have significant fission cross sections at high energies, and they can be used as high-energy threshold detectors.

Wollenberg and Smith (1969) have used Makrofol in contact with foils of Ta, Au, Bi, and U to determine the energy distribution of high-energy nucleons at the CERN proton synchrotron. At energies of several GeV the sensitivity is $\sim 10^{-6}$ tracks/nucleon. Heinzelman and Schuren (1970) have used bismuth foils, with a threshold at 50 MeV, to detect 90 MeV neutrons. With a detector area of ~ 120 cm^2 and spark-counting, they achieve a sensitivity of 0.4 track/mrem, corresponding to a detection limit of ~ 0.1 mrem.

In similar experiments at the Stanford Linear Accelerator, Svensson (1970) has used foils of ^{238}U and ^{232}Th. The advantage of ^{232}Th is that its nucleon fission cross section stays at a constant level of ~ 0.6 barn up to energies of at least ~ 350 MeV. At electron accelerators such as SLAC, high-energy gamma rays are produced and corrections for photofission must be made. These corrections can become substantial at energies above ~10 GeV; the photofission cross section of ^{232}Th at 13 GeV is ~0.3 barn.

High-energy nucleons induce spallation reactions in heavy nuclei causing heavy fragments to be ejected with a very broad range distribution, peaked at very short ranges but extending out to 20 microns or greater. The yield of observable tracks depends both on the Z of the target nuclei and on the detector sensitivity. It is best determined experimentally. Debeauvais et al. (1967) have used spallation tracks from Pb and Au to monitor the 18 GeV proton beam at CERN.

Though he used direct interactions of nucleons with C and O nuclei in Makrofol E instead of in a fissile foil, for completeness we should mention the work of Dutrannois (1971), who showed that neutrons and protons of energy above a few hundred MeV produce tracks at a constant rate of ~10^{-5} tracks/nucleon in the plastic itself. His fluence values agree very well with those determined by ion chambers and by activation of an Al foil.

Similarly, Annoni et al. (1970) have been able to determine the profile and absolute intensity of high-energy proton beams at levels of ~10^{11} to ~2×10^{12} particles/cm^2 with an error of ~10%. They used a microdensitometer to map out the density of spallation recoil tracks in etched cellulose nitrate.

9.3. NEUTRON DETECTION BY TRACKS FROM NONFISSION REACTIONS

In this category we consider primarily the detection of charged particle tracks from (n,α) and (n,p) reactions, recoil tracks from elastic collisions with hydrogen and heavier elements, and multipronged tracks from reactions in which several charged particles are emitted. Each of these reactions has certain attractive features for dosimetry, and each has its problems.

9.3.1. Alpha Tracks from (n,α) Reactions

In the previous chapter we saw that very low concentrations of isotopes such as ^{6}Li and ^{10}B with high cross sections for (n,α) reactions could be measured using plastics that are sensitive to low-energy alpha particles. Conversely, low fluxes of thermal and epithermal neutrons can be measured by using alpha-sensitive plastics in contact with a foil containing ^{6}Li or ^{10}B, or by using an alpha-sensitive plastic in which ^{6}Li or ^{10}B has been uniformly dispersed (Schultz, 1970). At low energies their cross sections for (n,α) reactions are inversely proportional to velocity. For a thermal distribution the cross sections are 3840 barns for ^{10}B and 950 barns for ^{6}Li, both of which are significantly higher than the thermal neutron fission cross section of ^{235}U. The sensitivity of an (n,α) dosimeter depends on the concentration of ^{6}Li or ^{10}B in the radiator and on the recording characteristics of the plastic detector. If ^{10}B is used, both the alpha and the recoil ^{7}Li fragment will record in most cellulosic polymers, but if ^{6}Li is used the alpha will record whereas the recoil ^{3}H will have too low an ionization to record until it has moved several microns beyond the point of emission. Roberts et al. (1970) have attained a sensitivity of 0.013 tracks/neutron, using a thick ($\sim$0.2 cm) ^{10}B disc with a cellulose nitrate or cellulose acetobutyrate detector to detect neutrons distributed over a 2π solid angle. Because of the complete absorption of thermal neutrons from the radiator side, isotropic neutrons would be detected with only half that sensitivity.

Johnson et al. (1970) have shown that the spark-counting method works well on alpha particle tracks in cellulose nitrate films 10 to 20 μm thick. Films that thin are not available commercially, but those cast from solution as described in Chapter 2 have excellent etching and spark-counting characteristics. Johnson et al. found that cast thin films of other cellulosics such as cellulose triacetate were unsatisfactory for spark-counting.

In personnel dosimetry the (n, α) method has several advantages over the (n,f) method:

1. No expensive and hazardous fissionable material is required.
2. Neither the skin near the dosimeter nor the gamma-sensitive detectors in the dosimeter are exposed to as much gamma radiation as would be emitted by fissionable materials.
3. There is no background track production from spontaneous fission.
4. There is no change in the etching characteristics of the detector because of an alpha radiation damage of its surface.

The practical importance of (n,α) dosimeters is, however, limited by the fact that significant thermal neutron exposures to personnel occur only very rarely. Various suggested methods of measuring fast neutron exposures by using the human body to thermalize and back-scatter neutrons are seriously limited by the strong $(\text{velocity})^{-1}$ dependence of (n,α) reactions. Of course, fast neutrons also initiate (n,α) reactions with heavier elements such as the Fe-group, but the cross sections are rather low.

Because of its extremely high sensitivity, the (n,α) method of thermal neutron dosimetry has found several special applications. Roberts et al. (1970) have proposed using a large area of cellulose acetobutyrate in contact with a 0.2 cm thick ^{10}B radiator to measure the distribution of low fluxes of thermal neutrons over land and water produced by cosmic rays.

Woolum et al. (1973) have made the first direct measurement of the distribution of thermal neutrons with depth in the lunar soil. Astronauts Cernan and Schmitt implanted their 2 m cylindrical rod into the hole left in the soil at the Apollo 17 site after the removal of a deep drill core. After the probe was in place it was activated by rotating an inner cylinder so that strips of ^{10}B and ^{235}U were brought next to strips of cellulose triacetate and mica that lined the adjacent, outer cylinder.

An on-off mechanism was necessary to prevent accumulation, during the flight to the moon, of background events produced by neutrons from the ^{238}Pu used as a power generator for lunar surface experiments and from cosmic ray neutrons produced in the spacecraft. Fig. 6-14 compares the resulting depth profile of thermal neutrons with the theoretical calculation by Lingenfelter et al. (1972). A knowledge of this profile is important because it permits one to infer the chronology and depth dependence of dynamical processes on the lunar surface from measurements of certain isotopes such as ^{157}Gd and ^{149}Sm, which are produced with extremely high cross sections in (n,γ) reactions with thermal neutrons.

9.3.2. Detection of Fast Neutrons by Means of Recoil Tracks

Fast neutrons can produce tracks by causing nuclei to recoil elastically either within the detector or from an external radiator foil. Provided their tracks are detectable, light nuclei are the most favorable because they have higher elastic scattering cross sections, they carry away a greater fraction of the energy from a collision with a neutron, and they have a greater range in the detector. Thus, the sensitive thickness of a radiator for recoil emission can be greater for lighter particles.

Proton recoil tracks will be detectable only in the most sensitive materials and then usually only in the last 1 to 2 microns of their range. Helium has been successfully used (Becker, 1969b) as an external radiator. The sensitivity of cellulose nitrate to fission neutrons is ~12 times higher in a helium atmosphere than when only the C, N, and O nuclei in the detector are the sources of tracks. Of course, a gaseous radiator is cumbersome and unsatisfactory for personnel use. Frank and Benton (1972) have systematically studied recoil track production as a function of Z of the radiator and the sensitivity of the plastic detector. They prefer beryllium, which is available in foils and has an elastic scattering cross section that slowly decreases from ~4.2 barns at 0.5 MeV to ~0.9 barn at ~20 MeV. Though Lexan is an adequate detector, they prefer using their own cellulose nitrate that has been doped with a plasticizer to reduce its sensitivity so that it does not record alpha particles from competing (n,α) reactions. Cronar is a commercially available plastic detector that is also insensitive to alpha particles (Fleischer and Lovett, 1968).

Many groups have used polycarbonate, cellulose nitrate, or cellulose acetate as both radiator and detector of recoil tracks (Medveczky and Somogyi, 1966; Becker, 1969a,b; Tuyn and Broerse, 1970a,b; Frank and Benton, 1970a; Nishiwaki et al., 1971). The sensitivity to fast neutrons ranges from $\sim 3 \times 10^{-6}$ to $\sim 2 \times 10^{-5}$ tracks/neutron, depending on the plastic and the neutron energy. This is comparable to the sensitivity of a plastic in contact with natural U and enclosed in Cd.

For some materials such as cellulose acetate the dependence on neutron energy follows the first-collision dose quite well down to ~0.5 MeV (Fig. 9-10). The

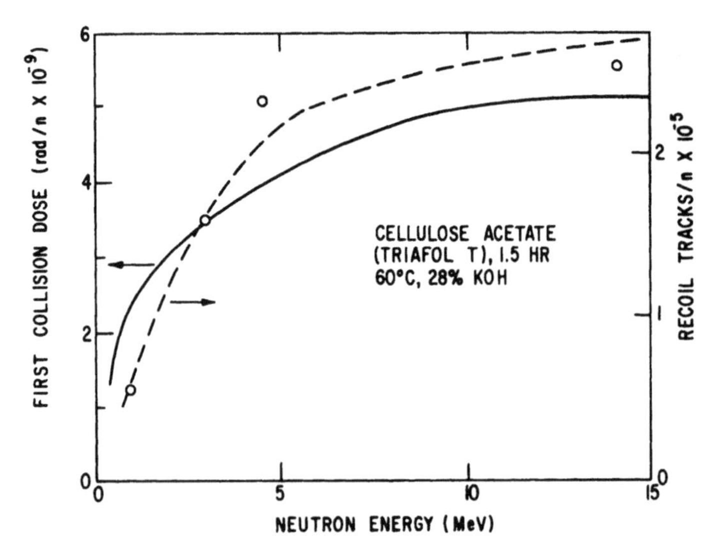

Fig. 9-10. *First-collision dose and recoil track density in cellulose triacetate as a function of neutron energy.* (*After Becker, 1969b.*)

minimum neutron energy for recoil track detection depends on such factors as the radiator and detector used, the etching procedures, and the scanning procedure. It ranges from ~0.2 to ~1 MeV (Jozefowicz, 1971).

Because the average range of a recoil nucleus increases with the energy of the colliding neutron, the average depth and diameter of the etch pits are a measure of neutron energy. In practice, the diameter distribution also depends on the direction of neutron incidence (Tuyn and Broerse, 1970b), which makes it somewhat difficult to use pit size to study the neutron energy distribution.

In an attempt to measure both the fluence and the energy spectrum of neutrons with $5 \lesssim E \lesssim 20$ MeV, Frank and Benton (1972) have introduced several thicknesses of a degrading foil between their radiator (Be foil) and their track recorder (insensitive cellulose nitrate). The degrader is Au, which has a very low cross section for (n,α) reactions and does not give rise to detectable elastic recoils in this energy range. A computer is used to unfold the neutron spectrum from measurements of track densities beneath various thicknesses of absorber.

The main drawback of recoil-track dosimeters is that the tracks etch more slowly and are smaller than fission fragment tracks. They are more difficult to count visually than fission tracks and are not long enough to be spark-counted (Johnson and Becker, 1970), except perhaps in the special case of high-energy ($E \gtrsim 5$ MeV) neutrons.

Recoil-track dosimeters have been used for personnel accident neutron dosimetry (in at least two major nuclear installations, ORNL and Karlsruhe, polymer foils have been incorporated in dosimetry belts worn by personnel) and for neutron depth-dose studies in phantoms (Becker, 1969a; Tuyn and Broerse, 1970a). In such applications tissue-equivalent composition is an important feature of plastic detectors.

9.3.3. Detection of Neutrons with $E \gtrsim 10$ MeV by Multiprong Tracks

In our 1965 review article (Fleischer et al., 1965a) we pointed out that fluxes of fast neutrons in various energy intervals above ~10 MeV could be measured and easily distinguished from fluxes of lower energy neutrons by virtue of nuclear breakup reactions such as $^{12}C(n,n')3\alpha$ that give rise to characteristic multipronged events. The cross section for the carbon breakup ranges from ~0.2 to ~0.3 barns in the interval from ~12 to somewhat more than 20 MeV, which leads to a sensitivity of $\sim 4 \times 10^{-8}$ tracks/neutron when carbon in the polymer itself is used as the target (Frank and Benton, 1970b). The three-pronged tracks have to be scanned microscopically, in order to be picked out of the ~100-fold higher background of single tracks from elastic recoils and other (n,α) reactions. This background imposes an upper limit of $\sim 10^{12}$ neutrons/cm^2. The energy window that can be studied, ~10 to ~20 MeV, is quite important for the monitoring of neutrons from the thermonuclear reaction $^3H(d,n)^4He$.

9.3.4. Proton Tracks from (n,p) Reactions

Cellulose nitrate (Fleischer et al., 1965b), cellulose acetate (Varnagy et al., 1970), and UV-sensitized Lexan (Stern and Price, 1972) are sensitive to protons of energy up to ~100 keV. The fraction of atoms that are hydrogen is 32% in cellulose nitrate and 45% in Lexan. In a neutron energy interval from perhaps 20 to 100 keV, tracks of recoil protons produced in elastic collisions with hydrogen in the plastic would be detectable. In this energy interval the cross section for (n,α) reactions with ^6Li and ^{10}B is too low to be useful, and recoils of heavier nuclei such as C and O in the detector would have too short a range to be detectable. More recently Lück (1974) has described a cellulose nitrate that will directly reveal protons with energies up to 700 keV.

9.4. ALPHA PARTICLE DOSIMETRY

At least two sources of naturally occurring alpha particles represent hazards to mankind: cosmic radiation, which is a potential hazard only at very high altitude and will be discussed in Section 9.5, and airborne radon gas and aerosol particles containing radon daughters. ^{222}Rn has a 3.8 day half-life and emits a 5.5 MeV alpha particle. Because radon is a gas, it emanates from the earth, and is present at trace levels even at high altitudes. We will see in Chapter 10 how subsurface uranium ores can be detected by means of alpha tracks in cellulose nitrate detectors mounted in inverted cups that have been planted in soil.

We are concerned here with the monitoring of highly toxic alpha emitters present at levels that represent a hazard. Of utmost concern is the risk to uranium miners, who regularly ingest radon gas and aerosols to which the nongaseous radon daughters become attached. Two of these, ^{218}Po and ^{214}Po, are alpha emitters (6.0 MeV and 7.68 MeV, respectively) and cause serious damage to bronchial tissues and lungs; the other two, ^{214}Pb and ^{214}Bi, are beta emitters. Prolonged exposure to alpha emitters may account for the increasing incidence of lung cancer among underground uranium miners (Federal Radiation Council, 1967).

At a maximum permissible concentration of radon in equilibrium with its daughters (3×10^{-14} curie/ml) for occupational workers for a 40-hour week, the flux of alpha particles is small and difficult to measure electronically without elaborate means of concentrating a large volume of gas. Because of the warm, humid atmosphere in mines and the necessity for an opaque wrapping, nuclear track emulsions are unsatisfactory as dosimeters. Alpha-sensitive plastics are obviously very attractive, and a number of groups have attempted to build working personnel dosimeters (Becker, 1969a; Rock et al., 1969; Lovett, 1969; Anno et al., 1970; Auxier et al., 1971; Liniecki et al., 1972; Frank and Benton, 1973a). Early tests in cooperation with the U.S. Bureau of Mines were encouraging (Lovett, 1967), but even now no completely satisfactory dosimeter exists.

Sensitivity is not a serious problem for plastic detectors, nor are humidity and temperature. At the so-called working level (1 WL = 100 pCi/l of air for ^{222}Rn in equilibrium with its daughters) the measured track-production rate is $\sim$28 tracks/cm^2 WL-hour or $\sim$1120 tracks/cm^2 WL-week for cellulose nitrate detectors exposed only to radon gas in an actual mine (Rock et al., 1969). The problems are mainly practical ones. In laboratory conditions and in passive tests where dosimeters were left in mines, several models give track counts that increase linearly with WL-hours of exposure as measured by a technique in which air is periodically pumped through a filter, which is then counted electronically. A dosimeter that is worn by a mine worker tends to become caked with dirt, which changes its response to daughters that plate out on solid surfaces, increasing its sensitivity if the deposit is thin enough to slow down the alpha particles enough to produce tracks, or blocking the detector if the deposit is thicker than the range of alpha particles.

Two approaches have been tried. One is to design the dosimeter so that only radon gas can produce etchable tracks. The General Electric dosimeter (Lovett, 1969) employs a cellulose nitrate sheet mounted on a miner's helmet in an orientation such that no alphas from daughters that have plated out on nearby surfaces will reach the detector. In principle, alphas from daughters plated out on the detector will have too high an energy, and too low an ionization rate, to record. In practice, accumulating surface deposits alter the recording rate. Frank and Benton (1973a) have attempted to circumvent the plate-out problem with a dosimeter consisting of a hollow cylinder with a cellulose nitrate sheet at one end and a filter at the other. Alphas from daughters that plate out on the filter cannot penetrate the several centimeter air path to the detector. Only ^{222}Rn gets through the filter; this gas and its daughters produce tracks at a rate that can be calibrated. One problem with a detector that measures only the radon concentration is that the dose from radon daughters (which account for most of the lung doses for uranium miners) may be overestimated by a large factor if the air is strongly ventilated or if fresh ^{222}Rn is released during rock chipping, for then the daughters will not have built up to equilibrium. At an atmospheric "age" of only 1 minute, the overestimate is by a factor of ten (Lovett, 1969).

A second approach is to monitor both radon and its daughters, using suitable absorbers to reduce the energy of the 6.0 and 7.68 MeV alphas so that they can produce detectable tracks. Auxier et al. (1971) have built a compact dosimeter with a tiny, motor-driven fan that sucks air through a filter in contact with an absorber and a plastic detector. Such a device is prone to clogging in actual mine use. Frank and Benton (1973b) have constructed a helmet-mounted dosimeter with a moving brush whose function is to wipe off accumulating dirt.

The acuteness of the need, and the extent of developmental activity (strongly supported by the Bureau of Mines), are such that it appears likely that an acceptable track-etch dosimeter may soon be put into use.

An unusual, hazardous situation exists in Grand Junction, Colorado, where alpha particle dosimeters are extremely important. Tailings from uranium-rich ore deposits were used in the construction of many of the houses. The Environmental Protection Agency, the Atomic Energy Commission, and the State of Colorado are concerned with the level of radioactivity in the air in these homes and have been using track-etch films as dosimeters. The results appear to be reliable because Rn and its daughters are in equilibrium and the films do not become caked with dirt.

Anno et al. (1970) have built a sensitive balloon-borne radon gas monitoring system for studying the distribution of radon in the atmosphere. With its 3.8 day half-life, radon can be used as a tracer in hydrology, aeronomy, and meteorology (for example, to predict monsoon patterns in India).

9.5. DOSIMETRY OF HIGH-*Z* NUCLEI

We first discuss the straightforward use of dielectric detectors in fluence measurement and then go into the more difficult problem of determining biological dose.

9.5.1. Heavy Ion Fluences in Accelerators

In Chapter 5 we discussed in detail the problem of determining the flux of heavy ions in the solar wind, solar flares, and galactic cosmic rays as a function of Z and energy. That is normally considered an astrophysical problem rather than a problem in dosimetry. Here we concern ourselves first with the task of measuring fluences of heavy ions in accelerators, perhaps amid a background of light ions.

Considerable effort at existing accelerators is being devoted to obtaining heavy ion beams. During the early stages of beam development when fluxes are extremely feeble, dielectric detectors are useful in tuning and aligning the beam and in establishing that the desired ions have been successfully accelerated. Sometimes a measurement of total track length is enough to prove that an ion of the desired Z has been accelerated. If several species with nearly the same range are believed to have passed through the system, measurements of etch rate as well as total range must be made, using criteria discussed in Chapter 3. Dielectric detectors have been used to verify the acceleration of weak beams of Xe ions at the 310-centimeter cyclotron Dubna (Flerov, 1972) and 0.29 GeV/nucleon ions of ^{14}N and ^{40}Ar at the Princeton-Penn accelerator (Schimmerling et al., 1971; Isaila et al., 1972). The problem of deciding whether heavy ions can be successfully carried along with the electromagnetic field of an intense "smoke ring" of electrons at the Berkeley Electron Ring Accelerator can be solved by using an insensitive dielectric detector as the target.

Plastic detectors can be calibrated once with ionization chambers so that they will provide a quick, integrated, two-dimensional map of the distribution of relativistic heavy ions across a target. The beam ions themselves do not record tracks, but the density and spatial distribution of tracks of highly ionizing fragments from nuclear interactions in the plastic will be proportional to fluence.

9.5.2. Biological Effects of Heavy Ions

The conventional dosimetric system based on the mean absorbed energy per unit volume of tissue breaks down when applied to highly ionizing heavy nuclei. The reason is that the radial distribution of energy density deposited along the path of the particle changes extremely rapidly on a scale much smaller even than the diameter of a single cell. As an example, in Fig. 9-11 one can see that the radial distribution of dose for an ^{16}O nucleus of velocity $0.1c$ and $0.4c$ changes by several orders of magnitude within a few microns of the trajectory of the ion. Kinematics limits the maximum energy that can be imparted by a fast particle to electrons (of mass m_e) in its path to $2m_ec^2\beta^2\gamma^2$, where $\gamma = (1 - \beta^2)^{-1/2}$. At the lower velocity, $0.1c$, the absorbed dose is localized within a cylinder of radius $\sim 2.5\ \mu$. At the higher velocity, $0.4c$, the kinematic limitation is weaker, and some of the energy is transferred to a much greater distance. Numerous experiments have shown that a single

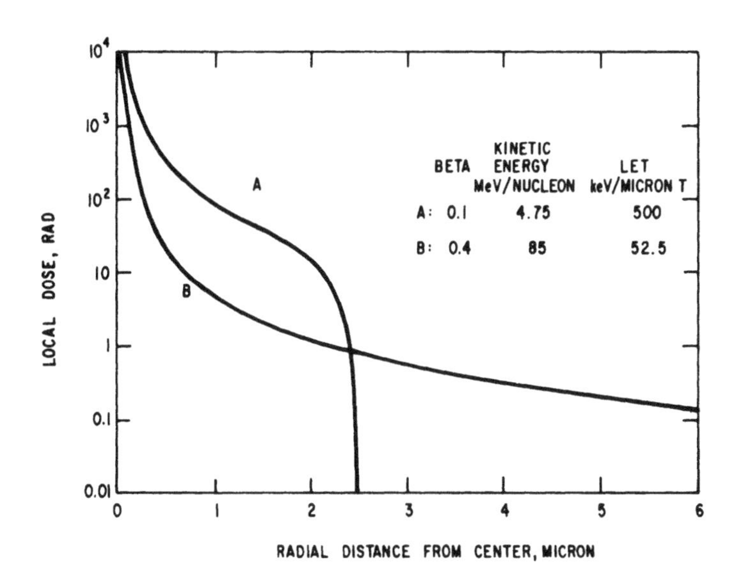

Fig. 9-11. *Calculated radial spread of energy dissipation around an oxygen track in tissue. Beta is the velocity relative to the velocity of light; LET is the energy loss per unit path length.*

ion with a sufficiently high ionization rate can inactivate with unit probability a cell through whose nucleus it passes. The interested reader should see the papers by Katz et al. (1971) and Schaefer (1972) for discussions of experiments and concepts. Of course, an ion with high β will not deposit all of its energy within a cylinder with cross-sectional area of cellular dimensions, and thus some of the total energy deposited contributes to radiation damage of cells not directly in its path. In an attempt to apply dosimetric concepts to radiation damage by high-Z ions, the traditional terminology of LET (= linear energy transfer in MeV/μ) has been replaced with the concept of a restricted LET, which applies only to that portion of the energy transfer deposited by secondary electrons with less than a specified range.

At present the biological effects of high-Z particles on an entire organism are not understood, nor have internationally agreed upon exposure limits been established. The problem is clearly an experimental one. Because of their similarity in density and composition to living tissue, and because they respond only to the very heavily ionizing particles that can kill individual cells with unit efficiency, track-recording plastics make ideal dosimeters for astronauts, who until now have been the best examples of persons exposed to naturally occurring high-Z particles. Since 1966 Benton and his coworkers have been using cellulose nitrate, cellulose triacetate, and Lexan to monitor doses of stopping high-Z particles received by the astronauts (see, for example, Benton and Collver, 1967; Schaeffer et al., 1972). They find that in satellites orbiting the earth at low inclination and at typical altitudes of ~200 nautical miles the main contribution of high-Z particles is from heavy recoils produced in nuclear reactions between energetic protons from the trapped radiation and nuclei in the plastic. Only very high-energy cosmic rays are permitted by the geomagnetic field to reach the spacecraft, and few of them come to rest in the body. Using a minimum track length criterion of 10 μm, Benton et al. (1972a) measured a heavy recoil track density of $\sim 10^3/\text{cm}^3$ day within Biosatellite III.

In high-inclination orbiting satellites such as Skylab, and in space probes that leave the earth's magnetosphere, such as the Apollo spacecrafts, the major contribution is from low-energy heavy nuclei in the galactic cosmic radiation (or occasionally in solar flares) that come to rest in tissue. Fig. 9-12 summarizes the data of Benton and Henke (1972a,b,c; 1973) for all the Apollo missions. To see if there was any correlation between the flux of stopping high-Z nuclei and the degree of modulation of the galactic cosmic ray intensity by solar magnetic fields, we have used as the abscissa the neutron monitor counting rate at Deep River. A definite correlation appears to exist.

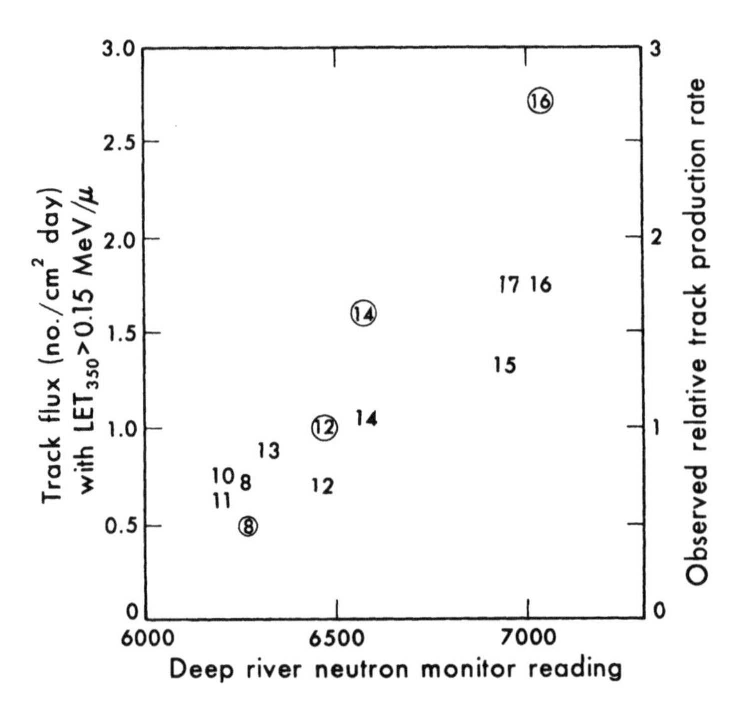

Fig. 9-12. *Correlation of flux of heavily damaging cosmic ray nuclei with solar activity. The numbers on the graph refer to Apollo missions 8 through 17. A high neutron monitor reading indicates a high flux of cosmic rays reaching the earth and thus implies weak solar modulation. Data without circles are from Benton and Henke (1972a,b,c; 1973) and correspond to left-hand ordinate. Circled data are from Fleischer et al. (1973) and give observed track production rates, corrected for shielding thicknesses and normalized to unity for Apollo 12.*

Detectors in different regions of the spacecraft show variations in fluence comparable in magnitude to those resulting from solar modulation. Benton and Henke (1972d) found that the heavy particle fluence at the heavily shielded Microbial Ecology Evaluation Device in Apollo 16 was only about one-fourth that recorded on the astronauts' dosimeters. Comstock et al. (1971) and Fleischer et al. (1973) have reported fluences in Lexan materials located in various parts of the Apollo 8, 12, 14, and 16 spacecrafts, the raw data from which appear uncorrelated with the solar cycle. However, after correcting for the large differences in shielding thicknesses at different locations, a strong correlation with solar cycle exists, as is shown by the circled points in Fig. 9-12. The materials examined by Comstock et al. (1971) were the Lexan helmets on the Apollo 8 and 12 missions. When processed without UV-sensitization, Lexan responds only to particles with ionization rates comparable to those that deactivate certain human cells with unit efficiency. The fluences reported by Comstock et al. were about half those reported by Benton and Henke (1972a), who increased the sensitivity of their Lexan with UV. From the Apollo 14 mission Fleischer et al. used the Lexan parts of the Electrophoresis Demonstration experiment (McKannan et al., 1971) that was stowed in the Command Module, and from the Apollo 16 mission they analyzed tracks produced when the Cosmic Ray Experiment (see Chapter 5) was inside the Command Module. The large differences between the fluences in the latter two devices and those in the astronaut dosimeters on the same missions correlated with the known differences in shielding and led Fleischer et al. (1973) to suggest that appropriate redistribution of shielding could be used to reduce the radiation hazard on long missions.

Using the data of Todd (1967) for the cross section for irreversible inactivation of human kidney cells (measured *in vitro*), both Comstock et al. (1971) and Benton and Henke (1972a,b,c) have calculated fractional cell losses of $\sim 10^{-4}$ for Apollo missions. The general feeling, privately expressed by a number of radiobiologists, is that problems from prolonged weightlessness will probably be more serious than the loss of such a small fraction of nonregenerative cells. However, the loss of 1% of the giant motor-control cells on a longer mission such as one to Mars might have much more serious biological effects and should not be dismissed lightly.

Supersonic transports are designed to carry passengers at altitudes of 60,000 to 70,000 feet. In polar flights at these altitudes there is a finite flux of stopping primary cosmic rays, mainly protons and alpha particles (Fukui and Young, 1971), but also including nuclei up to Fe (Allkofer et al., 1971). Fig. 9-13 shows the flux of the nuclei with $Z \geq 10$ that come to rest at various atmospheric depths, as measured in Lexan stacks (Allkofer et al., 1971) and in emulsion stacks (Fukui et al., 1969). The atmospheric depth corresponding to 60,000 feet is ~ 75 g/cm². Plastic dosimeters will probably be worn by the pilots of SST planes. Though the flux of high-Z primaries does not appear to pose a threat, Young and Fukui (1973) estimate that dark-adapted passengers on polar flights should see ~ 10 light flashes

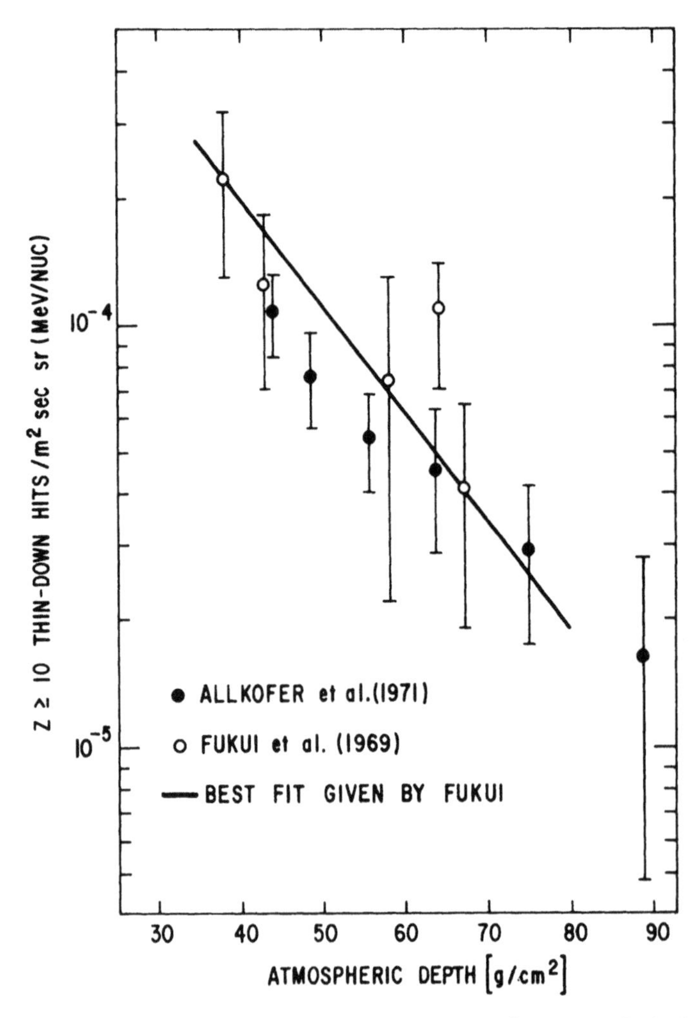

Fig. 9-13. *Flux of heavy nuclei coming to rest at various atmospheric depths in polar latitudes. The exposures of Allkofer et al. (1971) were made in 1970, near solar minimum; those of Fukui et al. (1969) were made in 1965 and 1966, near solar maximum.*

per hour due to alpha particles stopping in their retinas during the high-altitude part of a flight.

Beams of negative pions and heavy ions may be used for cancer therapy in the future, and tissue-equivalent plastic dosimeters should prove useful. Studies of high-LET fragments induced by such beams are underway (Benton et al., 1970a; 1972b). Plastics are also being used to follow the trajectories of heavy ions into individual cells (Benton et al., 1970b).

Two biological experiments involving plastic detectors have been carried on recent Apollo missions. The Biostack experiment on Apollo 16 (Bücker et al., 1972) consisted of ~200 layers of biological material, plastic detectors, and emulsions. Spores, seeds, and eggs of various sizes were embedded in inert polyvinyl alcohol foil permanently fixed to one side of cellulose nitrate sheets. The etching was done from one side of the plastic so that the biological objects were not damaged. The alignment of the stack was sufficiently good that the sites of direct hits by heavy ions could be traced from etched tracks into the biological objects. On the Apollo 17 mission the astronauts were accompanied by a group of desert mice into whose brains small Lexan detectors had been surgically inserted. Following the mission the mice were examined for linear arrays of microlesions (i.e., columns of dead cells) that might have been caused by heavy cosmic rays. The Lexan was etched to see if heavy ion tracks were aligned with the microlesions. In neither of these experiments have final results been published as of this writing.

Chapter 9 References

E. T. Agard, R. E. Jervis, and K. G. McNeill (1971), "Neutron Dosimetry with Nuclear Track Detectors Applied to *in vivo* Neutron Activation Analysis," *Health Phys.* **21,** 625–629.

O. C. Allkofer, W. Enge, W. Heinrich, and H. Rohrs (1971), "Preliminary Results of Measurements of Heavy Primaries in the Region of Supersonic Transport Using Plastic Stacks," paper at *Int. Conf. on Protection Against Accelerator and Space Radiation*, Geneva, April 1971.

J. Anno, D. Blanc, and J. L. Teyssier (1970), "Collection of Rn Daughters on a Filter," *Proc. Seventh Inter. Colloq. Corpuscular Photography and Visual Solid Detectors*, Barcelona, 543–551.

H. Annoni, R. Oppel, A. Cordaillat, A. J. Herz, and O. Mendola (1970), "Methode de Mesure du Profil d'un Faisceau pour des Intensites Elevees de Particules ($10^{12}/cm^2$)," *Proc. Seventh Inter. Colloq. Corpuscular Photography and Visual Solid Detectors*, Barcelona, 239–258.

R. J. Armani, R. Gold, R. P. Larsen, and J. H. Roberts (1970), "Application of Solid-State Track Recorders in Absolute Fission Product Yield Measurements," *Proc. Seventh Inter. Colloq. Corpuscular Photography and Visual Solid Detectors*, Barcelona, 495–502.

J. A. Auxier, K. Becker, E. M. Robinson, D. R. Johnson, R. H. Boyett, and C. H. Abner (1971), "A New Radon Progeny Personnel Dosimeter," *Health Phys.* **21,** 126–128.

G. Balducci, L. Bozzi, M. Mangialjo, and F. Rossini (1969), "Measurement of δ^{28} by Track Counting of Fission Fragments in Polycarbonates," *Energia Nucleare* **16,** 781–786.

W. V. Baumgartner and L. W. Brackenbush (1966), "Neutron Dosimetry Using the Fission Fragment Damage Principle," Report #BNWL-332, Battelle-Northwest, Richland, Wash.

W. V. Baumgartner, L. W. Brackenbush, and C. M. Unruh (1966), "A New Neutron and High Energy Particle Dosimeter for Medical Dosimetry Applications," *Symposium on Solid State and Chemical Radiation Dosimetry in Medicine and Biology*, International Atomic Energy Agency, Vienna.

K. Becker (1966), "Nuclear Track Registration in Dosimeter Glasses for Neutron Dosimetry in Mixed Radiation Fields," *Health Phys.* **12,** 769–785.

K. Becker (1967), "Neutron Personnel Dosimetry by Non-Photographic Nuclear Track Registration," *ENEA Symposium on Radiation Dose Measurements*, Stockholm, June 12–16.

K. Becker (1968), "Nuclear Track Registration in Solids by Etching," *Biophysik* **5,** 207–222.

K. Becker (1969a), "Alpha Particle Registration in Plastics and Its Applications for Radon and Neutron Personnel Dosimetry," *Health Phys.* **16,** 113–123.

K. Becker (1969b), "Direct Fast Neutron Interactions with Polymers," Oak Ridge National Laboratory Rept. 4446, p. 226.

K. Becker (1972), "Dosimetric Applications of Track Etching," *Topics in Radiation Dosimetry*, Suppl. 1, F. H. Attix (ed.), 79–142. New York: Academic Press.

F. Běhounek, J. Novotny, and Z. Spurny (1968), "Personnel Dosimetry of Slow Neutrons by Solid-State Nuclear Track Detector," *Czech. J. Phys.* **B18,** 743–748.

R. J. Benjamin, K. W. MacMurdo, and J. D. Spencer (1972), "Fission Cross-sections for Five Isotopes of Curium and ^{249}Cf," *Nucl. Sci. Engr.* **47,** 203–208.

E. V. Benton (1968), "A Study of Charged Particle Tracks in Cellulose Nitrate," U.S. Naval Rad. Defense Laboratory, Tech. Rept. 68-14.

E. V. Benton and M. Collver (1967), "Registration of Heavy Ions During the Flight of Gemini VI," *Health Phys.* **13,** 495–500.

E. V. Benton and R. P. Henke (1972a), "High-*Z* Particle Cosmic Ray Exposure of Apollo 8-14 Astronauts," Rept. AFWL-TR-72-5, Air Force Weapons Laboratory, Kirtland Air Force Base, N.M.

E. V. Benton and R. P. Henke (1972b), "Heavy Cosmic Ray Exposure of Apollo 15 Astronauts," Tech. Rept. 16, Dept. of Physics, Univ. of San Francisco.

E. V. Benton and R. P. Henke (1972c), "Heavy Cosmic Ray Exposure of Apollo 16 Astronauts," Tech. Rept. 20, Dept. of Physics, Univ. of San Francisco.

E. V. Benton and R. P. Henke (1972d), "The HZE Particle Exposure of the Microbial Ecology Evaluation Device (MEED) Aboard Apollo 16," Tech. Rept. 24, Dept. of Physics, Univ. of San Francisco.

E. V. Benton and R. P. Henke (1973), "Heavy Cosmic-Ray Exposure of Apollo 17 Astronauts," Tech. Rept. 26, Dept. of Physics, Univ. of San Francisco.

E. V. Benton, S. B. Curtis, M. R. Raju, and C. A. Tobias (1970a), "Studies of the Negative Pion Beams by Means of Plastic Nuclear Track Detectors," *Proc. Seventh Inter. Colloq. Corpuscular Photography and Visual Solid Detectors*, Barcelona, 423–428.

E. V. Benton, S. B. Curtis, J. T. Lyman, C. A. Tobias, and T. C. H. Yang (1970b), "Heavy Ion Monitoring of Biological Targets by Means of Nuclear Track Detectors," *Proc. Seventh Inter. Colloq. Corpuscular Photography and Visual Solid Detectors*, Barcelona, 585–590.

E. V. Benton, S. B. Curtis, R. P. Henke, and C. A. Tobias (1972a), "Measurement of High-LET Particles on Biosatellite III," *Health Phys.* **23,** 149–157.

E. V. Benton, M. R. Cruty, and C. Chao (1972b), "^{14}N Induced High-LET Fragments in Plastic Detectors," Tech. Rept. 23, Dept. of Physics, Univ. of San Francisco.

C. B. Besant (1970), "A High Spatial Resolution Fission Detector," *Proc. Seventh Inter. Colloq. Corpuscular Photography and Visual Solid Detectors*, Barcelona, 737–744.

C. B. Besant and S. S. Ipson (1970), "Measurement of Fission Rates in Zero Power Reactors Using Solid-State Track Recorders," *J. Nucl. Energy* **24,** 59–69.

R. C. Bhatt (1966), "Measurement of Neutron Fluence by Fission Fragment Damage Track Detection," Atomic Energy of Canada Ltd., Report AECL 2657, Chalk River, Ontario.

H. Bücker, G. Horneck, E. Reinholz, W. Scheuermann, W. Rüther, E. H. Graul, H. Planel, J. P. Soleilhavoup, P. Cüer, R. Kaiser, J. P. Massue, R. Pfohl, R. Schmitt, W. Enge, K. P. Bartholoma, R. Beaujean, K. Fukui, O. C. Allkofer, W. Heinrich, H. Francois, G. Portal, H. Kuhn, H. Wollenhaupt, and G. H. Bowman (1972), "Biostack Experiment," *Apollo 16 Preliminary Science Report*, NASA Special Pub. 315, Chap. 27.

K. Buijs, J. P. Vaane, B. Burghardt, and E. Priesch (1972), "Operational Experience with a Finger Dosimeter for Fast Neutrons," paper SM-167/2 in *Symp. on Neutron Monitoring for Radiation Protection Purposes*, IAEA, Vienna.

G. Burger, F. Grunauer, and H. Paretzke (1970), "The Applicability of Track Detectors in Neutron Dosimetry," paper SM-143.17 in *Proc. Symp. on New Radiation Detectors*, IAEA, Vienna.

M. M. Collver, E. V. Benton, and R. P. Henke (1965), "Delineation and Measurement of Sub-surface Tracks," U.S. Naval Rad. Defense Laboratory, Tech. Rept. 917.

G. M. Comstock, R. L. Fleischer, W. R. Giard, H. R. Hart, Jr., G. E. Nichols, and P. B. Price (1971), "Cosmic Ray Tracks in Plastics: Apollo Helmet Dosimetry Experiment," *Science* **172,** 154–157.

W. G. Cross and L. Tommasino (1970), "Rapid Reading Technique for Nuclear Particle Tracks in Thin Foils," *Rad. Effects* **5,** 85–89.

J. H. Davies and R. W. Darmitzel (1965), "Alpha Autoradiographic Technique for Irradiated Fuel," *Nucleonics* **22,** 86–87.

M. Debeauvais, M. Maurette, J. Mory, and R. Walker (1964), "Registration of Fission Fragment Tracks in Several Substances and Their Use in Neutron Detection," *Int. J. Appl. Rad. Isotopes* **15,** 289–299.

M. Debeauvais, R. Stein, J. Ralarosy, and P. Cüer (1967), "Spallation and Fission Fragments of Heavy Nuclei Induced by 18 GeV Protons Registered by Means of Solid Plastic Detectors," *Nucl. Phys.* **A90,** 186–198.

J. Debrue, M. de Coster, C. de Raedt, and A. Fabry (1968), "Techniques de Determination des Caracteristiques Nucleaires de la Boucle Refroidie au Sodium Utilisee dans BR2 pour des Essaies de Combustibles de Reacteurs Rapides," in *Fast Reactor Physics,* **1,** 413–432, IAEA, Vienna.

M. de Coster and D. Langela (1970), "Accurate Absolute Determination of Fission Densities in Fuel Rods by Means of Solid State Track Recorders," *Nucl. Applic. Technol.* **9,** 229–232.

M. N. de Meverngnies and P. del Marmol (1968), "Fission Cross Section of ^{232}Th for Thermal Neutrons," National Bureau of Standards, Special Pub. 299.

J. Dutrannois (1971), "Utilisation de Detecteurs Solides Visuels en Dosimetrie de Rayonnement de Haut Energie," paper at *Int. Conf. on Protection Against Accelerator and Space Radiations,* Geneva.

A. Fabry, M. de Coster, G. Minsart, J. C. Schepers, and P. Vandeplas (1970), "Implications of Fundamental Integral Measurements on High Energy Nuclear Data for Reactor Physics," *Proc. Conf. on Nuclear Data for Reactors,* IAEA, Helsinki, 15 June.

Federal Radiation Council (1967), "Guidance for the Control of Radiation Hazards in Uranium Mining," Staff Rept. 8, Revised.

R. L. Fleischer and D. B. Lovett (1968), "Uranium and Boron Content of Water by Particle Track Etching," *Geochim. Cosmochim. Acta* **32,** 1126–1128.

R. L. Fleischer, P. B. Price, and R. M. Walker (1965a), "Solid State Track Detectors: Applications to Nuclear Science and Geophysics," *Ann. Rev. Nuc. Sci.* **15,** 1–28.

R. L. Fleischer, P. B. Price, and R. M. Walker (1965b), "Tracks of Charged Particles in Solids," *Science* **149,** 383–393.

R. L. Fleischer, P. B. Price, and R. M. Walker (1965c), "Neutron Flux Measurements by Fission Tracks in Solids," *Nucl. Sci. Eng.* **22,** 153–156.

R. L. Fleischer, H. R. Hart, G. M. Comstock, M. Carter, A. Renshaw, and A. Hardy (1973), "Apollo 14 and 16 Heavy Particle Dosimetry Experiments," *Science* **181,** 436–438.

G. N. Flerov (1972), "Heavy Ion Research at Dubna," *IEEE Trans. Nucl. Sci.,* NS-19 **2,** 9–15.

A. L. Frank and E. V. Benton (1970a), "Measurements of ^{4}He Particles and Recoil Nuclei Produced by High Energy Neutrons in Plastics," *Proc. Seventh Inter. Colloq. Corpuscular Photography and Visual Solid Detectors,* Barcelona, 441–446.

A. L. Frank and E. V. Benton (1970b), "High Energy Neutron Flux Detection with Dielectric Plastics," *Rad. Effects* **3**, 33–37.
A. L. Frank and E. V. Benton (1972), "Development of a High Energy Neutron Detector," Defense Nuclear Agency Rept. 2918F.
A. L. Frank and E. V. Benton (1973a), "A Diffusion Chamber Radon Dosimeter for Use in Mine Environment," *Nucl. Instr. Methods* **109**, 537–539.
A. L. Frank and E. V. Benton (1973b), unpublished results.
K. Fukui and P. S. Young (1971), "Examination of the Density of Cosmic Ray Enders in Human Tissue from Solar Minimum to Maximum at the Level of SST Flight," paper at *Int. Conf. on Protection Against Accelerator and Space Radiation*, Geneva.
K. Fukui, Y. K. Lim, and P. S. Young (1969), "Cosmic-Ray Heavy-Nucleus Enders at Various Atmospheric Depths," *Nuovo Cim.* **61B**, 210–219.
H. Geisler and M. Heinzelman (1970), "Directional Dependence of Fission and Recoil Track Dosimeters in Plastic," Rept. JUL-670-ST, Atomic Research Establishment of North Rhein, W. Germany.
R. Gold, R. J. Armani, and J. H. Roberts (1968), "Absolute Fission-Rate Measurements with Solid-State Track Recorders," *Nucl. Sci. Engr.* **34**, 13–32.
M. A. Gomaa, A. M. Eid, and A. M. Sayed (1972), "Neutron Personnel Dosimetry Using Film Badge and Fission Fragment Track Dosimeters," paper SM-167/57 in *Symp. on Neutron Monitoring for Radiation Protection Purposes*, IAEA, Vienna.
M. Heinzelmann and H. Schuren (1970), "Spaltfragmentdosimeter für Hohe Neutronenenergien," Rept. JUL-670-ST, p. 147, Atomic Research Establishment of North Rhein, W. Germany.
M. V. Isaila, W. Schimmerling, K. G. Vosburgh, M. G. White, R. C. Filz, and P. J. McNulty (1972), "Acceleration of Argon Ions to 1.17×10^{10} Electron Volts," *Science* **177**, 424–425.
D. R. Johnson and K. Becker (1970), "Mechanism and Applications of Nuclear Track Etching in Polymers," Oak Ridge National Laboratory Rept. ORNL-TM-2826.
D. R. Johnson, R. H. Boyett, and K. Becker (1970), "Sensitive Automatic Counting of Alpha Particle Tracks in Polymers and Its Applications in Dosimetry," *Health Phys.* **18**, 424.
D. Jowitt (1971), "Measurement of $^{238}U/^{235}U$ Fission Ratios in Zebra Using Solid State Track Recorders," *Nucl. Instr. Methods* **92**, 37–44.
K. Jozefowicz (1971), "Energy Threshold for Neutron Detection in a Makrofol Dielectric Track Detector," *Nucl. Instr. Methods* **93**, 369–370.
R. Katz, B. Ackerson, M. Homayoonfar, and S. C. Sharma (1971), "Inactivation of Cells by Heavy Ion Bombardment," *Rad. Research* **47**, 402–425.
G. D. Kerr and T. D. Strickler (1966), "The Application of Solid-State Nuclear

Track Detectors to the Hurst Threshold Detector System," *Health Phys.* **12,** 1141–1142.
W. Köhler (1970), "Application of Solid State Track Recorders for Neutron Spectra Measurements in a Reactor Irradiation Facility," *Rad. Effects* **3,** 231.
H. D. Kosanke (1972), "Track Etch Registrants for High Temperature Applications," *Trans. Amer. Nucl. Soc.* **15,** 124.
P. Limkilde and G. Sletten (1973), "A Subnanosecond and a Nanosecond Fission Isomer in ^{238}Pu," *Nucl. Phys.* **A199,** 504–512.
R. E. Lingenfelter, E. H. Canfield, and V. E. Hampel (1972), "The Lunar Neutron Flux Revisited," *Earth Planet. Sci. Lett.* **16,** 355–369.
J. Liniecki, T. Domanski, and W. Chruscielewski (1972), "Studies on the Detection and Measurement Technique for Radon and Its Daughters Applying Triacetate-Cellulose Foils," *Second European Congress on Radiation Protection; Health Physics Problems of Internal Contamination*, Budapest, 88.
T. C. Liu and C. S. Su (1971), "The Application of Fission Track Detectors as Reactor Neutron Temperature Monitor," *Int. J. Appl. Rad. Isotopes* **22,** 227–232.
D. B. Lovett (1967), "Rifle Mine Field Test of Track Etch Dosimeters to Measure Radon Daughters Exposure," General Electric Rept. APED-5391.
D. B. Lovett (1969), "Track Etch Detectors for Alpha Exposure Estimation," *Health Phys.* **16,** 623–628.
H. B. Lück (1974), "Diameter Evolution of Proton Tracks in a Cellulose Nitrate Detector," *Nucl. Instr. Methods* **116,** 613–614.
A. P. Malykhin, I. V. Zhuk, Yu. I. Churkin, and O. I. Yaroshevich (1972), "Measurement of the Effective Fission Cross-section Ratio $\sigma_f^{235}/\sigma_f^{238}$ Using Solid State Track Detectors," *Vestsi Akad. Nauk BSSR Ser. Fiz. Energ. Nauk,* USSR **2,** 5–10.
M. Maurette and J. Mory (1968), "Contribution à la Dosimetrie des Fluences de Neutron Thermiques," *Rev. Phys. Appl.* **3,** 209–215.
E. C. McKannan, A. C. Kruysnick, R. N. Griffin, and L. R. McCreight (1971), "Electrophoresis Separation in Space—Apollo 14," NASA Tech. Memo X-64611.
L. Medveczky and G. Somogyi (1966), "Fast Neutron Flux Measurement by Means of Plastics," *Atomki Kozlem* **8,** 226–231.
J. Mory and R. M. Walker (1964), "Uses of Particle Track Detectors in Neutron Dosimetry," *Proc. Fifth Inter. Conf. Nuclear Photography*, CERN, Geneva, pp. I-12 to I-19.
T. Nakanishi and M. Sakanoue (1969), "Measurement of 14 MeV Neutron Flux Density by Fission Track Method," *Radiochem. Radioanal. Lett.* **2,** 313–319.
Y. Nishiwaki, T. Tsuruta, and K. Yamazaki (1971), "Detection of Fast Neutrons by Etch-Pit Method of Nuclear Track Registration in Plastics," *J. Nucl. Sci. Technol.* **8,** 162–166.

E. Piesch (1970), "Development of New Neutron Detectors for Accidental Dosimetry," paper SM-143/31 in *Proc. Symp. on New Radiation Detectors*, IAEA, Vienna.
P. Popa, M. de Coster, and P. H. M. Van Assche (1970a), "Determination of Fission Density in ^{235}U and Implications from Its Mass Distribution," *Nucl. Sci. Engr.* **39,** 50–55.
P. Popa, M. de Coster, and D. Langela (1970b), "Burnup Determination of Nuclear Fuels by High Resolution Gamma Spectrometry, Track Formation in Solid State Detectors and Neutron Dose Measurements," *Nucl. Applic. Technol.* **9,** 755–761.
H. E. Preston and C. O. Peabody (1972), "The Measurement of Personnel Neutron Dose in Reactor and Associated Areas," paper SM-167/37 in *Symp. on Neutron Monitoring for Radiation Protection Purposes*, IAEA, Vienna.
S. B. Pretre (1970), "Controle des Doses Neutroniques du Personnel par des Traces de Fragments de Fission dans Certains Plastiques," *Rad. Effects* **5,** 103–110.
S. Pretre, E. Tochilin, and N. Goldstein (1968), "A Standardized Method for Making Neutron Fluence Measurements by Fission Fragment Tracks in Plastics. A Suggestion for an Emergency Dosimeter with Rad Response," *Proc. First Inter. Congr. IRPA*, Rome, 1966, Part L, 491. New York: Pergamon Press.
P. Prevo, R. E. Dahl, and H. H. Yoshikawa (1964), "Thermal and Fast Neutron Detection by Fission-Track Production in Mica," *J. Appl. Phys.* **35,** 2636–38.
P. Rago and N. Goldstein (1967), "Fission Cross-section of ^{232}Th for Neutrons from 12.5 to 18 MeV," *Health Phys.* **13,** 654–655.
P. Rago, N. Goldstein, and E. Tochilin (1970), "Reactor Neutron Measurements with Fission Foil-Lexan Detectors," *Nucl. Applic.* **8,** 302–309.
G. Remy, J. Ralarosy, J. Tripier, M. DeBeauvais, and R. Stein (1970), "Dosimetrie et Spectrometrie Approchees de Neutrons de Fission et Thermonucleaires à l'Aide de Detecteurs Visuels Plastiques," *Rad. Effects* **5,** 221–225.
J. H. Roberts, S. T. Huang, R. J. Armani, and R. Gold (1968), "Fission Rate Measurements in Low-Power Fast Critical Assemblies with Solid-State Track Recorders," *Nucl. Applic.* **5,** 247–252.
J. H. Roberts, R. A. Parker, F. J. Congel, J. Kastner, and B. G. Oltman (1970), "Environmental Neutron Measurements with Solid State Track Recorders," *Rad. Effects* **3,** 283–285.
R. L. Rock, D. B. Lovett, and S. C. Nelson (1969), "Radon-Daughter Exposure Measurement with Track Etch Films," *Health Phys.* **16,** 617–621.
M. Sakanoue and T. Nakanishi (1969a), "Measurement of Neutron Flux in Nuclear Reactor by Means of Fission Track Method," *J. At. Energy Soc. Japan* **11,** 332–339.
M. Sakanoue and T. Nakanishi (1969b), "Measurement of Fission Cross-sections

by Fission-Track Method," *Proc. Ninth Japan Conf. on Radioisotopes*, Tokyo, May 13, 1969, 126–127.

H. J. Schaefer (1972), "Dosimeter Characteristics of HZE Particles in Space," Rept. NAMRL-1172, Naval Aerospace Medical Research Laboratory, Pensacola, Fla.

H. J. Schaefer, E. V. Benton, R. P. Henke, and J. J. Sullivan (1972), "Nuclear Track Recordings of the Astronauts' Radiation Exposure on the First Lunar Landing Mission Apollo XI," *Rad. Research* **49,** 245–271.

W. Schimmerling, K. G. Vosburgh, and P. W. Todd (1971), "Interaction of 3.9-GeV Nitrogen Ions with Matter," *Science* **174,** 1123–1125.

J. W. H. Schreurs, A. M. Friedman, D. J. Rokop, M. W. Hair, and R. M. Walker (1971), "Calibrated U-Th Glasses for Neutron Dosimetry and Determination of Uranium and Thorium Concentrations by the Fission Track Method," *Rad. Effects* **7,** 231–233.

W. W. Schultz (1966), "Neutron Measurement by Track Imaging in Insulating Materials," *IEEE Trans. Nucl. Sci.* **13,** 407–423.

W. W. Schultz (1970), "1/v Dielectric Track Detectors for Slow Neutrons," *J. Appl. Phys.* **41,** 5260–5269.

M. G. Seitz, R. M. Walker, and B. S. Carpenter (1973), "Improved Methods for Measurement of Thermal Neutron Dose by the Fission Track Technique," *J. Appl. Phys.* **44,** 510–512.

M. Sohrabi and K. Becker (1972), "Fast Neutron Personnel Monitoring by Fission Fragment Registration from ^{237}Np," *Nucl. Instr. Methods* **104,** 409–411.

R. A. Stern and P. B. Price (1972), "Charge and Energy Information from Heavy Ion Tracks in Lexan," *Nature Phys. Sci.* **240,** 82–83.

G. K. Svensson (1970), "A Feasibility Study of the Use of Track Registration from Fission Fragments for Neutron Personnel Dosimetry at the 20 GeV SLAC," paper at Second Inter. Congress IRPA, Brighton.

H. Tatsuta and K. Bingo (1970), "Evaluation of Dose Equivalent by Fission Fragment Detectors," paper 122 at Second Inter. Congress IRPA, Brighton.

P. Todd (1967), "Heavy-Ion Irradiation of Cultured Human Cells," *Rad. Research* (Suppl.) **7,** 196–207.

J. Trousil, J. Singer, and J. Marsal (1972), "Track Detectors in Czechoslovak National Personnel Dosimetry Service," paper SM-167/5 in *Symp. on Neutron Monitoring for Radiation Protection Purposes*, IAEA, Vienna.

J. W. N. Tuyn (1967), "Solid State Nuclear Track Detectors in Reactor Physics Experiments," *Nucl. Applic.* **3,** 372–374.

J. W. N. Tuyn (1969), "Measurement of Fission Density Distribution Within Fuel Elements in the Vicinity of Grey Absorber Control Elements," *Nukleonik* **12,** 183–185.

J. W. N. Tuyn (1970), "On the Use of Solid State Nuclear Track Detectors in Reactor Physics Experiments," *Rad Effects* **5,** 75–84.

J. W. N. Tuyn and J. J. Broerse (1970a), "On the Use of Makrofol Polycarbonate Foils for the Measurement of the Fast Neutron Dose Distribution Inside a Human Phantom," *Proc. Seventh Inter. Colloq. Corpuscular Photography and Visual Solid Detectors*, Barcelona, 527–531.

J. W. N. Tuyn and J. J. Broerse (1970b), "Analysis of the Etch Pit Size Distribution in Makrofol Polycarbonate Foil After Fast Neutron Exposure," *Proc. Seventh Inter. Colloq. Corpuscular Photography and Visual Solid Detectors*, Barcelona, 521–526.

M. Varnagy, J. Csikai, S. Szegedi, and S. Nagy (1970), "Observation of Proton Tracks by a Plastic Detector," *Nucl. Instr. Methods* **89**, 27–28.

R. M. Walker, P. B. Price, and R. L. Fleischer (1963), "A Versatile Disposable Dosimeter for Slow and Fast Neutrons," *Appl Phys. Letters* **3**, 28–29.

R. Weidenbaum, D. B. Lovett, and H. D. Kosanke (1970), "Flux Monitor Utilizing Track-Etch Film for Unattended Safeguards Application," *Amer. Nucl. Soc. Trans.* **13**, 524–526.

H. A. Wollenberg and A. R. Smith (1969), "Energy and Flux Determinations of High-Energy Neutrons," *Proc. Second Inter. Conf. Accelerator Dosimetry*, Stanford.

D. S. Woolum, D. S. Burnett, and C. A. Bauman (1973), "The Lunar Neutron Probe Experiment," *Apollo 17 Preliminary Science Report*, NASA Special Pub. 330, Chap. 18, pp. 1–12.

R. Yokota, S. Nakajima, and Y. Muto (1968), "Registration of Fission-Fragment Tracks in the Th- and U-doped Phosphate Glasses and Its Possible Application to Neutron Dosimetry," *Nucl. Instr. Methods* **61**, 119–120.

P. S. Young and K. Fukui (1973), "Predicting Light Flashes Due to α-Particle Flux on SST Planes," *Nature* **241**, 112–113.

Chapter 10

Diverse Applications in Science and Technology

The great fascination of particle tracks in solids comes in large measure from the wide-ranging diversity of scientific and technological areas in which track techniques can be applied, as indicated by the more than two dozen areas listed in Table 10-1. In this chapter we wish to summarize the applications in science areas that are not discussed in Chapters 1 through 9 and to emphasize particularly the technological applications of particle track etching, i.e., the uses that are of potential economic or public benefit.

Such applications of science provide one of the traditional justifications for the support of basic research. We believe that the case of particle track etching is an unusual example of essentially unpredictable practical results being derived from basic science. The applications range from semiconductor electronics to aerosol sampling, identification of microbiological particles, nuclear engineering, mine safety, neutron radiography, uranium exploration, sewage disposal, the color of gem diamonds, improvement of beer and wine, and securing nuclear reactors for peaceful uses. We categorize the applications in this chapter under the topics Holes and Controlled Geometry, Chemical Technology, Nuclear Technology, Imaging, and Miscellaneous.

Technological applications of track etching were reviewed recently (Fleischer et al., 1972a) and have been included in various other reviews (Alter, 1970; Fleischer et al., 1965a,b; 1969; Furman et al., 1966; Havens, 1968; Monnin and Isabelle, 1970; Sakanoue, 1968; 1970; Somogyi et al., 1971; Suits and Bueche, 1967; Tuyn, 1967; Weidenbaum et al., 1970a). Griffith (1973) lists thirty-seven "areas of interest" in track work and surveys the types of work that were in progress as of early 1972.

Table 10-1. Diverse Applications of Etched Particle Tracks

Science	Application	Chapter
Anthropology	Age of Bed I, Olduvai Gorge	4
	Age of man-made glasses	4
	Uranium in fossil bone	8
Archaeology	Uranium content of ancient glasses	8
Astrophysics	Ultraheavy cosmic rays: abundance measurements	5
	Detection of ^{244}Pu in early solar system	6
Avionics	Flight times and altitudes of birds	10
Biophysics	Size, count, and identify bacteria, viruses, and other small particles	10
Botany	Distribution of boron in leaves	10
Cosmic Ray Physics	Identification of heavy cosmic rays	5 and 6
	Energy spectrum of heavy solar flare particles	5 and 6
Chemistry	Distribution and quantity of fissionable elements (or those with (n,α) reactions) [U, Th, Pu, Li, B, N, O]	8
Cytology	Isolation of malignant cells	10
Elementary Particle Physics	Magnetic monopole searches	7
Geochemistry	Distribution in nature of U, B, Li	8
Geochronology	Fission track dating	4
Geophysics	Measure rate of ocean bottom spreading in N. Atlantic	4
Membrane Physics	Flow behavior in fine holes	10
Metallurgy	Locate impurities and boron-rich precipitates in metallurgical structures	8
Meteoritics and Selenology	Erosion and ablation of meteorites	6
	Early chronology and thermal history	6
	Surface exposure times on moon	6
	Accretion, erosion, and stirring on the moon	6
Nuclear Physics	High-resolution particle identification	3
	Study heavy particle reactions	7
	Measure cross sections, lifetimes	7
	Spallation studies	7
Radiation Biology	Heavy particle exposure of astronauts	9
Solid State Physics	Neutron-sensitive plate for neutron diffraction studies	10
	Defect identification, channeling and blocking studies	10
	Nature of particle tracks in solids	2
Superfluidity	Test understanding of superfluid flow through small, uniform holes	10
Superconductivity	Behavior of fine wires of known geometry	10
Solar Physics	Energy and composition of solar nuclei	5

Table 10-1 (*continued*)

Technology	Application	Chapter
Accelerator engineering	Beam attenuator for Superhilac	10
Alloy development	Structure of new superconducting alloys	10
Altimetry	Integrating barometer	10
Dosimetry	Count neutrons, protons, heavy particles	9
Environmental studies and radiation protection	Aerosol filtration	9
	Personnel radiation dosimetry	9
	Personnel radiation dosimetry in space	9
	Nuclear detonation detection	10
	^{239}Pu detection in bone, urine, etc.	8
	Uranium measurements in water supplies	8
	Tracer studies on sewage distribution in ocean waters	10
Filtration	Nuclepore filters for cytology, bacteria removal, etc.	10
Gem diamond manufacture	Boron and nitrogen distributions mapped in relation to color	8
Imaging	Neutron, proton and heavy ion radiography	10
Medical technology	DeBlois-Bean virus counter; diagnostic radiography	10
Nuclear engineering	Leak detection in fuel rods	10
	Neutron dosimetry	9
	α-Radiography to give Pu-U distribution in fuel rods	10
	^{239}Pu particle size analysis	10
Uranium exploration	^{222}Rn emanation mapping	10

10.1. HOLES AND CONTROLLED GEOMETRY

Many of the uses of particle tracks derive from the simple fact that an etched particle track is a hole. By proper control of the spatial distribution of the track-forming particles and of the etching conditions, structures can be produced that have unique, useful geometries.

10.1.1. Porous Membranes and Filters

The idea of a molecular sieve produced by etched tracks goes back more than a decade (Fleischer et al., 1963; 1964; 1972b; Moore, 1969; Porter and Schneider, 1973). One simply exposes a thin detector material to a collimated beam of particles that produce tracks across the entire thickness of the sheet, and one then etches to produce holes that perforate the membrane. The hole diameters can be controlled by specifying the etching time, and the number of holes is controlled by the particle dose. Holes with diameters ranging from 50 Å (Bean et al., 1970; Price and Walker, 1962) on up (Fleischer et al., 1963) can be produced in mica and various plastic materials.

In the commercial embodiment—Nuclepore® filters (made by the Nuclepore Corporation, 7035 Commerce Circle, Pleasanton, Calif.; and also available through Nomura Micro Science Co. Ltd., Japan)—polycarbonate film is irradiated with ^{235}U fission fragments in a nuclear reactor and later etched in a warm sodium hydroxide solution to give holes such as are shown in Figs. 10-1 to 10-4. The recent history of the formation of Nuclepore Corporation is given by Sabin (1973).

Commercially available filters have holes of uniform, specified sizes that range in steps from 300 Å to 8 μm and are routinely available in sheets of up to 50 cm width and essentially any length. Documentation of hole sizes and etching behavior is given by Bean et al. (1970) and Desorbo and Humphrey (1970) in Chapter 1 and also by Frank et al. (1970), Stamm (1971) and Bean (1972). Continuously adjustable holes can be obtained by perforating a stretchy plastic, as shown in Fig. 10-2 (Fleischer et al., 1972b). By placing such a perforated sheet on an adjustable support such as a camera iris, the hole size can be changed by small increments in order to separate groups of spherical objects having very nearly the same diame-

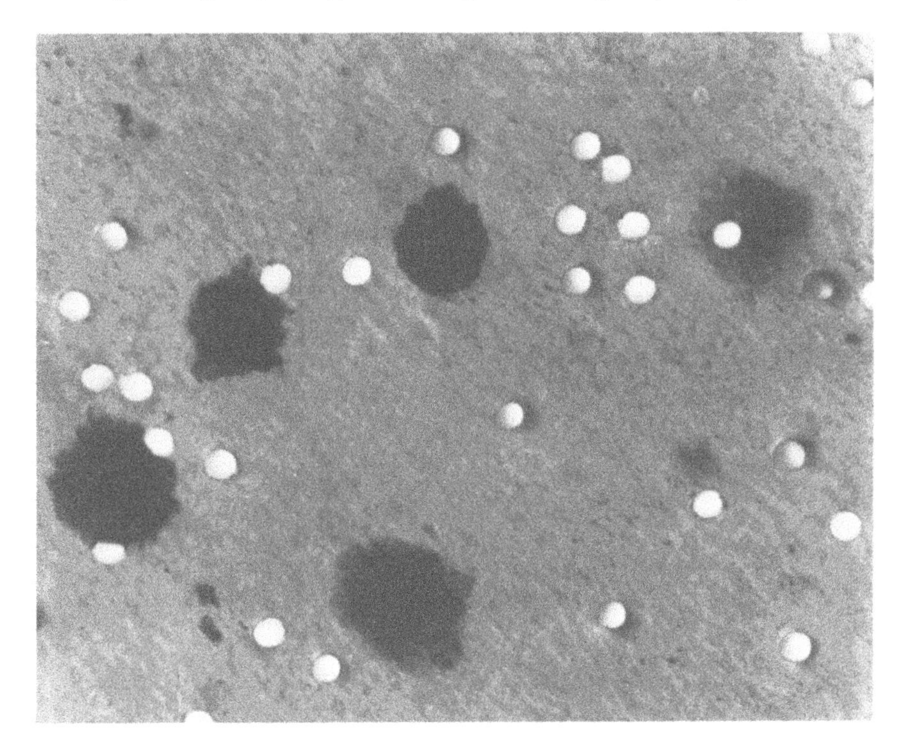

Fig. 10-1. *Polycarbonate film irradiated with fission fragments at normal incidence and etched with sodium hydroxide to produce circular holes of uniform size, about 5 μm in diameter. The large, dark blobs are cancer cells that have been filtered out of a large quantity of blood by virtue of being larger and more rigid than most normal cells in the blood.* (*After Fleischer et al., 1965a.*)

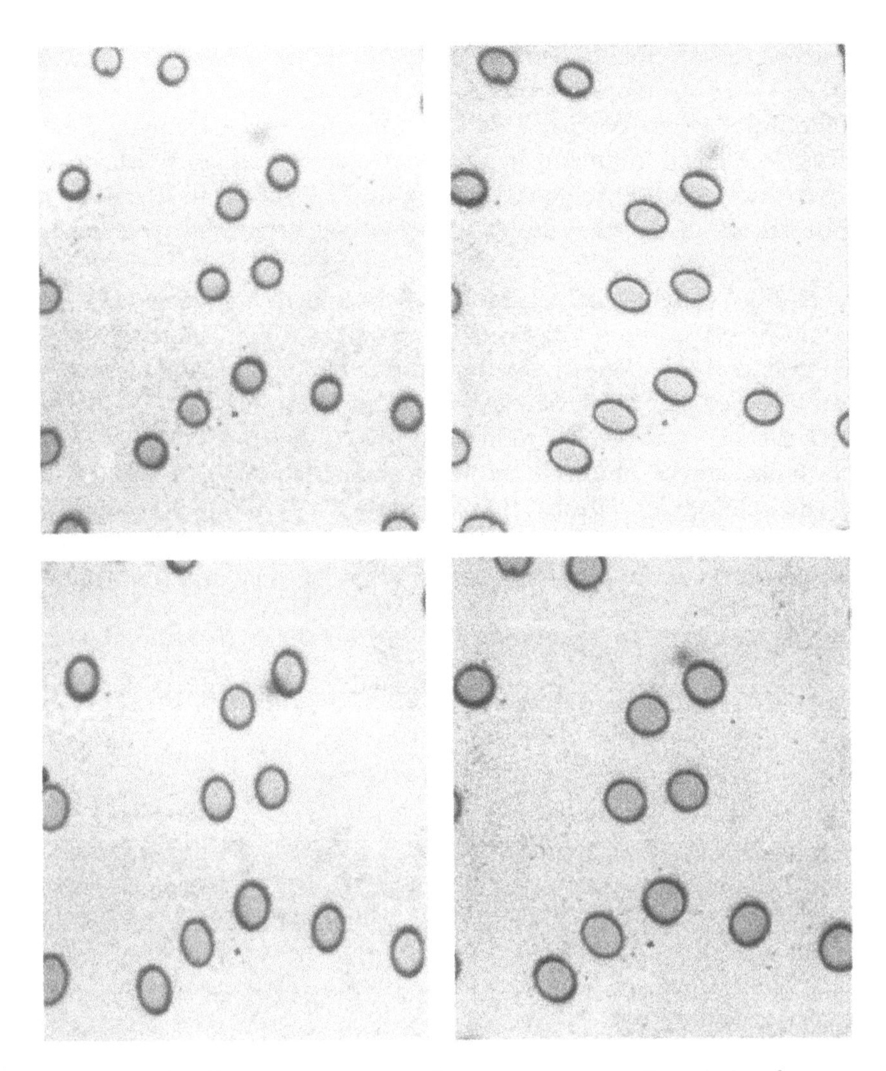

Fig. 10-2. *Silicone-polycarbonate filter produced by etching fission fragment tracks. Shown in nearly relaxed condition, horizontally stretched, vertically stretched, and biaxially stretched. The hole diameters in the relaxed state are 14 μm.* (*After Fleischer et al., 1972b.*)

ters. Also, as shown in Fig. 10-2, elliptical holes can be made so that special separations involving oblate spheroids are possible in principle.

Etched filters have been put to diverse uses. The earliest of these (see Fig. 10-1) (Fleischer et al., 1965a) was to separate circulating cancer cells from blood (Seal, 1964), making use of the fact that cancer cells are both larger and more rigid than normal blood cells. Seal demonstrated that when the malignant cells were present they could be separated and recognized; he did not establish whether such cells are present in the blood early enough in the course of the disease that locating them will allow their source to be identified and treated effectively. Song et al. (1971) have, however, examined blood samples from a series of cancer patients and have found that more than two-thirds of those with cancer of the breast, colon, or rectum had detectable cancer cells in the blood in cases where the malignancy had not metastasized, in contrast to less than 1 in 12 having such free-floating cells once the cancer had spread to the lymph nodes. Although unexplained, the direction of the effect is encouraging in that a malignancy is detectable in its earlier stage. Nuclepore has also been used to detect cancer cells in spinal fluid (Wertlake, 1972; Rich, 1969) and lung material sampled by needle aspiration biopsy (King and Russell, 1967).

Experimental procedures for filtering, staining, and mounting cells for optical examination have been given by Reynaud and King (1967) and McAlpine and Ellsworth (1969), and Todd and Kerr (1972) describe processing for cytological examination by scanning electron microscopy. Special properties of the filters are the cylindrical holes, which give a well-defined size separation and simultaneously add an internal size calibration during examination; and the uniform thickness of the clear material, which allows transmitted light examination to be carried out with the cells fixed *in situ*.

Gregersen et al. (1967) and Chien et al. (1971) showed that the size of the smallest hole that human red cells can squirm through can be understood by a model of the cell as a flexible sack of fixed volume. Thus, cells of average diameter 5.8 μm can pass through cylindrical pores less than 3 μm in diameter. Nuclepore filters have also been used in immunology (Horwitz and Garrett, 1971) and in the study of metabolic interactions (Batzdorf et al., 1969). In both studies, a membrane partition served as a barrier to cell migration but allowed free flow of metabolites, antibodies, and antigens.

Etched track filters have proved useful in environmental studies of particulates in the atmospheres and oceans. Spurny et al. (1969) have described the collection of airborne particles for optical and electron microscopy and Spurny and Lodge (1972a) have given the theory of collection efficiency along with elaborate tables. Anno et al. (1970) and Spurny and Lodge (1972b) describe the use of Nuclepore to sample radioactive aerosols. Twomey (1972) has used Nuclepore to sample cloud nuclei down to $\sim$100 Å diameter and to infer the residence time of these aerosols. Sampling of aerosol particles that are smaller than the holes in the filter depends on impaction of particles which do not follow the air flow perfectly and which therefore contact the exposed surface of the filter and stick by surface forces. Fig. 10-3, for example, shows salt particles caught in this manner from marine air on a filter with holes whose diameters are many times those of the crystals. P. Biscaye of Lamont-Doherty Geophysical Observatory (personal communication) has been using Nuclepore to strain fine particulate matter from suspension in ocean water

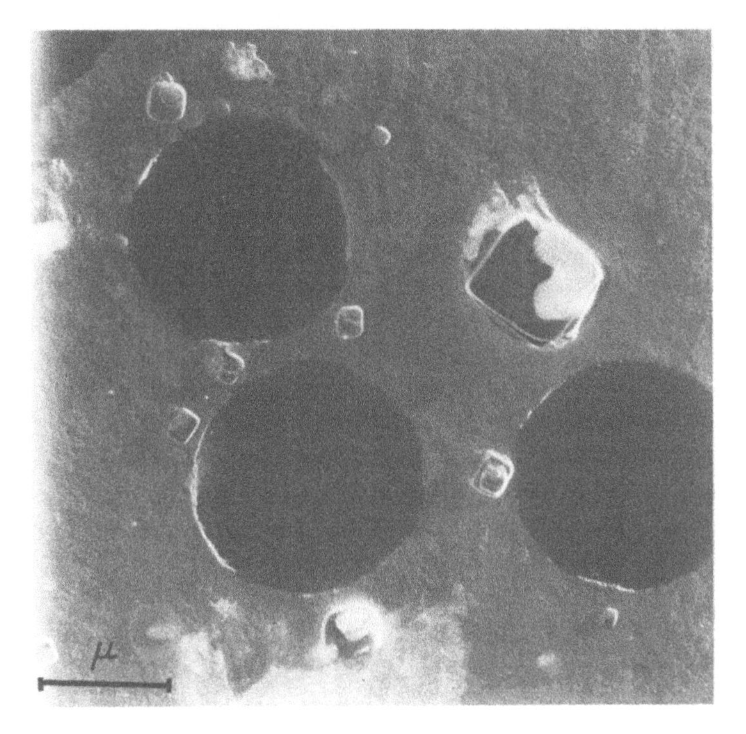

Fig. 10-3. *Airborne salt particles from marine air near Hawaii, caught by Nuclepore filter. The scale bar is 1 μm long. Photograph taken by E. R. Ackerman of the National Center for Atmospheric Research.*

samples in order to determine the quantity of suspended matter and to retrieve particles for further study.

Two of the more pleasurable qualities of Nuclepore are its demonstrated ability to clarify and stabilize wine and beer by removing bacteria, sediment, and yeast. Fig. 10-4 shows extracted material from such separation. Such stabilization allows draft beer to be stored safely at room temperature, with no need for the heating that otherwise would be required to destroy the bacteria in bottled beer, a heating that is responsible for the differences in taste between draft beer and the conventional bottled variety (*GE International*, 1971). Track-etch filters are also used for cleaning gases that are used for dusting electronic components prior to encapsulation.

Another novel use is as a heavy ion beam attenuator. The idea here is to decrease the beam current without transmitting particles of lowered energy. By placing in the beam a piece of mica with its holes aligned along the beam direction and continuously translating it normal to the beam direction, A. Ghiorso is able to transmit a desired fraction of the heavy ion beam at the Superhilac at Lawrence Berkeley Laboratory. This fraction is given by the porosity of the sheet.

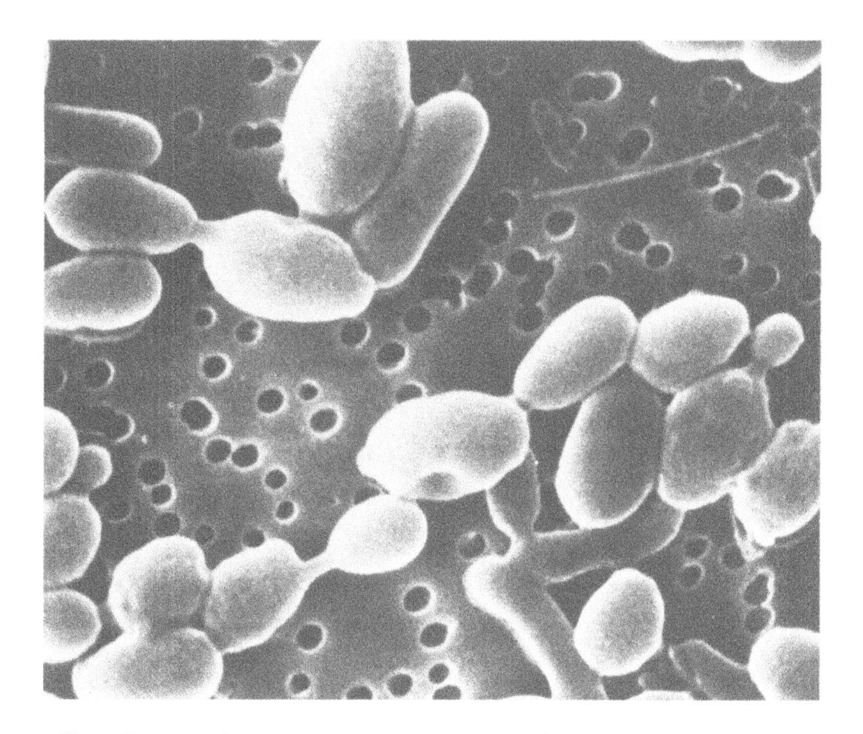

Fig. 10-4. *Nuclepore filter used to extract yeast from beer. Photograph supplied by H. J. Schneider of Nuclepore Corporation.*

10.1.2. Virus and Bacteria Counter

One of the most ingenious and economical uses of etched tracks is the *DeBlois-Bean Counter* (DeBlois and Bean, 1970; 1973), which employs a single etched track to count and size small particles in an electrolyte. It extends the principle and uses of the Coulter Counter (Coulter, 1953). As shown in Fig. 10-5, the two sides of an electrochemical cell are separated by a membrane with a track etched through it. The resistance between the electrodes depends primarily on the conducting path through the hole. As a charged, insulating particle enters, the resistance increases by an amount proportional to the volume of the particle. In the case of a spherical particle of diameter d, $\Delta R = 4\rho d^3/\pi D^4$, where ρ is the resistivity of the solution and D is the hole diameter. At the same time, the velocity of the particle through the hole is a measure of its charge, so that these two quantities—size and charge (or mobility)—can be used to characterize and identify individual particles. The oscilloscope traces show, for example, how clearly different are the signals of polystyrene spheres of 910 Å diameter from those of T-2 viruses of effective average diameter 1010 Å.

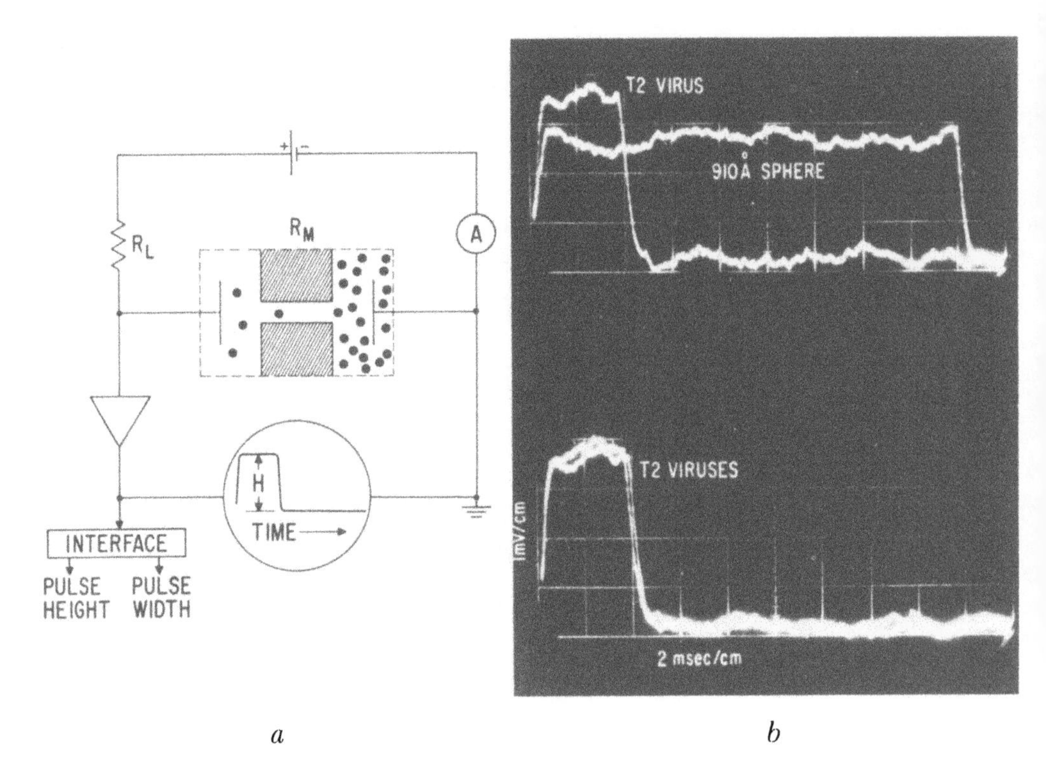

a *b*

Fig. 10-5. (*a*) *Principle of the DeBlois-Bean Counter* (*after DeBlois and Bean, 1970*); (*b*) *oscilloscope traces resulting from passage of polystyrene spheres and T-2 viruses through the hole in the counter. Individual particles are identified by their size* (*given by the pulse height*) *and charge* (*inferred from the transit time*). (*After DeBlois and Bean, 1970.*)

The special excitement which this instrument arouses comes from its ability to characterize virus particles. Viruses stand on the border of the animate and the inanimate (Frankel-Conrat, 1969). Although they are believed to have no intrinsic metabolic activity, viruses possess genetic information and the capacity to mobilize living cells to make exact duplicates of the viruses. More recently a second role of the virus has been noted: its ability to transform the host cell, presumably by permanently altering the genetic makeup and character of the cell by injecting genetic material of the virus.

In the study of animal cancer viruses and the search for those that may play a role in some human cancers, it is important to characterize different viruses and to measure their multiplication in given biochemical environments. Since viruses range from 150 Å to 4500 Å in dimension, none can be usefully examined in the optical microscope, but most can be counted and characterized by the new counter, which is in fact the only instrument capable of observing single viruses in liquid suspension.

Already the measurement of the T-2 virus shown in Fig. 10-5 has given useful information. The observed diameter is appreciably larger than that determined by electron microscopy of dried viruses, a result that is interpreted as implying that, in the "live" T-2 virus, water makes up roughly 30% of its volume. This inference is further supported by sizes estimated from electron micrographs of freeze-etched T-2 viruses (DeBlois and Bean, 1973).

It is expected that the present equipment will make sizes down to 300 Å accessible, with automatic readout of both the size and mobility distribution of particles. Fig. 10-6 indicates the impressive resolution of the instrument. In examining a mixture of polystyrene spheres with different size distributions, DeBlois et al. (1973) found that the mean size difference was 114 Å in contrast to the manufacturer's quoted 190 Å. The widths of the two distributions are real, not the result of instrumental limitations. These same workers have been applying the technique to characterizing, counting, and identifying different cancer viruses.

One of the fascinating characteristics of the device is that individual pulses with harmonic oscillation are sometimes observed when elongated particles tumble through the hole. From the distribution of amplitudes it is possible to infer the shape of such particles (Golibersuch, 1973).

The resistive pulse technique is useful not only for studying the contents of the

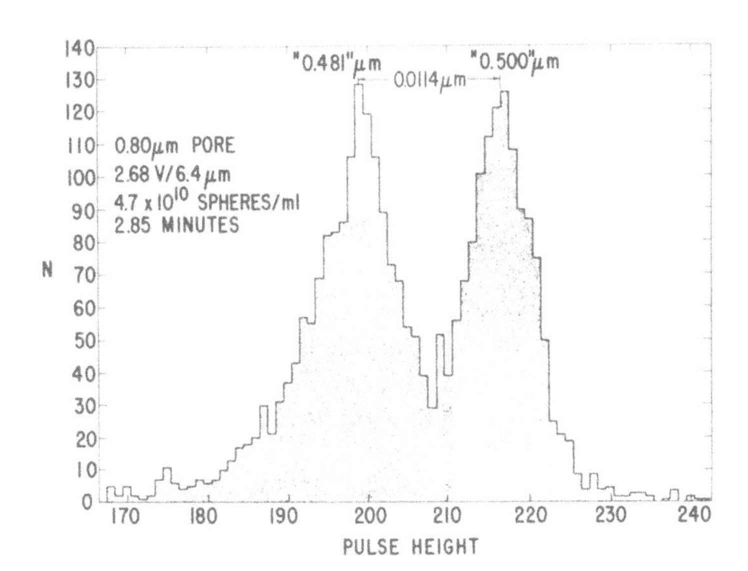

Fig. 10-6. *Volume distribution function of two mixed samples of polystyrene spheres. The horizontal axis is a measure of the pulse height with the zero displaced by 168 units to allow increased precision. The vertical axis plots the number of pulses of a given height in 2.85 minutes. The pore is 0.8 μm in diameter, and the particles have nominal diameters of 4810 Å and 5000 Å, but an observed true difference in diameter of 114 Å.*

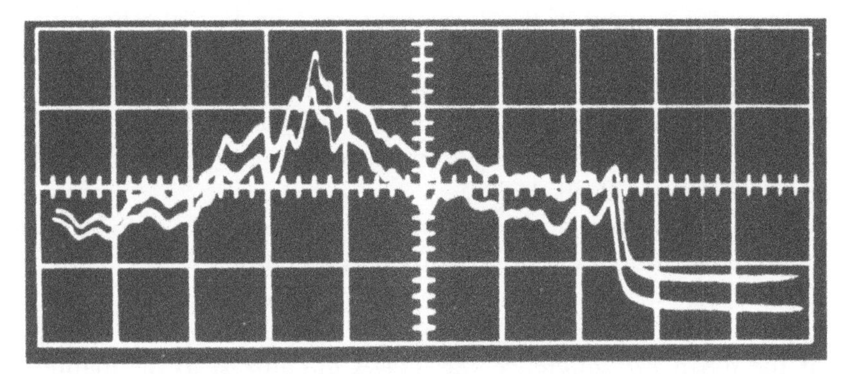

Fig. 10-7. *Resistive pulses as two nearly identical particles transit the same hole under an applied field. The ripples are duplicated, showing they are real, small-scale, diameter variations. The increase toward the center represents a narrowing of the hole and measures the taper (or cone angle) of the etched track. (Golibersuch and Fleischer, unpublished.)*

etched hole, but also the etched hole itself. As Fig. 10-7 shows (Golibersuch and Fleischer, unpublished), repeated passages through the holes by duplicate test particles reproduce the curve, complete with identical small-scale wiggles. These wiggles, consequently, represent the actual roughness of the interior of the hole. Because the resistance through the hole is highly sensitive to the geometry, being proportional to (d^3/D^4), the $\sim$10% roughness in the oscilloscope trace represents only a $\sim$2.5% variability in the regular shape of the hole. The overall tentlike shape of the trace reflects the finite taper to the track, constricted near the halfway point. Quantitative use of this curve shape would provide high sensitivity for measuring v_T as a function of position, using the geometrical relations discussed in Chapter 3 (Fleischer et al., 1970).

10.1.3. Special Devices

By other methods track-etched holes in a thin dielectric sheet such as mica or a plastic can be filled with solid plugs having a variety of possible properties of practical importance. One of us (PBP) once demonstrated that hot, liquid gallium (melting point = 30°C) could be forced into micron-sized holes in mica and retained when the temperature was lowered until the metal solidified. The necessary hydrostatic pressure is equal to $2\gamma/r$, where γ, the surface energy, is usually 10^2 to 10^3 erg/cm^2, and r is the hole radius. Another method of filling holes with solid plugs is to place solutions of ionic salts on either side of the sheet so that a chemical reaction occurs within the sheet, precipitating an insoluble product. Examples of such reactions are $AgNO_3 + KCl \rightarrow AgCl(ppt) + KNO_3$, where the AgCl precipitate is photosensitive, and $Pb(NO_3)_2 + 2KI \rightarrow PbI_2 + 2KNO_3$, where the

PbI_2 precipitate can be reduced to superconducting metallic Pb by exposing the PbI_2 to light or to x rays or to electrons when the sheet is at a temperature of 180°C. To our knowledge, no imaging or other devices based on arrays of plugs in dielectric sheets have been made.

Porous muscovite mica produced by track etching (Fleischer et al., 1963; Bean et al., 1970; and Quinn et al., 1972) has been used as a model system to test theories of flow and ideas of the structure of water and of superfluid flow. Bean (1969, 1972) calculated the combined effects of pressure and diffusion on solute flow through fine holes and found that the theory fit his observations on mica well. Some of the ideas (Derjaguin, 1970) that "anomalous" water exists as an unusual icelike structure on the interior of fine capillaries have been laid to rest with convincing experiments on fine holes (Bean, 1969; 1972; Beck and Schultz, 1970; 1972; Anderson and Quinn, 1972). In each case it was found that no unusual effect appeared even down to 56 Å radii. Petzny and Quinn (1969) have also shown how it is possible to attach organic molecules to the insides of holes in mica, building up to layers of $\sim$30 Å thickness one at a time. In this manner they have produced smaller holes with altered surface characteristics that may be closer to biological membranes in surface character and yet retain the well-defined geometry of track-etched holes.

10.1.4. Superfluidity

Notarys and Andelin (1967) and Notarys (1968) used etched mica to provide a connection between two liquid helium reservoirs. In the former study, sound was used to pump superfluid helium through the holes. It was found that in order for such pumping to occur, the sound frequency needed to be low enough to pass a vortex completely through a pore during a single cycle. In the latter case, an electric field suppression of the superfluid transition temperature was found to occur at sufficiently high fields. Gamota (1973) has observed single vortex rings in a similar experiment.

10.1.5. Superconductivity

Possin (1971) studied superconductivity in fine wires which he produced by electroplating tin into fine holes in etched mica. He describes how wires down to 400 Å diameter and 15 μm long can be produced in this manner from indium and zinc, as well as tin (Possin, 1970).

10.2. CHEMICAL TECHNOLOGY

The applications we note in this section make use of the high sensitivity and micromapping capability of the activation analysis and dosimetry procedures that we described in detail in Chapter 8. In these techniques useful information can be derived in an assortment of fields.

10.2.1. Metallurgy

Figs. 10-8 and 10-9 illustrate the metallurgical importance of tracks as means of understanding the structure of alloy steels (Rosenbaum and Armijo, 1967) and as a way to measure diffusion distances. On the boron map for the silicon-iron in Fig. 10-8, it is evident that the boron is distributed preferentially along crystal boundaries as well as in inclusions (R. Scanlon, personal communication). Similarly, the uranium map of taper-sectioned, Nb-Sn wires shown in Fig. 10-9 allows the thickness of the reaction zone containing uranium to be measured and to be related to the different phases observed by conventional metallographic examination of the wires. This result gives clear evidence that the uranium was uniformly dispersed in the superconducting phase Nb_3Sn (Fleischer et al., 1972a). The same principles have been used quantitatively by de Keroulas et al. (1967) to measure diffusion of trace amounts ($<10^{-7}$ atomic fraction) of uranium in titanium and face-centered cubic iron.

In an environmental context Hamilton (1970) measured uranium in aerosols filtered from various natural sites around the world. He found roughly five times more uranium at Sutton, U.K. than in marine and antarctic air.

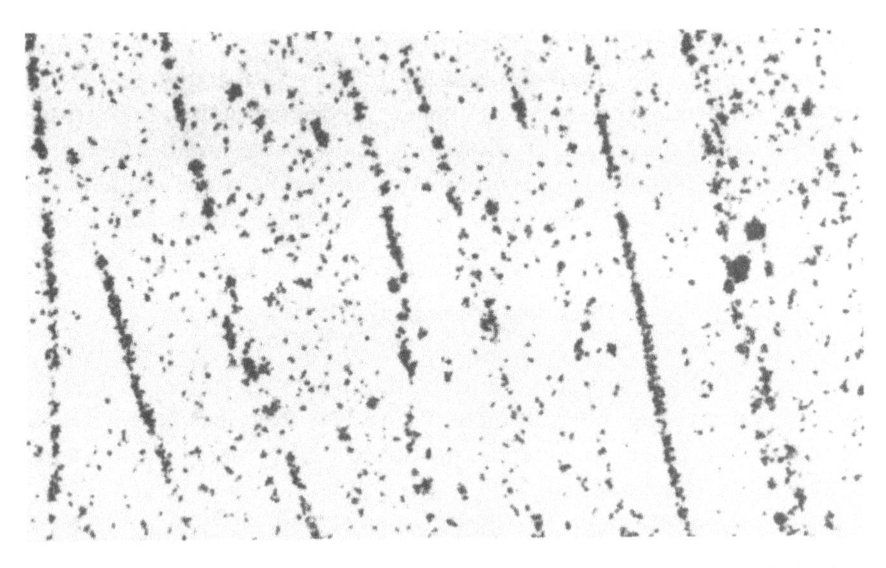

Fig. 10-8. *Mapping of boron, showing its segregation to grain boundaries in silicon-iron, which stabilizes the preferred orientations of crystals that lead to useful magnetic properties (R. Scanlon, unpublished).*

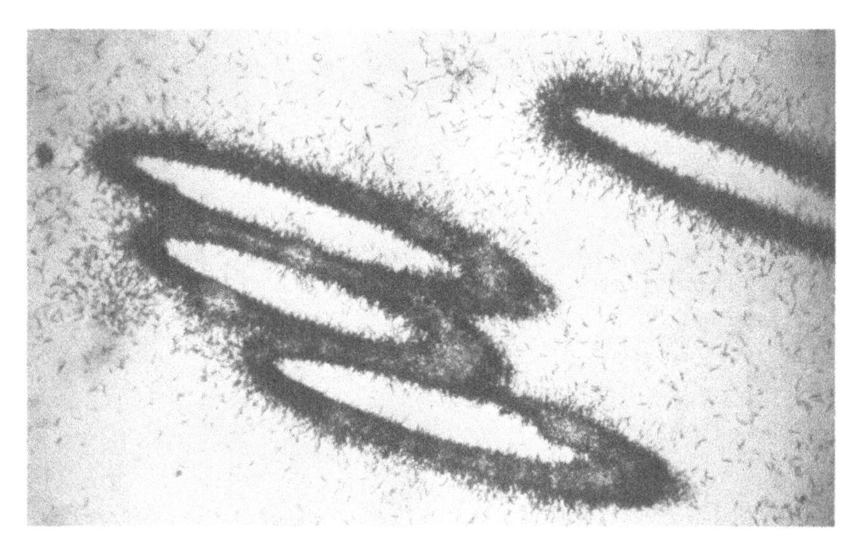

Fig. 10-9. *Micrograph showing uranium distribution in taper-sectioned 0.002 inch superconducting Nb_3Sn wires with uranium-enriched tin diffused into a thick outer section of the wire. Tracks from induced fission of uranium clearly delineate the thickness of the uranium-rich layer. (After Fleischer et al., 1972a.)*

10.2.2. Gemology

Boron mapping in diamond has shown a correlation between coloring—a major factor in the value of gems—and the presence of boron (Chrenko, 1971). The high-boron regions displayed by the lighter color of regions with many tracks in the cellulose nitrate pictured in Fig. 10-10 are also the most intensely blue. Chrenko noted that work is continuing to relate the boron distributions to the growth history and electrical and optical properties of diamonds.

10.2.3. Fluid Tracers

Small particles that are revealable by track techniques make it possible to trace the flow of a fluid in which they are suspended (Weidenbaum et al., 1970a; Weidenbaum and Lovett, 1971; Lovett, 1972). In preliminary studies uranium dioxide particles of 0.5 μm average diameter were added to water to study its flow. Samples taken at various dilutions were filtered to remove particulates, pressed against a fission track detector, and then neutron-irradiated to produce characteristic stars, similar to the one on the Frontispiece of this book (from the paper by Price and Walker, 1963) and to those shown in Fig. 4-9. Carpenter et al. (1973) showed that ^{17}O-enriched substances could be used as tracers by means of the (n,α) reaction.

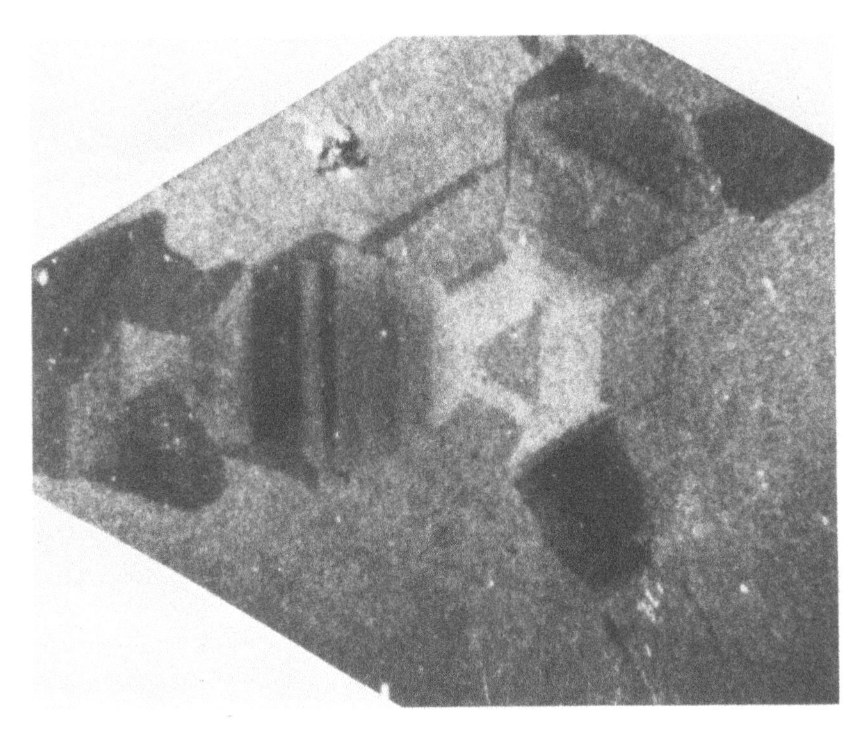

Fig. 10-10. *Photograph of particle track profile in plastic due to boron in a diamond. The lightest areas denote* $\sim 1.25 \times 10^{-4}$ *atom fraction of boron, and the darkest areas* $\lesssim 4 \times 10^{-6}$ *atom fraction boron. The diamond is* ~ 2.5 *mm wide. (After Chrenko, 1971.)*

Lovett found that 0.5 μm particles were readily visible after a dose of 10^{15} neutrons/cm^2 but settled from suspension with a half life of ~12 hours, i.e., too rapidly for universal usefulness in tracing bulk water flow. Use of Stokes law implies that by using 0.1 μm particles, the settling velocity could be usefully lowered by a factor of 25. At the desirable dose for these smaller particles, 10^{17} neutrons/cm^2, cellulosic or polycarbonate detectors are likely to have an annoying background of recoil tracks from fast neutron reactions; however, that problem can be easily avoided by using a less sensitive detector such as Cronar (polyethylene terephthalate) or muscovite mica.

The general procedure is by no means limited to uranium-rich particles. Other heavy elements that present lesser environmental questions, such as thorium, gold, or platinum, could be used and later caused to fission by high energy particles (as described in Chapter 8). It is also possible to use particles that are rich in ^{6}Li or ^{10}B. These would be detected by the products of (n,α) reactions, for which the cross sections are high—950 and 3,840 barns, respectively. H. W. Alter (personal communication) has, in fact, noted that particles could be individually coded by specifying, for example, the uranium/boron ratio. After recovery they would be irradiated while sandwiched between two detectors of different sensitivities. One would be a material such as mica, which records only fission fragments, and the other would be a detector such as cellulose triacetate, which would record both alpha particles and fission fragments. For a given particle the ratio of the track count in the mica to the excess above that number seen in the cellulose triacetate would determine the uranium/boron ratio.

Tracer experiments have many facets that need to be examined in each individual case: questions about settling time, chemical stability over the duration of the experiment, affinity for other surfaces in the medium being examined, magnitude of the background of other particles, and so on. The tracer technology area remains an important one for further exploration.

10.3. NUCLEAR TECHNOLOGY

10.3.1. Radiobiology

Since the dosimetry work was considered in detail in the previous chapter, we merely remind the reader of the variety of these applications, which include personnel dosimetry to measure neutron fluxes at nuclear reactors (Pretre, 1970) and to monitor the doses of alpha decays from ^{222}Rn and its daughters to which uranium miners are exposed (Auxier et al., 1971; Becker, 1969; Lovett, 1969); alpha monitoring of ^{222}Rn and its daughters in residential housing in parts of Colorado where uranium mine tailings were used as sand in concrete for construction purposes (D. Lovett, personal communication); and detectors to establish the heavy particle doses to which Apollo astronauts were exposed (Benton and Henke, 1972; Comstock et al., 1971; Fleischer et al., 1970; 1973).

10.3.2. Biological Microchemistry

These studies, reviewed in Chapter 8, include mapping and/or bulk measurements of uranium and plutonium in biological systems such as blood, bone, and sea weed. Fig. 10-11 shows a recent example of mapping in a botanical system. The light regions in the boron (n,α) map indicate strong depletions of boron along the veins of a tomato leaf. Stinson (1972) used the Kodak Pathé red cellulose nitrate described in Chapter 2 to enhance contrast and produce an etched sheet that could serve directly as a photographic negative to provide the print used for Fig. 10-11. Jee et al. (1972) simultaneously mapped both ^{239}Pu and (probably) boron in bone using Lexan detectors.

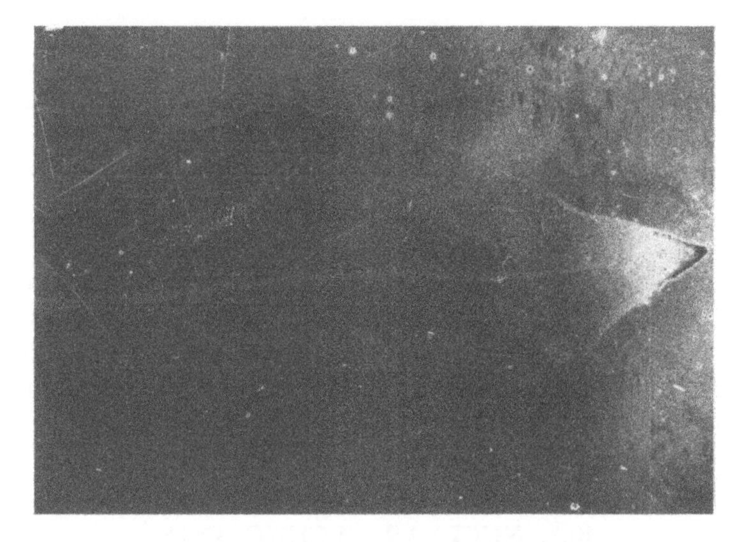

Fig. 10-11. *Distribution of boron in mature tomato leaflet. Light regions show areas of lowest concentration. This figure was made by using high contrast film and a point source of light in a photographic enlarger in which original autoradiograph is used as a negative. Diffraction effects produce undesired white areas near the tip in this photograph but do not hinder microscopic examination. (After Stinson, 1972.)*

10.3.3. Nuclear Fuel Development and Safeguards

The alpha autoradiography technique described in Chapter 8 and illustrated in Fig. 10-12 (Davies and Darmitzel, 1965; Davies et al., 1966; Kegley, 1970) has proven to be quite valuable. One difficult problem has been to map the Pu and U distribution in a mixed fuel rod after use under extreme conditions in a reactor. As shown by a conventional radiograph using photographic film, the β and γ radiation from fission products were so intense that fogging of the plate obscured details. By placing the sectioned fuel rod on an alpha-sensitive plastic, which is insensitive to the ambient β and γ radiation, only the plutonium-rich regions are imaged, plutonium being much more alpha-radioactive than is uranium. In this case, clear evidence was obtained for convective mixing of molten PuO_2 and UO_2. In another example, given by Fleischer et al. (1972a), the same technique was used to reveal the crack pattern produced in another mixed oxide fuel element by thermal stresses. Further examples are given by Weidenbaum et al. (1967).

Nuclear safeguards applications of track etching (Weidenbaum et al., 1970b; Gingrich et al., 1972a,b) have led to the preparation of a series of devices, such as the one shown in Fig. 9-9, which can be left unattended to monitor and record the neutron energy spectrum at sites around a nuclear reactor. By later periodic inspection the duration and time of use can be established along with the power level and possibly some details of the loading of the reactor.

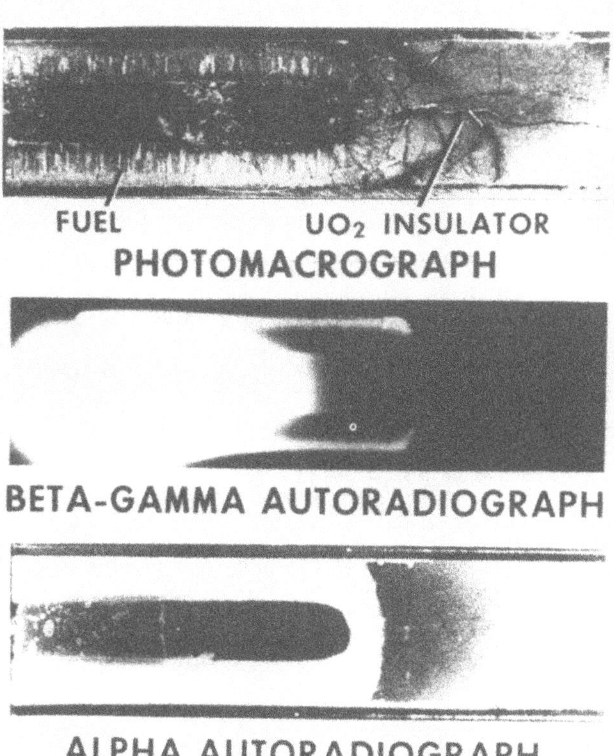

Fig. 10-12. *Photos of an intensely radioactive mixed fuel rod from an experiment in nuclear fuel development. Top: ordinary metallographic photograph of sectioned rod. Center: β-γ radiograph is fogged by radioactivity. Bottom: α-radiograph by track etching shows clearly the plutonium-rich regions as light-scattering regions that make clear where the fuel was melted.* (*After Davies and Darmitzel, 1965.*)

10.3.4. Reactor Applications

Fission rate, neutron flux, and nuclear fission cross section measurements are all simplified by using track detectors that do not register the effects of a background of β and γ radiation. The variety of the work in these areas is indicated by references given in Chapter 9. Neutron energy spectra can be measured by a time-of-flight technique described by Ewing (1973). A uranium foil, followed by a collimating slit and a rotating track detector, cause the energy spectrum to be displayed as the circumferential variation of the track density.

10.3.5. Uranium Exploration

Alpha-sensitive plastic detectors are being used to search large areas of land for sites of high radon concentration in the soil (Weidenbaum et al., 1970a; Gingrich and Lovett, 1972), as described in detail by Gingrich (1973). Such sites are then recognized as desirable locations for exploratory drilling for subsurface ore. The method depends on the fact that one member of the ^{238}U decay chain, ^{222}Rn, is a gas with a half life of 3.8 days, which is long enough for radon to be transported to the earth's surface from ores buried at considerable depths, provided the permeability of the overlying rock and soil is sufficiently high. The field procedure (now employed by Terradex Corporation, 1900 Olympic Blvd., Walnut Creek, Calif. 94596) is described by the sketch in Fig. 10-13. Plastic cups, each containing a piece of cellulose nitrate, are inverted and buried at shallow depths where they record alpha decays from ^{222}Rn and its daughters throughout a burial time of two to four weeks. The results for a large grid pattern of such measurements may be

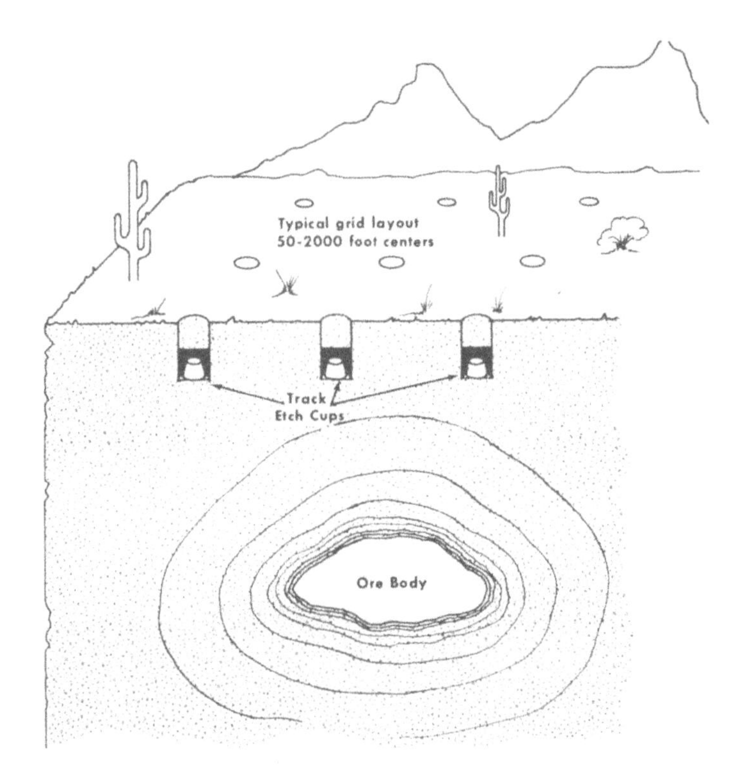

Fig. 10-13. *Uranium exploration technique. Radon-222, moving upward from uranium deposits, decays by alpha emission in buried, inverted cups, leaving tracks in cellulose nitrate detectors. Contours of track density reveal likely sites for more detailed mining exploration. (After Fleischer et al., 1972a.)*

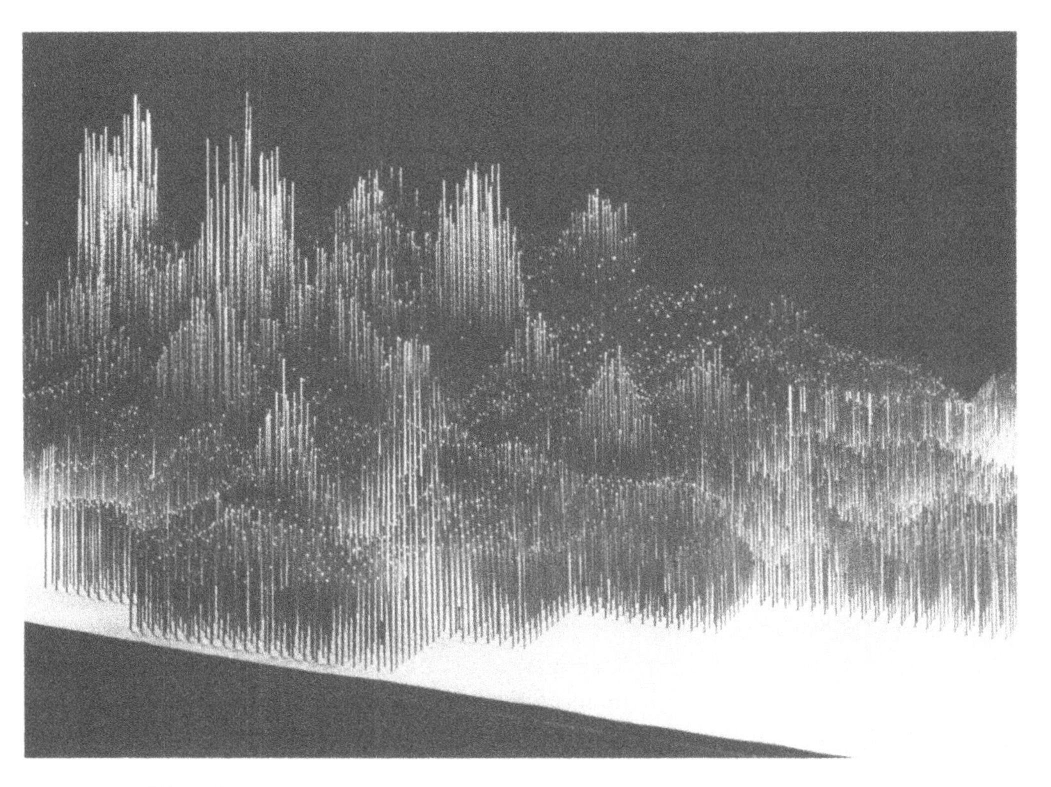

Fig. 10-14. *Radon emanation plot in which each pin represents the track density at a given position on a* ^{222}Rn *survey. Drilling at sites of high* ^{222}Rn *has led to the discovery of uranium at depths as great as 300 feet. (After Gingrich and Lovett, 1972.)*

represented by a pin plot such as that in Fig. 10-14, with the points of high radon concentration selected for exploratory drilling. This procedure has now located previously unknown ore bodies at depths of up to 300 feet and obtained clear signals from known deposits at 450 foot depths.

These depths are particularly remarkable because they are incompatible with simple diffusion models of the distance radon could move in a porous medium before being lost by alpha decay. The results support the idea that forced subsurface flow of the radon-bearing air occurs, very likely associated with changes of barometric pressure and soil moisture, and definitely associated with earth stresses that lead to seismic activity (Sadovsky et al., 1972). The likely association with barometric changes which have a time scale of a few hours or days is a plausible explanation as to why prompt-reading counters often do not locate radon anomalies that the track technique does recognize because it integrates over a time period that is much longer than that of the barometric fluctuations.

10.4. IMAGING

There are two main areas of application of track techniques: those aimed at solid state physics and metallurgy whereby some aspects of the internal structures of crystalline solids are imaged, and the various kinds of radiography aimed at revealing the internal structure of multicomponent objects including liquids and amorphous as well as crystalline solids.

10.4.1. Internal Structure of Crystalline Solids

Metallurgical uses of heavy ions and track detectors to examine internal structures have been developed and reviewed by Quere, Mory, and their associates (Couve, 1969; Delsarte et al., 1970; Jousset et al., 1969; Leteurtre et al., 1971; Mory, 1969; Mory and Quere, 1972; Quere, 1968; 1970; Quere and Mory, 1971) and by Medveczky and Somogyi, 1970. Fig. 10-15 shows a typical arrangement in which alpha particles from, for example, an ^{241}Am source are slowed down and effectively collimated by an amorphous absorber and then transmitted through the thin foil being examined. The distribution of emerging particles is recorded on a track detector such as cellulose nitrate or mica. Scattering in the foil allows various defects to be observed, including grain boundaries (Quere et al., 1966), twin boundaries (Quere and Couve, 1968; Mory and Delsarte, 1969), and the accumulation of dislocations from cold work (Quere and Couve, 1968; Mory et al., 1970). In Fig. 10-16, taken of a cold-worked Pt foil that was annealed 20 h at 950°C, twin and grain boundaries appear as white ribbons, and recrystallized areas appear dark because they have few defects that attenuate alpha particles.

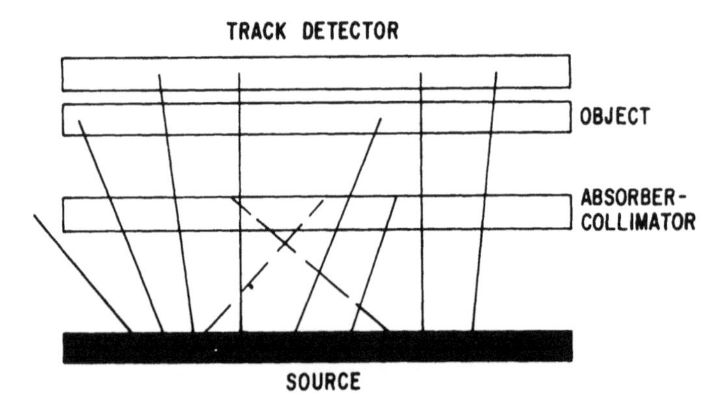

Fig. 10-15. *Experimental arrangement for alpha radiography of thin objects. A source of alpha particles passes through an absorber of the proper thickness to allow most of the particles to just pass through the object and register in the detector. Defects in the object scatter alpha particles and reduce the flux detected, giving maps such as are seen in Fig. 10-16. (After Quere et al., 1966.)*

Fig. 10-16. *Comparison between a reflected light micrograph (below) and an alpha radiograph (above) of a platinum foil. (After Quere and Couve, 1968.) Such radiographs permit sensitive detection of imperfections in metal foils. Width of micrographs is 700 μm.*

Quantitative measurements of the transmission through a sample after various stages of reheating to reduce the concentration of lattice defects allowed Quere and Couve (1968), Quere (1968) and Delsarte et al. (1970) to measure the activation energies for "recovery" of platinum and aluminum. Somogyi and Srivastava (1970; 1971) have discussed some procedures for alpha radiography. Mory and Delsarte (1969) have shown that fission fragments allow clearer definition of detail when examining platinum and tungsten than do alpha particles.

Channeling is one especially striking phenomenon that appears on many of the transmission photographs (called *channelographs* by Quere). In this process, the

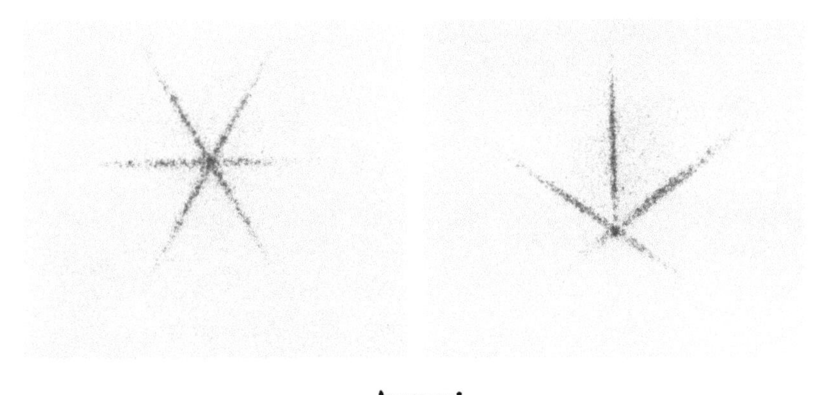

Fig. 10-17. *Channeling through small molybdenum crystals serves to display their orientations. Am-241 alpha particles penetrate the crystals preferentially along low-index planes and directions and display the crystal symmetries in a cellulose nitrate detector. Left side shows a ⟨111⟩ orientation and right side shows a ⟨210⟩ orientation.* (*Technique of Quere and Couve, 1968; photograph supplied by J. Mory.*)

charged particles move between atomic planes or atomic rows with less obstruction than in other directions. Such particles have an unusually long range and give patterns such as those in Fig. 10-17, which clearly reflect the crystal symmetry. Mory (1971) has studied the channeling process and Delsarte et al. (1971) have described how to use channeling patterns in much the same way as an x-ray Laue photograph is employed to determine crystal orientations.

Blocking is a closely related phenomenon in which transmission of particles is prevented along occupied atomic rows (but facilitated between rows or planes), as was illustrated so beautifully by Fig. 7-3. Patterns such as this are also useful for determining crystal orientation (Marsden et al., 1969) and, as described in Chapter 7, for determining or setting limits on nuclear lifetimes (Melikov et al., 1969). In such cases, if excited nuclei are ejected from an atomic row and move a mean distance of half an atomic spacing into the adjacent channel between rows, they can fission in positions where they are channeled rather than blocked. Conversely, atomic defects in specific interstitial positions along certain crystal directions can be located if blocking occurs along particular sets of directions where channeling would be the normal behavior in the absence of the point defects.

10.4.2. Neutron Radiography

Track-based neutron radiography can be performed using a detector assembly consisting of a boron-rich or uranium-rich plate pressed against an alpha-sensitive plastic detector such as cellulose nitrate or one of the cellulose acetates (Furman et al., 1966; Berger and Kraska, 1967). In this technique a beam of thermal neutrons is passed through the material to be radiographed and is then imaged by the particle tracks resulting from the ^{10}B (n,α) or ^{235}U (n,f) reactions. When etched, these tracks are light scatterers, so that the plastic sheet itself gives a direct visual image of the neutron transmission through the object. Neutron radiography is a complementary technique to normal x radiography, being in general more sensitive to light elements, particularly hydrogen, and it is therefore especially useful where organic material is to be examined—be it explosives or biological tissue. The neutron radiograph of the rubber date stamper in Fig. 10-18 illustrates the effective scattering by hydrogen. Strong neutron scattering by organic materials such as the wooden handle and the rubber belt provide good contrast for materials that are rich in hydrogen.

Berger (1973) has surveyed work in track-etch radiography, and earlier surveys of neutron radiography that discuss the track-etching technique are given by Berger (1968) and Hawkesworth and Walker (1969). Four special qualities of track-etch plates for neutron radiography are (1) their usefulness in high radiation environments—where other types are readily fogged; (2) their high resolution (since the limiting detail available is set by track lengths, the resolution is typically a few microns); (3) the fact that radiographs can be taken with as few as 2×10^7 neutrons/cm^2; and (4) their sensitivity to thickness differences as low as 1% (Berger and Lapinski, 1972; Berger, 1973). Contrast tends to be low, but a number

Fig. 10-18. *Neutron radiograph of a rubber date stamper recorded by alpha particle tracks in a cellulose nitrate sheet with an adjacent boron activation plate that supplies alpha particles from (n,α) reactions. (Taken by C. R. Porter, Vallecitos Nuclear Center; after Fleischer et al., 1972a.)*

of ways to enhance contrast have been described, including optical procedures (Berger and Lapinski, 1972; Berger, 1973) as well as optimal etching times and use of ozone treatments (Somogyi and Gulyas, 1972). Use of spark enhancement of etched holes in aluminized detector foils (as described in Chapter 2) has also been suggested (Somogyi and Gulyas, 1972).

In France G. Farny has begun using neutron radiography to inspect irradiated reactor fuel elements on a large-scale basis. With a 10 min exposure time and 20 min development time he is able to get high-resolution images of excellent quality, using a cassette camera containing a Kodak-Pathé CA 80-15 cellulose nitrate 10 m long and 12 cm wide.

Alpha particles from (n,α) reactions and fission fragments from ^{235}U (n,f) reactions are most commonly used for imaging thermal neutrons. Stinson et al. (1973) describe a neutron camera that images alpha particles from the $^{6}Li(n,\alpha)^{3}H$ reaction induced by slow neutrons. Fast neutrons can be imaged by a variety of reactions including recoil reactions within the plastic detector itself (Army Materials and Mechanics Research Center, 1972, 1973), carbon breakup into alpha particles (i.e., $^{12}C(n,n')\ 3\alpha$ reactions) (Fleischer et al., 1965b; Argonne National Laboratory, 1972), and fission reactions of various heavy nuclides (Berger, 1970). By using threshold detectors such as ^{237}Np or ^{232}Th, only neutrons of energy above 0.2 MeV or 1.3 MeV, respectively, would be imaged (Berger, 1970).

10.4.3. Proton and Heavy Ion Radiography

Whereas neutron radiography is sensitive to variations in the concentration of certain chemical elements, proton and heavy ion radiography are sensitive to small changes in density or mass thickness (g/cm^2) along the beam. The general principle is to shoot a monoenergetic beam of charged particles through the object to be imaged and to bring the beam to rest in a stack of plastic sheets. If the object were a uniform tank of water, the distribution of ion ranges would be tightly bunched in one or a few sheets of plastic, and if the plastic were sensitive only to ionization rates such as occur at the ionization peak for those particles, etch pits would be found uniformly distributed over the area struck by the beam as it stopped in the last sheet of plastic. If an object with variable density is immersed in the water, the particles travel either a greater or lesser distance before stopping, depending on whether the density is lower or higher. Various kinds of contrast can be achieved depending on the location of the etched sheet with respect to the position of the ionization peak for the beam in the pure water.

Fig. 10-19 shows ^{16}O ion radiographs in four sheets from within a stack of cellulose nitrate compared with an x ray; the object is a 14-day-old fertilized chicken egg. Benton et al. (1973), who took these radiographs, find that mass-thickness changes as small as 0.03 g/cm^2 in regions less than 1 cm in size can easily be detected, using about 10^6 ^{16}O ions, corresponding to an absorbed dose of about 1 rad.

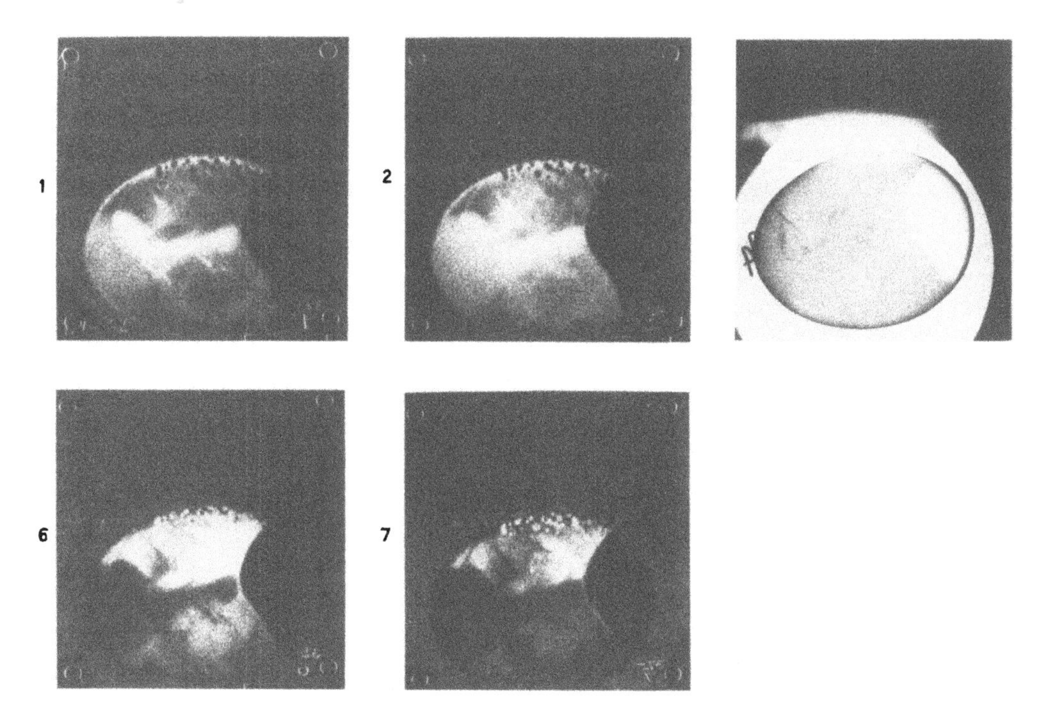

Fig. 10-19. *Four-layer CN radiographic sequence, produced by* ^{16}O *ions, of a 14-day-old fertilized chicken egg, compared with 35 kV x ray taken under optimum conditions. The head of the embryo is located on the left. The spots along the top of the egg are due to air bubbles trapped in a piece of foam rubber used to hold the egg under water.* (*After Benton et al., 1973.*)

The technique seems capable of being developed into a practical tool in diagnostic human radiography (possibly for tumor diagnosis) and is clearly the most sensitive tool presently available for the measurement of internal density distributions of objects. At present the Bevatron is the only accelerator that can produce heavy ion beams with ranges great enough to penetrate humans. Because both range-straggling and scattering decrease with increasing mass, radiographs of higher resolution can be produced with heavy ions than with protons. The greater availability of proton beams makes it important to continue development work on proton radiography such as has been begun by Berger et al. (1971). Somogyi and Srivastava (1970, 1971) have discussed experimental procedures for alpha radiography.

10.5. MISCELLANEOUS

A track use that evades simple classification is bird altimetry using an integrating barometer. Gustafson et al. (1973) placed a ^{210}Po alpha particle source in a thin-walled tube that also contained cellulose acetate detectors tilted at 45° to the particles from the source. The particles that record at the greatest distance from the source indicate the lowest atmospheric pressure to which the device is exposed. With such altimeters, which weigh one gram, Gustafson et al. determined that different, individual homing pigeons fly at different maximum altitudes—ranging from 300 to 1,700 meters—and that swifts fly at heights of 1,400 to 3,600 meters.

Chapter 10 References

H. W. Alter (1970), "Useful Applications of Charged Particle Tracks in Solids," paper presented at CIC/ACS Joint Conf., Toronto, May 24–29.

J. L. Anderson and J. A. Quinn (1972), "Ionic Mobility in Microcapillaries," *J. Chem. Soc. Faraday Trans. I*, **68,** Part 4, 744–748.

J. Anno, D. Blanc, and J. L. Teyssier (1970), "Collection of Ra Daughters on a Filter," *Proc. Seventh Inter. Colloq. Corpuscular Photography and Visual Solid Detectors*, Barcelona, 543–551.

Argonne National Laboratory (1972), "Report on Neutron Radiography," *Californium-252 Progress*, No. 12, 57–58.

Army Materials and Mechanics Research Center (1972), "Fast Neutron Imaging," *Californium-252 Progress*, No. 10, 18–20.

Army Materials and Mechanics Research Center (1973), "Fission Neutron Radiography," *Californium-252 Progress*, No. 14, 19–21.

J. A. Auxier, Klaus Becker, E. M. Robinson, D. R. Johnson, R. H. Boyett, and C. H. Abner (1971), "A New Radon Progeny Personnel Dosimeter," *Health Phys.* **21,** 126–128.

U. Batzdorf, R. S. Knox, S. M. Pokress, and J. C. Kennady (1969), "Membrane Partitioning of the Rose-Type Chamber for the Study of Metabolic Interaction Between Different Cultures," *Stain Technology* **44,** 71–74.

C. P. Bean (1969), "Characterization of Cellulose Acetate Membranes and Ultrathin Films for Reverse Osmosis," prepared for Office of Saline Water, Res. and Dev. Prog. Rept. 465.

C. P. Bean (1972), "The Physics of Porous Membranes—Neutral Pores," in *Membranes—A Series of Advances*, G. Eisenman (ed.), **1,** 1–54.

C. P. Bean, M. V. Doyle, and G. Entine (1970), "Etching of Submicron Pores in Irradiated Mica," *J. Appl. Phys.* **41,** 1454–1459.

R. E. Beck and J. S. Schultz (1970), "Hindered Diffusion in Microporous Membranes with Known Pore Geometry," *Science* **170,** 1302–1305.

R. E. Beck and J. S. Schultz (1972), "Hindrance of Solute Diffusion within Membranes as Measured with Microporous Membranes of Known Pore Geometry," *Biochem. Biophys. Acta.* **255,** 273–303.

K. Becker (1969), "Alpha Particle Registration in Plastics and its Applications for Radon, Thermal, and Fast Neutron Dosimetry," *Health Phys.* **16,** 113–123.

E. V. Benton and R. P. Henke (1972), "Heavy Cosmic Ray Exposure of Apollo 16 Astronauts," Tech. Rept. 20, Univ. of San Francisco.

E. V. Benton, R. P. Henke, and C. A. Tobias (1973), "Heavy Particle Radiography," *Science* **182,** 474–476.

H. Berger (1968), "Recent Progress in Neutron Imaging," *British J. of Non-Destructive Testing,* **10,** 26–33.

H. Berger (1970), "Image Detection Methods for 14.5 MeV Neutrons: Techniques and Applications," *Int. J. Appl. Rad. Isotopes* **21,** 59–70.

H. Berger (1973), "Track-Etch Radiography: Alpha, Proton, and Neutron," *Nuclear Technology* **19,** 188–198.

H. Berger and I. R. Kraska (1967), "A Track Etch, Plastic Film Technique for Neutron Imaging," *Trans. Am. Nucl. Soc.* **10,** 72–73.

H. Berger and N. P. Lapinski (1972), "Improved Sensitivity and Contrast, Track-Etch Thermal Neutron Radiography," *Trans. Am. Nucl. Soc.* **15,** 123–124.

H. Berger, N. P. Lapinski, and N. S. Beyer (1971), "Proton Radiography: A Preliminary Report," *Proc. 8th Symp. on Nondestructive Evaluation in Aerospace, Weapons Systems, Nuclear Applications,* Southwest Research Institute, San Antonio, Texas.

B. Carpenter, D. Samuel, and I. Wasserman (1973), "Quantitative Applications of ^{17}O Tracer," *Rad. Effects,* **19,** 59.

S. Chien, S. A. Luse, and C. A. Bryant (1971), "Hemolysis During Filtration Through Micropores: A Scanning Electron Microscopic and Hemorheologic Correlation," *Microvascular Research* **3,** 183–203.

R. M. Chrenko (1971), "Boron Content and Profiles in Large Laboratory Diamonds," *Nature* **229,** 165–167.

G. M. Comstock, R. L. Fleischer, W. R. Giard, H. R. Hart, Jr., G. E. Nichols, and P. B. Price (1971), "Cosmic Ray Tracks in Plastics: Apollo Helmet Dosimetry Experiment," *Science* **172,** 154–157.

W. H. Coulter (1953), U.S. Patent No. 2,656,509.

H. Couve (1969), "Some Applications of 'Channelography'," Rept. CEA-R-3741, C.E.N. Saclay, France.

J. H. Davies and R. W. Darmitzel (1965), "Alpha Autoradiographic Technique for Irradiated Fuel," *Nucleonics* **22,** 86–87.

J. H. Davies, R. F. Boyle, and J. F. Hanus (1966), "Fission Product and Pu Redistribution in High Performance UO_2 Fuel Rods," GE Rept. GEAP-5100-4, Atomic Energy Commission, avail. C.F.S.T.I.

R. W. DeBlois and C. P. Bean (1970), "Counting and Sizing of Submicron Particles by the Resistive Pulse Technique," *Rev. Sci. Instr.* **41,** 909–916.

R. W. DeBlois and C. P. Bean (1973), "Virus Detection and Characterization by the Resistive Pulse Technique," paper for presentation to the Ministry of

Electrotechnical Industry, Moscow, July 1 to 7, General Electric Company, preprint 73-CRD-188.
R. W. DeBlois, J. S. Wolff, and G. Schidlovsky (1973), "Viral Analysis by the Resistive Pulse Technique," abstract FPM-F11, the Biophysical Society, Feb. 27–March 2, Columbus, Ohio.
F. deKeroulas, J. Mory, and Y. Quere (1967), "La Fissiographie: Diffusion de L'Uranium en Dilution Quasi-Infinie dans le Titane et le Fer," *J. Nucl. Mat.* **22,** 276–284.
G. Delsarte, J. C. Jousset, J. Mory, and Y. Quere (1970), "Dechanneling of Fast Transmitted Particles by Lattice Defects," in *Atomic Collision Phenomena in Solids*, D. W. Palmer, M. W. Thompson, and P. D. Townsend (eds.), 50.
G. Delsarte, G. Desarmot, and J. Mory (1971), "Determination par Canaligraphie de l'Orientation de Microcristaux," *Phys. Stat. Sol.* **5,** 683–686.
B. V. Derjaguin (1970), "Superdense Water," *Sci. Amer.* **223** (11), 52–71.
W. Desorbo and J. S. Humphrey, Jr. (1970), "Studies of Environmental Effects on Track Etching Rates in Charged Particle Irradiated Polycarbonate Film," *Rad. Effects* **3,** 281–282.
I. Ewing (1973), "A Time-of-Flight Neutron Spectrometer Using Track-Etch Techniques," *Trans. Amer. Nucl. Soc.* **17,** 92–93.
R. L. Fleischer, P. B. Price, and R. M. Walker (1963), "Method of Forming Fine Holes of Near Atomic Dimensions," *Rev. Sci. Instr.* **34,** 510–512.
R. L. Fleischer, P. B. Price, and E. M. Symes (1964), "A Novel Filter for Biological Studies," *Science* **143,** 249–250.
R. L. Fleischer, P. B. Price, and R. M. Walker (1965a), "Tracks of Charged Particles in Solids," *Science* **149,** 383–393.
R. L. Fleischer, P. B. Price, and R. M. Walker (1965b), "Solid State Track Detectors: Applications to Nuclear Science and Geophysics," *Ann. Rev. Nucl. Sci.* **15,** 1–28.
R. L. Fleischer, P. B. Price, and R. M. Walker (1969), "Nuclear Tracks in Solids," *Sci. Amer.* **220,** 30–39, June.
R. L. Fleischer, H. R. Hart, Jr., and W. R. Giard (1970), "Particle Track Identification: Application of a New Technique to Apollo Helmets," *Science* **170,** 1189–1191.
R. L. Fleischer, H. W. Alter, S. C. Furman, P. B. Price, and R. M. Walker (1972a), "Technological Applications of Science: The Case of Particle Track Etching," *Science* **178,** 255–263.
R. L. Fleischer, J. R. M. Viertl, and P. B. Price (1972b), "Biological Filters with Continuously Adjustable Hole Size," *Rev. Sci. Instr.* **43,** 1708–1709.
R. L. Fleischer, H. R. Hart, Jr., G. M. Comstock, M. Carter, and A. Renshaw (1973), "Apollo 14 and 16 Heavy Particle Dosimetry Experiments," *Science* **181,** 436–438.

R. Frank, K. R. Spurny, D. C. Sheesley, and J. P. Lodge, Jr. (1970), "The Use of Nuclepore Filters in Light and Electron Microscopy of Aerosols," *J. de Microscopie* **9**, 735–740.

H. Frankel-Conrat (1969), *The Chemistry and Biology of Viruses*. New York: Academic Press.

S. C. Furman, R. W. Darmitzel, C. R. Porter, and D. W. Wilson (1966), "Track Etching—Some Novel Applications and Uses," *Trans. Am. Nucl. Soc.* **9**, 598–599.

G. Gamota (1973), "Creation of Quantized Vortex Rings in Superfluid Helium," *Phys. Rev. Letters* **31**, 517–520.

G. E. International (1971), **7**, No. 3, Oct., pp. 3–5.

J. E. Gingrich (1973), "Uranium Exploration Made Easy," *Power Eng.*, Aug., pp. 48–50.

J. E. Gingrich and D. B. Lovett (1972), "A Track Etch Technique for Uranium Exploration," *Trans. Am. Nucl. Soc.* **15**, 118.

J. E. Gingrich, H. D. Kosanke, D. B. Lovett, and R. W. Caputi (1972a), "Development of Neutron Flux Monitors for Unattended Safeguards Applications," Summary Rept. Contract ACDA/ST-209, prepared for the U.S. Arms Control and Disarmament Agency.

J. E. Gingrich, H. D. Kosanke, D. B. Lovett, and R. W. Caputi (1972b), "Development of Neutron Flux Monitors for Unattended Safeguards Applications," Final Rept. Contract ACDA/ST-209, prepared for the U.S. Arms Control and Disarmament Agency.

D. C. Golibersuch (1973), "Observation of Aspherical Particle Rotation in Poiseuille Flow via the Resistance Pulse Technique," *Biophys. J.* **13**, 265–280.

M. I. Gregersen, C. A. Bryant, W. E. Hammerle, S. Usami, and S. Chien (1967), "Flow Characteristics of Human Erythrocytes through Polycarbonate Sieves," *Science* **157**, 825–827.

R. V. Griffith (1973), "Result of the 1972 Survey on Track Registration," Lawrence Livermore Rept. UCRL-51362, March 15.

T. Gustafson, B. Kindkvist, and K. Kristiansson (1973), "New Method for Measuring the Flight Altitude of Birds," *Nature* **244**, 112–113.

E. I. Hamilton (1970), "The Concentration of Uranium in Air from Contrasted Natural Environments," *Health Phys.* **19**, 511–520.

W. W. Havens, Jr. (1968), "Nuclear Research as a Source of Technology," *Physics Today*, Sept., 47.

M. R. Hawkesworth and J. Walker (1969), "Review: Radiography and Neutrons," *J. Mat. Sci.* **4**, 817–835.

D. A. Horwitz and M. A. Garrett (1971), "Use of Leukocyte Chemotaxis *in Vitro* to Assay Mediators Generated by Immune Reactions," *J. Immunology* **103**, 649–655.

W. S. S. Jee, R. B. Dell, and L. G. Miller (1972), "High Resolution Neutron-Induced Autoradiography of Bone Containing ^{239}Pu," *Health Phys.* **22,** 761–763.

J. C. Jousset, J. Mory, and J. J. Quillico (1969), "Etude des Figures de Canalisation dans les Cristaux," *Proc. Inter. Conf. Nucl. Track Registration in Insulating Solids,* Clermont-Ferrand, Sect. IX, 47–51.

T. M. Kegley, Jr. (1970), "Use of Cellulose Nitrate for Alpha Autoradiography in Metallography," *Trans. Am. Nucl. Soc.* **13,** 528.

E. B. King and W. M. Russell (1967), "Needle Aspiration Biopsy of the Lung—Technique and Cytologic Morphology," *Acta Cytologica* **11,** 319–324.

J. Leteurtre, N. Housseau, and Y. Quere (1971), "Decanalisation des Particules α par les Dislocations dans l'Aluminum," *J. de Physique* **32,** 205–209.

D. B. Lovett (1969), "Track Etch Detectors for Alpha Exposure Estimation," *Health Phys.* **16,** 623–628.

D. B. Lovett (1972), "Track Etch Analysis of Uranium Particles as a Tracer System," *Trans. Am. Nucl. Soc.* **15,** No. 1, 121–122.

D. A. Marsden, N. G. E. Johansson, and G. R. Bellavance (1969), "The Use of Cellulose Nitrate Plastic in the Study of Blocking in Single Crystals," *Nucl. Instr. Methods* **70,** 291–297.

L. L. McAlpine and B. Ellsworth (1969), "A Modified Membrane Filter Technique for Cytodiagnosis," *Amer. J. Clinical Pathology* **52,** 242–244.

L. Medveczky and G. Somogyi (1970), "Induzierte Radiographie," *Atomki Kozlem* **12,** 191–194.

Yu. V. Melikov, Yu. D. Otstavnov, and A. F. Tulinov (1969), "The Blocking Effect in Uranium Fission," *Sov. Phys. JETP* **29,** 968–969.

M. Monnin and D. B. Isabelle (1970), "Les Detecteurs Solides de Traces et Leurs Applications en Biologie," *Annales de Phys., Biol. et Med.* **4,** 95–113.

J. G. Moore (1969), "The Development of a New Membrane Filter," *American Laboratory,* October.

J. Mory (1969), "Charged Particle Tracks in Insulating Solids," in *Solid State Dosimetry,* S. Amelinckx, B. Batz, and R. Strumane (eds.), 393–445. New York: Gordon and Breach.

J. Mory (1971), "Mesure du Coefficient de Decanalisation par un Defaut d'Empilement dans l'Or," *Rad. Effects* **8,** 139–141.

J. Mory and G. Delsarte (1969), "La Canalisation des Fragments de Fission," *Rad. Effects* **1,** 1–4.

J. Mory and Y. Quere (1972), "Dechanneling by Stacking Faults and Dislocations," *Rad. Effects* **13,** 57–66.

J. Mory, D. DeGuillebon, and G. Delsarte (1970), "Measure of Mean Path Length of Fission Fragments with the Mica Detector—Influence of the Crystalline Texture," *Rad. Effects* **5,** 37–40.

H. A. Notarys (1968), "Electric Field Suppression of the Lambda Point in Liquid Helium," *Phys. Rev. Letters* **20,** 1131.
H. A. Notarys and J. Andelin (1967), "Sound Stimulated Superfluid Flow," *Bull. Amer. Phys. Soc.* **12,** 1131.
W. J. Petzny and J. A. Quinn (1969), "Calibrated Membranes with Coated Pore Walls," *Science* **166,** 751–752.
M. C. Porter and H. J. Schneider (1973), "Nuclepore Membranes for Air and Liquid Filtration," *Filtration Engineering*, Jan./Feb., 8–16.
G. E. Possin (1970), "A Method for Forming Very Small Diameter Wires," *Rev. Sci. Instr.* **41,** 772–774.
G. E. Possin (1971), "Superconductivity in Nearly One-Dimensional Tin Wires," *Proc. Int. Conf. on the Science of Superconductivity*, F. Chilton (ed.), 339–343.
S. Pretre (1970), "Measurement of Personnel Neutron Doses by Fission Fragment Track Etching in Certain Plastics," *Rad. Effects* **5,** 103–110.
P. B. Price and R. M. Walker (1962), "Chemical Etching of Charged Particle Tracks," *J. Appl. Phys.* **33,** 3407–3412.
P. B. Price and R. M. Walker (1963), "A Simple Method of Measuring Low Uranium Concentrations in Natural Crystals," *Appl. Phys. Letters* **2,** 23–25.
Y. Quere (1968), "La Canaligraphie, Methode d'Etude des Defauts Cristallins," *J. de Physique* **29,** 215–219.
Y. Quere (1970), "Etude des Defauts Cristallins par Canalisation," *Ann. Phys.* **5,** 105–138.
Y. Quere and H. Couve (1968), "Radiography of Platinum by Means of Channeled Alpha Particles," *J. Appl. Phys.* **39,** 4012–4014.
Y. Quere and J. Mory (1971), "Etudes sur la Canalisation des Particules," *B.I.S.T. Commissariat a l'Energie Atomique*, No. 164, 3–16.
Y. Quere, J. C. Resneau, and J. Mory (1966), "Physique des Solides-Effets d'Obstruction sur la Canalisation dans l'Or," *Compt. rend.*, Paris **262,** 1528–1531.
J. A. Quinn, J. L. Anderson, W. S. Ho, and W. J. Petzny (1972), "Model Pores of Molecular Dimension; The Preparation and Characterization of Track-Etched Membranes," *Biophys. J.* **12,** 990–1007.
A. J. Reynaud and E. B. King (1967), "A New Filter for Diagnostic Cytology," *Acta Cytologia* **11,** 289–294.
J. R. Rich (1969), "A Survey of Cerebrospinal Fluid Cytology," *Bull. Los Angeles Neurological Societies* **34,** 115–131.
H. S. Rosenbaum and J. S. Armijo (1967), "Fission Track Etching as a Metallographic Tool," *J. Nuc. Mat.* **22,** 115–116.
S. Sabin (1973), "At Nuclepore They Don't Work for G.E. Anymore," *Fortune*, December, 144–153.
M. A. Sadovsky, I. L. Nersesov, S. K. Nigmatullaev, L. A. Latynina, A. A. Lukk,

A. N. Semenov, I. G. Simbireva, and V. I. Ulomov (1972), "The Processes Preceding Strong Earthquakes in Some Regions of Middle Asia," *Tectonophysics* **14,** 295–307.

M. Sakanoue (1968), "Charged Particle Tracks in Solids and Their Applications," *Radioisotopes* **17,** 212–229.

M. Sakanoue (1970), "Nuclear Particle Tracks in Solids and Their Development," *Kagaku no Ryoiki* **24,** 124–138, 207–221.

S. Seal (1964), "A Sieve for the Isolation of Cancer Cells and Other Large Cells from the Blood," *Cancer* **17,** 637–643.

G. Somogyi and J. Gulyas (1972), "Methods for Improving Radiograms in Plastic Detectors," *Radioisotopy* **13,** 549–568.

G. Somogyi and D. S. Srivastava (1970), "Investigations on Alpha Radiography with Plastic Track Detectors," *Atomki Kozlem* **12,** 101–117.

G. Somogyi and D. S. Srivastava (1971), "Alpha Radiography with Plastic Track Detectors," *Int. J. Appl. Rad. Isotopes* **22,** 289–299.

G. Somogyi, L. Medveczky, and M. Nagy (1971), "Solid State Track Detectors in Education," *Fizikai Szemle* **21,** 344–353 (Hungarian).

J. Song, P. From, W. Morrissey, and J. Sams (1971), "Circulating Cancer Cells: Pre- and Post-Chemotherapy Observations," *Cancer* **28,** 553–561.

K. R. Spurny and J. P. Lodge, Jr. (1972a), "Collection Efficiency Tables for Membrane Filters Used in the Sampling and Analysis of Aerosols and Hydrosols," NCAR Technical Note STR77, **1** (32 pages), "Techniques and Discussion"; **2** (282 pages), "Tables"; and **3** (402 pages), "Tables."

K. R. Spurny and J. P. Lodge, Jr. (1972b), "A Note on the Measurement of Radioactive Aerosols," *Aerosol Sci.* **3,** 407–409.

K. R. Spurny, J. P. Lodge, Jr., E. R. Frank, and D. C. Sheesley (1969), "Aerosol Filtration by Means of Nuclepore Filters: Aerosol Sampling and Measurement," *Environmental Sci. and Technol.* **3,** 464–468.

J. Stamm (1971), "Maximum Effective Pore Sizes of Nuclepore Membrane Filters," *Tappi* **54,** 1909.

R. H. Stinson (1972), "Boron Autoradiography of Botanical Specimens," *Can. J. Botany* **50,** 245–246.

R. H. Stinson, T. M. Holden, and P. A. Egelstaff (1973), "A High-Resolution Neutron Camera for Diffraction Experiments," Univ. of Guelph preprint.

C. G. Suits and A. M. Bueche (1967), "Cases of Research and Development in a Diversified Company," *Applied Science and Technol. Progress*, Rept. to the House of Representatives by Nat. Acad. Sci., 297–346.

R. L. Todd and T. J. Kerr (1972), "Scanning Electron Microscopy of Microbial Cells on Membrane Filters," *Applied Microbiology* **23,** 1160–1162.

J. W. N. Tuyn (1967), "On the Use of Solid State Nuclear Track Detectors in Reactor Physics Experiments," *Nucl. Applic.* **3,** 372–374.

S. Twomey (1972), "Measurements of the Size of Natural Cloud Nuclei by Means of Nuclepore Filters," *J. Atmos. Sci.* **29,** 318–321.
B. Weidenbaum and D. B. Lovett (1971), "Development of a New Technique for Bulk Water Tracing," Vallecitos Nuclear Center (Nucleonics Laboratory), Rept. NEDC-12015-5, Prepared for Contract 14-12-436 for EPA.
B. Weidenbaum, D. H. Coplin, and M. F. Lyons (1967), "Alpha and Beta-Gamma Autoradiography of High Performance Pellet and Powder Fuel," GEAP-5100-7, General Electric Company.
B. Weidenbaum, D. B. Lovett, and R. W. Caputi (1970a), "Track-Etch Research and Development at GE Vallecitos Nuclear Center," *Trans. Amer. Nucl. Soc.* **13,** 528–530.
B. Weidenbaum, D. B. Lovett, and H. D. Kosanke (1970b), "Flux Monitor Utilizing Track-Etch Film for Unattended Safeguards Application," *Trans. Amer. Nucl. Soc.* **13,** 524–526.
P. T. Werttake, B. A. Markovets, and S. Stellar (1972), "Cytology Evaluation of Cerebraspinal Fluid with Chemical and Histology Correlation," *Acta Cytologica* **16,** 224–239.

INDEX

Lightning Source UK Ltd.
Milton Keynes UK
UKHW041142140822
407235UK00001B/2